Membrane and Desalination Technologies

For further volumes:
http://www.springer.com/series/7645

VOLUME 13
HANDBOOK OF ENVIRONMENTAL ENGINEERING

Membrane and Desalination Technologies

Edited by

Lawrence K. Wang, PhD, PE, DEE
Zorex Corporation, Newtonville, NY, USA
Lenox Institute of Water Technology, Lenox, MA, USA
Krofta Engineering Corporation, Lenox, MA, USA

Jiaping Paul Chen, PhD
Division of Environmental Science and Engineering,
National University of Singapore, Kent Ridge, Singapore

Yung-Tse Hung, PhD, PE, DEE
Department of Civil and Environmental Engineering,
Cleveland State University, Cleveland, OH, USA

Nazih K. Shammas, PhD
Lenox Institute of Water Technology, Lenox, MA, USA
Krofta Engineering Corporation, Lenox, MA, USA

Humana Press

Editors
Dr. Lawrence K. Wang
Zorex Corporation, Newtonville, NY, USA
Lenox Institute of Water Technology, Lenox, MA, USA
Krofta Engineering Corporation, Lenox, MA, USA
larrykwang@juno.com

Dr. Jiaping Paul Chen
Division of Environmental Science and Engineering,
National University of Singapore, Kent Ridge, Singapore
paulchen@nus.edu.sg

Dr. Yung-Tse Hung
Department of Civil and Environmental Engineering,
Cleveland State University, Cleveland, OH, USA
y.hung@csuohio.edu

Dr. Nazih K. Shammas
Lenox Institute of Water Technology, Lenox, MA, USA
Krofta Engineering Corporation, Lenox, MA, USA
shammas@shammasconsult.com

ISBN: 978-1-4939-6085-9 ISBN: 978-1-59745-278-6 (eBook)
DOI 10.1007/978-1-59745-278-6
Springer New York Dordrecht Heidelberg London

Softcover re-print of the Hardcover 1st edition 2011

Printed on acid-free paper

Humana Press is part of Springer Science + Business Media (www.springer.com)

Preface

The past 35 years have seen the emergence of a growing desire worldwide that positive actions be taken to restore and protect the environment from the degrading effects of all forms of pollution – air, water, soil, thermal, radioactive, and noise. Since pollution is a direct or indirect consequence of waste, the seemingly idealistic demand for "zero discharge" can be construed as an unrealistic demand for zero waste. However, as long as waste continues to exist, we can only attempt to abate the subsequent pollution by converting it to a less noxious form. Three major questions usually arise when a particular type of pollution has been identified: (1) How serious is the pollution? (2) Is the technology to abate it available? and (3) Do the costs of abatement justify the degree of abatement achieved? This book is one of the volumes of the *Handbook of Environmental Engineering* series. The principal intention of this series is to help readers formulate answers to the above three questions.

The traditional approach of applying tried-and-true solutions to specific pollution problems has been a major contributing factor to the success of environmental engineering, and has accounted in large measure for the establishment of a "methodology of pollution control". However, the realization of the ever-increasing complexity and interrelated nature of current environmental problems renders it imperative that intelligent planning of pollution abatement systems be undertaken. Prerequisite to such planning is an understanding of the performance, potential, and limitations of the various methods of pollution abatement available for environmental scientists and engineers. In this series of handbooks, we review at a tutorial level a broad spectrum of engineering systems (processes, operations, and methods) currently being utilized, or of potential utility, for pollution abatement. We believe that the unified interdisciplinary approach presented in these handbooks is a logical step in the evolution of environmental engineering.

Treatment of the various engineering systems presented shows how an engineering formulation of the subject flows naturally from the fundamental principles and theories of chemistry, microbiology, physics, and mathematics. This emphasis on fundamental science recognizes that engineering practice has in recent years become more firmly based on scientific principles rather than on its earlier dependency on empirical accumulation of facts. It is not intended, though, to neglect empiricism where such data lead quickly to the most economic design; certain engineering systems are not readily amenable to fundamental scientific analysis, and in these instances we have resorted to less science in favor of more art and empiricism.

Since an environmental engineer must understand science within the context of application, we first present the development of the scientific basis of a particular subject, followed by the exposition of the pertinent design concepts and operations, and detailed explanations of their applications to environmental quality control or remediation. Throughout the series, methods of practical design and calculation are illustrated by numerical examples. These examples clearly demonstrate how organized, analytical reasoning leads to the most direct and clear solutions. Wherever possible, pertinent cost data have been provided.

Our treatment of pollution abatement engineering is offered in the belief that a trained engineer should more firmly understand fundamental principles, be more aware of the similarities and/or differences among many of the engineering systems, and exhibit greater flexibility and originality in the definition and innovative solution of environmental pollution problems. In short, an environmental engineer should by conviction and practice be more readily adaptable to change and progress.

Coverage of the unusually broad field of environmental engineering has demanded an expertise that could only be provided through multiple authorships. Each author (or group of authors) was permitted to employ, within reasonable limits, the customary personal style in organizing and presenting a particular subject area; consequently, it has been difficult to treat all subject materials in a homogeneous manner. Moreover, owing to limitations of space, some of the authors' favored topics could not be treated in great detail, and many less important topics had to be merely mentioned or commented on briefly. All authors have provided an excellent list of references at the end of each chapter for the benefit of the interested readers. As each chapter is meant to be self-contained, some mild repetition among the various texts was unavoidable. In each case, all omissions or repetitions are the responsibility of the editors and not the individual authors. With the current trend toward metrication, the question of using a consistent system of units has been a problem. Wherever possible, the authors have used the British system (fps) along with the metric equivalent (mks, cgs, or SIU) or vice versa. The editors sincerely hope that this redundancy of units' usage proves to be useful rather than being disruptive to the readers.

The goals of the *Handbook of Environmental Engineering* series are: (1) to cover entire environmental fields, including air and noise pollution control, solid waste processing and resource recovery, physicochemical treatment processes, biological treatment processes, biosolids management, flotation technology, membrane technology, desalination technology, water resources, natural control processes, radioactive waste disposal, and thermal pollution control, and (2) to employ a multimedia approach to environmental pollution control since air, water, soil, and energy are all interrelated.

This book is Vol. 13 of the *Handbook of Environmental Engineering* series, which has been designed to serve as an advanced chemical and environmental engineering textbook as well as a comprehensive reference book. We hope and expect that it proves of equal high value to advanced undergraduate and graduate students, to designers of water and wastewater treatment systems, and to scientists and researchers. The editors welcome comments from readers in all of these categories. It is our hope that the book not only provides information on membrane and desalination technologies, but also serves as a basis for advanced study or specialized investigation of the theory and practice of various membrane processes and systems.

This book, Membrane and Desalination Technologies, covers topics on principles of membrane technology, desalination requirements, historical developments, membrane fouling characterization, drinking water disinfection, regulations and determination of log removals, membrane systems planning and design, industrial waste treatment, municipal waste treatment, wastewater reclamation, food industry material separation, resource recovery, adsorption desalination, biofiltration, membrane bioreactor, thermal distillation, electrodialysis desalination, reverse osmosis desalination, point of use membrane applications, oil–water separation applications, and future membrane and desalination developments.

The editors are pleased to acknowledge the encouragement and support received from their colleagues and the publisher during the conceptual stages of this endeavor. We wish to thank the contributing authors for their time and effort, and for having patiently borne our reviews and numerous queries and comments. We are very grateful to our respective families for their patience and understanding during some rather trying times.

Lawrence K. Wang, Lenox, MA, USA
Jiaping Paul Chen, Kent Ridge, Singapore
Yung-Tse Hung, Cleveland, OH, USA
Nazih K. Shammas, Lenox, MA, USA

Contents

Contributors

JIRAPAT ANANPATTARACHAI, PhD • *Lecturer, Department of Civil Engineering, Naresuan University, Phitsanulok, Thailand*

ANUTOSH CHAKRABORTY, PhD • *Assistant Professor, School of Mechanical and Aerospace Engineering, Nanyang Technological University, Singapore*

JIAPING PAUL CHEN, PhD • *Associate Professor, Division of Environmental Science and Engineering, National University of Singapore, Singapore*

MUNIR CHERYAN, PhD • *Professor, Agricultural Bioprocess Laboratory, University of Illinois at Urbana–Champaign, Urbana, IL, USA*

EDWARD S.K. CHIAN, PhD • *Retired Professor, School of Civil and Environmental Engineering, Georgia Institute of Technology, Atlanta, GA, USA*

IBRAHIM I. EL-SHARKAWY, PhD • *Lecturer, Mechanical Power Engineering Department, Mansoura University, El-Mansoura, Egypt*

ANTHONY GORDON FANE, PhD • *Professor, Director, Singapore Membrane Technology Centre, Nanyang Technological University, Singapore*

RAJA GHOSH, PhD • *Associate Professor, Canada Research Chair in Bioseparations Engineering, Department of Chemical Engineering, McMaster University, Hamilton, ON, Canada*

TAY KWEE GUAN, BS • *Process Engineer, Keppel Seghers Engineering Singapore Pte Ltd, Singapore*

YUNG-TSE HUNG, PhD • *Professor, Department of Civil and Environmental Engineering, Cleveland State University, Cleveland, OH, USA*

YUE JIA, PhD • *Associate Professor, Institute of Biological and Chemical Engineering, Tianjin Polytechnic University, Tianjin, PR China*

PUANGRAT KAJITYICHYANUKU, PhD • *Associate Professor, Department of Civil Engineering, Naresuan University, Phitsanulok, Thailand*

KIRAN A KEKRE, MSc • *Vice President, Centre for Advanced Water Technology, PUB Consultants Pte Ltd, Singapore*

TAKESHI MATSUURA, PhD • *Emeritus Professor, Department of Chemical Engineering, University of Ottawa, Ottawa, ON, Canada*

RAVINDER MENON, BS • *Former Senior Principal Engineer, Industrial Biological Systems ONDEO Degremont, Inc. (formerly Infilco Degremont, Inc.), Richmond, VA, USA*

KAUSTUBHA MOHANTY, PhD • *Assistant Professor, Department of Chemical Engineering, Indian Institute of Technology Guwahati, Guwahati, Assam, India*

HONGHUI MOU, MENG • *Process Engineer, Chartered Semiconductor Manufacturing Ltd, Singapore*

K.G. NADEESHANI NANAYAKKARA, BS • *Research Scholar, Division of Environmental Science and Engineering, National University of Singapore, Singapore*

KIM CHOON NG, PhD • *Professor, Department of Mechanical Engineering, National University of Singapore, Singapore*

JIANJUN QIN, PhD • *Senior Research Scientist, Centre for Advanced Water Technology, PUB Consultants Pte Ltd, Singapore*

JIZHONG REN, PhD • *Professor, Director of Research and Development, National Engineering Research Center of Membrane Technology, Dalian Institute of Chemical Physics, Dalian, PR China*

BIDYUT BARAN SAHA, PhD • *Professor, Mechanical Engineering Department, Kyushu University, Fukuoka, Japan*

NAZIH K. SHAMMAS, PhD • *Professor and Environmental Engineering Consultant, Ex-Dean and Director, Lenox Institute of Water Technology, Lenox, MA, USA* • *Krofta Engineering Corporation, Lenox, MA, USA*

PING-XIN SHENG, PhD • *Facilities Engineer, Siltronic Singapore Pte Ltd, Singapore*

LIANFA SONG, PhD • *Associate Professor, Department of Civil and Environmental Engineering, Texas Technological University, Lubbock, TX, USA*

KWEE GUAN TAY, BS • *Research Fellow, Environmental Engineering Laboratory, National University of Singapore, Singapore*

STEPHEN TIONG-LEE TAY, PhD • *Associate Professor, School of Civil and Environmental Engineering, Nanyang Technological University, Singapore*

YEN-PENG TING, PhD • *Associate Professor, Department of Chemical and Biomolecular Engineering, National University of Singapore, Singapore*

LAWRENCE K. WANG, PhD • *Ex-Dean and Director (retired), Lenox Institute of Water Technology, Lenox, MA, USA* • *Assistant to the President (retired), Krofta Engineering Corporation, Lenox, MA, USA* • *Vice President (retired), Zorex Corporation, Newtonville, New York, NY, USA*

RONG WANG, PhD • *Associate Professor, School of Civil and Environmental Engineering, Deputy Director, Singapore Membrane Technology Centre, Nanyang Technological University, Singapore*

YUTING WEI, MENG • *Research Scholar, Division of Environmental Science and Engineering, National University of Singapore, Singapore*

LEI YANG, PhD • *Senior Engineer, Areate Pte Ltd, Singapore*

YU-MING ZHENG, PhD • *Research Fellow, Division of Environmental Science and Engineering, National University of Singapore, Singapore*

SHUAI-WEN ZOU, MENG • *Research Fellow, Division of Environmental Science and Engineering, National University of Singapore, Singapore*

1

Membrane Technology: Past, Present and Future

A.G. (Tony) Fane, Rong Wang, and Yue Jia

Contents

Abstract Membrane science and technology have experienced a long historical development in laboratory study before realizing their first significant industrial application in the 1960s. With nearly 50 years of rapid advancement, today, membrane-based processes enjoy numerous industrial applications and have brought great benefits to improve human life. In this chapter a general introduction is given to membrane technology in terms of the historical development, current status and future prospects. It begins with a description of the basic terms such as membrane, membrane structures, membrane classifications and membrane configurations. Membrane processes based on the different driving forces applied, the operation modes for filtration and membrane fouling are also briefly introduced. Section 2 is an overview of the historical development of membranes and membrane processes, including reverse osmosis, ultrafiltration, nanofiltration, microfiltration, gas separation, pervaporation and membrane bioreactors. Section 3 describes major applications and commercial relevance of the above-mentioned processes. In Sect. 4, future market development trends for membrane technology are indicated and critical technical challenges for further growth of the membrane industry are addressed. In addition, some promising novel applications of membrane technology are pointed out in the final section.

Key Words Membrane technology • historical development • current industrial applications • commercialization • future prospects • challenges.

From: *Handbook of Environmental Engineering, Volume 13: Membrane and Desalination Technologies*
Edited by: L.K. Wang et al., DOI: 10.1007/978-1-59745-278-6_1

1. INTRODUCTION

Membrane technology is an emerging technology and has become increasingly important in our life. A significant breakthrough for industrial applications of synthetic membranes started in the 1960s, although the earliest recorded study of membrane phenomena can be traced back to the middle of the eighteenth century (1). With nearly 50 years of rapid development, today, various membrane processes have found numerous industrial applications, which cover water and dairy purification, sea and brackish water desalination, wastewater reclamation, food and beverage production, gas and vapor separation, energy conversion and storage, air pollution control and hazardous industrial waste treatment, hemodialysis, proteins and microorganisms separation, etc. Membrane technology has greatly enhanced our capabilities to restructure production processes, protect the environment and public health, and provide new technologies for sustainable growth. The scope of the applications of membrane technology is still extending stimulated by the developments of novel or improved membrane materials and membranes with better chemical, thermal and mechanical properties or better permeability and selectivity characteristics, as well as by the decrease of capital and operation costs. This chapter provides a general review of membrane technology in terms of historical development, current status and future prospects. It can be seen that membrane science and technology have experienced a long period of development in laboratory studies. Although having enjoyed numerous industrial applications since 1960s, membrane technology still requires steady improvement to satisfy future broader applications.

1.1. Membranes, Membrane Classifications and Membrane Configurations

For a better understanding of the development of membrane technology, the fundamentals of membranes and membrane processes will be briefly reviewed. The word "membrane" is derived from the Latin word "membrana" and was first used in popular English media sometime before 1321 (Webster's Online Dictionary, http://www.websters-online-dictionary.org/). "Membrane" has different meanings in different domains. In association with separation, concentration or purification processes, a membrane can be essentially defined as a barrier to separate two phases and be able to restrict the transport of various components in a selective manner, as shown schematically in Fig. 1.1. A conventional filter also meets the definition of a membrane; however, the term "filter" is usually limited to structures that separate particulate suspensions larger than 1–10 μm.

There are many ways to classify synthetic membranes. They can be classified by the nature of the membrane material, the membrane morphology, geometry, preparation methods, separation regime and processes, etc. For instance, synthetic membranes can be organic (polymeric) or inorganic (ceramic/metal), solid or liquid, electrically charged or neutral in nature; they can be homogeneous or heterogeneous, symmetric or asymmetric in structure. Grouped by membrane geometric shapes, synthetic membranes can be flat, tubular or hollow fiber membranes. There are separation membranes to change the composition of mixtures, packaging membranes to prevent permeation, ion-exchange and biofunctional membranes to physically/chemically modify the permeating components, proton conducting membranes to conduct electric current, or non-selective membranes to control the permeation rate (2).

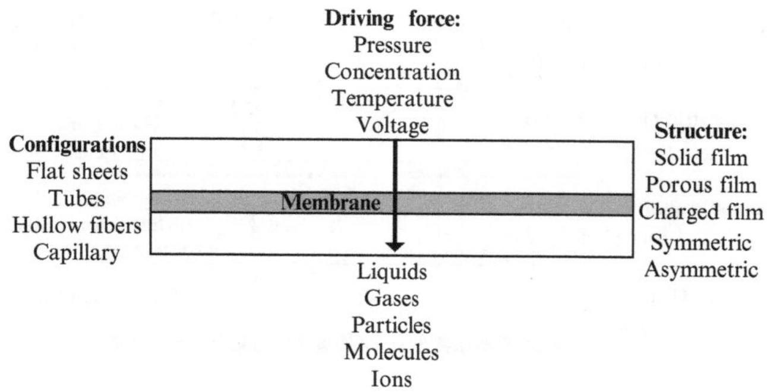

Fig. 1.1. Fundamentals of membrane and membrane processes.

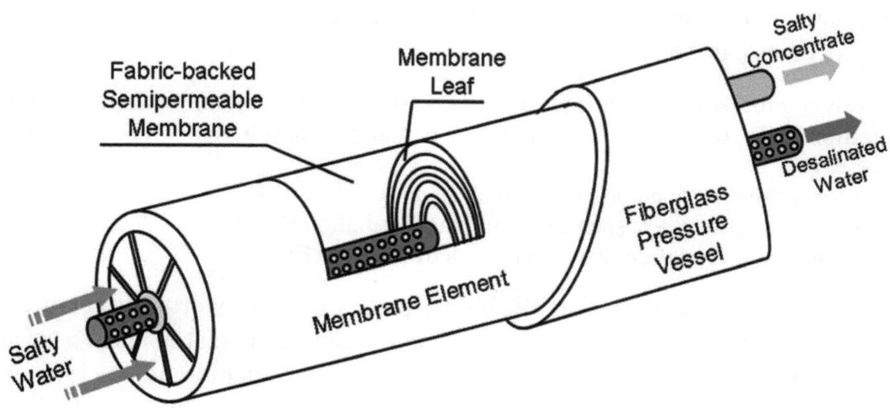

Fig. 1.2. Diagram of spiral-wound module.

Membranes have to be configured into membrane modules for practical applications (3) (Fig. 1.2). Membrane modules are the core elements in membrane-based separation and purification systems. In general, membrane modules are classified into three types: plate & frame, spiral wound and hollow fiber modules. Normally, the membrane surface to volume ratio is about 328–492 m^2/m^3 (100–150 ft^2/ft^3) for plate and frame modules, 656–820 m^2/m^3 (200–250 ft^2/ft^3) for spiral wound modules and 6,562–13,123 (2,000–4,000 ft^2/ft^3) for hollow fiber modules. The key properties of efficient membrane modules are high packing density, good control of concentration polarization and membrane fouling, low operating and maintenance costs, and cost-efficient production (4).

1.2. Membrane Processes, Operation Modes and Membrane Fouling

Membrane-based processes enjoy numerous industrial applications as they potentially offer the advantages of highly selective separation, continuous, automatic and economical operation at ambient temperature, and simple integration into existing production processes,

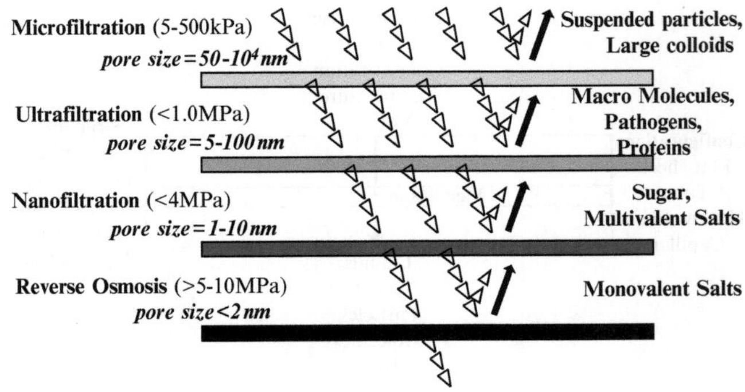

Fig. 1.3. Pressure-driven membranes for water and wastewater treatments.

as well as appreciable energy savings (2, 5). According to the driving force applied, the membrane processes can be classified as:

- Pressure-driven processes, such as reverse osmosis (RO), nanofiltration (NF), ultrafiltration (UF), microfiltration (MF), or gas separation (GS); or partial-pressure-driven processes, such as pervaporation (PV).
- Concentration-gradient-driven processes, such as dialysis.
- Temperature-driven processes, such as membrane distillation (MD).
- Electrical-potential-driven processes, such as elctrodialysis (ED).

Figure 1.3 shows the ranges of applications, pore sizes and working pressures for the pressure driven membrane processes for water and wastewater treatments. There are other processes or hybrid processes such as membrane contactors, membrane reactors and membrane bioreactors (MBRs), in which the function of the membrane is integrated with conventional processes to provide highly efficient performance. Different membrane processes can also be combined together to overcome the limitations of individual systems, and to maximize the productivity of the target separation processes.

Membrane filtration can be operated basically in two modes: dead-end and cross flow (Fig. 1.4). In dead-end mode, the entire feed flow transports towards the membrane perpendicularly so that the retained particles and other components accumulate and deposit on the membrane surface. In a cross-flow operation, the feed stream moves parallel to the membrane surface and only a portion of the feed stream passes through the membranes under the driving pressure. Moving the feed flow tangentially to the membrane surface can result in much higher permeation fluxes as the stream continuously removes retained material. In addition, back washing, ultrasonic vibration or periodic flow can also be used to remove the deposited material from the membrane surface. The membrane life-span may be prolonged (6). However, more complex equipment is required by the cross-flow mode compared to the dead-end mode, and also, the operating cost of the cross-flow mode is higher than the dead-end mode because of the energy needed to circulate the feed flow (7). The dead-end mode tends to be used in the water/wastewater industry for dilute feeds such as surface waters or secondary effluents. Periodic backwash is required to control the cake formation and fouling.

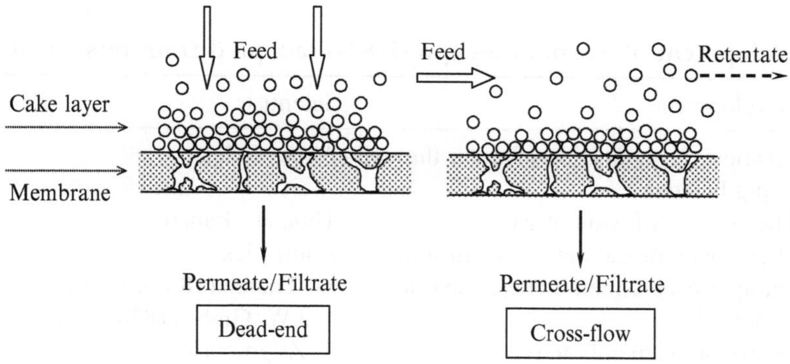

Fig. 1.4. Schematics of dead-end and cross-flow filtrations.

In membrane filtration applications, the main concerns are the loss of performance due to membrane fouling and additional costs from membrane cleaning and replacement. Membrane fouling is the deposition of material on or within the structure of membranes that is not readily reversed by simply releasing the pressure or by backwashing. The consequence of membrane fouling is to decrease membrane permeability (flux declines at constant pressure, or pressure rises at constant flux), and to alter solute retention (tends to increase for MF and UF, and decrease for NF and RO). Different fouling mechanisms may occur depending on the specific process, which may include pore blocking, pore narrowing or adsorption, irreversible cake layer formation, etc. Membrane fouling is a very complex phenomenon, which is related to feed solution properties (concentration, pH, ionic strength and component interactions), the nature of the membrane (hydrophobicity, charge, roughness, pore size, pore size distribution and porosity) and operating conditions (temperature, transmembrane pressure (TMP) and cross-flow velocity). Control of fouling can be achieved, for instance, by modifying the membrane surface and optimizing operating conditions (8, 9). Pretreatment is frequently used to modify the fouling tendency of the feed.

2. HISTRORICAL DEVELOPMENTS OF MEMBRANES AND MEMBRANE PROCESSES

2.1. *Historical Background (Pre-1980s)*

The development of membrane science and technology experienced a long history in laboratory study before realizing its first significant industrial application in the 1960s. The important events are listed in Table 1.1 (1, 10–12); references 10 and 11 are particularly illuminating. The earliest recorded membrane phenomenon, "osmosis", was discovered accidentally by Abbé Nollet in 1748. The stronger permeability of water through a piece of pig's bladder covering a jar containing "spirits of wine" demonstrated the concept of semi-permeability for the first time. Eighty years later, the observation of hydrogen gas escaping through cracked jars by Doebereiner in 1823 stimulated Thomas Graham to study gas diffusion in 1833 and later he proposed Graham's law of gas diffusion in 1866. Thomas

Table 1.1
Historical development of membranes (pre-1980s) (adapted from refs. (1, 10–12))

Year	Development	Scientists
1748	"Osmosis", permeation of water through pig bladders	Abbé Nollet
1833	The law of diffusion of gases	Thomas Graham
1855	Phenomenological laws of diffusion	Adolf Fick
1860–1880s	Semipermeable membranes: osmotic pressure	M. Traube (1867), W. Pfeffer (1877), J.W. Gibbs (1878), J.H. van't Hoff (1887)
1907–1920	Microporous membranes	R. Zsigmondy
1920s	Prototype of reverse osmosis	L. Michaelis (1926), E. Manegod (1929), J.W. McBain (1931)
1930s	Electrodialysis membranes, modern membrane electrodes	T. Teorell (1935), K.H. Meyer and J.F. Sievers (1936)
1950s	Electrodialysis, microfiltration, and hemodialysis, ion-exchange membranes	Many
1963	Defect-free, high-flux, anisotropic reverse osmosis membranes	S. Loeb and S. Sourirajan
1968	Basics of pervaporation	P. Aptel, and J. Neel
	Spiral-wound RO module	J. Westmorland
1977	Thinfilm composite membrane	J. Cadotte
1970–1980	Reverse osmosis, ultrafiltration, microfiltration, electrodialysis	Many
1980s	Industrial membrane gas separation processes	J.M.S. Henis and M.K. Tripodi (1980)
1989	Submerged membrane (bioreactor)	K. Yamamoto

Graham was the pioneer in the quantitative measurement of the rate of gas permeation through natural rubber films. Adolf Fick formulated the phenomenological laws of diffusion in 1855, which soon became the important rule describing diffusional transport in membranes. Semi-permeable membranes were first made by Traube in 1867 and used to measure solution osmotic pressure by Pfeffer (1877), which led to the classical theory of solutions – van't Hoff's osmotic pressure equation (1887). This milestone work was awarded the first Nobel Prize for Chemistry in 1901. The Donnan equilibrium involving charged species was worked out in 1911. In 1907–1920, Zsigmondy developed extremely fine particle filters or molecular filters which were early microfilters and ultrafilters. A little later in the 1920s, the prototype of RO was observed and studied by Michaelis (1926), Manegod (1929) and McBain (1931) with cellophane or cellulose-nitrate membranes. The basis for electrodialysis membranes and modern membrane electrodes was established by Teorell and Meyer in 1930s from their study on transport across neutral and fixed-charge membranes. Willem Johan Kolff achieved the first clinically successful hemodialysis with a human patient in 1945. Later in the 1950s, electrodialysis, MF and ion-exchange membranes were explored and used in lab-scale and in testing drinking water safety in Europe. The first commercial plant for the production of potable water from a saline source using electrodialysis and ion-exchange membranes was put into operation in 1954 (13).

One of the milestones in the industrial application of membranes was the fabrication of defect-free, high-flux, anisotropic RO, asymmetric cellulose acetate (CA) membranes by Loeb and Sourirajan in 1962 (14). With the original Loeb–Sourirajan technique, other membrane formation processes and membrane packaging methods were developed in the next two decades. By 1980, RO, UF, MF and electrodialysis (ED) had found worldwide applications in large scale plants. In the 1980s, commercial PV systems and industrial membrane GS processes were developed. Generally, it was widely agreed that the broad industrial applications of membrane processes began from the year 1960; therefore, the development of various types of membrane separation processes in the period of 1960–1980s are discussed in their order of emergence for a large scale application in the following sections.

2.2. *Reverse Osmosis*

Osmosis, a natural phenomenon discovered as early as 1748, is diffusion of fluid (usually water) through a semipermeable membrane from a solution with a low solute concentration to a solution with a higher solute concentration until an equilibrium of fluid concentration on both sides of the membrane is reached. The pressure difference between the two sides of the membrane equals to osmotic pressure of the solution. The chemical potential gradient across the membrane acts as the driving force for the water transport. By applying a pressure in excess of the osmotic pressure, pure water flows from the high solute concentration side through a membrane to the low solute concentration side and thus, the separation of water from the solution is achieved. This is the reverse of the normal osmosis process and termed as "reverse osmosis" (Fig. 1.5). The membrane between the two liquid phases is semipermeable which allows the passage of water but not the solutes. It should be noted that the term "hyperfiltration" is also used to describe the process, but the common parlance is "reverse osmosis" or RO.

Osmosis and RO have been studied over two centuries; however, it was not until the early 1960s the use of RO as a feasible separation process was demonstrated. In 1959, the work of Reid and Breton showed that CA films were capable of separating salt from water, but the water fluxes obtained were too small to be practical (15). Then, in 1962, Loeb and Sourirajan reported the first synthetic RO membrane from CA polymer which was capable of rejecting the salt with relatively high water fluxes (14). Cadotte and Petersen later made the first really efficient thin-film composite membranes, which was a key breakthrough to make large-scale application of RO membranes economically feasible (16). The key developments of RO membranes between 1960 and 1990 are summarized in Fig. 1.6.

The surface of polymeric RO membranes is considered as being nonporous, consisting of a polymer network in which the "solvent" can be dissolved. In reality, the RO membrane has a "dynamic" polymer network which allows water molecules to enter. In effect the "pores" are transient channels in the dense film. The transport process is controlled by solution and

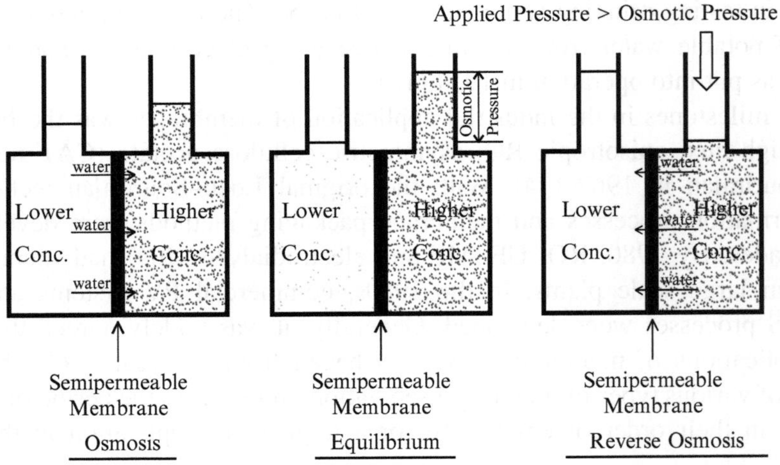

Fig. 1.5. Osmosis and reverse osmosis.

diffusion, which retains almost all ionic species. RO membranes are generally categorized into asymmetric membranes and thin-film, composite membranes. An asymmetric RO membrane contains one polymer material and has a thin, permselective skin layer supported by a more porous sublayer in structure. The dense skin layer (from 0.1 to 1 μm) determines the flux and selectivity while the porous sublayer has little effect on the membrane separation properties. Asymmetric membranes are most commonly formed by a phase inversion (polymer precipitation) process based on the Loeb–Sourirajan technique. Since the invention of the asymmetric cellulose acetate RO membranes in 1962, significant improvements were achieved in the following 20 years and cellulosic polymers (CA, cellulose triacetate, etc.) and linear aromatic polyamide membranes are found to be the most widely used examples of asymmetric membranes (17–19). A thin-film composite membrane is made of two or more polymer materials and consists of a thin polymer barrier layer formed on one or more porous support layers (different from the surface layer) in structure. The extremely thin surface layer (in the order of 0.1 μm or less) allows high water fluxes. The most important thin-film, composite membranes are made from cross-linked aromatic polyamide on a polymer (e.g., polysulfone or aryl-alkyl polyetherurea) support layer by the interfacial polymerization method (17, 19–20). Comparison between the asymmetric membranes and the thin-film composite membranes in terms of the characteristics which are relevant to their applications in seawater and brackish water desalination is given in Table 1.2 (21).

A typical efficient RO process is designed to achieve higher water flux with relatively lower energy expenditure. In addition to the membrane materials, the packaging of membranes is also a very important factor in the feasibility of the RO process. The commercially available membrane modules are plate-and-frame, tubular, spiral-wound and hollow fiber elements. RO systems can be arranged either in single-pass or in multi-pass configurations (22). A typical single-pass RO system is depicted in Fig. 1.7. Limitation in product recovery (defined as product flow divided by feed flow) is governed by the concentration of the feed,

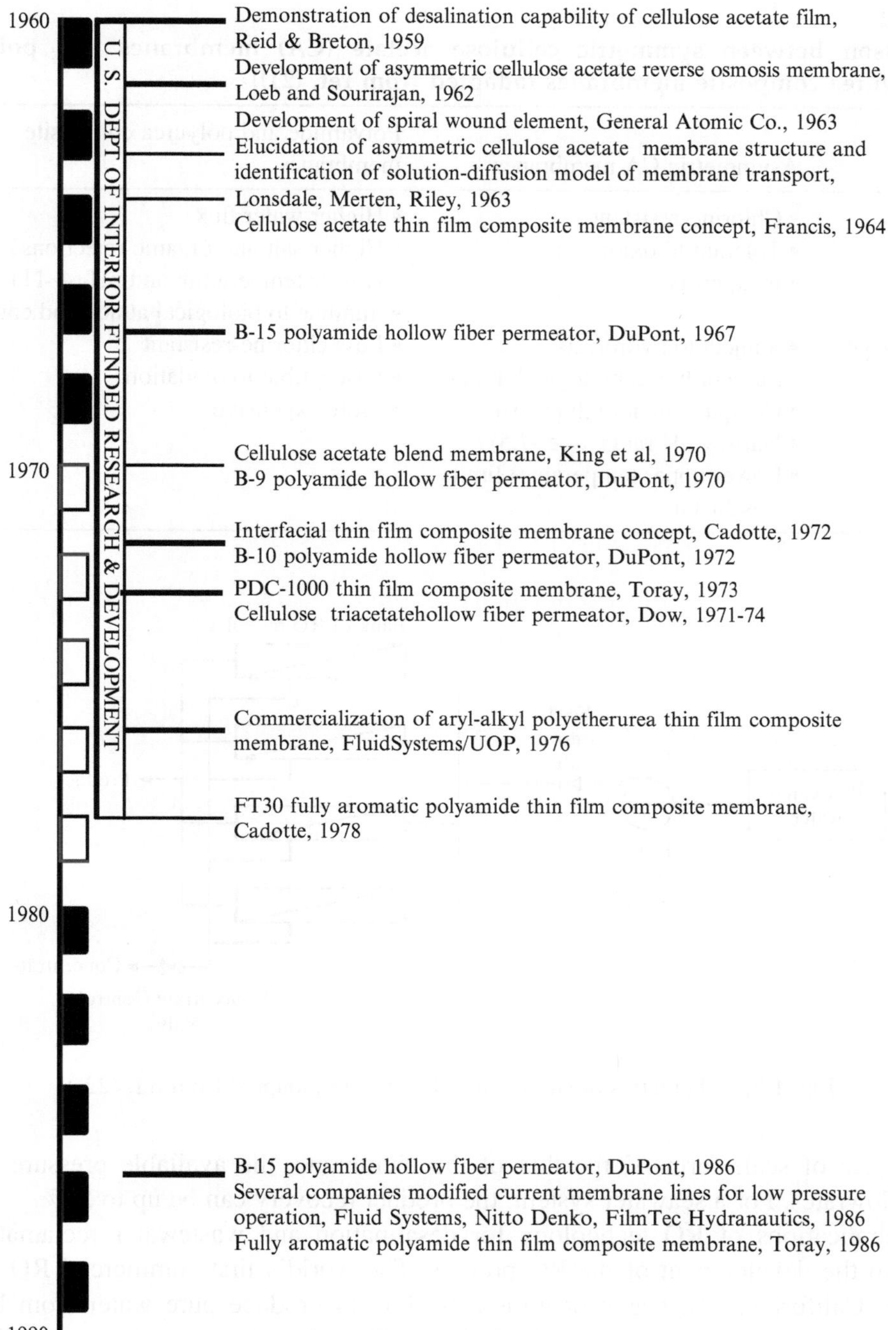

Fig. 1.6. Key developments of RO process by 1990 (adapted from ref. (12)).

Table 1.2
Comparison between symmetric cellulose acetate (CA) membranes and polyamide and polyurea composite membranes (adapted from ref. (21))

	Asymmetric CA membranes	Polyamide and polyurea composite membranes
Advantages	• Chlorine-resistant • Tolerant to oxidation • Inexpensive	• Higher water flux • Higher salt and organic rejections • Higher temperature and pH (4–11) • Immune to biological attack and compaction
Disadvantages	• Subject to hydrolysis • Susceptible to biological attack • Compaction at high pressures • Narrow pH range (4.5–7.5) • Lower upper temperature limits (~35°C)	• Less chlorine-resistant • Susceptible to oxidation • More expensive

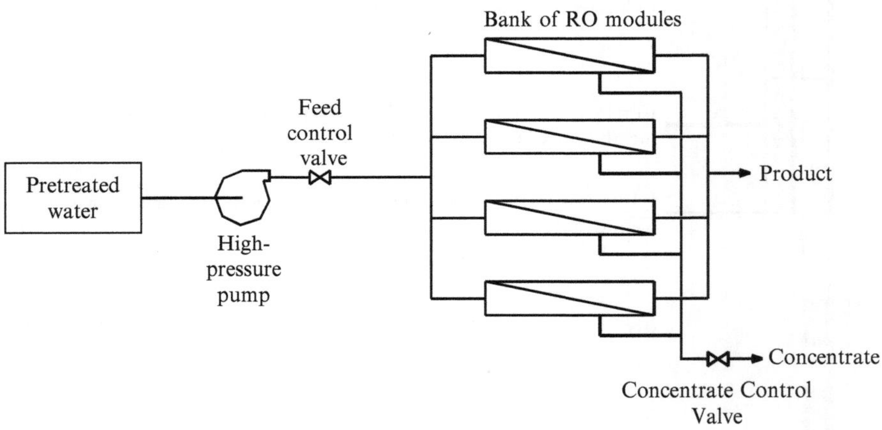

Fig. 1.7. Schematics of single-pass RO system (adapted from ref. (22)).

the presence of scale-formers, the flux of the membrane, the available pressure and the desired flowrates. For a seawater system, the product recovery can be up to 50%.

The applications of RO technology for desalination and wastewater reclamation run parallel to the development of the RO process. The world's first commercial RO plant in Coalinga, California, US, began operation in 1965 to produce pure water from brackish groundwater at a capacity of up to 22.7 kL/day (6,000 gallon/day). A subsequent pilot plant was constructed at La Jolla to extract fresh water from the sea (http://www.engineer.ucla.edu/history/osmosis.html/). The first large RO plant with 18.9 MLD (5 MGD) of product capacity to treat municipal wastewater, which was a part of the "Water Factory 21" in Orange County, California, commenced operation in the late 1970s (23). By the end of 1990, the RO technique had found its possible applications in almost all the industries in addition to the seawater and brackish water desalination processes (Table 1.3) (12). For instance, in the metal finishing industry, RO membranes are used for rinse water recovery and reuse of plating salts from plating waste streams (24). High purity water is a necessity for manufacturing miniature electronic components. The large volumes of spent ultrapure water can be purified for reuse by RO systems (25). RO membranes are also employed in the photographic and printing industries to solve their discharge problems while simultaneously reducing water expense by recovery (26). Additionally, RO technology can be combined

Table 1.3
Reverse osmosis market applications (adapted from ref. (12))

Status	Industry	Applications
Mature	Desalination	• Potable water production
		• Seawater
		• Brackish water
		• Municipal wastewater reclamation
	Ultrapure water	• Semiconductor manufacturing
		• Pharmaceuticals
		• Medical uses
Growing	Utilities and power generation	• Boiler feedwater
		• Cooling tower blowdown recycle
	Point of use	• Home reverse osmosis
Emerging	Chemical process industries	• Process water production and reuse
		• Effluent disposal and water reuse
		• Water/organic liquid separation
		• Organic liquid mixtures separation
	Metals and metal finishing	• Mining effluent treatment
		• Plating rinse water reuse
		• Recovery of metals
	Food processing	• Dairy processing
		• Sweeteners concentration
		• Juice and beverage processing
		• Production of light beer and wine
		• Waste stream processing
	Textiles	• Dyeing and finishing
		• Chemical recovery
		• Water reuse
	Pulp and paper	• Effluent disposal and water reuse
	Biotechnology/medical	• Fermentation products recovery purification
	Analytical	• Isolation, concentration and identification of solutes and particles
	Hazardous substance removal	• Removal of environmental pollutants from surface and ground waters

with UF, PV, distillation and other separation techniques to provide highly efficient and selective separations.

2.3. Ultrafiltration

UF membranes were studied at the laboratory scale or small scale as early as 1907 (27), but they were firstly prepared with an initial goal of producing high-flux RO membranes. The first commercial UF membranes were introduced in the mid-1960s by Millipore and Amicon as a spin-off of the development of asymmetric RO membranes. UF membranes have their roots in commercial cellulose acetate RO membranes and thereafter were steadily developed in parallel with RO membranes. A few milestones are shown in Fig. 1.8 (28).

Different from RO membranes, no significant osmotic pressure is generated across the UF membranes because the porous structure (pore size 1–100 nm) allows the permeability of microsolutes (MWs < 300) through the membranes (29). UF membranes, practically, are used as a barrier to separate macromolecules, colloids and solutes with molecular weight larger than 10,000 from low molecular weight species. Although these species can produce an osmotic pressure, it is usually only a few bar. Thus, the hydrostatic pressure difference used as driving force in UF is in the range of 1–10 bars. The selectivity of UF membranes is based on the difference in the size and surface charge of the components to be separated, the properties of the membrane as well as the hydrodynamic conditions.

Most UF membranes have an asymmetric porous structure and are often prepared by the phase-inversion process. CA was the main membrane material in the first decade of UF. However, the chemical and thermal stabilities of CA are low and it has a relatively narrow range of pH tolerance and is highly biodegradable. Therefore, other polymers or polymer blends were employed to produce UF membranes, such as polyacrylonitrile, aromatic polyamides,

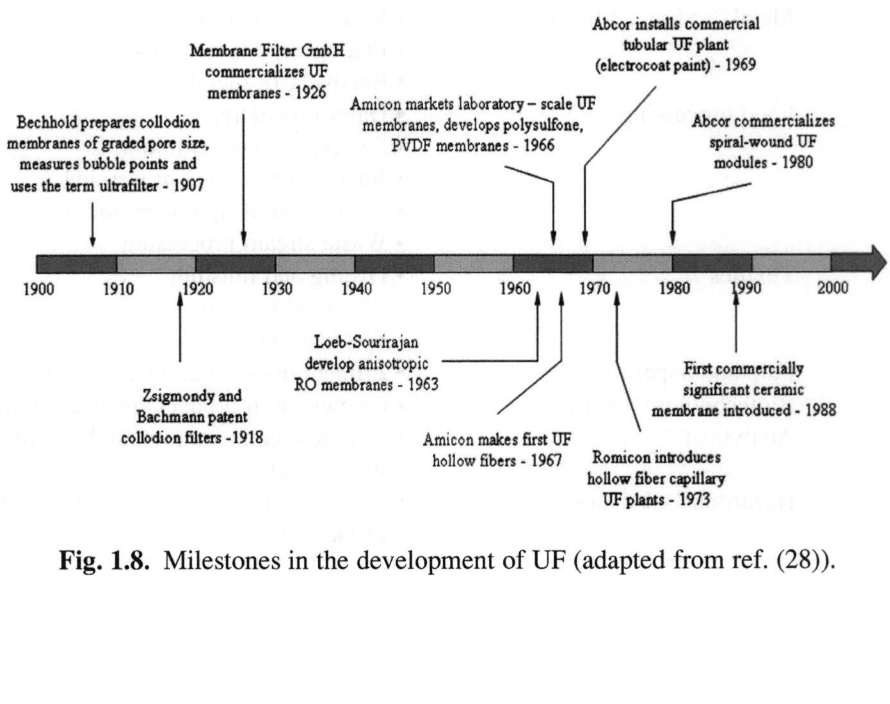

Fig. 1.8. Milestones in the development of UF (adapted from ref. (28)).

polysulfone, polyethersulfone, polyvinylchloride and polyvinylidene fluoride. UF membranes prepared from these materials show a wide range of pH and temperature resistances, and are fairly resistant to chlorine, which significantly broaden their applications. The UF membrane configurations also varied from tubular to hollow fibers, spiral-wound and capillary modules by the end of 1980s.

Thirty years after their emergence, UF membranes had successfully found a wide range of industrial applications (Table 1.4) which created a world wide market up to US$60 million for water treatment, US$44 million for food industry, US$130 million for medical devices and US$15 million for chemical industry (30). Two key industries were involved in the commercial development of UF. UF membranes were adopted for the recovery of electrophoretic paints from rinse water. Considerable saving of paints and water was achieved by using tubular CA membranes without requirement of thermal or chemical stability (29), and without UF the electrophoretic paint industry would have struggled. Another important application of UF membranes has been the dairy industry. Not only used for recovering proteins, lactose and lactic acid from whey, the membranes were also used to concentrate dietary milk or produce cheese. The double benefits in water pollution treatment and valuable products recovery made the dairy industry a large market area for UF membranes (31).

Table 1.4
Industrial applications of UF membranes

Industries	Applications
Automobile and household appliances industry	Recovery of electrophoretic paints from rinse waters
Metal-processing industry	Recycling emulsions used in metal forming
Food processing	Dairy:
	• Recovery of proteins from whey
	• Dietary milk production
	Recovery of Proteins from meat processing effluents
	Concentration of egg white
	Sterilization and clarification of beverages, especially wine
	Recovery of valuable constituents from starch and yeast processing effluents
Pulp and paper industry	Waste paper mill effluent treatment
Pharmaceutical industry	Sterile filtration of water or solutions
	Isolation, concentration and purification of biologically active substances (enzymes, viruses, nucleic acids, specific proteins)
	Fractionation of blood
Semiconductor industry	Production of ultra pure water
Water industry	Pretreatment before NF and RO

2.4. Nanofiltration

The term NF was introduced by FilmTec in the second half of the 1980s to describe a "RO process" that selectively and purposely allows some ionic solutes in a feed water to permeate through. It was derived from the membrane's selectivity towards noncharged solutes of approximately 1 nm cutoff (32). Within a few years, other membrane scientists began using the word and Eriksson (33) was one of the first who used it explicitly in 1988. As a matter of fact, such membranes with the selectivity between the RO and UF regions already existed in the 1960s and were categorized as open, loose, low-pressure RO, intermediate RO/UF, or tight UF membranes (34–35).

Different from RO membranes which has nonporous structure and a transport mechanism of solution-diffusion, NF membranes operate at the interface of porous and nonporous membranes with both sieving and diffusion transport mechanisms. Bhattacharyya, Jitsuhara and Tsuru (36–38) observed that charged UF membranes performed same features as NF in the case of salt retention. Therefore, it was acknowledged that NF performed an intermediate capability as "loose" RO (nonporous, diffusion) or "tight" UF (porous, sieving). This "loose" property of NF membranes enables them to be operated at higher water fluxes (water recoveries) with much lower pressure compared to RO membranes, that results in significant energy saving. In addition, most NF membranes are surface charged so that electric interactions also add to the transport and selective rejection behavior of NF membranes.

NF membranes have a high permeability for monovalent salts (e.g., NaCl, KCl), but they are able to eliminate multivalent salts near completely and remove relatively small organic compounds (39). NF was used as an alternative to RO for the concentration and demineralization of whey in dairy industry. Till 1990, NF membranes had found their more applications in seawater softening, food, textile and mining industries (40–42). The NF-50 membrane was the first example of RO membrane capable of operation at UF membrane pressures (20). Characteristics of this membrane include 30–40% NaCl rejection, 85–90% $MgSO_4$ rejection, 98% sucrose rejection and 99% raffinose rejection. However, the first generation NF (loose RO) membranes made by CA possessed poor biological and chemical stability. They did not allow application in organic solvent due to the lack of chemical resistance of the membrane polymers against the solvents (42). The most important drawback of NF membranes is the problem to control the reproducibility of the membrane pore size and the pore size distribution. Moreover, NF membranes are liable to fouling, possibly resulting in important flux decline.

2.5. Microfiltration

MF membranes are used for separation of impurities (particles, viruses, and bacteria) with a size range of 0.1–10 μm from a solvent or other low molecular weight components. The separation mechanism is based on a sieving effect and particles are separated according to their dimensions although some charge or adsorptive separation is possible. The applied pressure in MF is quite low (<2 bars) compared with other filtration processes (43).

Systemic studies on MF membranes started at the beginning of the twentieth century by Bechold, who made membranes from different compositions of the casting solution and found that pore size could be changed. He also applied bubble point measurement to determine the maximum pore size of membrane filters and this technique is still used today. Around 1918, Zsigmondy and Bachmann developed the first method to produce nitro-cellulose membranes that were applicable on a commercial scale. The term "membrane filter" was used by them for the first time. The membrane filter technique became important just after World War II as it was applied to the bacteriological examination of water supply systems, which had inspired extensive studies on the membrane filtration technology. In 1950, Goetz developed a new method capable of producing membranes with improved performance and Millipore Co. was then setup based on it in 1954 (44).

The first commercialized application of MF membranes was in biological and pharmaceutical manufacturing in the 1960s. In the following 20 years, MF membranes were mainly used in sterile filtration (microorganisms removal) in the pharmaceutical industry, and final filtration (particle removal) of the rinse water in the semiconductor industry. Not so stringent as in the pharmaceutical industry, MF was also used in sterilization of beer and wine, as well as clarification of cider and other juices, easily and economically. It was not until the mid-1980s, that MF was introduced to water treatment industry because of its cost effective benefit. The application of MF (and UF) in water treatment gained in momentum following the cryptosporidium outbreak in the USA in 1992. Stricter regulation on pathogen removal for water supply have seen a major shift to the low pressure membrane processes. Before 1963, membrane microfilters were predominantly made of nitro-cellulose or cellulose esters. The development of highly resistant membranes such as polypropylene, polyamide and polysulfone enables the MF technique to be used as a large-scale separation tool.

2.6. *Gas Separation*

The first important scientific discovery of GS membranes was made by Mitchell who found that different gases penetrated natural rubbers at different rates in 1831 (45). A great contribution to GS membranes can be attributed to Thomas Graham, who proposed the concept of solution-diffusion being the gas transport mechanism for the first time, conducted the first membrane GS to separate oxygen from air, and made the first effort to prepare composite membrane materials as well as to design experiments on hydrogen permeation across inorganic membranes in the 1860s (46, 47). In the 1930s and 1940s, R.M. Barrer also conducted many important studies related to gas permeation (48, 49), and he was later acknowledged by defining the units of the permeability coefficient widely used today as: 1 Barrer = 10^{-10} cm^3(STP) cm/(cm^2 s cmHg) and the corresponding SI unit is mol m/m^2 s Pa (1 Barrer = 0.33×10^{-15} mol m/m^2 s Pa).

It was not until 1945, that the first large-scale application of microporous membranes in GS was initiated by using inorganic membranes to separate uranium isotopes for military purpose and nuclear applications (50). Even though the selectivity of nonporous polymeric membranes is much higher than that of microporous membranes, the nonporous polymeric GS membranes only became economically competitive in the 1970s. Unsatisfactory low fluxes resulting from the membrane thickness had limited their commercial use in the early days. This issue was then overcome with the invention of integrally skinned asymmetric membranes made by Loeb and Sourirajan in 1962. For GS purposes, this membrane invention was modified while maintaining the integrity between the thin dense selective layer and the porous layer. The asymmetric structure was first made into flat sheets and eventually developed into hollow fibers (51). The Loeb-Sourirajan technique not only laid a foundation for the development of RO, NF, UF, MF membranes for liquid separation, but also paved the way for the commercialization of GS membranes.

In the 1980s, membrane technology in GS made a large stride forward. The revolutionary composite membranes were invented by Henis and Tripodi who coated a thin layer of silicone-rubber on the asymmetric substrate to achieve high hydrogen/carbon monoxide selectivity with a low cost module system (52). This major breakthrough led to the success of the first large-scale membrane GS system, Prism®, which was installed by the Monsanto Co. in 1980 for hydrogen recovery from the purge gas in an ammonia plant (50). Monsanto's success has stimulated intensive research in novel membranes and separation processes. Membrane engineers in Permea introduced Lewis acid:base complex dopes to fabricate asymmetric hollow fibers with a graded density skin (53, 54). Dow produced systems to separate nitrogen from air, and Cynara and Separex developed systems to separate carbon dioxide from natural gas. In 1990, Pinnau and Koros invented defect-free ultra-thin asymmetric membranes by a dry/wet phase inversion method, which enhanced gas permeation through the membrane tremendously to meet the need for large-scale industrial applications (55). Later intensive studies mainly focused on improving the fabrication of high performance asymmetric hollow fibers. One of the examples is that of Chung et al. who had successfully modified Permea's technology to produce 6FDA-polyimide membranes for air separation (56).

A simple GS membrane process setup is not significantly different from that used for solid–liquid separations except for the different phases in the streams. A gas mixture is fed into the membrane module at high pressure. One component diffuses faster through the membrane and is enriched in the permeate stream, the rest of the gas is concentrated in the residue stream. For a more complete separation, recycling of some of the permeate or residue stream may be required.

2.7. *Pervaporation*

PV is a relatively new membrane process used for the separation of liquid mixtures. In PV, a liquid mixture is brought into contact with one side of a membrane and the permeate is removed as low pressure vapor from the other side (Fig. 1.9). Transport through the membrane is induced by the vapor pressure difference between the feed solution and the permeate vapor (57). Separation is achieved based on the sorption and diffusion differences between the feed components, which are mainly controlled by the complex interactions between the feed components, the membrane materials, and the permeate. PV has certain common elements with both RO and GS. Nevertheless, it is different from RO, UF, or MF, as PV involves a change of the permeating species from liquid to vapor phase, and the driving force of this process is provided by lowering the chemical potential of the permeate stream.

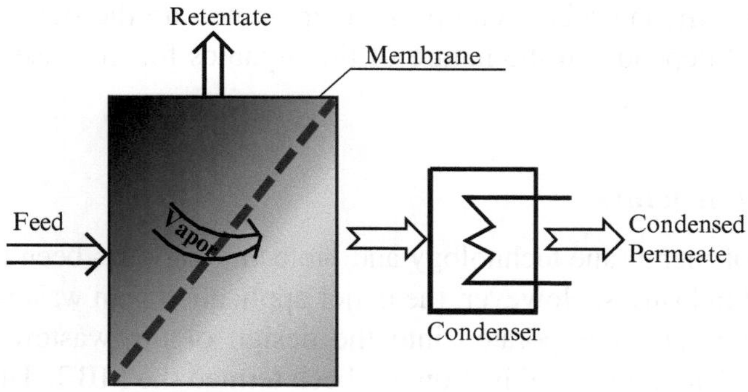

Fig. 1.9. Schematics of pervaporation process.

As early as 1906, L. Kahlenberg reported a qualitative study on the separation of a mixture of a hydrocarbon and an alcohol through a rubber membrane (58). In 1917, Kober introduced the term "PV" in his report about the selective permeation of water from aqueous solutions of albumin and toluene through collodion (cellulose nitrate) films (59). Between 1958 and 1962, Binning and coworkers at American Oil established the principles of PV and proposed the potential of this technology for the separation of a liquid-liquid mixture into a vapor mixture using a nonporous polymeric film (60). However, the low permeation flow rate through homogenous dense films hindered the large-scale industrial applications of PV. Commercialization of PV systems was not economically viable until the fabrication techniques of asymmetric membranes and composite membranes were invented. The real breakthrough was achieved in Gesellchaft fur Trenntechnik mbH (GFT) of Germany by developing a polyvinyl alcohol and polyacrylonitrile composite membrane to dehydrate alcohol solutions and the first pilot plant was installed in 1982 (61), which heralded a surge of commercial applications of PV for dehydration of alcohol and other solvents. The other important application of PV is the removal of small amounts of organic compounds from contaminated water. These two main applications of PV take advantage of the different polarities of water and organic solvents to achieve easy and economical separations. In addition, the utilization of this process were extended to other industrial fields in the following 10 years, including pollution control, solvent recovery and organic–organic separations (62).

Compared with the traditional distillation process, where the vapor–liquid equilibrium determines the separation characteristic, PV provides a potentially more efficient approach to separate a given mixture. The energy consumption of PV is mainly involved in the vaporization of the permeate stream so it is attractive when the concentration of the permeating species is low. In general, PV systems typically use plate-and-frame modules and spiral-wound modules which consist of flat membranes made from glassy, rubbery or ionic polymers (63). Most PV membranes are composites formed by solution-coating the selective layer onto a microporous support. For dehydration of organic solvents, several excellent membrane materials are available such as crosslinked poly(vinyl alcohol) coated on a microporous polyacrylonitrile support, Chitosan and Nafion. For separating volatile organic compounds

(VOCs) from water, silicone rubber composite membranes are the state-of-the-art material. The membrane used depends on the nature of the organics for the case of organic/organic separation (57).

2.8. Membrane Bioreactors

The integration of membrane technology and biotechnology has been applied in the food and pharmaceutical industries. However, the major application is in wastewater treatment. In this case, membranes are incorporated into the design of the wastewater bioreactors to separate solids/liquid, and this combination has been termed the MBR. The first descriptions of this technique dated from the late 1960s. A bench-scale membrane separation system linked with an activated sludge process was reported, in which a UF membrane module was placed externally to an aerated biological treatment basin in place of the gravity clarifier for the separation of mixed liquor (64, 65) (Fig. 1.10a). At around the same time, the first MBRs were developed commercially by Dorr–Oliver (66). Later in that period, similar external MBRs were used in fermentation of ethanol or acetone/butanol from low-grade food processing waste, such as cheese whey (67) and, in producing protein antibodies by genetically engineered mammalian cells (68, 69). This "first generation" MBR system has benefits in its continuous operation, the high cell densities achievable and easy control of the flux of different molecular weight molecules. However, the power costs associated with the operation of the external membrane MBR system limited its application to smaller wastewater flows.

In the 1980s, the Japanese government invested in the development of MBR with a low footprint that would be suitable for water recycling. In 1989, Yamamoto (70) demonstrated the application of a vacuum-driven membrane immersed directly in the active sludge for solid-liquid separation, which was later termed the submerged MBR (Fig. 1.10b). In the submerged MBRs, the membranes are self-supported (no pressure vessel) and the modules can be fabricated simply and economically. The submerged MBR operates at a much lower TMP and at a lower liquid cross-flow velocity. This has paved the way towards a significant reduction in capital and operating costs. The Kubota plate and frame submerged membrane system, developed as a result of the Japanese Government initiative, has revolutionized wastewater treatment in Japan. The other type of submerged membrane system uses hollow fibers in bundles or curtains, either horizontally or vertically aligned. More details of submerged membrane systems can be found elsewhere (71).

In MBRs, solids can accumulate at the membrane surface; thus, a back transport of solids is necessary to sustain membrane permeability. In the external MBR, solids back transport is provided by cross flow produced through recycling the mixed liquor at high flow rates. In the submerged MBR, coarse bubble aeration is used to prevent solids accumulation. By using the MBR, the conventional multi-stage activated sludge process, which includes primary settlement, secondary biological treatment and possible tertiary disinfection/polishing, can be simplified to a single filtration stage. The main advantages of MBRs lie in their high-quality effluents (BOD/TSS <5 ppm), a very low excess sludge production, high biodegradation efficiency and a relatively small footprint (72). However, membrane fouling and marginally

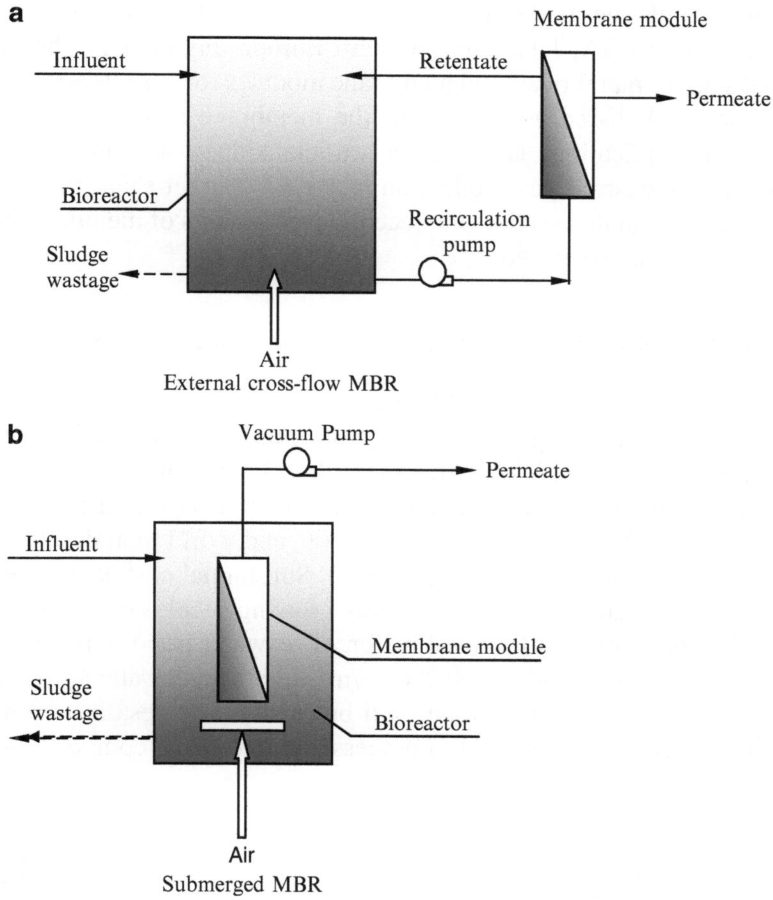

Fig. 1.10. Schematics of MBR systems.

higher capital and operating costs are considered as the disadvantages of MBRs compared to other biological processes (73).

3. CURRENT STATUS OF MEMBRANE TECHNOLOGY

Membrane technology has come a long way since its origins in the 1960s and is now experiencing a blossoming initiated in the early 1990s. Membranes are widely accepted as the best available technology for water and wastewater treatment. Rising concerns for increasing population growth and scarcity of freshwater resources, as well as more stringent water quality regulations are the major reasons for the adoption of membrane technology. Membranes are also used in application areas such as chemical and pharmaceutical processing, energy production, environmental monitoring and quality control, as well as food and beverage processing. Besides this, membranes are key components in fuel cells and bioseparation systems.

The worldwide membrane market grows at nearly 8–10% per year and 75% of the total market share belongs to USA, Japan and Western Europe. Figure 1.11 shows the statistic results on annual sales of membranes and membrane modules for various applications (72). In 1998, MF, UF, and RO shared up to 41% of the membrane market value for water and wastewater treatment applications. Dialysis (the artificial kidney) accounted for another 43%. GS for chemical and industrial gas production had a 5% market share but had the highest growth rate of 15% (74). In this section, the recent developments of membrane technology in several important fields are reviewed.

3.1. *RO for Seawater and Brackish Water Desalination and Water Reclamation*

3.1.1. *Seawater and Brackish Water Desalination*

RO process is currently the fastest growing and the most promising technique for seawater and brackish water desalination. Worldwide more than 15,000 industrial scale desalination plants had been installed by the year 2002, and RO made up to 44% of the total capacity (75). This has increased to levels approaching 60% in 2006, and world total desalination capacity presently is estimated as 10–30 million m^3/day (76). Substantial cost decline of desalination by RO process has contributed much to the rapidly growing success of RO technology. Over the last decade, the operating cost for desalination of seawater has dropped from USD1.5 to USD0.5–0.8/m^3; the cost is around USD0.2–0.35/m^3 for brackish water (77, 78) (Table 1.5).

The other parameters of RO in seawater and brackish water desalination are also compared in Table 1.5. Applying one stage RO process and interfacial composite membranes in

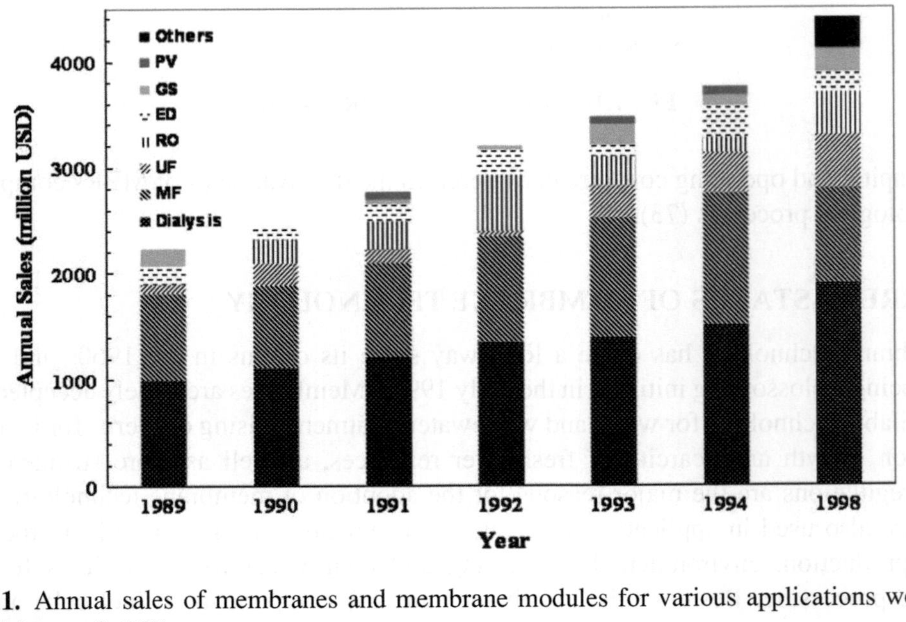

Fig. 1.11. Annual sales of membranes and membrane modules for various applications worldwide (adapted from ref. (72)).

Table 1.5
Comparison between recent seawater and brackish water desalination using RO (adapted from refs. (77, 78))

Parameter	Seawater	Brackish water
Operating pressure (psi)	800–1000	150–400
Solution concentration (%)	1–5	0.2–0.5
One-stage NaCl rejection (%)	>99.3	96–98
Target salt concentration (ppm)	<500	200–500
Membrane used	Interfacial composite membranes	CA membranes/composite membranes
Cost (per m^3)	USD 0.5–0.8	USD 0.2–0.35

Table 1.6
Advances in spiral-wound module RO performance (adapted from ref. (78))

Year	Cost normalized (1980 USD)	Productivity normalized (to 1980)	Reciprocal salt Passage normalized (to 1980)
1980	1.00	1.00	1.00
1985	0.65	1.10	1.56
1990	0.34	1.32	2.01
1995	0.19	1.66	3.52
2000	0.14	1.94	7.04

seawater desalination plants can achieve a salt rejection greater than 99.3%. For brackish water desalination, membranes with NaCl rejection greater than 95% are typically considered adequate. Thus, CA membranes with NaCl rejection of 96–98% can be applied despite some disadvantages (see Table 1.2). Among all available module configurations, the spiral-wound module is preferably used in seawater desalination plants because of its good balance between ease of operation, fouling control, permeation rate, packing density, as well as initial, operating and replacement costs (79). A typical spiral-wound module with a dimension of 8-in.-diameter, 40-in-long is the most commonly used for RO to produce permeate of 30–40 kL/day (8,000–10,000 gallon/day) (78). Table 1.6 shows the improved performance and costs of spiral-wound modules over the period 1980–2000. Since the introduction of the first interfacial composite membranes in 1980, the cost of spiral-wound membrane modules on a per square meter basis has decreased sevenfold till year 2000. At the same time the water flux has doubled while the salt permeability has decreased sevenfold. Today, the RO membranes are almost 100 times better than in the early 1980s. This type of incremental improvement is likely to continue for some time (78).

Compared to other desalination methods, RO requires much less energy consumption which ensures RO a good market worldwide. For example, for large-scale desalination plants with a same capacity of 32 MLD, the specific energy consumption of RO was reported as

6.7–8 kWh/m^3, which is much lower than that of multi-stage flush distillation (51.7 kWh/m^3) and multi-effect distillation (45 kWh/m^3) (80). The energy consumption keeps decreasing to 3–5 kWh/m^3 by installation of energy recovery systems (81), and by the application of high rejection and high flow membranes to RO plants (82). As recently reported, the specific energy consumption can even drop down to as low as 1.6 kWh/m^3 in large-scale seawater RO plants with energy recovery devices (83). Whereas, the energy consumption of small-scale RO desalination plants (<100 kL/day) without energy recovery devices is normally above 10 kWh/m^3 and special pumps are required for energy saving (84).

In practice, due to the complexity of the feed water and the membranes' pronounced sensitivity to their environment, a careful pretreatment system is required to protect RO membranes from contamination and fouling with colloids, humic substances, microorganisms and heavy metals, and thus, extend the RO lifetime. MF and UF are being introduced, as a replacement for conventional pretreatment processes, such as coagulation, adsorption, sedimentation, flotation, sand filtration, disinfection and antiscalant addition (85). With the rapid development of membrane technology in the last decade, pretreatment with membrane processes has shown great advantages over conventional pretreatment processes. For instance, UF pretreatment can be 20% more aggressive in terms of flux. In addition, both the capital cost and the operating cost of RO system can be reduced by UF pretreatment in either seawater or brackish water desalination (86).

Table 1.7 gives a list of large-scale seawater RO desalination (SWRO) plants constructed in the last 10 years up to 2004 (87). The first large-scale SWRO plant with 95 MLD capacity in the U.S. began operation in Tampa, Florida, in 2003. Singapore's first SWRO plant at Tuas, which is the largest of its kind in Asia, commenced operation in 2005 at a capacity of 110 MLD. It ranks among the most energy efficient ever constructed, enabling it to achieve the lowest desalinated seawater price in the world. Completed in 2005, Ashkelon SWRO plant, built by IDE (Israel) and VEOLIA (France), is the world's largest and most advanced desalination plant with a production capacity of 320 MLD of drinking water. It received the prestigious Global Water Awards 2006 organized by Global Water Intelligence (http://www.water-technology.net/projects/).

3.1.2. Contaminated Aquifer Water and Municipal Wastewater Reclamations

During the last decade, water reclamation and reuse have increased in popularity as more communities and industries have realized that water reclamation and reuse are effective approaches conserving limited, high-quality freshwater supplies. Water from underground aquifers contaminated with industrial or agricultural drainage and municipal wastewater can be further treated to produce water with higher quality that can meet the criteria for safe recharge to natural water bodies, indirect potable reuse or for industrial purposes such as boiler make-up and process water applications. RO is the key treatment process used in these areas.

RO processing of contaminated aquifers is usually similar to the conventional brackish RO application. Since this type of raw water usually contains low concentrations of suspended solids and low levels of biological activity, it requires a relatively simple pretreatment process. For instance, the Arlington Desalter, which is located in Riverside County, California, and has started operation since 1990, processes agricultural drainage water of about

Table 1.7
Large-scale RO seawater desalination plants in the last 10 years (adapted from ref. (87))

Plant name/location	Capacity MLD (MGD)	In operation since
Tianjin (Dagang), China	98 (26)	2006
Kwinana (Perth), Australia	125 (33)	2006
Hamma, Algeria	200 (53)	2006
Cartagena, Mauricia, Spain	64 (17)	2004
Tuas, Singapore	136 (36)	2005
Ashkelon, Israel	394 (104)	2004
Fujairah, UAE	170 (45)	2003
Tampa Bay, USA	95 (25)	2003
Alikante, Spain	53 (14)	2003
Carboneras, Almeria, Spain	121 (32)	2003
Point Lisas, Trinidad	114 (30)	2002
Las Palmas, Telde	38 (10)	2002
Larnaca, Cyprus	53 (14)	2001
Al Jubail III, Saudi Arabia	91 (24)	2000
Muricia, Spain	64 (17)	1999
The Bay of Palma, Palma de Mallorca	64 (17)	1998
Dhekelia, Cyprus	42 (11)	1997
Marbella, Mallaga,	57 (15)	1997
Okinawa, Japan	42 (11)	1996

1,000 ppm total dissolved solids (TDS) salinity, and produces up to 6 MGD (23 MLD) of desalinized water by blending 4 MGD (15 MLD) of RO permeate with 2 MGD (7.5 MLD) of ground water (88) (http://www.sawpa.org/arlington/).

The application to municipal wastewater treatment requires extensive pretreatment prior to the RO process. A typical process for municipal wastewater consists of primary, secondary and tertiary treatments. This process produces an effluent with low turbidity that can be disinfected for discharge. However, the effluent still contains dissolved solids such as colloids, organic matter and bacterial activity. As a result, the water is not generally suitable for reuse. RO membranes are able to produce water with a quality much higher than tertiary effluent, but early attempts to treat this water with RO membranes were hindered by rapid fouling caused by the dissolved solids. Frequent membrane cleaning was thus required and it not only shortened membrane life, but also greatly increased operating costs. Therefore, a series of pretreatments are performed before the effluent enters the RO units. Until recently these pretreatment steps were chemical and physical. An example of such a system is the RO plant known as "Water Factory 21" in Orange County, California. The influent was received from the Sanitation District and passed through the treatment stages of lime clarification, recarbonation, chlorination and media filtration before it was directed to the RO system (89) (Fig. 1.12).

These conventional pretreatment processes are relatively complicated and costly. Recently, a new pretreatment technology consisting of backwashable MF and UF membrane elements has been introduced into the RO processing of municipal effluents. In this integrated "dual-membrane" process, an RO membrane system is preceded by a low-pressure MF or UF membrane for removal of those solids that may otherwise foul the RO membranes to ensure more reliable and efficient operation of the RO systems. The utilization of MF pretreatment successfully reduces the RO cleanings from once per month in a typical plant to no cleanings in over 1.5 years with an excellent permeate quality produced (86). This "dual membrane" process is now being used at "Water Factory 21". Other examples for municipal wastewater reclamation include the Singapore "NeWater" plants which has been installed at Kranji and Bedok since the end of 2002. Bedok plant applies MF in their first phase and submersible UF in their second phase, while Kranji plant takes use of submersible MF as the pretreatment of RO (90).

Numerous large-scale commercial membrane plants are now being used to reclaim municipal wastewater. Table 1.8 gives a list of some wastewater reclamation plants that repurify secondary or tertiary effluent with MF/UF and RO (91). As an example, the configuration of the NEWater plants in Singapore is shown schematically in Fig. 1.13. Figure 1.13 illustrates the "dual membrane" process and shows its simplicity compared with conventional pretreatment (Fig. 1.12). These plants utilize specifically developed low fouling composite membranes that have low affinity for dissolved organics. The reduced fouling rates lead to lower energy consumption, lower chemical use, longer membrane life, less maintenance labor and greater system online time. The cumulative effect of these benefits makes the

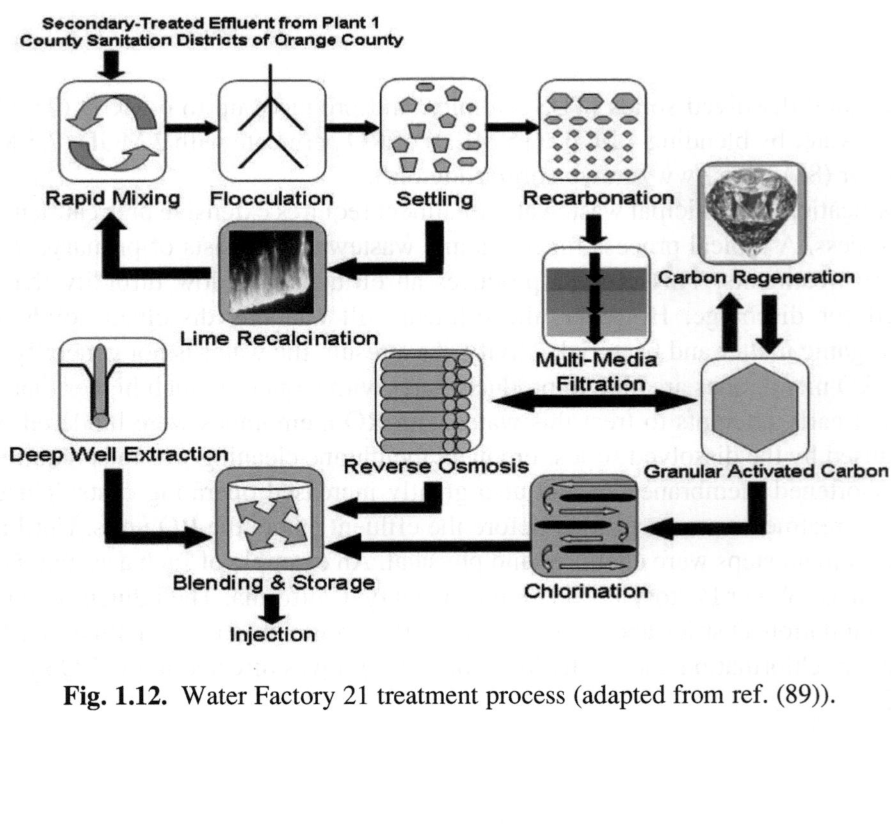

Fig. 1.12. Water Factory 21 treatment process (adapted from ref. (89)).

Table 1.8
Wastewater repurification plants (adapted from ref. (91))

Plant	Location	Capacity MLD (MGD)	MF/RO feed source	Product use	Start date
West Basin, CA WRP – Phase 2	California	9.5 (2.5)	Secondary	Barrier	1997
West Basin Mobil Oil Refinery	California	12 (3.2)	Tertiary	Industrial	1998
Scottsdale WC	Arizona	70 (18.5)	Tertiary	Indirect potable	1998/1999
West Basin, CA WRP – Phase 3	California	19 (5.0)	Tertiary	Industrial	2001
West Basin, CA Carson RWRP	California	19 (5.0)	Secondary	Industrial	2002
Bedok Water Reclamation Plant	Singapore	43 (11.4)	Secondary	Indirect Potable/Indus	2002
Kranji Water Reclamation Plant	Singapore	38 (10.0)	Secondary	Indirect potable/Indus	2002
Honoluilui Reclamation Plant	Hawaii	7.6 (2.0)	Secondary	Industrial	2000
Orange County WF21 – GWRS	California	265 (70)	Secondary	Indirect potable	2007

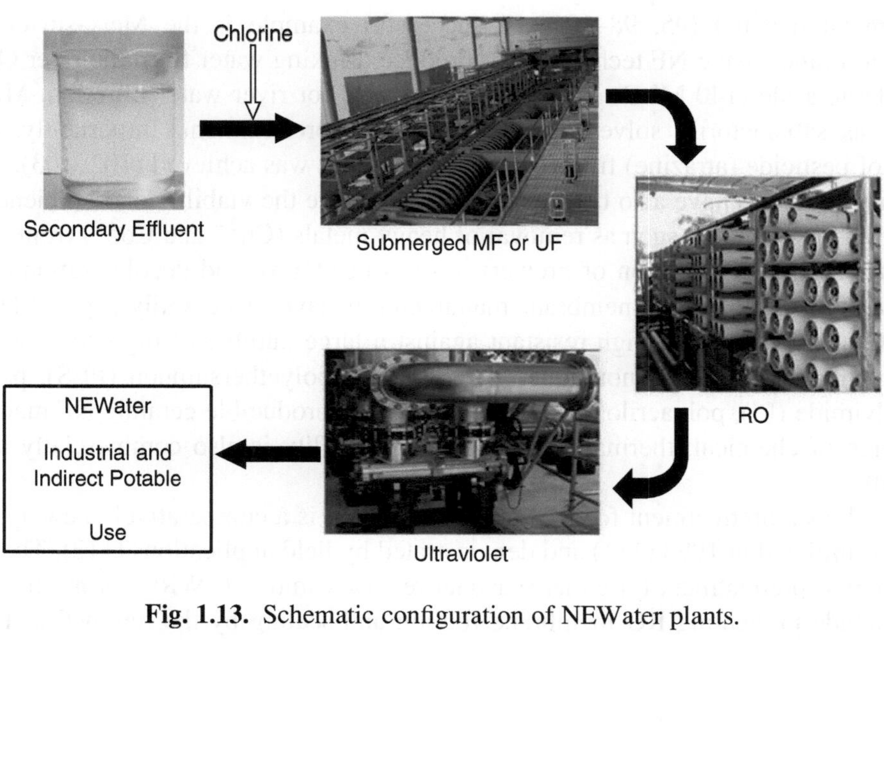

Fig. 1.13. Schematic configuration of NEWater plants.

overall system less expensive despite the UF/MF equipments being more expensive than the conventional pretreatments. By far the largest facility in the world to use RO and UF membrane-based water purification is the Sulaibiya Wastewater Treatment and Reclamation Plant completed in November 2004. With an initial capacity of up to 375 MLD and designed for extension to 600 MLD in the future, the Sulaibiya plant treats wastewater to potable quality for non-potable uses in agriculture, industry and aquifer recharge. Other larger plants include the 270 MLD "Water Factory" plant in Orange County, California, USA, and the 170 MLD Ulu Pandan plant in Singapore. It is evident that RO technology has gained acceptance worldwide (http://www.water-technology.net/projects/).

3.2. *Applications of NF, UF and MF Membranes*

3.2.1. NF Membranes

Although the early applications of NF were in the dairy industry to separate lactose and salts, it was soon recognized to have a role in water. At first the applications of NF were limited to the softening of drinking water and the removal of pesticides from surface water. Nowadays, NF membranes have found broad applications well beyond their roots as "loose RO" membranes. The development of new generations of NF membranes and the expansion of operation knowledge are attributed to this growth. A comprehensive book on NF is now available (32).

Water softening is still one of the major purposes of current NF (92–94), whereas new trends for removing nitrates (95, 96) fluoride and aluminum (97) are also in pilot scale investigation. Utilization of NF membranes for natural organic matter (NOM) and pesticides removal from both surface water and ground water has attracted increasing interest during the last 15 years, due to their capability to achieve more than 90% rejection, high recovery with low chemical demand (95, 98–101). A successful example is the Mery-sur-Oise plant installed in France using NF technology to produce drinking water from the river Oise. It is the first large-scale (140 MLD) NF plant in the world for river water filtration. Membrane fouling was satisfactorily solved with an integrated pretreatment. Importantly, efficient removal of pesticide (atrazine) from 1 to below 0.1 μg/L was achieved (102, 103).

Recently, interests have also been extended to evaluate the viability and efficiency of NF for wastewater treatment, such as removal of heavy metals (Cu^{2+} and Cd^{2+}) from synthetic wastewater (104), regeneration of brewery wastewater (105), and decolorization of textile wastewater (106). Moreover, membrane manufacturers have successfully exploited thin-film composite membranes with high resistant against a large number of organic solvents. The specialty polymers used for non-aqueous NF can be polyethersulphon (PES), polyamide (PA), polyimide (PI), polyacrilonitrile (PAN) (107). Reproducible ceramic NF made with a high degree of chemical, thermal and mechanical stability is also commercially available (108–110).

Using NF as a pretreatment for seawater desalination is a comparatively new application which was initiated in 1998 (111) and demonstrated by field applications (112). The benefits of using NF as pretreatment for either seawater reverse osmosis (SWRO) or multistage flash (MSF) include preventing RO membrane fouling and scaling by the removal of turbidity, bacteria and scale forming hardness ions, and lowering the required operating pressure by reducing the seawater feed TDS by 30–60%, depending on the type of NF membrane and operating conditions (113).

At the same time, NF membrane fouling due to the concurrence of NOM and calcium ions has also been actively studied to control the flux decline and thus, to reduce the additional costs for membrane cleaning and replacement (114–119). The developments in NF technology raise the possibility of a one-step treatment for both organic and inorganic compounds from water, and thus replacing complex multi-step treatment processes.

3.2.2. MF and UF Membranes for Water Purifications

The low-pressure membranes including MF and UF are primarily used for turbidity, bacteria and protozoa removal as stand-alone processes, retrofit of existing conventional treatment plants, or as pretreatment to advanced processes such as NF and RO. Like all other technologies, MF and UF membranes have evolved from an R&D phase through relatively high cost initial applications to widespread utilizations.

Over the past 10 years, MF and UF membrane technologies have gained considerable acceptance in the drinking water industry. Addressing chlorine-resistant pathogens while controlling disinfection by-product formation is a main regulatory objective and challenge faced by the water industry. MF and UF membrane technologies have consistently proven effective for the removal of larger pathogens such as Cryptosporidium and Giardia in surface water (120, 121). Virus removal can also be achieved by UF membranes and many MF/UF filtration plants have been developed for this purpose. The types of water sources treated include reservoirs, lakes, rivers, surface water impoundments and aquifers under the influence of surface water. The manufacturers for installation and MF/UF membrane production include Siemens-Memcor, GE-Zenon, Aquasource, Pall, Koch, Ionics, Hydranautics and Norit. All commercially available MF and UF membranes currently used for drinking water treatment are constructed in hollow fiber configuration (122).

The early market for the low-pressure membranes was dominated by small-scale facilities with a treatment capacity less than 2 MLD. As the use of membranes increased, the number of larger installations has also increased. An example is the MF plant located in the City of Kenosha, Wisconsin, that began operation in 1998 at a capacity of 53 MLD. Appleton Water Treatment Plant in Wisconsin, which commenced operation in 2001, is another example that employs UF membranes to meet challenges posed by its raw water supply. Most recently, Asahi Kasei Chemicals announced that its Microza™ hollow-fiber membrane filtration system has been selected for a new water treatment facility located in Fridley, Minnesota, with a full-capacity of 360 MLD operation scheduled to begin in January 2011. This will make it one of the largest drinking water plants in the US, and among the largest plants anywhere to utilize a pressurized membrane filtration system for water purification (http://www.asahi-kasei.co.jp/asahi/en/news/2006/e070117.html). Today, more than 3000 MLD of drinking water is being produced worldwide using low-pressure membranes, and UF applications represent about 74% of the total capacity (123). Furthermore, MF membrane costs have dropped more than 80% since the 1990s as a result of increased competition, better system design, and lower manufacturing costs (123).

However, the application of MF/UF for NOM removal has been hindered due to serious membrane fouling by NOM and organic aggregation in the presence of calcium (124, 125). NOMs consist of a broad spectrum of molecular weights and size distributions, functional groups and sub-structures, and have been shown to be the major foulant in the membrane filtration of natural waters (126, 127). Substantial removal of NOM requires NF or low pressure MF/UF coupled with coagulants or adsorbents such as powdered activated carbon (PAC), which are the common practice in water and wastewater treatments (128–130).

Trace organics in surface or ground water, such as pesticides, herbicides, as well as taste-and-odor causing materials, are not effectively removed by MF or UF processes because the larger membrane pore size cannot retain the low molecular weight materials. Removal of these hazardous contaminants from drinking water can be fulfilled by hybrid systems combining either PAC adsorption or photocatalytic oxidation with membrane processes. The trace organics are effectively separated from water by PAC (131, 132) or degraded by oxidation using UV irradiation and TiO_2 photocatalyst (133, 134). The fine PAC or TiO_2 particles are efficiently retained by MF or UF membranes.

In many cases one membrane process is followed by another with the purpose of enhancing the function of the other to meet goals ranging from disposal of wastewater to production of drinking water from various sources. The application of MF or UF as a pretreatment for the RO process is one example. This type of membrane application has been described in Sect.3.1.2.

3.3. MBRs for Wastewater Treatments

MBR technology for wastewater treatment has experienced rapid development from the early 1990s onwards. It has emerged as an effective solution to transform various wastewaters into high quality effluents suitable for discharge or for reclamation by RO (135). Today, the MBR market is predominated by submerged membrane modules with configurations of the flat-sheet (or plate & frame) and hollow fibers, such as the Kubota Submerged Membrane Unit, and GE-Zenon's ZeeWeed®, ZenoGem® systems. The main suppliers of MBR systems for wastewater treatment are Kubota (Japan), GE-Zenon (USA), Siemens-Memcor (USA) and Mitsubishi Rayon (Japan). Other suppliers include Degremont (France), Wehrle Werk (Germany), Hans Huber (Germany), Orelis Mitsui (Japan) and Koch (USA).

In Japan, since the first wastewater treatment plant with Kubota submerged MBR was installed in 1990, there have been total 1,149 Kubota submerged membrane units applied in wastewater treatment plants by 2003. The markets in South-East Asia and Japan are dominated by the Kubota and Mitsubishi Royon. In the US, the first large-scale external MBR system for treatment of industrial wastewater was constructed in 1991. Zenon Environmental introduced the first ZenoGem® MBR process in the early 1990s. The first large-scale internal MBR system for treatment of industrial wastewater was installed in 1998. The MBR market in North America is currently dominated by Zenon. In Europe, The first pilot-scale submerged MBR plant for municipal wastewater treatment was built at Kingston Seymour, UK in 1996, followed by the construction of full-scale wastewater treatment plants at Porlock, UK in 1998, Büchel and Rödingen, Germany in 1999, and Perthes-en-Gâtinais, France in 1999. The largest MBR plant worldwide was commissioned with a capacity of 50 MLD in Kaarst, Germany in 2004. By 2006, around 100 municipal full-scale plants with a capacity greater than 500 P.E. or population equivalent were in operation in Europe, and there were around 250 large industrial plants with a capacity greater than 20 m^3/d (0.02 MLD) (http://www.mbr-network.eu/mbr-projects/index.php).

3.4. *Gas Separation*

Since Monsanto launched its first hydrogen separation Prism® system in 1980, membrane-based GS has grown into a $150 million/year business in 2000 (136). Current applications of GS membranes can be categorized into two groups (1) large scale applications including enrichment of oxygen and enrichment of nitrogen, recovery of hydrogen from various process streams in ammonia production and syngas, and removal of carbon dioxide from natural gas; (2) emerging large scale separations such as olefin-paraffin separations and natural gas dehydration. Table 1.9 shows the industrial applications of GS membranes (137) and the main players in 2002 and their products are listed in Table 1.10 (136).

Oxygen or oxygen-enriched air has a great demand in medical applications and other processes such as in refineries or in combustion systems. For example, oxygen or oxygen-enriched air can reduce energy consumption of furnace operators by 25–60% and reduce environmental emissions. Lowering the cost of oxygen would dramatically improve manufacturing processes, resulting in less pollution, higher productivity and lower production costs. Thus, many world-leading gas companies have conducted intensive R&D on novel membranes that preferentially permeate oxygen with a high flux. Nitrogen enrichment using polymeric membranes has also gained a promising position in the field today followed by the success of economically obtaining nitrogen with purity up to 99.5% (138). Nitrogen

Table 1.9
Industrial applications of gas separation membranes (adapted from ref. (137))

Gas separation	Applications
O_2/N_2	Nitrogen generation, oxygen enrichment
H_2/hydrocarbons	Refinery, hydrogen recovery
H_2/CO	Syngas ratio adjustment
H_2/N_2	Ammonia purge gas
CO_2/hydrocarbons	Acid gas treating, enhanced oil recovery, landfill gas upgrading
H_2S/hydrocarbons	Sour gas treating
H_2O/hydrocarbons	Natural gas dehydration
H_2O/air	Air dehydration
Hydrocarbons/air	Pollution control, hydrocarbon recovery
Hydrocarbons from process streams	Organic solvent recovery, monomer recovery

Table 1.10
Major gas separation membrane producer and membrane systems (adapted from ref. (136))

Producer	Major membrane material used	Module type
Permea (Air products)	Polysulfone	Hollow fiber
Medal (Air Liquid)	Polyimide/polyaramide	Hollow fiber
Generon (MG Industries)	Tetrabromo polycarbonate	Hollow fiber
IMS (Praxair)	Polyimide	Hollow fiber
Kvaerner	Cellulose acetate	Spiral wound
Separex (UOP)	Cellulose acetate	Spiral wound
Cynara (Natco)	Cellulose acetate	Hollow fiber
Parker Hannifin	Polyimide	Hollow fiber
Ube	Polyimide	Hollow fiber
GKSS licensees	Silicone rubber	Plate and frame
MTR	Silicone rubber	Spiral wound

membranes are widely commercialized in many inert gas applications. Examples include the OBIGGS (On Board Inert Gas Generating Systems) in aerospace, purging and blanketing in chemical processing, chemical tanker inerting in marine systems, as well as in packaging of food and beverages.

Separation of hydrogen is an ideal application for membranes as hydrogen is a fast gas with a high selectivity over other gases. Membrane separation technology has been widely used in hydrogen recovery from purge gas streams in ammonia plants. A similar application is hydrogen/carbon monoxide ratio adjustment in syngas plants to meet the requirement of carbonylation processes. Several hundred hydrogen separation plants to treat these gas streams have been installed. In addition, the demand for hydrogen in refineries is increasing because of the increased environmental regulation and heavier crude oils. Hydrogen recovery from refinery fuel gas streams, PSA tail gas, and hydrocracker/hydrotreater off-gas, which contains 30–80% hydrogen mixed with light hydrocarbons (C1–C5), is another large application of hydrogen-permeable membranes (136).

Natural gas treatment is one of the presently growing application areas of membranes. Except for the major component of methane, which is typically 75–90% of the total, natural gas also contains undesirable impurities such as water, carbon dioxide, nitrogen, and hydrogen sulfide. Removal of these undesirable impurities to meet the gas pipeline specifications is the main purpose of natural gas treatment, and the total worldwide market for new natural GS equipment is probably $5 billion/year, making it by far the largest application of industrial GS. Currently, membrane processes have less than 1% of this market, almost all for the removal of carbon dioxide (136). The main opportunities and challenges for membrane technology are reducing the high cost of the membrane process and minimizing the main product (methane) loss to make membrane process more competitive with the conventional processes.

For vapor/gas mixtures, rubbery polymers are commonly used for condensable vapor separation, while the glassy polymers for permeating the smaller gas. More recently, membranes have begun to be used to recover hydrocarbons and processing solvents from petrochemical purge gas with a recovery value up to \$1–2 million/year (136). Separation of vapor mixtures, such as ethylene/ethane, propylene/propane, *n*-butane/isobutene, and olefin/paraffin, is a potential market for membrane processes. Therefore, development of appropriate membrane materials is currently under investigation for commercialized applications in this area.

3.5. *PV and its Hybrid Systems*

The commercialization of PV process has been gradually realized on a large scale as an attractive alternative to other separation methods since the installation of the first PV pilot plant in 1982. The first pilot-plant results for the separation of methanol from methyl *tert*-butyl ether/isobutene mixtures was reported by Separex in 1988 (57). The first commercial plant for small amounts of VOCs removal from contaminated water was developed by Membrane Technology and Research in 1996.

Table 1.11 lists some industrial suppliers of PV systems, their main applications and membranes (62). The major applications of PV can be roughly classified into three categories. One is the dehydration of organic solutions; the second is the removal of small amount of organic components from a mixture; the third is the separation of organic mixtures. Among these, the dehydration of organic solutions, in particular, the dehydration of 90–95% ethanol solutions, a difficult separation problem because of the ethanol–water azeotrope at 95% ethanol, remains the most important application in the industry. Table 1.12 summarizes 63 PV separation systems installed by GFT and associates between year 1984 and 1996 (139).

Table 1.11
Industrial suppliers of pervaporation systems (adapted from ref. (62))

Company	Applications	Membranes/modules
GFT	Solvent dehydration	Corsslinked PVA composite membranes. Plate and frame modules
Lurgi	Solvent dehydration	GFT PVA membranes. Plate and frame modules
Membrane technology and research (MTR)	Solvent recovery	Composite membranes. Spiral wound modules
British Petroleum (Kalsep)	Solvent dehydration	Composite membranes based on ion exchange polymers. Tubular and plate and frame modules
Air products (Separex)	Organic-organic separation, methanol/MTBE	Cellulose acetate membranes. Spiral wound modules
Texaco	Solvent dehydration especially ethylene glycol, isopropyl alcohol, organic/organic separation	Their own composite membranes in GFT plate and frame modules

Table 1.12
Pervaporation membrane separation systems installed by GFT and associates between 1984 and 1996 (adapted from ref. (139))

For organic solvent dehydration (62 units)
Dehydration of alcohols forming azeotropes with water:
Ethanol (22 units): Germany, Ostrich, France, Italy, Hungray, Japan, Malaysia, Slovakia, Taiwan
Isopropanol (16 units): Germany, France, Ireland, Japan, The Netherlands, UAS
Multifunctional systems (12 units):
Germany, Canada, France, Great Britain, Japan, UAS
Dehydration of other solvents or complex reaction mixtures:
Esters (4 units): Germany, India
Ethers (4 units): France, Japan, Switzerland, UAS
Solvent mixtures (3 units): Germany, Japan
Triethylamine (1 unit): Japan
Extraction of volatile organic compounds (VOC) from water (1 unit)
Recovery and recycling of VOC
Tetrachloroethylene (1 unit): VOC recovery in a dry-cleaning unit

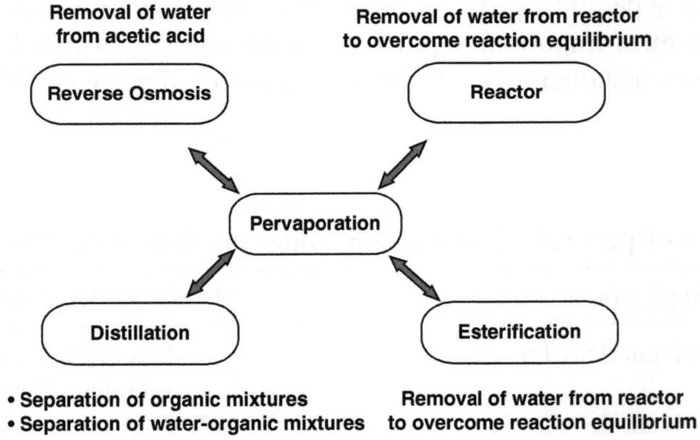

Fig. 1.14. Diagram of pervaporation-based hybrid systems and their applications (adapted from ref. (140)).

Although pevaporation has established itself as one of the most promising technologies for dehydration of organic compounds and separation of organic mixtures in recent years, a stand-alone process may not be economical in industrial applications due to the purity required in production and the limitation of flux (140). To overcome these drawbacks, an integrated process package combining PV and another separation process or a reactor shows a high potential in its low energy requirement and reduced use of additives. Existing and potential PV-based hybrid systems are shown in Fig. 1.14 (140), in which PV–distillation is the most commonly used hybrid system for both dehydration and organics separation. In the hybrid PV–distillation processes, PV performs a first, crude, low-energy and low-cost separation, leaving the polishing operation for distillation. These hybrid systems should establish a long-term market perspective for PV with their technological viability and economic benefits.

4. FUTURE PROSPECTS

Membrane technology has become an important and versatile process in our daily life. In the twenty-first century, the world is facing more severe challenges than ever in terms of meeting increasing demands for energy, providing adequate supplies of water in developed and developing countries, controlling the impact of global warming and protecting our environment. The demand for the use of membranes to address these issues in a variety of applications will continue to increase.

4.1. Membranes for Water, Food and Biopharmaceutical Industries

Water treatment is one of the great success stories of membrane technology. It will continue to play a significant role in resolving water issues in the world, especially in arid coastal regions and other areas wherever saline waters are available but good-quality fresh water is limited. RO, UF and MF membranes could indeed be the technology to take the drinking-water and sanitation industries into the twenty-first century. One emerging area is the development of decentralized system where the modulus nature of membranes is well suited. The challenge is to see whether these membrane technologies can offer a long-term solution to the water needs by supplying an inexpensive process and offering a high degree of reliability. Within pharmaceuticals and biotech, membrane filtration is also increasingly adopted as a choice for separating amino acids, enzymes, yeast, glycoproteins and biopesticides from proteins. Membranes will play a major role in the recovery, isolation and concentration of these proteins. As for MBRs, due to more stringent regulations and water reuse strategies, it is expected that a significant increase in MBR plant capacity and widening of application areas will occur in the future.

A detailed view of market size, growth and trends for the RO, UF, MF, MBR processes as well as for the membrane systems used in the food and biopharmaceutical processes has been provided by BCC Research in a series of reports on Membrane and Separation Technology. The followings are some key points (http://www.bccresearch.com/mst/):

- The global market for RO membranes and system components is expected to rise at an average annual growth rate of 10.3% from $1.9 billion in 2005 to more than $3 billion in 2010.
- The market for UF is rising at an average annual growth rate of 9.7% from $458 million in 2002 and reaching $727 million in 2007.
- Estimated at $792 million in 2005, the global market for MF membranes is expected to rise at an average annual growth rate of 9.4% to more than $1.2 billion.
- The global MBR market is rising at an average annual growth rate of 10.9% from $216.6 million in 2005 and is expected to approach $363 million in 2010.
- Membrane sales to food and beverage processors is forecast to rise at an average annual growth rate of about 4.6% during the next 5 years, from $185 million in 2006 to $230 million.

- Estimated at $740 million dollars in 2004, the market for membrane technology used in biopharmaceutical discovery, development and commercial production is expected to rise at an average annual growth rate of 10.7% to $1.23 billion in 2009.

Overall, the application of RO, UF, MF and MBR processes are growing at 10% or more per annum to a market of about $8 billion by 2010.

4.2. *Membranes for Refinery, Petrochemical and Natural Gas Industries*

With global goals of cleaner energy and shortening supply of fossil fuels, the market for novel energy efficient separation technologies is expected to see sustained growth over the foreseeable future, and membrane-based separation technologies are particularly well positioned for high demand. As shown in Table 1.13, The total membrane market for GS is expected to reach $350 million by 2010 and to double by 2020 to $760 million with an average growth rate of 7–8%/year in the refinery, petrochemical and natural gas industries (136).

Specifically, the development of improved membrane separation and purification technology for use in the production of hydrogen will be one of the focuses. Various membranes including ceramic ionic transport membranes, microporous membranes and palladium based membranes are expected to be used for a wide range of production capacities, from large central production facilities to small-distributed production units. Offshore natural gas processing is another area where membrane technology exhibits great potential. Conventional gas treatment technologies are increasingly unattractive for deep water operations or for remote sites where gas pipelining is uneconomical. Membrane-based processes could offer the benefits of operating at well-head conditions and without the introduction of additional components. The ongoing progress in membrane technology will also enable sequestration of carbon dioxide from the process of coal gasification for the reduction of greenhouse gases (141). Given the growing concerns about greenhouse gas emission, this application could become the most significant application of membrane technology. However, it presents major

Table 1.13
Future membrane market for gas separation (adapted from ref. (136))

Separation	Membrane market (USD million)		
	2000	2010	2020
Nitrogen from air	75	100	125
Oxygen from air	<1	10	30
Hydrogen	25	60	100
Natural gas	30	90	220
Vapor/nitrogen	10	30	60
Vapor/vapor	0	30	125
Other	10	30	100
Total	150	350	760

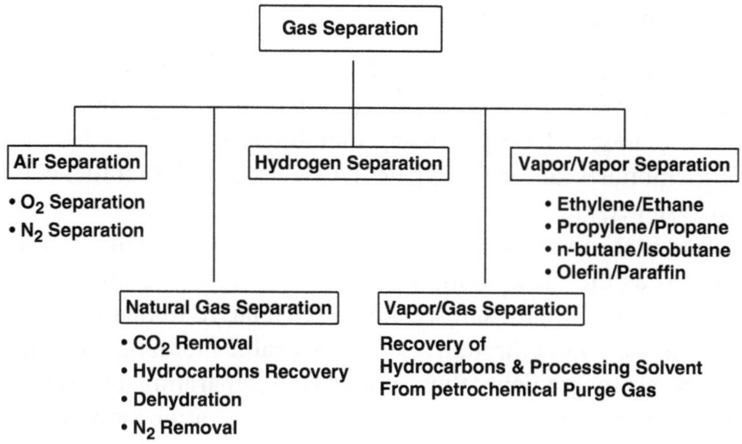

Fig. 1.15. Applications of gas separation membranes (adapted from ref. (136)).

challenges in scale-up, cost and energy minimization. Figure 1.15 summarizes various applications of GS membranes (136).

4.3. Challenges for the Membrane Industry

Despite the successes and advancements, there is a clear need for further improvements of both membranes and processes. Current applications of membrane technology for liquid separation and purification are almost entirely restricted to the processing of relatively benign water-based systems. Although an increasing variety of membranes are available for water production and wastewater treatment, there are very few membranes that are able to tolerate harsh conditions posed by oils, solvents and organics. A similar challenge also exists in the case of gas/vapor separations. The diversity and operating range of actual large-scale membrane-based separations remain limited, as the selectivity of current membrane materials cannot be maintained even under moderately aggressive feed conditions. These challenges must be overcome to ensure a sustainable industrial growth of existing membrane processes and to identify opportunities for new applications. The technical barriers include fouling, instability, low flux, low separation factors and poor durability. Advancements on new generations of organic, inorganic and ceramic membranes, and mixed matrix polymer–inorganic and polymer–carbon composite membranes are urgently needed. In addition, economical processes to form large modules are important and innovative ways to incorporate membranes in industrial processes are also required. The other major challenge and driver for change and innovation is the need to reduce the energy used in separation processes, such as membrane technology. For example, in 2007, RO desalination world wide consumes $<1\%$ of energy used in separations, but with the predicted growth in desalination and the proposed cuts in greenhouse gas emissions by 2050, RO could be a major energy user and greenhouse gas emitter ($>50\%$ separations) on a "business-as-usual" basis. Clearly, RO desalination faces a major energy challenge. An assessment of the challenges and applications for membrane technology has been provided by Fane (142).

4.4. Promising Membrane Systems

With the development of new generations of membrane materials and the improvement in membrane processes, some membrane applications may be highly attractive in the future for energy-saving consideration. These promising membrane systems include forward osmosis (FO), biomimetic membranes, anaerobic MBRs (AMBRs), microbial fuel cell (MFC), decentralized sanitation systems, etc.

Forward osmosis (FO), also referred to as "osmosis", is actually a traditional process which was discovered as the background phenomena of RO (see also Fig. 1.5). In the FO process, a "draw" solution of high osmostic pressure (relative to that of the feed solution) is used to induce a net flow of water through a semi-permeable membrane into the draw solution under an osmotic pressure gradient, thus, effectively separating the feed water from its solutes. If fresh water is the desired product, a second separation step is required to extract the fresh water from the less concentrated draw solution. The first separation step of FO, driven by the osmotic pressure gradient, does not require a significant energy input. The main energy consumption is involved in the production of fresh water and reconcentration or regeneration of the draw solution. Forward osmosis has been studied for a range of applications including water and wastewater treatments, seawater/brackish water desalination, food processing, drug delivery, and electric power production, but commercial applications are still limited. It is worth a further exploitation as a promising energy-effective process. The breakthroughs for FO process require the development of robust and efficient membranes and modules, and the development of draw solutions that can induce high osmotic pressure with low energy requirements for regeneration or reconcentration (143).

Nature has developed a most efficient way for water transport across the cell membranes of microorganisms. Aquaporins or water channel proteins, which were discovered in 1990s by Peter Agre (2003's Nobel Laureate for the discovery), are highly permeable to water but highly retentive to dissolved solutes. They are typically bound in phospholipid cellular membranes and responsible for water transport across the membrane. The ideal property of the water channel proteins makes water delivery highly efficient. An artificial membrane may be developed to mimic the natural cellular membranes by incorporating and immobilizing aquaporins into an ultrathin amphiphilic polymeric layer to form a similar rejection film (144). With a robust and carefully engineered microporous support structure, the biomimetic membrane could be applied to ultrapure water production, water reuse and seawater desalination. Compared to the conventional RO membranes, the biomimetic membranes offer potential advantages including better selectivity, improved water permeability, reduced energy consumption, and improved product water quality because of its remarkable selectivity and water permeability. However, the biomimetic membranes are only at the stage of conceptual demonstration for water reuse and desalination. The technical challenges include the preparation of a suitable membrane support, the incorporation of aquaporin into a copolymer film and its interaction with the support.

The AMBR uses anaerobic bacteria to degrade organic substrate in the absence of oxygen. AMBR systems have similar configurations to aerobic MBR systems except for the application of biogas instead of air in the submerged reactor (Fig. 1.16) (145). Compared to

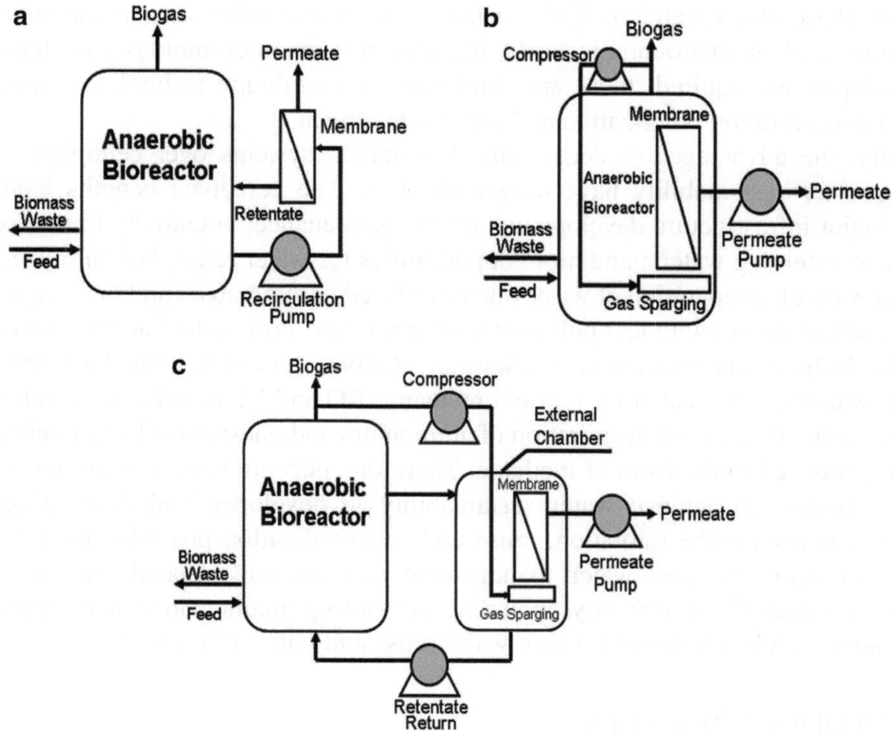

Fig. 1.16. Schematics of AMBR configurations (adapted from ref. (145)). (**a**) Pressure-driven external cross-flow membrane. (**b**) Vacuum-driven submerged membrane with the membrane immersed directly into the reactor. (**c**) Vacuum-driven submerged membrane with the membrane immersed in an external chamber.

conventional activated sludge systems, the AMBR systems should consume lower energy or be energy neutral. The process energy requirement of the AMBR can be offset by utilization of the product methane, which is produced by the process. Thus, AMBR offers a major advantage over the aerobic MBR based on the net energy consumption. The disadvantage of the AMBR is that it may not be able to achieve the removal efficiencies required for discharge. To date, AMBR has established its application in the treatment of synthetic wastewaters, food processing wastewaters, industrial wastewaters, high-solids-content waste streams, municipal wastewater, etc., but the commercial applications have been very limited. The AMBR process is a prospective system for the future due to its advantages in sustainability of high organic load, maintenance of high biomass concentration, recovery of energy and organic acid, as well as less sludge production (145).

A MFC is a device in which microorganisms oxidize organic matter to produce electrical power. This intriguing process has been demonstrated using wastewater as substrate for the direct generation of small amounts of electricity (146). The dual activity could significantly reduce the costs associated with current water treatment. The MFC is considered as a novel type of MBR, as the anode compartment is separated from the cathode by a proton conducting membrane. Presently, research on MFCs using wastewater as substrate is in the initial stages of laboratory evaluation around the world. In addition to the microbiology and electrochemistry developments required, there are challenges in membrane technology, scale-up for control of concentration polarization and feed pretreatment.

Recently, the advantages of decentralized treatment systems over centralized ones in achieving water sustainability have drawn attention. The perceived benefits include less need for major infrastructure development and/or maintenance; potentially lower costs; less discharge to receiving waters; and more opportunities for water reuse. Decentralized sanitation systems involve separation of wastewater produced within houses and buildings into two streams: a larger stream with less pollutant load termed as "gray water" and a smaller stream from toilet flushing water termed as "black water". Gray water can be treated for further reuse and black water can be treated for nutrient recovery. RO and NF membranes can be used in the anaerobic treatment for concentration of ammonium and phosphate (147). Energy, in this system, is produced in the form of methane. Therefore, decentralized sanitation system is a promising process to achieve water sustainability in developed and developing worlds because membranes make sanitation, reuse and decentralization possible (148). However, effective decentralized wastewater management systems will depend on the technical resources of a centralized authority, including monitoring, maintenance, and technical services. More research is required for safe water reuse and nutrient recovery.

5. CONCLUDING REMARKS

The development of membrane science and technology experienced a long history in laboratory studies before realization of its industrial applications. Today, membranes are widely accepted as the best available technology for water and wastewater treatments (149–151). Membrane technology also offers a viable alternative to conventional separations and purifications in chemicals and pharmaceutical processing, energy production, environmental monitoring and quality control, food and beverage processing, as well as fuel cells and bioseparation systems (152–154). Many successful large-scale real applications have brought great benefits to better human life (152). In the twenty-first century, the world is facing great challenges in the aspects of energy, water and the environment. Membrane technology is envisaged to play continuously a significant role to address these issues through extensive R&D on the membranes and process improvements.

REFERENCES

1. Baker RW (2004) Overview of membrane science and technology. In: Membrane technology and applications, 2nd edn., Wiley, England, pp 1–14
2. Mulder M (1996) Basic principles of membrane technology, Kluwer, Netherlands
3. Brennan MB (2001) Research accelerates on advanced water-treatment technologies as their use in purification grows. Sci Technol 79(15):32–38
4. Strathmann H (2001) Membrane separation processes: current relevance and future opportunities. AIChE J 47(5):1077–1087
5. Ferry JD (1936) Ultrafilter membranes and ultrafiltration. Chem Rev 18(3):373–455

6. Milisic V (1986) Anti-fouling techniques in cross-flow microfiltration. Filtration & Separation 23:347–349
7. Henry JD (1972) Cross flow filtration. In: Li NN (ed) Recent developments in separation science, vol 2. CRC Press, Cleveland, OH, pp 205–225
8. Fane AG, Fell CJD (1987) A review of fouling and fouling control in ultrafiltration. Desalination 62:117–136
9. Wei X, Wang R, Fane AG (2006) Development of a novel electrophoresis-UV grafting method to modify PES UF membranes used for NOM removal. J Memb Sci 273:47–57
10. Lonsdale HK (1982) Review: the growth of membrane technology. J Memb Sci 10:81–181
11. Mason EA (1991) Review: from pig bladders and cracked jars to polysulfones: an historical perspective on membrane transport. J Memb Sci 60:125–145
12. Riley RL (1991) Reverse osmosis. In: Baker RW, Cussler EL, Eykamp W, Koros WJ, Riley RL, Strathmann H (eds) Membrane separation systems, vol. 2. Noyes Data Corporation, Park Ridge, NJ, pp 276–328
13. Powell JH, Guild ME (1961) Salinity problems in the arid zones: Field operation of electric membrane equipment for water desalinating. Proceedings of the Teheran Symposium, UNESCO, 363–370
14. Loeb S, Sourirajan S (1962) Sea water demineralization by means of an osmotic membrane. Adv Chem Ser 38:117–132
15. Reid C, Breton E (1959) Water and ion flow across cellulosic membranes. J Appl Polym Sci 1:133
16. Cadotte JE, Petersen RI (1981) Thin film reverse osmosis membranes: origin, development, and recent advances. In: Turbak AF (ed) Synthetic membranes, ACS Symposium Series 153, vol. I: Desalination. American Chemical Society, Washington, DC, pp 305–325
17. Cabasso I (1987) Membranes. In: Encyclopedia of Polymer Science and Engineering, 9. Wiley, New York, pp 509–579
18. Kesting R (1985) Synthetic polymeric membranes: a structural perspective, Wiley, New York
19. Strathmann H (1990) Synthetic membranes and their preparation. In: Porter M (ed) Handbook of industrial membrane technology. Noyes, Park Ridge, NJ, pp 1–60
20. Petersen RI, Cadotte JE (1990) Thin film composite reverse osmosis membranes. In: Porter M (ed) Handbook of industrial membrane technology. Noyes, Park Ridge, NJ, pp 307–348
21. Bhattacharyya D, Williams M, Ray R, McCray S (1992) Reverse osmosis. In: Ho W, Sirkar K (eds) Membrane handbook. Van Nostrand Reinhold, New York, pp 263–390
22. Montgomery JM (1985) Facilities design. In: Water treatment principles and design. Wiley, New York
23. Argo DG et al. (1979) Evaluation of membrane processes and their role in wastewater reclamation, Report to US department of Interior, Office of Water Research an Technology Vol. 1: November
24. Schrantz J (1973) Rock island arsenal's zero discharge, Industrial Finishing June
25. Haralson RN, Jondahl KE (1983) Ultrafiltration aids in production of high quality water for electronics manufacture, A case history, Proceedings of the 44th International Water Conference, Pittsburgh, PA, October 24–26
26. Kohn PM (1978) Photo-processing facility achieves zero discharge. Chemical engineering, McGraw-Hill, New York, December 4
27. Bechhold H (1907) Kolloidstudien mit Filtrationsmethode. Z Phys Chem 60:257–318

28. Baker RW (2004) Ultrafiltration. In: Membrane technology and applications, 2nd edn. Wiley, England, pp 237–274
29. Strathmann H (1981) Membrane separation processes. J Memb Sci 9:121–189
30. Strathmann H (1992) Economic assessment of membrane processes. In: Li NN, Calo JM (eds) Separation and purification technology. Marcel Dekker, New York
31. Gekas V, Hallstrom B, TrBgardh G (1985) Food and dairy applications: the state of the art. Desalination 53:95–127
32. Schafer AI, Fane AG, Waite TD (eds) (2005) Nanofiltration – principles and applications. Elsevier, Oxford
33. Eriksson P (1988) Nanofiltration extends the range of membrane filtration. Environ Prog 7(1): 58–62
34. Conlon WJ (1985) Pilot field test data for prototype ultra low pressure reverse osmosis elements. Desalination 56:203–226
35. Kurihara M, Uemura T, Nakagawa Y, Tonomura T (1985) The thin-film composite low-pressure reverse osmosis membranes. Desalination 54:75–88
36. Bhattacharyya D, McCarthy JM, Grieves RB (1974) Charged membrane ultrafiltration of inorganic ion in single and multi-salt systems. AIChE J 20:1206–1212
37. Jitsuhara I, Kimura S (1983) Structure and properties of charged ultrafiltration membranes made of sulfonated polysulfone. J Chem Eng Jpn 16:389–393
38. Tsuru T, Nakao SI, Kimura S (1990) Effective charge-density and pore structure of charged ultrafiltration membranes. J Chem Eng Jpn 23:604–610
39. Rautenbach R, Gröschl A (1990) Separation potential of nanofiltration membranes. Desalination 77:73–84
40. Bhattacharyya D, Adams R, Williams M (1989) Separation of selected organic and inorganic solutes by low pressure reverse osmosis membranes. Prog Clin Biol Res 292:153–167
41. Watson BM, Hornburg CD (1989) Low-energy membrane nanofiltration for removal of color organics and hardness from drinking water supplies. Desalination 72:11–22
42. Reddy KK, Kawakatsu T, Snape JB, Nakajima M (1996) Membrane concentration and separation of L-aspartic acid and L-phenylalanine derivates in organic solvents. Sep Sci Technol 31 (8):1161–1178
43. Baker RW (2004) Microfiltration. In: Membrane technology and applications, 2nd edn. Wiley, England, pp 275–300
44. Gelman C (1965) Microporous membrane technology, Part 1: historical development and development. Anal Chem 37:29
45. Mitchell JK (1833) On the penetration of gases. Am J Med Sci 25:100–112
46. Graham T (1866) On the absorption and dialytic separation of gases by colloid septa. Part I. action of a septum of cauotchouc. Philos Mag 32:401
47. Graham T (1866) Philos Mag 32:503
48. Barrer RM (1939) Permeation, diffusion and solution of gases in organic polymers. Trans Faraday Soc 35:628–656
49. Barrer RM (1943) The zone of activation in rate processes. Trans Faraday Soc 39:237–241
50. Ismail AF, David LIB (2001) A review on the latest development of carbon membranes for gas separation. J Memb Sci 193:1–18
51. Stern SA (1994) Polymers for gas separations: the next decade. J Memb Sci 94:1–65
52. Henis JMS, Tripodi MK (1980) U.S. Patent 4230463
53. Kesting RE, Fritzsche AK (1993) Polymeric gas separation membranes. Wiley, New York

54. Kesting RE, Fritzsche AK, Murphy MK, Handermann AC, Cruse CA, Malon RF (1989) U.S. Patent 4,871,494
55. Pinnau I, Koros WJ (1990) U.S. Patent 4902422
56. Chung TS, Kafchinski ER, Vora RH (1995) U.S. Patent 5,413,852
57. Baker RW (2004) Pervaporation. In: Membrane technology and applications, 2nd edn. Wiley, England, pp 355–392
58. Kahlenberg L (1906) On the nature of the process of osmosis and osmotic pressure with. observations concerning dialysis. J Phys Chem 10:141–209
59. Kober PA (1917) Pervaporation, perstillation and percrystallization. J Am Chem Soc 39:944
60. Binning RC (1961) Separation of mixtures. U.S. Patent 2,981,680, April 25
61. Ballweg AH, Brüschke HEA, Schneider WH, Tüsel GF, Böddeker KW (1982) Pervaporation membranes. In: Proceedings of Fifth International Alcohol Fuel Technology Symposium, Auckland, New Zealand, pp. 97–106
62. Baker RW (1991) Pervaporation. In: Baker RW, Cussler EL, Eykamp W, Koros WJ, Riley RL, Strathmann H (eds) Membrane separation systems, vol. 2. Noyes Data Corporation, Park Ridge, NJ, pp 151–188
63. Feng X, Huang RYM (1997) Liquid separation by membrane pervaporation: a review. Ind Eng Chem Res 36:1048–1066
64. Smith CV, Gregorio DO, Talcott RM (1969) The use of ultrafiltration membrane for activated sludge separation, Proceedings of 24th Industrial Waste Conference, Purdue, IN
65. Hardt FW, Cleseri LS, Nemerow N, Washington D (1970) Solids separation by ultrafiltration for concentrated activated sludge. J Water Pollut Control Fed 42:2135–2148
66. Bemberis I, Hubbard PJ, Leonard FB (1971) Membrane sewage treatment systems – potential for complete wastewater treatment. Am Soc Agric Engineers Winter Meeting 71–878:1–28
67. Cheryan M, Mehaia MA (1985) Membrane bioreactors for high-performance fermentations. In: Sourirajan S, Matsura T (eds) Reverse osmosis and ultrafiltration. ACS Symposium Series Number 281. American Chemical Society, Washington, DC, pp 231–246
68. Kanazek RA, Gullino PM, Kohler PO, Dedrick RC (1972) Cell culture on artificial capillaries: an approach to tissue growth in vitro. Science 178:65–72
69. Hu WS, Dodge TC (1985) Cultivation of mammalian cells in bioreactors. Biotechnol Prog 1:209–215
70. Yamamoto K, Hiasa M, Mahmood T, Matsuo T (1989) Direct solid-liquid separation using hollow fiber membrane in an activated-sludge aeration tank. Water Sci Technol 21:43–54
71. Fane AG (2008) Chapter 10. Submerged membranes. In: Li N, Fane AG, Ho W, Matsuura T (eds) Advanced Membrane Technology and Applications. John Wiley & Sons, Inc. NY pp 239–270
72. Wenten IG (2002) Recent development in membrane science and its industrial applications. J Sci Technol Membrane Sci Technol 24(Suppl):1010–1024
73. Ye Y, Le CP, Chen V, Fane AG (2005) Evolution of fouling during crossflow filtration of model EPS solutions. J Memb Sci 264(1–2):190–199
74. Strathmann H (2001) Membrane separation processes: current relevance and future opportunities. AIChE J 47:1077–1087
75. Wangnick K (2002) 2002 IDA Worldwide Desalting Plants Inventory Report, No. 17
76. Fell CJD (2003) Membrane technology: past successes – futureopportunities, IMSTEC'03 International Membrane Science & Technology Conference, Sydney, November

77. Semiat R (2000) Desalination: present and future, International Water Resources Association. Water International 25(1):54–65
78. Baker RW (2004) Reverse osmosis. In: Membrane technology and applications, 2nd edn. Wiley, England, pp 191–235
79. Schwinge J, Neal PR, Wiley DE, Fletcher DF, Fane AG (2004) Spiral wound modules and spacers: Review and analysis. J Memb Sci 242:129–153
80. Wade NM (2001) Desalination plant development and cost update. Desalination 136:3–12
81. Leitner GF (1998) Breaking the cost barrier for seawater desalting. Desalination Water Reuse 81(1):15–20
82. Redondo JH, Lanzarote IV (2001) A new concept for two pass SWRO at low O&M cost using the new high flow FilmTec SW 30–380. Desalination 138:231–236
83. MacHarg JP (2002) The evaluation of SWRO energy recovery systems. Energy Recovery Inc, USA. www.energy-recovery.com
84. Avlonitis SA, Kouroumbas K, Vlachakis N (2003) Energy consumption and membrane replacement cost for seawater RO desalination plants. Desalination 157:151–158
85. Hilal N, Al-Abri M (2006) Enhanced membrane pre-treatment processes using macromolecular adsorption and coagulation in desalination plants: a review. Sep Sci Technol 41:403–453
86. Truby R (2001) Desalination processes enhanced by multiple membrane systems, *European Conference On Desalination And The Environment*, EDS Newsletter Issue 12, htttp://www.edsoc.com, May
87. Voutchkov NV (2004) The ocean – a new resource for drinking water. Public Works 30–33, June
88. Dunivin W, Lange P, Sudak R, Wilf M (1991) Reclamation of ground water using RO technology, Proceedings of IDA World Conference on Desalination and Water Reuse, Washington, August
89. Overview of WATER FACTORY 21 (2001) Orange County Water District, http://www.Ocwd.com
90. Thompson M, Powell D (2003) Case study – Kranji high grade water reclamation plant, Singapore, IMSTEC'03, Sydney, Australia, September
91. Guendert D Comparing MR/RO performance on secondary and tertiary effluents in reclamation/reuse applications, IDS-Water – White Paper, http://www.idswater.com/water/us/conference.html
92. Duran FE, Dunkelberger GW (1995) A comparison of membrane softening on three South Florida groundwaters. Desalination 102(1):27–34
93. Tanninen J, Nyström M (2002) Separation of ions in acidic conditions using NF. Desalination 147:295–299
94. Schaep J, Van der Bruggen B, Uytterhoeven S, Croux R, Vandecasteele C, Wilms D, Van Houtte E, Vanlerberghe F (1998) Removal of hardness from groundwater by nanofiltration. Desalination 119:295–302
95. Van der Bruggen B, Everaert K, Wilms D, Vandecasteele C (2001) Application of nanofiltration for the removal of pesticides, nitrate and hardness from ground water: retention properties and economic evaluation. J Memb Sci 193(2):239–248
96. Santafé-Moros A, Gozálvez-Zafrilla JM, Lora-García J (2005) Performance of commercial nanofiltration membranes in the removal of nitrate ions. Desalination 185:281–287
97. Kettunen R, Keskitalo P (2000) Combination of membrane technology and limestone filtration to control drinking water quality. Desalination 131:271–283
98. Alborzfar M, Jonsson G, Grén C (1998) Removal of natural organic matter from two types of humic ground waters by nanofiltration. Water Res 32(10):2983–2994
99. Fu P, Ruiz H, Lozier J, Thompson K, Spangenberg C (1995) A pilot study on groundwater natural organics removal by low-pressure membranes. Desalination 102:47–56

100. Gorenflo A, Veliizquez-Padrón D, Frimmel FH (2002) Nanofiltration of a German groundwater of high hardness and NOM content: performance and costs. Desalination 151:253–265
101. Berg P, Hagmeyer G, Gimbel R (1997) Removal of pesticides and other micropollutants by nanofiltration. Desalination 113(2–3):205–208
102. Cyna B, Chagneau G, Bablon G, Tanghe N (2002) Two years of nanofiltration at the Mery-sur-Oise plant, France. Desalination 147:69–75
103. Ventresque C, Gisclon V, Bablon G, Chagneau G (2000) An outstanding feat of modem technology: the Mery-sur-Oise nanofiltration treatment plant (340,000 ml/day). Desalination 131:1–16
104. Qdais HA, Moussa H (2004) Removal of heavy metals from wastewater by membrane processes: a comparative study. Desalination 164:105–110
105. Braeken L, Van der Bruggen B, Vandecasteele C (2004) Regeneration of brewery waste water using nanofiltration. Water Res 38:3075–3082
106. Voigt I, Stahn M, Wohner St, Junghans A, Rost J, Voigt W (2001) Integrated cleaning of colored waste water by ceramic NF membranes. Sep Purif Technol 25:509–512
107. Cuperus P, Ebert K (2002) Non-aqueous applications of NF. In: Nanofiltration – principles and applications. Elsevier, Oxford, pp 521–536
108. Van Gestel T, Van der Bruggen B, Buekenhoudt A, Dotremont C, Luyten J, Vandecasteele C, Maes G (2003) Surface modification of γ-Al2O3/TiO2 multilayer membranes for applications in nonpolar solvents. J Memb Sci 224:3–10
109. Tsuru T, Sudoh T, Yoshioka T, Asaeda M (2001) Nanofiltration in non-aqueous solutions by porous silica-zirconia membranes. J Memb Sci 185:253–261
110. Webar R, Chmiel H, Mavrov V (2003) Characteristics and application of new ceramic nanofiltration membrane. Desalination 157:113–125
111. Hassan A, Al-So M, Al-Amoudi A, Jamaluddin A, Farooque A, Rowaili A, Dalvi A, Kither N, Mustafa G, AI-Tisan I (1998) A new approach to thermal seawater desalination processes using nanofiltration membranes (Part 1). Desalination 118:35–51
112. Hassan AM, Farooque A, Jamaluddin A, Al-Amoudi A, Al-Sofi M, AI-Rubaian A, Kither N, Al-Tisan I, Rowaili A (2000) A demonstration plant based on the new NF-SWRO process. Desalination 131:157–171
113. Hilal N, Al-Zoubi H, Darwish NA, Mohammad AW, Abu Arabi M (2004) A comprehensive review of nanofiltration membranes: Treatment, pretreatment, modeling, and atomic force microscopy. Desalination 170:281–308
114. Nystrom M, Kaipia L, Luque S (1995) Fouling and retention of nanofiltration membranes. J Memb Sci 98(3):249–262
115. Hong S, Elimelech M (1997) Chemical and physical aspects of natural organic matter (NOM) fouling of nanoflltration membranes. J Memb Sci 132:159–181
116. Seungkwan H, Menachem E (1997) Chemical and physical aspects of natural organic matter (NOM) fouling of nanofiltration membranes. J Memb Sci 132(2):159–181
117. Schafer AI, Fane AG, Waite TD (1998) Nanofiltration of natural organic matter: removal, fouling and the influence of multivalent ions. Desalination 118(1–3):109–122
118. Shaalan HF (2002) Development of fouling control strategies pertinent to nanofiltration membranes. Desalination 153:125–131
119. Laine JM, Vial D (2000) Pierre Moulart, Status after 10 years of operation – overview of UF technology today. Desalination 131:17–25

120. Wohlsen T, Bates J, Gray B, Katouli M (2004) Evaluation of five membrane filtration methods for recovery of Cryptosporidium and Giardia isolates from water samples. Appl Environ Microbiol April:2318–2322
121. Falk CC, Karanis P, Schoenen D, Seitz HM (1998) Bench scale experiments for the evaluation of a membrane filtration method for the recovery efficiency of Giarida nad Cryptosporidium from water. Water Res 32(3):565–568
122. EPA (2001) Low-pressure membrane filtration for pathogen removal: application, implementation, and regulatory issues, EPA 815-C-01–001, April
123. Hanft S (2006) The membrane microfiltrations market, BCC Research Pub ID: WA1271760, http://www.bccresearch.com/mst/, March 1
124. Schäfer AI, Schwicker U, Fischer MM, Fane AG, Waite TD (2000) Microfiltration of colloids and natural organic matter. J Memb Sci 171:151–172
125. Amy G, Cho J (1999) Interactions between natural organic matter (NOM) and membranes: rejection and fouling. Water Sci Technol 40(9):131–139
126. Bersillon JL (1998) Fouling analysis and control in future industrial prospects of membrane processes. Elsevier, Oxford
127. Fane AG, Wei X, Wang R (2006) Membrane filtration processes and fouling. In: Newcombe G, Dixon D (eds) Interface science in drinking water treatment: theory and applications. Elsevier, USA, .ISBN: 0120883805, September, pp 107–132
128. Pirbazari M, Badriyha BN (1992) MF-PAC for treating waters contaminated with natural and synthetic organics. J Am Water Works Assoc 84(12):95–103
129. Guigui C, Rouch JC, Durand-Bourlier L, Bonnelye V, Aptel P (2002) Impact of coagulation conditions on the in-line coagulation/UF process for drinking water production. Desalination 147:95–100
130. Abdessemed D, Nezzal G (2002) Treatment of primary effluent by coagulation-adsorption-ultrafiltration for reuse. Desalination 152:367–373
131. Campos C, Mariñas BJ, Snoeyink VL, Baudin I, Laîné JM (1998) Adsorption of trace organic compounds in CRISTAIJ® processes. Desalination 117:265–271
132. Chang S, Waite TD, Ong PEA, Schäfer AI, Fane AG (2004) Assessment of trace estrogenic contaminants removal by coagulant addition, powdered activated carbon adsorption and powdered activated carbon/microfiltration processes. J Environ Eng 130(7):736–742
133. Fu J, Ji M, Wang Z, Jin L, An D (2006) A new submerged membrane photocatalysis reactor (SMPR) for fulvic acid removal using a nano-structured photocatalyst. J Hazard Mater B 131:238–242
134. Lee J, Kwon T, Ramesh T, Moon I (2006) Adsorption and photocatalytic degradation of bisphenol A using TiO2 and its separation by submerged hollowfiber ultrafiltration membrane. J Environ Sci 18(1):193–200
135. Judd S, Judd C (2006) The MBR book: principles and applications of membrane bioreactors in water and wastewater treatment. Elsevier, Amsterdam
136. Baker RW (2002) Future directions of membrane gas separation technology. Ind Eng Chem Res 41:1393–1411
137. Noble RD, Stern SA (1995) Membranes separation technology. Principles and applications. Elsevier, Amsterdam
138. Prasad R, Notaro F, Thompson DR (1994) Evolution of membranes in commercial air separation. J Memb Sci 94:225–248

139. Jonquières A, Clément R, Lochon P, Néel J, Dresch M, Chrétien B (2002) Industrial state-of-the-art of pervaporation and vapour permeation in the western countries. J Memb Sci 206:87–117
140. Lipnizki F, Field RW, Ten PK (1999) Pervaporation-based hybrid process: a review of process design, applications and economics. J Memb Sci 153:183–210
141. Fell CJD (2003) Membrane technology: past successes – future opportunities, IMSTEC'03 International Membrane Science & Technology Conference, Sydney, November
142. Fane AG (2007) Membranes achievements, challenges and the future. Proceedings of IWA, Leading Edge Technology Conference, Singapore, June 4–6
143. Cath TY, Childress AE, Elimelech M (2006) Forward osmosis: principles, applications, and recent developments. J Memb Sci 281:70–87
144. Kumar M, Grzelakowski M, Zilles J, Clark M, Meier W (2007) Highly permeable polymeric membranes based on the incorporation of the functional water channel protein Aquaporin Z. Proc Natl Acad Sci U S A 104:20719–20724
145. Liao BQ, Kraemer JT, Bagley DM (2006) Anaerobic membrane bioreactors: applications and research direction. Crit Rev Environ Sci Technol 36:489–530
146. Logan BE, Hamelers B, Rozendal R, Schrorder U, Keller J, Freguia S, Aelterman P, Verstraete W, Rabaey K (2006) Microbial fuel cells: methodology and technology. Environ Sci Technol 40 (17):5181–5192
147. van Voorthuizen EM, Zwijnenburg A, Wessling M (2005) Nutrient removal by NF and RO membranes in a decentralized sanitation system. Water Res 39:3657–3667
148. Fane AG (2004) Exploration of the potential of membrane technology for sustainable decentralized sanitation (Editorial). J Memb Sci 228:127–128
149. Wang LK, Wang MHS (2008) Development and applications of membrane bioreactor technologies. 2008 National Engineers Week Conference, Albany Marriott, Albany, NY. Feb. 14–15
150. Wang LK, Wang MHS (2008) Application of Membrane Filtration Technologies in Food Processing Industry. 2008 National Engineers Week Conference, Albany Marriott, Albany, NY. Feb. 14–15
151. Wang LK, Wang MHS, Suozzo T, Dixon RA, Wright TL, Sarraino S (2009) Chemical and Biochemical Technologies for Environmental Infrastructure Sustainability. 2009 National Engineers Week Conference, Albany Marriott, Albany, NY. Feb. 5–6
152. Levesque S, Wallis-Large C, Hemken B, Bontrager S, Kreuzwiesner S (2009) Plan ahead with MBRs. Water Environ Technol 21(1):34–37
153. Wang LK, Ivanov V, Tay JH, Hung YT (eds) (2010) Environmental biotechnology. Humana, Totowa, NJ
154. Nadeeshani Nanayakkara KG, Wei Y, Zheng YM, Chen JP (2010) Food Industry wastewater treatment. In: Wang LK, Hung YT, Shammas NK (eds) Handbook of advanced industrial and hazardous wastes treatment. CRC Press, Boca Raton, FL, pp 1233–1254

2
Preparation of Polymeric Membranes

Jizhong Ren and Rong Wang

CONTENTS

Abstract This chapter mainly describes the principles of membrane formation process for polymeric membranes. With a brief introduction of relevant background information such as various membranes and membrane processes, a comprehensive list of polymer materials, which are suitable for making membranes, has been given. The most common technique used to prepare polymeric membranes – phase inversion process, including thermally induced phase separation (TIPS) and diffusion induced phase separation (DIPS), is discussed in detail. The thermodynamic behavior of the casting polymer solution, the process of membrane formation, and the fabrication of hollow fiber and flat sheet membranes are involved.

The thermodynamic description of the polymer solution is based on the concepts of spinodal, binodal, vitrification boundary, gelation boundary et al. in the phase diagram. The linearized cloud point curve correlation is presented. In addition, two important parameters, the approaching ratio of the polymer solution and the approaching coagulant ratio, are discussed in association with membrane formation. For the membrane formation process, the delay time and gelation time are two macroscopic time scales, which influence the membrane morphologies simultaneously. The formation of nascent porous membranes, the vitrification of the membrane morphology and the membrane surface formation are described. The macrovoid formation is related to the viscous fingering phenomenon. In the fabrication of hollow fiber membranes, the shear flow of the polymer solution inside the spinneret and the elongation flow in the air gap strongly influence the performance of resultant membranes. Thus, the shear flow and elongation flow are discussed in great detail. The hydrodynamics of the polymer solution at the casting window in the process of preparing flat sheet membranes is also concerned.

From: *Handbook of Environmental Engineering, Volume 13: Membrane and Desalination Technologies*
Edited by: L.K. Wang et al., DOI: 10.1007/978-1-59745-278-6_2

Key Words Polymeric membranes • phase inversion process • thermodynamics • hydrodynamics • hollow fiber membranes • flat sheet membranes.

1. INTRODUCTION

Simply, a membrane is a selective barrier between two phases (1), and a membrane process should be always associated with its application according to this definition, which can range from gas separation, pervaporation and vapor permeation, desalination, nanofiltration, ultrafiltration, microfiltration, membrane extraction, membrane distillation, supported liquid membrane, and so on. The term of membrane is different from filter, which is usually limited to the structures that separate particulate suspensions larger than 1–10 μm (2). Since Loeb and Sourirajan (3) developed the first anisotropic membranes by phase inversion technique using cellulose acetate in the 1960s, membrane separation technology has been becoming one of the most innovative and rapidly growing fields across science and engineering, and has been used in a variety of applications. Depending on different applications, a large number of new polymers are employed to make different membranes with specific morphologies and transport properties suitable to the desired separation processes. Except for the Loeb-Sourirajan's phase inversion technique, some new membrane formation processes such as interfacial polymerization, composite coating, multilayer composite casting, stretching, etc., were also developed to make high performance membranes in the past decades. To make various membranes with controlled morphology and desired transport properties, it is vital to understand the principles of membrane formation. Some critical reviews have been made to address relevant issues (2, 4–10).

This chapter mainly describes the principles of polymeric membrane formation process with focus on the phase inversion technique. A variety of polymer materials suitable for membrane fabrication are briefly introduced. The thermodynamic behavior of the casting polymer solution, the membrane formation process and the fabrication of hollow fiber and flat sheet membranes.are involved. Based on the thermodynamic description of the polymer solution, the linearized cloud point curve correlation is presented. As for the membrane formation process, the vitrification of the membrane morphology, the membrane surface formation, as well as the formation of nascent porous membranes are all associated with two macroscopic time scales of delay time and gelation time. In addition, two important parameters for membrane fabrications, the approaching ratio of the casting solution and the approaching coagulant ratio, are also described and discussed.

2. MEMBRANE CLASSIFICATION

Membranes are generally classified by the nature of the materials, the membrane morphology, geometry, preparation method, separation regime and processes et al. (8, 10). A representative membrane classification is shown in Fig. 2.1.

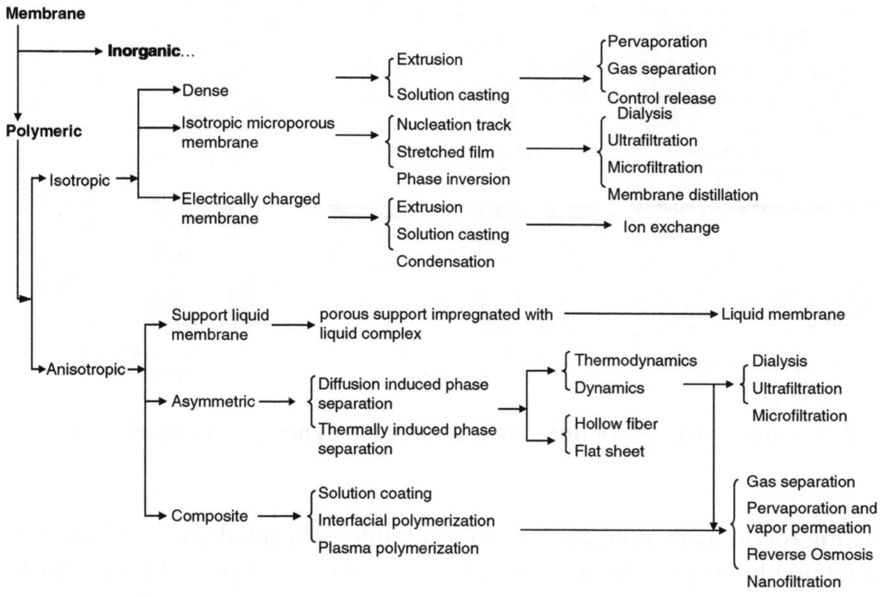

Fig. 2.1 Schematic diagram of membrane classification.

Based on the membrane materials, membranes are grouped into polymeric and inorganic membranes. For polymeric membranes, many organic polymers including crystalline and amorphous, glassy and rubbery, are suitable for the membrane fabrication. The preparation methods involve phase inversion, interfacial reaction, coating, stretching, etc. Among these preparing methods, the phase inversion is the main approach to prepare most of commercial membranes, which will be described in greater detail in the following section. As for inorganic membranes, oxides, ceramics, metals, carbon, etc., are commonly used, which are not discussed in this chapter.

Polymeric membranes are usually classified into isotropic and anisotropic membranes in terms of membrane morphologies (2). Isotropic membranes consist of nonporous dense membranes, microporous membranes, and electrically charged membranes, while anisotropic membranes include integrated asymmetric membranes, composite membranes, and supported liquid membranes, which are shown in Fig. 2.2.

2.1. *Isotropic Membranes*

Dense membranes are rarely used in practical membrane separation processes because of its low flux caused by its high membrane thickness, but the intrinsic properties of polymers will determine the membrane performance and separation characteristics. Dense membranes are mainly used in laboratory to characterize the intrinsic membrane properties for control release, gas separation, pervaporation, nanofiltration, and reverse osmosis membranes for material screening. They are prepared by solution casting and thermal melting extrusion approaches.

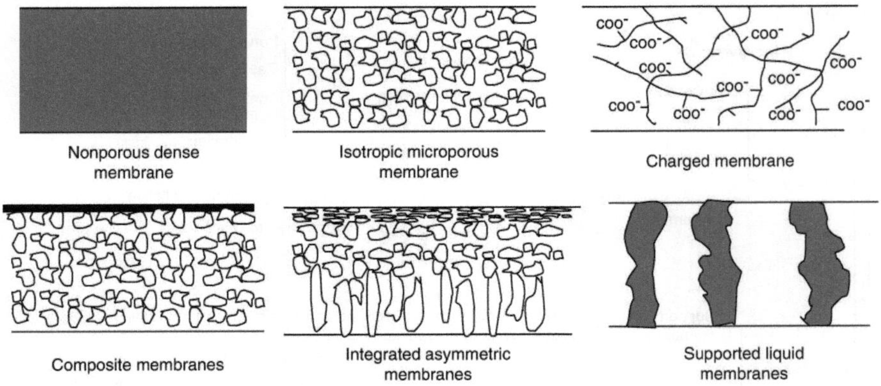

Fig. 2.2 Schematic diagram of different membrane morphologies [adapted from ref. (2)].

Isotropic microporous membranes have a rigid, interconnected pore, and voided structure distributed randomly; its pore diameter is in the order of 0.01–10 μm. The particles larger than the largest pores are completely rejected and the particles smaller than the largest pores, will be partially rejected by deep filtration and screen filtration according to the hydrodynamic and hindered transport theory. The separation process is controlled by the pore size distribution of microporous membranes and the hydrodynamic conditions. The microporous membranes are prepared by phase separation, tracked etch, stretching, or leaching, etc. The phase separation is the most important method for the isotropic microporous membrane preparation.

Electrically charged membranes are referred to either anion-exchange membranes or cation-exchange membranes, which possess dense/microporous structures carrying fixed positive or negatively charged ions. The charge and the concentration of the ions in solution are the key factors to control the separation process.

2.2. *Anisotropic Membranes*

Anisotropic membranes are layer structures, changing the porosity and pore size over the whole membrane wall. The anisotropic membranes usually have a very thin surface layer supported on a thick microporous substrate. The thin skin layer is the selective layer to perform separation, while the microporous substrate mainly provides the mechanical strength. Because of the very thin selective layer, the membrane fluxes are very high. Integrally asymmetric membranes, composite membranes and supported liquid membranes are in the category of anisotropic membranes.

Integrally asymmetric membranes were developed by Loeb and Sourirajan in 1960s (3), which were prepared by the phase inversion method using a single membrane material. The membrane porosity and pore size change in different layers. A selective thin layer exists on the top surface followed by more and more loose layers. The thin skin layer and porous substrate are coupled with each other as they are formed simultaneously in the phase inversion processphase inversion process.

Compared with integrally asymmetric membranes, composite membranes usually contain two separated layers with different separation functions and different membrane materials. The porous substrate acts as mechanical support and the skin layer is mainly used for the selective purpose. They are not coupled with each other in the preparation process. The skin layer can be prepared separately by interface polymerization, plasma polymerization, solution coating, etc.

Supported liquid membranes are porous polymer matrixes filled with liquid carriers in pores. The separation process is mainly carried out by the carrier facilitated transport. And the microporous substrate mainly provides pores holding the liquid carrier phase by the capillarity.

2.3. *Membrane Processes*

Different preparation methods result in various isotropic and anisotropic membranes, which are related to different membrane processes. Based on the pore size range of the membrane and the separated species, the membrane processes can be classified into gas separation, pervaporation, reverse osmosis, nanofiltration, ultrafiltration, microfiltration, and so on, which may correspond to either nonporous, mesoporous or macroporous membranes. The relationship between the pore size of the membrane and the separated species is shown in Fig. 2.3.

2.3.1. Gas Separation and Pervaporation

Gas separation and pervaporation use nonporous membranes. And the transport process is controlled by solution and diffusion. The nonporous membranes for gas separation and

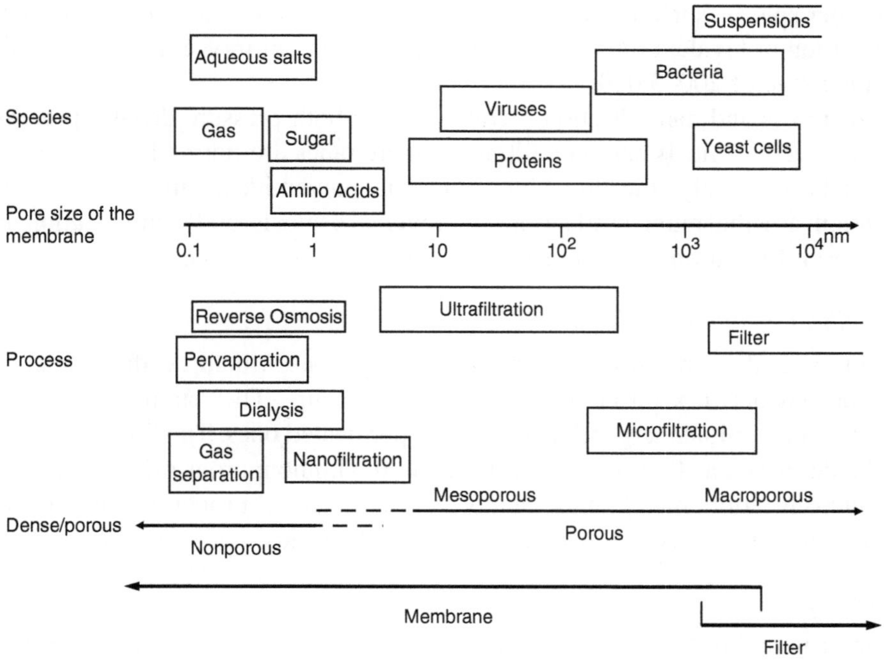

Fig. 2.3 Relationship between membrane pore size and separated species.

pervaporation possess integrally asymmetric structure or composite structure. For some commercial gas separation membranes (N_2/H_2, CO_2/CH_4, etc.), a thin layer is coated on the top surface of the integrally asymmetric membranes to block the defects on the active skin layer of the original membranes, which cannot contribute to the intrinsic separation properties of the membranes. However, for other gas separations such as O_2/N_2, volatile organic compounds (VOCs)/N_2, pervaporation and vapor permeation, etc., pervaporation and vapor permeation, the composite coating mainly acts as the selective layer, which contributes to the intrinsic separation properties of the membranes, and the original microporous membranes only provide the mechanical strength.

2.3.2. *Reverse Osmosis and Nanofiltration*

Generally, polymeric reverse osmosis membranes have a nonporous structure on the membrane surface, but consist of a polymer network in which solutes can be dissolved. The transport process is controlled by solution and diffusion, which retains almost all ionic species. Nanofiltration membranes can be divided into porous membranes and nonporous membranes with a swollen network structure. Though the porous and network swollen nanofiltration membranes may have the same separation performance, their transport mechanisms are different. For a porous membrane, uncharged solutes will be separated by a sieving mechanism, whereas for a nonporous membrane, solution-diffusion mechanism will determine the transport phenomena in the swollen network. Compared with reverse osmosis, nanofiltration membranes have a high flux and relative low retention of monovalent ionic species.

Reverse osmosis membranes only permit some small molecules such as water to transport and reject other species such as monovalent ions Na^+, K^+ et al. But nanofiltration is mainly for the separation of bivalent ions such as Ca^{2+}, Mg^{2+}, etc., or molecules with the molecular weight in the range of 200–5,000.

Reverse osmosis and nanofiltration membranes are both pressure driving processes. The commercial reverse osmosis and nanofiltration membranes are normally prepared using the same materials. Currently, the thin-film composite polyamide membranes via interfacial polymerization are the main productions of reverse osmosis and nanofiltration by using slightly different formation conditions to produce a more or less open polymer structure.

2.3.3. *Ultrafiltration and Microfiltration*

Ultrafiltration and microfiltration membranes are porous structures, they have a distinct, permanent porous network through which transport occurs. The retain species are usually several orders of magnitude larger than that of the permeated ones. The flow in the membrane pores can be described as Pouiseille flow. Compared with nonporous membranes, ultrafiltration and microfiltration membranes cannot reveal intrinsic properties of the polymeric materials and the intrinsic selectivity for the transport species.

2.3.4. *Filter*

Filters are usually limited to the structures that separate particulate suspensions larger than 1–10 μm. From Fig. 2.3, the dividing line between different membrane processes cannot be distinguished precisely and they are overlapped partially sometimes.

3. MEMBRANE MATERIALS

Most of materials such as polymer, ceramic, metal, carbon and glass can be used to make membranes. At present, the commercial membranes are mainly fabricated from polymeric materials. Being membrane materials, the polymers should demonstrate thermal stability over a wide range of temperature and a chemical stability over a range of pH, and possess strong mechanical strength. Besides, they can also be processed into flat sheet or hollow fiber membranes easily. The polymers that are suitable for making membranes include cellulose, cellulose acetates, cellulose acetate butyrate, cellulosic ester, cellulose nitrates, polycarbonate esters, ethyl celluloses, polyacetylenes, polyacrylonitrile, polyamides, polyamide esters, polyamide-hydrazide, polyamideimides, polyaryletherketone, polyetherketones, polycarbonates, poly(phenylene oxide) polyesters, polyestercarbonate, polyethers, polyetherimides, polyetherketones, polyethersulfones, polyethylenes, polyhydrazides, polyimides, polyphenylene oxides, polyphenylene sulfide, poly(phthalazine ether sulfone ketone), sulfonated poly (phthalazine ether sulfone ketone), polypropylene, polysiloxanes, polysulfones, sulfonated polysulfones, polytetrafluoroethylene, poly(trialkylsilylacetylenes), poly(trimethylsilylpropyne), polyureas, polyurethanes, polyvinylalcohol, polyvinylchloride, polyvinylidene fluorides, etc. The early commercial membranes were based on cellulose acetate and polysulfone, etc. Afterward, other polymers such as polyethersulfones, polyacrylonitrile, polyvinylidene fluorides, polycarbonate, polyamide, polyimide, polyetherimide, etc., were used one after the other. Some commercial and representative polymeric membrane materials are introduced further in the following sections.

3.1. Cellulose and Cellulose Acetate

Cellulose is a polysaccharide with a regular repeating structure (Fig. 2.4), which can form strong intermolecular hydrogen-bonds between the hydroxy groups and gives rise to a high level of crystallinity. And the crystallinity also prevents cellulose form dissolving in water although the polymers are highly hydrophilic. Cellulose can dissolve in *N*-Methylmorpholine-*N*-Oxide (NMMO), *N,N*-dimethylacetamide (DMAC) and *N*-methylpyrrolidone (NMP) with lithium chloride (LiCl) added. Regenerated cellulose has been used extensively to fabricate dialysis membranes. Cellulose acetate, diacetate, triacetate and their blends are widely used to form microfiltration, ultrafiltration and reverse osmosis membranes, which are strongly hydrophilic and thus demonstrate lower fouling characteristics. The disadvantages of cellulose and cellulose acetate membranes are their stability over a very narrow pH range between 4 and 6.5 and their poor resistance to microbial attack.

3.2. Polysulfone

Polysulfone (PS) is an amorphous high performance polymer and demonstrates a high degree of chemical and thermal stability, which chemical structure is shown in Fig. 2.5. Polysulfones are mainly used to form ultrafiltration, microfiltration, and gas separation membranes, but they are also used to form the porous support layer of many reverse osmosis, nanofiltration, and some gas separation membranes. Compared with cellulose acetate, polysulfone membranes can

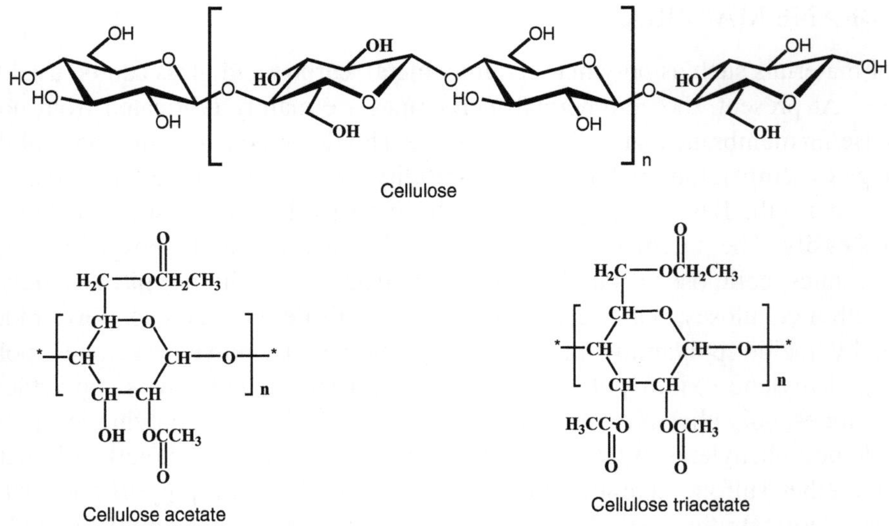

Fig. 2.4 Chemical structures of cellulose, cellulose acetate and triacetate.

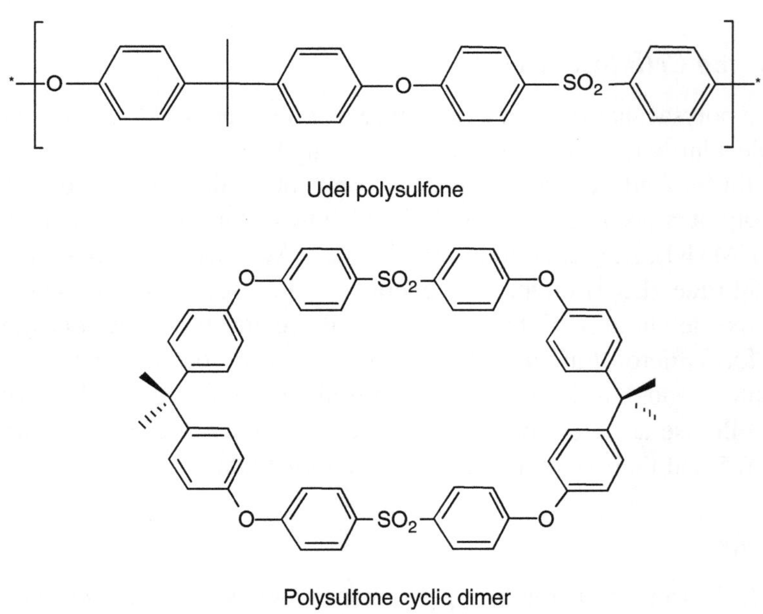

Fig. 2.5 Chemical structure of polysulfone.

tolerate a wider range of pH, which can be cleaned with acids or alkalis. For some membrane preparations, the level of cyclic dimer is an important parameter for the processing efficiency and fiber quality (11). Too much cyclic dimer can result in cloudy dope solution. UDEL polysulfone only contains 1.20% by weight of cyclic dimer, which is very suitable for membrane making.

3.3. *Polyethersulfone*

Polyethersulfone (PES) membranes also demonstrate very high chemical and thermal stability, which chemical structure is shown in Fig. 2.6. Compared with Polysulfone's glass transition temperature (T_g) of 195°C, PES has a higher T_g of 230°C. Like PS material, PES membranes also demonstrate a low tolerance of aromatic hydrocarbons or ketones. PES membranes are mainly used in ultrafiltration, microfiltration, and dialysis membranes.

3.4. *Polyacrylonitrile*

Polyacrylonitrile (PAN) is mainly used to prepare ultrafiltration membranes and porous supports of composite membranes due to its superior resistance to hydrolysis and oxidation, which chemical structure is shown in Fig. 2.7. PAN is hydrophobic in its pure state though a polar group is presented in the backbone of the polymer. It is usually copolymerized with more hydrophilic monomers to improve its hydrophilicity and processability to make it less brittle.

3.5. *Polyvinylidene Fluoride*

Polyvinylidene fluoride (PVDF) is a very good material for microfiltration membranes due to its chemical resistance and thermal stability; the chemical structure is shown in Fig. 2.8. It is resistant to most inorganic and organic acids and tolerant to a wide pH range. PVDF is soluble in dimethyl formamide (DMF), DMAC, NMP, dimethylsulfoxide (DMSO) and triethyl phosphate (TEP). It is a hydrophobic and semi-crystalline polymer and commonly used for making microfiltration membranes by the phase inversion process.

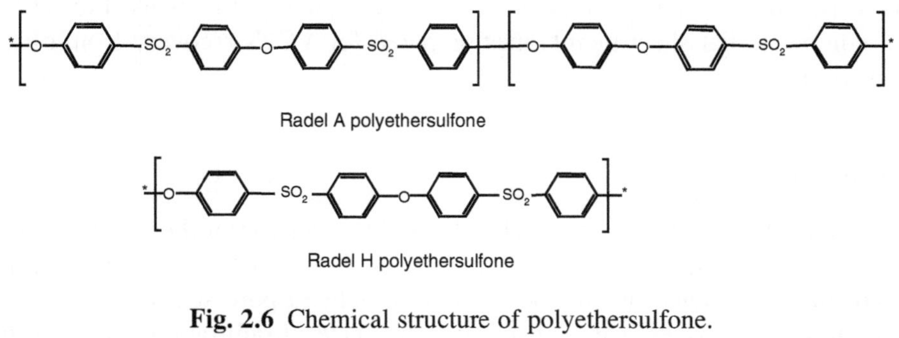

Fig. 2.6 Chemical structure of polyethersulfone.

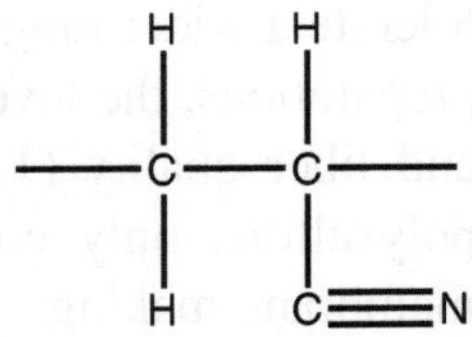

Fig. 2.7 Chemical structure of polyacrylonitrile.

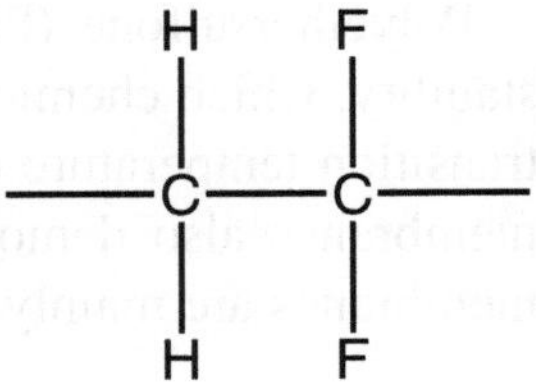

Fig. 2.8 Chemical structure of polyvinylidene fluoride.

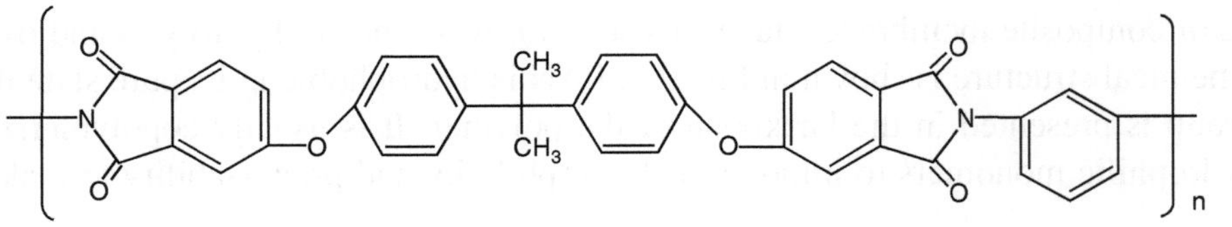

Fig. 2.9 Chemical structure of polyetherimide.

3.6. Polyetherimide

Polyetherimide (PEI) is a clear amorphous high performance polymer with a glass transition temperature of close to 216°C, the chemical structure of which is shown in Fig. 2.9. Its chemical stability is much higher than cellulose acetate, but less than PVDF and PSF at a high pH. PEI cannot be used in contact with chloroform and dichloromethane. It is a good material for the fabrication of the integrally asymmetric membranes in gas separation, especially for the recovery of helium (12). Usually, PEI is used to fabricate the support of composite flat sheet membranes prepared for VOCs recovery from off-gases in petrochemical industries.

3.7. Polycarbonate

Polycarbonate (PC) is a tough, dimensionally stable, transparent thermoplastic with high performance properties, which chemical structure is shown in Fig. 2.10. Polycarbonate is mainly used for track-etched membranes with well-defined pore structures and very good mechanical strength. Now PC can also be used to make ultrafiltration and microfiltration membranes by the phase inversion process (13).

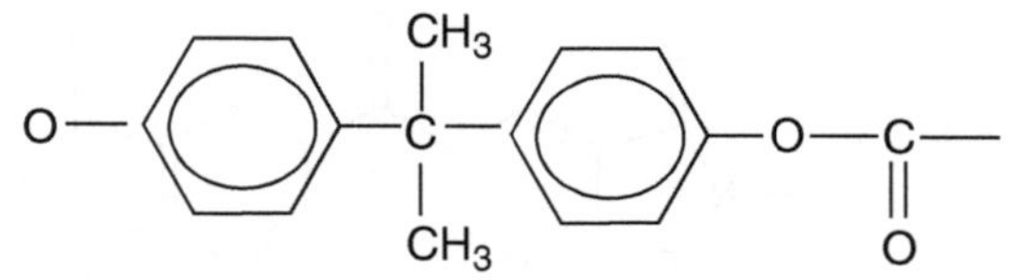

Fig. 2.10 Chemical structure of polycarbonate.

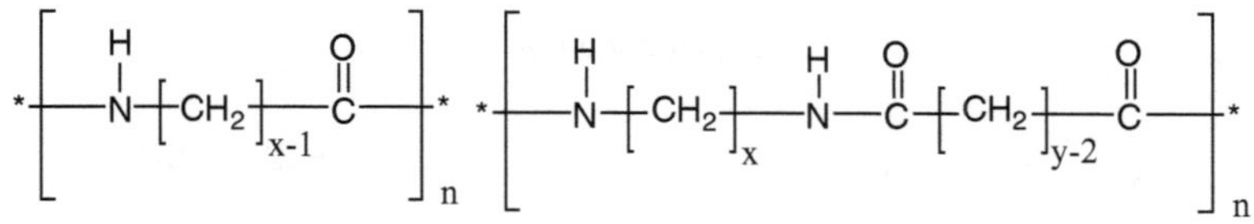

Fig. 2.11 Chemical structures of basic aliphatic polyamides (nylon x or nylon x,y).

3.8. Polyamide

Polyamides (PA) are synthetic amides in which the amide group (–NHCO–) is an integral group of the main polymer chain, which include aliphatic and aromatic polyamides. The basic aliphatic polyamides are referred as nylons, which chemical structures are shown in Fig. 2.11. Nylon 6, nylon 4,6 and nylon 6,6 are semicrystalline polymers, they are important engineering thermoplastics and possess good thermal stability and mechanical strength (14). The porous polyamide membranes have been commercialized for many years by immersion precipitation. Nylons are resistant to many organic solvents such as aliphatic, aromatic hydrocarbons as well as some linear alcohols and aprotic liquids such as NMP and DMF.

Now polyamides are frequently used as the thin dense layer in composite membranes for reverse osmosis and nanofiltration, which are formed by interfacial polymerization of diamines and diacylchlorides. Polyamide membranes overcome some of the pH limitations of cellulosic membranes. However, the chlorine tolerance is lower that cellulose based membranes.

3.9. Polyimide

Polyimides (PIs) are generally known as polymers with excellent thermal stability because of their high glass transition temperature, which demonstrate high thermal and chemical stability. Polycondensation of aromatic dianhydrides with aromatic diamines is to form a soluble poly (amic acid). The poly(amic acid) is shaped into the polyimide product by dehydration step. P84 copolyimide is an amorphous commercial copolyimide from Lenzing. Its solvent resistance to many organic solvents is excellent, thus it is mainly used for the preparation of solvent-resistant nanofiltration membranes. Whereas it is well soluble in amides and it is also not resistant to aqueous alkaline solutions. The T_g of P84 copolyimide is 315°C, which can be used at a high temperature (above 200°C). Matrimide 5218 is another commercial polyimide, which is manufactured by Ciba-Geigy. It is originally developed for use in microelectric industry. Now it is also used as a material for making gas separation and nanofiltration membranes. The chemical structures of copolyimide P84 and Matrimid 5218 are shown in Fig. 2.12.

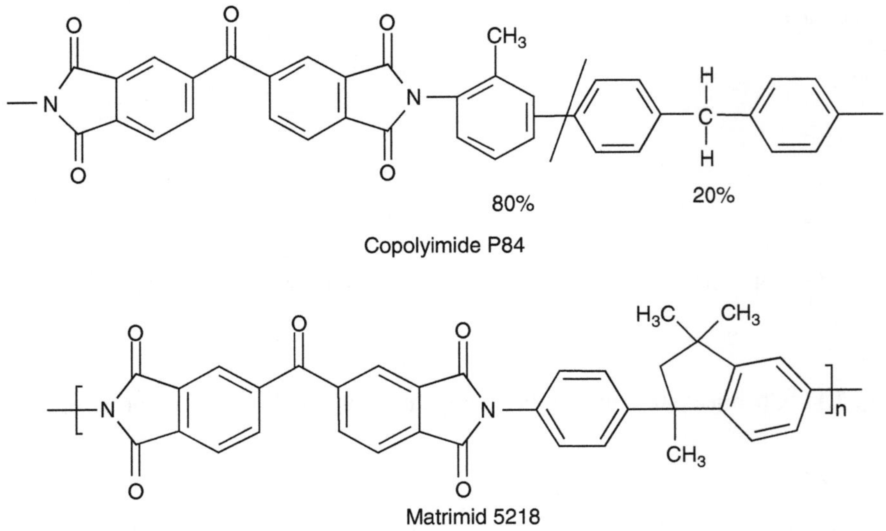

Fig. 2.12 Chemical structures of copolyimide P84 and matrimid 5218.

Fig. 2.13 Chemical structures of PEEK and SPEEK.

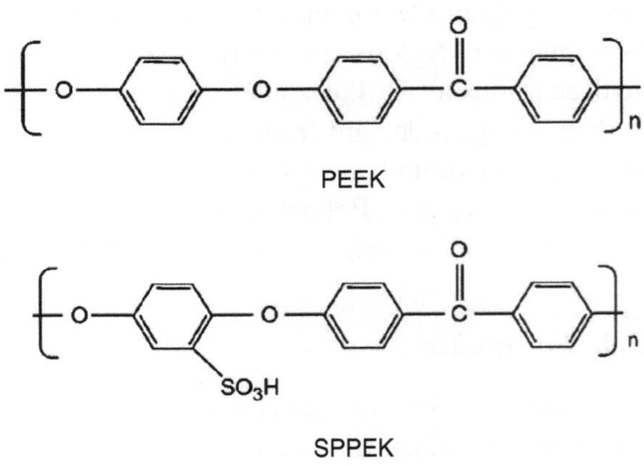

3.10. Polyether Ether Ketones

Polyether ether ketones (PEEK) are engineering thermoplastics with an exceptional combination of heat and chemical stability. The high insolubility in common solvents makes PEEK membranes successfully being used in chemical processes as a solvent resistant membrane. But the PEEK membrane fabrication is limited by the conventional solvent based methods for membrane casting. The sulfonated PEEK(SPEEK) is soluble in common solvents, which can be used for the preparation of hydrophilic membranes or ion-exchange membranes. The chemical structure of PEEK and SPEEK are shown in Fig. 2.13.

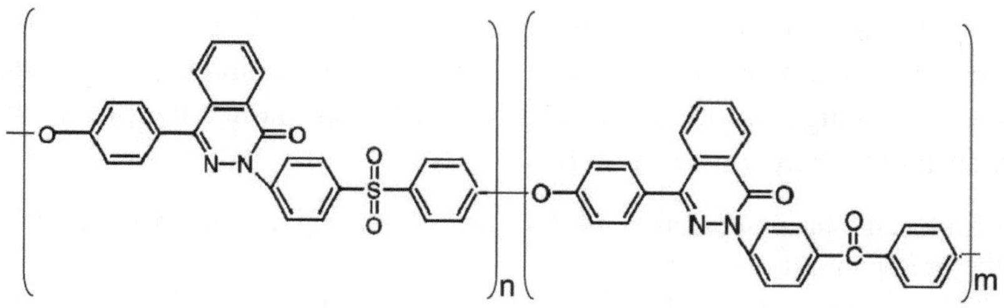

Fig. 2.14 Chemical structure of PPESK.

Fig. 2.15 Chemical structure of PEBAX.

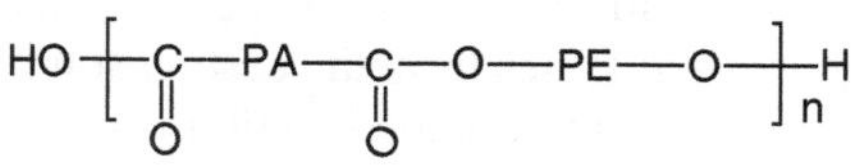

3.11. *Poly(phthalazine ether sulfone ketone)*

Poly(phthalazine ether sulfone ketone) (PPESK) contains rigid aromatic rings and has shown very good mechanical strength and chemical resistance, which chemical structure is shown in Fig. 2.14. Its glass transition temperature is near 300°C. PPESK is a potential membrane material for high temperature gas separation. It is also used to prepare ultrafiltration and nanofiltration membranes for water treatment.

3.12. *Polyether Block Amide*

Polyether block amide (PE-b-PA) copolymer resin is a thermoplastic elastomer combining linear chains of rigid polyamide segments interspaced with flexible polyether segments. The commercial PE-b-PA copolymer resin is under the trademark PEBAX by Elf-Atochem, which has the general chemical structure shown in Fig. 2.15.

Where PA is an aliphatic polyamide "hard block" (i.e. Nylon 6 [PA6] or Nylon 12 [PA-12]) and PE is an amorphous polyether "soft block". The soft segment is either polyether oxide (PEO) or polytetramethylene oxide (PTMEO). This crystalline/amorphous structure demonstrates properties of thermoplastics and elastic rubbers. Usually, the hard amide block provides the mechanical strength of the prepared membranes, and the soft ether blocks mainly favor the transport and separation characteristics.

4. PHASE INVERSION MEMBRANES

Following the development of the first anisotropic membrane via precipitating a cellulose acetate casting solution in a water bath by Loeb and Sourirajan in 1960s, different techniques such as solute casting, interfacial polymerization and plasma polymerization have been proposed to make selective, permeable anisotropic membranes during the past three decades. Nevertheless, the phase inversion technique is still the most popular and important preparation method, especially for commercial membrane productions. In the process of the phase inversion, the polymer solutions may be precipitated by different approaches, including cooling, immersion in a nonsolvent coagulant bath, evaporation and vapor adsorption as well.

According to the change of the operating parameters that induce the phase inversion, two different separation mechanisms are involved:

1. Thermally induced phase separation (TIPS): by decreasing the temperature of the polymer solution, the precipitation is induced.
2. Diffusion induced phase separation (DIPS): by contacting a polymer solution to a vapor or liquid, diffusional mass exchange leads to a change in the local composition of the polymer film and the precipitation is induced. Three types of techniques have been developed to reach DIPS, which are shown in Fig. 2.16.
   (a) Immersion precipitation: The casting solution is immersed into a nonsolvent bath, and the nonsolvent from the coagulation bath diffuses into the casting solution and the solvent diffuses from the casting solution into the nonsolvent bath, which cause a rapid precipitation of the casting solution from the top surface downward.
   (b) Vapor adsorption: The casting solution is subjected to a vapor containing nonsolvent and some gases (such as air, nitrogen, etc.). The adsorption of the nonsolvent will cause the precipitation of the casting solution.
   (c) Solvent evaporation: The casting solutions are prepared with a volatile solvent, a less or no volatile nonsolvent and a polymer. Preferential evaporation of the volatile solvent will generate meta- or unstable compositions and precipitation will be induced.

In the process of membrane fabrication, the solvent evaporation, vapor adsorption, and immersion precipitation may occur simultaneously. For example, in the spinning of hollow fiber membranes, the nascent hollow fiber membranes spun from the spinneret will go through an air gap and then immerse into a water bath. At the air gap, the solvent in the casting solution will evaporate into the air, and the humidity in the air gap will be adsorbed into the nascent hollow fiber membranes, then the immersion precipitation occurs when the nascent hollow fiber membranes is immersed into the water bath.

In the phase inversion process, the formation of asymmetric membranes is controlled by both the thermodynamics of the casting solution and the kinetics of transport process. The thermodynamics of the casting solution is related to the phase equilibrium between the

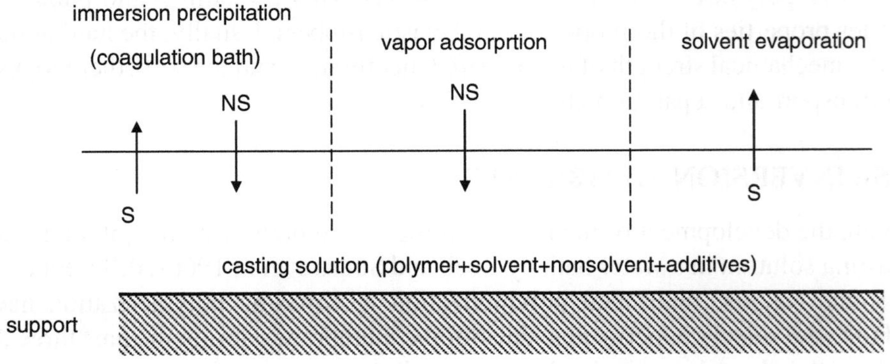

Fig. 2.16 Schematic representation of three DIPS processes (S: solvent; NS: nonsolvent).

components of the casting solution, while the kinetics of transport process can be described by the mutual diffusion and transport of the components. In all the phase inversion processes, a liquid polymer solution is precipitated into two phases: a polymer-rich phase that will form the matrix of the membrane and a polymer-poor phase that will form the membrane pores in an unstable nascent membrane structure. The porous asymmetric membrane morphology is then fixed according to subsequent solidification process.

4.1. Thermodynamics of the Polymer Solution

The thermodynamic condition of the polymer solution is very important for the membrane preparation, which involves the formula of the polymer solution, the choice of the solvent, nonsolvent and additives, etc., for a specific casting polymer system. The phase inversion process is induced by either TIPS or DIPS to change the thermodynamic properties of the dope solution.

4.1.1. Thermally Induced Phase Separation

TIPS is one of main approaches to prepare microporous membranes. In the membrane preparation by TIPS, a polymer is dissolved into a poor solvent or a mixture of solvent and nonsolvent at a high temperature. By cooling down the homogeneous solution, the phase separation is induced. After the polymer-rich phase is solidified by glass transition or crystallization, the porous membrane structure can be created by removing the solvent via extraction. Many studies reported on the membrane formation by TIPS were mainly from the basic standpoints of thermodynamics and kinetics (15–33). The phase diagrams and structure growth mechanisms were clarified in many systems, which concerned liquid–liquid demixing, solid–liquid demixing, crystallization, gelation, vitrification (glass transition), etc. Fig. 2.17 shows a phase diagram of TIPS with an amorphous polymer, where the liquid–liquid demixing and glass transition processes are involved. According to Berghmans' mechanism (34, 35), gelation results from the liquid–liquid phase separation arrested by the vitrification of the polymer-rich phase. From the phase diagram, three different phase areas of homogeneous solution phase I, the liquid–liquid demixing (metastable areas II, III, and unstable area IV) and one phase glass V exhibit, which are separated by binodal curve, spinodal curve and glass transition curve.

For a solvent (component 1)-polymer (component 2) system, its thermodynamic behavior can be described by Gibbs free energy of mixing:

$$\frac{\Delta G_{\mathrm{m}}}{RT} = n_1 \ln \phi_1 + n_2 \ln \phi_2 + n_1 \phi_2 \chi, \tag{1}$$

where the subscripts 1 and 2 denote the solvent and the polymer, respectively. ΔG_{m}, R, T, n, and ϕ are Gibbs free energy of mixing, gas constant, temperature, the molar number and volume fraction of the polymer and solvent, respectively. χ is the interaction parameter reflecting the non-ideality of the system. The chemical potential of the components can be described as:

$$\Delta \mu_i = \left(\frac{\partial \Delta G_{\mathrm{m}}}{\partial n_i} \right)_{P,T,n_{j,j \neq i}} \quad i,j = 1,2\,, \tag{2}$$

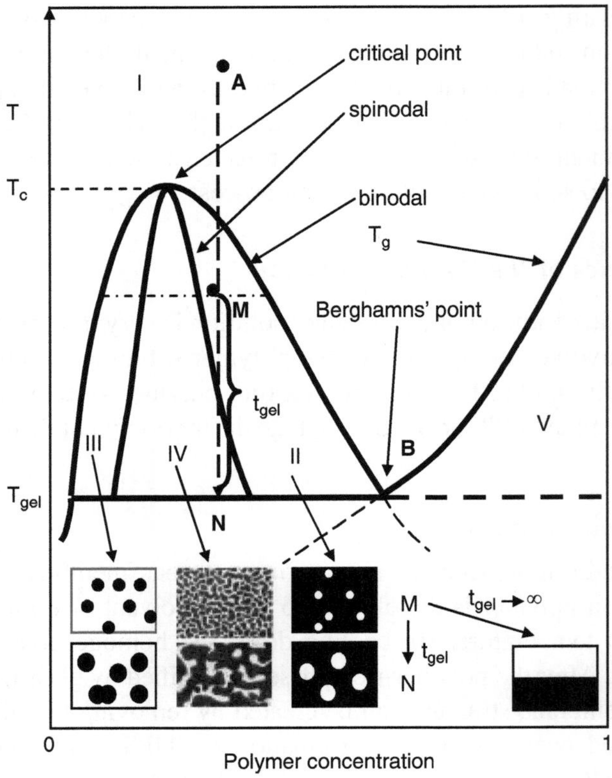

Fig. 2.17 Schematic phase diagram of thermally induced phase separation.

where $\Delta\mu_i$ and P are the chemical potential of component i and the system pressure, respectively. For the binodal curve, the polymer-rich and polymer-poor phases present the following relation:

$$\Delta\mu_i^{(\text{polymer-poor})} = \Delta\mu_i^{(\text{polymer-rich})}. \tag{3}$$

And the spinodal curve can be obtained by (1)

$$\frac{\partial^2 \Delta G_{\mathrm{m}}}{\partial \phi_i^2} = 0. \tag{4}$$

The critical temperature and composition can be obtained via Eqs. (4) and (5) (1):

$$\frac{\partial^3 \Delta G_{\mathrm{m}}}{\partial \phi_i^3} = 0. \tag{5}$$

Glass Transition

The glass transition temperature depression of a polymer by a low molecular weight diluent was proposed by different theories in literature. For a membrane forming system, it was reported that the glass transition temperature depression can be described well using the Kelley–Bueche equation (36). The Kelley–Bueche equation was based on the free volume concept (36):

$$T_g = \frac{R\phi_1 T_{g1} + \phi_2 T_{g2}}{R\phi_1 + \phi_2}, \tag{6}$$

where the subscripts 1 and 2 denote the diluent and the polymer, respectively. T_g is the glass transition temperature of diluted polymer. T_{gi} and ϕ_i are the glass transition temperature and the volume fraction of the component i. The glass transition temperature of a polymer can be found in literature. The estimate of a solvent can be obtained according to an empirical equation (37):

$$\gamma = \frac{T_m + T_b}{T_g + T_b}, \tag{7}$$

where T_m, T_b, and T_g are the melting temperature, boiling temperature and glass transition temperature of one solvent, respectively. The coefficient of γ in Eq. (7) is almost a constant of 1.15, which changes from 1.09 to 1.30 for some solvents (37). A modified formula is given by (38),

$$\gamma^* = \frac{T_m + T_b}{T_g + T_b} + 0.6\frac{T_b - T_m}{T_b + T_m}. \tag{8}$$

The coefficient of γ^* is 1.36, which deviates from 1.30 to 1.41.

Gelation Point

The gelation mechanism for a binary solution of an amorphous polymer was defined by Berghmans (34). The gelation arises in the system characterized by the intersection of the binodal and the glass transition boundary in a temperature–concentration phase diagram. The Berghmans point (B point) is the intersection between the binodal curve and the glass transition boundary of the system. So the gelation results from the liquid–liquid phase separation arrested by the vitrification of the polymer-rich phase. By cooling down a homogeneous solution noted as A in Fig. 2.17, the solution separates into two phases at point M after it crosses the binodal curve. Upon further cooling, the composition of the polymer-rich phase finally reaches point B, and the polymer-rich phase is vitrified. Consequently the structure of the demixed solution is fixed. The time from the original phase separation (point M) to the final vitrification of the polymer-rich phase (point B) is defined as the gelation time (t_{gel}). If the cooling speed is not infinitely slow, a porous glass will be formed. The original porous structure is created by the liquid–liquid demixing and the final porous structure is related to the vitrification of the polymer solution. The liquid–liquid demixing will cease as soon as gelation starts.

Vitrification and Coarsening Phenomenon

When the phase separation (point M) occurs in the metastable region (II, III) between the binodal and spinodal curves, the mechanism of the phase separation is commonly referred to as nucleation and growth (NG). At the metastable region II, only the structure of the polymer-rich phase is interpenetrating and the polymer-poor phases is discrete. Closed pores can be observed in the membrane structure, which is shown in Fig. 2.17. The similar situation takes place in metastable region III, and the system is a suspension solution in which the polymer particles are dispensed in a polymer-poor phase.

When the phase separation (point M) occurs inside the unstable region IV, the system is unstable. Any infinitesimal concentration fluctuation can induce phase separation and the mechanism of phase separation in this region is called spinodal decomposition (SD). It is a spontaneous process without the need of a nucleus. For the spinodal decomposition process, an interpenetrating (bicontinuous) three-dimensional network is formed, which is also shown in Fig. 2.17.

Since the polymer solution cannot be vitrified in the initial stage of the phase separation (M point), the original phase separation structures based on NG and SD (shown in Fig. 2.17) are not stable. They will grow and coarsen within a limited gelation time (shown in Fig. 2.17) and two fully separated phases may be obtained with infinite gelation time (39–42). Due to the coarsening phenomenon, it is almost impossible to attribute obtained membrane structures to the demixing processes (4). For a SD system within a short gelation time, the coalescence process is "frozen" early enough by vitrification and a morphology with high interconnectivity is obtained. With the increase of the gelation time, the interpenetrating structure coarsens and the interconnectivity declines (43). For a long gelation time, the phase coalescence may lead to a closed cell structure (44). Thus, the coarsening processes are very important in the phase separation process, which can determine the final size and interconnectivity of the porous structure of the membranes. The coarsening processes can be controlled or adjusted according to the gelation time. Different membrane morphologies can be obtained by controlling the different gelation times.

4.1.2. Diffusion Induced Phase Separation

Phase Diagram

The thermodynamic behavior of different spinning systems in the process of DIPS can be also described in a phase diagram, which was first used by Strathmann in 1971 (45) and then advanced by Smolders et al. (46, 47). For an immersion precipitation process, at least three components of polymer, solvent and nonsolvent are involved. Usually, other components such as blended polymers, mixed solvents, and nonsolvent additives are used to get desired membrane morphology with good performance. Nevertheless, a ternary combination of polymer, solvent and nonsolvent is discussed here for the simplicity. Figure 2.18 is a schematic illustration of the phase behavior in a polymer, a solvent and a nonsolvent system. From the phase diagram, four different regions are found.

Region I: one phase solution
Region II: liquid–liquid two-phase solution
Region III: liquid–solid two-phase swelling
Region IV: one phase glass and swelling

In order to distinguish different regions mentioned above, Berghmans' point B, the spinodal curve, the binodal curve, the vitrification boundary, the swelling boundary, and

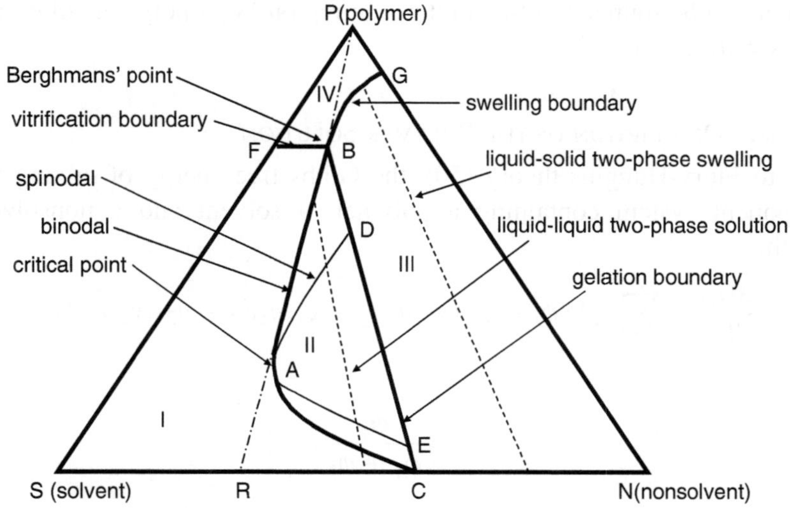

Fig. 2.18 Schematic phase diagram of a ternary system for DIPS.

the gelation boundary are depicted. Berghmans' point is defined as the intersection between the cloud point curve and the glass transition boundary. Although it is originally derived from a binary system, a generalization was made by Li et al. (38, 47) to extend to a ternary membrane forming system consisting of a polymer, a solvent and a nonsolvent.

From this diagram, the binodal curve BAC and the spinodal curve DAE exist in the liquid–liquid two-phase solutions. One metastable area (BAD) exists between the spinodal and binodal curves at high polymer concentrations, and the other metastable area (CAE) is at lower polymer concentrations. The unstable area (DAE) is also formed by the spinodal curve. Point "B" is the Berghmans' point. At this point, the porous structure of the polymer-rich phase will be vitrified. The gelation boundary BC and the solvent–nonsolvent axis intersects at point C, which denotes the maximum solvent concentration in a coagulation bath that can be used to prepare a solidified membrane. Point C is defined as the "critical coagulation point", which is abbreviated to "C-point" (48, 49). C-point is in the equilibrium with the Berghmans' point, which is also an extension of polymer-poor phase at solvent–nonsolvent axis, so it is a mixture of only solvent and nonsolvent, and no polymer is included. When the nonsolvent concentration in the mixed solvent is higher than C-point, the polymer can only be swollen. For membrane making, the membrane structure becomes more rigid and the mechanical strength will be enhanced at the high nonsolvent concentration above C-point in coagulant bath. C-point is a clear boundary between the liquid–liquid phase separation and polymer swelling.

The solution in region I is homogeneous with one phase liquid state. In region II, the solution will separate into two liquid phases: one polymer-rich phase and one polymer-poor phase. In region III, there are a swollen polymer phase and a mixed liquid phase. In this region, the polymer cannot be dissolved but swollen in the mixture of solvent and nonsolvent, and no polymer can be founded in the mixture. In region IV, a polymer solution reaches its glass state or swelling state.

Thermodynamic Description of the Polymer Solution

According to Flory-Huggins theory (50), the Gibbs free energy of mixing (ΔG_m) for a ternary component system containing a polymer, a solvent and a nonsolvent, can be expressed with:

$$\frac{\Delta G_m}{RT} = \sum_{i=1}^{3} n_i \ln \phi_i + g_{12}(u_2)n_1\phi_2 + g_{13}n_1\phi_3 + g_{23}(v_2)n_2\phi_3, \quad (9)$$

with

$$u_2 = \frac{\phi_1}{\phi_1 + \phi_2}, \quad (9a)$$

$$v_2 = \frac{\phi_2}{\phi_2 + \phi_3}, \quad (9b)$$

where the subscripts 1, 2, and 3 denote nonsolvent, solvent, and polymer, respectively, n_i and Φ_i are the number of mole and the volume fraction of component i, g_{ij} is binary interaction parameter between component i and j.

The chemical potential of the individual component i is expressed as:

$$\frac{\Delta\mu_1}{RT} = \ln \phi_1 - s\phi_2 - r\phi_3 + (1 + g_{12}\phi_2 + g_{13}\phi_3)(1 - \phi_1) - sg_{23}\phi_2\phi_3 - g'_{12}\phi_2, \quad (10)$$

$$\frac{s\Delta\mu_2}{RT} = s\ln \phi_2 - \phi_1 - r\phi_3 + (s + g_{12}\phi_1 + sg_{23}\phi_3)(1 - \phi_2) - g_{13}\phi_1\phi_3 + g'_{12}\phi_1 + sg'_{23}\phi_3, \quad (11)$$

$$\frac{r\Delta\mu_3}{RT} = r\ln \phi_3 - \phi_1 - s\phi_2 + (r + g_{13}\phi_1 + sg_{23}\phi_2)(1 - \phi_3) - g_{12}\phi_1\phi_2 - rg'_{23}\phi_2 \quad (12)$$

with g'_{ij} being as follows:

$$g'_{12} = u_2(1 - u_2)\frac{\partial g_{12}}{\partial u_2}, \quad (10a)$$

$$g'_{23} = v_2(1 - v_2)\frac{\partial g_{23}}{\partial v_2}, \quad (11a)$$

where $\Delta\mu_i$ is the chemical potential of component i and $s = \bar{v}_1/\bar{v}_2$, $r = \bar{v}_1/\bar{v}_3$. $\bar{v}_i$ is the partial molar volume of component i.

The requirements for the phase equilibrium (the binodal curve) are:

$$\Delta\mu_i^{(\text{polymer-rich})} = \Delta\mu_i^{(\text{polymer-poor})}, \quad i = 1, 2, 3, \quad (13)$$

The objective function F is defined as:

$$F = \sum_{i=1}^{3} \left(\frac{\Delta\mu_i^{(\text{polymer-rich})}}{RT} - \frac{\Delta\mu_i^{(\text{polymer-poor})}}{RT} \right)^2. \quad (14)$$

The phase equilibrium (the binodal curve) can be obtained when the objective function F is in minimum. The system within the binodal curve is unstable. It will decompose into two equilibrium phases, which are related by tie lines in the phase diagram.

For the swelling curve BG, since there is no polymer dissolved in the mixture of solvent and nonsolvent, the objective function F_1 is defined as:

$$F_1 = \sum_{i=1}^{2} \left(\frac{\Delta\mu_i^{(\text{swelling})}}{RT} - \frac{\Delta\mu_i^{(\text{mixture/solvent-nonsolvent})}}{RT} \right)^2. \tag{15}$$

The unstable liquid–liquid two-phase region is divided into two regions by the spinodal curve. The metastable region is formed between the spinodal and binodal curves. In this metastable region, small perturbations will decay, and the phase decomposition can only be caused by a large perturbation. Within the spinodal curve, any small perturbation will induce the phase separation of the system. The spinodal curve in a ternary system is defined as (51):

$$G_{22}G_{33} = G_{23}^2, \tag{16}$$

G_{22}, G_{33} and G_{23} are the derivatives of ΔG_m with respect to ϕ_i, ϕ_j

$$G_{ij} = \left\{ \frac{\partial \Delta G_\mathrm{m}/(VRT)}{\partial\phi_i \partial\phi_j} \right\} \bar{v}_1. \tag{17}$$

Here, V is the total volume. The spinodal curve for a ternary system with constant interaction parameters can be expressed as follows:

$$\left(\frac{1}{\phi_1} + \left(\frac{\bar{v}_1}{\bar{v}_2}\right)\left(\frac{1}{\phi_2}\right) - 2g_{12}\right)\left(\frac{1}{\phi_2} + \left(\frac{\bar{v}_1}{\bar{v}_3}\right)\left(\frac{1}{\phi_3}\right) - 2g_{13}\right) = \left(\frac{1}{\varphi_1} - g_{12} - g_{13} + \left(\frac{\bar{v}_1}{\bar{v}_2}\right)g_{23}\right)^2. \tag{18}$$

Linearized Cloud Point Curve Correlation

The phase behavior of a ternary polymer solution can be determined by cloud point measurements and the compositions of the cloud points refer to the position of the binodal. The correlation of the linearized cloud point curve was obtained for polymer solutions in the mixture of one solvent and one nonsolvent (52–54). With this correlation, the information about the thermodynamics of ternary systems can be obtained obviously, which is helpful for preparing membrane forming systems. The linearized cloud point curve correlation describes a linear relationship between the nonsolvent concentration relative to the polymer concentration and the solvent concentration relative to the polymer concentration at the condition of phase separation, which is shown as (54):

$$\ln\frac{\phi_1}{\phi_3} = b\ln\frac{\phi_2}{\phi_3} + a, \tag{19}$$

where

$$b = \frac{\bar{v}_1 - \bar{v}_3}{\bar{v}_3 - \bar{v}_3}, \tag{19a}$$

$$\begin{aligned} a = {} & (g_{12} + g'_{12})(-\bar{v}_2 b\phi_1 + \bar{v}_1\phi_2) + g_{13}(\bar{v}_1\phi_3 - \bar{v}_3(1-b)\phi_1) \\ & +(g_{23} + g'_{23})(-\bar{v}_2 b\phi_3 + \bar{v}_3(1-b)\phi_2) \end{aligned}. \tag{19b}$$

The conditions for the validity of this correlation function appear to be that the polymer is strongly incompatible with the nonsolvent, and that only the liquid–liquid demixing occurs (53). The slope of the linearized cloud point seems to be only dependent on the molar volumes of the components. Information about the binary Flory–Huggins interaction parameters and their concentration dependence can be obtained from the intercept of the linearized curve.

In an earlier published paper by Li et al. (52), an empirical rule for cloud point curves was given for the membrane forming system of PS–DMAC–water:

$$\frac{\phi_1}{\phi_2} = \beta. \tag{20}$$

From this description, the ratio of volume fractions of the nonsolvent and solvent is a constant of β, which is a simplification of the linearized cloud point curve correlation [Eq. (19)] at the following condition:

$$\phi_3 >> \exp\left(\frac{a}{1-b}\right), \tag{21}$$

β in the empirical rule [Eq. (20)] can be represented as:

$$\beta = \exp(a). \tag{22}$$

Form the ternary system phase diagram in Fig. 2.18, the composition of the binodal curve above the critical point is extended to one straight line PR. And R point in the phase diagram with volume fraction can be determined by:

$$\frac{\overline{SR}}{\overline{RN}} = \beta, \tag{23}$$

where $\overline{SR}$ and $\overline{RN}$ are the Length of line SR and RN in Fig. 2.18, which are equal to the nonsolvent and solvent concentration at point R, respectively. For a phase diagram with weight fraction, R point can be determined by:

$$\frac{\overline{SR}}{\overline{RN}} = \beta\left(\frac{\bar{v}_2}{\bar{v}_1}\right)\left(\frac{M_1}{M_2}\right), \tag{24}$$

M_1 and M_2 are the molar masses of nonsolvent and solvent, respectively. The empirical rule [Eq. (20)] is important for the preparation of membrane casting system.

Approaching Ratio

From the linearized cloud point curve correlation of Eq. (20), it is easy to determine the maximum ratio of nonsolvent and solvent in the membrane casting system, which is independent of the concentration of the polymer concentration. And it is also clear that the weight ratio (or volume ratio) of nonsolvent and solvent at the cloud point curve (above critical point) was constant. This suggests that the binodal curve is strongly dominated by the weight ratio of nonsolvent and solvent in the casting solution (above critical point), and is almost not influenced by the polymer concentration. In order to describe the thermodynamic state of the casting solution, the approaching ratio α (55–57), is defined as follows:

$$\alpha = \frac{\left(\frac{C_{\text{nonsolvent}}}{C_{\text{solvent}}}\right)_{\text{dope}}}{\left(\frac{C_{\text{nonsolvent}}}{C_{\text{solvent}}}\right)_{\text{cloud point}}}, \tag{25}$$

where the numerator is the weight ratio of nonsolvent and solvent in the casting solution, and denominator is the weight ratio of nonsolvent and solvent at the cloud point. The dope and cloud point in Eq. (25) refer to the casting solution and the linearized cloud point curve, respectively. The approaching ratio α reflects the quality of the solvent mixture and the thermodynamic characteristics, which has a significant influence on the membrane morphologies.

Approaching Coagulation Ratio

Similar to the approaching ratio, the solvent content in the coagulation bath is characterized by the approaching coagulation ratio γ, which is defined as the ratio of the solvent concentration in the coagulation bath to that at the C-point (48, 49):

$$\gamma = \frac{(C_{\text{solvent}})_{\text{Bath}}}{(C_{\text{solvent}})_{C-\text{point}}}, \tag{26}$$

where $(C_{\text{solvent}})_{\text{Bath}}$ is the mass concentration of the solvent in the coagulant bath, and $(C_{\text{solvent}})_{C\text{-point}}$ is the mass concentration of the solvent in the coagulant bath at *C*-point shown in Fig. 2.18. In the case of $\gamma = 0$, the coagulant is a 100% pure nonsolvent. If $\gamma = 1$, the bath has the same composition as in the *C*-point where the polymer solution just can be solidified in such a bath. With $\gamma > 1$, a solidified membrane structure cannot be obtained. For the immersion precipitation, the approaching coagulant ratio of the coagulant bath is usually controlled between 0 and 1 so that the exchange rate of solvent and nonsolvent between polymer solution and coagulant bath can be controlled to influence the final membrane morphologies. The approaching coagulation ratio is also a very important parameter in membrane fabrication, especially for the selection of a bore fluid in spinning hollow fiber membranes.

4.2. Membrane Formation Processes

The membrane formation process can be illustrated with a ternary phase diagram as shown in Fig. 2.19. Point A represents the initial composition of a casting solution. Paths 1 and 2

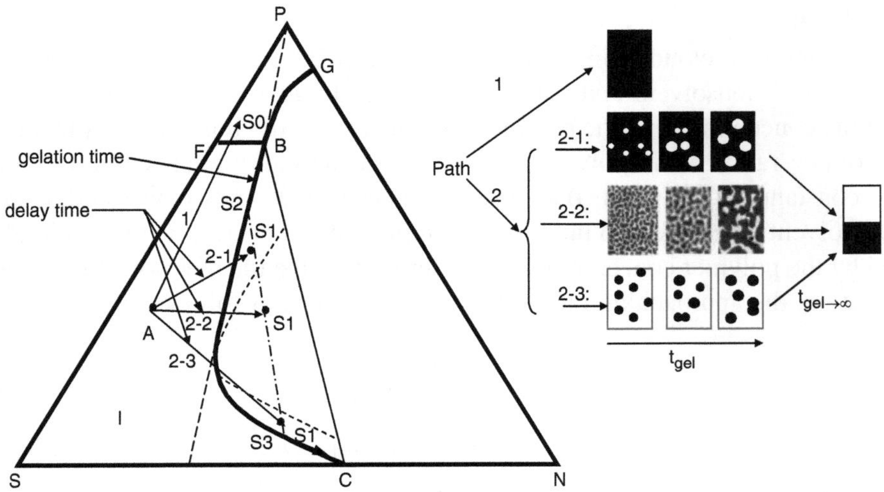

Fig. 2.19 Schematic diagram of membrane formation processes for DIPS.

(2-1, 2-2, 2-3) represent two possible composition paths if a polymer casting solution with composition A is immersed into a nonsolvent bath. For path 1, the polymer solution undergoes a glass transition and goes into the glass and swelling phase IV directly. Consequently, the solution becomes a homogeneous glass film. For path 2, the gelation dominates the formation of the porous membrane morphologies.

4.2.1. Delay Time

For path 2 (2-1, 2-2, 2-3), when the composition path crosses the binodal and reaches point S1, liquid–liquid phase separation (S1→S2 + S3) occurs and nascent porous membrane structure is formed. The time interval between the moment of the immersion of a polymer solution into the coagulant and the onset of the liquid–liquid demixing (A→S1) is defined as delay time (58, 59). Based on different delay times, two different mechanisms occur in the phase separation by immersion precipitation: instantaneous demixing and delayed demixing. Instantaneous demixing occurs when the phase separation begins immediately after immersion. The precipitation path crosses the binodal and two distinct phases are formed. For delayed demixing, the composition of the entire solution remains in the homogeneous region of the phase diagram for a certain period of time.

4.2.2. Gelation Time

When the composition path reaches the gelation boundary, i.e., the rich phase of the polymer solution reaches the Berghmans' point B, where the polymer-rich phase vitrifies and the structure of the demixed solution is fixed. The time interval between the onset of the demixing and the solidification of the polymer solution (S2→B) is defined as the gelation time (47). The delay time and gelation time, two macroscopic time scales, strongly influence the final membrane morphology (47). However, the membrane formation is a competitive process between demixing and gelation (60).

4.2.3. Formation of Nascent Porous Membrane Morphologies

For the path 2, according to different locations of liquid–liquid phase separation point S1, three different nascent membrane morphologies are formed by "the nucleation and growth of the polymer-poor phase" (path 2-1),"the spinodal decomposition" (path 2-2), and "the nucleation and growth of the polymer-rich phase" (path 2-3).

For the path 2-1, the solvent–nonsolvent exchange brings the system first to a metastable condition above critical point, NG mechanism is favored (8, 61). Under this condition, the free energy associated with the phase separation decreases and nuclei less than the critical size will simply "re-dissolve" into the homogeneous polymer solution. The nuclei of polymer-poor phase larger than the critical size grow, forming dispersed nuclei and becoming stable if the activation energy for the nucleus formation is higher than their surface free energy. NG is usually a slow process and leads to a morphology with dispersed pores.

For the path 2-2, the demixing path crosses the critical point or the metastable region and goes directly into the unstable region, SD mechanism predominates. A concentration fluctuation appears in the initially homogeneous system and progresses with increasing amplitude, leading to a bicontinuous network of the polymer-poor and polymer-rich phases without any nucleation and growth due to instantaneous demixing (8). The polymer-poor phase will also form the pores. A truly bicontinuous network formed via SD promotes an open porous membrane support without selective function ideally.

For the path 2-3, when the liquid–liquid demixing starts somewhere below the critical point, the nucleation and NG mechanism is also favored. The polymer-rich phase nucleates and grows, leading to low-integrity powdery agglomerates. Such a polymer morphology is not practical and it thus rarely happens in the membrane formation.

4.2.4. Vitrification of Membrane Morphology

The membrane formation is a competitive process between liquid–liquid demixing and gelation process (38, 48, 49, 60). In order to understand the formation process of porous membranes, it is important to know how long it takes to start the liquid–liquid demixing, how fast the demixing takes place and when the demixing can be stopped, which are associated with the delay time (62–66) and the gelation time (38, 49), respectively. According to the exchange rate of solvent and nonsolvent, the nascent membrane morphology will be formed at the phase separation point S1 based on NG or SD mechanisms. Compared with NG and SD demixing (S1→S2 + S3), the gelation process (S2→B) also has a significant influence on the final membrane morphology after the liquid–liquid phase inversion, as both NG and SD usually progress to a phase coalescence and the final structure can be controlled by the following gelation process.

Nunes et al. (9) also considered that the point where the developing structure is fixed is at least as important as the starting mechanism of phase separation. After the liquid–liquid demixing, the concentration of the polymer-rich solution increases (along S2→B) by the solvent–nonsolvent exchange. As the solvent concentration decreases, the membrane structure will be solidified due to the vitrification of the polymer-rich phase at point B. If the coagulation concentration approaches the point C (in Fig. 2.19) from the pure nonsolvent, the gelation time will also increase. The influence of the gelation time on the membrane morphology is also shown in Fig. 2.19. At the very short gelation time, the system (S2→B) solidifies quickly after the first step of phase separation (A→S1), the membrane will have a fine pore structure and the original characteristics given by the initial demixing mechanism can be kept. For the NG process, a membrane morphology of closed cells would be formed because of rapid stop of the demixing during the initial stages. With the time going on, the nuclei would grow and touch each other forming interconnected pores. The SD demixing would induce the formation of an interconnected pore structure from the beginning, which might evolve into closed cell structure overtime (44). Because of the coarsening phenomenon, the final membrane morphology cannot be completely attributed to the original demixing process. For the NG and SD demixing processes, when the concentration of the coagulation bath is or lower than point C shown in Fig. 2.19, the nascent membrane morphology cannot be solidified and eventually two fully separated layers may be obtained.

4.2.5. Membrane Surface Formation of Porous Membranes

The formation process of the membrane surface for porous membranes can be illustrated with a ternary phase diagram as shown in Fig. 2.20. According to solvent evaporation or solvent–nonsolvent exchange, the surface polymer composition will reach the vitrification boundary directly (path 1) and consequently a dense membrane surface will be formed. When the polymer solution under the skin layer crossed the binodal curves, a porous membrane morphology under the skin layer is formed. The resultant membranes will show intrinsic

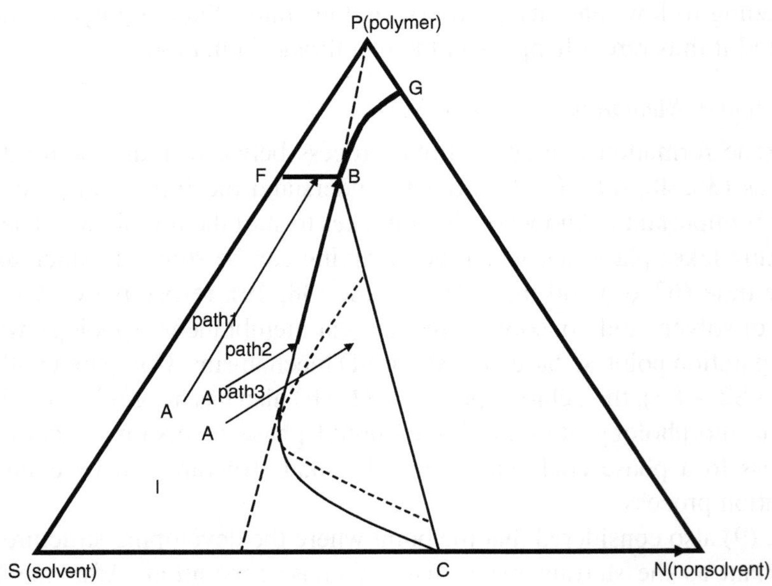

Fig. 2.20 Schematic diagram of membrane surface formation process for DIPS.

polymer properties for gas separation, pervaporation, reverse osmosis, et al. For the second path (path 2), the decomposition path of the membrane surface will go to the binodal curve. And the polymer concentration at the membrane surface will be always on the binodal position, which is in equilibrium with solvent concentration of the coagulant bath at the interface between the membrane and coagulant bath (38, 65, 67). With the time going on, the polymer concentration will change along the binodal curve, but no liquid–liquid demixing occurs on the surface layer because no polymer-poor phase is formed. At this period the skin layer cannot be solidified. The liquid–liquid demixing occurring at the sublayer underneath the membrane surface will influence the skin layer, and cause different pore size distributions and different skin thicknesses. The surface structure with different pore size distributions will be solidified until it reaches point B. The resulting membranes may be ultrafiltration, microfiltration and some gas separation membranes et al. Actually, for a dry-phase inversion process induced by forced-convective evaporation, the polymer concentration on the membrane surface sometimes crosses the binodal curve and goes into the spinodal decomposition (path 3) (68). Because of relative long gelation time, the coarsening of the bicontinuous structure occurs at a later stage of the phase separation process (before solidification) to form spherical nodules. Because of an additional physical process such as capillary pressure arising from the curvature of the interface of polymer-poor phase contacting with an external gas phase, an essentially homogeneous nascent skin layer will be formed by the coalescence of highly plasticized nodule network (43, 68, 69), shown in Fig. 2.21.

4.2.6. Macrovoid Formation

Macrovoids are very large elongated pores which can extend over the membrane thickness. It may have different shapes such as tear-like, pin-like, finger-like and so on. Macrovoids are undesirable because they cause poor mechanical strength in the membrane. The formation of macrovoids in the phase inversion process has been studied extensively in the literature (1, 4, 70–77). And a variety of mechanisms have been proposed and used to describe the formation of macrovoids such as the shrinkage of polymer matrix (71, 72), surface tension gradient-induced convective flow (75), osmotic pressure (73), instantaneous phase demixing (70), and concentration gradient-induced instability (76). However, a lot of controversy exists concerning the original of the macrovoids.

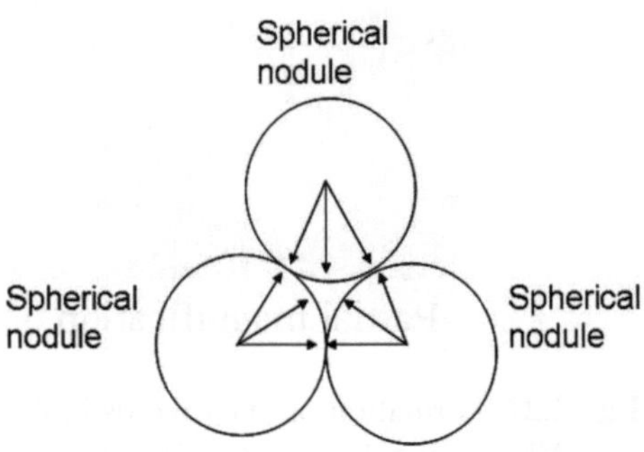

Fig. 2.21 Contracting forces of polymer particles resulting from capillary pressure in interstitial spaces of film [adapted from ref. (68)].

Viscous fingering phenomenon is a kind of far from equilibrium pattern formation, which is evolved from the flow instability resulting from a less viscous fluid displacing a more viscous fluid (78). The phenomenon of viscous fingering in the polymer and gelation systems has been studied in recent years (79–81), which is used to explain the macrovoid formation recently (57). Generally, for an immersion precipitation process especially for an instantaneous phase demixing, a nascent porous membrane structure is formed in the sub-layer with a nascent skin layer. Due to osmotic pressure (73), the nonsolvent diffused into the polymer poor phase formed in the nascent porous membrane structure with linearly unstable condition and initiated the formation of some macrovoids (invading phase, Fig. 2.22a, c). There existed a clear boundary between the macrovoid and its side or front casting solution. However, the base of the macrovoid was open, which mounted onto the membrane top layer like a pedicel, which can also be found from some membrane structures after solidification shown in Fig. 2.23. It was clear that the bases of macrovoids were relatively loose. These loose structures enhanced nonsolvent's diffusion into the polymer poor phase and initiated the formation of macrovoids (invading phase). The viscosity of invading phase (polymer-poor phase, macrovoids) was much lower than that of defending phase (casting solution). Thus the polymer poor phase would displace casting solution (defending phase) and formed viscous

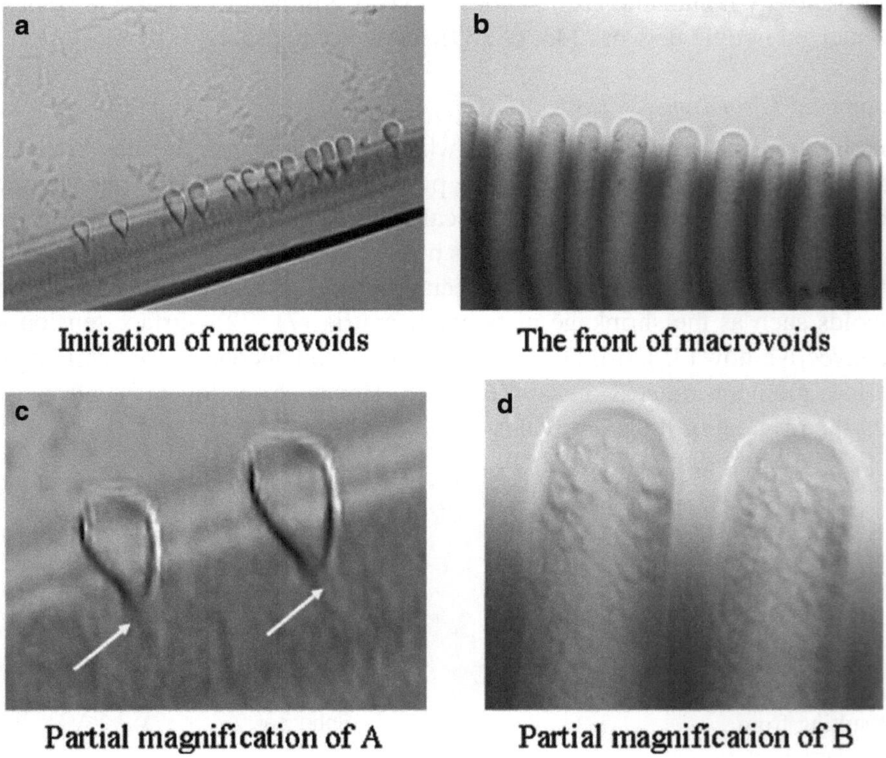

Fig. 2.22 Formation of macrovoids in the phase inversion process for casting solution (21 wt% P84, pure NMP; coagulant: water; 25°C).

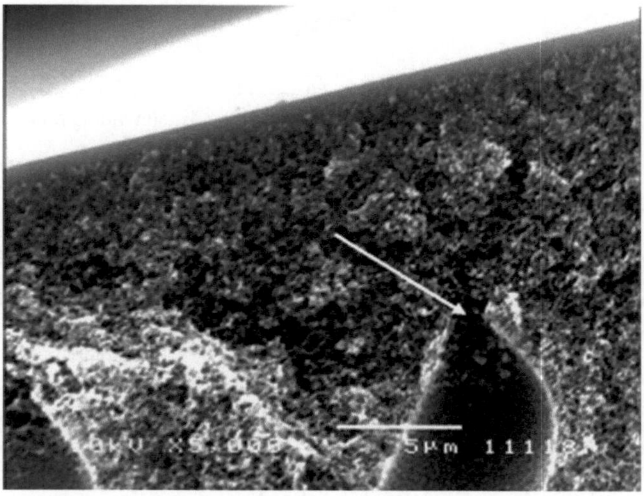

Fig. 2.23 Membrane structures for P84 copolyimide asymmetric membranes (15 wt% P84, H_2O/NMP: 4.8/95.2 (w/w); coagulant: water; casting temperature: 25°C).

fingering, which could advance quickly into the casting solution (Fig. 2.22b, d). Additionally, the loose bases of macrovoids were also the main transport channels for nonsolvent and polymer-lean phase to go into the viscous fingering and displace the casting solution. Hence, the finger-like structure of porous membranes was a result of hydrodynamically unstable viscous fingering that was developed when the casting solution was displaced by a polymer poor phase. The schematic diagram of macrovoid formation is shown in Fig. 2.24. Analogue to the viscous fingering in the polymer system in the Hele–Shaw cells, the advance of viscous fingering for a non-Newtonian polymer solution in the Hele–Shaw cells can be described with Darcy's Law (80):

$$v = \frac{b^2}{12\eta_{\text{eff}}} \nabla P, \tag{27}$$

where v is the finger tip velocity, b is the plate gap of the Hele–Shaw cell, η_{eff} is the shear dependent viscosity of the viscous polymer solution and ∇P is the pressure gradient.

Because of the phase inversion process, the viscous finger phenomenon in the immersion precipitation is different from traditional viscous fingering for two immiscible phases. Compared with the viscous fingering in the polymer system in the Hele–Shaw cells, the advance of viscous fingering in the immersion precipitation is controlled by three factors: the osmosis pressure between the polymer-poor phase and the coagulant, the viscosity of the dope solution and the approaching ratio (which can directly reflect the competition between the phase inversion process and viscous fingering). When poor coagulants are used or some solvents were added to the coagulation bath, the exchange rate of solvent and nonsolvent between the coagulation bath and the polymeric system will be very low and the delay demixing process occurs. Compared with a pure strong coagulation bath, its osmosis pressure

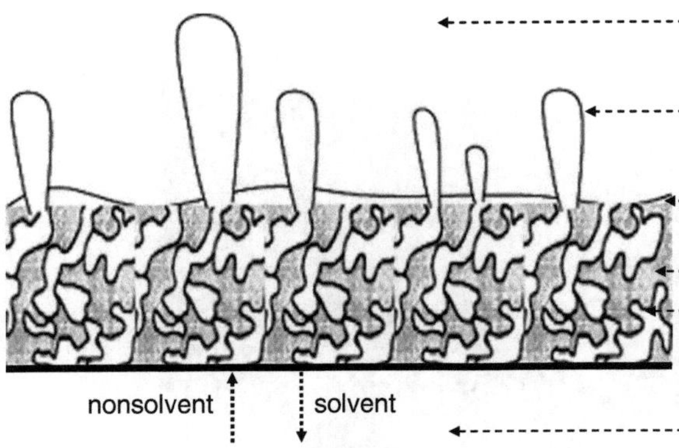

Fig. 2.24 Schematic diagram of macrovoid formation in the phase inversion process.

will be decreased. For the polymeric system with a low osmosis pressure and high viscosity, the advance of the viscous finger will be compressed. The tear-like or sponge-like membrane structure can be obtained.

Due to the exchange of the solvent and nonsolvent between the viscous fingering and the casting solution, a condition existed where there was much solvent inside the viscous fingering. This resulted in a delayed demixing and a sponge-like structure with porous surface was formed on the viscous finger walls. The phase inversion fronts induced by the delayed demixing on the viscous finger walls moved lower than the viscous fingering fronts (Fig. 2.22b, d). The competition between viscous fingering and phase inversion determined the membrane structure. When the rate of viscous fingering was faster than that of the phase inversion front, a finger-like structure would be obtained, as shown in Fig. 22.2. Conversely, macrovoids would not be formed and a sponge-like structure was observed when the rate of viscous fingering was slower than that of the phase inversion front. For the casting membrane system with a high approaching ratio, the composition of casting solution moved closer to the binodal curve and the phase inversion process could occur more easily. Even if some small macrovoids were irritated, the viscous fingering was also not formed and because the front interface between the irritating macrovoids and the casting solution would be solidified rapidly and easily. Compared with traditional viscous fingering phenomenon in liquid–liquid displacement, "viscous fingering" phenomenon in membrane making are more difficult to be described due to osmosis pressure, the exchange of different components and the competition between the viscous fingering and the phase inversion process. The formation of macrovoids may consist of three steps: instantaneous demixing (nascent skin layer with porous sub-layer) → initiation and growth of viscous fingering (a finger-like supporting structure) → delayed demixing (a sponge-like structure on the walls of viscous fingering).

Influence of Approaching Ratio on Membrane Morphology

The approaching ratio describes the quality of the solvent mixture and the thermodynamic characteristic, and represents the approaching degree of the dope solution to the phase demixing, which can influence the membrane morphology strongly. The membrane morphology changed from finger-like to sponge-like structure with an increase in the approaching ratio (57). In order to describe the change in the membrane structure more accurately, the relative sponge-like thickness β was defined as the ratio of the thickness of the sponge-like layer and that of the total membrane. The relationship between the approaching ratio α and relative sponge-like thickness β is shown in Fig. 2.25. (57). The critical approaching ratio α^* reflected the sharp change of membrane structure from finger-like to sponge-like. For the P84/NMP/H_2O casting solution (15 wt% P84; coagulant: water; casting temperature: 25°C), the critical approaching ratio α^* was about 0.67. In this figure, the membrane structures were classified into three parts: Part I: $0 \leq \alpha < \alpha^*$, the membrane was mainly a finger-like structure; Part II: the membrane would change from finger-like to sponge-like structure at the short period including α^*. Part III: $\alpha^* < \alpha < 1.0$, the membrane was mainly a sponge-like structure. When the approaching ratio was below α^*, it was suggested that the advancing rate of viscous fingering was faster than that of the phase-inversion front, hence a finger-like membrane structure was formed. When the approaching ratio was above α^*, the composition of casting solution moved closer to the binodal curve and the phase inversion process could occur more easily in the water bath. As a result, the mechanism of phase inversion process changed and the viscous fingering could not be initiated, hence only a sponge-like membrane structure could be formed.

Influence of Membrane Thickness on Membrane Morphology

The influence of membrane thickness on the membrane morphologies was first studied by Vogrin et al. (82) with a ternary cellulose acetate/acetone/water system. In their study, macrovoids appeared on the membranes prepared from a 12.5 wt % casting solution at the

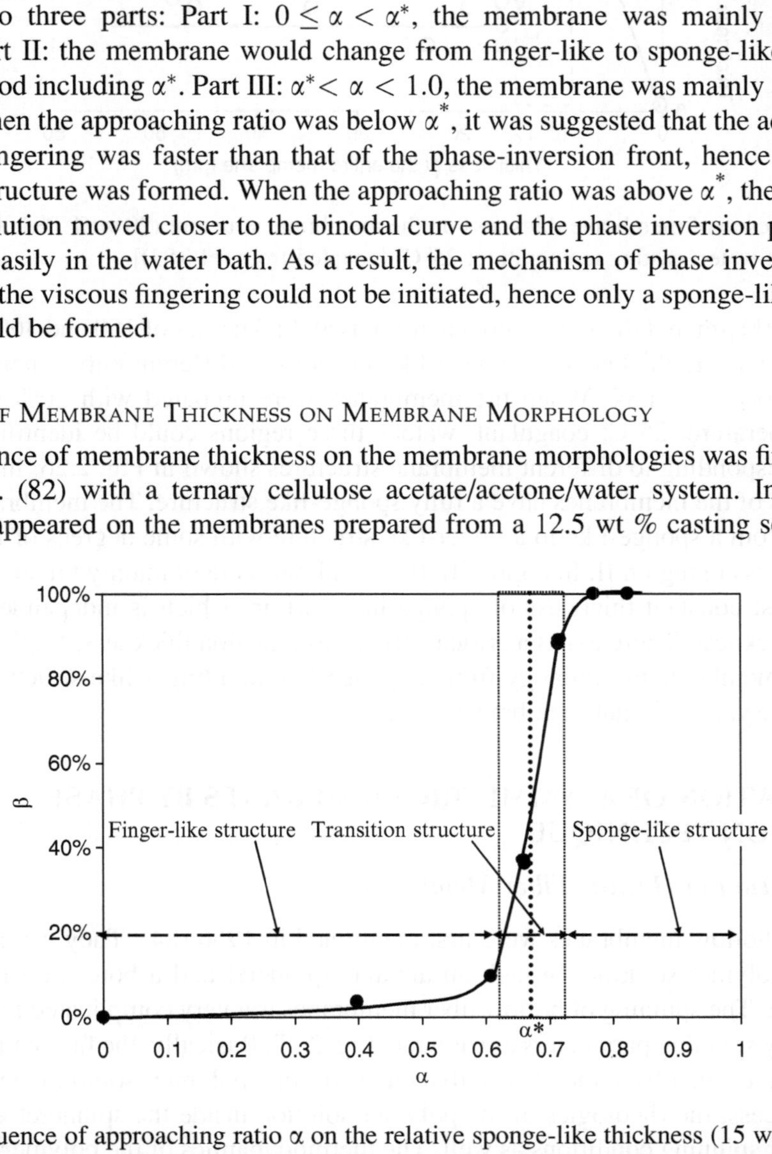

Fig. 2.25 Influence of approaching ratio α on the relative sponge-like thickness (15 wt% P84; mixed solvent: NMP/H_2O; coagulant bath: water; casting temperature: 25°C) [adapted from ref. (57)].

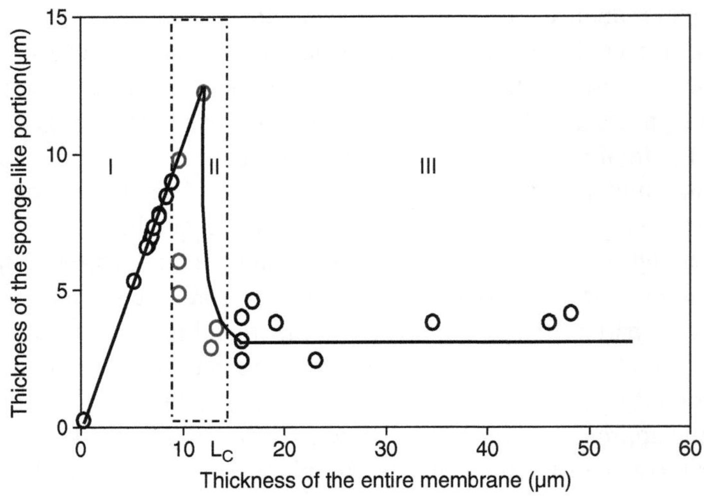

Fig. 2.26 Influence of membrane thickness on the membrane structure (20 wt% P84; solvent: NMP; coagulant bath: water; casting temperature: 25°C) [adapted from ref. (83)].

thickness of 500 μm, but there were no macrovoids at thicknesses of 150 and 300 μm. Li et al. (83) investigated the thickness of sponge-like portions at different entire membranes with PES and copolyimide P84. When the membranes were prepared with 20% P84 in NMP (casting temperature, 25°C; coagulant, water), three regions could be identified along the abscissa corresponding to different membrane structures shown in Fig. 2.26. In region I, the cross sections of the membranes have a fully sponge-like structure. The membrane morphology transits from a sponge-like to a finger-like structure with some degrees of fluctuation in sponge thickness in region II. In region III, the membranes are of mainly finger-like structure with an almost constant thickness of sponge-like portion, which is independent of overall membrane thickness. There exists a critical structure-transition thickness, L_c, that reflects the transition of membrane morphology from a sponge-like to a finger-like structure during the formation of asymmetric flat membranes.

5. PREPARATION OF ASYMMETRIC MEMBRANES BY PHASE INVERSION TECHNIQUE

5.1. Preparation of Hollow Fiber Membranes

Polymeric hollow membranes were first mentioned in 1966 (84). They were prepared by extruding a polymer solution through an annular spinneret and a bore fluid flowed in the annular centre. The spinning of hollow fiber membranes is a very complicated process, which involves many spinning parameters as shown in Fig. 2.27. Basically, the fabrication of hollow fiber membranes involves the thermodynamics of the polymer solution and the phase inversion process, the rheologies of the polymer solution inside the spinneret and at the air gap, and other spinning conditions as well. The thermodynamics of the polymer solution and the phase inversion process have been discussed in the previous sections. As for the spinning conditions, Mckelvey et al. (85) gave a detailed guidance on how to control some macroscopic properties of hollow fiber membranes using dominant process parameters (such as spinneret design, dope extrusion rate, drawing ratio, air gap distance, bore fluid extrusion rate, solvent concentration in the bore fluid, vitrification kinetics, and spin line guides) and how to determine the optimal macroscopic properties. The importance of dope rheology in the spinning process has received much attention recently, and some works have been done to study the effect of shear rate on the performance and morphology of hollow fiber membranes (86–96).

The fabrication of hollow fiber membranes can be carried out using dry-jet spinning or wet spinning or dry-jet wet spinning methods. Being an extremely complex process, the fabrication of hollow fiber membranes entails highly sophisticated mechanical, thermodynamic and kinetic considerations. For the dry-jet spinning process, the spun hollow fiber membranes are only exposed to a gaseous environment prior to vitrification, whereas the spun hollow finer membranes are exposed to a liquid environment prior to vitrification for the wet-spinning process. As for the dry-jet wet spinning process, there exists an air gap between the spinneret and the coagulation bath. The nascent hollow fiber membranes are spun into an air gap prior to quenching in a liquid nonsolvent bath, shown in Fig. 2.27.

The dry-jet wet spinning process mainly concerns three steps: the rheology of the dope solution in the spinneret, the formation of nascent hollow fiber membranes in the air gap and the solidification or vitrification of hollow fiber membranes in the coagulation bath. Two different flow patterns of shear flow and elongation flow exhibit inside the spinneret and in the air gap, respectively (96).

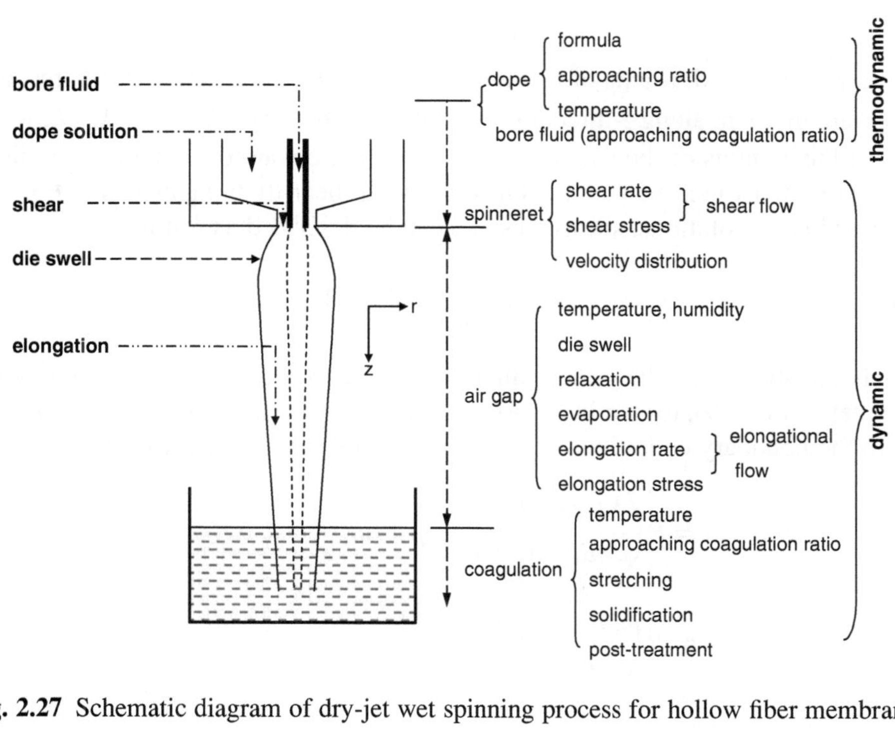

Fig. 2.27 Schematic diagram of dry-jet wet spinning process for hollow fiber membranes.

5.1.1. Rheology of the Polymer Solution Inside the Spinneret

For the shear flow inside the annular channel of the spinneret in Fig. 2.27, the velocity changes in r direction perpendicular to the direction of the motion (z direction), but it remains constant along the z direction of the flow as long as one stays on the same streamline. In contrast, for the nascent hollow fiber membranes in the air gap, a uniaxial elongational field exists, the fluid velocity changes in the z direction of the flow, but it remains constant in the perpendicular r direction. The shear flow generates a rotational flow component by the perpendicular velocity gradient. And the macromolecules of the dope solution are partly compressed and extended, which causes some orientations imparted on the parts of the molecules. But for an uniaxial elongational flow field, only a velocity gradient in the flow direction exists and no rotational flow component is generated. And the macromolecules only experience elongation. Compared with the shear flow, the elongation flow is more efficient in imparting molecular orientation provided that the velocity gradients and resident times in the elongational and shear fields are comparable, especially for hollow fiber membranes spun at a very high drawing ratio.

When a dope solution is extruded through a spinneret, its rheological properties could be correlated with shear flow induced by the shear stress within the spinneret, which has an effect on chain conformation and induces molecular orientation in the skin layer during the phase inversion process of hollow fiber formation.

During the hollow fiber spinning, when a non-Newtonian polymer solution is pumped through an annular spinneret, the z-component of momentum equation may be written in cylindrical coordinates in terms of shear rate as follows (89, 97):

$$-\frac{1}{r}\frac{\mathrm{d}}{\mathrm{d}r}(r\tau_{rz}) = \frac{\mathrm{d}p}{\mathrm{d}z}, \tag{28}$$

where r and $\tau_{rz}$are the radius axis of the annular spinneret and the shear stress, respectively. $\frac{\mathrm{d}p}{\mathrm{d}z}$ is a pressure gradient along z-direction, and thus, can be replaced by $-\Delta P/L$, where L is the length of the annulus of the spinneret. Polymer solutions used for spinning hollow fiber membranes are usually non-Newtonian fluids, showing shear-thinning behavior. For a power-law fluid of spinning solutions, shear stress τ_{rz} can be described as follows:

$$\tau_{rz} = -m\left|\frac{\mathrm{d}v_z}{\mathrm{d}r}\right|^{n-1}\left(\frac{\mathrm{d}v_z}{\mathrm{d}r}\right), \tag{29}$$

where $\frac{\mathrm{d}v_z}{\mathrm{d}r}$ is the shear rate along the r direction, m and n are the viscosity coefficient and power number of Eq. (29), respectively. At $r = \lambda R$, v_z is maximum and $\tau_{rz} = 0$. The velocity profile with the boundary conditions of $v_z = 0$ at $r = kR$ and $r = R$ is as follows:

$$v_z = R\left(\frac{\Delta PR}{2mL}\right)^s \int_k^\rho \left(\frac{\lambda^2}{\rho} - \rho\right)^s \mathrm{d}\rho, \quad k \le \rho \le \lambda \tag{30}$$

$$v_z = R\left(\frac{\Delta PR}{2mL}\right)^s \int_\rho^1 \left(\rho - \frac{\lambda^2}{\rho}\right)^s \mathrm{d}\rho, \quad k \le \rho \le \lambda \tag{31}$$

where $s = 1/n$, $\rho = r/R$, λ can be obtained by solving Eq. (32) numerically:

$$\int_k^\lambda \left(\frac{\lambda^2}{\rho} - \rho\right)^s \mathrm{d}\rho = \int_\rho^1 \left(\rho - \frac{\lambda^2}{\rho}\right)^s \mathrm{d}\rho. \tag{32}$$

The volumetric flow rate of the dope solution in an annulus can be calculated by

$$Q_\mathrm{p} = \pi R^3 \left(\frac{\Delta PR}{2mL}\right)^s \int_k^1 \left|\lambda^2 - \rho^2\right|^{s+1} \rho^{-s} \mathrm{d}\rho, \tag{33}$$

where Q_p is the volumetric flux of the dope solution and R is the outer radius of the spinneret. The shear rate induced at the outer wall of a spinneret is expressed as:

$$\begin{aligned} \left|\frac{\mathrm{d}v_z}{\mathrm{d}r}\right|_{r=R} &= \left(\frac{\Delta PR}{2mL}\right)(1-\lambda^2)^s, \\ &= \frac{Q_\mathrm{p}}{\pi R^3} \frac{(1-\lambda^2)^s}{\int_k^1 \left|\lambda^2-\rho^2\right|^{s+1} \rho^{-s} \mathrm{d}\rho}. \end{aligned} \tag{34}$$

The velocity distribution, shear rate of a dope solution with 21% copolyimide P84 are illustrated in Fig. 2.28 (98). The maximum shear rates were obtained at the outer and inner edges of the annular dope flow in the spinneret. When the outer surface of the hollow fiber membranes was the selective skin layer, the shear rate of the dope solution on the outer edge of the annular dope flow in the spinneret represented the shear rate of dope solution within

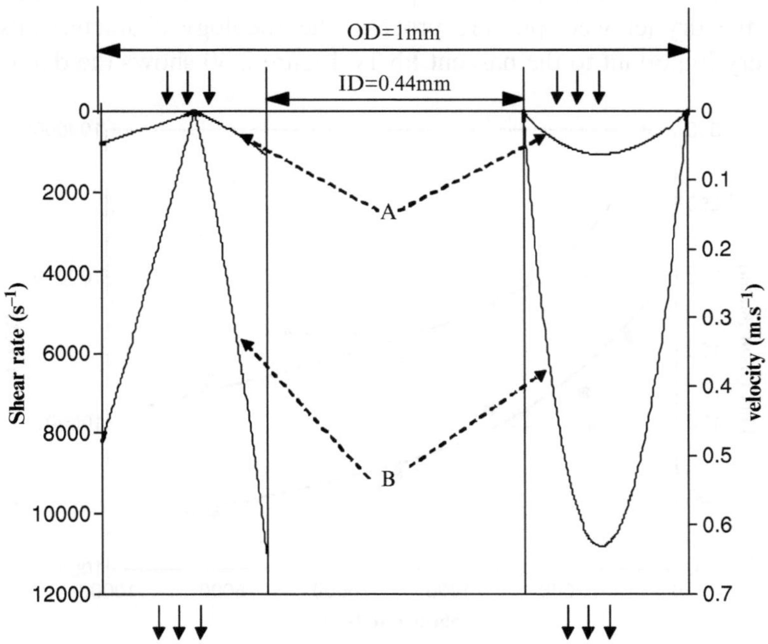

Fig. 2.28 Velocity and shear rate distributions of dope solution within the spinneret for different spinning conditions (polymer solution: 21 wt% P84; H_2O/NMP: 3.6/96.4 (wt/wt); $\alpha = 0.45$. Spinneret: OD = 1 mm; ID = 0.44 mm; extruding velocity: A: 1.6 mL/min; B: 16 mL/min. Power law; σ: shear stress, $\dot{\gamma}$: shear rate) [adapted from ref. (98)].

the spinneret. For some membranes with the inner skin layer due to the strong coagulant in the bore side, the shear rate of the dope solution on the inner edge of the annular dope flow in the spinneret was representative.

The shear rate of the dope solution within the spinneret strongly influenced the membrane performance, which was shown in Fig. 2.29. (98). With an increase in shear rate of the dope solution, the molecular weight cut-off (MWCO) changed from 110,000 to 19,000 Da, and the flux decreased greatly from 159 to 38 L/m^2 h atm. The shear rate of the dope solution within the spinneret also strongly influenced the membrane morphology (98). With an increase in shear rate, the defect on the surface of hollow fiber membranes could be suppressed and a low flux and a good selectivity were obtained (98). With an increase in shear rate, the inner surface of hollow fiber membranes became much tighter, and the outer surface of hollow fiber membranes became smoother and denser.

5.1.2. Nascent Hollow Fiber Membranes in the Air Gap

When the hollow fibers are spun using the dry-jet wet spinning process, the nascent hollow fiber membranes are strong influenced by the viscoelastic properties of the polymer dope. A viscoelastic fluid exhibits both energy dissipation and energy storage in its mechanical behavior, which can be reflected according to creep recovery test and dynamic oscillation test.

Viscoelasticity and Relaxation Time

Usually the dope solution prepared for spinning hollow fiber membranes is a viscoelastic fluid. During the dry-jet wet spinning process, the rheology characteristics of the dope solution are very important to the nascent fibers. Figure 2.30 shows the dynamic frequency

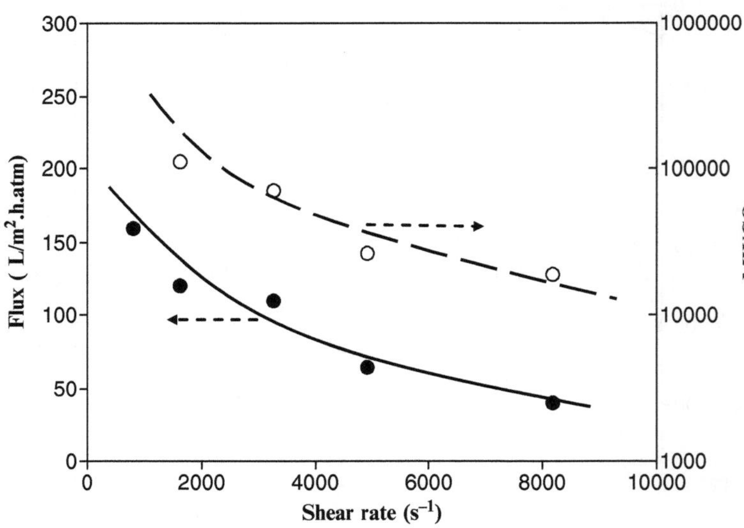

Fig. 2.29 Influence of shear rate of dope solution within the spinneret on the performance of hollow fiber membranes (polymer solution: 21 wt% P84; H_2O/NMP: 3.6/96.4 (wt/wt); $\alpha = 0.45$. Spinneret: OD = 1 mm; ID = 0.44 mm) [adapted from ref. (98)].

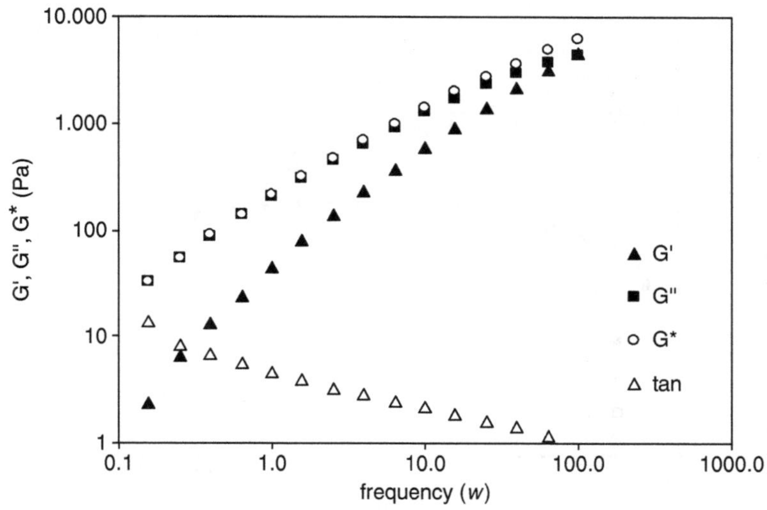

Fig. 2.30 Dynamic frequency sweep test of a dope solution at the same strain (polymer solution: 24 wt % 6FDA-ODA/NDA; 3 wt% EtOH; 22°C; strain = 20%) [adapted from ref. (99)].

test of a spinning polymer solution at 20% strain at room temperature (99). At a low frequency, the loss shear modulus G'' is almost the same as the complex modulus $G^*$and the storage shear modulus G' is much lower than G''. It can be concluded that the rheology of the dope solution is mainly influenced by viscosity in low shear rates. However, with an increase in frequency, the influence of elasticity of the dope solution on the rheology of the dope solution will increase. This can be reflected according to the decrease of loss tangent, $\tan\delta$. The influence of frequency on the dynamic viscosity and complex viscosity is indicated in Fig. 2.31 (99). At a low frequency, the dynamic shear viscosity η' is a constant, η_0. The dynamic relaxation viscosity η'' is much lower as compared to the dynamic shear viscosity, η'. With the increase of frequency, η' experiences the transition from η_0 to a power-law behavior and decreases. But η'' increases with the increase of frequency and then is kept almost the same at a high frequency. So it can be concluded that the rheological characteristics of the dope solution also change from viscosity to elasticity with the increase of shear rate. Viscoelasticity of the polymer solution related to the relaxation time, which is the characteristic time of the exponential strain decay curve and indicates the level of viscoelasticity of polymer solutions.

The relaxation time is an important parameter in the membrane fabrication. It indicates the level of molecular orientation of the polymer dope induced under shear, which will finally influence the membrane performance (68, 100). With the progress of the phase separation process of the polymeric solution, the relaxation time of the polymer chains will actually increase due to the onset of solidification. In the instantaneous phase separation process, the shear-induced molecular orientation will be frozen quickly into the nascent skin layer of the membrane because the polymer solution has no chance of relaxing (101, 102). Even for some delayed demixing processes, certain level of orientation would still be kept because the relaxation times become progressively longer at low levels of residual strain. Generally, the

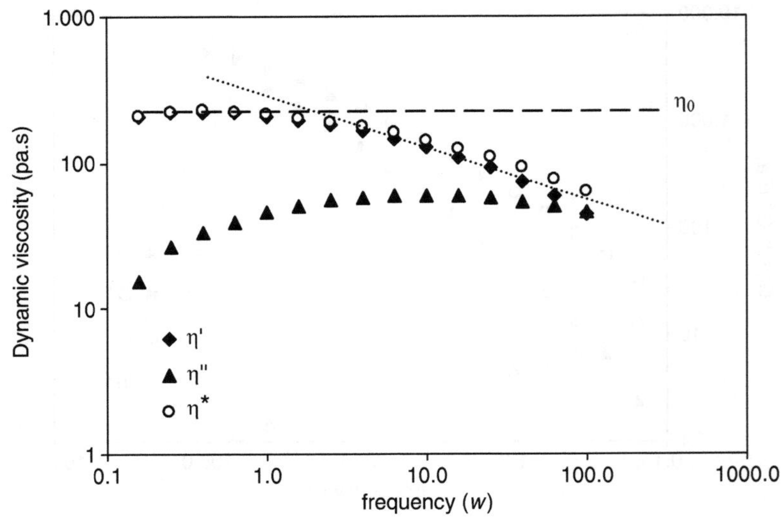

Fig. 2.31 Influence of dynamic frequency on the dynamic viscosity (polymer solution: 24 wt% 6FDA-ODA/NDA; 3 wt% EtOH; 22°C; strain = 20%) [adapted from ref. (99)].

approaching ratio also strongly influences the relaxation time and molecular orientation. With the increase of the approaching ratio, the dope formulation is approached to the cloud point curve and close to the thermodynamic instability limit, which speeds up the phase separation process and the relaxation effects on the membrane orientation will be reduce.

Elongation Flow in the Air Gap

Because of the viscoelasticity of the dope solution, the nascent hollow fiber at the exit of the spinneret would exhibit die-swell, where the outside diameter of the nascent fibers in the air gap is in the maximum and the velocity is the smallest. The nascent hollow fiber membranes accelerate from the velocity at the die-swell to the take up velocity by free-falling or drawing process. The drawing ratio (DR) is defined as (85):

$$\mathrm{DR} = \frac{V_{\mathrm{F}}}{V_{\mathrm{D}}} = \frac{V_{\mathrm{F}}\pi(D_0^2 - d_0^2)}{4Q_{\mathrm{P}}}, \tag{35}$$

where V_{F} and V_{D} are the drawing velocity and average extrusion rate at the spinneret exit (die-swell), respectively. Q_{p} is the volumetric flux of the dope solution. D_0 and d_0 define the dope annular cross-section leaving the spinneret. If there is no die swell, then the velocity at the spinneret exit is equal to the average extrusion velocity in the annular hollow channel. However, for a viscoelastical dope solution, the rheology of the dope solution inside the spinneret influences the die-swell of the nascent membranes in the air gap. At the higher shear rate of the dope solution inside the spinneret, the dope solution also change from viscosity to elasticity, which causes a bigger outer diameter of the nascent hollow fiber membranes, shown in Fig. 2.27.

In the air gap, the tensile stress τ_{zz} is described as (96):

$$\tau_{zz}(z) = \eta_E \left(\frac{dv_z(z)}{dz} \right), \tag{36}$$

where η_E is the elongation viscosity. $v_z(z)$ and $\frac{dv_z(z)}{dz}$ are the velocity and the elongation rate of the nascent hollow fiber membrane along z direction in the air gap, respectively. The ratio of elongation viscosity to shear viscosity is defined as Trouton ratio. For a Newtonian fluid, the elongation viscosity is three times of its shear viscosity (96). However, a viscoelastical fluid exhibits a Trouton ratio of great than 3. A higher trouton ratio indicates greater viscoelasticity. Normally, the shear viscosity decreases with increasing shear rate, the elongation viscosity may decrease or increase as the rate of elongation increases. The elongation viscosity is described as a power law function of elongation rate (96):

$$\eta_E = k \left(\frac{dv_z(z)}{dz} \right)^{n-1}, \tag{37}$$

where k and n are the elongation viscosity coefficient and power number in Eq. (37). Then the elongation stress can be described as

$$\tau_{zz}(z) = k \left(\frac{dv_z(z)}{dz} \right)^{n}. \tag{38}$$

In the spinning process of hollow fiber membranes, the extruded polymer solution is accelerated to a higher velocity by two different ways: gravity or external tensile force. In order to obtain small diameter hollow fiber membranes, usually a higher draw ratio of 5 to 15 can be used, which is associated with a high tensile stress and increased macromolecular chain orientation (85). The external force is provided by a godet or a take-up drum. The process of accelerating the extrudate or filament is called drawing. When the fiber in the air gap is free-falling, the external tensile stress is zero, the extruded polymer solution is only accelerated by its own gravity, which is called the free-fall spinning.

External tensile force

When the hollow fiber membranes are spun with a drawing ratio due to external tensile force, the spinning velocity is significantly greater than that corresponding to the free-fall spinning. It is assumed that the spinline in the air gap is isothermal and the tension or tensile force is the same at all the points in the air gap. Furthermore, the temperature change, solvent evaporation and the exchange between the polymer solution and the bore fluid are not considered here.

Referring to Eq. (38), if $n = 1$, the extension viscosity is constant. If $n > 1$, the extension viscosity increases with the increase of the extension rate, showing extension hardening such as the polyaramide solution prepared with DMAC (96). If $n < 1$, the dope solution exhibits extension softening, such as the PES formation contained propionic acid/NMP (96). And the tensile force F is given by:

$$F = \tau_{zz} A, \tag{39}$$

where A is the cross-section area of nascent hollow fiber membrane in the air gap, which can be described as:

$$A = \frac{Q_P}{v_z(z)}, \tag{40}$$

combining with the following equation

$$\frac{\tau_{zz}(z)}{v_z(z)} = \frac{\tau_{zz}(0)}{v_z(0)}, \tag{41}$$

here $\tau_{zz}(0)$ and $v_z(0)$ are the tension stress and the velocity of the die-swell at the exit of the spinneret.

The tensile force F can be obtained at any point in the air gap (z) or at the air gap distance L:

$$\begin{aligned} F &= k\left(\tfrac{n}{n-1}\right)^n \left(\left(\tfrac{v_z(z)}{v_z(0)}\right)^{\frac{n-1}{n}} - 1\right)^n \left(\tfrac{v_z(0)}{z}\right)^n A_0, \\ &= k\tfrac{\pi(D_0^2-d_0^2)}{4}\left(\tfrac{n}{n-1}\right)^n \left(DR^{\frac{n-1}{n}} - 1\right)^n \left(\tfrac{v_z(0)}{L}\right)^n. \end{aligned} \tag{42}$$

where, A_0 is the cross-section area of the nascent hollow fiber membranes at the die-swell. DR, D_0 and d_0 are defined in Eq. (35). The velocity distribution of the nascent hollow fiber in the air gap can be described as:

$$v_z(z) = v_z(0)\left[\left(DR^{\frac{n-1}{n}} - 1\right)\left(\frac{z}{L}\right) + 1\right]^{\frac{n}{n-1}}. \tag{43}$$

And the elongation rate inside the air gap can be written as:

$$\frac{v_z(z)}{dt} = v_z(0)\frac{n}{n-1}\left[\left(DR^{\frac{n-1}{n}} - 1\right)\left(\frac{z}{L}\right) + 1\right]^{\frac{1}{n-1}} \left(\frac{DR^{\frac{n-1}{n}} - 1}{L}\right), \tag{44}$$

where $\dfrac{v_z(z)}{dt}$ is the elongation rate at the point z in the air gap. And the tensile stress at the bottom of the air gap ($\tau_{zz}(L)$) is described as:

$$\tau_{zz}(L) = kDR\left(\frac{n}{n-1}\right)^n \left(DR^{\frac{n-1}{n}} - 1\right)^n \left(\frac{v_z(0)}{L}\right)^n. \tag{45}$$

The elongation rate inside the air gap and the tensile stress at the bottom of the air gap are strongly controlled by DR, the air gap distance (L), and the velocity of the polymer solution at the die-swell.

The theoretical velocity distribution in the air gap at different power numbers (with $n < 1$, $n = 1$ and $n > 1$) is shown in Fig. 2.32. In fact, when the hollow fibers are spun into the air gap from the spinneret, the solvent evaporation, temperature change, the exchange between the polymer solution and the bore fluid, the phase inversion and solidification strongly influence the spinline dynamics in the air gap. The actual spinline dynamics can be divided into three regions, which is shown in Fig. 2.32:

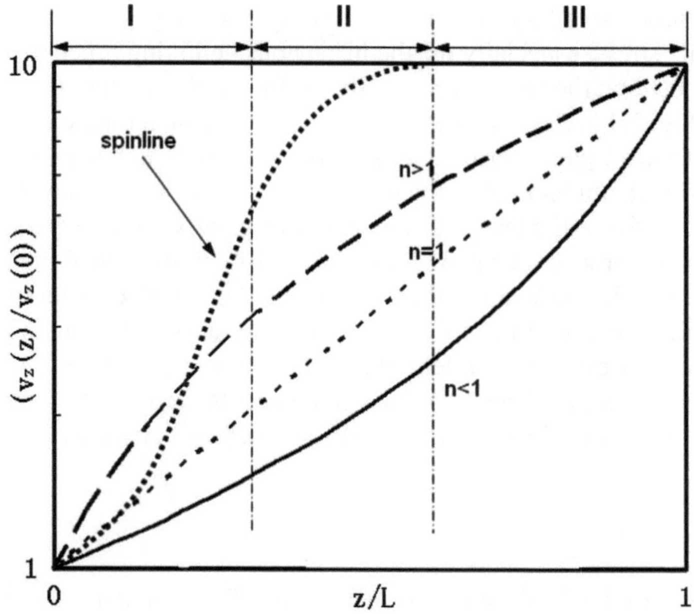

Fig. 2.32 Velocity distribution of nascent hollow fiber membranes verse distance at the air gap.

I. relative low extension viscosity
II. relative high extension viscosity
III. solidification (high extension viscosity)

At region I, the nascent hollow fiber membranes are just spun from the exit of the spinneret, they have the same rheological characteristics as the dope solution inside the spinneret. When the nascent hollow fiber membranes goes into region II, the elongation viscosity increases greatly with the time because of the solvent exchange, temperature change and the phase inversion induced by the exchange of solvent and nonsolvent between the nascent membrane and the bore fluid. Due to the high elongation viscosity, the elongation rate will decrease, which causes a lower slope of spinline in region II. When the vitrifiction or solidification occurs in region III, the polymer structure is solidified and the velocity can be kept at the drawing rate.

For different polymer solutions, bore fluids and spinning conditions, the different spinline velocities can be achieved at the same draw ratio. When the instantaneous demixing occurs inside the air gap, three distinct regions exist in the air gap and the drawing speed is mainly accelerated in the beginning part of the air gap. For the delay demixing in the air gap, the region I and II may describe its spinline velocity in the air gap, whereas the region II may be extended into the coagulation bath. When the nascent polymer cannot be solidified because of the poor coagulant or short air gap, the region III will not exist and the region II is extended into the coagulation below the water surface and the final velocity is attained under the water surface (96).

In the spinning process of hollow finer membranes, the air gap can also be used for the release of the tensile stress, especially for the instaneous demixing process. The influence of air gap distance and *DR* on the tensile stress at the bottom of the air gap for the same spinning speed is shown in Fig. 2.33. The tensile stress increases with the increase of the drawing ratio and the decrease of the air gap. A critical tensile stress exists, which may cause the breakage of the nascent hollow fiber membranes, the formation of some defects, the poor performance, irregular shapes, et al. Since the tensile stress increases greatly with a very small increase of DR in the wet spinning process (air gap is zero), it is difficult to control the tensile stress. In order to release the tensile stress, a small air gap should exist between the spinneret exit and the coagulation bath. Because of the air gap, the drawing ratio will be mainly attained at the air gap and the tensile stress will be released greatly due to the relative low viscosity of the polymer solution in the air gap. Thus, it is easy to control the tensile stress below the critical tensile stress. And the hollow fiber breakage and some defects caused by a high tensile stress can be overcome.

Free fall spinning

When the take-up speed is the same as that of the free-gravitational flow dictated at the bottom of the air gap, the elongation stress at the bottom of the air gap is zero. In this case, the gravitational force component at the distance *z* (from the exit of the spinneret) is equal to the elongation force

$$F = \rho g \int_{z}^{L} \frac{Q_{\mathrm{P}}}{v_z(z)} \mathrm{d}z, \tag{46}$$

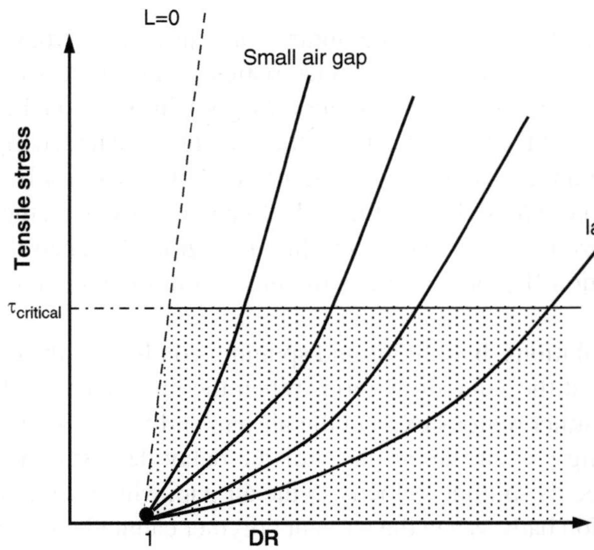

Fig. 2.33 Schematic diagram of the relationship between tensile stress and drawing ratio at different air gaps.

where F, ρ and g are the elongation force, density of the polymer solution and the gravity constant, respectively. And the force in the air gap decreases from the exit of the spinneret ($z = 0$) to the bottom of the air gap ($z = L$). The elongation stress ($\tau_{zz}(z)$) at point z can be described as:

$$\tau_{zz}(z) = \frac{F}{A} = \rho g v_z(z) \int_z^L \frac{1}{v_z(z)} \mathrm{d}z = k \left(\frac{\mathrm{d}v_z(z)}{\mathrm{d}z} \right)^n, \tag{47}$$

where A is the cross-section area of the nascent hollow fiber membrane. With the low deformation at air gap, the elongation viscosity can be assumed to be a constant. And $n = 1$, the rheology equation can be obtained in the air gap:

$$\frac{\mathrm{d}^2 v_z(z)}{\mathrm{d}z^2} - \frac{1}{v_z(z)} \frac{\mathrm{d}v_z(z)}{\mathrm{d}z} + \frac{\rho g}{k} = 0. \tag{48}$$

The boundary conditions are as follows:

$$v_z(z) = v_0 \quad \text{at} \quad z = 0, \tag{48a}$$

$$\frac{\mathrm{d}v_z(z)}{\mathrm{d}z} = 0 \quad \text{at} \quad z = L. \tag{48b}$$

To spin ultrafiltration and microfiltration hollow fiber membranes with a small air gap, the spinline velocity does not change greatly, and the extension stress can be described as:

$$\tau_{zz}(z) = \frac{F}{A} \approx \rho g (L - z). \tag{49}$$

And the velocity distribution in the air gap is written as:

$$v_z(z) = v_z(0) + \left(\frac{\rho g}{k} \right)^{\frac{1}{n}} L^{\frac{n+1}{n}} \left[1 - \left(1 - \frac{z}{L} \right)^{\frac{n+1}{n}} \right]. \tag{50}$$

The velocity at the bottom of the air gap:

$$v_z(L) = v_z(0) + \left(\frac{\rho g}{k} \right)^{\frac{1}{n}} L^{\frac{n+1}{n}}. \tag{51}$$

For the free-fall spinning process at the same air gap, with the increase of power number n, the velocity of the nascent hollow fiber membranes in the air gap will decrease. And the velocity at the bottom of the air gap increases as a power number ($n + 1/n$) of the air gap distance (L). That is to say, for the free-falling process, the spinning velocity and the air gap are coupled each other, which is different from the drawing process. When the factors of solvent evaporation, temperature decrease, solvent and nonsolvent exchange between the polymeric solution and the bore fluid, the phase inversion and solidification are considered, the velocity of the nascent fibers will decrease.

For the free-fall wet spinning process, there is almost no elongation stress existing and the spinning velocity is almost the same as that of the spinneret exit (die-swell). A small drawing velocity change caused by the spinning roll or drum will cause a greatly increase of the tensile stress, especially for the instantaneous demixing process, which may induce some defects on the membrane surface and cause poor performance of hollow fiber membranes. Compared with the dry-jet wet spinning process, it is more important to control the free-fall spinning velocity of the wet spinning process.

Usually, the elongation rate induced by elongation stress (gravity and drawing force) in the air gap and/or the coagulant bath causes chain conformation and induce molecular orientation. Ekiner et al. (96) observed that an increase in the drawing ratio or elongation strain resulted in a denser morphology and higher gas selectivity. The macrovoids inside the hollow fiber membranes can also be eliminated at a high elongation rate (103). With the increase of the drawing rate to six times of the free-falling velocity, the membrane morphology changed from original finger-like structure to teardrop-like structure, to complete sponge-like structure (103). From the thermodynamic point of view, the approaching ratio can control the membrane structure transition form finger-like to sponge-like structure. From the spinning dynamic point of view, the shear rate and elongation rate can also cause the membrane structure transition from finger-like to sponge-like structure.

5.1.3. Dimension of the Hollow Fiber Membranes

To make hollow fiber membranes, the ratio of membrane thickness to the fiber outer diameter has also to be considered, as it is associated with membrane performance, membrane mechanical strength and so on. In the spinning process, when the dope solution is extruded into the air gap, because of the die swell and the bore fluid, the dimension of the nascent hollow fiber membrane will be re-arranged according to the volume ratio of the dope solution to the bore fluid (ω). The relationship between the outer and inner diameter of the nascent hollow fiber membranes (D, d) at the air gap is correlated as follows:

$$\frac{Q_\mathrm{p}}{Q_\mathrm{b}} = \frac{\pi(D^2 - d^2)v_\mathrm{p}}{\pi d^2 v_\mathrm{b}} = \omega, \tag{52}$$

where Q_p and Q_b are the volumetric flow rate of the dope solution and the bore fluid, respectively. D and d are the outer and inner diameter of the nascent hollow fiber membranes in the air gap. In fact, the velocity of dope solution (v_p) is almost the same as that of bore fluid (v_b) in the air gap. The relationship between the thickness (δ) of nascent hollow fiber membranes and the membrane outer diameter in the air gap can be concluded as follows:

$$\frac{\delta}{D} = \frac{D - d}{2D} = \frac{1}{2}\left(1 - \frac{1}{\sqrt{1+\omega}}\right). \tag{53}$$

Due to the exchange of solvent and nonsolvent between nascent hollow fiber membranes and the coagulants (water and bore fluid), the membrane dimension changed and a revised factor ε was introduced into the Eq. (53) (98):

$$\frac{\delta}{D} = \frac{1}{2}\left(1 - \frac{1}{\sqrt{1+\varepsilon\omega}}\right). \tag{54}$$

where δ is the membrane thickness. The membrane thickness (δ) was a function of the volume ratio of the dope solution to the bore fluid (ω), the outer diameter of hollow fiber membranes (*D*) and the revised factor (ε). The revised factor ε should be less than 1, which reflected the shrinkage degree of nascent hollow fiber membranes due to the exchange of solvent and nonsolvent in the coagulant bath. Because of the exchange of solvent and nonsolvent between the dope solution and the coagulant (water and bore fluid), the revised factor ε for P84/NMP/water spinning system was approximately 0.8–0.9 (98). In the spinning process, the ratio of the membrane thickness to membrane outer diameter can be controlled by the volume ratio of the dope solution to the bore fluid (ω) and the revised factor ε.

5.2. *Preparation of Flat Sheet Membranes*

The original Loeb-Sourirajan membranes were made by casting 22–25% cellulose acetate solutions on glass plates. After solvent evaporation, the cast film was immersed in a water bath to precipitate and form the membrane. Rozelle et al. (104) also used the same method to prepare supported membranes with polysulfone and polycarbonate et al. for composite membranes. This method has been using for the preparation of flat sheet membranes in the laboratory till now, but for commercial production, a casting machine is usually used. Figure 2.34 shows a typical schematic diagram of the fabrication of flat sheet membranes. The nascent films were cast onto a moving nonwoven fabric web by the casting blade, and then exposed to a gaseous environment for solvent evaporation prior to entering the coagulant bath, finally precipitated by quenching in the coagulant bath as shown in Fig. 2.34.

The fabrication of flat sheet membranes is also a complicated process, which involves the dope preparation, the rheology of the casting solution, air gap and immersion precipitation in the coagulant bath et al. The process mainly concerns three steps: the rheology of the dope

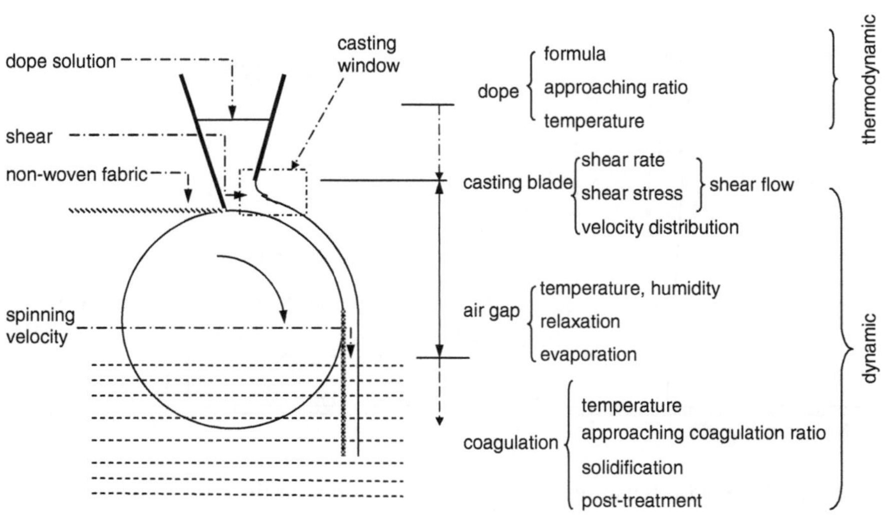

Fig. 2.34 Schematic diagram of flat sheet membrane fabrication.

solution at the casting window, the formation of nascent flat sheet membranes in the air gap and solidification or vertification of flat sheet membranes in the coagulation bath.

5.2.1. Hydrodynamics of the Polymer Solution at the Casting Window

When the nonwoven fabric web was drawn at a speed of V_0, the polymer solution was cast onto the thin support through a casting window with a casting gap H. The hydrodynamics of the polymer solution at the casting window is illustrated in Fig. 2.35. The flow field inside the casting window is assumed to be laminal and the velocity V_z along the y direction is a linear distribution. And the shear rate at the casting window can be described as:

$$\frac{dV_z}{dy} = \frac{V_0}{H}, \tag{55}$$

where H is the height of the casting window.

A casting die exists at the exit of the casting window, where a very complicated flow field exists. According to the transformation of flow field, the linear velocity distribution at the casting window will be changed into an equal velocity distribution (V_0) with a nascent membrane thickness L, which is immersed into the coagulant bath for precipitation.

The shear rate at the casting window strongly influences the membrane morphologies and performance of flat sheet membranes (105, 106). For the fabrication of flat sheet membranes, only shear flow exists in the casting window and no elongation flow is involved at the air gap.

He and Ulbricht (107) have reported a synergist immobilization method for surface selective photo-grafting on porous polymer membranes. Practical applications of polymeric membranes for air pollution control, portable water treatment, wastewater reclamation, and hazardous industrial waste treatment can be found from the literature (108–110).

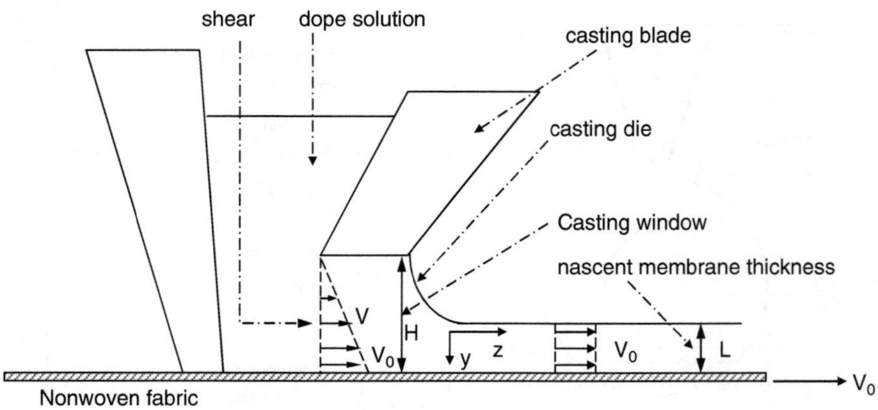

Fig. 2.35 Hydrodynamics of the polymer solution at the casting window.

6. ACRONYMS

B = Berghmans' point
C-point = Critical coagulation point
D = Dialysis
DIPS = Diffusion induced phase separation
DMAC = *N,N*-dimethylacetamide
DMF = Dimethylformamide
DMSO = Dimethylsulfoxide
DR = Drawing ratio
EtOH = Ethanol
6FDA-ODA/NDA = Poly(2,6-toluene-2,2bis(3,4-dicarboxyphenyl)hexafluoropropane diimide)
GS = Gas separation
ID = Inner diameter of the spinneret
LiCl = Lithium chloride
MF = Microfiltration
MWCO = Molecular weight cut-off
NMMO = *N*-Methylmorpholine-*N*-Oxide
NF = Nanofiltration
NG = Nucleation and growth mechanism
NMP = *N*-methylpyrrolidone
NS = Nonsolvent
OD = Outer diameter of the spinneret
P = Polymer
P84 = The trademark of copolyimide from Lenzing
PA = Polyamide
PAN = Polyacrylonitrile
PC = Polycarbonate
PEBAX = The trademark of PE-b-PA from Elf-Atochem
PE-b-PA = Polyether block amide
PEEK = Polyether ether ketones
PEI = Polyetherimide
PEO = Polyether oxide
PES = Polyethersulfone
PI = Polyimide
PPESK = Poly(phthalazine ether sulfone ketone)
PS = Polysulfone
PTMEO = Polytetramethylene
PV = Pervaporation
PVDF = Polyvinylidene fluoride
RO = Reverse osmosis
S = Solvent
SPEEK = Sulfonated PEEK
SD = Spinodal decomposition
TEP = Triethyl phosphate

TIPS = Thermally induced phase separation
UF = Ultrafiltration
VOCs = Volatile organic compounds

7. NOMENCLATURE

a = A parameter in Eqs. (19) and (19b).
A = The cross-section area of nascent hollow fiber membrane in the air gap (m^2).
A_0 = The cross-section area of nascent hollow fiber membranes at die-swell (m^2).
B = A parameter in Eqs. (19) and (19a); the plate gap in Eq. (27) (m).
C = Concentration (kg/m^3).
d = Inner diameter of the nascent hollow fiber in the air gap (m).
d_0 = Inner diameter of the nascent hollow fiber at die-swell (m).
D = Outer diameter of the nascent hollow fiber in the air gap (m).
D_0 = Outer diameter of the nascent hollow fiber at die-swell (m).
F = Tensile force, N; objective function in Eq. (14).
F_1 = Objective function in Eq. (15).
g = Gravity constant, 9.8 N/kg.
g_{ij} = The binary interaction parameter between component i and j.
G' = Storage shear module (Pa).
G'' = Loss shear module (Pa).
G^* = Complex module (Pa).
G_{ij} = The derivative of G_m with respect to $\phi_i$and ϕ_j.
H = The height of the casting window for flat sheet membrane (m).
i, j = Component i, j.
k = Coefficient in Eq. (37).
L = The annular length of the spinneret, the length of the air gap, the thickness of the flat sheet membranes (m).
L_c = Critical structure-transition thickness (m).
m = Viscosity coefficient in Eq. (29).
M_i = The molar mass of the component i (kg/mole).
n = The molar number, the power number in Eqs. (29) and (37).
p = Pressure (Pa).
Q_b = The volumetric flux of the bore fluid (m^3/s).
Q_p = The volumetric flux of the dope solution in the spinneret (m^3/s).
r = r axis in radius direction of the spinneret, m; a parameter in Eqs. (10)–(12).
R = Gas constant, 8.314 m^3 Pa/(mol K), the outer radium of the spinneret, m, a parameter in Eq. (6).
s = A parameter in Eqs. (10)–(12), a parameter in Eq. (32).
t_{gel} = Gelation time (s).
T = Temperature (K).
T_g = Glass transition temperature (K).
T_m = Melting temperature (K).
T_b = Boiling temperature (K).

v = Velocity (m/s).
$\bar{v}_i$ = The partial molar volume of component i (m^3/mol).
v_b = The velocity of bore fluid in the air gap (m/s).
v_p = The velocity of the dope solution extruded from the spinneret in the air gap (m/s).
v_z = The velocity of the polymer solution in z-direction in the air gap in Fig. 2.27. (m/s).
V = Total volume (m^3).
V_0 = The velocity of the drawing nonwoven for flat sheet membrane fabrication in Fig. 2.35 (m/s).
V_D = Average extrusion rate at die swell (m/s).
V_F = Drawing velocity (m/s).
V_z = The velocity of the polymer solution in the casting window in Eq. (55) (m/s).
w = Frequency (s^{-1}).
y = y axis in casting window for flat sheet membrane (m).
z = The axis in z direction in Fig. 2.27 (m).
α = The approaching ratio.
α^* = Critical approaching ratio.
β = The relative sponge-like thickness the volume ratio of nonsolvent and solvent at cloud point.
β' = The weight ratio of nonsolvent and solvent at cloud point.
χ = Interaction parameter between two components.
δ = Membrane thickness (m).
ε = Revised factor
ϕ_i = Volume fraction of component i.
γ = Coefficient in Eq. (7), the ratio of the solvent concentration in the coagulation bath to that at the C-point.
γ^* = Coefficient in Eq. (8).
η = Viscosity (Pa s).
η' = Dynamic shear viscosity (Pa s).
η'' = Dynamic relaxation viscosity (Pa s).
η_E = Elongational viscosity (Pa s).
η_{eff} = Shear dependent viscosity (Pa s).
λ = The ratio of r and R for the spinneret at the maximum velocity.
ρ = Density (kg/m^3); the ratio of r and R for the spinneret in Eq. (30)–(34).
τ = Shear stress (Pa).
ω = The volume fraction of dope solution to bore fluid.
ΔG_m = Gibbs free energy of mixing (kJ).
ΔP = Pressure difference (Pa).
∇P = Pressure gradient (Pa/m).
$\Delta\mu_i$ = Chemical potential of component i (kJ/mol).

REFERENCES

1. Mulder MHV (1991) Basic principals of membrane technology. Kluwer, Dordrecht
2. Baker RW (2004) Membrane technology and applications, 2nd ed., Wiley, New York
3. Loeb S, Sourirajan S (1962) Sea water demineralization by means of an osmotic membranes. Adv Chem Ser 38:117
4. van de Witte P, Dijkstra PJ, van de Berg JWA, Feijn BJ (1996) Phase separation processes in polymer solutions in relation to membrane formation. J Memb Sci 117:1

5. Kesting RE (1985) Synthesis polymeric membranes, a structural perspective. Wiley, New York
6. Strathmann H (1981) Review – membrane separation processes. J Memb Sci 9:121
7. Matsuura T (1994) Synthetic membranes and membrane separation process. CRC Press, Boca Raton
8. Pinnau I, Freeman B (eds) (2000) Membrane formation and modification. American Chemical Society, Washington, DC.
9. Nunes SP, Peinemann KV (eds) (2001) Membrane technology in the chemical industry. Wiley-VCH, Verlag GmbH, Weinheim
10. Ulbricht M (2006) Advanced functional polymer membranes. Polymer 47:2217
11. http://www.solvaymembranes.com/
12. Peinemann KV, Fink K, Witt P (1986) Asymmetric polyetherimide membranes for helium separation. J Memb Sci 27:215
13. Bodzek M, Bohdziewicz J (1991) Porous polycarbonate phase-inversion membranes. J Memb Sci 60:25
14. Mark HF, Bikales NM, Overbeger CG, Menges G (1985) Encyclopedia of polymer science and engineering. 2nd ed, vol 11, Polyamides, Wiley, NY
15. Castro AJ (1981) Methods for making microporous products, US Patent 4247498.
16. Caneba GT, Soong DS (1985) Polymer membrane formation through the thermal inversion process: 1. Experimental study of membrane structure formation. Macromolecules 18:2538
17. Lloyd DR, Kinzer KE, Tseng HS (1990) Microporous membrane formation via thermally induced phase separation. I. Solid–liquid phase separation. J Memb Sci 52:239
18. Tsai FJ, Torkelson JM (1990) Microporous poly(methyl methacrylate) membranes: effect of a low-viscosity solvent on the formation mechanism. Macromolecules 23:4983
19. Lloyd DR, Kim S, Kinzer KE (1991) Microporous membrane formation via thermally-induced phase separation. II. Liquid–liquid phase separation. J Memb Sci 64:1
20. Vadalia HC, Lee HK, Myerson AS, Levon K (1994) Thermally induced phase separation in ternary crystallizable polymer solutions. J Memb Sci 89:37
21. Mehta RH, Madsen DA, Kalika DS (1995) Microporous membranes based on poly(ether ether ketone) via thermally-induced phase separation. J Memb Sci 107:93
22. Cha BJ, Char K, Kim JJ, Kim SS, Kim CK (1995) The effects of diluent molecular weight on the structure of thermally induced phase separation membrane. J Memb Sci 108:219
23. McGuire KS, Laxminarayan A, Martula DS, Lloyd DR (1996) Kinetics of droplet growth in liquid–liquid phase separation of polymer-diluent systems: model development. J Colloid Interface Sci 182:46
24. Matsuyama H, Berghmans S, Batarseh MT, Lloyd DR (1998) Effects of thermal history on anisotropic and asymmetric membranes formed by thermally induced phase separation. J Memb Sci 142:27
25. Matsuyama H, Berghmans S, Lloyd DR (1999) Formation of anisotropic membranes via thermally induced phase separation. Polymer 40:2289
26. Matsuyama H, Kudari S, Kiyofuji H, Kitamura Y (2000) Kinetic studies of thermally induced phase separation in polymer diluent system. J Appl Polym Sci 76:1028
27. Liu B, Du Q, Yang Y (2000) The phase diagrams of mixtures of EVAL and PEG in relation to membrane formation. J Memb Sci 180:81
28. Atkinson PM, Lloyd DR (2000) Anisotropic flat sheet membrane formation via TIPS: atmospheric convection and polymer molecular weight effects. J Memb Sci 175:225

29. Kim JJ, Hwang JR, Kim UY, Kim SS (1995) Operation parameters of melt spinning of polypropylene hollow fiber membranes. J Memb Sci 108:25
30. Sun H, Rhee KB, Kitano T, Mah SI (2000) Hollow-fiber membrane via thermally induced phase separation HDPE II. Factors affecting the water permeability of the membrane. J Appl Polym Sci 75:1235
31. Matsuyama H, Maki T, Teramoto M, Asano K (2002) Effect of polypropylene molecular weight on porous membrane formation by thermally induced phase separation. J Memb Sci 204:323
32. Matsuyama H, Yuasa M, Kitamura Y, Teramoto M, Lloyd DR (2000) Structure control of anisotropic and asymmetric polypropylene membrane prepared by thermally induced phase separation. J Memb Sci 179:91
33. Yang MC, Perng JS (2001) Microporous polypropylene tubular membranes via thermally induced phase separation using a novel solvent–camphene. J Memb Sci 187:13
34. Arnauts J, Berghmans H (1987) Amorphous thermoreversible gels of atactic polystyrene. Polym Commun 28:66
35. Frank FC, Keller A (1988) 2-fluid phase-separation – modified by a glass-transition. Polym Commun 29:186
36. Kelley FN, Bueche F (1961) Viscosity and glass temperature relations for polymer-diluent systems. J Polym Sci 50:549
37. Fedors RF (1979) Universal reduced glass-transition temperature for liquids. J Polym Sci Part C Polym Lett 17:719
38. Li SG (1994) Preparation of HollowFiber Membranes for Gas Separation, Ph.D. thesis, University of Twente, NL
39. Siggia ED (1979) Late stages of spinodal decomposition in binary mixtures. Phys Rev A 20:595
40. Caneba GT, Soong DS (1985) Polymer membrane formation through the thermal inversion process. 2. Mathematical modeling of membrane structure formation. Macromolecules 18:2545
41. Tsai FJ, Torkelson JM (1990) Roles of phase separation mechanism and coarsening in the formation of poly(methylmethacrylate) asymmetric membranes. Macromolecules 23:775
42. Laxminarayan A, McGuire KS, Kim SS, Lloyd DR (1994) Effect of initial composition, phase separation temperature and polymer crystallization on the formation of microcellular structure via thermally induced phase separation. Polymer 35:3060
43. Koros WJ, Pinnau I (1994) Membrane Formation for gas separation processes. In: Paul DR, Yampol'skii YP (eds) Polymeric gas separation membranes. CRC Press, Boca Raton
44. Guo HF, Laxminarayan HFA, Caneba GT, Solc K (1995) Morphological studies of late-stage spinodal decomposition in polystyrene-cyclohexanol system. J Appl Polym Sci 55:753
45. Strathmann H, Scheible P, Baker RW (1971) A rational for the preparation of Lob-Sourirajan type cellulose acetate. J Appl Poly Sci 15:811
46. Altena FW, Smolders CA (1982) Calculation of liquid-liquid phase separation in a ternary system of a polymer in a mixture of a solvent and a nonsolvent. Macromolecules 15:1491
47. Li SG, van den Boomgaard T, Smolders CA, and Strathmann H (1996) Physical gelation of amorphous polymers in a mixture of solvent and nonsolvent. Macromolecules 29:2053
48. Wang L (1999) Studies on gelation time and formation mechanism of support layer of PES membrane-making system, Master thesis, Dalian Institute of Chemical Physics, Chinese Academy of Sciences, China
49. Wang L, Li Z, Ren J, Li SG, Jiang C (2006) Preliminary studies on the gelation time of poly(ether sulfones) membrane-forming system with an elongation method. J Memb Sci 275:46
50. Flory PJ (1953) Principles of Polymer Chemistry, Cornell University Press, Ithaca, NY

51. Matsuyama H, Nishiguchi M, Kitamura Y (2000) Phase separation mechanism during membrane formation by dry-cast process. J Appl Polym Sci 77:776
52. Li S, Jiang C, Zhang Y (1987) The investigation of solution thermodynamics for the polysufone – DMAC – water system. Desalination 62:79
53. Boom RM, van den Boomgaard T, van den Berg JWA, Smolders CA (1993) Linearized cloudpoint curve correlation for ternary systems consisting of one polymer, one solvent and one non-solvent. Polymer 34:2348
54. Kools WFC, van den Boomgaard T, strathmann H (1998) Considerations and restrictions on the theoretical validity of the linearized cloudpoint correlation. Polymer 39:4835
55. He T, Jiang CZ (1998) Effects of nonsolvent additives on performance of poly(ether sulfone) microporous membranes. Memb Sci Technol 18:43
56. Li Z, Jiang C (2001) Investigation into the rheological properties of PES/NMP/nonsolvent membrane-forming systems. J Appl Polym Sci 82:283
57. Ren J, Li Z, Wong FS (2004) Membrane structure control of asymmetric BTDA-TDI/MDI (P84) co-polyimide membranes by phase inversion process. J Memb Sci 241:305
58. Reuvers AJ and Smolders CA (1987) Formation of membranes by means of immersion precipitation: Part II. the mechanism of formation of membranes prepared from the system cellulose acetate-acetone-water. J Memb Sci 34:67
59. Reuvers AJ (1987) Membrane Formation-diffusion Induced Demixing Processes in Ternary Polymeric Systems, Ph.D thesis, University of Twente, Enschede, The Netherlands
60. Tanaka F and Stockmayer WH (1994) Thermoreversible gelation with junctions of variable multiplicity. Macromolecules 27:3943
61. McKelvey SA (1997) Formation and characterization of hollow fiber membranes for gas separation. The University of Texas at Austin, Ph.D. thesis
62. Van't Hof JA, Reuvers AJ, Boom RM, Rolevink HHM, Smolders CA (1992) Preparation of asymmetric gas separation membranes with high selectivity by a dual-bath coagulation method. J Memb Sci 70:17
63. Radovanovic P, Thiel SW, Hwang ST (1992) Formation of asymmetric polysulfone membranes by immersion precipitation. 1. Modeling mass-transport during gelation. J Memb Sci 65:213
64. Radovanovic P, Thiel SW, Hwang ST (1992) Formation of asymmetric polysulfone membranes by immersion precipitation. 2. The effects of casting solution and gelation bath compositions on membrane-structure and skin formation. J Memb Sci 65:231
65. Boom RM, van den Boomgaard T, Smolders CA (1994) Mass-transfer and thermodynamics during immersion precipitation for a 2-polymer system-evaluation with the system PES-PVP-NMP-water. J Memb Sci 90:231
66. Li SG, Koops GH, Mulder MHV, van den Boomgaard T, Smolders CA (1994) Wet spinning of integrally skinned hollow-fiber membranes by a modified dual-bath coagulation method using a triple orifice spinneret. J Memb Sci 94:329
67. Kools WFC (1998) Membrane Formation by Phase Inversion In Multicomponent Polymer Systems, Mechanism and Morphologies, Ph.D. thesis, University of Twente, NL
68. Ismail AF, Yean LP (2003) Review on the development of defect-free and ultrathin-skinned asymmetric membranes for gas separation through manipulation of phase inversion and rheological factors. J Appl Poly Sci 88:442
69. Pinnau I, Koros WJ (1993) A qualitative skin layer formation mechanism for membranes made by dry wet phase inversion. J Polym Sci Part B Polym Phys 31:419
70. Smolders CA, Reuvers AJ, Boom RM, Wienk IM (1992) Microstructures in phase-inversion membranes. Part 1: Formation of macrovoids. J Memb Sci 73:259

71. Strathmann H, Kock K, Amar P, Baker RW (1975) The formation mechanism of asymmetric membranes. Desalination 16:179
72. Strathmann H, Kock K (1977) The formation mechanism of phase inversion membranes. Desalination 21:241
73. McKelvey SA, Koros WJ (1996) Phase separation, vitrification, and the manifestation of macrovoids in polymeric asymmetric membranes. J Memb Sci 112:29
74. Pekny MR, Zartman J, Krantz WB, Greenberg AR, Todd P (2003) Flow-visualization during macrovoid pore formation in dry-cast cellulose acetate membranes. J Memb Sci 211:71
75. Matz R (1972) The structure of cellulose acetate membranes 1. The development of porous structures in anisotropic membranes. Desalination 10:1
76. Ray RJ, Krantz WB, Sani RL (1985) Linear stability theory for finger formation in asymmetric membranes. J Memb Sci 23:155
77. Frommer MA, Messalem FM (1973) Mechanism of membrane formation. VI. Convective flows and large void formation during membrane precipitation. Ind Eng Chem Prod Res Dev 12:328
78. Meakin P (1998) Fractals, scaling, and growth of far from equilibrium. Cambridge University Press, Cambridge
79. Kawaguchi M, Makino K, Kato T (1997) Viscous fingering patterns in polymer solutions. Physica D 109:325
80. Kawaguchi M (2001) Viscous fingering patterns in polymer systems. Nonlinear Anal 47:907
81. Lal J, Bansil R (1992) Effect of gelation on morphology of spinodal decomposition and viscous finger. Physica A 186:88
82. Vogrin N, Stropnik C, Musil V, Brumen M (2002) The wet phase separation: the effect of cast solution thickness on the appearance of macrovoids in the membrane forming ternary cellulose acetate/acetone/water system. J Memb Sci 207:139
83. Li D, Chung TS, Ren J, Wang R (2004) Thickness dependence of macrovoid evolution in wet-phase inversion asymmetric membranes. Ind Eng Chem Res 43:1553
84. . Mahon HI (1966) Permeability separatory apparatus and membrane element, method of making the same and process ultilizing the same, US patent 3228876, Dow chemical
85. Mckelvey SA, Clausi DT, Koros WJ (1997) A guide to establishing hollow fiber macroscopic properties for membrane applications. J Memb Sci 124:223
86. Chung TS, Lin WH, Vora RH (2000) The effect of shear rates on gas separation performance of 6FDA-Durene polyimide hollow fibers. J Memb Sci 167:55
87. Chung TS, Teoh SK, Lau WWY, Srinivasan MP (1998) Effect of shear stress within the spinneret on hollow fiber membrane morphology and separation performance. Ind Eng Chem Res 37:3930
88. Chung TS, Qin JJ, Gu J (2000) Effect of shear rate within the spinneret on morphology, separation performance and mechanical properties of ultrafiltration polyethersolfone hollow fiber membranes. Chem Eng Sci 55:1077
89. Qin JJ, Wang R, Chung TS (2001) Investigation of shear stress effect within a spinneret on flux, separation and thermomechanical properties of hollow fiber ultrafiltration membranes. J Memb Sci 175:197
90. Ismail AF, Shilton SJ, Dunkin IR, Gallivan SL (1997) Direct measurement of rheologically induced molecular orientation in gas separation hollow fiber membranes and effects on selectivity. J Memb Sci 126:133
91. Shilton SJ, Ismail AF, Gough PJ (1997) Molecular orientation and the performance of synthetic polymeric membranes for gas separation. Polymer 38:2215

92. Ren J, Chung TS, Li D, Wang R, Liu Y (2002) Development of asymmetric 6FDA-2,6DAT hollow fiber membranes for CO_2/CH_4 separation 1. The influence of dope composition and rheology on membrane morphology and separation performance. J Memb Sci 207:227
93. Idris A, Ismail AF, Gordeyev SA, Shilton SJ (2003) Rheology assessment of cellulose acetate spinning solution and its influence on reverse osmosis hollow fiber membrane performance. Polym Test 22:319
94. Ismail AF, Ng BC, Rahman WA (2003) Effects of shear rate and forced convection residence time on asymmetric polysulfone membranes structure and gas separation performance. Sep Purf Technol 33:255
95. Gordeyev SA, Lees GB, Dunkin IR, Shilton SJ (2001) Super-selective polysulfone hollow fiber membranes for gas separation: rheological assessment of the spinning solution. Polymer 42:4747
96. Ekiner OM, Vassilatos G (2001) Polyaramide hollow fibers for H_2/CH_4 separation II. Spinning and properties. J Memb Sci 186:71
97. Bird RB, Armstrong RC, Hasager O (1987) Dynamics of polymeric liquids, vol. 1. Fluid mechanics, 2nd ed. Wiley, New York
98. Ren J, Li Z, Wong FS, Li D (2005) Development of asymmetric BTDA-TDI/MDI (P84) co-polyimide hollow fiber membranes for ultrafiltration: The influence of shear rate and approaching ratio on membrane morphology and performance. J Memb Sci 248:177
99. Ren J, Wang R, Chung TS, Li D, Liu Y (2003) The effect of chemical modifications on morphology and performance of 6FDA-ODA/NDA hollow fiber membranes for CO_2/CH_4 Separation. J Memb Sci 222:133
100. Hagler GE (1981) Qualitative prediction of the effects of changes in spinning conditions on spun fiber orientation. Polym Eng Sci 21:121
101. Shilton SJ, Ismail AF, Gough PJ, Dunkin IR, Gallivan S (1997) Molecular orientation and the performance of synthetic polymeric membranes for gas separation. Polymer 38:2215
102. Shilton SJ, Bell G, Ferguson J (1994) The rheology of fiber spinning and the properties of hollow-fibre membranes for gas separation. Polymer 35:5327
103. Wang KY, Li DF, Chung TS, Chen SB (2004) The observation of elongation dependent macrovoid evolution in single and dual-layer asymmetric hollow fiber membranes. Chem Eng Sci 59:4657
104. Rozelle LT, Cadotte JE, Corneliussen RD, Erickson EE (1968) Final report on development of new reverse osmosis membranes, NTIS Report No.PB.206329, June
105. Ismail AF (1997) Novel studies of molecular orientation in synthetic polymeric membranes for gas separation, University of Strathclyde, Ph.D thesis
106. Ismail AF, Lai PY (2003) Effects of phase inversion and rheological factors on the defect-free and ultrathin-skinned asymmetric polysulfone membranes for gas separation. Sep Purf Technol 33:127
107. He D, Ulbricht M (2006) Surface selective photo-grafting on porous polymer membranes via a synergist immobilization method. J Mater Chem 16:1860
108. Wang LK, Pereira NC, Hung YT (eds) (2004) Air pollution control engineering. Humana, Totowa, NJ, pp 471–475
109. Wang LK, Hung YT, Shammas NK (eds) (2005) Physicochemical treatment processes. Humana, Totowa, NJ, pp 526–671
110. Wang LK, Hung YT, Lo HH, Yapijakis C (eds) (2006) Hazardous industrial waste treatment. CRC Press/Taylor & Francis, New York, pp 468–487

3
Advanced Membrane Fouling Characterization in Full-Scale Reverse Osmosis Processes

Lianfa Song and Kwee Guan Tay

CONTENTS

Abstract Fouling alleviation and control in full-scale reverse osmosis (RO) processes can be seriously hindered by ineffective fouling characterization. Fouling characterization is currently done primarily by measuring the silt density index (SDI) of feed water and monitoring the average permeate flux of full-scale RO processes. However, the SDI or related fouling indices have been known not capable to catch all possible foulants to RO membranes, and the average permeate flux may fail to correctly reflect membrane fouling in full-scale RO processes under certain circumstances.

In this chapter, an advanced characterization method is introduced for membrane fouling in full-scale RO processes. An inclusive fouling potential indicator is proposed for a better characterization of the fouling strength of feed water, which is defined in a way that it can be easily determined with a small lab-scale RO membrane device and that it is readily usable to predict fouling development in the full-scale RO processes. A model based on basic membrane filtration and mass conservation principles is presented to calculate the average permeate flux in a full-scale RO process as fouling is progressing. The main intention of the advanced characterization method is to provide a reliable assessment of membrane fouling in the full-scale RO processes, so that the need for pilot tests can be significantly reduced.

From: *Handbook of Environmental Engineering, Volume 13: Membrane and Desalination Technologies*
Edited by: L.K. Wang et al., DOI: 10.1007/978-1-59745-278-6_3 © Springer Science+Business Media, LLC 2011

Key Words Fouling characterization • reverse osmosis • full-scale processes • fouling potential • thermodynamic equilibrium.

1. INTRODUCTION

Industrial applications using (RO) membranes as separation medium began in 1960s when Loeb and Sourirajan (1) successfully developed high flux cellulose acetate RO membrane. Today, RO has been widely used for separation and concentration of solutes in many fields, such as chemical and biomedical industry, food and beverage processing, and water and wastewater treatment. The most common uses of RO membrane are in the seawater desalination and water reclamation industries. There are estimated 12,500 desalination plants around the world, supplying 20 million cubic meters per day or 1% of world's production of drinking water. With the continuous decrease in cost of RO desalination, the number of RO desalination plants and its share in world drinking water supply is expected to rapidly increase in the near future (2, 3).

The increasing acceptance of RO technology in seawater desalination and water reclamation is no coincidence. Shortage of potable water around the world and more stringent water quality requirements by the regulating bodies (4, 5) are two major driving factors that make RO an attractive option for separation process in water treatment and reclamation. In addition, RO has many advantages over conventional separation processes such as distillation and other physical operations. Energy consumption in terms of electricity use by RO is only about one third to that of distillation (6). RO does not require clarification tanks and disinfection units; hence, it usually requires smaller space and is easy to operate. Compared with conventional separation processes, RO produces less sludge as it does not or seldom involves the use of chemicals such as coagulants or polymers (6).

Despite the attractive attributes of RO, the separation process is unfortunately plagued by the single most critical problem – membrane fouling (7, 8). Fouling is the accumulation of contaminants or foulants on the membrane surface, which results in the reduction of water production, product water quality, and membrane lifespan (9). Membrane fouling can be significantly reduced by appropriate pretreatment of the feed water to the RO process. The fouled membranes can be cleaned to restore its permeability. As a guide, cleaning is usually done when there is a 10% decrease in water production (10) at constant operating conditions, or a 10% increase in the driving pressure to maintain the same production at constant temperature, or an increase of 15–20% in the pressure differential between feed and reject flows. However, the costs that associate with fouling control and membrane cleaning represent a significant proportion in the total operating cost. Studies indicated that scale inhibition and cleaning chemicals account for about 5–20% of the total cost for a typical seawater RO system (11, 12) and the pretreatment cost in RO systems in Middle East to range between 10 and 25% of the total cost (13). If membrane fouling is not effectively controlled and the extent of fouling is so severe that membrane cleaning fails to reverse the decline in water production, as in the case of Yuma Desalting Plant in Arizona (14), plant has to be shut down and membrane replacement has to be carried out.

The ascertainment of fouling behavior or extent of fouling in any full-scale RO process is termed as *fouling characterization*. Fouling characterization is an important aspect in RO processes as the appropriate actions administered to mitigate membrane fouling, such as pretreatment and membrane cleaning, are largely dependent on the results concluded from fouling characterization. Information for fouling characterization is commonly derived from (a) of feed water and (b) average permeate flux. In actual full-scale RO processes, the variation of both variables with time means the fouling characterization is a continuous effort that demands constant monitoring.

In this chapter, discussions are focused mainly on the primary means of fouling characterization, that is, fouling potential of feed water and average permeate flux. Membrane fouling is first defined and basic information on fouling types and fouling control will be given. Current methods to quantify fouling potential and measure membrane fouling are examined, and attempts will be made to provide a more effective fouling characterization methodology for full-scale RO processes. The development of membrane fouling in a long membrane channel will be simulated with model using a more effective fouling potential. Efforts are also made to study various factors that may influence the fouling behavior in full-scale RO processes.

2. MEMBRANE FOULING AND CONTROL

Owing to its ability to reject most of the particles, colloids, solutes, and ions, RO process is susceptible to membrane fouling. Generally, membrane fouling refers to the attachment, accumulation, or adsorption of foulants onto membrane surfaces and/or within the membrane, deteriorating the performance of the membranes over time. In RO processes, pore plugging is seldom the main contributor to fouling because of the "nonporous" nature of RO membranes. The main fouling mechanism in RO processes is often associated with the formation of dense fouling layer on the membrane surface. Foulants are usually brought onto the membrane surface by the downward drag of fluid (permeate) as shown in Fig. 3.1. As more foulants accumulate on the membrane surface, a layer of foulant is formed. The fouling layer is dense and compacted because of the high pressure usually applied in RO processes, and therefore provides significant resistance to the permeate flow. In this study, membrane fouling in RO processes is defined precisely as the increase in membrane resistance caused by the formation of fouling layer on the membrane surface.

2.1. *Factors Affecting Membrane Fouling*

The ease of formation of fouling layer on membrane surface is dependent on numerous factors, which can be categorically divided into three main groups: driving pressure, dimensions and physicochemical properties of membrane channel, and feed water characteristics. It is mentioned earlier that foulants are brought upon the membrane surface by the downward fluid flow due to water permeating through the membrane. As the permeate is primarily driven by the driving pressure, an increase in driving pressure will increase the accumulation of foulant on membrane surface. Driving pressure also affects the porosity of the fouling

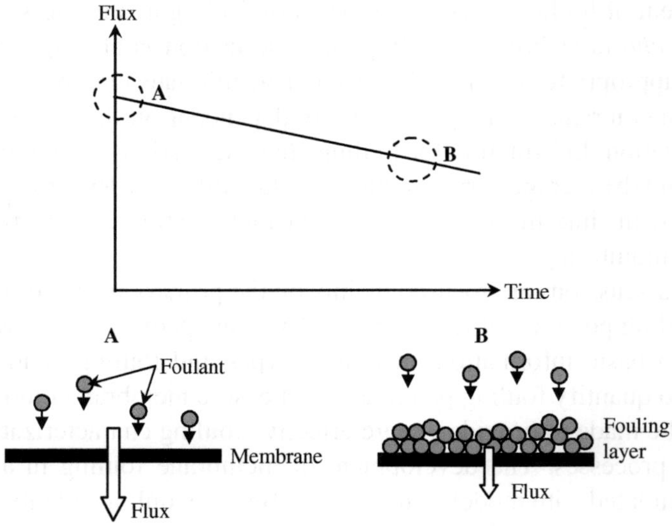

Fig. 3.1. Decline of permeate flux with the development of fouling layer on membrane surface.

layer. A more compact fouling layer reduces the amount of water that is allowed to reach the membrane surface and increases the rate of fouling.

The dimensions of membrane channel can affect the rate of membrane fouling by creating hydrodynamics that does not favor the deposition of foulant on membrane surface. One popular method, commonly found in spiral-wound and hollow-fiber membrane modules, is to pass the feed water through a narrow membrane channel, so that the crossflow velocity generates sufficient wall shear to remove the foulants from the membrane surface. To enhance the shearing effect and promote mixing, the membrane channel is constructed in a way such that the feed water is forced to travel through a series of bends within the channel, generating turbulence that hinders foulant deposition. This method to develop unstable feed water flow can be achieved with the use of feed spacers commonly found in spiral-wound membrane modules. However, it is important to note that the use of narrow channel and generation of turbulence can significantly reduce the downstream driving pressure. In addition, the presence of feed spacers may trap or capture the suspended foulants, leading to "clogging" problem. Hydrophobicity of RO membranes is also recognized to affect membrane fouling in RO processes. It is believed that hydrophobic membranes attract organic foulants more easily than hydrophilic ones. This is why some hydrophobic RO membranes are pretreated to give them a hydrophilic property.

Feed water characteristics – such as types and concentration of foulants, ionic strength, and pH – can have huge impact on membrane fouling in RO processes. A higher foulant concentration naturally poses a greater fouling problem with the availability of larger pool of foulants for deposition. The foulant type and composition influence the porosity of the fouling layer, which in turn determines the hydraulic resistance of the fouling layer. For example, a fouling layer consisting of organic matters and colloidal particles can have higher resistance than that made up of same colloidal particles only. The organic matters can fill up the voids between the colloidal particles, narrowing and reducing the pathways the feed water can take to reach the membrane surface. Ionic strength and pH of feed water affect the membrane–foulant and foulant–foulant interactions by altering the surface properties of membrane and foulants. Generally, a high ionic strength and low pH enhance the accumulation of foulants on membrane surface.

2.2. *Types of Fouling in RO Processes*

Membrane fouling is a complicated process and fouling behavior varies according to the types of feed water and operating conditions. According to the nature of foulants, membrane fouling in RO processes can be commonly classified as colloidal fouling, organic fouling, biological fouling, and scaling.

Colloidal particles are major foulants in all kinds of membrane processes. The size of colloidal particles ranges from a few nanometers to a few micrometers. They are ubiquitous in natural waters and examples of inorganic colloids include metal oxides, clay minerals, colloidal silica, and silicon. There are also plenty of colloidal particles of organic and biological origins. Most colloids carry negative surface charge in pH range of natural waters. Under the drag force of permeate flux, these colloidal particles will accumulate on the RO membrane surface to form a cake layer. The formation of this cake layer of the deposited colloidal particles adds on an additional resistance to the membrane resistance. This type of fouling is known as *colloidal fouling*.

Organic fouling is another prevalent problem in water treatment and reclamation plants. A common organic foulant is the natural organic matter (NOM), which is a complex heterogeneous mixture of different organic macromolecules from degradation and decomposition of living organisms. Another organic foulant is the residual organic matter in the effluent of wastewater treatment plants. The ability of organic foulants to foul membranes includes their affinity for RO membrane, molecular weight, and functionality. Negative functional groups on organic polyelectrolytes may be repulsed by the negatively charged membrane surfaces of RO membranes. Greater charge density on the membrane surface is often associated with higher hydrophilicity. Because most RO membranes are made of hydrophobic polymers, organic matters in the feed water usually tend to be preferentially adsorbed onto the membrane surfaces.

Biological fouling or biofouling occurs when living microorganisms in the feed water are transported onto the membrane surface where they absorb or adhere, forming a thin fouling layer. Once attached, microorganisms may grow and multiply at the expense of nutrients in the feed, forming a biological film or biofilm on membrane surface. Biofilm can be detected directly with microscopic inspection of fouled membrane surfaces, or indirectly with measurements of consequences of biofilm on membrane properties and system performance, such as solute rejection and permeate flux. Biofouling not only increases the membrane resistance, but also biodegrades the membranes through hydrolysis of membrane polymers.

Fouling by precipitation of the sparingly insoluble minerals to form an impermeable layer on the membrane surface is commonly known as *scaling*. As the salt concentration in the feed increases downstream due to the loss of water through permeation, the dissolved inorganic minerals such as carbonates and sulfates of calcium, magnesium, and iron, and some silica minerals may reach their solubility limits and precipitate on the membrane surface. Another possible reason that leads to precipitation is concentration polarization. Scaling can be reduced or avoided with the addition of acid to reduce anion species in feed water, pretreatment with lime softening to remove scale-forming metals, and addition of antiscaling agents such as hexametaphosphate.

2.3. *Silt Density Index*

One of the major components of membrane fouling in RO systems is deposition of foulants on membrane surface. These foulants are usually not retained by microfiltration or other filtering methods that treat feed water prior to RO processes. The extent of membrane fouling in RO system is very much dependent on the concentration, physical and chemical properties of foulants in the feed water that can be represented by a single entity known as *fouling potential*. The physical characteristics of the feed channel, such as channel height and the presence of feed spacers, are also believed to be factors influencing membrane fouling.

Silt density index (SDI) is the most widely used fouling index to quantify fouling potential of colloidal particles in feed water. This empirical test, shown in Fig. 3.2, involves filtering the feed water through a 0.45-μm filter membrane at constant pressure in dead-end flow. As shown in Eq. (1), the index is calculated based on the time taken to filter a fixed volume of feed water of a clean membrane and the time required to filter the same volume of feed water after the membrane has been used for a defined length of time:

$$\mathrm{SDI} = \frac{100(t_\mathrm{i}/t_\mathrm{f})}{t_\mathrm{total}}, \tag{1}$$

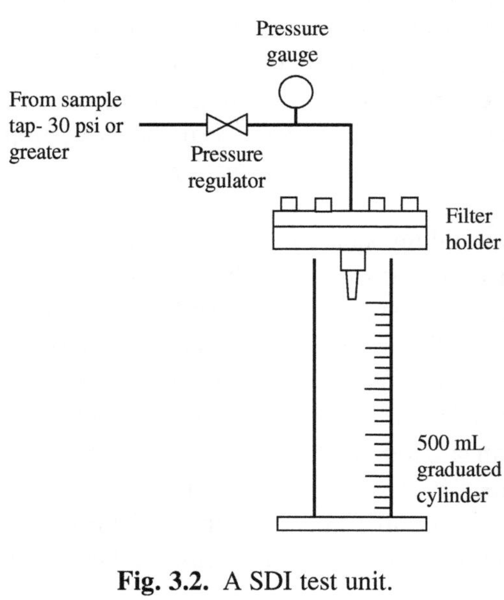

Fig. 3.2. A SDI test unit.

where t_i and t_f are the times (min) to collect initial and final 500 mL of sample, respectively, and t_{total} is the total running time (min) of test.

Apparently, the fouling potential determined with this method is not suitable for RO applications, considering the vast physical and operating differences between the SDI and RO units. Most notably, the SDI is only measuring the fouling rate associated with particles larger than 0.45 μm, since it uses a 0.45-μm filter membrane. In many RO processes where feed water is prefiltered before reaching RO systems, membrane fouling is usually caused by particles much smaller than 0.45 μm. In addition, the SDI adopts dead-end flow mode, which is different from the crossflow mode commonly employed in conventional full-scale RO processes. As the operating flow mode affects the fouling behavior of the particles on the membrane surface, fouling potentials measured when operating in these flow modes may be different. The accuracy of SDI also became questionable when it was reported that the index given by SDI is not proportional to the colloidal concentration (15). Other indices, such as modified fouling index (MFI), have been proposed in view of the deficiencies in SDI. Unlike the SDI that is empirically derived, MFI is based on the theory of cake filtration and is observed to be proportional to the colloidal concentration in the reported tests. Unfortunately, this method to measure colloidal fouling potential adopts the same type of filter membrane and operating flow mode as the SDI. Hence, the index may not be suitable for use in RO processes.

2.4. *Pretreatment*

Membrane cleaning reduces production time and increases operating costs. In addition, excessive cleaning may damage the membranes, resulting in more frequent membrane replacement. To reduce the cleaning frequency and duration, it is critical to slow down the rate of membrane fouling by keeping the fouling potential of the feed water to a minimum. This can be achieved by pretreating the feed water before it enters into the RO membrane modules. Some examples of pretreatments are as follows:

(a) Removal of large particles using coarse strainer. For smaller particles, microfiltration is used
(b) Water disinfections with chlorine
(c) Clarification and hardness removal using lime
(d) Addition of scale inhibitor
(e) Water sterilization using UV radiation

To ensure optimized performance of the full-scale RO system, the pretreated feed water must satisfy the requirements set by the plant operators. For example, to reduce the potential of organic fouling, designers can limit the total organic carbon (TOC) at 3 mg/L, biochemical oxygen demand (BOD) at 6 mg/L, and chemical oxygen demand (COD) at 8 mg/L. SDI is widely used to determine the potential of colloidal/suspended fouling. Generally, the maximum allowable SDI of the feed water is about 5 or less.

2.5. *Membrane Cleaning*

Membrane cleaning is a direct method to alleviate membrane fouling by restoring the permeability of RO membranes. Membrane cleaning is an important part of full-scale RO

Table 3.1 General categories of cleaning agents

Category	Chemicals	Adjusted pH
Weak acidic	Citric acid	4.0
Weak basic	Sodium tripolyphosphate (STPP) and sodium ethylenediaminetetraacetate (Na–EDTA)	10.0
Strong acidic	Hydrochloric acid (HCl)	2.5
Strong basic	STPP and sodium dodecylbenzene sulfonate (Na–DDBS)	10
	Sodium hydroxide (NaOH)	11.5

From Hydranautics Technical Service Bulletin 107.10.

operation, without which an early plant failure may occur or frequent replacement of membranes may be required. The membrane cleaning efficiency is affected by numerous factors, including the types of cleaning agents, types of foulant, chemical dosages, frequency of cleaning, and contact time. The efficiency of membrane cleaning and its effects on the performance of full-scale RO process are usually evaluated from past experiences or through pilot studies.

As a guide, cleaning is usually done when there is a 10% decrease in water production at constant operating conditions, or a 10% increase in the driving pressure to maintain the same production at constant temperature, or an increase of 15–20% in the pressure differential between feed and reject flows. In full-scale RO processes where spiral-wound membrane modules are the predominant membrane configuration, membrane cleaning is commonly done with chemicals.

Chemical cleaning is generally divided into low and high pH cleanings, which are used to remove inorganic and organic foulants, respectively. Cleaning agents are categorized into strong/weak acids and bases, as shown in Table 3.1. In practice, both inorganic and organic fouling occur together and membrane cleaning usually starts off with acid cleaning to remove inorganic scale or soluble colloidal materials, before the membranes are subjected to high pH cleaning to remove any remaining insoluble inorganic colloidal material, organic material, and/or biological organisms.

2.6. *Challenges*

Since RO membranes were first used in water and wastewater treatment, membrane fouling has emerged as one of the most serious obstacles to the technology. Scientists and engineers in this field have worked for decades to mitigate membrane fouling. Although many pretreatment processes and membrane cleaning techniques are invented, membrane fouling has not been totally eliminated and still remained as the major threat of RO processes. One of the reasons for ineffective efforts on membrane mitigation and control is the lack of effective fouling characterization methods. There are two obvious shortcomings in current fouling characterization (a) the commonly used SDI and related indices are inadequate to include all possible foulants and (b) fouling development in a full-scale RO process is not quantitatively related to the fouling indices. Therefore, SDI and related indices cannot be used as a rigorous parameter in process design. Pilot tests usually have to be conducted for observation of fouling development in the full-scale RO processes to generate the needed design parameters in fouling mitigation and control. The duration of pilot test may take months or over a year to obtain the meaningful information. However, because pilot tests are usually costly and time consuming, only limited scenarios can be tested and evaluated. Hence, the information obtained for fouling characterization is quite limited and incomplete. A quick, cost-effective, and reliable method for fouling characterization is therefore the key to a successful fouling control.

3. QUANTIFICATION OF FOULING POTENTIAL OF FEED WATER

3.1. Desirable Attributes for Fouling Potential Parameter

A more effective index or indicator for fouling potential of feed water is needed for a better characterization. It is preferred that the water fouling potential is determined with RO membrane and preferably under the conditions similar to the designed working conditions for the full-scale RO processes. Because of the use of RO membrane, all possible foulants in the feed water are considered in the measurements. In this way, the fouling index determined is "inclusive," which is a big advantage over most existing indices for fouling potential measurement of feed water.

Another desirable attribute of fouling index is that it is indicative to membrane fouling in full-scale RO processes. A quantitative relationship between the index and fouling development in the full-scale RO processes can be effectively and accurately established with a simulation model. When the fouling index of feed water with different pretreatments is determined, the fouling behavior of the water in the can be immediately simulated and assessed. The effectiveness of a pretreatment alternative can be evaluated in a few hours or at most a few days. It becomes possible to evaluate more pretreatment methods and operating conditions compared to conventional pilot testing. The need for pilot tests will be significantly reduced or eliminated with the new fouling characterization and modeling method developed (see Fig. 3.3).

In Sect. 3.2, an inclusive parameter for water fouling potential will be introduced. In Sect. 4, the associated mathematical model for the linkage of the parameter to fouling development in the full-scale RO processes will be introduced.

3.2. An Inclusive Parameter for Fouling Potential

Under the driving pressure, permeate flux moves perpendicularly to the membrane surface from the bulk solution and passes through the membrane. The foulants containing in the water will be retained on the membrane surface. The foulant deposition rate onto the RO membrane surface induced by the permeate flow is

$$j = vc_{f0}, \tag{2}$$

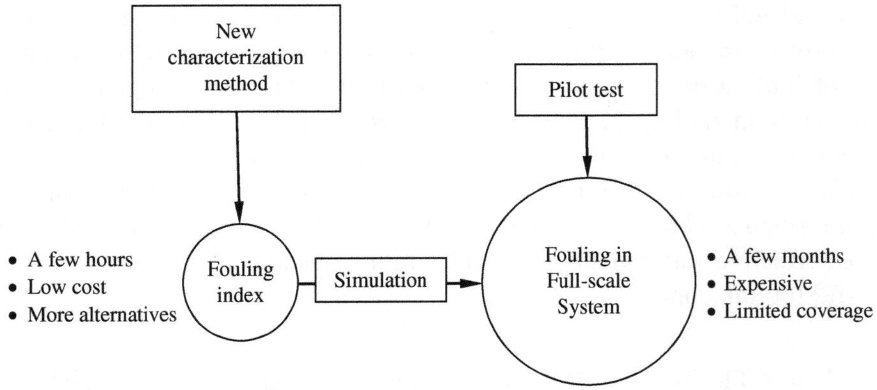

Fig. 3.3. Comparison of the new characterization method and the pilot test.

where j is the foulant deposition rate onto the membrane surface (g/m^2 s), v is the permeate flux (m^3/m^2 s or m/s), and c_{f0} is the foulant concentration (mg/L) in the bulk flow. The foulant amount M (g/m^2) accumulated over time t is calculated by

$$M = \int_0^t j\,\mathrm{d}t = c_{f0}\int_0^t v\,\mathrm{d}t. \tag{3}$$

Resistance of the fouling layer at any point of time is related to the amount of accumulated foulants with

$$R_f = r_s M = r_s c_{f0}\int_0^t v\,\mathrm{d}t = k_f\int_0^t v\,\mathrm{d}t, \tag{4}$$

where R_f is the resistance of the fouling layer (Pa s/m), r_s is the specific resistance of the fouling layer (Pa s m/g), and k_f (= $r_s c_{f0}$) is a new parameter (Pa s/m^2) introduced and is termed as the *fouling potential of the feed water*. Rearranging Eq. (4) gives (15, 16)

$$k_f = \frac{R_f}{\int_0^t v\,\mathrm{d}t}. \tag{5}$$

Because the integration term is the volume of permeate collected per unit area of membrane over time period t, Eq. (5) shows that the fouling potential of the feed water is defined as the resistance of the fouling layer caused by a unit volume of permeate collected per unit area of membrane.

In the above derivation, the fouling potential is calculated as the product of the specific cake resistance (r_s) and bulk foulant concentration (c_{f0}). However, the calculation proves to be difficult with real feed water, primarily because it is really impossible to establish the specific cake resistance of all foulants in the feed water. Fortunately, the fouling potential of feed water can be easily determined with a simple fouling test using a small piece of RO membrane. The total hydraulic resistances at the beginning (R_0) and the end (R_t) of the fouling test are determined and the difference between these two hydraulic resistances gives the resistance of the fouling layer:

$$R_{\mathrm{f}} = R_{\mathrm{t}} - R_0. \tag{6}$$

Substituting Eq. (6) into Eq. (5) results in

$$k_{\mathrm{f}} = \frac{R_{\mathrm{t}} - R_0}{\int_0^t v\,\mathrm{d}t}. \tag{7}$$

The hydraulic resistances at the beginning and the end of fouling test are related to the permeate fluxes with the basic membrane transport principle:

$$R_0 = \frac{\Delta p - \Delta \pi}{v_0}, \tag{8}$$

$$R_{\mathrm{t}} = \frac{\Delta p - \Delta \pi}{v_{\mathrm{t}}}, \tag{9}$$

where v_{t} and v_0 represent the measured permeate flux at the beginning and the end of the fouling experiment, respectively, $\Delta\pi$ is the osmotic pressure (Pa), and Δp is the driving pressure (Pa). Unlike in most ultrafiltration and microfiltration processes, the permeate flux in RO processes usually declines with time at a near constant rate. This is because the resistance of fouling layer is much smaller than the resistance of RO membranes and nearly constant permeate flux can be maintained during the fouling experiment. In this case, the integral term in Eq. (7) can be well approximated to

$$\int_0^t v\,\mathrm{d}t = \frac{t}{2}(v_0 + v_{\mathrm{t}}). \tag{10}$$

Substituting Eqs. (8)–(10) into Eq. (7) results in

$$k_{\mathrm{f}} = \frac{2(\Delta p - \Delta\pi)}{t}\,\frac{(v_0 - v_{\mathrm{t}})}{v_0 v_{\mathrm{t}}(v_0 + v_{\mathrm{t}})}. \tag{11}$$

Fouling potential in Eq. (11) can be easily determined since all the terms on the right-hand side of the expression can be either measured from a crossflow RO test cell unit or predetermined by the investigator.

3.3. *Membrane Device for Fouling Potential Measurement*

The RO membrane cell for fouling potential measurement is schematically shown in Fig. 3.4. A small piece of RO membrane is housed in a stainless steel case that can sustain high pressure. The membrane cell is operated under crossflow mode in order to simulate fouling under similar hydraulic condition as in the full-scale RO processes. The feed water under test is pumped into the cell at the testing crossflow velocity and pressure with a high-pressure pump. Permeate from the cell is collected in the glass container, weighed and recorded at preset interval. The measurement usually last for a few hours, depending on the fouling potential of the feed water. The duration is shorter for feed water with higher fouling potential. The permeate fluxes at the beginning and the end of measurement are determined and the fouling potential is calculated from Eq. (11).

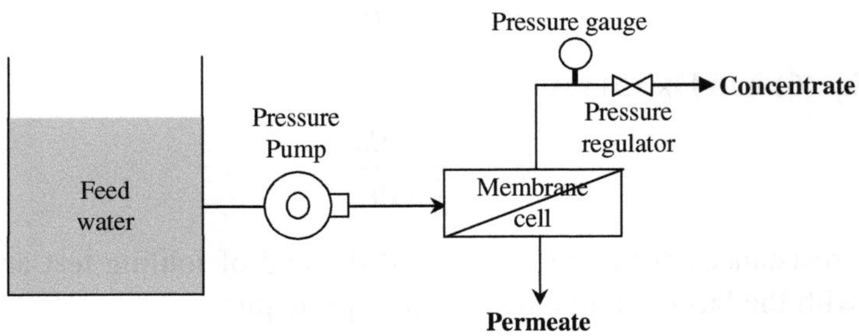

Fig. 3.4. The membrane cell setup for fouling potential measurement. Permeate and concentrate from the membrane cell either discarded or recirculated back into feed tank.

The resistance of some new RO membranes may increase slightly in the initial period under high driving pressure. This phenomenon is commonly known as *membrane compaction*. Membrane compaction will affect the accuracy of the measurement of fouling potential. To avoid the effect of membrane compaction, new membranes are recommended to be precompacted under high pressure with deionized (DI) water for at least 24 h before use.

It is preferred that large volume feed water be stored for the measurement, so that permeate and concentrate from the membrane cell can be discarded. The recirculation of permeate and concentrate back to the feed tank has obvious disadvantages. First, it will cause change in feed water fouling property due to gradual detainment of foulants by the membrane. Second, the high-pressure pump tends to increase the water temperature. Changes in both foulant concentration and temperature will introduce substantial errors in the measurement of fouling potential. When measurement has to be conducted with permeate and concentration recirculation (e.g., with a small feed tank), automatic temperature control is almost mandatory to keep the temperature at a constant value.

3.4. *Properties of Fouling Potential of Feed Water*

In this section, the properties of fouling potential are examined and discussed with some examples. With the membrane cell described in Sect. 3.2, synthetic feed water consisting of sodium chloride (provide osmotic pressure) and different concentrations of silica colloids is measured for their fouling potentials. The silica colloids (Snowtex ZL, Nissan Chemical Industries Ltd, Tokyo, Japan) used to make the synthetic feed water has an average diameter of 130 nm. Operating conditions are given in the figure captions. Permeate fluxes from the measurements are plotted with time in Fig. 3.5.

From the measurements, the permeate fluxes at the beginning and the end of the test period are identified and used with Eq. (11) to calculate the fouling potentials at each colloidal concentration. When there are fluctuations in the permeate fluxes, the data points can be fitted with straight lines and the permeate fluxes at the beginning and end are determined from the fitted lines. In this way, the experimental errors are significantly reduced and much more accurate fouling potential can be obtained. Table 3.2 presents the fouling potentials of the

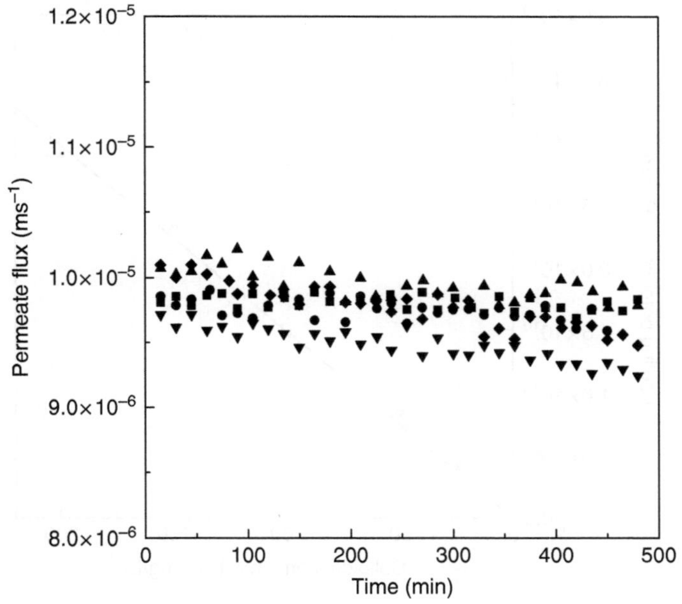

Fig. 3.5. Decline of permeate flux with time in fouling experiments with different colloidal concentrations of 25 mg/L (*filled square*), 50 mg/L (*filled circle*), 100 mg/L (*filled triangle*), 150 mg/L (*filled inverted triangle*), and 200 mg/L (*filled diamond*). Driving pressure = 1.5 × 10^6 Pa, salt concentration = 1,000 mg/L NaCl, and feed crossflow velocity = 0.15 m/s.

Table 3.2 Properties of the fitted lines at different colloidal concentrations

Colloidal concentration (mg/L)	Gradient	Initial flux (m/s)	k_f (Pa s/m^2)
25	-1.55×10^{-10}	9.84×10^{-6}	3.97×10^{9}
50	-3.19×10^{-10}	9.83×10^{-6}	8.32×10^{9}
100	-5.99×10^{-10}	1.01×10^{-5}	1.47×10^{10}
150	-8.22×10^{-10}	9.69×10^{-6}	2.33×10^{10}
200	-1.11×10^{-9}	1.01×10^{-5}	2.87×10^{10}

synthetic feed water with different colloidal concentrations, determined from the lines best fitted to the flux data shown in Fig. 3.5.

Table 3.2 shows that the fouling potential of the synthetic water increases with colloidal concentration. This phenomenon is reasonable because more colloidal particles are brought onto membrane surface at higher particle concentration. The fouling potentials are plotted against colloidal concentrations and shown in Fig. 3.6. The graph shows that all the data points can be well fitted with a straight line. The strong linear relationship between fouling potential and colloidal concentration indicates that fouling potential is an excellent indicator of the amount of foulants in the feed water.

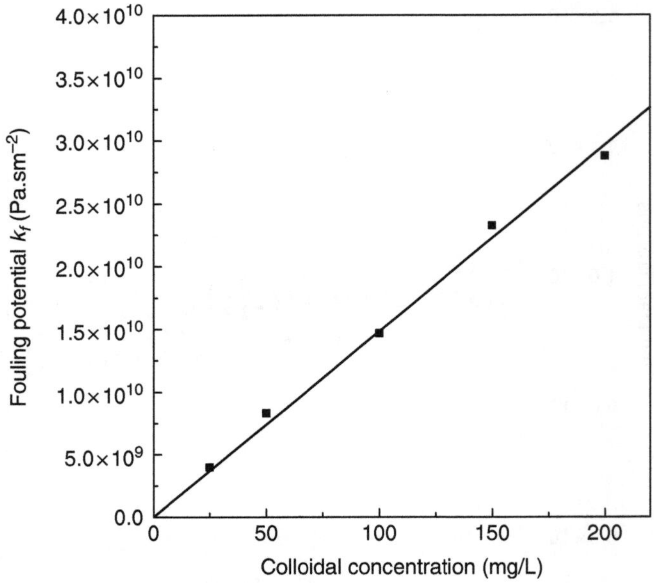

Fig. 3.6. Linear relationship between fouling potentials and colloidal concentrations.

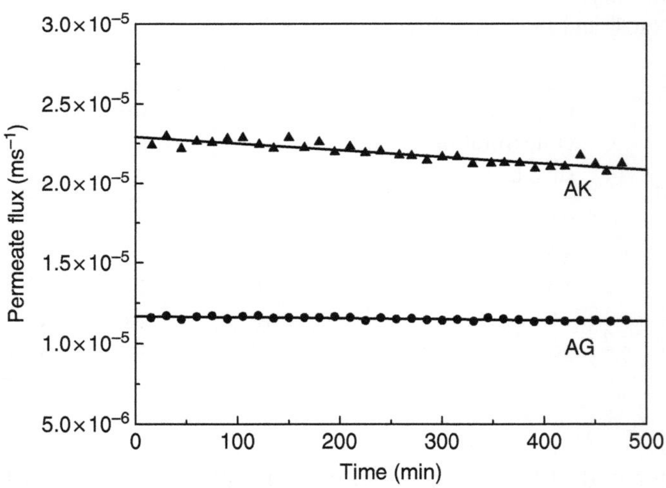

Fig. 3.7. Permeate flux decline with time on two different RO membranes. Driving pressure = 1.38×10^6 Pa, salt concentration = 1,000 mg/L NaCl, feed crossflow velocity = 0.15 m/s, and colloidal concentration = 50 mg/L.

Figure 3.7 shows permeate fluxes in the membrane device as a function of time on two different RO membranes (AG and AK) from Osmonics (Minnetonka, MN). Colloidal concentration and salt concentration in the synthetic feed water under test are 50 and 1,000 mg/L, respectively. The initial permeate flux of AK membrane is twice the value of AG membrane and this indicates that the clean membrane resistance of AG membrane is two times of AK membrane. Although the permeate flux is observed to decline at a much higher rate with AK membrane compared with AG membrane, fouling potentials determined with AG and AK membranes are similar (AK 8.66×10^9 Pa s/m^2 and AG 8.51×10^9 Pa s/m^2). The difference of the two values is about 2% and well within the experimental errors. It is clear that the fouling potential is the intrinsic property of the feed water and is independent of the membrane used in the measurement (17). However, precaution has to be taken when there is strong interaction between the foulant components and the RO membrane. For example, if a particular component in the feed water has a strong affinity to certain type of RO membrane, higher fouling potential may be obtained with this membrane compared to other types of membranes. This problem can be avoided by using the same type of RO membrane used in full-scale RO process.

In actual RO practices, the fouling potential of the feed water may increase with pressure due to compaction of the cake or fouling layer at high pressure. A more compact fouling layer leads to a higher rate of flux decline as less water passes through the fouling layer, resulting in a higher fouling potential measured, even though the colloidal concentration is unchanged. Fouling potential is studied with AG membrane under different driving pressures and presented in Fig. 3.8. Colloidal concentration and salt concentration in the synthetic feed water under test are 50 and 1,000 mg/L, respectively. It is observed that the rate of flux decline is more rapid at higher pressure, even though the colloidal concentration remains unchanged in

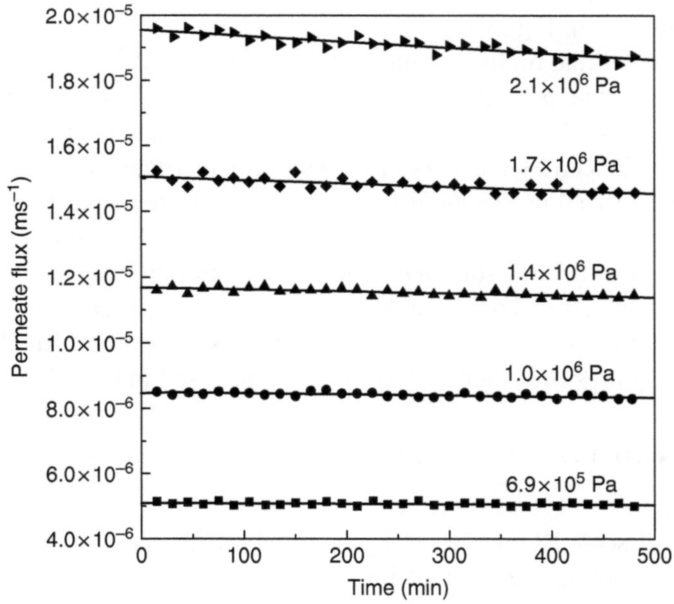

Fig. 3.8. Permeate fluxes in fouling experiments under different pressures. Salt concentration = 1,000 mg/L NaCl, feed crossflow velocity = 0.15 m/s, and colloidal concentration = 50 mg/L.

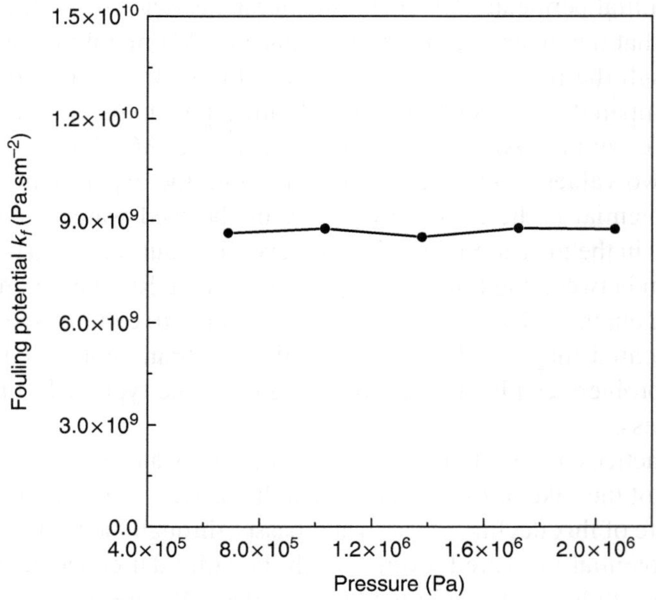

Fig. 3.9. Fouling potentials at different driving pressures measured with AG membrane.

the experiments. The rapid flux decline can be attributed to the increase in particle deposition, which is caused by the increase in permeate flux that carries the colloidal particles onto the membrane surface when pressure is increased. Fouling potential at different pressures is calculated and presented in Fig. 3.9.

It is shown in Fig. 3.9 that fouling potential determined with AG membrane is not sensitive to driving pressure in the pressure range tested. The compaction of fouling layer is not noticeable in the measured fouling potentials in this case. Increase in the permeate flux at higher pressures is balanced by the same increase in the resistance of the fouling layer. This nonchalant response of fouling potential to increasing pressure suggests that the duration of fouling tests to determine fouling potential can be shortened by increasing the driving pressure, so that permeate flux can decline sufficiently to give meaningful results. However, the lack of method to determine the critical pressure that gives rise to compaction of fouling layer means precaution has to be taken when operating the fouling tests at high pressure. For most accurate measurement, driving pressure should be identical to that used in the full-scale RO process.

4. PREDICTION OF FOULING IN FULL-SCALE REVERSE OSMOSIS PROCESSES

Fouling in the full-scale RO processes is commonly indicated by the decline of permeate flux under constant driving pressure or increase in driving pressure to maintain constant permeate flux. It has been known for a long time that the flux decline in the full-scale RO processes is quite different from that observed from the fouling test in the laboratory-scale RO device. This fact makes it impossible to quantitatively predict flux decline in the full-scale

RO processes directly from the fouling potential determined from the laboratory membrane device without a well-correlated mathematical model.

In this section, a model based on fundamental mass conservation and membrane transfer principles is first presented to link the fouling behavior of full-scale RO processes to the fouling potential of feed water. Simulations are also presented to illustrate fouling development in full-scale RO processes under various operating scenarios. In this section, the impact of fouling potential of the feed water on fouling behavior in the full-scale RO processes is particularly emphasized.

4.1. Model Development

The membrane channel in a full-scale RO processes is predominantly formed by several spiral-wound modules connected in series in a pressure vessel. The channel of spiral-wound membrane modules is schematically shown in Fig. 3.10. The spacers in the channel enhance mixing, so that complete mixing in the transverse direction (along the channel height) of the membrane channel can be reasonably assumed.

The membrane channel in the full-scale RO processes is typically a long narrow channel, which cannot be treated as a homogeneous system when highly permeable RO membranes are used. Under normal operating conditions, the parameters – such as crossflow velocity, salt concentration, and permeate flux – change substantially along the membrane channel. Localized parameters must be considered to accurately describe the membrane system (18–20). Furthermore, the membrane resistance is a function of time because membrane fouling is a continuous process. Therefore, most parameters of the membrane system are time dependent even though the operating conditions (i.e., feed driving pressure and feed crossflow velocity) are kept unchanged.

Permeate flux at any point of the membrane channel is commonly accepted to linearly relate to the net driving pressure:

$$v(x,t) = \frac{\Delta p(x,t) - \Delta\pi(x,t)}{R_{\mathrm{m}}(x,t)}. \tag{12}$$

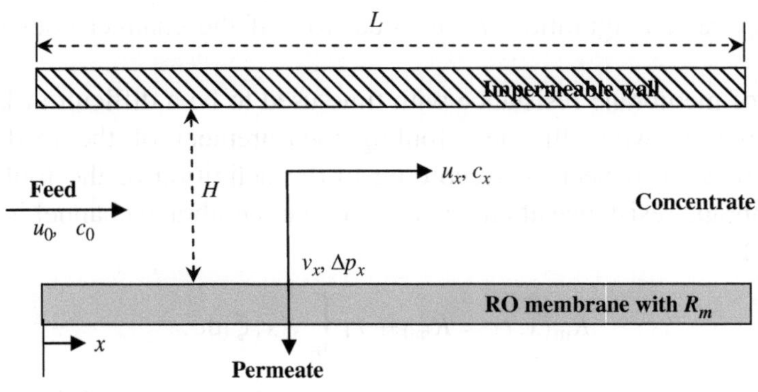

Fig. 3.10. A long RO membrane channel.

The symbols x and t in the brackets indicate that the variables are dependent on location and time (this point will not be mentioned later). The osmotic pressure of the feed water is usually linearly related to salt concentration with the following equation:

$$\Delta\pi(x,t) = f_{os}c(x,t), \tag{13}$$

where f_{os} is the osmotic coefficient (Pa L/mg) and c is the salt concentration (mg/L).

Salt concentration in the membrane channel can be determined by the following mass conservation principle:

$$c(x,t)u(x,t)H = c_0u_0H - (1 - r_{rej})\int_0^x c(\xi,t)v(\xi,t)\mathrm{d}\xi, \tag{14}$$

where u is the crossflow velocity (m/s), H is the height of the membrane channel (m), u_0 is the feed flow velocity (m/s), c_0 is the feed salt concentration (mg/L), ξ is the integration variable, and r_{rej} is the salt rejection of the membrane. The term on the left-hand side of the equation is the total amount of salt remaining in the feed channel at location x. The first term on the right-hand side of the equation is the total amount of salt in the feed flow. The second term on the right-hand side of the equation is the total amount of salt in the permeate collected from the inlet to location x in the membrane channel.

Similarly, applying the mass conservation principle on water results in the following expression for crossflow velocity along the long channel:

$$u(x,t)H = u_0H - \int_0^x v(\xi,t)\mathrm{d}\xi. \tag{15}$$

As the water flows down the membrane channel, the friction of the feed water with channel walls and the spacers in the channel causes pressure loss. The pressure loss is related to the channel height and spacer configuration by

$$\Delta p(x,t) = \Delta p_0 - \frac{12K_s\eta}{H^2}\int_0^x u(\xi,t)\mathrm{d}\xi, \tag{16}$$

where Δp_0 is the feed driving pressure (Pa), η is the water viscosity (Pa s), and K_s (≥ 1) is a coefficient for spacer configuration. K_s is equal to 1 if the channel does not contain any spacer.

The tracking of membrane resistance with time used to be a difficult task. This problem, however, is overcome when the new fouling measurement of the feed water (fouling potential) is introduced in Sect. 3. According to the definition of the fouling potential of feed water, membrane resistance at any location in the membrane channel is mathematically expressed as (21)

$$R_m(x,t) = R_{m0} + k_f\int_0^t v(x,\xi)\mathrm{d}\xi, \tag{17}$$

where $R_m(x, t)$ is the membrane resistance (Pa s/m) at location x and time t and R_{m0} is the clean membrane resistance (Pa s/m).

The average permeate flux of the membrane channel is then calculated with

$$\bar{v}(t) = \frac{1}{L}\int_0^L v(\xi, t)\mathrm{d}\xi, \tag{18}$$

where $\bar{v}(t)$ is the average permeate flux (m/s).

4.2. Fouling Development in a Long Membrane Channel

With Eqs. (12)–(18), the change in average permeate flux of a long membrane channel can be simulated with time for different fouling potentials of feed water and operating conditions. A simulation example of fouling development in a 6-m long membrane channel is shown in Fig. 3.11 with operating conditions given in Table 3.3.

Figure 3.11 shows the average permeate flux in the membrane channel over 100 days for the filtration of feed water with a fouling potential of 1.0×10^9 Pa s/m^2. The simulated average permeate flux remains constant in the first 20 days of operation even membrane fouling has been occurring in the membrane channel from the beginning. This fouling behavior is entirely different from that in the small laboratory-scale RO devices, where the average permeate flux declines immediately from the start of the fouling tests. The constant average permeate flux of the full-scale RO processes in the initial period of operation is observed and reported in literature (22, 23). The use of homogeneous membrane transfer equation ($v = (\Delta p - \Delta\pi)/R_{\mathrm{m}}$) to interpret the invariant average flux will mislead to the conclusion that no fouling is occurring in the membrane channel.

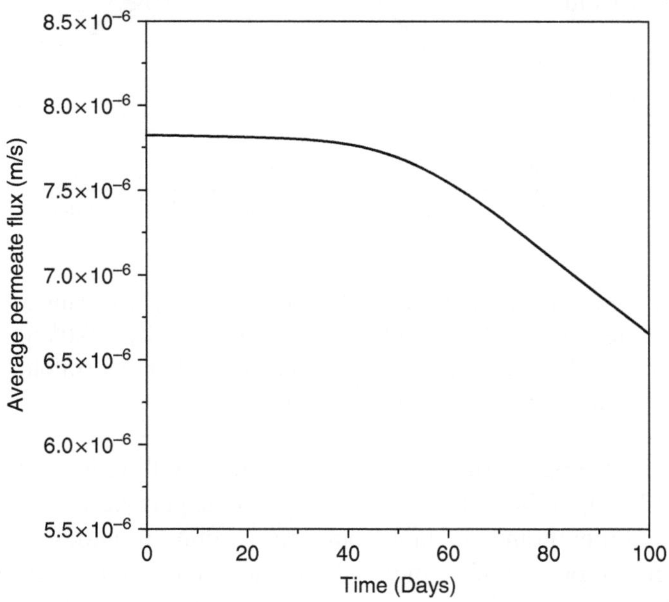

Fig. 3.11. Fouling development in a 6-m long membrane channel with operating conditions given in Table 3.3.

Table 3.3 Operating parameters used in the simulation

Operating parameters	Value
Channel length (m)	6
Channel height (m)	0.5×10^{-3}
Feed driving pressure (Pa)	1.21×10^{6}
Feed salt concentration (mg/L)	1,000
Osmotic coefficient	68.95
Feed crossflow velocity (m/s)	0.1
Clean membrane resistance (Pa·s/m)	8×10^{10}
Solute rejection	0.995
Operating period (day)	100
Fouling potential (Pa s/m^2)	1×10^{9}

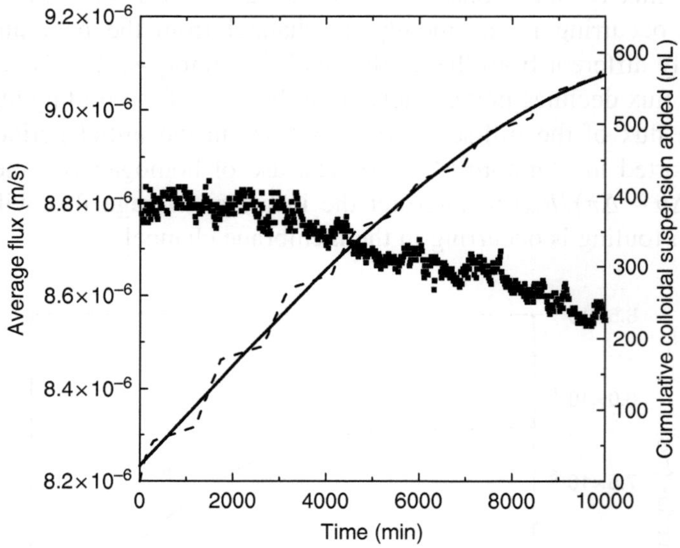

Fig. 3.12. Decline in average permeate fluxes with time in a fouling test using a 4-m long membrane channel. Driving pressure = 1.3×10^6 Pa and feed crossflow velocity = 0.05 m/s. *Scatter points* for measured average permeate fluxes, *dash curve* for cumulative volume of colloidal suspension added, *solid curve* for Gaussian fit of dash curve with $R^2 = 0.995$.

The insensitivity of average permeate flux to membrane fouling is experimentally demonstrated with Fig. 3.12, which shows the variation of average permeate flux with time of an RO system with 4-m long membrane channel (four spiral-wound modules connected in series). Synthetic feed water comprising of 30 mg/L of commercial colloidal particles (Snowtex ZL) and 1,000 mg/L of sodium chloride is used in the experiment. An initial period of constant average flux, which lasts for about 1,500 min, was observed before obvious average permeate flux decline occurs. During this period of constant average flux, the colloidal concentration in the feed water is measured with a UV spectrophotometer and colloidal suspension is regularly added into the feed tank to maintain the colloidal concentration. The cumulative volume of colloidal suspension added is shown as the dash curve in Fig. 3.12. The addition of colloidal suspension is a clear indication that colloidal fouling is taking place from the start of the experiment. But the deposition of the colloids on membrane surface is clearly not reflected in the average permeate flux.

Using the operating conditions in Table 3.3, the changes in membrane resistance profiles of the membrane channel with time are simulated and shown in Fig. 3.13. It obviously indicates that the membrane resistance increases with time right from the beginning of the operation, which means that there is membrane fouling in the membrane channel. Corresponding localized permeate flux profiles along the membrane channel at different times are presented in Fig. 3.14. It can be seen from Figs. 3.13 and 3.14 that the localized permeate flux is sensitive to the change in the localized membrane resistance.

A more evenly distributed local permeate flux profile evolves as membrane fouling progresses. This phenomenon is attributed to the strong interplay between the local permeate flux and membrane resistance. At the beginning of membrane filtration, permeate flux usually has its maximum (peak) value at the channel entrance. The membrane resistance will increase faster due to a higher rate of foulant deposition, which in turn causes a more rapid decline in permeate flux. In the downstream region with a lower permeate flux, a lower fouling rate will induce a slower flux decline. Eventually, an evenly distributed flux profile will form along the membrane channel.

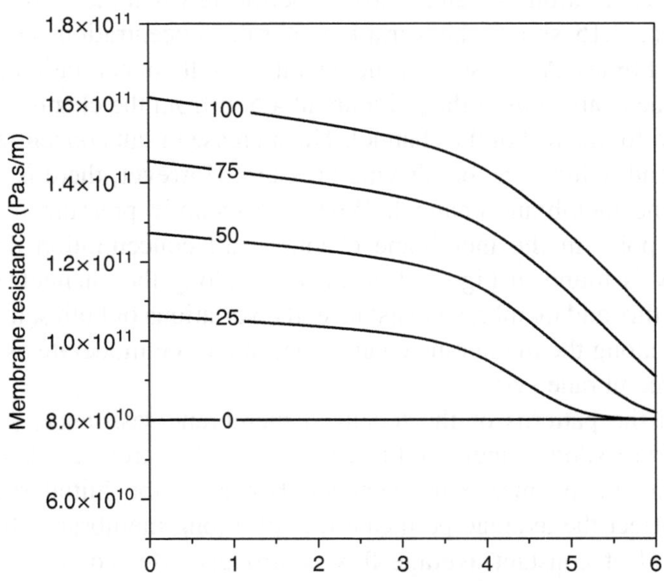

Fig. 3.13. Changes in membrane resistance profiles of the membrane channel with time.

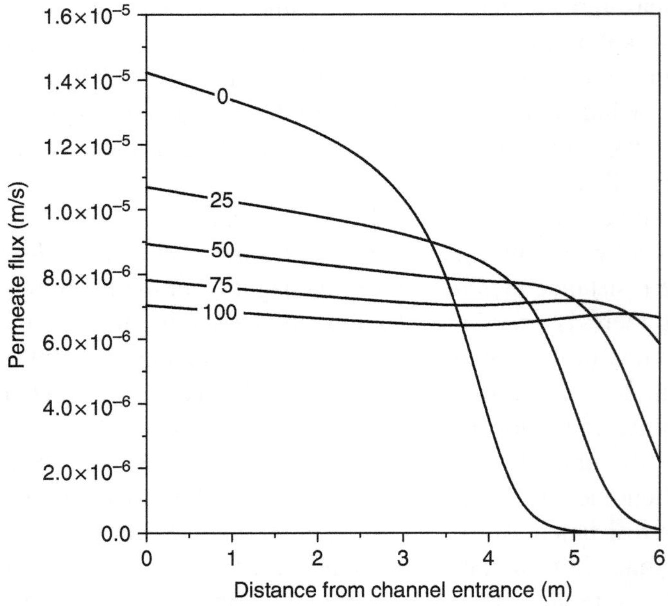

Fig. 3.14. Evolvement of evenly distributed permeate flux profile of the membrane channel with time.

The reason for the decline in permeate flux along the membrane channel in Fig. 3.14 is the build-up of salt concentration as water is lost as permeate when the feed water travels along the channel. Figure 3.15 shows the variation of salt concentration along the membrane channel at different times. At the start of the operation, salt concentration is seen to increase along the membrane channel until the point about 4.5 m from the channel entrance, where it remains unchanged for the rest of the channel. The increase in salt concentration increases the osmotic pressure and reduces the net driving pressure. However, there is a limit for the salt concentration in the membrane channel. When the osmotic pressure reaches the driving pressure at some point in the membrane channel, salt concentration ceases to increase. Crossflow velocity is found in Fig. 3.16 to decrease along the channel due to water loss. Like the permeate flux and membrane resistance, the variations of both salt concentration and crossflow velocity along the membrane channel became more moderate as fouling continues to develop in the membrane system.

The changes in the patterns of the localized membrane resistance, permeate flux, salt concentration, and crossflow velocity in Figs. 3.13–3.16 give strong evidence that membrane fouling is occurring in the membrane channel. However, the initial stage of membrane fouling does not affect the average permeate flux of a long membrane channel. The length of the initial period of constant average flux is strongly affected by the driving pressure employed in the membrane process. The average permeate flux of the same membrane channel (using parameters given in Table 3.3) is simulated for two driving pressures of 1.21×10^8 and 1.38×10^8 and the results are shown in Fig. 3.17. It can be seen that the 15%

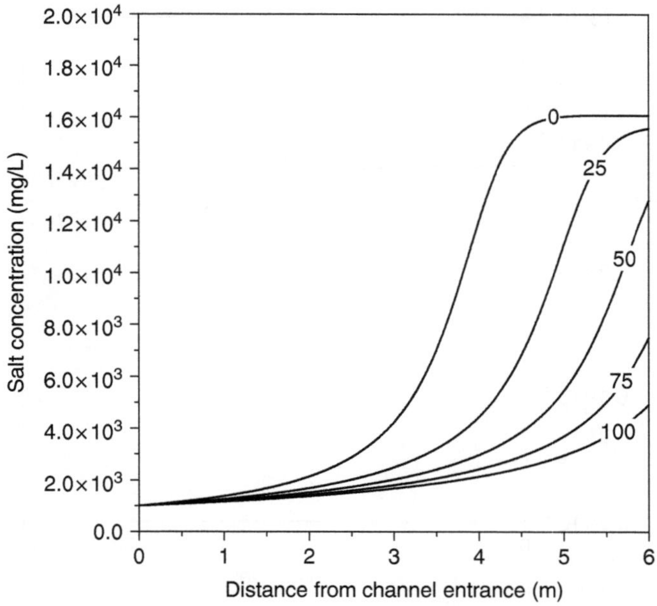

Fig. 3.15. Variation of salt concentration along the membrane channel with time.

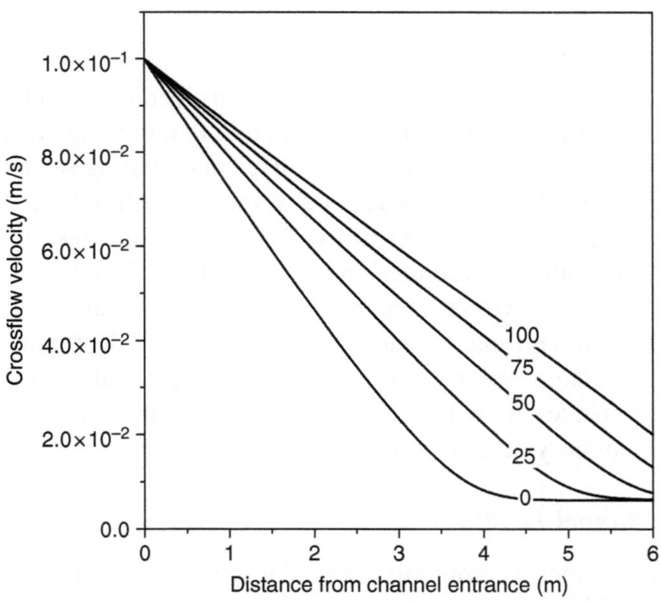

Fig. 3.16. Crossflow velocity profiles along the membrane channel at different operating times.

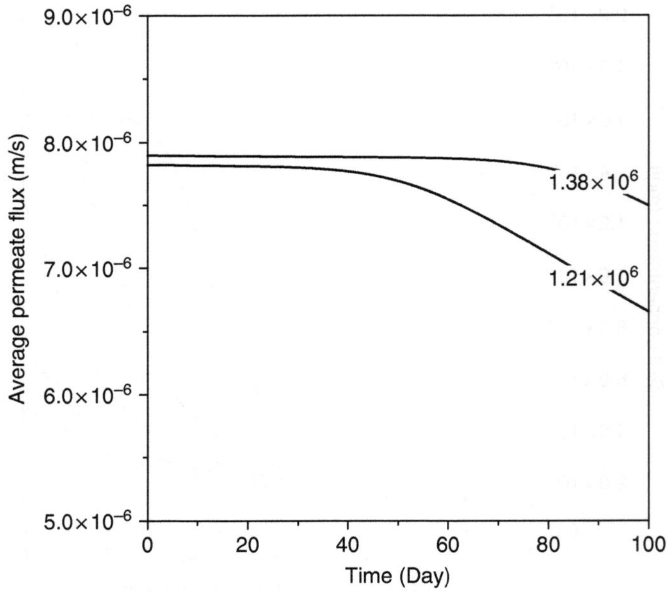

Fig. 3.17. The duration of constant average permeate flux greatly affected by driving pressure.

increase in driving pressure doubled the duration of the constant average flux with only marginal increase in the magnitude of the average flux.

4.3. *Influence of Feed Water Fouling Potential*

With the introduction of the new fouling potential of the feed water and the development of the mathematical model for fouling development in a long membrane channel, the effect of membrane fouling on the system performance at different operating scenarios can be quantitatively predicted. Figure 3.18 shows the fouling development in a long membrane channel (indicated by the average permeate flux) for feed waters with different fouling potentials. Operating conditions in Table 3.3 are used in the simulations. The initial average permeate fluxes are identical for all feed waters since the same clean membrane resistance is used in the simulations. The period of constant average flux increases with decreasing fouling potential of the feed water. It is also noted that feed water with higher fouling potential has a larger decline in average permeate flux with time. This phenomenon is similar to that observed in fouling tests using laboratory-scale membrane cells.

4.4. *Influence of Channel Length*

A long membrane channel is commonly adopted in most full-scale RO systems to increase the permeate recovery. This is achieved by connecting several membrane modules in series in a pressure vessel. The influence of channel length on the system performance is simulated with operating parameters in Table 3.3 using feed water with a fouling potential of 10^9 Pa s/m^2 and

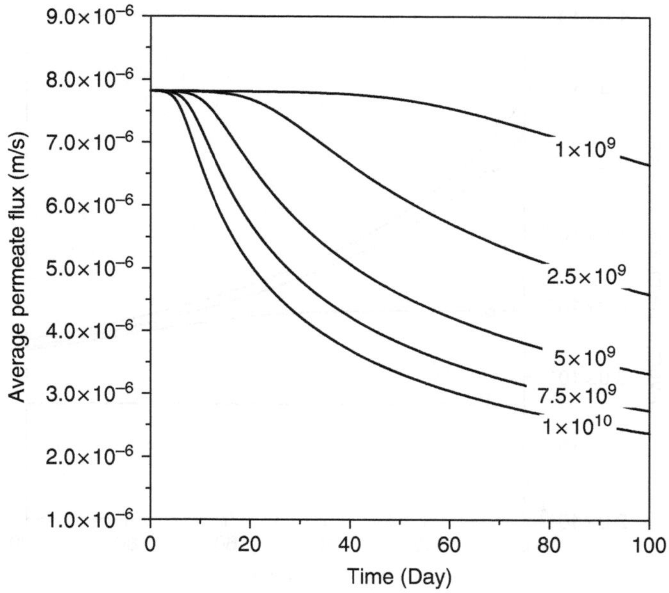

Fig. 3.18. Fouling developments in a 6-m long membrane channel for feed water of different fouling potentials. Other operating parameters given in Table 3.3.

the results are shown in Fig. 3.19. Channel length is found to play a key role on the behavior of average flux. The period of constant average flux does not occur in 1- and 3-m channel as the average permeate fluxes of both channels decline immediately from the start of the filtration operation. The profiles of these average permeate fluxes are similar to that obtained from the laboratory-scale RO systems. An initial period of constant flux is observed in 6- and 9-m channels and the duration of the constant flux increases markedly with channel length.

It is also observed that longer membrane channels have lower initial average permeate fluxes, even though the operating parameters used are identical. This finding indicates that the maximal recovery of the feed water has been reached and, therefore, it cannot be further increased by increasing channel length. Because the permeate rate is fixed, a lower average permeate flux is obtained for a longer membrane channel. Interestingly, this behavior is not observed or reported in the older generation of RO membranes. These RO membranes have higher membrane resistance that justifies the increase of permeate rate with longer membrane channel. In a long membrane channel of highly permeable RO membranes, it is possible that the permeate rate is not influenced by the membrane area, but by the thermodynamic properties of the system.

4.5. *Influence of Clean Membrane Resistance*

The advancement in membrane technology has led to the development of highly permeable membranes with high salt rejection property. Currently used RO membranes are at least one order more permeable than those in the 1980s (24–28). It is interesting to know how the

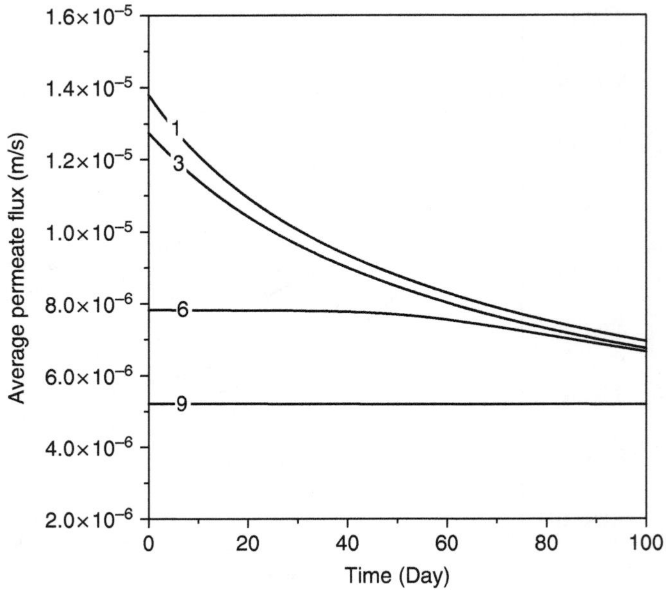

Fig. 3.19. Effect of channel length on fouling development in a long membrane channel using feed water with fouling potential of 10^9 Pa s/m^2.

change in clean membrane resistance affects the fouling behavior in a long membrane channel. Figure 3.20 shows the variation of average permeate flux with time under two different clean membrane resistances. Other operating parameters are given in Table 3.3. An initial period of constant average permeate flux is observed when clean membrane resistance of 8×10^{10} Pa s/m is used in the simulation. In contrast, average permeate flux is observed to decline immediately from the start of operation at clean membrane resistance of 1.5×10^{11} Pa s/m with identical fouling potential. The absence of initial period of constant average permeate flux in RO membranes with very high resistances is attributed to the lower localized permeate flux along the membrane channel. Since the entire membrane channel is fully utilized to produce permeate, any increase in membrane resistance due to fouling will reduce the average permeate flux.

4.6. Characteristic Pressure of a Long Membrane Channel

The above simulations indicate that there are two distinguished regimes in the operation of a long membrane channel. Membrane fouling can only be indicated by average permeate flux in one of the regimes. Further study reveals that there exists a specific pressure for a given membrane channel:

$$\Delta p^* = \frac{u_0 H R_m}{L}, \tag{19}$$

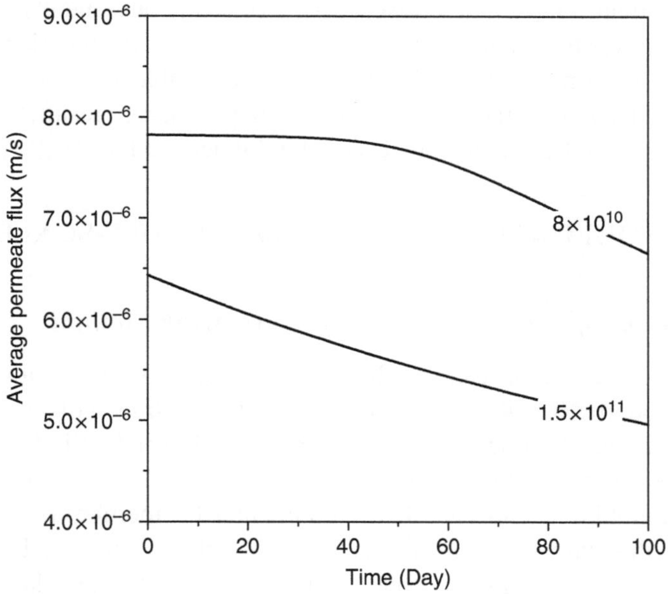

Fig. 3.20. Different fouling behaviors for 6-m long membrane channels with different membrane resistances using operating parameters given in Table 3.3.

Table 3.4
Typical characteristic pressures for a 6-m membrane channel of various RO membranes[a]

Membranes	Membrane resistance (Pa s/m)	Characteristic pressure (Pa) (ψ)
Low-pressure membranes	8.00×10^{10}	9.33×10^{5} (135)
Seawater membranes	3.00×10^{11}	3.50×10^{6} (507)
Old membranes	2.00×10^{12}	2.33×10^{7} (3,384)

[a]Channel height $H = 7 \times 10^{-4}$ m and feed velocity $u_0 = 0.1$ m/s.

When the driving pressure Δp is smaller than Δp^*, the average permeate flux is controlled by mass transfer and fouling is immediately reflected by the average flux. However, when the driving pressure is greater than Δp^*, the average permeate flux is controlled by and is not affected by membrane fouling. It clearly shows that Δp^* is one of the most important characteristics of a membrane filtration system. It is termed the *characteristic pressure* (Pa) of the membrane channel.

As shown by Eq. (19), the characteristic pressure of an RO process is a function of the membrane resistance. Table 3.4 lists the typical characteristic pressures of some commercial RO membranes. It can be seen that the typical characteristic pressures of the old RO membranes in the common configuration (6-m long channel) are very high. The driving pressures were usually way below their characteristic pressures. This explains that membrane fouling in the early RO processes can be adequately indicated by the decline in the average permeate flux. The new generation RO membranes have much lower resistances and consequently their characteristic pressures are much lower than their predecessors. In fact, the characteristic pressures of a 6-m long channel of highly permeable RO membranes are comparable to or lower than the common driving pressures used in many practical water treatment plants. As a result, the initial periods of constant average permeate fluxes may be observed in modern full-scale RO processes.

5. MEMBRANE FOULING QUANTIFICATION IN FULL-SCALE REVERSE OSMOSIS PROCESSES

5.1. *The Need for an Effective Fouling Characterization Method*

An initial period of constant average flux can be observed before obvious flux decline occurs in many modern RO plants. The eventual flux decline indicates the presence of membrane fouling in the RO processes. The observation of initial period of constant average flux reveals that membrane fouling in the initial stage cannot affect the average flux, that is, the average flux cannot be taken as an effective indicator for membrane fouling in this case. The conditions for this case to occur are discussed in detail with model simulations in the previous section. The main reason is that the RO membranes nowadays are much more permeable than their previous generations. The typical clean membrane resistances of modern and older generation membranes are listed in Table 3.5. When modern highly permeable membranes are used to form long membrane channel similar to the configurations of their predecessors, there would likely be some surplus membrane area at the end of the channel that is "unused" at the beginning of operation. No permeate flux is produced in the "unused" membrane area. As fouling progresses from the entrance of the channel, the surplus membrane area is gradually utilized. However, before all the surplus membrane area is used up, the permeate production rate (i.e., average flux) will not be affected by membrane fouling. This fouling development in a long membrane channel has been simulated and discussed in Sect. 4.

Table 3.5
Clean membrane resistances of old and new generation RO membranes

Types of RO membrane	Resistance (Pa s/m)
1980s membrane	
Homopolar cellulose acetate (CA)	2.07×10^{12}
Cellulose triacetate (CTA)	1.83×10^{12}
Cellulose acetate propionate (CAP)	2.71×10^{12}
Cellulose acetate butyrate (CAB)	2.26×10^{12}
Currently used membrane	
Hydranautics SWC1-4040 (CP)	4.11×10^{11}
Hydranautics LFC2 (CP)	1.13×10^{11}
Hydranautics ESPA (CP)	8.43×10^{10}
Osmonics SG4040F (TFC)	1.35×10^{11}
Osmonics SC8040F (TFC)	6.64×10^{11}

CP composite polyamide, *TFC* thin-film composite.

For similar reason, the use of average permeate flux is not appropriate when assessing the efficiency of membrane cleaning in the full-scale RO processes. Because of the existence of the surplus membrane area in the long channel, partial removal of the foulants from the channel can fully restore the average permeate flux. The failure of the average flux in assessing cleaning efficiency is often experienced by many plant operators, who observe that this fully restored average flux cannot be maintained for as long as when new RO membranes are initially used (23). The duration of fully restored flux becomes shorter in subsequent cleanings until the initial average flux cannot be fully restored. These observations suggest that some foulants still remain on the membrane surface after cleaning even when average flux is fully restored. The use of flux restoration as a measurement of the effectiveness of membrane cleaning will lead to inaccurate and most likely overestimated conclusions.

5.2. *Filtration Coefficient of a Long Membrane Channel*

By definition, membrane fouling always increases the total resistance of the RO membrane regardless of decline in the average permeate flux. Therefore, the basic requirement of an effective membrane fouling measurement technique is to reflect the increase in membrane resistance because it is the most fundamental feature of membrane fouling. For this purpose, a lumped parameter, the filtration coefficient F (m/Pa s), is introduced to reflect the overall membrane resistance as

$$F = \frac{1}{L}\int_0^L \frac{1}{R_\mathrm{m}}\mathrm{d}x. \tag{20}$$

It can be seen from Eq. (20) that the increase of membrane resistance on any point of the channel length will cause corresponding decrease in the filtration coefficient F. When there is membrane fouling in a membrane channel, the value of the filtration coefficient of the membrane channel will decrease with time. Therefore, the filtration coefficient is an intrinsic indicator of fouling in a long membrane channel that is independent of average flux. Only in cases when the membrane channel is controlled by mass transfer, the average permeate flux can be related to the filtration coefficient by the following equation:

$$\bar{v} = F(\Delta p - \Delta\pi). \tag{21}$$

Although the filtration coefficient is virtually a different way to represent the membrane resistance, its advantages are obvious in fouling characterization of a long membrane channel over the direct use of the membrane resistance. First, filtration coefficient is a collective parameter that describes the whole system with a single value while the membrane resistance is a distributive parameter that varies along the membrane channel. Secondly, it is practically impossible to measure or determine membrane resistance distribution in the membrane channel. On contrary, the filtration coefficient of a membrane channel can be easily determined with a simple filtration test. Conducting filtration experiment at a pressure under the characteristic pressure of the clean membrane channel, the filtration coefficient can be readily calculated from the measured average flux and the net driving pressure with Eq. (21).

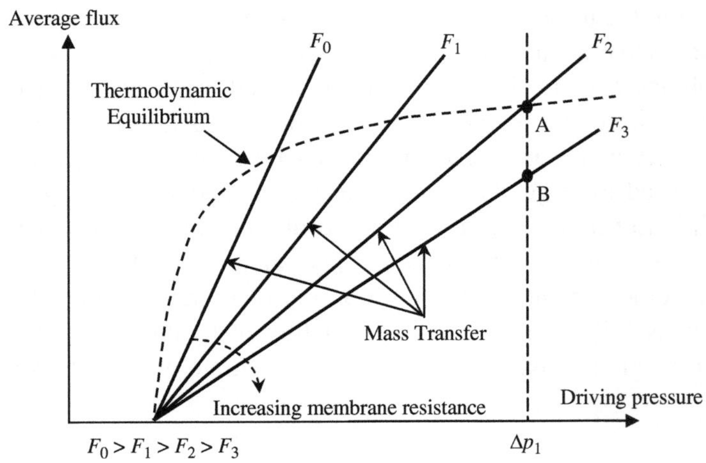

Fig. 3.21. A presentation of fouling development in a membrane channel with the use of filtration coefficients.

The use of filtration coefficient to describe membrane fouling in a full-scale RO process is illustrated in Fig. 3.21. The driving pressure for the RO process is indicated on the figure by Δp_1. The straight lines, indicated with filtration coefficients F_0 to F_3, are the average permeate fluxes at different times under mass transfer restriction. The filtration coefficients are also the slopes of the straight lines. The thermodynamic equilibrium-controlled average permeate flux is plotted with dotted line. The line F_0 with the largest slope describes the average permeate fluxes of clean RO membrane at the start of operation. At the start of the operation under pressure Δp_1, the average permeate flux of the membrane channel is indicated by point A, the intersection of the thermodynamic equilibrium curve with the driving pressure. Thermodynamic equilibrium restriction is obviously the limiting mechanism because it has smaller average permeate flux than that of mass transfer restriction at driving pressure Δp_1. The filtration coefficient decreases gradually with time as fouling develops in the membrane system. When the mass transfer flux line passes F_2, the flux-controlling mechanism shifts from thermodynamic equilibrium to mass transfer restriction, as the mass transfer-controlled flux now gives a lower value. Observed average permeate flux is now indicated by point B, where flux line F_3 intersects with the driving pressure.

5.3. Fouling Index for a Long Membrane Channel

Filtration coefficient is equivalent to average permeate flux for fouling characterization in the mass transfer regime, because in this regime any increment in membrane resistance is immediately reflected in the average permeate flux. Filtration coefficient remains an effective fouling indicator in the thermodynamic equilibrium regime as well because it is related to the increase in membrane resistance. Membrane resistance increases as long as there is membrane fouling. The remaining problem is that the filtration coefficient cannot be determined in the thermodynamic equilibrium regime. This problem, however, can be easily overcome by temporarily reducing the driving pressure to the mass transfer regime for the measurement of the filtration coefficient. By comparing the filtration coefficient at any time with the initial filtration coefficient, membrane fouling can be quantitatively expressed by a fouling index I_f defined as

$$I_f = \frac{F_i - F_t}{F_i}, \tag{22}$$

where F_i and F_t are the values of filtration coefficient at times 0 and t, respectively. The value of 0 indicates the absence of membrane fouling and the value of 1 indicates fouling is so severe that the membrane becomes impermeable. The fouling index can also be used to effectively evaluate the efficiency of membrane cleaning through the comparison of fouling indexes before and after membrane cleaning. If the fouling index is reduced to zero, the membrane cleaning method is said to be capable of removing the foulants completely from the RO membrane.

6. CONCLUSIONS

The single most critical problem faced in RO processes is membrane fouling, a process of foulants' accumulation on membrane surface that reduces water production and shortens membrane lifespan. Fouling alleviation and control are seriously hindered by ineffective fouling characterization. It is well known that the most widely used SDI is not a good indicator of the fouling potential of feed water for RO separation. Furthermore, fouling in full-scale RO processes is customarily indicated by the decline in the average flux. Unfortunately, many evidences demonstrate that membrane fouling in the RO processes is a very complex phenomenon that cannot be sufficiently or adequately characterized by flux decline. In this chapter, an advanced fouling characterization method is introduced that is capable of tracking fouling under different operating conditions and provides accurate feed water fouling potential that can be used for accurate prediction of membrane fouling in full-scale RO processes.

A new fouling indicator, fouling potential, is proposed for the fouling strength of feed water. The new fouling potential can be conveniently determined with a crossflow RO membrane cell under conditions similar to the designed working conditions for the RO processes. Because of the use of RO membrane in the measurement, all possible foulants to the RO process are captured and contributed to the fouling potential. Experiments show that the new fouling potential is linearly related to the foulant concentration and is independent from membrane resistance, driving pressure, or other operating parameters. More importantly, the fouling potential can be readily used to predict fouling development in RO processes.

Fouling development in a long membrane channel is well predicted by incorporating the fouling potential of the feed water into the model for the performance of the system. The model is able to trace the variation in resistance along the membrane channel due to accumulation of foulant with time and predict the corresponding average flux of the channel. Simulations of fouling development in a long membrane channel reveal an interesting behavior of the average flux. Under certain conditions (thermodynamic restriction), the average flux remains constant for an initial period of operation before decline in flux can be observed. The behavior has been reported in the literature and verified with fouling experiments on a 4-m long membrane channel in the laboratory. This interesting behavior highlights that the decline in average flux cannot be used as an effective indicator of membrane fouling in a long membrane channel that is potentially operated under thermodynamic equilibrium regime.

A more effective fouling characterization method is developed to overcome the inadequacy of using average flux to track fouling in full-scale RO processes. The central idea of the new characterization method is based on the intrinsic feature of membrane fouling, which is the increase in the total membrane resistance. The total resistance of the membrane increases whenever there is membrane fouling. With new fouling characterization method, fouling development in a long membrane channel can be accurately accounted for, even when no obvious flux decline is observed in the process. Furthermore, the new characterization method can also be used to more accurately assess the effectiveness of membrane cleaning.

7. ACRONYMS

BOD = Biochemical oxygen demand
COD = Chemical oxygen demand
MFI = Modified fouling index
NOM = Natural organic matter
RO = Reverse osmosis
SDI = Silt density index
TOC = Total organic carbon

8. NOMENCLATURE

c = Salt concentration (mg/L or g/m^3)
c_0 = Feed salt concentration (mg/L or g/m^3)
c_{f0} = Foulant concentration in the bulk flow (mg/L or g/m^3)
F = Filtration coefficient (m/Pa s)
f_{os} = Osmotic coefficient (Pa L/mg)
H = Height of the membrane channel (m)
η = Water viscosity (Pa s)
I_f = Fouling index
j = Foulant deposition rate onto the membrane surface (g/m^2 s)
k_f = Fouling potential of the feed water (Pa s/m^2)
K_s = Coefficient for spacer configuration (≥ 1)
M = Foulant amount accumulated over time (g/m^2)
Δp = Driving pressure (Pa)
Δp^* = Characteristic pressure (Pa)
Δp_0 = Feed driving pressure (Pa)
$\Delta \pi$ = Osmotic pressure (Pa)
R_0 = Total hydraulic resistances at the beginning of the fouling test (Pa s/m)

R_f = Resistance of the fouling layer (Pa s/m)
R_m = Membrane resistance (Pa s/m)
R_{m0} = Clean membrane resistance (Pa s/m)
r_{rej} = Salt rejection of the membrane
r_s = Specific resistance of the fouling layer (Pa s m/g)
R_t = Total hydraulic resistances at the end of the fouling test (Pa s/m)
t_f = Time to collect final 500 mL of sample (min)
t_i = Time to collect initial 500 mL sample (min)
t_{total} = Total running time of SDI test (min)
u = Crossflow velocity (m/s)
u_0 = Feed flow velocity (m/s)
v = Permeate flux (m/s)
v_0 = Measured permeate flux at the beginning of the fouling experiment (m/s)
v_t = Measured permeate flux at the end of the fouling experiment (m/s)
$\bar{v}(t)$ = Average permeate flux at time t (m/s)
ξ = Integration variable

REFERENCES

1. Loeb S, Sourirajan S (1962) Sea water demineralisation by means of an osmotic membrane. Adv Chem Ser 38:117–132
2. Martin-Lagardette JL (2000) Competing seawater desalination technologies help to meet demand for fresh water. Memb Technol 124:11–12
3. Committee on Advancing Desalination Technology (NRC) (2008) Desalination: a national perspective. National Academies Press, Washington DC
4. Mallevialle J, Odendaal PE, Wiesner MR (1996) The emergence of membranes in water and wastewater treatment. In: Mallevialle J et al (eds) Water treatment membrane processes. McGraw Hill, New York
5. Shannon MA, Bohn PW, Elimelech M, Georgiadis JG, Marinas BJ, Mayes AM (2008) Science and technology for water purification in the coming decades. Nature 452:301–310
6. Pantell SE (1993) Seawater desalination in California. California Coastal Commission, California
7. Bryne W (1995) Reverse osmosis – a practical guide for industrial users. Tall Oaks, Littleton
8. Mulder M (1996) Basic principles of membrane technology, 2nd edn. Kluwer, London
9. AWWA Membrane Technology Research Committee (1992) Committee report: membrane processes in potable water treatment. J Am Water Works Assoc 84(1):59–64
10. Avlonitis SA, Kouroumbas K, Vlachakis N (2003) Energy consumption and membrane replacement cost for seawater RO desalination plants. Desalination 157:151–158
11. Dudley L, Vigo Pisano F, Fazel M (2000) Optimizing membrane performance – practical experiences. Membr Technol 121:5–8
12. Madaeni SS, Mohamamdi T, Moghadam MK (2001) Chemical cleaning of reverse osmosis membranes. Desalination 134:77–82
13. Shahalam AM, Al-Harthy A, Al-Zawhry A (2002) Feed water pretreatment in RO systems: unit processes in the Middle East. Desalination 150:235–245

14. Kaakinen JW, Moody CD (1985) Characteristics of reverse osmosis membrane fouling at the Yuma desalting test facility. In: Sourirajan S, Matsuura T (eds) Reverse osmosis and ultrafiltration. American Chemical Society, Washington DC
15. Song L, Chen KL, Ong SL, Ng WJ (2004) A new normalization method for determination of colloidal fouling. J Colloid Interface Sci 271:426–433
16. Singh G, Song L (2005) Quantifying the effect of ionic strength on colloidal fouling potential in membrane filtration. J Colloid Interface Sci 284(2):630–638
17. Singh G, Song L (2005) Influence of various monovalent cations and calcium ion on the colloidal fouling potential. J Colloid Interface Sci 289(2):479–487
18. Song L, Hong S, Hu JY, Ong SL, Ng WJ (2002) Simulations of full-scale reverse osmosis membrane process. J Environ Eng 128:960–966
19. Zhou W, Song L, Tay KG (2006) A numerical study on concentration polarization and system performance of spiral wound RO membrane modules. J Memb Sci 271:38–46
20. Tay KG, Song L, Ong SL, Ng WJ (2005) Nonlinear behavior of permeate flux in full-scale reverse osmosis processes. J Environ Eng 131:1481–1487
21. Chen KL, Song L, Ong SL, Ng WJ (2004) Prediction of membrane fouling in full-scale RO process. J Memb Sci 232:63–72
22. Rico PD, Arias MFC (2001) A reverse osmosis potable water plant at alicante university: first year of operation. Desalination 137:91–102
23. Saad MA (1999) Optimize water cost by early prediction of membrane system fouling trends. In: IDA World Congress on desalination and water reuse, San Diego, CA
24. Keith S (1995) Handbook of industrial membranes, 1st edn. Elsevier Advanced Technology, Oxford
25. Lennitech BV (2010) Membrane fouling. Delft, The Netherlands. http://www.lenntech.com/membrane-fouling.htm
26. Water Treatment Guide (2010) Solutions to membrane fouling. Technical Database for the Water Treatment Industry. http://www.watertreatmentguide.com/membrane_fouling_solutions.htm
27. Bartels C, Lai K, Wilf M (2007) New generation of low fouling nanofiltration membranes. EDS Conference, Halkidiki, Greece, April, 2007
28. Wiesner Research Group, Membranes Fouling (2010) Investigation of the nature of adhesion of organic foulants to membrane surfaces and strategies for cleaning. http://wiesner.cee.duke.edu/?q=node/70

4
Membrane Filtration Regulations and Determination of Log Removal Value

Nazih K. Shammas and Lawrence K. Wang

CONTENTS

Abstract The United States Environmental Protection Agency (US EPA) promulgated the *Long Term 2 Enhanced Surface Water Treatment Rule*, which has identified membrane filtration as a treatment technology that may be used to achieve the required level of *Cryptosporidium* treatment. This rule, along with the companion: *Stage 2 Disinfectants and Disinfection By-products Rule*, constitutes the principal US regulations for the application of the membrane technology in potable water treatment.

The primary elements of the regulatory requirements for membrane filtration including the definition of membrane filtration, as well as challenge testing, direct integrity testing, and continuous indirect integrity monitoring, are summarized in this chapter. The requirements and procedure for challenge testing, which is required to demonstrate the ability of the treatment process to remove a specific target organism, are explained in detail. The removal efficiency demonstrated during challenge testing establishes the log removal value (LRV) or

From: *Handbook of Environmental Engineering, Volume 13: Membrane and Desalination Technologies*
Edited by: L.K. Wang et al., DOI: 10.1007/978-1-59745-278-6_4

removal credit that a membrane process would be eligible to receive. The core requirements of direct integrity testing and continuous indirect integrity monitoring are fully discussed. Design example and case studies are presented. Various water and wastewater physicochemical processes (such as, conventional sand filtration, direct filtration, contact filtration, slow sand filtration, cartridge filtration, diatomaceous earth filtration, dissolved air flotation, sedimentation, lime softening, coagulation, bank filtration, second stage filtration, continuous backwash upflow dual sand filtration, UV, chlorination, chloramination, ozonation, chlorine dioxide oxidation, etc.) are compared with membrane filtration for removal and/or inactivation of *Cryptosporidium*, *Giardia*, and virus.

Key Words Membrane filtration • regulations • *Cryptosporidium* removal • *Giardia* removal • removal efficiency • log removal value (LRV) • challenge testing, direct integrity testing • continuous indirect integrity monitoring • process comparison • Long Term 2 Enhanced Surface Water Treatment Rule • Disinfectants and Disinfection By-Products Rule.

1. INTRODUCTION

Currently, the most common form of drinking water treatment for surface water sources involves the physicochemical removal of particulate matter by coagulation, flocculation, sedimentation, and filtration processes, along with disinfection to inactivate any remaining pathogenic microorganisms. Filtration remains the cornerstone of drinking water treatment, conventionally in the form of granular media depth filters. Although granular media filters can produce high-quality water, they represent a probabilistic rather than an absolute barrier; consequently, pathogens can still pass through the filters and pose a health risk. The disinfection process provides an additional measure of public health protection by inactivating these microorganisms. However, some microorganisms, such as *Cryptosporidium*, are resistant to common primary disinfection practices such as chlorination and chloramination. Furthermore, drinking water have established Maximum Contaminant Levels (MCLs) for disinfection by-products (DBPs) that may create incentive for potable water utilities to minimize the application of some disinfectants. As a result of the concern over chlorine-resistant microorganisms and DBP formation, the potable water industry is increasingly utilizing alternative treatment technologies in effort to balance the often-competing objectives of disinfection and DBP control. One such alternative technology that has gained broad acceptance is membrane filtration.

Although the use of membrane processes has increased rapidly in recent years, the application of membranes for water treatment extends back several decades. Reverse osmosis (RO) membranes have been used for the desalination of water since the 1960s, with more widespread use of nanofiltration (NF) for softening and the removal of total organic carbon (TOC), dating to the late 1980s (1, 2). However, the commercialization of backwashable hollow-fiber microfiltration (MF) and ultrafiltration (UF) membrane processes for the removal of particulate matter (i.e., turbidity and microorganisms) in the early 1990s has had the most profound impact on the use, acceptance, and regulation of all types of membrane processes for drinking water treatment.

The United States Environmental Protection Agency (US EPA) developed a *Membrane Filtration Guidance Manual* (3) in support of the *Long Term 2 Enhanced Surface Water Treatment Rule*. (LT2ESTWR), which has identified membrane filtration (i.e., MF, UF, NF, RO, and certain types of cartridge filters) as one treatment technology in a "toolbox" of options that may be used to achieve the required level of *Cryptosporidium* treatment. This rule, along with the companion: *Stage 2 Disinfectants and Disinfection By-products Rule* (DBPR), constitutes the principal US regulations for the application of the membrane technology in potable water treatment.

2. MEMBRANES FOR THE POTABLE WATER INDUSTRY

Driven by the need to protect public health from waterborne pathogens, US EPA has progressively developed regulations that require higher standards for filtered water quality to prevent the passage of infectious pathogens through the treatment process and into the finished drinking water supply. In 1986, the *Surface Water Treatment Rule* (SWTR) required surface water systems to provide treatment equivalent to 3 log *Giardia* and 4 log virus reduction via a combination of removal and inactivation. US EPA estimated conventional filtration plants (i.e., including coagulation and sedimentation) meeting the filter effluent turbidity requirements provided a minimum of 2.5 log *Giardia* and 2 log virus reduction, while direct filtration plants (i.e., without sedimentation) provided 2 log *Giardia* and 1 log virus removal. The *Interim Enhanced Surface Water Treatment Rule* (IESWTR) focused on filter effluent turbidity control, setting a combined filter effluent turbidity limit of 0.3 NTU and requiring individual filter turbidity monitoring for systems serving at least 10,000 people. The rule also introduced a requirement for 2 log . The IESWTR also required that turbidity monitoring be conducted "continuously" (i.e., every 15 min) on individual filters. Based on available data, US EPA believed conventional and direct filtration plants meeting the IESWTR filter effluent turbidity requirements provided a minimum 2 log removal of *Cryptosporidium*.

The *Long Term 1 Enhanced Surface Water Treatment Rule* (LT1ESWTR) promulgated the requirements of the IESWTR, with some modifications, for all water systems serving less than 10,000 people. Under the existing surface water treatment rules, log removal credits for alternative filtration technologies (AFTs), such as membrane processes and bag and cartridge filters, were not explicitly addressed, but instead covered under a special state primacy requirement. For compliance with the SWTR, IESWTR, and LT1ESWTR, many states grant removal credits to membrane processes based on the guidelines for AFTs in the *Guidance Manual for Compliance With the Filtration and Disinfection Requirements for Public Water Systems Using Surface Water Sources* (commonly called the SWTR Guidance Manual) (4). However, there is significant variability in the manner in which states regulate membrane processes, as summarized for MF/UF systems in the US EPA report *Low-Pressure Membrane Filtration for Pathogen Removal: Application, Implementation, and Regulatory Issues* (5).

The *Long Term 2 Enhanced Surface Water Treatment Rule* (LT2ESWTR) builds on the previous surface water treatment rules by requiring additional treatment for those systems with elevated influent *Cryptosporidium* levels. The rule identifies a number of "toolbox" technologies that may be employed to achieve additional *Cryptosporidium* treatment requirements. The range of removal or inactivation credits allocated to each of the various toolbox options under the rule varies based on the capabilities of the particular treatment technology. Because the various types of membrane filtration processes represent one of these toolbox alternatives, utilities have the option of using membrane filtration for compliance with the rule requirements as a distinct technology rather than simply as a general AFT. Consequently, US EPA has developed specific regulatory requirements and associated guidance, as contained in the *Membrane Filtration Guidance Manual* (3), for membrane filtration processes used for compliance with the LT2ESWTR.

3. LONG TERM 2 ESWTR AND STAGE 2 DBPR REGULATIONS

3.1. Long Term 2 Enhanced Surface Water Treatment Rule

Under the LT2ESWTR, systems that use either a surface water or ground water under the direct influence of surface water (GWUDI) source (collectively referred to as surface water systems) are required to conduct source water monitoring to determine average *Cryptosporidium* concentrations. Based on the average *Cryptosporidium* concentration, a system is classified in one of four possible bins, as shown in Table 4.1. The bin assignment dictates the supplemental level of *Cryptosporidium* treatment required in addition to the existing requirements of the SWTR, IESWTR, and LT1ESWTR. Utilities may comply with additional treatment requirements by implementing one or more management or treatment techniques from a toolbox of options that includes membrane filtration. A guidance manual has also been

Table 4.1
LT2ESWTR additional treatment requirements for filtered systems

If source water *Cryptosporidium* concentration for filtered systems is in oocyst/L	And if the system uses the following filtration treatments, then additional treatment requirements are...				
	Bin classification is...	Conventional Filtration Treatment	Direct Filtration	Slow sand or diatomaceous earth	Alternative filtration technologies
<0.075	1	No additional treatment	No additional treatment	No additional treatment	No additional treatment
≥0.075 and <1.0	2	1.0 log treatment	1.5 log treatment	1.0 log treatment	At least 4.0 log
≥1.0 and <3.0	3	2.0 log treatment	2.5 log treatment	2.0 log treatment	At least 5.0 log
≥3.0	4	2.5 log treatment	3.0 log treatment	2.5 log treatment	At least 5.5 log

developed by the US EPA for the use of ultraviolet (UV) disinfection for LT2ESWTR compliance (6). Guidance for the use of all other toolbox options is given in the *Toolbox Guidance Manual for the Long Term 2 Enhanced Surface Water Treatment Rule* (7).

3.2. Stage 2 Disinfectants and Disinfection Byproducts Rule

The Stage 2 DBPR is designed to reduce DBP occurrence peaks in the distribution system by changing compliance monitoring requirements. The requirements of the Stage 2 DBPR apply to all community water systems (CWSs) and nontransient noncommunity water systems (NTNCWSs) – both ground and surface water systems – that either add a chemical disinfectant (i.e., a disinfectant other than UV light) or deliver water that has been treated with a chemical disinfectant. The primary components of the rule are summarized as follows (3).

3.2.1. Initial Distribution System Evaluations

Under the Stage 2 DBPR, systems are required to conduct an initial distribution system evaluation (IDSE) to identify compliance monitoring locations with high total trihalomethane (TTHM) and haloacetic acid (HAA5) levels. The IDSE consists of either a standard monitoring program (SMP) or a system-specific study (SSS). NTNCWSs serving fewer than 10,000 people are not required to conduct an IDSE, and other systems may be eligible to receive waivers from the IDSE requirement.

3.2.2. Compliance Determination and Schedule

The Stage 2 DBPR changes the way DBP sampling results are averaged to determine compliance, implementing a methodology based on a locational running annual average (LRAA) rather than the system-wide running annual average (RAA) used under the Stage 1 DBPR. This new methodology introduced under the Stage 2 DBPR is introduced in two phases: Stage 2A and Stage 2B. Under Stage 2A, all systems must comply with TTHM and HAA5 MCLs of 120 and 100 mg/L, respectively, measured as LRAAs at each Stage 1 DBPR monitoring site, while continuing to comply with the respective TTHM and HAA5 Stage 1 DBPR MCLs of 80 and 60 μg/L, measured as RAAs. Subsequently, under Stage 2B, systems must comply with respective TTHM and HAA5 MCLs of 80 and 60 μg/L at the sampling locations identified under the IDSE.

3.2.3. Compliance Monitoring

Stage 2B compliance monitoring requirements (in terms of number of sites and frequency of sampling) are expected to be similar to the Stage 1 DBPR requirements for most, but not all, systems. Some small systems will have to add an additional monitoring location if the highest TTHM and highest HAA5 site do not occur at the same location.

3.2.4. Significant Excursion Evaluations

Because Stage 2 DBPR MCL compliance is based on an annual average of DBP measurements, a system could from time to time have DBP levels significantly higher than the MCL (referred to as a significant excursion) while still maintaining compliance. If a significant excursion occurs, a system must conduct a significant excursion evaluation and discuss the evaluation with the state prior to the next sanitary survey.

3.3. Requirements for Membrane Filtration under the LT2ESWTR

In order to receive removal credit for *Cryptosporidium* removal under the LT2ESWTR, a membrane filtration system must meet the following three criteria:

1. The process must comply with the definition of membrane filtration as stipulated by the rule.
2. The of a membrane filtration process must be established through a product-specific challenge test and direct integrity testing.
3. The membrane filtration system must undergo periodic direct integrity testing and during operation.

The rule does not prescribe a specific removal credit for membrane filtration processes. Instead, removal credit is based on system performance as determined by challenge testing and verified by the direct integrity testing. Thus, the maximum removal credit that a membrane filtration process may receive is the lower value of either of the following:

1. The removal efficiency demonstrated during challenge testing.
2. The maximum log removal value (LRV) that can be verified by the direct integrity test used to monitor the membrane filtration process.

Based on this framework, a membrane filtration process could potentially meet the Bin 4 *Cryptosporidium* treatment requirements, as shown in Table 4.1. Additionally, if a membrane filtration system has been previously approved for 5.5 log *Cryptosporidium* removal by the state, the utility would not be required to conduct source monitoring under the LT2ESWTR.

These primary elements of the regulatory requirements for membrane filtration under the LT2ESWTR, including the definition of membrane filtration, as well as challenge testing, direct integrity testing, and continuous indirect integrity monitoring, are summarized in the following subsections.

3.3.1. *Definition of a Membrane Filtration Process*

For the purposes of compliance with the LT2ESWTR, membrane filtration is defined as a pressure- or vacuum-driven separation process in which particulate matter larger than 1 μm is rejected by a nonfibrous engineered barrier, primarily through a size exclusion mechanism and which has a measurable removal efficiency of a target organism that can be verified through the application of a direct integrity test. This definition is intended to include the common membrane technology classifications: MF, UF, NF, and RO. In addition, any cartridge filtration device that meets the definition of membrane filtration and that can be subject to direct integrity testing in accordance with rule requirements would also be eligible for *Cryptosporidium* removal credit as a membrane filtration process under the LT2ESWTR. In the guidance manual (3), such processes are called membrane cartridge filtration (MCF). Filtration processes that are reliant on mechanisms such as adhesion to filter media or accumulation of a fouling layer to remove particulate matter are excluded from the definition of membrane filtration. Examples of processes that would not be considered membrane filtration devices for the purposes of LT2ESWTR compliance include bag filters and cartridge filters using a fibrous filtration medium.

3.3.2. Challenge Testing

Since there are no uniform design criteria that can be used to assure the removal efficiency of a membrane process, challenge testing is required to demonstrate the ability of a treatment process to remove a specific target organism. The removal efficiency demonstrated during challenge testing establishes the *maximum* removal credit that a membrane process would be eligible to receive, provided that this value is less than or equal to the maximum LRV that can be verified by the direct integrity test, as described in the following section. The LT2ESWTR only requires product-specific challenge testing; once the removal efficiency has been demonstrated, additional testing is not required unless the product is significantly modified. Data from challenge studies conducted prior to promulgation of this regulation can be considered in lieu of additional testing at the discretion of the state. However, the prior testing must have been conducted in a manner that demonstrates removal efficiency for *Cryptosporidium* equivalent to or greater than the treatment credit awarded to the process. Challenge testing is discussed in detail in Sect. 4.

3.3.3. Direct Integrity Testing

While challenge testing can demonstrate the ability of an integral membrane process to remove the target organism, integrity breaches can develop in the membrane during routine operation that could allow the passage of microorganisms. In order to verify the removal efficiency of a membrane process during operation, direct integrity testing is required for all membrane filtration processes used to comply with LT2ESWTR. A direct integrity test is defined as a physical test applied to a membrane unit to identify and isolate integrity breaches (8–11). The rule does not mandate the use of a specific type of direct integrity test, but rather performance criteria that any direct integrity test must meet. These criteria include requirements for resolution, sensitivity, and frequency (3):

1. *Resolution*: The direct integrity test must be applied in a manner such that a 3-μm hole contributes to the response from the test.
2. *Sensitivity*: The direct integrity test must be capable of verifying the LRV awarded to the membrane process.
3. *Frequency*: The direct integrity test must be applied at a frequency of at least once per day.

A control limit must also be established for a direct integrity test, representing a threshold response which, if exceeded, indicates a potential integrity problem and triggers subsequent corrective action. For the purposes of LT2ESWTR compliance, this threshold response must be indicative of an integral membrane unit capable of achieving the *Cryptosporidium* removal credit awarded by the state.

3.3.4. Continuous Indirect Integrity Monitoring

Because currently available direct integrity test methods require the membrane unit to be temporarily taken out of service, or are either too costly or infeasible to apply continuously, direct testing is only conducted periodically. Thus, in the absence of continuous direct integrity test that meets the resolution and sensitivity requirements of the LT2ESWTR, continuous indirect integrity monitoring is required. Although the indirect monitoring methods are typically not as sensitive as direct tests for detecting losses of membrane integrity, the indirect methods do provide some measure of performance assessment between applications of direct testing (12). For the purposes of the LT2ESWTR, continuous indirect integrity monitoring is defined as monitoring some filtrate water parameter that is indicative of the removal of particulate matter at a frequency of no less than once every 15 min. Although turbidity monitoring (13) is specified as the default method of continuous indirect integrity monitoring under the rule, other methods, such as particle counting (14, 15) or particle monitoring (16, 17), may be used in lieu of turbidity monitoring at the discretion of the state. For any indirect method used, a control limit must be established that is indicative of acceptable performance. Monitoring results exceeding the control limit for a period of more than 15 min must trigger direct integrity testing.

3.4. Considering Existing Membrane Facilities under the LT2ESWTR

As shown in Table 4.1, the LT2ESWTR only requires additional treatment measure for those drinking water systems with source water *Cryptosporidium* levels greater than or equal to 0.075 oocysts/L – bins 2, 3, or 4. Thus, existing systems utilizing membrane filtration that fall into bin 1 may continue to operate under the previous surface water treatment rules (i.e., the SWTR and either the IESWTR or the LT1ESWTR) as administered by the state.

Utilities with existing membrane filtration facilities will be affected by the LT2ESWTR in one of five ways. These cases are summarized as follows (3):

- *Case 1*: The utility has previously been awarded 5.5 log *Cryptosporidium* treatment credit via a combination of physical removal (which may include membrane filtration) and chemical inactivation. In this case, the utility is not required to conduct source water monitoring.
- *Case 2*: The utility conducts source water monitoring for *Cryptosporidium* and determines that it is in bin 1 (i.e., concentrations less than 0.075 oocysts/L). In this case, the system may continue to operate under the previous surface water treatment rules (i.e., the SWTR and either the IESWTR or the LT1ESWTR) as administered by the state. No additional action is required under the LT2ESWTR.
- *Case 3*: The utility conducts source water monitoring for *Cryptosporidium* and determines that it is in either bins 2, 3, or 4 (i.e., concentrations greater than or equal to 0.075 oocysts/L). The utility then successfully demonstrates to the state that its membrane filtration system can achieve the total required *Cryptosporidium* treatment credit, as listed in the right-hand column of Table 4.1, up to a maximum of 5.5 log credit.
- *Case 4*: The utility conducts source water monitoring for *Cryptosporidium* and determines that it is in either bins 2, 3, or 4 (i.e., concentrations greater than or equal to 0.075 oocysts/L). The utility then successfully demonstrates to the state that its membrane filtration system can achieve part of the required *Cryptosporidium* treatment credit, as listed in the right-hand column (i.e., under "Type of Existing Filtration – Alternative Filtration Technologies") of Table 4.1. In this case, the utility would be required to use other toolbox options to obtain the balance of *Cryptosporidium* treatment credit required under the rule.
- *Case 5*: The utility conducts source water monitoring for *Cryptosporidium* and determines that it is in either bins 2, 3 or 4 (i.e., concentrations greater than or equal to 0.075 oocysts/L). The utility then

successfully demonstrates to the state that its membrane filtration system can achieve part of the required *Cryptosporidium* treatment credit, as listed in the right-hand column (i.e., under "Type of Existing Filtration – Alternative Filtration Technologies") of Table 4.1. However, the utility opts not to use its membrane filtration system for the purposes of LT2ESWTR compliance. In this case, the utility would be required to use other toolbox options to obtain all of the *Cryptosporidium* treatment credit required under the rule.

Under the LT2ESTWR, the regulatory basis for a membrane filtration process to receive treatment credit for *Cryptosporidium* is the demonstration of removal efficiency through challenge testing and the demonstration of membrane system integrity through routine direct integrity testing and continuous indirect integrity monitoring. These criteria form the basis for the potential for membrane filtration systems to be awarded up to a maximum of 5.5 log *Cryptosporidium* removal credit for complying with the requirements of the previously promulgated surface water treatment rules in combination with any additional *Cryptosporidium* treatment credit that may be required under the LT2ESWTR.

With respect to the challenge testing requirements of the rule, the two most likely options available to utilities with existing membrane filtration systems are to grandfather data generated during the pilot testing conducted as part of the permitting process (if applicable) or to use data from challenge testing conducted by the membrane manufacturer in its effort to qualify its product(s) for *Cryptosporidium* removal credit under LT2ESWTR (since challenge testing is required on a product-specific and not a site-specific basis).

Existing membrane facilities will also have to meet the direct integrity testing and continuous indirect integrity monitoring requirements of the LT2ESWTR to qualify for treatment credit under the rule. This may necessitate that some facilities implement new integrity verification practices, since state requirements vary widely, and some may not require direct integrity testing at all (5, 18, 19). However, this is not anticipated to be problematic for many existing facilities, since most membrane filtration systems applied to surface water are equipped with the ability to conduct some form of direct integrity testing. In addition, although states may not have explicit indirect integrity monitoring requirements, turbidity monitoring (the default method of continuous indirect integrity monitoring under the LT2ESWTR) is nonetheless required for compliance with the various existing surface water treatment rules. In some cases, utilities may need to purchase additional equipment to comply with the integrity verification requirements of the rule.

Another consideration for existing membrane facilities required to meet the LT2ESWTR criteria is replacement membrane modules. When replacement modules are installed, it is necessary to verify that the specific modules used meet the quality control release value of the nondestructive performance test as a means of indirectly verifying removal efficiency.

The regulatory framework developed for membrane filtration under the LT2ESWTR addresses many of the specific capabilities and requirements of the technology and thus may introduce new concepts that might not be included in a given state's current regulatory approach for membrane processes, particularly if the state currently considers membrane filtration as an AFT, as described in the US EPA *Guidance Manual for Compliance With the Filtration and Disinfection Requirements for Public Water Systems Using Surface Water Sources* (4). Although states may choose to adopt aspects of the LT2ESWTR framework for broader regulation of membrane filtration systems, US EPA only requires that this regulatory framework be applied to systems that utilize membrane filtration to meet the additional *Cryptosporidium* treatment requirements of the LT2ESWTR.

3.5. Membrane Terminology Used in the Guidance Manual

In the development of the regulatory language and associated guidance for LT2ESWTR, it was necessary to select the most appropriate terminology for various aspects of membrane treatment, with the understanding that use of such terminology can vary widely throughout the industry. The purpose of this section is to clarify the use of membrane treatment terminology associated with the LT2ESWTR and to note generally synonymous terms that are also in common use, where applicable. This section also presents some new terms defined under the rule that are critical to the regulatory framework.

There are 12 terms formally defined under the language for the LT2ESWTR, as follows:

- *Membrane filtration* – a pressure- or vacuum-driven separation process in which particulate matter larger than 1 μm is rejected by a nonfibrous engineered barrier, primarily through a size exclusion mechanism and which has a measurable removal efficiency of a target organism that can be verified through the application of a direct integrity test.
- *Module* – the smallest component of a membrane unit in which a specific membrane surface area is housed in a device with a filtrate outlet structure. For the purposes of the LT2ESWTR, this term encompasses hollow-fiber modules and cassettes, spiral-wound elements, cartridge filter elements, plate-and-frame modules, and tubular modules, among other membrane devices of similar scope and purpose.
- *Membrane unit* – a group of membrane modules that share common valving, which allows the unit to be isolated from the rest of the system for the purpose of integrity testing or other maintenance. For the purposes of the LT2ESWTR, "membrane unit" is intended to include the commonly used synonymous terms rack, train, and skid.
- *Challenge test* – a study conducted to determine the removal efficiency (i.e., LRV) of the membrane filtration media.
- *Flux* – flow per unit of membrane area.
- *Recovery* – the ratio of filtrate volume produced by a membrane to feed water volume applied to a membrane over the course of an uninterrupted operating cycle. In the context of the LT2ESWTR, the term recovery does not consider losses that occur due to the use of filtrate in backwashing or cleaning operations.
- *Direct integrity test* – a physical test applied to a membrane unit to identify and isolate integrity breaches.
- *Integrity breach* – one or more leaks (in a membrane filtration system) that could result in the contamination of the filtrate.
- *Resolution* – the smallest leak that (i.e., integrity breach) contributes to a response from a direct integrity test.
- *Sensitivity* – the maximum LRV that can be reliably verified by the direct integrity test associated with a membrane filtration system.
- *Control limit* – an integrity test response that, if exceeded, indicates a potential problem with the system and triggers a response. In this context, the terms upper control limit (UCL) and lower control limit (LCL) are also used. The term UCL is always used in reference to the control limit that is mandated under the LT2ESWTR, or to the last control limit that could be exceeded before the

unit must be taken off-line for diagnostic testing and repair. The term LCL was established to distinguish any more conservative voluntary or additional state-mandated control limits that may trigger increased monitoring or other action short of taking the unit off-line.

- *Indirect integrity monitoring* – monitoring some aspect of filtrate water quality that is indicative of the removal of particulate matter. In the context of indirect integrity monitoring, "continuous" is defined as a frequency of no less than once every 15 min.

In addition to these terms, it is important to note that the term filtrate, as used in the rule language, includes the synonymous term permeate, which is commonly used in the industry in association with the treated water from NF and RO semipermeable membrane processes.

3.6. *Summary of US EPA Regulatory Framework*

Under the US Environmental Protection Agency (US EPA) Long Term 2 Enhanced Surface Water Treatment Rule (LT2ESWTR), water systems may be required to achieve as much as 3 log additional *Cryptosporidium* removal and/or inactivation credit depending on the results of source water quality monitoring and the subsequent Bin assignment. Thus, when combined with the prescribed *Cryptosporidium* treatment credit awarded to a system in compliance with the US EPA Interim Enhanced Surface Water Treatment Rule (IESWTR) or the Long Term 1 Enhanced Surface Water Treatment Rule (LT1ESWTR), as applicable, the total *Cryptosporidium* treatment credit required for a system in Bins 2, 3, and 4 is 4 log, 5 log, and 5.5 log, respectively. Membrane filtration is one of several toolbox options that has been determined to be capable of achieving the maximum required credit as a stand-alone process. In order to receive *Cryptosporidium* removal credit under the rule, a membrane filtration system must meet the following three criteria.

3.6.1. Compliance with the Definition of Membrane Filtration

The first of three criteria is that the process to be tested must comply with the definition of membrane filtration as stipulated by the rule. Membrane filtration is defined under the rule as a pressure- or vacuum-driven separation process in which particulate matter larger than 1 μm is rejected by an engineered barrier, primarily through a size exclusion mechanism and which has a measurable removal efficiency of a target organism that can be verified through the application of a direct integrity test. This definition includes the following membrane processes commonly used in drinking water treatment: (a) MF, (b) UF, (c) NF. and (d) RO.

In addition, any cartridge filtration device that meets the definition of membrane filtration and which can be subject to direct integrity testing in accordance with rule requirements would also be eligible for *Cryptosporidium* removal credit as a membrane filtration process under the LT2ESWTR. The Membrane Filtration Guidance Manual (MFGM) refers to these processes as MCF (3).

3.6.2. Establishment of Membrane Filtration Process's Removal Efficiency Through a Product-Specific Challenge Test and Direct Integrity Testing

Another rule is that the removal efficiency of a membrane filtration process must be established through a product-specific challenge test and direct integrity testing. This rule does not prescribe a specific removal credit for membrane filtration processes. Instead, removal credit is based on system performance as determined by challenge testing and verified by direct integrity testing. Thus, the maximum removal credit that a membrane filtration process may receive is the lower value of either of the following: (a) the removal efficiency demonstrated during challenge testing (b) the maximum LRV that can be verified by the direct integrity test used to monitor the membrane filtration process.

3.6.3. Requirement of Periodic Direct Integrity Testing and Continuous Indirect Integrity Monitoring During Operation

The third rule is that the membrane filtration system must undergo periodic direct integrity testing and continuous indirect integrity monitoring during operation. The LT2ESWTR requires that the *Cryptosporidium* log removal credit awarded to the membrane filtration process be verified on an ongoing basis during operation. This verification is accomplished by the use of direct integrity testing. Currently available direct integrity test methods represent the most sensitive means of detecting integrity breaches, but these tests cannot be conducted on a continuous basis while the membrane filtration system is in operation. Thus, direct integrity testing is implemented at regular intervals and complemented by indirect integrity monitoring, which is generally less sensitive but can be conducted continuously during filtration. This continuous indirect integrity monitoring allows for a coarser assessment of membrane integrity in between periodic applications of a more sensitive direct integrity test.

3.6.4. Summary of Rule Requirements

The LT2ESWTR specifies requirements for three critical aspects of implementing membrane filtration for the removal of *Cryptosporidium* in compliance with the rule: (a) Challenge Testing; (b) Direct Integrity Testing, and (c) Continuous Indirect Integrity Monitoring.

As a whole, these rule requirements are designed to first establish what *Cryptosporidium* removal credit a membrane product is able to achieve and subsequently how the allocated removal credit for a site-specific system (as determined by the State) is verified on an ongoing basis during operation. The more detailed requirements for challenge testing, direct integrity testing, and continuous indirect integrity monitoring are addressed in the MFGM (3).

4. CHALLENGE TESTING: DETERMINATION OF LRV

The Long Term 2 Enhanced Surface Water Treatment Rule (LT2ESWTR) requires that any membrane filtration system used to comply with the *Cryptosporidium* treatment requirements of the rule undergo challenge testing. The primary purpose of this challenge testing is to establish the LRV that an integral membrane can achieve. Under the LT2ESWTR, the maximum removal credit that a membrane filtration system is eligible to receive is the lower of the two values established by either of the following:

1. The removal efficiency demonstrated during challenge testing.
2. The maximum LRV that can be verified by the particular direct integrity test used during the course of normal operation.

The requirement for challenge testing under the LT2ESWTR is intended to be product-specific such that site-specific demonstration of *Cryptosporidium* removal efficiency is not necessary. Once the LRV of a membrane has been established through a challenge test that meets the requirements of LT2ESWTR, additional challenge testing is not required unless significant modifications are made to the membrane process. The rule specifies criteria for the following aspects of challenge testing:

1. Full-scale versus small-scale module testing.
2. Appropriate challenge particulates.
3. Challenge particulate concentrations.
4. Test operating conditions.
5. Calculation of removal efficiency.
6. Verifying characteristic removal efficiency for untested modules.
7. Module modifications.

The discussion of challenge testing in this chapter applies similarly to MF, UF, NF, RO, and MCF, except as otherwise noted. Although the primary focus of challenge testing as required under the LT2ESTWR is demonstration of *Cryptosporidium* removal, the general framework for challenge testing may be adapted for use in establishing removal efficiencies for other microbial pathogens of concern, including bacteria, viruses, and other protozoa such as *Giardia*.

4.1. Core Requirements for Challenge Testing

The LT2ESWTR specifies the core requirements that a challenge test must meet to demonstrate the removal efficiency of a membrane filtration system with respect to *Cryptosporidium*. These requirements are summarized as follows:

1. *Full-scale vs. small-scale module testing*: Challenge testing must be conducted on a full-scale membrane module identical in material and construction to the membrane modules proposed for use in full-scale treatment facilities. Alternatively, challenge testing may be conducted on a smaller scale module that is identical in material and similar in construction to the full-scale modules.
2. *Appropriate challenge particulates*: Challenge testing must be conducted using *Cryptosporidium* oocysts or a surrogate that has been determined to be removed no more efficiently than *Cryptosporidium* oocysts. The organism or surrogate used during challenge testing is referred to as the "challenge particulate". The concentration of the challenge particulate must be determined using a method capable of discretely quantifying the specific challenge particulate used in the test; indirect water quality measurements such as turbidity, particle counting, or conductivity cannot be used for this purpose.
3. *Challenge particulate concentrations*: The maximum allowable feed water concentration used during a challenge test is based on the detection limit of the challenge particulate in the filtrate and is determined as follows:

$$\text{Maximum feed concentration} = \left(3.16 \times 10^6\right) \times (\text{Filtrate detection limit}). \tag{1}$$

This expression allows for the demonstration of up to 6.5 log removal during challenge testing if the challenge particulate is removed to the detection limit.

4. *Test operating conditions*: Challenge testing must be conducted under representative hydraulic conditions at the maximum design flux and maximum design system recovery specified by the manufacturer.
5. *Calculation of removal efficiency*: The removal efficiency of a membrane filtration process as determined from the results of the challenge test is expressed in terms of a LRV according to the following relationship:

$$\mathrm{LRV} = \log(C_\mathrm{f}) - \log(C_\mathrm{p}), \tag{2}$$

where LRV is the log removal value demonstrated during challenge testing, C_f is the feed concentration used during challenge testing (number or mass/volume), C_p is the filtrate concentration observed during challenge testing (number or mass/volume).
6. *Verifying characteristic removal efficiency for untested modules*: Because the LT2ESWTR does not require that every membrane module be subject to challenge testing, a nondestructive performance test (NDPT) must be applied to each production membrane module that did not undergo challenge testing to verify removal efficiency. A quality control release value (QCRV) must be established for the NDPT that is directly related to the removal efficiency of the membrane filtration process as demonstrated during challenge testing. Membrane modules that do not meet the established QCRV are not eligible for the removal credit demonstrated during challenge testing.
7. *Module modifications*: Any significant modification to the membrane media (e.g., a change in the polymer chemistry), hydraulic configuration (e.g., changing from suspension to deposition mode), or any other modification that could potentially affect removal efficiency or NDPT parameters would require additional challenge testing to both demonstrate the removal efficiency of the modified module and define a new QCRV for the NDPT.

4.2. *Test Organization Qualification*

The LT2ESWTR does not specify any requirements with respect to the qualifications of an organization conducting a challenge test, as long as the test is performed according to the criteria mandated under the rule. Each state has discretion in approving the results from a challenge test conducted by any organization. However, since challenge testing is intended to be product-specific, it is important that there be some consensus regarding what constitutes an acceptable test. The US EPA Guidance Manual (3) provides an outline of the skills and capabilities that a test organization should possess to produce quality data. In general, conducting a successful challenge test necessitates that the testing organization demonstrate effective knowledge of the following:

1. The various membrane processes commonly used in drinking water treatment.
2. Operation of membrane filtration equipment and related system processes, including pretreatment, posttreatment, backwashing, chemical cleaning, and integrity testing.
3. Proper challenge particulate seeding and sampling techniques.
4. Analytical techniques for the enumeration of the challenge particulate used in the challenge test, including analyses of microorganisms, inert particulate markers, or molecular markers (as applicable).
5. Adequate quality assurance (QA)/quality control (QC) procedures to ensure that data quality objectives are achieved.

6. Basic statistical procedures that may be used in data analysis.
7. Preparation of reports for regulatory agencies.

Historically, many utilities and states have used independent, third party organizations to conduct verification testing (i.e., challenge testing) to ensure an unbiased evaluation of the process. While there are advantages to this approach, a membrane manufacturer may have the ability to conduct an acceptable challenge test if it can demonstrate that appropriate QA/QC procedures are used.

4.3. *General Procedure for Designing a Challenge Test Protocol*

The core challenge test requirements of the LT2ESWTR should be incorporated into a detailed protocol for implementing the test that documents the necessary equipment, procedures, and analyses. Due to the variety of membrane systems available from numerous suppliers, it is not possible to develop a single comprehensive protocol. However, the following general list of procedures describes the basic steps in the development of such a protocol (3).

1. Document basic membrane module specifications, including:
   - Maximum design flux.
   - Mode of operation (i.e., inside out or outside in).
   - Hydraulic configuration (i.e., deposition or suspension).
   - Module dimensions and filtration area.
   - Operating constraints (e.g., maximum feed pressure, temperature, pH range, oxidant tolerance, etc.).
   - Backwash and chemical cleaning procedures.

   Although not all of the information provided with any particular product specification sheet may be necessary for developing a challenge test, it is nevertheless prudent to compile as much available information about the product as possible not only for its potential use in challenge testing but also for long-term use during operation of the full-scale facility.
2. Document the manufacturer's procedure for conducting nondestructive performance testing (NDPT) and ensure that the associated quality control release value (i.e., the minimal result from the NDPT that constitutes an acceptable product) is indicative of a NDPT resolution of 3 μm to demonstrate *Cryptosporidium* removal capability. If available, the statistical distribution of the NDPT results for the product line may also be useful.
3. Determine the number of modules that will be evaluated during the challenge test and the method or criteria that will be used to select specific modules for testing.
4. Determine whether or not small-scale module testing is an option.
5. Identify the target organism or contaminant for the test. For the purposes of compliance with the LT2ESWTR, the target organism is *Cryptosporidium*.
6. Establish the target LRV for the challenge test. Because challenge testing is intended to be product-specific under the LT2ESWTR, it is generally advantageous for a manufacturer to set this target at the maximum LRV for which it is anticipated that the system will qualify.
7. Select the challenge particulate to be used for testing. If it is not feasible or desirable to use the target organism as the challenge particulate, it is necessary to identify an acceptable surrogate that is removed on an equivalent or more conservative basis.

8. Select an analytical method that will be used to discretely quantify (i.e., enumerate) the challenge particulate and collect information relevant to the methodology for use in developing a sampling plan. Determine the detection limit for the challenge particulate in the filtrate based on the method capabilities and filtrate sample volume.
9. Design the challenge test solution and establish the method for seeding the challenge particulates into the solution (e.g., continuous or batch seeding).
10. Design and construct the testing apparatus, and select appropriate operational parameters.
11. Develop a sampling and monitoring plan that specifies the following:
    - The number of feed and filtrate samples to be collected and analyzed.
    - The frequency of feed and filtrate sample collection.
    - The feed and filtrate sample volumes.
    - Procedures for sample collection.
    - Additional operating and water quality parameters to be monitored and associated monitoring frequency.

After completing the steps outlined above, the specific protocol for conducting challenge testing should be documented and submitted for state approval, if required. Note that the LT2ESWTR does not require the challenge test protocol to be reviewed or approved by the US EPA; however, each state may exercise its discretion regarding whether approval of the protocol is required before results of the challenge testing are accepted in that state.

4.4. *Nondestructive Performance Testing*

While challenge testing is used to establish the LRV of an integral module of a particular product type, it does not necessarily guarantee that all such modules produced will achieve the same level of performance due to variability in the manufacturing process. In order to address this issue, a NDPT is applied to all subsequently manufactured modules that are not subject to challenge testing to ensure that these modules comply with the minimum standards for *Cryptosporidium* removal under the LT2ESWTR. A NDPT is a physical test applied to the membrane module with the objective of characterizing some aspect of process performance and which does not alter or damage the membrane.

In order to be utilized in a membrane filtration system that is applied for the purpose of receiving *Cryptosporidium* removal credit under the LT2ESWTR, each module must pass a NDPT that is consistent with the 3-μm resolution requirement of the rule. For example, one commonly used type of NDPT is the bubble point test, which characterizes the largest pore (or defect) in a membrane module. In a bubble point test, a pressure is applied to a fully wetted membrane module and gradually increased. The pressure at which water is first evacuated from the pores represents the bubble point of the membrane associated with a particular module. If the established bubble point of the membrane is sufficient to demonstrate that there are no pores (or defects) larger than 3 μm, then the NDPT is consistent with resolution specified in the rule for the removal of *Cryptosporidium*.

The minimum passing test result for a NDPT is known as the quality control release value (QCRV). In the context of the LT2ESWTR, a test result that surpasses the QCRV indicates both that a module can adequately remove *Cryptosporidium* and is sufficiently similar in quality to the modules subjected to challenge testing to demonstrate the ability of the module of interest (which would not have been subject to challenge testing) to achieve the same LRV. After a group of modules has been subject to challenge testing, the NDPT is applied to those modules to determine an appropriate QCRV associated with the removal efficiency observed during the test. Subsequently, all modules that are not subjected to challenge testing must pass the same NDPT by exceeding the established QCRV applicable to *Cryptosporidium* removal under the LT2ESWTR. Modules that do not pass the NDPT at the QCRV would not be eligible for *Cryptosporidium* removal credit under the rule and could not be used in any membrane filtration systems applied for this purpose.

The LT2ESWTR does not specify a particular procedure for determining the QCRV from the various modules that are subjected to challenge testing. Thus, the manufacturer or independent testing organization may exercise its discretion in selecting an appropriate methodology. The QCRV may be selected as the average result among the various modules tested, the most conservative result (to establish the most stringent QA/QC standards), or the least conservative result (to maximize the number of modules eligible for removal credit under the LT2ESWTR). Alternatively, a methodology similar to that required by the LT2ESWTR for determining the overall removal efficiency based on the number of LRV observations could be applied to the various NDPT results to yield an appropriate QCRV.

Note that the rule does not specify the manner in which the QCRV is determined from the challenge test data; however, the methodology must be acceptable to each state in which the product line is applied for the purpose of receiving *Cryptosporidium* removal credit under the LT2ESWTR. It is recommended that each module subjected to challenge testing also undergo subsequent nondestructive performance testing for the purpose of establishing a QCRV. The manner in which the NDPT results are used to determine the QCRV should reflect the manner in which modules are selected, as well as the range of LRVs observed during the test. If modules are selected in a conservative manner and if the range of LRVs observed during challenge testing is small, the average of the respective NDPT results from the modules subjected to challenge testing might be selected to represent the QCRV. However, if a statistical distribution of modules is selected for challenge testing or if the range of LRVs observed during challenge testing is significant, a more conservative value is recommended for the QCRV, such as the minimum of the NDPT results observed among the tested modules.

Because it is common for manufacturers to conduct some type of NDPT on every module as a routine component of its QA/QC program, the NDPT requirements of the LT2ESWTR simply ensure that the QCRV used by the manufacturer is sufficient to justify the LRV for *Cryptosporidium* demonstrated via challenge testing. Note that because different NDPTs may be used by the various membrane module manufacturers, the rule does not specify a particular type of NDPT. However, the NDPT used must be consistent with the resolution requirements of the LT2ESWTR for a module to be eligible for *Cryptosporidium* removal credit.

4.5. Selection of Modules for Challenge Testing

The intent of challenge testing under the LT2ESWTR is to characterize the removal efficiency of a specific membrane product without requiring challenge testing for all production modules. In addition, the rule does not specify a particular number of modules that are required to undergo challenge testing to demonstrate *Cryptosporidium* removal efficiency. However, because it is important that manufacturing variability in the product line be considered in the development of an appropriate challenge test, the number of modules subject to challenge testing, as well as the particular modules chosen, should be carefully selected on a rational and scientific basis. Although manufacturers or independent testing organizations may develop any number of different procedures for module selection, two common approaches are discussed as illustrative examples (3):

1. Selection of modules based on previously collected QC data for the product line.
2. Random sampling of membrane modules from several manufactured lots according to a statistical sample design.

Use of the first approach listed above is predicated on the existence of significant QC data for the product line accumulated over time by the manufacturer. Since manufacturers typically conduct some kind of NDPT on all modules produced to ensure quality and characterize the variability of a product line independent of the link established between nondestructive performance testing and challenge testing under the LT2ESWTR, such data should generally be available. Because the modules subject to challenge testing will be subsequently recharacterized with the NDPT to establish an acceptable QCRV required for all modules to be eligible for *Cryptosporidium* removal credit under the LT2ESWTR, it may be most advantageous for a manufacturer to select modules for challenge testing that are near the lower end of the statistical distribution of acceptable (i.e., under the manufacturer's in-house QC procedures) NDPT results based on historical data. If these tested modules yield a QCRV that is consistent with the resolution requirement of the rule, then it is likely that the majority of production modules will also meet the established QCRV and thus be eligible for *Cryptosporidium* removal credit. Using this approach, the number of modules selected for challenge testing is generally at the discretion of the manufacturer or independent testing organization.

If historical QC data for the product line is not available from the manufacturer, the second approach listed above may represent an appropriate option. This method involves the evaluation of a statistically significant random sample of modules from a number of production lots. Because the modules at the lower end of the QC data are not artificially selected (as with the first approach described above), it is likely that this method will result in a higher QCRV, resulting in a somewhat higher rejection rate of modules eligible for *Cryptosporidium* removal credit under LT2ESWTR. This number of modules selected for challenge testing using this approach will likely be dictated by the particular statistical sampling technique used. Either of these two approaches or other rationale approach developed by the manufacturer or independent testing organization could be utilized to select modules for challenge testing. Regardless of the method used, it is suggested that at least five membrane modules from different manufactured lots be evaluated during a challenge test.

4.6. Small-Scale Module Testing

The evaluation of small-scale (as opposed to full-scale) modules during a challenge test is permitted under the LT2ESWTR to allow for cases in which it may not be feasible or practical to test a full-scale module. For example, if it is desirable to conduct challenge testing using the target organism (i.e., *Cryptosporidium* for the purposes of the LT2ESWTR) rather than a surrogate, the use of a small-scale module may be the only economically viable alternative.

All challenge testing requirements under the LT2ESWTR are equally applicable to both full-scale and small-scale modules. However, any small-scale module tested must be similar in design to the full-scale modules of the product of interest such that it can be operated under a hydraulic configuration and at a maximum design flux and recovery that are representative of the full-scale modules. Simulating the full-scale recovery and hydraulic configuration are important considerations for small-scale challenge testing, since both of these parameters affect the concentration of suspended solids on the feed side of the membrane.

Although the decision to allow the use of small-scale module testing is left to the discretion of the state, the option is permitted under the LT2ESWTR since it is considered a valid approach for characterizing removal efficiencies. For the purposes of consistency, it is recommended that manufacturers or independent testing agencies that opt to subject a product line to challenge testing using small-scale modules utilize a protocol that has been accepted by a wide range of stakeholders. Such a protocol has been proposed for use under the National Sanitation Foundation (NSF) Environmental Technology Verification (ETV) program (20).

4.7. Target Organisms and Challenge Particulates

The purpose of a challenge test is to determine the removal efficiency of a membrane module for one or more target organisms or pathogens. Challenge testing can be conducted using either the target organism itself or an appropriate surrogate; the organism or surrogate used in the test is referred to as the *challenge particulate*. The selection of a suitable challenge particulate is critical to the design of a challenge test.

This section provides guidance for selecting an appropriate challenge particulate, including the selection of a target organism for the test and characteristics of suitable surrogates for the target organism. A more detailed discussion of particular surrogates for *Cryptosporidium* is also provided.

4.7.1. Selecting a Target Organism

The target organism or pathogen of interest for the purposes of challenge testing is selected based on the treatment objectives for the membrane filtration system. For example, *Cryptosporidium* would be the target organism in a challenge test conducted to demonstrate the ability of a membrane filtration system to comply with the treatment requirements of the LT2ESWTR. However, in some cases, it may be desirable to determine the removal efficiency of a system for multiple target organisms. In such cases, the most conservative target organism should be selected for the purpose of designing a challenge test. For example, if the

Table 4.2 Potential target organisms for challenge testing

Target organism	Size range (μm)
Enteric viruses	0.03–0.1
Fecal coliform	1–4
Cryptosporidium	3–7
Giardia	7–15

challenge test is designed to evaluate the removal efficiency of a system for both *Cryptosporidium* and *Giardia*, then the smaller of these two pathogens should be used as the target organism.

Although the approximate size ranges for these two organisms overlap to some degree, as shown in Table 4.2, *Cryptosporidium* has the smaller lower bound. Since membrane filtration is a barrier technology based primarily on the principle of size exclusion, the removal efficiency for the smallest organism of interest should be conservative for larger pathogens.

4.7.2. Surrogate Characteristics

Although use of the target organism as the challenge particulate offers the advantages of directly measuring removal efficiency for the pathogen of interest and eliminates issues regarding the appropriateness of a surrogate, it may not be practical or feasible as a result of economic considerations or concerns about working directly with the pathogen. Thus, the use of surrogates may be the most viable option for challenge testing. An ideal surrogate should have characteristics that are likely to affect removal efficiency which are similar to those of the target organism, while a conservative surrogate would have characteristics that may result in a lower removal efficiency relative to the target organism. In general, it is necessary to use a conservative surrogate unless there are data to support the use of an ideal surrogate.

As a result of the cost and potential health concerns associated with conducting a challenge test using *Cryptosporidium* oocysts, the LT2ESWTR allows challenge testing to be administered with a surrogate that has been verified to be removed no more efficiently than *Cryptosporidium* oocysts (i.e., an ideal or conservative surrogate). The most direct means of demonstrating that a surrogate is ideal or conservative is through a comparative test in which removal of the surrogate and the target organisms are evaluated side by side. However, if the characteristics of a surrogate are sufficiently conservative, direct verification may not be necessary. Key physical characteristics to consider when evaluating the suitability of a surrogate for *Cryptosporidium* removal using membrane filtration include size, shape, and surface charge. Other important considerations include ease of use and measurement, as well as cost.

1. *Particle size and shape*: The effective size of an appropriate surrogate should be equivalent to or smaller than the lower bound of the size range of the target organism. Furthermore, the effective size of the surrogate should be characterized using an upper bound of its size distribution such the 99th or 99.9th percentile rather than the median. Ideally, a surrogate would have a relatively narrow size distribution and a high uniformity coefficient. For example, the lower size range of

Cryptosporidium is approximately 3 μm, and thus, a conservative surrogate might be the one in which 99% of particles have a diameter of 1 μm or less.

Generally, it is desirable to use a surrogate that is of the same shape as the target organism. In the case of *Cryptosporidium*, an appropriate surrogate would have a spherical shape, although in some cases a nonspherical surrogate might be considered. If a nonspherical surrogate is used, it is recommended that the smallest dimension be considered as the effective size since particles can interact with a membrane barrier at any orientation.

Another consideration is the surface structure of the proposed surrogate. A particle that has a highly irregular surface structure may be removed more efficiently than a similarly sized particle that has a smooth surface. While it may be difficult to completely characterize the surface of a potential surrogate, those with rough surfaces that are known to exhibit a high degree of adherence may be removed through mechanisms other than size exclusion, and thus may not provide a conservative estimate of removal efficiency.

The manner in which the surrogate disperses in the challenge test solution has a significant impact on the effective size and shape of the challenge particulate. Some may agglomerate or become attached to other particles while in solution which would yield larger effective particle sizes. For example, organisms such as *Staphylococci* exist as clumps and *Streptococci* exist as chains. In its aggregate form, each of these organisms is too large to be considered conservative surrogates for *Cryptosporidium*. Surface structure also impacts the tendency of particles to agglomerate, and in general, particles with a smooth surface are more likely to be monodispersed in solution.

2. *Particle surface charge*: A conservative challenge particulate should have a neutral surface charge, since charged particles may interact with other particles and surfaces, thus enhancing removal. The solution pH can also affect the charge of some surrogates and thus should be considered in the preparation of a test solution. If there is a concern regarding the charge of the surrogate such that mechanisms of particle retention other than size exclusion may be responsible for surrogate removal in a MF or UF system, a nonionic surfactant could be used in the challenge test solution to significantly reduce the impact of charge-related removal mechanisms.
3. *Ease of use and measurement:* Although factors such as ease of handling and measurement are not critical in determining the appropriateness of a surrogate, these nevertheless may be important factors to consider. Handling the surrogate could expose personnel to the challenge particulate, and thus, the surrogate should be selected to minimize unacceptable risk to the technicians conducting the test. The material should also be easy to work with, in a dose-accurate way, since repeated tests may be conducted in which reproducibility is desirable. Surrogates that could degrade during the test, resulting in an inconsistent challenge concentration, should be avoided.

   It is also desirable to use a surrogate that is easy to enumerate through established analytical techniques. Furthermore, the LT2ESWTR requires that the concentration of challenge particles determined using a discrete measuring technique such that gross measurements such as turbidity are not acceptable.
4. *Cost*: The cost of seeding and analysis may preclude the use of some surrogates. Both the cost of the surrogate itself and cost of the required analytical techniques should be considered, as well as any other miscellaneous costs associated with the surrogate.

4.7.3. Surrogates for Cryptosporidium

In the absence of an acceptable surrogate, formalin- or heat-fixed *Cryptosporidium parvum* could be used as the challenge particulate for compliance with the requirements of the LT2ESWTR. However, the rule does permit surrogates for the purpose of challenge testing, and several different surrogates have been successfully used in studies evaluating physical removal of *Cryptosporidium* (21–23). There are three general classifications of surrogates (3):

1. Alternate microorganisms.
2. Inert particles.
3. Molecular markers.

It is important to note that not all of these classes of surrogates are appropriate for each type of membrane filtration system, and it is critical that these be compatible for the purposes of challenge testing. Generally, particulate surrogates such as alternate microorganisms and inert particles are appropriate for MF, UF, and MCF systems, while molecular markers would not be removed by these types of membranes. It may be necessary to use molecular makers with NF and RO membrane systems that can remove dissolved substances and that are not designed to accommodate large particulate concentrations. Some of the potential advantages and disadvantages associated with each class are summarized in Table 4.3.

1. *Alternate microorganisms*: Numerous organisms that have a history of use in filter evaluation studies are smaller than 1 μm (when monodispersed in solution), and these could be considered conservative surrogates for *Cryptosporidium*. A number of these organisms and an appropriate enumeration method are listed in Table 4.4, including both bacteria and viruses. Table 4.4 also includes common surrogates for *Giardia* and enteric viruses. *Serratia marcescens* and *Pseudomonas diminuta* have been widely used as surrogates within the membrane filtration industry, and the use of MS2 bacteriophage has generally been accepted as a surrogate for enteric viruses, since it is similar in size and shape to the poliovirus and hepatitis virus.

   Although *Bacillus subtilis* has been used as a surrogate for *Cryptosporidium* for testing the removal efficiencies of conventional treatment processes, it is not necessarily a suitable *Cryptosporidium* surrogate for challenge testing membrane filtration devices. Because there is limited data currently available regarding the use of *B. subtilis* in membrane challenge studies, a rigorous characterization of this organism would be necessary to determine whether it could be used as a *Cryptosporidium* surrogate for the purposes of challenge testing under the LT2ESWTR. Based on

Table 4.3
Comparative summary of *Cryptosporidium* and potential surrogates

Challenge particulate	Size range	Advantages	Disadvantages
Cryptosporidium parvum	3–5 μm	• No verification of surrogate required	• High cost • Difficult to measure
Alternate microorganisms	0.01–1 μm	• Low cost • Easy to measure • Accepted use	• Difficult to handle • Potential clumping
Inert particles	<1 μm	• Moderate cost • High uniformity • Easy to use	• Difficult to measure accurately
Molecular markers	<100,000 Da	• Low cost • Easy to measure	• Inappropriate for some applications

Table 4.4
Potential microbiological surrogates for *Cryptosporidium*

Microorganism	Size range (μm)	Target organism	Enumeration method
Micrococcus l.	7–12	*Giardia*	Standard Methods (24)
Bacillus subtilis	~ 1	*Cryptosporidium*	Barbeau et al. (25)
Escherichia coli	1–4	*Cryptosporidium*	Standard Methods (24)
Pseudomonas diminuta	0.3	*Cryptosporidium*	Standard Methods (24)
Serratia marcescens	0.5	*Cryptosporidium*	Standard Methods (24)
MS2 bacteriophage	0.01	Enteric virus	Adams (26)

the size range cited in Table 4.4, *B. subtilis* could potentially be considered an ideal surrogate for *Cryptosporidium*, pending a rigorous comparison of other characteristics (e.g., shape, surface charge, etc.) between these two organisms. However, because of this same size range overlap, *B. subtilis* could not be considered a conservative surrogate for *Cryptosporidium*.

The primary advantage of many microbial surrogates is that enumeration is fairly simple and inexpensive, typically involving culturing the test organisms present in the feed and filtrate samples. The ease with which these organisms can be cultured allows many to be grown in a laboratory to produce a stock for use in challenge testing. Bacteria can be cultured to yield stock concentrations in the range of 10^5 to 10^9 organisms per 100 mL, while MS2 bacteriophage can be grown at concentrations in the range of 10^7–10^{12} organisms per 100 mL. Any microbial stock used for the purpose of seeding during a challenge test should be enumerated prior to conducting the challenge test to facilitate seeding at the target level.

2. *Inert particles*: Inert particles may also be used as a surrogate for *Cryptosporidium* under the LT2ESWTR. For example, polystyrene latex microspheres (i.e., latex beads) have been used as a surrogate for *Cryptosporidium* in a number of studies. Historically, microspheres have been used in the calibration of particle counters and similar optical equipment in which a challenge particle of a known size and geometry is required by the investigator. Microspheres can be manufactured with very high particle uniformity and a smooth surface, both of which are important considerations when selecting a conservative surrogate. Microspheres are chemically inert, easy to handle, and relatively inexpensive. Furthermore, microspheres without a significant surface charge can be produced to minimize the potential for adsorption and interaction with either other particles or the membrane surface. Microspheres are also readily available with particle concentrations ranging from 10^7 to 10^9 particles per mL.

The primary difficulty associated with the use of microspheres is particulate enumeration. Although particle counting is a simple means of enumeration, this technique may not meet the rule requirement that the challenge particulate be discretely quantified as a result of the potential for background particles other than the microspheres to affect the results. Furthermore, other problems such as coincidence error and the dynamic range of most particle counting instruments may also skew the results. Any clumping of microspheres may also complicate particulate enumeration. A more reliable, albeit more expensive, means of enumerating microspheres is through capture (normally on a laboratory-grade membrane filter) and direct examination. The use of fluorescent microspheres is recommended to facilitate particulate identification. Methods for microscopic analysis of fluorescent microspheres are reported in the literature (27, 28).

The appropriateness of microspheres as a surrogate for *Cryptosporidium* could be directly verified through a comparative study; however, microspheres that meet certain criteria might be deemed conservative surrogates that would not require direct verification. For example, neutral, spherical-shaped microspheres with a maximum diameter of 1 μm and which are completely monodispersed in solution might constitute a conservative surrogate for *Cryptosporidium* that would not require direct verification.

3. *Molecular markers*: The suitability of molecular markers as surrogates for *Cryptosporidium* should be considered on a case-by-case basis. While the justification for using microorganisms and inert particles as surrogates for *Cryptosporidium* is more straightforward given that all are particulates, molecular markers are dissolved substances that are fundamentally different from particulate contaminants. As such, the removal mechanisms for molecular markers may be different than for those associated with discrete particles in many cases. However, semipermeable membranes that are capable of achieving very high removal efficiencies for dissolved substances may be capable of achieving similar removal of particulates such as *Cryptosporidium*. In addition, porous membranes with very fine pore sizes may be able to remove large macromolecules via mechanisms similar to those that filter discrete particles. Thus, the use of molecular challenge particulates is permitted for the purposes of challenge testing under the LT2ESWTR if the molecular marker used is determined to be conservative for *Cryptosporidium* and is discretely quantifiable.

A variety of molecular markers have been historically used to characterize the pore size or removal capabilities of membrane processes. For example, macromolecular protein compounds are used to determine the molecular weight cut-off (MWCO) for many UF membranes. In addition, fluorescent dyes such as Rhodamine WT and FDC Red #40 are used to characterize NF and RO membranes. These substances have high spectrophotometric absorbance characteristics that allow measurement and detection at the μg/L level (29). However, these low molecular weight (~500 Da) solutions could only be used with RO and less permeable NF membranes. If molecular markers are considered for challenge testing, it is desirable to use compounds that are more similar to discrete particles such as macromolecular proteins. It is also recommended that a mass balance be conducted on the feed, filtrate, and concentrate streams prior to challenge testing, to assess the efficacy of using a particular molecular marker.

With some molecular markers, it may be difficult to demonstrate removal in excess of 3 log unless sufficiently sensitive instrumentation is used. For challenge tests conducted with molecular markers, the feed and filtrate concentrations are typically quantified in terms of mass per unit volume. If the analytical method is specific for the molecular marker used in the test, use of a mass-based concentration is acceptable since the mass of a known substance can be related to moles, which is a discrete quantification. As is the case with any challenge particulate, gross measurements cannot be used for the purpose of quantification. This requirement would preclude the use of analytical techniques such as TOC monitoring and conductivity monitoring in most cases (3).

4.8. Challenge Test Solutions

Generation of an appropriate challenge test solution is an essential component of an effective test program. The purpose of the challenge test solution is to deliver the challenge particulate to the module of interest under the established test conditions. The design of the challenge test solution includes:

1. Establishing acceptable water quality of the solution.
2. Determining volume requirements.

3. Determining the challenge particulate concentration.
4. Selecting a seeding method.

4.8.1. Test Solution Water Quality

While the LT2ESWTR does not stipulate any constraints on the design of the challenge test solution, it is desirable to conduct the test in a manner that would be considered valid under any anticipated conditions to which the product undergoing testing might be applied in the field. Thus, the water quality of the test solution should be taken into consideration (3, 30). In designing the test solution, it is important to consider that the primary objective of challenge testing is to evaluate the removal efficiency of the challenge particulate during filtration. While water quality may not have a significant impact on the removal efficiency, particulate matter in the feed water can enhance removal of smaller contaminants. Thus, it is generally accepted that a test solution matrix comprised of high-quality water provides the most conservative estimate of removal efficiency. Note that challenge testing is not intended to yield meaningful information regarding membrane productivity and fouling potential, and thus, the use of a high-quality water matrix should not be considered inhibitory to testing these measures of membrane performance that are unrelated to the primary objective of challenge testing – to evaluate the removal efficiency of the challenge particulate. Productivity and fouling are addressed using site-specific pilot testing and not on a product-specific basis.

Some particular considerations regarding water quality characteristics and their implications for challenge testing are as follows:

1. High-quality water or "particle–free water" with a low concentration of suspended solids (e.g., membrane filtrate) should be used as the matrix for the challenge solution, minimizing the potential for formation of a fouling layer during the challenge test that would enhance removal of the challenge particulate.
2. No oxidants, disinfectants, or other pretreatment chemicals should be added to the test solution unless necessitated by process requirements (e.g., acid addition and/or scale inhibitor which may be necessary with NF/RO processes).
3. If the challenge particulate is a molecular marker, the water quality of the matrix for the test solution should not interfere with the introduction, dispersion, or measurement of the marker. Thus, the impact of water quality parameters, such as pH and ionic strength, on the chemical characteristics and speciation of molecular markers should be considered in the design of the test solution. This is particularly critical for NF or RO membrane modules, which can concentrate organic and inorganic solutes, potentially interfering with some molecular markers.
4. If a microbial challenge particulate is used, it may be necessary to add buffers or other materials to maintain the viability of the organisms. Any additives used must not interfere with any aspect of the test or result in a change in the concentration of the challenge particulate over the duration of the test. In addition, because water quality parameters such as pH and ionic strength can affect microbial aggregation, these solution characteristics should also be considered.
5. It is recommended that the challenge test solution be characterized with respect to basic water quality parameters, such as pH, turbidity, temperature, total dissolved solids (TDS), TOC, and any other water quality parameters that are critical to the test or interpretation of the results.

4.8.2. Test Solution Volume

The solution volume necessary to conduct a challenge test depends on several factors determined during the test design, including:

1. Filtrate flow.
2. Recovery.
3. Test duration.
4. Hold-up volume of the test system.
5. Equilibration time for the test solution.

Taking these factors into account, the volume of solution required for challenge testing may be calculated using Eq. (3):

$$V_{\text{test}} = \left(\frac{Q_{\text{p}}T_{\min}}{R} + V_{\text{hold}} + V_{\text{eq}}\right)\text{SF}, \tag{3}$$

where V_{test} is the minimum challenge test solution volume, gal; Q_{p} is the filtrate flow, gpm; $T_{\min}$ is the challenge test duration, min; R is the system recovery during test, decimal percent; V_{hold} is the hold-up volume of the test system, gal; V_{eq} is the system volume required to attain equilibrium feed concentration, gal; SF is the safety factor, dimensionless.

The safety factor in Eq. (3) accounts for unanticipated circumstances that might require additional solution volume. The value of the safety factor should be at least 1.0 and may be as high as 2.0 under conservative conditions.

The parameters Q_{p} and R are dictated by the system operating conditions. The LT2ESWTR requires that the flux and recovery used during the challenge test be set at the maximum design values for each parameter, as per manufacturer's specifications. The hold-up volume of the test system V_{hold} is the unfiltered test solution volume that would remain in the system on the feed side of the membrane barrier at the end of the test (i.e., after the system is shut down). At a minimum, this volume would include the feed side volume of the membrane module and associated piping. In general, it is desirable to design the system with a small hold-up volume, which could potentially allow V_{hold} to be ignored in Eq. (3). However, the hold-up volume can be significant in some systems, and in such cases, V_{hold} should be measured and taken into account. Another approach for dealing with the system hold-up volume is to finish the test with a "chaser" of clean water (i.e., without the challenge particulate) through the system allowing the entire test solution to be filtered. However, because this approach can dilute the feed concentration of the challenge particulate, the use of a clean water chaser is only recommended when the entire filtrate stream is sampled as a single composite.

The equilibrium volume (V_{eq}) is the quantity of the test solution that must pass through the membrane module(s) at the beginning of the test before the system stabilizes (i.e., the feed side concentration reaches an equilibrium value). In general, filtrate sampling cannot begin until at least this equilibrium volume has passed through the system. For most test apparatuses, a reasonable assumption is that a system achieves 90% of its equilibrium condition after three hold-up volumes (i.e., 3 V_{hold}) have passed through the system.

Table 4.5
Example challenge test solution volume for various types of modules

Module		Example membrane area [$ft^2(m^2)$]	Example maximum flux [$gal/ft^2/d$ ($L/m^2/h$)]	Test duration (min)	Recovery (%)	Volume (V_{test}) [gal (L)]
Cartridge filter		5 (0.46)	1,364 (2,316)	30	100	155 (586)
Spiral-wound	4" diameter	75 (7.0)	17.8 (30.2)	30	85	36.1 (137)
	8" diameter	350 (32.5)	17.8 (30.2)	30	85	168 (635)
Hollow-fiber	Outside-in	350 (32.5)	53.5 (90.8)	30	100	429 (1623)
	Inside-out	1,400 (130)	107 (182)	30	100	3,438 (13,013)

The duration of the challenge test, T_{min} as given in Eq. (3), does not include the time required for the test solution to come to equilibrium, as this is taken into account by the parameter V_{eq}. Thus, T_{min} represents the time necessary to implement the sampling program associated with the challenge test, which typically requires less than 1 h.

Table 4.5 provides examples of the challenge test solution volumes (i.e., V_{test}) required as calculated using Eq. (3) for various membrane configurations under the listed conditions, in these cases assuming the system hold-up volume (V_{hold}) and volume required to achieve equilibrium (V_{eq}) are negligible and a safety factor of 1.1. The filtrate flow (Q_p) is not shown specifically in the table but is calculated simply by multiplying the membrane area and maximum flux. Table 4.5 is intended to be illustrative only. Thus, it is recommended that the solution volume requirements for a specific challenge test be determined according to the procedure described above. Also, note that the values listed in Table 4.5 are examples only and that particular product specifications will vary by module manufacturer.

4.8.3. Test Solution Concentration

The concentration of the test solution is based on the target LRV to be demonstrated during the challenge test (LRV_t) and the detection limit for the challenge particulate in the filtrate samples. Since challenge testing is intended to be a one-time, product-specific requirement, it is generally advantageous to select a LRV_t at or near the maximum of 6.5 log removal that can be demonstrated under the LT2ESWTR. The detection limit is a function of the analytical technique used to enumerate the challenge particulate and the filtrate sample volume. For example, if the method can detect 1 particle in a sample, and the filtrate sample volume is 1 L, the detection limit is 1 particle/L. The detection limit and maximum 6.5 LRV are used to calculate the maximum feed concentration that can be used during a challenge test, as shown in Eq. (4)

$$C_{f-max} = \left(3.16 \times 10^6\right)\mathrm{DL}, \tag{4}$$

where $C_{f\text{-}max}$ is the maximum feed concentration, number or mass/volume, DL is the detection limit in the filtrate, number or mass/volume.

The coefficient in Eq. (4) represents the antilog 6.5, thus capping the maximum feed concentration in the test solution to allow a maximum of 6.5 log removal to be demonstrated if the challenge particulate is removed to the detection limit in the filtrate. The 6.5 log limit under the rule is intended to prevent excessive overseeding that can result in artificially high LRVs. The minimum required feed concentration can be calculated from the $\mathrm{LRV_t}$ and the detection limit using Eq. (5):

$$C_{\mathrm{f-min}} = 10^{\mathrm{LRV_t}}\mathrm{DL}, \tag{5}$$

where $C_{\text{f-min}}$ is the minimum feed concentration, number or mass/volume; $\mathrm{LRV_t}$ is the target log removal value for the challenge test, dimensionless; DL is the detection limit in the filtrate, number or mass/volume.

Equation (5) implicitly assumes complete removal of the challenge particulate by an integral membrane as a conservative means of estimating the minimum feed concentration. Note that Eqs. (4) and (5) are identical for a $\mathrm{LRV_t}$ of 6.5. Equations (4) and (5) also demonstrate that the requisite feed concentration is a function of the detection limit associated with the analytical technique used to enumerate the challenge particulate in the filtrate. The detection limit is typically expressed in terms of the number of challenge particulates per unit volume, or in the case of a molecular marker, mass per unit volume. For a given analytical method with a known sensitivity, the detection limit can be reduced by increasing the sample volume analyzed. For example, if a microbiological method is capable of detecting one organism in a sample, the detection limit can be improved by an order of magnitude by increasing the volume analyzed from 100 to 1,000 mL; however, many methods have limitations with respect to the maximum sample volume that can be analyzed. Also note that the feed and filtrate concentrations must be expressed in terms of equivalent volumes for the purposes of calculating log removal, even if different sample volumes are collected and analyzed during the test. If the $\mathrm{LRV_t}$ selected is less than the maximum of 6.5 permitted under the LT2ESWTR, the maximum and minimum feed concentrations will be different. Note that it is desirable to use a concentration greater than the minimum since use of the minimum feed concentration would only demonstrate the $\mathrm{LRV_t}$ if the challenge particulate were removed to the detection limit in the filtrate. After the maximum and minimum concentrations are established using Eqs. (4) and (5), respectively, a convenient value between these boundaries can be selected for the challenge particulate concentration (C_{test}).

Once the required test solution volume and concentration have been determined, the total number of challenge particulates required for the test can be calculated using Eq. (6). If a molecular marker is used, Eq. (4) would be used to determine the total mass of the marker required.

$$\mathrm{TCCP} = C_{\mathrm{test}} V_{\mathrm{test}}, \tag{6}$$

where TCPP is the total challenge particulate population, number or mass of particles; C_{test} is the feed concentration of challenge particulate, number or mass/volume; V_{test} is the challenge test solution volume, gal.

4.8.4. Challenge Particulate Seeding Method

There are two approaches commonly used to introduce the challenge particulate into the test solution: batch seeding and in-line injection. Batch seeding involves the introduction of the total challenge particle population (TCPP) into the entire volume of test solution, followed by complete mixing to a uniform concentration. In-line injection allows for continuous or intermittent introduction of challenge particulates into the feed stream entering the membrane filtration system. The specific method used may depend on the circumstances of the particular challenge test, although either is permitted under the LT2ESWTR.

Batch seeding is simpler and requires less equipment than in-line injection, but it is only feasible when the entire test solution volume is contained in a reservoir. Furthermore, mixing the challenge particulates into large volumes of water to create a uniform concentration can be logistically problematic. For these reasons, batch seeding is typically only used in challenge studies for small-scale modules with a relatively small membrane area, for which test solution volumes are easier to handle and mix.

In-line injection is the most common seeding approach used in challenge studies, particularly for those involving full-scale modules with greater membrane area. In-line injection allows the challenge particulate to be introduced into the feed on either a continuous or intermittent basis. In general, continuous seeding is advantageous for challenge testing, although intermittent seeding may be appropriate for long-term studies in which it is only necessary to seed organisms at key times during an operational cycle. If intermittent seeding is used, it is necessary to ensure that equilibrium is achieved during each seeding event prior to collection of any feed or filtrate samples.

In-line injection requires additional equipment, such as chemical feed pumps, injection ports, and in-line mixers. These components must be properly designed and integrated into the test apparatus to ensure a consistent challenge particulate concentration in the feed. A chemical metering pump that delivers an accurate and steady flow of challenge material is recommended, while pumps that create a pulsing action should be avoided. The injection port should introduce the challenge material directly into the bulk feed stream to aid in dispersion. Examples of acceptable and unacceptable injection ports are shown in Fig. 4.1. An in-line static mixer should be placed downstream of the injection port, and a feed sample tap should be located approximately ten pipe diameters downstream of the mixer.

The in-line injection method of seeding delivers challenge particulates from a stock solution with a known concentration as calculated using Eq. (7):

$$C_{\mathrm{ss}} = \mathrm{TCPP}/V_{\mathrm{ss}}, \tag{7}$$

where C_{ss} is the challenge particulate concentration in the stock solution, number or mass/volume; TCPP is the total challenge particulate population, number or mass of particles; V_{ss} is the challenge particulate stock solution volume, gal.

The TCPP is calculated according to Eq. (6), while the stock solution volume, V_{ss} can be selected for convenience. However, the V_{ss} should be between 0.5 and 2% of the total test solution volume, as determined using Eq. (3).

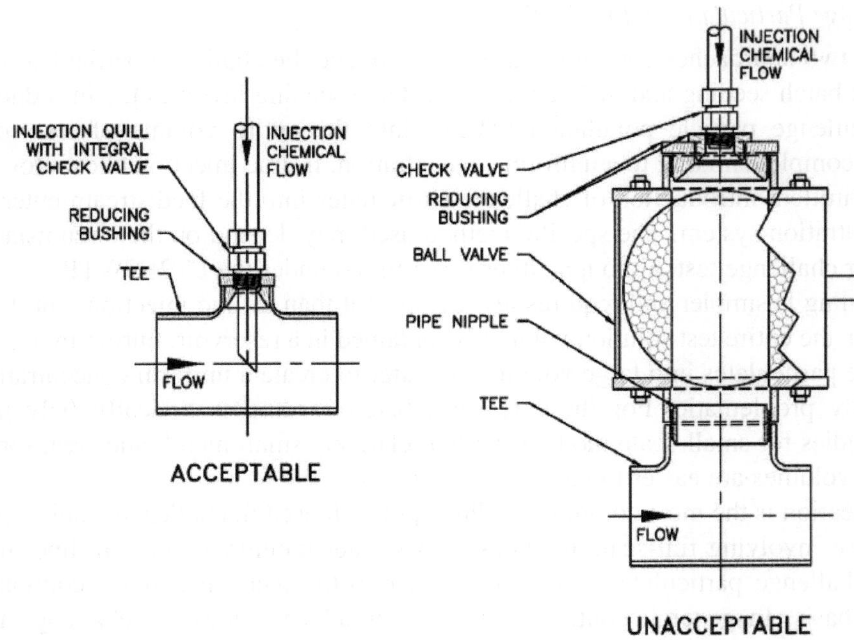

Fig. 4.1. Schematic of acceptable and unacceptable sample injection ports.

Once the concentration of the stock solution has been determined, the stock solution delivery rate (SSDR) for the in-line injection method of seeding can be calculated using Eq. (8):

$$\mathrm{SSDR} = C_{\mathrm{test}} Q_{\mathrm{p}} / C_{\mathrm{ss}} R, \quad (8)$$

where SSDR is the stock solution delivery rate, gpm; C_{test} is the feed concentration of challenge particulate, number or mass/volume; Q_{p} is the filtrate flow, gpm; C_{ss} is the challenge particulate concentration in the stock solution, number or mass/volume; and R is the system recovery during test, decimal percent.

The SSDR represents the setting for the chemical feed pump used to inject the challenge particulate into the feed stream.

4.9. Challenge Test Systems

A system used for challenge testing must be carefully designed to both meet the objectives of the test and simulate full-scale operation to the greatest practical extent. Guidance for both designing an appropriate test apparatus and determining operational parameters for challenge testing is provided in the following subsections.

4.9.1. Test Apparatus

The equipment used to conduct challenge testing is product-specific to some extent, although there are some basic components that are common to all systems. In many cases, a manufacturer may maintain a special test apparatus to check individual modules as a component of its QA/QC program. Such an apparatus may be suitable for conducting challenge testing and typically includes equipment such as pumps, valves, instrumentation, and controls necessary to evaluate full-scale modules. This same type of equipment would be used in the design of systems for testing small-scale modules.

Both the seeding and sampling methods selected for challenge testing, as well as the hydraulic configuration of the system, affect the design of the test apparatus. Batch seeding requires a feed tank and mixing equipment, while continuous seeding requires a stock solution reservoir, chemical metering pump, and in-line mixers. Sampling requirements may dictate the location and design of sample taps in the system. In addition, the test apparatus should be designed to mimic the hydraulic configuration of the full-scale system as much as practical; however, the test apparatus may alternatively utilize a more conservative recovery than the full-scale system. If 100% recovery (i.e., the most conservative scenario) is used, a cross-flow system must operate without a bleed stream such that all of the concentrate is recirculated, and a deposition mode system must filter the entire test solution volume. Note that a full-scale cross-flow system could not operate at 100% recovery on a sustained basis, since the feed would become increasingly concentrated. However, operation at 100% recovery is feasible for a short-term challenge test in the interest of generating conservative results. The test apparatus should allow the membrane module to undergo direct integrity testing both before and after the challenge test. Figure 4.2 through Fig. 4.6 are schematic representations of typical apparatuses for challenge testing under various conditions. Note that ancillary equipment and operational processes (e.g., backwash, chemical cleaning, and integrity testing) are not shown.

Figure 4.2 illustrates a pressure-driven apparatus operating in deposition mode with batch seeding and composite sampling. This type of system may be well suited for a MCF or other membrane module with limited surface area. With this apparatus, the test solution is prepared as a batch and a composite filtrate sample would be generated, yielding a single data pair (i.e., a feed sample and a composite filtrate sample) for the purposes of calculating the log removal efficiency for the challenge particulate.

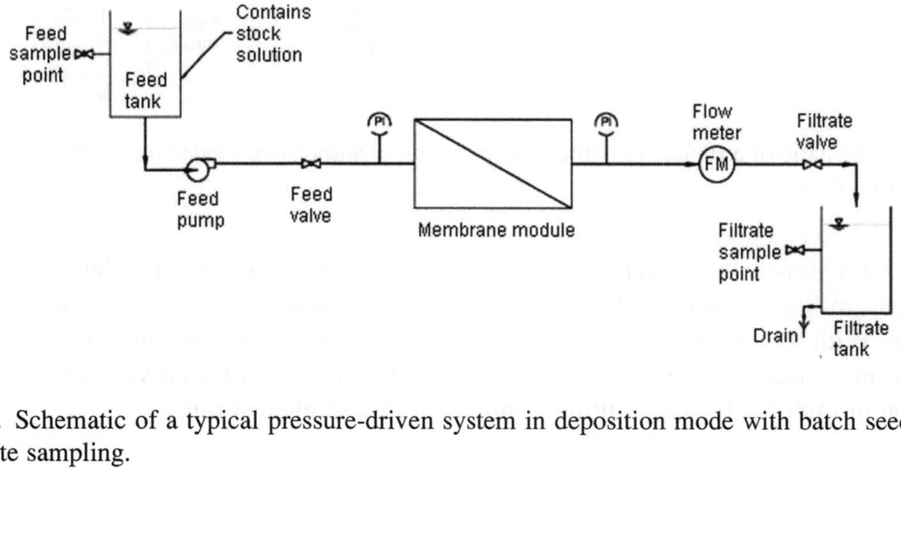

Fig. 4.2. Schematic of a typical pressure-driven system in deposition mode with batch seeding and composite sampling.

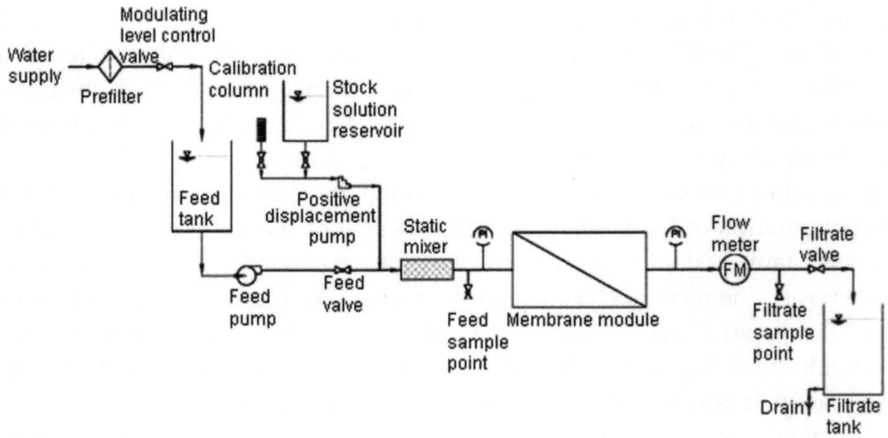

Fig. 4.3. Schematic of a typical pressure-driven system in deposition mode with continuous seeding and grab sampling.

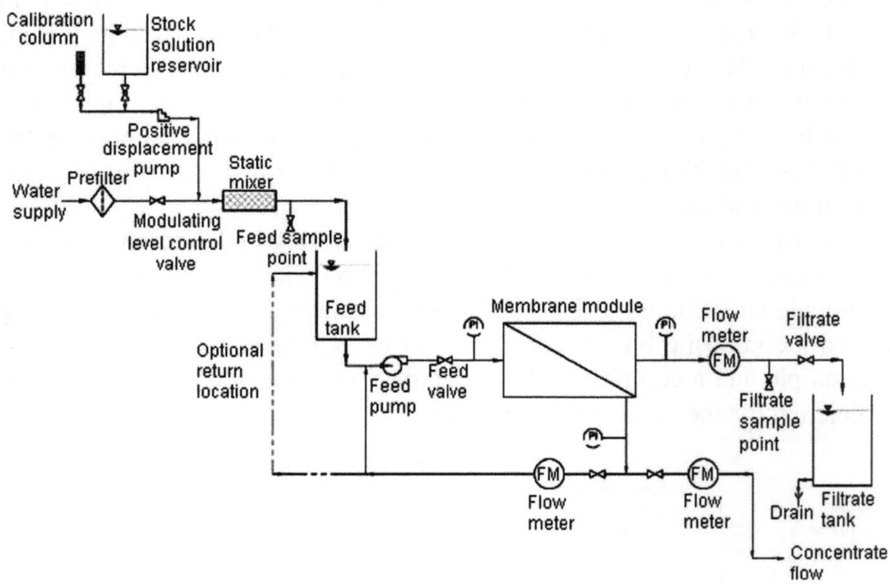

Fig. 4.4. Schematic of a typical pressure-driven system in suspension mode with continuous seeding and grab sampling.

Figure 4.3 illustrates an apparatus similar to that shown in Fig. 4.2, but designed for continuous in-line seeding and grab sampling rather than batch seeding and composite sampling. With this apparatus, the challenge particulate is introduced from a stock solution reservoir and mixed prior to the feed sampling point. The use of grab sampling allows the collection of multiple feed and filtrate samples from a single test run.

Figure 4.4 shows a schematic of a pressure-driven apparatus operated in suspension mode with continuous seeding and grab sampling. In systems that operate in suspension mode, the concentration of suspended solids increases on the feed side of the membrane. While it may not be practical to accurately replicate the solids concentration profile for a membrane system in which the feed side concentration of suspended solids varies as a function of filtration time, concentrate recirculation can produce conservative feed side conditions for the purpose of challenge testing, assuming appropriate operating conditions for recovery and recycle ratio are selected. Selection of an appropriate recovery for challenge testing can be complicated by the fact that system recoveries can vary significantly in some cases (particularly for NF/RO systems). The recycle ratio should be selected such that velocities across the membrane surface are high enough to keep particles in suspension. The manufacturer can typically recommend a minimum scour velocity for a cross-flow system.

For systems that utilize concentrate recycling, there are some additional considerations that are important to be taken into account regarding the feed side system volume and the location of the feed sample point. In general, larger feed side system volumes require longer system equilibration times. For example, Fig. 4.4 shows the concentrate return location at the feed tank rather than directly into the module feed line, thus increasing the effective feed side system volume significantly. If such an arrangement is necessary (e.g., to provide an air break in the recirculation system), then the feed tank volume should be minimized. In an apparatus utilizing concentrate recycling, the feed sample point must be located upstream of the return point, as shown in Fig. 4.4.

Figure 4.5 shows a typical vacuum-driven test apparatus operated in deposition mode with continuous seeding and grab sampling. Although the module is immersed in a tank, the feed water is not agitated, thus allowing particles to deposit on the membrane surface. With this apparatus, the filtrate sampling point must be located downstream of the vacuum pump.

Figure 4.6 illustrates an apparatus for a vacuum-driven system operated in suspension mode with continuous seeding and grab sampling. With this apparatus, the feed tank is mechanically agitated to keep particles in suspension and can be modeled as a continuous stirred tank reactor (CSTR). As with the vacuum-driven apparatus shown in Fig. 4.5, the filtrate sampling point must be located downstream of the vacuum pump.

4.9.2. Test Operating Conditions

The design of a challenge test includes specifications for the following operating conditions: flux, recovery, and hydraulic configuration. The LT2ESWTR requires the challenge test to be conducted at the maximum design flux and recovery and that the test apparatus be operated under representative or conservative hydraulic conditions. These requirements dictate the operating conditions for the test apparatus during challenge testing. Note that under the LT2ESWTR, recovery is defined as the volumetric percent of feed water that is converted to filtrate in the treatment process over the course of an uninterrupted operating cycle (i.e., excluding losses that occur due to the use of filtrate in backwashing or cleaning operations).

Testing at the maximum recovery is important to ensure that the volumetric concentration factor (VCF) simulated during challenge testing is representative of (or conservative for) full-scale system operation. For systems that operate in deposition (i.e., direct flow or "dead-end")

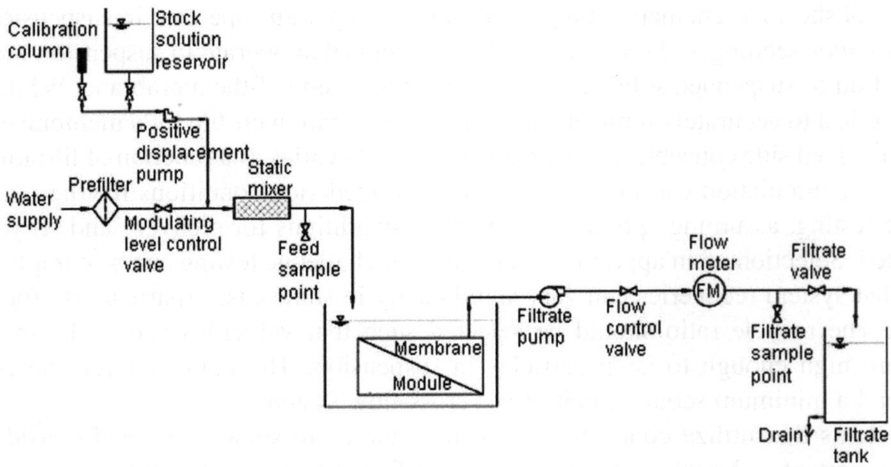

Fig. 4.5. Schematic of a typical vacuum-driven system in deposition mode with continuous seeding and grab sampling.

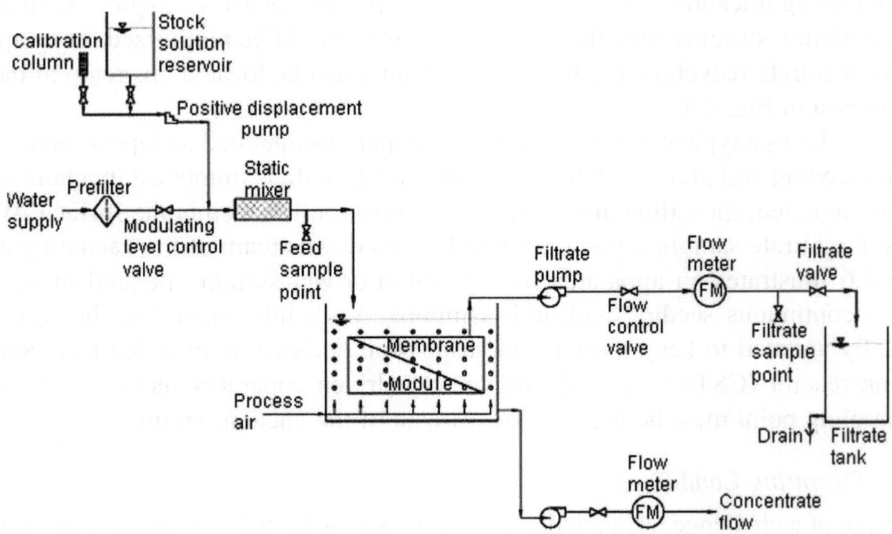

Fig. 4.6. Schematic of a typical vacuum-driven system in suspension mode with continuous seeding and grab sampling.

mode, such as most MF/UF and MCF systems, the value of the VCF is one, and thus, the recovery does not have a significant impact on the suspended solids concentration. However, it is still recommended that systems operating in deposition mode process at least 90% of the challenge test solution, thus resulting in an effective recovery of at least 90%. Similarly, for MF/UF systems that operate in suspension mode but without a concentrate waste stream (i.e., a bleed stream), it is also recommended that at least 90% of the challenge test solution be processed to generate an effective recovery of at least 90%. For MF/UF systems that operate with a concentrate waste stream, it is recommended that a recovery of at least 75% be utilized for challenge testing unless a more representative system recovery can be demonstrated by the manufacturer; a recovery of 100% would represent the most conservative case. NF/RO systems, for which the utilization of a concentrate waste stream is standard, should be operated at a recovery of at least 45% for the purposes of challenge testing, which is representative of a flow-weighted average recovery for a module in such a system operating at an overall recovery of greater than 90% for a single stage. If challenge testing is conducted on a small-scale NF/RO system (such as one using a single module), then concentrate recycling should be used to increase the recovery to at least 45%.

The membrane area is also typically given in the specifications, and the maximum filtrate flow can be calculated by multiplying the membrane area and the maximum flux (31–33). Although the LT2ESWTR stipulates specific requirements for challenge testing flux, the filtrate flow is also necessary for designing the challenge test solution. Table 4.6 summarizes some typical specifications for membrane area and maximum flux associated with various types of membrane modules, as well as the corresponding filtrate flow. For the purposes of challenge testing, the membrane area exposed to the feed (i.e., as opposed to that to the filtrate) should be used in all calculations. Note that the values listed in Table 4.6 are examples only and that particular product specifications will vary with module manufacturer.

A challenge test should be designed to simulate the hydraulic configuration of full-scale system operation, since it affects the concentration of suspended solids on the feed side of the membrane and thus the removal efficiency of the process. A membrane filtration system can be operated in either suspension or deposition mode. While it is relatively straightforward to simulate a system operating in deposition mode during a challenge test, it may not be possible to simulate all variations of suspension mode operation using a single-module challenge test apparatus. For example, it is not practical to simulate a plug flow reactor (PFR) configuration typical of full-scale RO systems. In such cases, the challenge test apparatus should be designed and operated as a CSTR, since a CSTR configuration generally results in the highest concentration of suspended solids on the feed side of the membrane. If the challenge test can successfully demonstrate the target LRV with a higher concentration of suspended solids on the feed side of the membrane than expected with the full-scale system, then the use of a

Table 4.6
Typical parameters for various types of modules

Module		Example membrane area [ft^2(m^2)]	Example maximum flux [gal/ft^2/d (L/m^2/h)]	Filtrate flow [gpm (L/min)]
Cartridge filter		5 (0.46)	1,364 (2,316)	4.8 (18.2)
Spiral-wound	4" diameter	75 (7.0)	17.8 (30.2)	0.9 (3.4)
	8" diameter	350 (32.5)	17.8 (30.2)	4.3 (16.3)
Hollow-fiber	Outside-in	350 (32.5)	53.5 (90.8)	13.1 (49.6)
	Inside-out	1,400 (130)	107 (182)	103 (390)

CSTR in place of a PFR configuration would represent more conservative, and thus acceptable challenge test conditions. A single RO module can be operated as a CSTR by including a concentrate recycling loop, as shown in Fig. 4.4.

4.10. Sampling

Although the LT2ESWTR does not stipulate any particular requirements for sampling, it is an important component of a challenge test. An effective sampling program is dependent on the development of detailed and thoroughly documented sampling plan, as well as the selection of appropriate sampling methods and locations. These critical aspects of sampling, along with the monitoring of operational parameters during the execution of the challenge test, are discussed in the following subsections (3).

4.10.1. Sampling Methods

The two most common approaches for sampling are the grab and composite methods. Grab sampling involves the collection of one or more aliquots from the feed or filtrate stream, while composite sampling involves collection of the entire process stream for processing and subsequent analysis. The concentration of challenge particulates in the feed solution is typically characterized through grab sampling, while the filtrate stream may be sampled using either grab or composite sampling. If grab sampling is used for both the feed and filtrate streams, the number of feed and filtrate samples does not need to be equivalent, and samples can be collected on different schedules during the challenge test. In many cases, it may be advantageous to collect more filtrate samples than feed samples, since filtrate concentrations are expected to be very low, and an error of just a few particles in a filtrate sample can have a significant impact on the demonstrated removal efficiency. Moreover, if batch seeding is used, the feed concentration should not vary significantly over the course of challenge testing, assuming appropriate feed stock mixing. However, if continuous seeding is used, paired sampling may be preferred for simplicity of data reduction.

Grab sampling typically involves the collection of a predetermined volume of water in an appropriate collection vessel at predetermined times, as documented in the sampling plan. In some cases, a composite sample is collected directly into a sampling vessel; however, it is more common to capture the challenge particulate during composite sampling. A composite sample is usually compiled by passing the entire filtrate stream through an absolute filter capable of 100% capture of the challenge particulate. The challenge particulate would then be enumerated either directly from the filter media or removed from the filter for subsequent analysis. Grab samples typically do not require this type of processing in the field, and any extraction or concentration steps for grab samples are typically conducted in the laboratory.

It is important that good sampling practices be employed during challenge testing, such as flushing taps prior to sample collection (if applicable) and isolating filtrate sampling locations from feed sampling locations to prevent cross-contamination. Furthermore, appropriate QA/QC measures should be implemented during sampling, such as collection of duplicate samples and blanks.

4.10.2. Sample Port Design and Location

As with the challenge particulate injection port, the design of the feed and filtrate sample withdraw ports should yield a uniform sample. Figure 4.7 illustrates examples of both acceptable and unacceptable sampling ports. The unacceptable apparatus requires the sample to be pulled from a large section of pipe and has an area where stagnation of sample flow may occur. By contrast, the acceptable apparatus has a quill that extends into the center of the pipe to obtain a more representative sample.

The feed sample tap should be located at least ten pipe diameters downstream of challenge particulate injection points and in-line mixers to ensure uniform concentration. For apparatuses that utilize concentrate recycling, as illustrated in Fig. 4.4, the feed sample tap should be located upstream of the T-connection where the concentrate is blended with the incoming feed water. A check valve may be used to prevent backflow of concentrate into the feed line. As a guideline, the feed sample rate should be no more than 1% of the flow to the membrane. Filtrate samples should be collected at a point after the filtrate passes through any filtrate side instrumentation such that any important measurements are not affected by the sampling event. In vacuum-driven apparatuses, the filtrate sample tap must be located downstream of the filtrate pump, as shown in Figs. 4.5 and 4.6. Note that the filtrate sampling valve should be positioned as close as possible to the filtrate port to minimize error due to any potential for adhesion of the challenge particulate to the piping prior to the sampling point. In addition, if a microorganism is used as the challenge particulate, it is prudent to use a metal or heat-resistant sampling valve to allow the tips of the valve to be flame-sterilized.

4.10.3. Process Monitoring

During the challenge test, it is important that the operational parameters be monitored to ensure that the test conditions remain constant. Continuous monitoring for flow (or flux) and pressure should be conducted if the apparatus is equipped with the appropriate instrumentation. If periodic monitoring must be utilized, it is important that operational parameters be

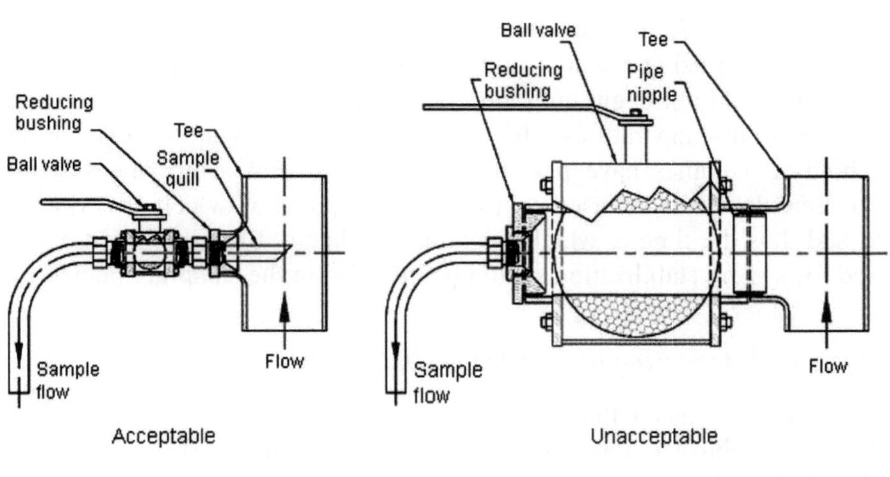

Fig. 4.7. Schematic of acceptable and unacceptable sampling ports.

checked at least both before beginning and after completing sampling for the challenge particulate. Operational data collected prior to initiating sampling should be used to verify that the flux and recovery are at the required levels. It is also important that the membrane module(s) undergo direct integrity testing both before and after challenge testing to verify that the modules were integral during the test.

If the water quality of the test solution matrix is a particular concern, other water quality parameters should be sampled accordingly. Some examples of other parameters that may be important in some cases include pH, temperature, turbidity, TDS, TOC, and ionic strength. The measurement of such water-quality parameters may be most relevant if a molecule marker is used as the challenge particulate.

All aspects of process monitoring, including what parameters to monitor, how often to monitor, and the range of acceptable results should be included in the sampling plan for the challenge test. If any results are outside acceptable tolerances, the challenge test should be restarted.

4.10.4. Sample Plan Development

The primary purpose of a sampling plan is to define the samples to be collected and provide an accompanying sampling schedule for the challenge test. A sampling plan should include:

1. Type of sample(s) (i.e., composite vs. grab).
2. Number of feed and filtrate samples to be collected.
3. Sample locations.
4. Sampling interval.
5. Estimate of time required to collect each sample.
6. Sampling equipment required.
7. Sample volume(s).
8. Process monitoring requirements.

The sampling plan should also specify any particular requirements associated with the analytical technique to be employed, as well as procedures for shipping the sample(s) if they are to be analyzed at an off-site laboratory. Samples should be collected, preserved, stored, prepared, and analyzed using methods and techniques appropriate for the challenge particulate. Sampling should not begin until the system has stabilized (i.e., reached equilibrium concentration). Most test apparatuses achieve greater than 90% of equilibrium concentrations after three hold-up volumes have passed through the system. Thus, the system hold-up volume and feed flow can be used to estimate the point at which the system is near equilibrium and thus the time at which sampling can begin. Both the hold-up volume and time required for system stabilization should be included in the sampling plan.

4.11. Analysis and Reporting of Challenge Test Results

After challenge testing is completed for a particular product, the results must be analyzed to determine the established removal efficiency of the module (i.e., $LRV_{C\text{-test}}$) for the purposes of LT2ESWTR compliance. The following subsections provide guidance regarding the calculation of removal efficiency under the rule, the statistical analysis of the challenge test results, and summarizing challenge testing in a report for state review.

4.11.1. Calculation of Removal Efficiency

The removal efficiency established during challenge testing – $LRV_{C\text{-test}}$ – is determined from the various LRVs generated during the testing process. The LT2ESWTR requires that a single LRV be generated for each module tested for the product line under evaluation. The LRVs for each respective module tested are then combined to yield a single value of $LRV_{C\text{-test}}$ that is representative of the product line.

Under the LT2ESWTR, the LRV is calculated according to Eq. (2):

$$\mathrm{LRV} = \log(C_{\mathrm{f}}) - \log(C_{\mathrm{p}}), \tag{2}$$

where: LRV is the log removal value demonstrated during a challenge test; C_{f} is the feed concentration of the challenge particulate, number or mass/volume; C_{p} is the filtrate concentration of the challenge particulate, number or mass/volume.

Note that the feed and filtrate concentrations must be expressed in identical units (i.e., based on equivalent volumes) for Eq. (2) to yield a valid LRV. If the challenge particulate is not detected in the filtrate, then the term C_{p} is set equal to the detection limit. The overall value of $LRV_{C\text{-test}}$ (i.e., the removal efficiency of the product) is based on the entire set of LRVs obtained during challenge testing, with one representative LRV established per module tested. The manner in which $LRV_{C\text{-test}}$ is determined from these individual LRVs depends on the number of modules tested. Under the LT2ESWTR, if fewer than 20 modules are tested, then the lowest representative LRV among the various modules tested is the $LRV_{C\text{-test}}$. If 20 or more modules are tested, then the 10th percentile of the representative LRVs is the $LRV_{C\text{-test}}$. The percentile is defined by $[i/(n + 1)]$ where "i" is the rank of "n" individual data points ordered from lowest to highest. It may be necessary to calculate the 10th percentile using linear interpolation.

Although the LT2ESWTR requires that one representative LRV be established per module tested, the rule does not restrict the manner in which the representative LRV for each module is calculated. Consequently, there are numerous methods that could be used to calculate the representative LRV for a module. If multiple feed/filtrate sample pairs are collected, a LRV can be calculated for each set of paired data, and the LRV for the tested module could be selected as the lowest LRV (more conservative) or the average of the LRVs (less conservative). Another approach is to average all the respective feed and filtrate concentrations from among the various samples collected and calculate a single LRV for a tested module using Eq. (2). A more conservative approach would be to use the average feed concentration but the maximum filtrate concentration sampled, which would result in a lower representative LRV for a tested module. Likewise, a still more conservative approach would be to use the minimum feed and maximum filtrate concentrations. Note that these methods simply represent potential options; other approaches may be used for calculating a representative LRV for each module tested.

Regardless of the particular method used, the range of data collected for a single module can provide some indication about the experimental error associated with the study (i.e., errors due to seeding, sampling, analysis, etc.). If a statistically valid sampling method was used to select the modules for challenge testing, a comparison of the LRVs across the different modules tested would provide an indication of variability within the product line.

4.11.2. Statistical Analysis

If a sufficient number of modules are evaluated in the course of a challenge test to be considered a statistically significant sample of the product line, it may be useful to conduct a formal statistical analysis of the data to make inferences about the entire population from the sample set. Note that such a statistical analysis is not considered a substitute for the methodology for determining $LRV_{C\text{-test}}$ required under the rule, but could provide the manufacturer with useful information. For example, if challenge testing were conducted using a random sampling of membranes from a production lot, a statistical analysis of the challenge test data (i.e., the LRV observed for each of the modules tested) would provide an estimate of the range of removal efficiency for the entire product line. This information could be used to infer the number of membrane modules in the product line that would be expected to achieve the $LRV_{C\text{-test}}$. If it is desirable to apply a statistical analysis to the LRV data generated during challenge testing, the particular method of analysis should be considered in the design of the challenge test protocol, since the method to be employed may dictate the number of modules selected for testing.

4.11.3. Reporting

The LT2ESWTR does not mandate any reporting requirements for challenge testing. However, the results of the challenge test must be acceptable to any state that permits a particular membrane module product line to be used for obtaining *Cryptosporidium* removal credit under the rule. In order to facilitate this acceptance by the various states, a sample outline of a challenge test report is provided as follows. Note that this outline should be customized to meet any particular state requirements.

1. Introduction
   - Description of testing organization
   - Test site
   - Description of membrane filtration product
   - Testing objectives (including target LRV)
2. Membrane Modules and Test Apparatus
   - Membrane module specifications for each module evaluated
   - Considerations for small-scale module testing (if applicable)
   - Nondestructive performance testing
   - Description of test apparatus
3. Challenge Test Protocol
   - Challenge particulate (including rationale for selection)
   - System operating conditions
   - Challenge test solution design
   - Seeding method
   - Process monitoring
   - Detailed sampling plan

- QA/QC procedures
- Data management

4. Results and Discussion
   - Summary of measured system operating conditions
   - Summary of LRV results for each module tested
   - Summary of system integrity evaluation
   - Determination of removal efficiency
   - Summary of NDPT results for each module tested
   - QCRV determination based on the results of nondestructive performance testing
   - Statistical evaluation of results (if applicable)
5. Summary and Conclusions
   - Summary description of membrane filtration product
   - Summary of challenge test protocol
   - LRV demonstrated during challenge testing
   - Quality control release value for nondestructive performance testing

4.12. Retesting of Modified Membrane Modules

As a component of ongoing innovation and product development, manufacturers may make changes to a particular product line or its associated manufacturing process (34). If such a change affects the fundamental characteristics of the module, the removal efficiency, and/or the NDPT results and associated QCRV, the LT2ESWTR requires the modified product to be resubjected to challenge testing.

Because it is not possible to develop a comprehensive listing of the potential modifications that would require retesting, the need to retest a modified product must be evaluated on a case-by-case basis and at the discretion of the state. Some examples of membrane properties, which if modified, might alter the fundamental removal characteristics of a module, include:

1. Membrane material (e.g., a change in the polymer or backing material).
2. Pore size (nominal and absolute).
3. Porosity.
4. Permeability.
5. Membrane symmetry (i.e., symmetric, asymmetric, or composite).

As shown in the examples given above, most of the changes that may necessitate retesting are modifications to the membrane material itself, such as changes to the polymer chemistry and/or pore size distribution. Even if the change to the membrane media is not intended to affect removal efficiency or NDPT parameters, it may be necessary to retest the module, since a modification to an intrinsic property of the membrane media could have an impact on one or both of these criteria.

Minor changes or changes that are not related to the fundamental removal characteristics of the membrane media or NDPT parameters would not require retesting. For example, a change in the membrane area within a module would not be expected to affect either the removal efficiency or nondestructive performance testing.

Although not directly related to the product itself, any modifications to the manufacturer's NDPT may also require that the product line be resubjected to challenge testing, since the

Table 4.7
Typical range of VCF values for various hydraulic configurations

Hydraulic configuration		VCF
Deposition mode	Dead-end	1
	PFR	3–20
Suspension mode	Cross-flow	4–20
	CSTR	4–20

results of the specific NDPT used ensure that the modules produced meet the minimum requirements for achieving the demonstrated LRV. However, if the modified NDPT can be correlated to the specific test associated with product line's challenge test, additional challenge testing may not need to be conducted, at the discretion of the state.

Modifications to the hydraulic configuration of a membrane filtration system might warrant retesting, since the concentration of suspended solids on the feed side of the membrane may be affected, in turn potentially affecting removal efficiency. In determining whether or not a modification to the hydraulic configuration would require retesting, one significant factor is whether the hydraulic conditions of the original challenge test are conservative for the new hydraulic configuration of the system. In general, testing performed under a hydraulic configuration with a higher VCF than that associated with the modification would be considered conservative. A summary of the various hydraulic configurations and associated VCFs is provided in Table 4.7.

In general, most modifications to the module itself (as opposed to the membrane media) are unlikely to affect either removal efficiency or nondestructive performance testing, although retesting might be warranted in some cases. Although the manufacturer can exercise its discretion in deciding the circumstances under which additional testing is necessary, the state has authority regarding whether or not to accept a particular change such that the product line would still be eligible to receive *Cryptosporidium* removal credit under the rule without retesting.

4.13. Grandfathering Challenge Test Data from Previous Studies

As a result of the commercialization of membrane treatment processes for drinking water production, a number of states, consulting engineers, and independent testing organizations have developed programs to demonstrate the removal efficiency of membrane systems. Under these existing programs, numerous challenge studies have been conducted in which complete removal of *Cryptosporidium* was demonstrated, independent of the challenge level (5). The state has the discretion to accept data from a previous study provided that the testing was conducted in a manner that demonstrates removal efficiency for *Cryptosporidium* commensurate with the treatment credit awarded to the process.

While a previously conducted challenge test may adequately demonstrate the required *Cryptosporidium* removal efficiency, it may be more difficult to correlate the results of a prior test to the NDPT currently used by the manufacturer. If the state determines that the grandfathered test does not meet the intent of the LT2ESWTR, a new challenge test that is consistent with the rule requirements must be conducted.

As a general guide, the following challenge test conditions have been identified as potentially yielding results that do not satisfy the intent of the rule:

1. Challenge testing conducted on obsolete products.
2. Challenge testing conducted on small-scale modules: Small-scale module testing is permitted under the LT2ESWTR only if certain criteria are met.
3. Challenge testing using unacceptable surrogates for *Cryptosporidium*: The challenge particulate used in a grandfathered test must provide equivalent or sufficiently conservative removal efficiency relative to *Cryptosporidium* oocysts.
4. Challenge particulate enumeration using unacceptable methodology: The challenge particulate must have been quantified using an acceptable method.
5. Unavailable QCRV: If nondestructive performance testing was not used to establish a suitable QCRV in a previous study, it may be difficult or impossible to relate the demonstrated removal efficiency to the NDPT results for untested modules that are produced.

There may also be cases in which deviations from challenge testing requirements under the LT2ESWTR may not be significant, such that additional testing would not be required. Some potential cases in which challenge test data that do not specifically comply with the rule might receive favorable consideration for grandfathering, at the discretion of the state, are listed below.

1. Removal efficiency determined using different method: It may be necessary to recalculate the removal efficiency (i.e., $LRV_{C\text{-test}}$) from a previous challenge test according to the requirements of the LT2ESWTR. If some moderate deficiencies exist in the data such that the $LRV_{C\text{-test}}$ cannot be calculated according to the procedure described in the guidance manual (3), the state may exercise discretion as to whether or not to approve the prior test. The state may also evaluate the impact of such deficiencies and consequently consider reducing the $LRV_{C\text{-test}}$ on the basis of this evaluation.
2. Elevated feed challenge particulate concentration: The LT2ESTR places a limit on the feed concentration of the challenge particulate that essentially establishes a maximum LRV of 6.5 to prevent excessive overseeding during challenge studies. However, exceeding this maximum feed concentration does not necessarily invalidate a prior test. In reviewing grandfathered data, it is most important to consider whether or not high LRVs were achieved through excessive overseeding. For example, if a removal study was conducted at a challenge level of 10^8 particles/L and all filtrate concentrations were less than 10 particles/L, the process could potentially qualify for the maximum removal credit of 6.5 log. By contrast, if the same challenge level was applied to a membrane that allowed 10^3 particles/L to pass into the filtrate, states may opt not to qualify the test for a LRV of 5, since the high LRV is likely a result of overseeding an otherwise inadequate or nonintegral membrane.
3. Challenge testing at a flux or recovery other than the maximum design value(s): Deviations from testing at the maximum flux and/or recovery might not exclude a prior test from consideration for grandfathering. Most available data suggests that flux and recovery will not significantly impact the removal efficiency of a membrane filtration process for *Cryptosporidium*-sized particles. If complete removal of the challenge particulate was achieved during the challenge test, then deviations from testing at maximum design flux and recovery values can probably be ignored.

4. Challenge testing using a poor-quality feed water matrix: Although it is recommended that challenge testing be conducted using clean make-up water for the test solution and unfouled membranes to provide a conservative estimate of removal efficiency, prior studies conducted with waters with a greater fouling potential (e.g., untreated surface waters) can potentially be considered to meet the LT2ESWTR challenge test requirement, since the removal of *Cryptosporidium*-sized particles is generally unaffected by the presence of foulants for membrane filtration processes as defined under the LT2ESWTR. However, recent research suggests that integrity defects on the order of 200 μm can be obscured by foulants in some cases, improving pathogen rejection (29, 39). Thus, prior studies conducted with poorer quality water should be considered for grandfathering on a case-by-case basis.

4.14. Summary of the US EPA Required Challenge Testing

The *Cryptosporidium* log removal that a membrane product is capable of achieving is determined via challenge testing. Thus, the objective of challenge testing is to demonstrate *Cryptosporidium* removal efficiency. Challenge testing is intended to be a one-time, product-specific test to establish the maximum *Cryptosporidium* log removal credit that the product is eligible to receive as applied for LT2ESWTR compliance (subject to state approval); however, the demonstrated sensitivity of the site- and system-specific direct integrity test utilized may limit the maximum log removal credit that can be awarded. Retesting for a specific membrane product may be required if the manufacturer makes significant changes to the product. Guidance to assist membrane manufacturers and state regulators with assessing what types of changes may require retesting is provided in the MFGM (3). A general overview of challenge testing under the LT2ESWTR is provided in Table 4.8.

Challenge testing involves: (a) seeding the feed water with *Cryptosporidium* or an acceptable surrogate (i.e., a "challenge particulate"), (b) determining the maximum feed concentration using Eq. (1), and (c) measuring the log reduction in the concentration of the challenge particulate between the feed and filtrate, as shown in Eq. (2).

The rule requirements associated with challenge testing are summarized in Table 4.9. A detailed explanation of these requirements, along with guidance for compliance, is provided in the MFGM (3).

Table 4.8
General overview of challenge testing (Source: US EPA)

Description	One-time, product-specific test event designed to demonstrate *Cryptosporidium* removal ability
Purpose	*Demonstrate Cryptosporidium* removal efficiency of an integral membrane product and establish the maximum removal credit that product is eligible to receive
Applicability	Membrane product
Frequency	Once

5. DIRECT INTEGRITY TESTING

5.1. Core Requirements of Direct Integrity Testing

After the *Cryptosporidium* removal capability of an integral membrane product is demonstrated via challenge testing, the LT2ESWTR requires that the removal efficiency of a membrane filtration system be verified on an ongoing basis during operation. This verification is accomplished by the use of direct integrity testing. Thus, the objective of direct integrity testing is to verify that the membrane has no integrity breaches (i.e., leaks) of a magnitude that would compromise the ability of the membrane to achieve the *Cryptosporidium* removal credit awarded by the state. It is important to note that direct integrity testing is not necessarily intended to validate the *Cryptosporidium* log removal demonstrated by challenge testing, but rather the log removal credit that has been awarded to the membrane filtration system by the state, even if the sensitivity of the direct integrity test allows for the validation of a greater LRV. A general overview of direct integrity testing under the LT2ESWTR is provided in Table 4.10.

The LT2ESWTR does not specify the use of a particular type of direct integrity test but rather allows for the utilization of any type of test meeting the requirements for test resolution, sensitivity, and frequency. The MFGM provides specific guidance for the use of both pressure-based tests (e.g., pressure or vacuum decay test) and marker-based tests (both particulate markers for MF/UF systems and molecular markers for NF/RO systems), the two types of direct integrity tests that are currently in most common use (3).

However, the rule does not preclude the use of other types of direct integrity tests that may be developed in the future, provided the basic requirements for test resolution, sensitivity, and frequency can be satisfied. It should also be noted that by definition a direct integrity test must directly test the membrane barrier by some physical means. For example, pressure-based tests apply pressurized air directly to the membrane to detect integrity breaches, and marker-based tests involve spiking the feed water to the membrane with a known concentration of a marker to directly challenge the membrane and demonstrate the log removal ability of the process.

5.2. Resolution and Sensitivity

In the context of membrane treatment technology, resolution and sensitivity are new terms introduced with the LT2ESWTR and defined as follows: (a) resolution: the size of the smallest integrity breach that contributes to a response from a direct integrity test and (b) sensitivity: the maximum LRV that can be reliably verified by the direct integrity test associated with a given membrane filtration system.

It is noted that resolution is expressed as a size of integrity breach. Because the LT2ESWTR is specifically concerned with *Cryptosporidium*, the rule requires that a direct integrity test must have a resolution of 3 μm, the lower bound of the *Cryptosporidium* size range. For marker-based tests, the resolution requirement dictates that the surrogate (i.e., the "marker") used must have an effective size of 3 μm or smaller to demonstrate *Cryptosporidium* removal ability. In order to meet the resolution requirement with a pressure-based test, the net pressure applied must be sufficient to overcome the capillary forces in a 3-μm breach,

Table 4.9
Summary of requirements of challenge testing (Source: US EPA)

Topic	*Requirement synopsis*
Scale of testing	• Testing must be conducted on a full-scale membrane module or small-scale module that is identical in material and similar in construction
Challenge particulates	• Testing must be conducted using *Cryptosporidium* oocysts or a suitable surrogate that is removed no more efficiently than *Cryptosporidium* • Challenge particulate concentration must be measured using a method capable of discrete quantification; gross measurements may not be used
Maximum feed concentration	• Maximum feed concentration = $(3.16\ 10^6) \times$ filtrate detection limit
Test operating conditions	• Testing must be conducted under representative hydraulic conditions at the maximum design flux and maximum design system recovery specified by the membrane module manufacturer
Removal efficiency equation	• LRV = $\log(C_f) - \log(C_p)$
Calculating removal efficiency	• Calculate a single LRV value for each module tested • Overall membrane *product* removal efficiency: – For sample size <20 modules: product LRV = lowest value in sample set – For sample size ≥20 modules: product LRV = 10th percentile value
Verifying removal efficiency for untested modules	• Apply a nondestructive performance test (NDPT) to all modules subjected to challenge testing process • Establish a quality control release value (QCRV) from NDPT results that is directly related to the LRV demonstrated during challenge testing • Apply identical NDPT to all modules of that product • Modules not meeting the QCRV are not eligible for the *Cryptosporidium* removal credit demonstrated during challenge testing
Module modifications	• Additional challenge testing must be conducted for a membrane product that is modified in manner that could affect the established removal efficiency or the applicability of the NDPT, and a new QCRV must be determined
Reporting	• Systems must report the results of challenge testing associated with the membrane filtration system to be used for rule compliance to the state
Grandfathered data	• All data submitted to the state for grandfathering consideration must originate from studies that can be demonstrated to have been conducted in a manner consistent with the challenge testing requirements of the rule

(*Continued*)

Table 4.9
(Continued)

Topic	*Requirement synopsis*
	• The maximum credit for which the process is eligible cannot exceed the removal efficiency demonstrated by the grandfathered data

Table 4.10
General overview of direct integrity testing (Source: US EPA)

Description	Physical testing applied directly to the pathogen barrier associated with a membrane unit (i.e., a rack, a skid, etc.) to identify and isolate integrity breaches
Purpose	*Verify* that the membrane pathogen barrier has no integrity breaches that would compromise the ability to achieve the *Cryptosporidium* removal credit awarded by the State on an ongoing basis during operation
Applicability	Membrane units in a site-specific membrane filtration system
Frequency	Once per day

thus ensuring that any breach large enough to pass *Cryptosporidium* oocysts would also pass air during the direct integrity test. Guidance for determining the resolution of both pressure- and marker-based tests is provided in the MFGM (3).

The sensitivity of a direct integrity test is expressed in terms of a LRV, which must be equal to or greater than the *Cryptosporidium* removal credit awarded to the system to achieve compliance with the LT2ESWTR. If the direct integrity test used is not sensitive enough to verify *Cryptosporidium* removal on the order of that demonstrated in challenge testing, the sensitivity dictates the maximum removal credit for which the process is eligible. Accordingly, the results of a direct integrity test must be correlated to a corresponding LRV. This correlation is straightforward for marker-based tests, which can use water quality monitoring instrumentation to quantify the concentration of the marker in both the spiked feed and the filtrate under fully integral conditions. The rule-specified expression for calculating the sensitivity of marker-based direct integrity tests is shown as Eq. (9).

Sensitivity of marker-based tests

$$\mathrm{LRV_{DIT}} = \log(C_\mathrm{f}) - \log(C_\mathrm{p}), \tag{9}$$

where $\mathrm{LRV_{DIT}}$ is the direct integrity test sensitivity in terms of LRV, C_f is the feed concentration, C_p is the filtrate concentration.

The correlation between the results of pressure-based tests and sensitivity (expressed as a LRV) is more complex, given that these types of tests typically yield results in terms of airflow (e.g., mL/min) or rate of pressure change per unit time (e.g., psi/min). The rule-specified expression for calculating the sensitivity of pressure-based direct integrity tests is shown as Eq. (10).

Sensitivity of pressure-based tests

$$LRV_{\text{DIT}} = \log\left(\frac{Q_{\text{p}}}{\text{VCF} \bullet Q_{\text{breach}}}\right), \tag{10}$$

where LRV_{DIT} is the direct integrity test sensitivity in terms of LRV, Q_{p} is the membrane unit design capacity filtrate flow, Q_{breach} is the flow from the breach associated with the smallest integrity test response that can be reliably measured, VCF is the volumetric concentration factor.

If the pressure-based test used does not yield results in terms of the flow of water through an integrity breach, these results must be converted for Eq. (10) to be utilized. Guidance for converting the rate of pressure change or the flow of air through an integrity breach to an equivalent flow of water is provided by the MFGM (3). Additional guidance for determining the volumetric concentration factor (VCF) for a site-specific membrane filtration system is also provided by the MFGM (3). This parameter accounts for the degree to which some systems concentrate particulate matter in the feed water just above the membrane surface.

5.3. *Summary of the US EPA Required Direct Integrity Testing*

A summary of the rule requirements associated with direct integrity testing is provided in Table 4.11.

6. CONTINUOUS INDIRECT INTEGRITY MONITORING

6.1. *Core Requirements of Continuous Indirect Integrity Monitoring*

Indirect methods do not assess the integrity of the membrane barrier directly but instead utilize water-quality parameters as a surrogate to infer information about membrane integrity based on the levels of the monitored parameters relative to the known baseline in a fully integral system. Although indirect integrity monitoring is generally not as sensitive for detecting integrity breaches as the various direct methods, the indirect methods do have the advantage of being able to be applied to continuously monitor membrane filtrate quality during production, thus providing some means of assessing integrity between direct integrity test applications. Consequently, the objective of continuous indirect integrity monitoring is to monitor a membrane filtrate system for significant integrity problems between direct integrity test applications. Note that the LT2ESWTR does allow the requirement for continuous indirect integrity monitoring to be waived if a continuous method of direct testing that meets the resolution and sensitivity requirements of the rule is used. A general overview of continuous indirect integrity monitoring under the LT2ESWTR is provided in Table 4.12.

The LT2ESWTR requires filtrate turbidity monitoring (for each membrane unit) as the default method for continuous indirect integrity monitoring. However, alternative methods such as particle counting, particle monitoring, conductivity monitoring (for NF/RO systems), or others may also be approved at the discretion of the State. Independent of the method used, "continuous" monitoring is defined as one reading at least every 15 min. The rule specifies a control limit of 0.15 NTU for turbidity monitoring, such that if the filtrate turbidity associated

Table 4.11
Summary of requirements of direct integrity testing (Source: US EPA)

Topic	*Requirement synopsis*
Scale of testing	• Testing must be conducted on each membrane unit (i.e., rack, skid, etc.) in service
Resolution	• The test method used must have a resolution of 3 μm or less
Sensitivity	• The test method used must have sensitivity sufficient to verify the ability of the membrane filtration system to remove *Cryptosporidium* at a level commensurate with the credit awarded by the state • Formulae for sensitivity calculation: – For pressure-based tests: $LRV_{DIT} = \log[Q_p/(VCF(Q_{breach})]$ – For marker-based tests: $LRV_{DIT} = \log(C_f) - \log(C_p)$
Control limit	• A control limit must be established within the sensitivity limits of the direct integrity test that is indicative of an integral membrane unit capable of achieving the log removal credit awarded by the state • If the direct integrity test results exceed the control limit for any membrane unit, that unit must be removed from service • Any unit taken out of service for exceeding a direct integrity test control limit cannot be returned to service until repairs are confirmed by subsequent direct integrity test results that are within the control limit
Frequency	• Direct integrity testing must be conducted on each membrane unit at a frequency of at least once per day that the unit is in operation • States may approve less frequent testing based on demonstrated process reliability, the use of multiple barriers effective for *Cryptosporidium*, or reliable process safeties
Reporting	• The sensitivity, resolution, and frequency of the direct integrity test proposed for use with the full-scale facility must be reported to the state • Any direct integrity test results exceeding the control limit, as well as the corrective action taken in response, must be reported to the state within 10 days of the end of the monthly monitoring cycle • All direct integrity test results must be retained for a minimum of 3 years

with any membrane unit exceeds 0.15 NTU for a period greater than 15 min (i.e., two consecutive 15-min readings higher than 0.15 NTU), that unit must immediately undergo direct integrity testing. Although control limits for alternative methods are determined at the discretion of the state, two consecutive 15-min readings exceeding the state-approved control limit for any alternate method would likewise trigger immediate direct integrity testing for the associated membrane unit.

6.2. Summary of the US EPA Required Continuous Indirect Integrity Monitoring

A summary of the rule requirements associated with direct integrity testing is provided in Table 4.13.

Table 4.12
General overview of continuous indirect integrity monitoring (Source: US EPA)

Description	Monitoring some aspect of filtrate water quality that is indicative of the removal of particulate matter
Purpose	*Monitor* a membrane filtration system for significant integrity problems between direct integrity test applications
Applicability	Membrane units in a site-specific membrane filtration system
Frequency	Continuous

7. DESIGN EXAMPLE: CHALLENGE TEST SOLUTION DESIGN SCENARIO

Design a challenge test solution using the following assumptions and parameters:

1. The target LRV for the challenge test is 4 log.
2. The membrane module has an area of 100 m^2.
3. The maximum flux for the module 85 Lmh, where Lmh = L/(m^2 – h).
4. The module operates in deposition mode.
5. A test duration of 30 min is required to conduct the required sampling.
6. A system hold-up volume of 200 L.
7. The filtrate sample volume used during the challenge test is 500 mm.
8. The detection limit for the filtrate sampling technique is 1 particle per 500 mL.
9. Challenge particulate seeding is conducted via continuous in-line injection.

Solution:

Step 1: Determine the required test solution volume.

$$V_{\text{test}} = \left(\frac{Q_{\text{p}} T_{\text{min}}}{R} + V_{\text{hold}} + V_{\text{eq}}\right)\text{SF}, \tag{3}$$

where T_{min} = 30 min (from given information), R = 100% (standard for deposition mode hydraulic configuration), V_{hold} = 200 L (from given information), V_{eq} = ? (to be determined), SF = ? (to be determined), Q_{p} = ? (to be determined).

Assume that the equilibrium volume is equal to three times the holdup volume,

$$V_{\text{eq}} = 3V_{\text{hold}},$$

$$V_{\text{eq}} = 3(200\text{L}),$$

$$V_{\text{eq}} = 600\text{L}.$$

A suitable safety factor is approximately in the range of 1.1–1.5. Since no other information is given in this example, a value for the safety factor is arbitrarily assumed.

$$\text{SF} = 1.2$$

The filtrate flow, Q_{p}, can be calculated simply by multiplying the given maximum flux and the membrane area (and converting to convenient units), as shown below:

Table 4.13
Summary of requirements of continuous indirect integrity monitoring (Source: US EPA)

Topic	*Requirement synopsis*
Scale of testing	Monitoring must be conducted separately on each membrane unit (i.e., rack, skid, etc.) in service
Monitoring method	Continuous indirect integrity monitoring must be conducted using turbidity monitoring unless the state approves an alternative method
Frequency	Continuous indirect integrity monitoring must be conducted at a frequency of at least one reading every 15 min
Control limit	If the continuous indirect integrity monitoring results exceed the specified control limit for any membrane unit for a period greater than 15 min (i.e., two consecutive readings at 15-min intervals), direct integrity testing must be immediately conducted on that unit The control limit for turbidity monitoring is 0.15 NTU Control limits for state-approved alternative methods must be established by the state
Reporting	Any continuous indirect integrity monitoring results triggering direct integrity testing, as well as any corrective action taken in response, must be reported to the state within 10 days of the end of the monthly monitoring cycle All continuous indirect integrity monitoring results must be retained for a minimum of 3 years

$$Q_\mathrm{p} = \frac{(85\mathrm{Lmh})100\mathrm{m}^2}{60\,\mathrm{min}/\mathrm{h}}$$

$$Q_\mathrm{p} = 142\mathrm{L/min}.$$

Therefore, the required test solution volume can be calculated as follows:

$$V_\mathrm{test} = (142\mathrm{L/min} \times 30\mathrm{min} + 200\mathrm{L} + 600\mathrm{L}) \times 1.2,$$

$$V_\mathrm{test} = 6,070\mathrm{L}.$$

Step 2: Determine feed challenge particulate concentration within the bounds established by Eqs. 4 and 5.

Minimum feed concentration:

$$C_{\mathrm{f-min}} = 10^{\mathrm{LRV_t}}\mathrm{DL},\ \ \mathrm{From}\ (5) \qquad (5)$$

where $\mathrm{LRV_t}$ = 4 log (from given information), DL = 1 particles per 500 mL (from given information), DL = 2 particles per L

$$C_{\mathrm{f-min}} = 10^4 \times 2\ \mathrm{particles/L},$$

$$C_{\mathrm{f-min}} = 2 \times 10^4\ \mathrm{particles/L}.$$

Maximum feed concentration:

$$C_{\mathrm{f-max}} = (3.16 \times 10^6)\ \mathrm{DL},\ \text{From (4)}$$

$$C_{\mathrm{f-max}} = \left(3.16 \times 10^6\right) \times 2\ \text{particles/L},$$

$$C_{\mathrm{f-max}} = 6.32 \times 10^4\ \text{particles/L}. \tag{4}$$

Select a feed challenge particle concentration between the minimum (2 × 10^4 particles/L) and the maximum (6.32 × 10^6 particles/L). Arbitrarily:

$$C_{\mathrm{test}} = 5 \times 10^4 \text{particles/L}.$$

Step 3: Determine the total challenge particulate population required for the test solution.

$$\mathrm{TCCP} = C_{\mathrm{test}} V_{\mathrm{test}},\ \text{From (6)} \tag{6}$$

where C_{test} = 5 × 10^4 particles/L (as calculated in step 2 above), V_{test} = 6,070 L (as calculated in step 1 above)

$$\mathrm{TCPP} = \left(5 \times 10^4 \text{particles/L}\right) 6,070\ \mathrm{L},$$

$$\mathrm{TCCP} = 3.04 \times 10^8 \text{particles}.$$

Step 4: Determine the challenge particulate stock solution concentration using Eq. (7).

$$\mathrm{C_{ss}} = \mathrm{TCPP}/V_{\mathrm{ss}},\ \text{From (7)} \tag{7}$$

where TCPP = 3.04 × 10^8 particles (as calculated in step 3 above), V_{ss} = ? (to be determined).

Select a stock solution volume that is between 0.5 and 2% of the total test solution volume of 6,07 L (as determined in step 1 above). Thus:

$$(0.005 \times 6,070\mathrm{L}) < \mathrm{V_{ss}} < (0.02 \times 6,070\mathrm{L})$$

$$(30.3\mathrm{L}) < \mathrm{V_{ss}} < (121\mathrm{L})$$

$\mathrm{V_{ss}}$ = 100 L arbitrary selection

Substituting values:

$$\mathrm{C_{ss}} = \mathrm{TCPP}/V_{\mathrm{ss}}, \tag{7}$$

$$C_{\mathrm{ss}} = 3.04 \times 10^8 \text{particles/100L},$$

$$C_{\mathrm{ss}} = 3.04 \times 10^6 \text{particles/L}.$$

Step 5: Determine the challenge particulate stock solution delivery rate for continuous, in-line injection.

$$\mathrm{SSDR} = C_{\mathrm{test}} Q_{\mathrm{p}} / C_{\mathrm{ss}} R, \tag{8}$$

where $C_{\mathrm{test}} = 5 \times 10^4$ particles/L (as calculated in step 2 above), Q_{p} = 142 L/min (as calculated in step 2 above), $C_{\mathrm{ss}} = 3.04 \times 10^6$ particles/L (as calculated in step 4 above), $R = 100\%$ (standard for deposition mode hydraulic).

Substituting values:

$$SSDR = \frac{5 \times 10^4 \text{particles/L}(142\text{L/min})}{3.04 \times 10^6 \text{particles/L}(1)},$$

$$SSDR = 2.34\text{L/min}.$$

8. GUIDELINES FOR COMPARING MEMBRANE FILTRATION WITH OTHER WATER AND WASTEWATER TREATMENT PROCESSES FOR GIARDIA CYSTS, CRYPTOSPORIDIUM OOCYSTS AND VIRUS REMOVAL

Giardiasis is a diarrheal disease caused by infection of the small intestine by the parasite *Giardia*. Infections are transmitted by contamination of the environment with resistant cysts excreted by infected hosts. Parasites have been detected in mammals, birds, and amphibians, but only one mammalian species has been associated with human disease. Many infections show no symptoms, while others can cause mild-to-moderate intermittent diarrhea which usually resolves itself. Infections may be prolonged in infants, children, and people with lowered immune systems. Effective antibiotics are available to treat this infection. Infections are usually spread from person to person by hand-to-mouth transfer of cysts, although studies overseas have occasionally implicated contaminated food or water. (40)

Cryptosporidiosis is a diarrheal disease caused by the parasite *Cryptosporidium*. All infections are transmitted by contamination of the environment with resistant oocysts excreted by infected hosts. The parasite infects a wide range of animals and has been associated with watery diarrhea in mammals, diarrhea and respiratory signs in birds, and gastritis in reptiles and possibly in fish. Many infections show no symptoms. At present, there is no effective treatment for cryptosporidiosis, and most infections resolve themselves. However, prolonged chronic infections may develop in high-risk patient groups, especially babies and people with lowered immune systems. Different parasite species occur in different animal classes, and only one mammalian species has been associated with human disease. Infections have been shown to be transmitted from person to person, from animal to person, and via contaminated food and water.

Cyst means the specialized cell produced either in response to environmental stress or as a natural part of the life cycle of a protozoan. *Giardia* cysts are round to oval (8 to 20 μm long by 5–15 μm wide).

Oocyst means the encrusted zoospore that results from completion of the protozoan life cycle. Cryptosporidium oocysts are ovoid to spherical (3–7 μm in diameter). (Oo)cyst means *Giardia* cyst or *Cryptosporidium* oocyst in this book.

Microfiltration (membrane filters of absolute pore size 1 μm or smaller) removes cysts and oocysts by size exclusion. Membrane filters act as an absolute barrier provided membrane integrity is maintained. The Queensland Government, Australia, has provided the following guide to the removal/inactivation of *Giardia* and *Cryptosporidium* by various physicochemical water treatment processes (40):

1. 2.5 log or 99.7% removal for conventional coagulation,clarification, and rapid filtration system.

2. 2 log or 99% removal for direct or contact filtration involving coagulation and rapid filtration.
3. 2 log or 99% removal for slow sand filtration.
4. 2 log or 99% removal for diatomaceous earth filtration.
5. greater than 2 log removal or greater than 99% removal for dissolved air flotation.
6. 6 log or 99.9999% removal for microfiltration.
7. 2 log or 99% removal for ozonation at neutral pH, 25°C and 5–10 CT value.
8. 90% removal for chlorine dioxide oxidation at neutral pH, 25°C and 78 CT value.
9. 90% removal for chlorination at neutral pH, 25°C and 7,200 CT value (40).

CT means the product of the disinfectant residual concentration (prior to the first user), C, in the water in mg/L, and the contact time, T, in minutes, that the water is in contact with the disinfectant. Equation (9) shows how CT value is calculated.

$$\mathrm{CT} = (C)(T). \tag{9}$$

Inactivation means the reproductive sterilization or the destruction of organisms through disinfection. Removal means the physical removal of organisms usually through filtration.

In the USA, the original Surface Water Treatment Rule (SWTR) that went into effect in 1993 required all public water supply systems using surface water or GWUDI (ground water under direct influence of surface water) sources to provide a 3 log or 99.9% –inactivation and a 4 log enteric virus removal/inactivation. The US EPA has recently promulgated the Long Term 2 Enhanced Surface Water Treatment Rule (LT2ESWTR), which requires all public water systems with surface water sources or GWYDI sources to provide treatment (filtration and/or disinfection) to adequately remove and/or inactivate various pathogenic organisms depending on the source water quality. Specifically, additional treatment may be required consisting of a 99.9% through 99.9995% removal/inactivation of *Cryptosporidium.* This is also defined as a 3 log through a 5.5 log removal/inactivation of *Cryptosporidium.* The additional level of treatment required under LT2ESWTR is dependent on the amount of *Cryptosporidium* in the source water, and those requirements are detailed in the LT2ESWTR bin classification table (Table 4.1). The US EPA has provided what is called the microbial tool boxes (Table 4.14) outlining what additional treatment is acceptable for meeting the LT2ESWTR *Cryptosporidium* treatment requirements. Specifically, Table 4.14 shows the applicable log removal–inactivation of *Cryptosporidium* credits for various common types of water treatment. Some of these filtration credits may generally apply to *Giardia lamblia* also. It is noted from Table 4.14 that the log credit equivalent to removal efficiency must be demonstrated in challenge test for membrane filtration if supported by direct integrity testing. Tables 4.15 and 4.16 present additional pathogenic microorganisms removal/inactivation data of UV, chlorine dioxide, and ozone treatments.

Under the LT2ESWTR, in addition to the *Cryptosporidium* treatment requirement, viruses are also required to be removed and/or inactivated to a level of 99.99% or a 4 log level as they were under the original SWTR. Table 4.17 summarizes the US EPA data on virus inactivation using chemical disinfectants, free chlorine, chloramines, chlorine dioxide, and ozone. Table 4.18 indicates the baffling factors which should be applied to any chemical treatment, such as chlorine, chlorine dioxide, chloramine, or ozone when CT value is involved in determining the microbial inactivation.

Table 4.14
Microbial toolbox summary: options, treatment credits and criteria (Source: LT2ESWTR; 71FR777)

Toolbox option	*Cryptosporidium treatment credit with design and implementation criteria*
Source protection and management toolbox options	
(1) Watershed control program	0.5-log credit for state-approved program comprising required elements, annual program status report to state, and regular watershed survey. Unfiltered systems are not eligible for credit.
(2) Alternative source/intake management	No prescribed credit. Systems may conduct simultaneous monitoring for treatment bin classification at alternative intake locations or under alternative intake management strategies.
Prefiltration toolbox options	
(3) Presedimentation basin with coagulation	0.5-log credit during any month that presedimentation basins achieve a monthly mean reduction of 0.5-log or greater in turbidity or alternative state-approved performance criteria. To be eligible, basins must be operated continuously with coagulant addition and all plant flow must pass through basins.
(4) Two-stage lime softening	0.5-log credit for two-stage softening where chemical addition and hardness precipitation occur in both stages. All plant flow must pass through both stages. Single-stage softening is credited as equivalent to conventional treatment.
(5) Bank filtration	0.5-log credit for 25-foot setback; 1.0-log credit for 50-foot setback; aquifer must be unconsolidated sand containing at least 10% fines; average turbidity in wells must be less than 1 NTU. Systems using wells followed by filtration when conducting source water monitoring must sample the well to determine bin classification and are not eligible for additional credit.
Treatment performance toolbox options	
(6) Combined filter performance	0.5-log credit for combined filter effluent turbidity less than or equal to 0.15 NTU in at least 95% of measurements each month Specific criteria are in § 141.718(a).
(7) Individual filter performance	0.5-log credit (in addition to 0.5-log combined filter performance credit) if individual filter effluent turbidity is less than or equal to 0.15 NTU in at least 95% of samples each month in each filter and is never greater than 0.3 NTU in two consecutive measurements in any filter.

(*Continued*)

Table 4.14
(Continued)

Toolbox option	*Cryptosporidium treatment credit with design and implementation criteria*
(8) Demonstration of performance	Credit awarded to unit process or treatment train based on a demonstration to the state with a state-approved protocol.
Additional filtration toolbox options	
(9) Bag or cartridge filters (individual filters)	Up to 2-log credit based on the removal efficiency demonstrated during challenge testing with a 1.0-log factor of safety.
(10) Bag or cartridge filters (in series) *Series may include any pretreatment filter if final filter is 1 μm absolute rated.*	Up to 2.5-log credit based on the removal efficiency demonstrated during challenge testing with a 0.5-log factor of safety.
(11) Membrane filtration	Log credit equivalent to removal efficiency demonstrated in challenge test for device if supported by direct integrity testing.
(12) Second stage filtration	0.5-log credit for second separate granular media filtration stage if treatment train includes coagulation prior to first filter.
(13) Slow sand filters	2.5-log credit as a secondary filtration step; 3.0-log credit as a primary filtration process. No prior chlorination for either option.
Inactivation Toolbox Options	
(14) Chlorine dioxide	Log credit based on measured CT in relation to CT table.
(15) Ozone	Log credit based on measured CT in relation to CT table.
(16) UV	Log credit based on validated UV dose in relation to UV dose table; reactor validation testing required to establish UV dose and associated operating conditions.

9. CASE STUDY OF CHALLENGE TESTING FOR COMPARING MICROFILTRATION AND CONTINUOUSLY BACKWASHED DUAL SAND FILTRATION TECHNOLOGIES

In May of 1997, the New York City Department of Environmental Protection (NYC DEP), USA, issued new rules and regulations governing activities in the city's watershed. Contained in the city's regulations is the requirement that all wastewater treatment plants (WWTPs) discharging in either the Catskill or Delaware or Croton watersheds be upgraded to incorporate advanced wastewater treatment. This upgrade, by virtue of the city's regulations, must include microfiltration (assumed to be the best available technology) or an approved

Table 4.15
***Cryptosporidium, Giardia,* and virus log inactivation values for UV (Source: US EPA)**

Log credit	*Cryptosporidium* dose (mj/cm^2)	*Giardia* dose (mj/cm^2)	Virus dose (mj/cm^2)
0.5	1.6	1.5	39
1.0	2.5	2.1	58
1.5	3.9	3.0	79
2.0	5.8	5.2	100
2.5	8.5	7.7	121
3.0	12	11	143
3.5	15	15	163
4.0	22	22	186

Table 4.16
Chlorine dioxide and ozone disinfection for *Giardia* inactivation (Source: US EPA)

CT Values ($CT_{99.9}$) for 99.9% inactivation of *Giardia lamblia* cysts by chlorine dioxide and ozone[a]

Degrees (°C)	$\leq$1	5	10	15	20	$\geq$25
Chlorine dioxide	63	26	23	19	15	11
Ozone	2.9	1.9	1.4	0.95	0.72	0.48

[a]These CT values achieve greater than a 99.99% inactivation of viruses. CT values between the indicated pH values may be determined by linear interpolation. If no interpolation is used, use the $CT_{99.9}$ value at the lower temperature for determining $CT_{99.9}$ values between indicated temperatures

equivalent. Moreover, the U.S. EPA Region 2 Filtration Avoidance Determination issued in May of 1997 also requires all WWTPs discharging into NYC's Catskill and Delaware watersheds utilize by May 1, 2002 either microfiltration or an approved equivalent.

Accordingly, NYC DEP developed and implemented protocols for a study that compares the pathogen removal capabilities of microfiltration (MF) and continuous-backwash-upflow, dual-sand (CBUDS) filtration by operating a pilot facility of each system, side by side, with the same influent and testing methods, shown in Fig. 4.8. The test units are located at the Stamford Wastewater Treatment Plant (WWTP), immediately downstream of the secondary clarifier. This plant treats mostly domestic wastewater. The microfiltration unit is Memtec America Corporation's Memcor 6M10C unit. The CBUDS test unit consists of two Parkson Corporation Dynasand filters operated in series. Both the CBUDS and the microfilter test units were operated using the manufacturer's standard operating procedures.

During each run of the pilot facility, the mean concentration in the spike solution tank, from which the spiking material was pumped into the influent stream, was approximately 1×10^8 units per 100 L for both *Giardia* cysts and *Cryptosporidium* oocysts. Approximately 42 L (11 gallons) of this solution was delivered to each effluent stream. Hence, the total numbers of *Giardia* cysts and *Cryptosporidium* oocysts delivered to each unit was approximately 4.2×10^7.

Table 4.17
Chemical disinfection for virus inactivation (Source: US EPA)

CT values for inactivation of viruses by free chlorine, pH 6–9						
Log inactivation	1°C	5°C	10°C	15°C	20°C	25°C
2	5.8	4.0	3.0	2.0	1.0	1.0
3	8.7	6.0	4.0	3.0	2.0	1.0
3.5	10.2	7.0	5.0	3.5	2.5	1.5
4	11.6	8.0	6.0	4.0	3.0	2.0
CT values for inactivation of viruses by chloramine, pH 6–9						
Log inactivation	1°C	5°C	10°C	15°C	20°C	25°C
2	1243	857	643	428	321	214
3	2063	1423	1067	712	534	356
4	2883	1988	1491	994	746	497
CT values for inactivation of viruses by chlorine dioxide, pH 6–9						
Log inactivation	1°C	5°C	10°C	15°C	20°C	25°C
2	8.4	5.6	4.2	2.8	2.1	1.4
3	25.6	17.1	12.8	8.6	6.4	4.3
4	50.1	33.4	25.1	16.7	12.5	8.4
CT values for inactivation of viruses by ozone, pH 6–9						
Log inactivation	1°C	5°C	10°C	15°C	20°C	25°C
2	0.9	0.6	0.5	0.3	0.25	0.15
3	1.4	0.9	0.8	0.5	0.40	0.25
4	1.8	1.2	1.0	0.6	0.50	0.30

Table 4.18
Baffling factors for chemical disinfection CT value calculations (Source: US EPA)

Baffling condition	T10/T	Baffling description
Unbaffled (mixed flow)	0.1	None, agitated basin, very low length-to-width ratio, high inlet and outlet flow velocities. Can be approximately achieved in flash mix tank
Poor	0.3	Single or multiple unbaffled inlets and outlets, no intrabasin Baffles
Average	0.5	Baffled inlet or outlet with some intrabasin baffles
Superior	0.7	Perforated inlet baffle, serpentine or perforated intrabasin baffles, outlet weir or perforated launders
Perfect (plug flow)	1.0	Very high length to width ratio (pipeline flow), perforated inlet, outlet, and intrabasin baffles

T10 = Detention time (min) of segment × baffling fact

After system stabilization (based on dye flow tests), three filters were used to monitor the effluent concentration from each system for 80 min. The effluent monitoring filters were all absolute pore size filters (Nuclepore # 641505 filters) designed to capture *all Giardia* cysts and *Cryptosporidium* oocysts.

The first run of the test facility was on Monday, June 2, 1997. Additional runs were conducted every other Monday through October 20, and the final run was on Monday, October 27, 1997. The engineering reports that document the experimental plants, experimental conditions, challenge testing procedures, results, and recommendations can be found from the refernces (37–39).

The following are the conclusions drawn from the case studies (37):

1. Both MF and CBUDSF systems produced finished water quality in compliance with existing regulatory standards for potable water (e.g., consistent effluent turbidities less than 0.1 NTIJ; complete coliform bacterial kill).
2. Both systems reduced influent BOD_5 and TSS to less than 1 mg/L.
3. The MF system achieved an average reduction of total phosphorus of 68%, resulting in an average effluent total phosphorus concentration of 0.099 mg/L. The reduction was achieved with the addition of PAC or PASS polymers as coagulants. Peak concentrations reached 0.41 mg/L. The use of coagulants increases the frequency of system CIP (clean-in-place caustic cleaning) from biweekly to weekly. Combined average and peak concentrations without coagulant addition were 0.412 mg/L and 2.4 mg/L, respectively.
4. The CBUDSF system achieved an average reduction of total phosphorus of 96%, resulting in an average effluent phosphorus concentration of 0.013 mg/L. The coagulant used to achieve this reduction was PASS polymer. Combined peak concentrations reached 0.29 mg/L. Coagulant addition is required for normal operation of this system.
5. There was no reduction in ammonia (typically less than 0.1 mg/L in the pilot influent) nor nitrate through either tertiary filtering system.
6. The extremely low levels of BOD_5, TSS, and particularly phosphorus in the final effluent(s) dictate extremely sensitive sampling and analytical procedures.
7. Twelve protozoan spike challenge events to both filtering systems were completed. The MF system average removal for *Cryptosporidium* oocysts was 6.47 log. The CBIJDSF system average removal for pathogens was 7.00 log.
8. Oxidants (e.g., sodium hypochlorite) will deteriorate the polypropylene hollow fiber membranes and cannot be introduced into the MF unit. The filter modules must be cleansed with a caustic solution as specified in the testing procedure.
9. The chlorination required by the CBIJDSF unit prevents biofilm accumulation and may provide additional wastewater disinfection due to the detention time afforded within each filter unit.
10. 80 amp of electrical service were supplied to the MF unit, whereas 10 amperes of electrical service were supplied to the CBUDSF system.
11. To accurately measure the filtrate particulates from the MF and CBUDSF systems, on-line turbidimeters and particle counters must be employed. Coordination of monitoring equipment results with visual observations of the process train must be performed by operations staff.
12. The net water production during this pilot study for the MF unit was 90%. The net water production for the CBUDSF unit was 95%.
13. System downtime during this pilot study for the MF unit was 16 days out of 159 days or 10%. For the CBUDSF system, downtime was 4 h out of 159 days or 0.1%. During this pilot study, on-site

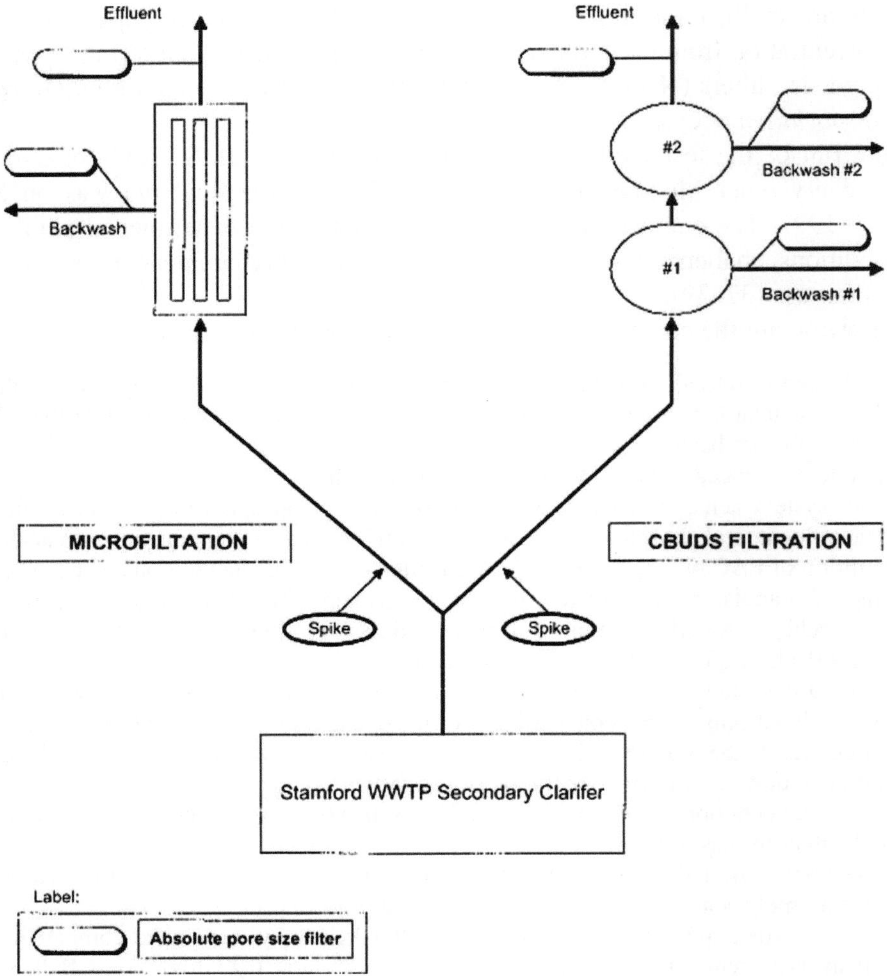

Fig. 4.8. Schematic of challenge testing facility (Source: ref. (37)).

operators accomplished all repairs to the CBUDSF system with readily available tools. For many of the repairs to the MF unit, factory-trained technicians were needed.

14. Results of the increased solids loading test indicate that both systems are capable of processing high solids source water with modified operational protocol (e.g., increased coagulant, chlorine turnover rate for CBIJDSF and increased backwash frequency for MF). The MF unit outperformed the CBUDSF unit during the tests; however, both units produced filtrate in compliance with existing regulatory standards for potable water (e.g., consistent effluent turbidities less than 0.1 NTU) throughout the test.
15. Both systems performed well as tertiary filtration units and in such a way as to confirm that both units are in compliance with the NYC Watershed Rules and Regulations.

Muhammad et al. (40) evaluated various surrogates for *Cryptosporidium* removal. The results are useful for future pilot plant challenge testing.

10. ACRONYMS

AFT	Alternate filtration technology
ALCR	Air-liquid conversion ratio
APHA	American Public Health Administration
ASTM	American Society for Testing and Materials
AWWA	American Water Works Association
AWWARF	American Water Works Association Research Foundation
BOD	Biochemical oxygen demand
CL	Control limit
CSTR	Continuous stirred tank reactor
DBP	Disinfection by-product
DBPR	Disinfectants and disinfection by-products rule
DOC	Dissolved organic carbon
ED	Electrodialysis
EDR	Electrodialysis reversal
ETV	Environmental technology verification
FR	Federal register
GWUDI	Ground water under the direct influence (of surface water)
HAA	Haloacetic acid
HAA5	(sum of five) Haloacetic acids
HFF	Hollow fine fiber
HPC	Heterotrophic plate count
IESWTR	Interim enhanced surface water treatment rule
IMS	Integrated membrane system
ISO	International organization for standardization
IVP	Integrity verification program
LCL	Lower control limit
LED	Light-emitting diode
LRC	Log removal credit
LRV	Log removal value
LT1ESWTR	Long term 1 enhanced surface water treatment rule
LT2ESWTR	Long term 2 enhanced surface water treatment rule
MCF	Membrane cartridge filtration
MCL	Maximum contaminant level
MF	Microfiltration
MWCO	Molecular weight cut-off
NA	Not applicable
NDP	Net driving pressure
NDPT	Nondestructive performance test
NF	Nanofiltration
NSF	National sanitation foundation
NTNCWS	Nontransient noncommunity water system
OEM	Original equipment manufacturer
PAC	Powdered activated carbon

PFR	Plug flow reactor
QA	Quality assurance
QC	Quality control
QCRV	Quality control release value
RAA	Running annual average
RO	Reverse osmosis
SDI	Silt density index
SEM	Scanning electron microscopy
SMP	Standard monitoring program
SSDR	Stock solution delivery rate
SUVA	Specific ultraviolet absorbance
SWTR	Surface water treatment rule
TCF	Temperature correction factor
TCPP	Total challenge particulate population
TDS	Total dissolved solids
TMP	Transmembrane pressure
TOC	Total organic carbon
TSS	Total suspended solids
TTHM	Total trihalomethanes
UCL	Upper control limit
UF	Ultrafiltration
US EPA	United States Environmental Protection Agency
UV	Ultraviolet (light)
VCF	Volumetric concentration factor

11. NOMENCLATURE

C_{f}	Feed concentration used during challenge testing mg/L or number/L
$C_{\mathrm{f-max}}$	Maximum feed concentration, mg/L or number/L
$C_{\mathrm{f-min}}$	Minimum feed concentration, mg/L or number/L
C_{p}	Filtrate concentration observed during challenge testing, mg/L or number/L
C_{ss}	Challenge particulate concentration in the stock solution, mg/L or number/L
C_{test}	Feed concentration of challenge particulate, mg/L or number/L
DL	Detection limit in the filtrate, number or mass/volume
i	The rank of n individual data points
LRV	Log removal value demonstrated during challenge testing, dimensionless
$\mathrm{LRV_t}$	The target log removal value for the challenge test, dimensionless
n	Individual data points
Q_{p}	Filtrate flow, gpm
R	System recovery during test, decimal percent
SF	Safety factor, dimensionless
SSDR	Stock solution delivery rate, gpm (L/min)
TCPP	Total challenge particulate population, number or lb (g)
T_{min}	Challenge test duration, min

V_{eq}	System volume required to attain equilibrium feed concentration, gal (L)
V_{hold}	Hold-up volume of the test system, gal (L)
V_{ss}	Challenge particulate stock solution volume, gal (L)
V_{test}	Challenge test solution volume, gal (L)

REFERENCES

1. Chen JP, Yang L, Ng W-J, Wang LK, Song SL (2006) Membrane filtration. In: Wang LK, Hung YT, Shammas NK (eds.) Advanced physicochemical treatment processes. Humana, Totowa, NJ
2. WEF (2006) Membrane systems for wastewater treatment, Water Environment Federation, McGraw-Hill, New York, NY
3. US EPA (2003) Membrane filtration guidance manual, US Environmental Protection Agency, Cincinnati, OH. Office of Water, EPA 815-D-03-008, June
4. US EPA (1991) Guidance manual for compliance with the filtration and disinfection requirements for public water systems using surface water sources, US Environmental Protection Agency, Cincinnati, OH. EPA 68-01-6989, March
5. US EPA (2001) Low-pressure membrane filtration for pathogen removal: application, implementation, and regulatory issues, US Environmental Protection Agency, Cincinnati, OH. EPA 815-C-01-001, April
6. US EPA (2003) UV disinfection guidance manual, Proposal Draft, US Environmental Protection Agency, Cincinnati, OH
7. US EPA (2003) Toolbox guidance manual for the long term 2 enhanced surface water treatment rule, Proposal Draft, US Environmental Protection Agency, Cincinnati, OH
8. Côté P, Sutherland I, Adams N, Cadera J (2003) Validation of membrane integrity methods in a challenge test with *Bacillus subtillis*. AWWA Membrane Technology Conference Proceedings, Atlanta, GA, March 2–5
9. Johnson WT (1997) Predicting log removal performance of membrane systems using in-situ integrity testing. AWWA Annual Conference Proceedings, Atlanta, GA, June 15–19
10. Adham SS, Jacangelo JG, Laine J-M (1995) Low-pressure membranes: assessing integrity. *J Am Water Works Assoc* 87, 3, 62
11. Farahbakhsh K, Smith D (2003) Estimating air diffusion contribution to pressure decay during membrane integrity tests. AWWA Membrane Technology Conference Proceedings, Atlanta, GA, March 2–5
12. Anerjee A, Carlson K, Lozier J (2000) Monitoring membrane integrity using ultra high sensitive laser light. AWWA Water Quality Technology Conference Proceedings, Salt Lake City, UT, November 5–9
13. US EPA (1999) Guidance Manual for Compliance with the Interim Enhanced Surface Water Treatment Rule: Turbidity Provisions, US Environmental Protection Agency, EPA 815-R-99-010, Cincinnati, OH, April
14. American Water Works Association Research Foundation (2000) Fundamentals of drinking water particle counting. McGraw-Hill, New York
15. American Water Works Association Research Foundation (2000) Particle count method development for concentration standards and sample stabilization. McGraw-Hill, New York

16. International Organization for Standardization (1999) Hydraulic fluid power – on-line automatic particle counting systems for liquids – methods of calibration and validation, 11943, Geneva, Switzerland
17. ASTM (2000) Standard Practice for Calibration of a Liquid-Borne Particle Counter Using an Optical System Based Upon Light Extinction, American Society for Testing and Materials, F 658–00a, West Conshohocken, PA
18. Degremont Technologies (2010) *Membranes*, http://www.degremont-technologies.com/dgtech.php?rubrique56
19. Schulze JC (2001) The Texas approach to regulating MF/UF drinking water applications, *AWWA Membrane Technology Conference Proceedings*, San Antonio, TX, March 4–7
20. NSF (2002) ETV Protocol for Equipment Verification Testing for Physical Removal of Microbiological and Particulate Contaminants, National Sanitation Foundation, Ann Arbor, MI
21. Shammas NK (2005) Coagulation and flocculation. In: LK Wang, YT Hung, NK Shammas (eds.) Physicochemical treatment processes, Humana, Totowa, NJ, p 103–140
22. Asano T, Metcalf & Eddy, Burton FL, Leverenz HL, Tsuchihashi R, Tchobanoglous G (2007) Water reuse, issues, technologies and applications, McGraw-Hill Professional, New York, NY, p 1570
23. Neemann J, Freeman S, Cooke T, Delphos P (2010) Filtration membranes for drinking water applications, http://ww.pennnet.com/display_article/322060/141/ARTCL/none/none/1/Filtration-Membranes-for-Drinking-Water-Applications, April
24. American Public Health Association, American Water Works Association and Water Environment Federation (1998) Standard Methods for the Examination of Water and Wastewater, 20th Edition, Baltimore, MD
25. Barbeau B, Boulos L, Desjardins R, Coallier J, Prévost M, Duchesne D (1997) A modified method for the enumeration of aerobic spore-forming bacteria. Can J Microbiol 43, 976–980
26. Adams BH (1959) Methods of study of bacterial viruses. In: Bacteriophages. Interscience, New York, NY, pp 443–522
27. Abbaszadegan M, Hansan MN, Gerba CP, Roessler PF, Wilson BR, Kuennen R and Van Dellen E The disinfection efficacy of a point-of-use water treatment system against bacterial, viral, and protozoan waterborne pathogens, *Water Research*, 31, 3, 574–582 (1997)
28. Li SY, Goodrich JA, Owens JH, Willeke GE, Schaefer FW III and Clark RM Reliability of non-hazardous surrogates for determining *Cryptosporidium* removal in bag filters, *Journal Am. Water Works Assoc.*, 89, 5, 90–99 (1997)
29. Lozier J, Colvin C, Kim J-H, Kitis M, Mariñas B, Mi B (2003) Microbial Removal and Integrity Monitoring of high-pressure membranes. American Water Works Association Research Foundation, Denver, CO
30. Childress A and Elimelech M (1996) Effect of solution chemistry on the surface charge of polymeric reverse osmosis and nanofiltration membranes. J Memb Sci 119, 253–268
31. GEA Process Engineering Inc. (2010) Membrane Filtration, http://www.niroinc.com/membrane_filtration.asp
32. AMTA (2008) Technology transfer workshop – membrane applications for all your thirsty needs, American Membrane Technology Association, Stuart, FL, http://www.amtaorg.com/publications.html, March 12–13
33. Pearce G (2007) Introduction to membranes: filtration for water and wastewater treatment. Filtr Separat 44, 2, 24–27
34. Energy Center of Wisconsin (2009) Membrane filtration fact sheet. Madison, WI, http://www.ecw.org/prod/425-1

35. Wang LK, Hung YT, Shammas NK (eds.) (2007) Advanced physicochemical treatment technologies. Humana, Totowa, NJ, p 710
36. Queensland Government (2007) Cryptosporidium and Giardia in Drinking Water. Management and Response Protocol. Queensland Government, Health Department, Australia, February
37. NYCDEP (2000) Village of Stamford/NYCDEP Tertiary Wastewater Treatment Demonstration Project comparing continuously backwashed upflow dual sand filtration and microfiltration technologies. New York City Department of Environmental Protection, NY, May
38. US EPA (1998) Testing equivalence of microfiltration and continuous-backwash-upflow, dual-sand filtration technologies, Report No. RTI/6922/371-04F, May 13
39. Wang LK (2009) Technologies for environmental infrastructural sustainability. National Engineers Week, Marriott, Albany, NY, February 5–6
40. Muhammad N, Sinha R, Krishnan ER, Piao H, Patterson CL, Cotruvo J, Cumberland SL, Nero VP, Delandra C (2008) Evaluating surrogates for Cryptosporidium removal in POU systems. JAWWA, 100 12, p. 98–107

5

Treatment of Industrial Effluents, Municipal Wastes, and Potable Water by Membrane Bioreactors

Lawrence K. Wang and Ravinder Menon

Contents

Abstract Membrane bioreactor (MBR) is a biochemical engineering process involving the use of both (a) a suspended growth bioreactor for biochemical reactions (such as fermentation, bio-oxidation, nitrification, and denitrification) and (b) a membrane separator for subsequent solids–liquid separation. In a chemical engineering fermentation process, the solids may be yeasts and the liquid may be an alcohol. In an environmental engineering process, the solids may be activated sludge and the liquid may be the biologically treated water or wastewater. Practically speaking, the membrane separator replaces clarifier, such as sedimentation or dissolved air flotation (DAF) in a conventional suspended growth biological process system. The membrane module can be either submerged in a suspended growth biological bioreactor, or situated outside the bioreactor. This chapter introduces historical development, engineering applications, various MBR process systems, design considerations, and practical environmental engineering applications, such as treatment of dairy industry wastes, landfill leachate, coffee industry wastes, cosmetics industry wastes, municipal waste, and potable water.

Key Words Membrane bioreactor • suspended growth bioreactor • fermentation • bio-oxidation • nitrification • denitrification • membrane process • activated sludge • design • dairy industry waste • landfill leachate • coffee industry waste • cosmetics industry waste • municipal waste • potable water • MBR

From: *Handbook of Environmental Engineering, Volume 13: Membrane and Desalination Technologies*
Edited by: L.K. Wang et al., DOI: 10.1007/978-1-59745-278-6_5

1. INTRODUCTION

1.1. General Introduction

With increasing pressures worldwide on existing water resources due to increases in human population and activity, reuse and conservation of water resources assume a very high priority. Membrane bioreactor (MBR) is a tool that has the potential to help industries (and municipalities) manage their water resources better. MBR is mainly an innovative wastewater treatment (WWT) technology, based on proven processes of activated sludge biological treatment and membrane separation (1–50), although it has also been applied to potable water treatment. The MBR system has been implemented in several full-scale industrial and municipal applications. The synergistic combination of enhanced biological treatment, and membrane filtration produces a treated effluent quality, which is not merely excellent, but very reliably so. This provides the opportunity to facilities to recycle/reuse part or all of the treated effluent, thereby reducing costs for fresh water and water treatment on the one hand, and reducing sewer surcharge (for pretreatment facilities), on the other. This chapter discusses some basic aspects of design of MBRs and presents some full-scale examples of its various applications.

1.2. Historical Development

The chemical engineering processes involving the use of membranes for phase separation are termed "membrane processes". The phases include solid phase (suspended solids, dissolved solids, etc.), liquid phase (water, ethanol, chloroform, etc.), and gas phase (air, nitrogen, oxygen, etc.). A membrane is a porous filtration medium, which can be cationic, anionic, or nonionic in nature, and acts as a barrier to prevent mass movement of selected phases, but allows passage of remaining phases. The main applications of the membrane processes are processing water and wastewater (WW) streams (1). Recently, membrane processes have been used for purification of gas streams (2) and potable water.

The membrane processes include at least five main subcategories for processing water and wastewater (1).

Microfiltration It is a pressure filtration process for the separation of mainly suspended solids in the particle size range of about 0.08–10 μm, such as chemical flocs, Giardia, algae, Cryptospordium, visible particles, pollen, yeast, TSS, activated sludge, etc. The hydraulic pressure (trans-membrane pressure) applied in Microfiltration (MF) is about 1–2 bars, or 15–20 psig, primarily for overcoming resistance of the "cake" (1 micron = 1 micrometer = 0.00004 in. = 10,000 Å units) (1).

Ultrafiltration It is another pressure filtration process for the separation of macromolecular solids in the particle size range of about 0.01–0.1 μm, such as colloids, turbidity, bacteria, fungi, activated. The primary factor affecting the separation of solids from water relies on the size of macromolecular solids. The hydraulic pressure required by ultrafiltration (UF) for overcoming hydraulic resistance of the polarized macromolecular layer on the membrane surface is about 1–7 bars (1).

1.2.1. Nanofiltration

Nanofiltration (NF) membranes are multiple-layer thin-film composites of polymer consisting of negatively charged chemical groups, and are used for retaining molecular solids (such as sugar) and certain multivalent salts (such as magnesium sulfate), but passing substantial amounts of most monovalent salts (such as sodium chloride), at an operating pressure of about 14 bars or 200 psig. The sizes of molecular solids and multivalent salts to be rejected by NF are normally in the range of 0.0005–0.007 μm (1), such as color, virus, hardness, calcium ion, magnesium ion, iron ion, permanganate ion, phosphate ion, sulfate ion, THM, HAA, VQC, SOOD etc.

1.2.2. Reverse Osmosis

Reverse osmosis (RO) membranes are mainly made of cellulose acetate with the pore sizes of about 5–20 Å units, and are used for rejecting salts (as high as 98%) and organics (as high as 100%), at an operating pressure of about 20–50 bars or 300–750 psig. The sizes of molecular solids and salts (multivalent as well as monovalent) to be rejected by RO are normally in the range of 0.00025–0.003 μm (1), including sodium ion, potassium ion, chloride ion, nitrate ion, organic etc.

Figure 5.1 illustrates the relationships among MF, UF, NF, RO, and conventional sand filtration process for separation of algae, pollen, yeast, bacteria, colloids, virus, organic compounds and dissolved salts. Figure 5.2 shows the separation capabilities of MF, UF, NF, and RO for water and wastewater treatment.

In this chapter, mainly MF and UF are related to MBRs. RO is an effective posttreatment unit to MBR (see Sect. 5.2).

1.3. Physical–Chemical Pretreatment Prior to Membrane Process

Theoretically and technically speaking, the membrane process alone will be feasible for purifying water or wastewater. For membrane process operation, build up of a layer on the surface of the membrane and deposition of foulants within the membrane pore structure due to high solution concentrations are the major mechanisms responsible for membrane flux decline (1). Membrane fouling requires frequent chemical cleanings and, in the worst case, membrane replacement. Accordingly pretreatment prior to membrane process is required.

Accordingly membrane process alone is mainly used for water purification when the quality of an influent raw water is good.

For potable water or industrial water treatment using membrane processes, physical–chemical pretreatment prior to treatment in a membrane reactor will significantly prolong the membrane life and, in turn, reduce the treatment cost (3). Typical physical–chemical pretreatment processes include pH adjustment, chemical coagulation, clarification, sand filtration, and cartridge cloth filtration (3). An ultra-low pressure drop oleophilic filter is also effective (41). Accordingly pretreatment prior to membrane process is required.

Again, a membrane process reactor with physical–chemical pretreatment is technically feasible but not economically feasible for treatment of industrial waste or domestic sewage, due to high organic load in the influent streams.

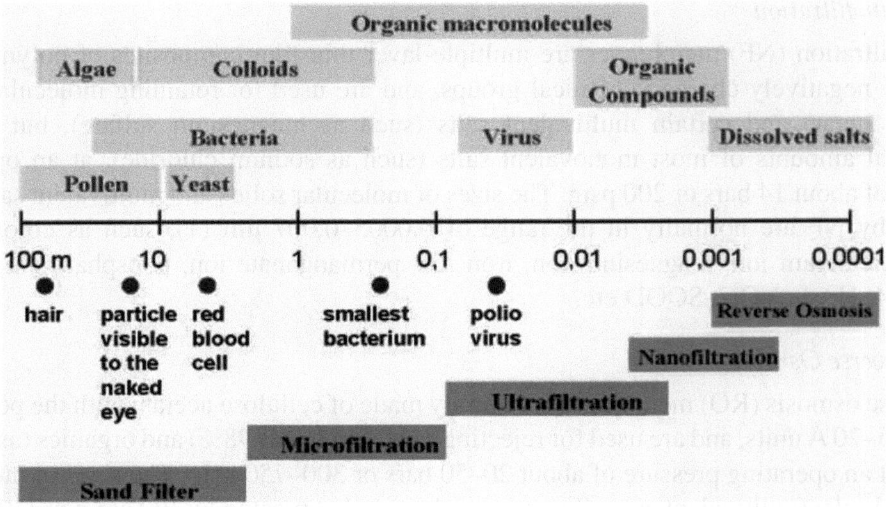

Fig. 5.1. Particle size and separation processes.

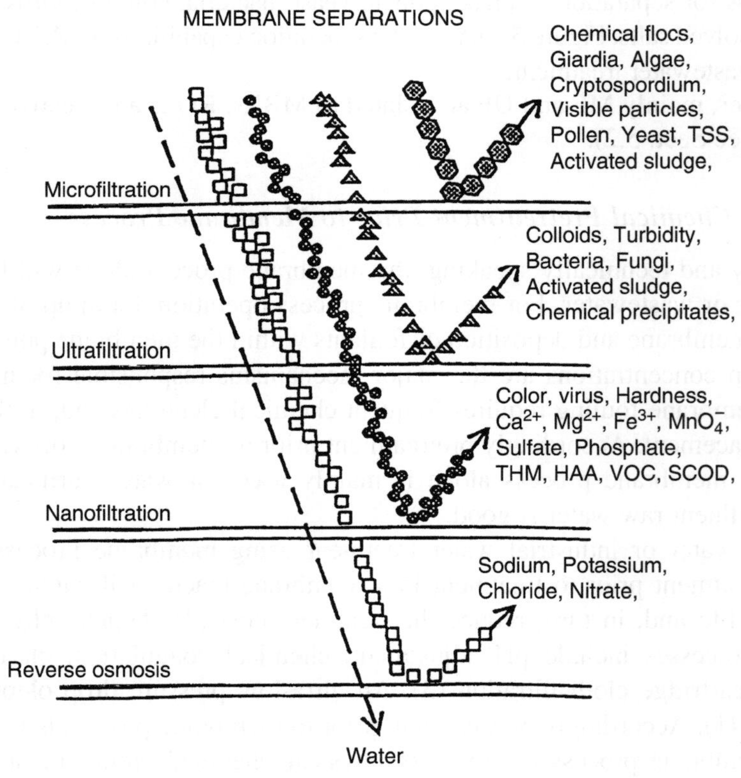

Fig. 5.2. Separation capabilities of MF, UF, NF, and RO for water and wastewater treatment.

1.3.1. Biological Pretreatment Prior to Membrane Process

Traditionally, biological treatment processes, such as activated sludge, tickling filters, lagoons, fluidized-bed reactors, rotating biological contactors, sequencing batch reactors (SBRs), adopt either sedimentation clarification or dissolved air flotation (DAF) clarification for solids–liquid separation and for microorganisms (sludge) return (4–11).

When a biological process is used in conjunction with a membrane reactor (either microfiltration or ultrafiltration), the entire process system becomes an MBR. The MBR (either microfiltration or ultrafiltration) will accomplish the following three tasks:

1. Solids–water separation (clarification)
2. Sludge return (microorganisms return)
3. Tertiary wastewater treatment (WWT), capable of disinfection, nutrients removal, metals removal, and toxic organics removal.

When treating wastewater streams at same flow, biological treatment processes are usually more cost effective than comparable physical–chemical processes.

Biological processes can (a) adopt suspended growth reactors, attached growth reactors, or both; (b) be operated as a continuous process system or an sequencing batch reactor (SBR) process system; and (c) be controlled under aerobic, anoxic, or anaerobic conditions for biochemical treatment of wastewater. The theory and principles of biological treatment processes can be found elsewhere (4–10, 49, 51, 52), and are beyond the scope of this chapter. In general, biological processes can accomplish carbonaceous oxidation (aerobic), nitrification (aerobic), denitrification (anoxic/anaerobic), digestion (aerobic or anaerobic), phosphorus removal (aerobic/anoxic), and methane production (anaerobic).

There are many kinds of membrane bioreactor (MBR) systems, which will be introduced in Sect. 1.3, MBR Research and Engineering Applications.

SBR can also adopt membrane modules, forming an SBR–MBR batch process system for WWT (51).

Only the common MBRs consisting of activated sludge aeration basins and microfiltration or ultrafiltration will be presented and discussed in this chapter in detail.

Although MBRs are mainly used for WWT, they can also be used for potable water treatment, aiming at nitrogen removal (50).

1.4. Physical–Chemical–Biological Pretreatment Prior to Membrane Process

It has been known that many non-biodegradable or toxic pollutants cannot be removed by biological processes. On the contrary, many pollutants cannot be chemically coagulated or adsorbed (10, 12).

For treatment of certain polluted water or wastewater streams, it may be necessary to adopt combined physical–chemical–biological pretreatment prior to a membrane process reactor treatment.

1.5. Membrane Bioreactors Research and Engineering Applications

In this section, various MBRs will be reviewed and discussed, although the remaining sections of this chapter will introduce and discuss only the most common MBRs, which are well-established and practically applied to the environmental engineering field for pollution control.

MBRs consisting of an aerobic or anaerobic reactor with suspended biomass and membranes for liquid–solid separation are now the mainstream environmental engineering processes for water treatment and WWT. MBRs have been used for treatment of municipal and industrial wastewater (13) and for reclamation of water for potential reuse in public water supplies (14).

With respect to treatment efficiency and system stability, MBRs have several advantages over conventional processes:

1. With complete solids–liquid separation by the membrane, high biomass concentrations and relatively short reaction times are possible (13–21).
2. MBRs can produce a clear final effluent regardless of hydraulic retention time (HRT) and without concerns of biomass settleability characteristics (15).
3. Biological nitrogen removal is also possible in an intermittently aerated single-stage MBR (16).

An MBR with powdered activated carbon (PAC) addition was applied for drinking water treatment to remove nitrate, natural organic matter, and pesticides and to disinfect the water (22). Also, the addition of PAC to the activated-sludge process with attached microbial growth on the PAC enhanced membrane permeability. The flux enhancement could be attributed to the development of dense floc particles around the PAC (23, 52).

Brindle and Stephenson (13) studied three generic membrane processes within bioreactors for WWT, solids separation and recycling, bubble-less aeration, and priority organic pollutants extraction. Commercial aerobic and anaerobic MBRs are already in use, producing a high-quality effluent at high organic loading rates. However, bubble-less aeration and extractive MBRs are still in development.

Dollorer and Wilderer (24) compared oxygenation by bubbling and via a silicone rubber, bubble-free membrane system in sequencing batch biofilm reactors (SBBRs). The clay-packed SBBRs achieved 68% dissolved organic compounds removal from hazardous waste with a 12-h cycle. The bubble-free SBBR emitted less biodegradable volatile organics than the bubbled system.

Livingston et al. (25) used an extractive membrane bioreactor (EMBR) to remove a range of toxic organic compounds from the chemical industry, achieving more than 99% removal with a wastewater reactor contact time of fewer than 30 min. The removal efficiency was modeled and a new EMBR configuration was discussed. Data on the effect of biofilms on membrane mass transfer were shown. In an additional work, this group demonstrated that the addition of sodium chloride to the biomedium increases the maintenance energy requirement of the degradative organisms and resulted, in a carbon-limited situation, in reduction of biofilm growth. Organic substrate flux remained high under reduced biofilm growth conditions (26).

Cote et al. (27) discovered that when a submerged membrane was placed in an aeration tank for municipal WWT with an anoxic–aerobic process, total Kjeldahl nitrogen (TKN) removal efficiencies were greater than 69 and 94% at mixed liquor suspended solids (MLSS) concentrations of 15,000 and 25,000 mg/L, respectively. Further studies on aeration strategies to optimize nitrogen removal designs are needed. The application of membranes to biological WWT is limited by membrane fouling and high energy consumption. Back-flushing with permeate or air in a crossflow membrane coupled to a biological reactor has been used to reduce membrane fouling (22, 28, 29). An improvement in flux rates compared to that for operations without back-flushing was reported.

Air contaminated with trichloroethylene (TCE) was passed through microporous hollow fibers in a hollow-fiber MBR, whereas an oxygen-free nutrient solution was recirculated through the shell side of the membrane module. A removal efficiency of 30% was achieved at inlet TCE concentrations of 20 ppmv and a 36-s gas phase residence time (30).

A bioreactor was developed by Clapp et al. (31) using silicone tubing with an attached methanotrophic biofilm to treat TCE-contaminated waste streams. The reactor was developed to overcome the low solubility of methane, competitions between methane and TCE, the lack of NADH regeneration in the presence of TCE only, and the cytotoxic products of TCE metabolism.

Many other techniques such as formation of a dynamic membrane, precoat, or hydrophobic skin layers atop the membrane have been introduced to reduce fouling in crossflow MBRs, but these are still in an early stage of evaluation. Some researchers using crossflow MBRs have reported that the pumping shear stress caused biological floc break-up, leading to severe flux decline in long-term operations caused by the small flocs forming a denser biomass cake layer on the membrane. Additionally, continuous recycling of mixed liquor in a crossflow MBR requires a relatively large amount of energy (32–35).

Yamamoto et al. (36) studied an alternative to a crossflow membrane operation using a submerged membrane with permeate removal by vacuum suction. Power consumption per unit volume of treated water was greatly reduced by eliminating the circulation pump, but the permeate flux was reduced to an impractical low level of less than 2 L/m^2 h. The energy consumption associated with filtration in these new submerged membrane reactors was at a substantially low level of 0.2–0.4 kW h/m^3 compared to the relatively high energy consumption (2–10 kW h/m^3) with circulation loops (27).

Performance of an SBR using a membrane for effluent filtration was investigated by Choo and Stensel (37) in terms of chemical oxygen demand (COD) removal, nitrogen removal, and membrane permeability during long-term continuous operation treating synthetic wastewater. The reactor was operated with six 4-h cycles per day consisting of 0.2, 2.0, and 1.5 h for fill, aeration, and effluent filtration-idle, respectively. Minimal solids wasting occurred for the first 10 months of operation, followed by an 8-day solids retention time (SRT) for the final 1.5 months. Membrane fouling was controlled by backwashing with aeration for 10 min during each cycle. A stable permeate flux of approximately 0.34 (L/m^{2}h)/kPa or 34 (L/m^2 h)/bar was achieved and was independent of MLSS concentrations of 700–10,000 mg/L. The reactor effluent turbidity averaged less than 0.20 NTU and more than 98% COD removal occurred. Nitrogen removal efficiency ranged from 87 to 93% through biological nitrification and denitrification. Most of the nitrate was removed during the mixed and unaerated fill period, but a significant amount of nitrogen was removed by simultaneous nitrification–denitrification (SNDN) during aeration at dissolved oxygen (DO) concentrations less than 3.0 mg/L.

A new membrane separation process developed by Osmotek of Corvallis, Oregon, USA, is moving out of the pilot testing phase and is available for a variety of applications, such as treating wastewater and landfill leachate, according to the US Department of Energy (DOE) which helped develop the technology (38).

The technology called direct osmosis concentration (DOC) is a cold temperature membrane process that separates waste streams in a low-pressure environment. DOC uses salt brine as an osmotic agent to treat wastewater on board US Navy vessels. The technology has also been shown to remove 95% of water from landfill leachate with little maintenance (38).

Anaerobic WWT processes have become increasingly popular for treating industrial effluents, especially those containing high levels of fermentative products. Because of their ability to withhold slow-growing bacteria, anaerobic MBRs have generated increased interest in recent years.

Beaubien et al. (20) used a 1.5-m^3 anaerobic MBR pilot plant to treat condensate from a distillery to evaluate the possibility of recycling and reusing the treated effluent in alcoholic s. Consisting mostly of acetate, propionate, and ethanol with a mean COD of 3,000 mg/L, distillery condensates are particularly suitable for anaerobic treatment. Following a 5-day biological anaerobic MBR start-up period, during which removal efficiency increased from 40 to 80%, satisfactory performance of the anaerobic MBR was obtained. More than 75% of the applied load was removed. The suitability of the treated effluent for reuse in alcoholic fermentations was evaluated by comparing the alcohol concentration obtained using treated and untreated effluents, and water as process-dilution fluid in fermentations. The results clearly show that the untreated effluent significantly inhibits the fermentative organisms, while the treated effluent does not induce a noticeable inhibition of alcoholic fermentation.

2. MBR PROCESS DESCRIPTION

2.1. Membrane Bioreactor with Membrane Module Submerged in the Bioreactor

This type of MBR process uses the same biological WWT as conventional activated sludge (CAS), but provides tertiary treatment with far fewer unit processes. Aeration, secondary clarification, and filtration (without the need for coagulation/flocculation) occur within a single bioreactor (shown in Fig. 5.3), rather than in separate basins.

The MBR process uses hollow-fiber microfiltration or ultrafiltration membranes. The membranes are bundled into "modules" and grouped together in "cassettes". The cassettes are connected by a header to a permeate (effluent) pump and are submerged in the bioreactor. In more recent configurations, the cassettes are submerged in separate tanks, for the ease of cleaning. The permeate pumps create a vacuum that pulls the effluent into the hollow-fiber membranes, but leaves the solids behind in the bioreactor. This eliminates the need for secondary clarification and return sludge pumping (18, 44).

Because activated sludge stays in the bioreactor, the concentration of MLSS is much higher (10,000–12,000 mg/L) than it would be in a Conventional activated sludge (CAS)

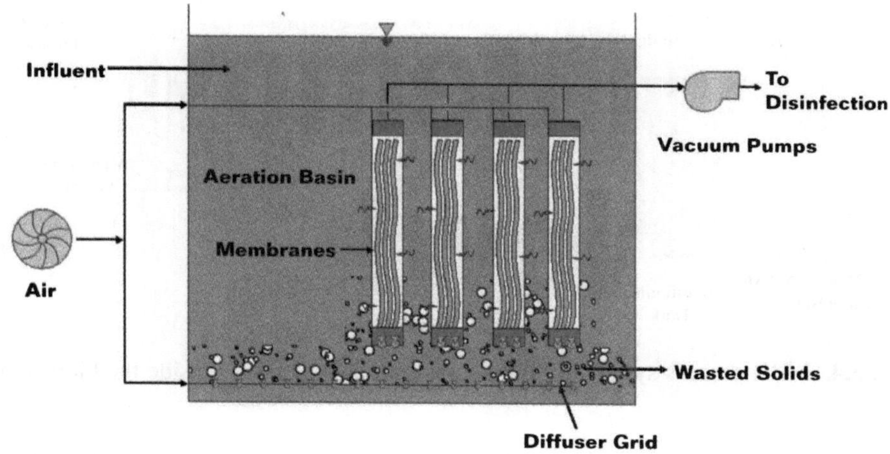

Fig. 5.3. MBR process system with membrane module submerged in the bioreactor.

process. The high MLSS concentrations facilitate treatment within a smaller bioreactor volume.

The hydraulic capacity of the MBR process is limited by the flow of water per unit area of the membrane surface. The average flow rate per unit area, or flux, for membranes that are used for wastewater treatment (WWT) is typically 10–15 gallons per square foot per day (gfd).

This type of MBR process can be built from the ground up, or retrofitted into an existing CAS aeration basin, as shown in Fig. 5.3. In operation, air is supplied through coarse bubble diffusers at the base of the membrane cassettes to agitate and scour the membranes for cleaning and to provide oxygen for biological treatment. At regular intervals, automatic backwash (backpulse) cycles clean and restore permeability to the membranes. The coarse bubble diffusers used for membrane cleaning do not transfer oxygen efficiently, so fine bubble diffusers (or other means of aeration) are added to supply more air for treatment.

2.2. *Membrane Bioreactor with Membrane Module Situated Outside the Bioreactor*

This type of MBR process system is schematically shown in Fig. 5.4. Screened influent enters the bioreactor, where it is oxidized to remove organic pollution, as well as ammonia, if any. The mixed liquor from the bioreactor at an MLSS concentration ranging from 10 to 20 g/L is withdrawn and pumped through a crossflow membrane filtration module. The permeate from the membranes constitutes the treated effluent. The retentate stream representing concentrated biosolids is returned to the bioreactor. Excess biosolids are wasted from the bioreactor or from the return line.

It may be noted that due to the membranes acting as absolute barrier for solids, it is possible to maintain the desired sludge age or SRT accurately. Also, the micro- or ultrafiltration

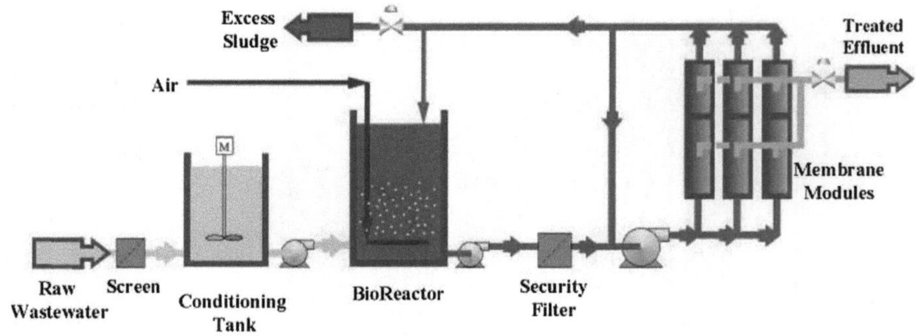

Fig. 5.4. MBR process system with membrane module situated outside the bioreactor.

membranes used for separation are capable of separating suspended and colloidal solids, organic macromolecules, and microorganisms from treated effluent. Figures 5.1 and 5.2 illustrate this point.

The MBR system introduced in this section is based on an external or in-series configuration, where the membrane units follow and are situated outside the bioreactor. This helps to keep the two processes separate, avoiding interferences and enabling individual optimization.

The complete separation of hydraulic and SRTs provides optimum control of biological reactions, and greater reliability and flexibility in use (15, 20, 21). The MBR system typically uses high SRTs in the range of 60–100 days. The high SRT used helps in the development of slow-growing microorganisms, such as nitrifying bacteria, and provides complete biodegradation of difficult-to-treat components such as organic macromolecules, which are retained by the membrane units and kept in the system until biodegradation.

2.3. *MBR System Features*

The two most common types of MBR system are introduced in Sects. 2.1 and 2.2 (see Figs. 5.3 and 5.4). Since both are very similar to each other, only one type (Sect. 2.2; Fig. 5.4) is described in detail in the remaining sections of this chapter. The entire MBR system shown in Fig. 5.4 includes, but is not limited to, screen, conditioning tank, bioreactor, pumps, and pipes. Special system features of the innovative MBR system (Fig. 5.4) include the following:

1. Very high-quality bacteria-free effluent
2. High organic loading (2–4 kg COD/m^3 day or 0.12–0.25 lb/ft^3 day)
3. High MLVSS (10,000–20,000 mg/L)
4. Efficient oxygen transfer
5. Very high sludge ages used (30–100 days)
6. 35–45% less excess biosolids (sludge) production
7. Promotes the growth of slow-growing bacteria
8. Immune to filamentous and other bulking

The working principle of the crossflow membrane filtration is illustrated in Fig. 5.5, based on one membrane module.

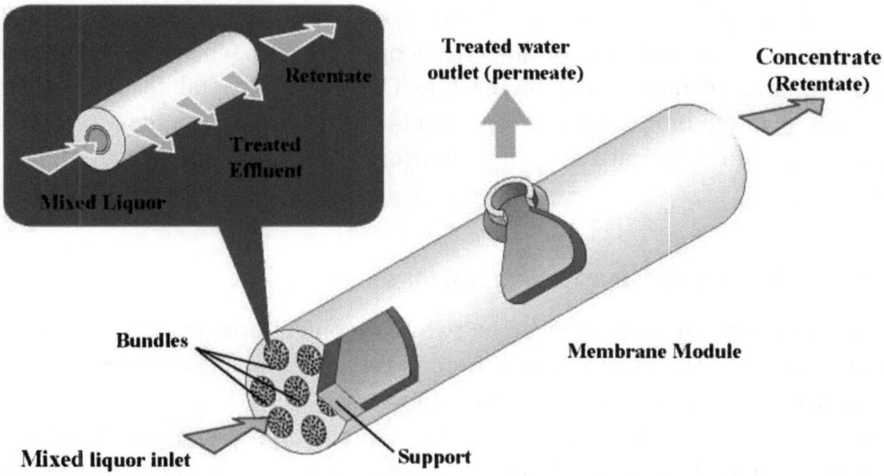

Fig. 5.5. Working principle of crossflow membrane filtration.

In operation, the mixed liquor from the bioreactor (see Fig. 5.4) passes a security filter and then enters the mixed liquor inlet of a membrane module, which consists of many bundles of membrane filters. Through the mixed liquor inlet, the mixed liquor enters the tube-type membrane filters where water–solid separation occurs under high pressure. The liquid portion of the mixed liquor is forced by pressure to pass through the tube-type membrane filter, becoming the treated effluent, while the suspended solids remain, becoming highly concentrated retentate (or concentrate). The suspended solids are mainly microorganisms from the bioreactor having particle sizes larger than that of the membrane filter's pores.

The treated effluent (or treated water), as shown in Figs. 5.4 and 5.5, is discharged to a receiving water or reused.

The retentate from the membrane modules is partially returned to the bioreactor as return activated sludge (RAS), and partially wasted as excess sludge (Fig. 5.4).

Since the direction of mixed liquor flow inside a tube-type membrane filter is perpendicular to the direction of treated effluent passing through the membrane, this flow pattern is called crossflow filtration – this is the first special membrane feature of the membrane filtration operation.

The second special membrane feature, created by ONDEO Degremont, involves the use of ceramic membranes, which have very high corrosion resistance, can be cleaned more efficiently, and are less prone to biofouling.

Availability of various pore sizes in ultrafiltration and microfiltration range (Figs. 5.1 and 5.2) is the third special membrane feature.

The fourth special membrane feature is that the membranes can be cleaned by clean-in-place (CIP) techniques, using crossflow filtration, and reverse backwash operation.

As discussed previously, a complete MBR system (Fig. 5.4), developed by ONDEO Degremont, is based on external or in-series configuration, where the membrane units follow, and are situated outside the bioreactor. This added special feature may keep the biological process in the bioreactor and the liquid–solid separation process (clarification) in the membrane module separate, avoiding interferences and enabling individual process optimization.

2.4. Membrane Module Design Considerations

Membrane processes are characterized by two basic process parameters: (a) flux, which is the rate of transport of solvent or solution through the membrane and (b) rejection, which is the degree of separation of a particular feed component.

There are five major variables that affect the two basic process parameters: (a) driving force in terms of applied trans-membrane pressure and/or electric voltage/current; (b) flow velocity, which affects turbulence and mass transfer coefficient; (c) process water temperature, which has effects on physical properties such as density, viscosity, diffusivity, osmotic pressure, surface tension, and others; (d) feed stream characteristics in terms of particle concentration, particle size, viscosity, molecular weight, molecular configuration, ionic charges, and fouling potential; and (e) membrane module in terms of materials, pore sizes, membrane configuration, membrane ionic charges, and feed compatibility (1).

There are basically six different designs of membrane modules: (a) tubular modules with channel diameters greater than 3 mm; (b) hollow-fiber or capillary modules made of self-supporting tubes, usually 2 mm or less in internal diameters; (c) plate modules; (d) spiral-wound modules; (e) pleated sheet modules; and (f) rotary modules. The latter four module designs use flat sheets of membrane in various configurations (1).

In selecting a particular membrane module and a particular membrane process, the major criteria are (a) feed stream characteristics, which affect the biocompatibility of the membranes; (b) flux requirements, which are controlled by the volumetric rate of a feed stream; (c) rejection requirements, which decide the process objectives and treatment efficiencies; and (d) cost requirements, which are affected by energy consumption, membrane replacement cost, and operating and cleaning costs.

Biocompatibility of the membrane relates to the interaction between the membrane module and the feed stream. Major biocompatibility factors include (a) stability to extremes in temperature, pressure, and pH, especially under cleaning and sanitizing conditions; (b) membrane–solute interactions, which affect the rate of fouling, cleaning, yields, and rejection of individual feed substances; and (c) acceptability of the membrane as a contact material for the final product, which essentially implies using membrane materials that are inert and do not leach out any toxic substances from the membrane into the final product. In this regard, there are new generations of membranes, made of expensive inorganic materials, such as ceramics, stainless steel, carbon–zirconia, etc.

MF membranes are made of a wide range of inorganic materials (such as alumina, zirconia–carbon composites, carbon–carbon composites, ceramics, stainless steel, silica, etc.) and natural and synthetic polymers (such as polypropylene, polycarbonates, polysulfone, polyvinylchloride, PVC copolymer, cellulose esters, cellulose acetate, etc.) (1).

UF membranes are mainly made of polysulfone-type materials (such as polyether sulfone, polyphenyl sulfone, sulfonated polysulfone, etc.), although they are also available in a wide range of organic materials (such as PVC copolymer, cellulose acetate, etc.) and inorganic materials (such as ceramic composites, stainless steel, etc.).

Most NF membranes are multiple-layer thin-film composites of synthetic polymers. The active NF membrane layer usually consists of negatively charged chemical groups. NF membranes are of porous filter media with an average pore diameter of 2 nm. The nominal molecular weight cutoff ranges from 100 to 200. The active NF membrane layer can be made of polyamide, polyvinyl alcohol, sulfonated polysulfone, and sulfonated polyether sulfone. Salt rejection by NF membranes is mainly due to electrostatic interaction between the ions and the NF membrane. Rejection of neutral substances is by size.

Cellulose acetate and derivatives are widely used as the RO membrane, despite their real and perceived limitations. Thin-film composite membranes containing a polyamide separating barrier on a polysulfone or polyethylene supporting layer generally give better performance for RO applications with regard to temperature and pH stability and cleanability, but have almost zero chlorine resistance. In general, these thin-film composite membranes will be the material of choice for RO applications, unless there is a specific fouling problem with these membranes.

There are four types of membrane equipment: tubular membrane modules, hollow-fiber membrane modules, plate membrane modules, and spiral-wound membrane modules. Each design has its own special applications, advantages, and disadvantages (1).

The large-bore tubular membrane modules are suitable for food streams with high concentrations of suspended solids, such as citrus juices and animal waste streams, even though the tubular membrane modules have the lowest packing densities and highest energy consumption among all the modules. The tubular designs with ceramic inorganic membranes are frequently used in the food processing industries.

The hollow-membrane modules have extremely high packing density (surface area to volume ratios) and comparatively low energy consumption, and are suitable for comparatively clean feed streams with low concentrations of suspended solids and macromolecules. Certain macromolecules display non-Newtonian behavior. Their viscosity will increase dramatically above certain concentrations, making pumping difficult and reducing mass transfer within the boundary layer. This will eliminate most hollow-fiber/capillary modules because they cannot withstand high pressure drops.

Membrane modules utilizing flat sheets (spiral-wound, plate, and pleated sheet modules) usually have a mesh-like spacer between sheets of membrane. This restricts their use to clear feed streams containing only fine SS.

Feed streams containing large SS would be treated poorly in spiral-wound modules, owing to the spacers in their feed channels. On the contrary, spiral-wound membrane modules are the lowest in capital costs and energy consumption. The trend in the food and beverage industries in recent years seems to be away from plate modules and toward spiral-wound modules, with ceramic tubular modules holding their own.

3. PROCESS COMPARISON

3.1. *Similarity*

Figure 5.6 and Table 5.1 show the similarity and dissimilarity of Conventional activated sludge (CAS) process system and innovated MBR process system.

Both CAS process system and MBR process system are similar to each other from biochemical engineering view points. The basic process system of either CAS or MBR includes the unit processes of influent feed, biological oxidation, final clarification, treated effluent discharge, RAS, and excess sludge discharge. Both the CAS and MBR require air supply to support the biological oxidation. Since theory and principles of process chemistry

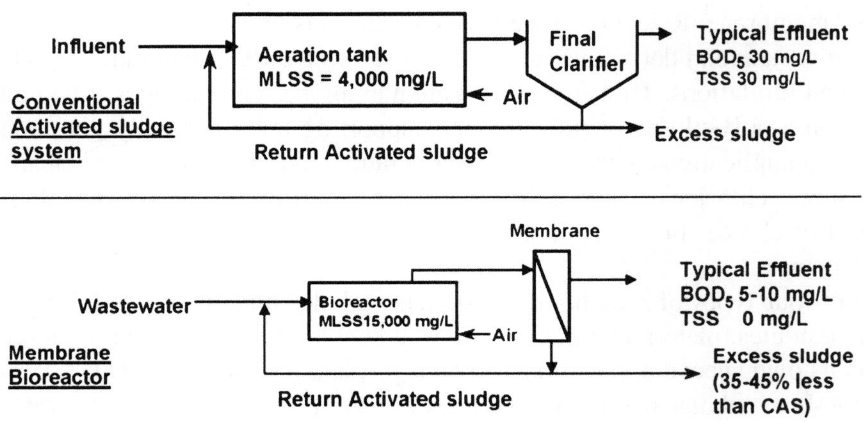

Fig. 5.6. Comparison between membrane bioreactor and conventional activated sludge system.

Table 5.1
Comparison of MBR and conventional activated sludge (CAS) systems

Dairy application	CAS	MBR
WW flow, m^3/d	600	600
Influent COD, mg/L	5,000	5,000
Influent BOD_5, mg/L	3,000	3,000
BOD_5, kg/d	1,800	1,800
Recycle of treated effluent, m^3/d	0	400
Aeration volume, m^3	4,500	600
Total floor space requirement, m^2	1,300	260
Effluent COD, mg/L	90	30
Effluent BOD_5, mg/L	30	5
Effluent TSS, mg/L	30	0

Based on a 5-month pilot study at the dairy site in France.

of both CAS and MBR are the same, the detailed process chemistry can be seen in the chapter on Activated Sludge Process, and will not be repeated here.

Both CAS and MBR can be operated for the desired purposes of carbonaceous oxidation, nitrification, and denitrification.

3.2. *Dissimilarity*

Then what are the differences between a CAS system and an MBR process system? Again, the readers are referred to Fig. 5.6 and Table 5.1.

3.2.1. *Reactor, MLSS, and Space Requirement Comparison*

Assuming a CAS process system and a comparable MBR process system will be treating the same wastewater influent flow of 600 m^3/d, the influent BOD_5 concentration and BOD_5 load are 3,000 mg/L and 1,800 kg/d, respectively (see Table 5.1). The CAS system will require a large 4,500 m^3 aeration tank (MLSS = 4,000 mg/L) for biological oxidation, while the MBR system will require a much smaller 600 m^3 bioreactor (MLSS = 15,000 mg/L) for biological oxidation. As shown in Fig. 5.6, the mixed liquor from the CAS aeration tank flows to a large sedimentation clarifier (2–4 h hydraulic detention time) for clarification (or liquid–solid separation), while the mixed liquor from the MBR bioreactor flows to a compact membrane modules system (less than 0.5-h hydraulic detention time) for clarification. Accordingly, the footprint of an MBR process system is much smaller. In this specific example, the total floor space requirements of the CAS and MBR systems are 1,300 m^2 and 260 m^2, respectively.

3.2.2. *Effluent Quality Comparison*

The most significant comparison has been done on the treated effluent quality. Table 5.1 indicates that the effluent COD, BOD_5, and TSS are 90, 30, and 30 mg/L, respectively, for the CAS process system. The MBR process system's performance is much better: the effluent COD, BOD_5, and TSS are 30 mg/L, 5 mg/L, and 0 mg/L, respectively. The above process comparison is based on a 5-month pilot study at a dairy processing plant in France.

3.2.3. *Cost Comparison and Water Recycle Considerations*

Based on limited cost data, the innovative MBR process system is cheaper to build, but more expensive to operate, in comparison with the CAS process system, if the treated effluent is not to be recycled for reuse.

Due to the high biomass concentration in the bioreactor (10,000–20,000 mg/L), the reactor can be made much more compact compared to the CAS process systems. Additionally, this facilitates the system to accept higher organic loads.

Another major advantage of the MBR process system is that the excess sludge production is lower than that in the conventional systems. This creates higher SRTs used in the process and is a function of the shear forces imparted to the biomass as they move through the crossflow membrane units (for external membrane MBR configurations).

Thus while the MBR system is an enhancement of the CAS system, it is quite different in space requirements, especially in effluent quality. Figure 5.7 further illustrates graphically a

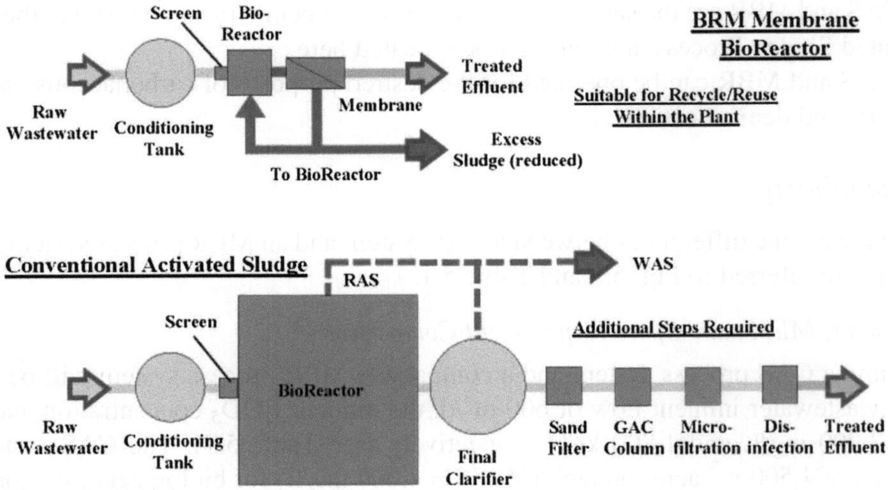

Fig. 5.7. Comparison between MBR and equivalent traditional WWTP.

comparison of the MBR and CAS systems, both designed to produce an effluent quality, suitable for recycle/reuse within and without the production facility.

In a water shortage region, such as California, the treated effluent should be recycled for reuse as much as possible. When water recycle is under consideration by environmental engineers, then both capital and O&M (Operations & Maintenance) costs of an MBR system will be much lower than that of a comparable CAS system. As shown in Table 5.1, 67% of the MBR treated effluent will meet the water quality requirements for direct nonpotable reuse, while the CAS treated effluent will not be suitable for recycle and reuse, unless tertiary treatment process units, such as sand filter (SF), activated carbon filter (ACF), and disinfection, are added for further effluent purification.

3.2.4. *Waste Treatment Consideration*

Finally, the average MLSS in a CAS aeration tank is around 4,000 mg/L, whereas that for an MBR system is approximately 15,000 mg/L (10,000–20,000 mg/L range), as shown in Fig. 5.6. Then, an MBR process system with much higher MLSS concentration is more suitable than a CAS process system when treating a high-strength wastewater stream.

3.2.5. *Summary*

In summation, the following are the advantages of an MBR process system over a CAS process system (15, 39, 40):

1. Excellent quality of treated effluent.
2. Possibility of recycle/reuse of treated effluent – better overall water economy.
3. Very compact installation: low construction costs.
4. Lower sludge production: lower sludge handling and nutrient costs.
5. Operating flexibility and simplicity; no sludge bulking problems, full automation possible.
6. Ideal preparation for the future; more stringent standards, rising costs of make-up water, etc.

7. Good esthetics – appearance, odor, etc.
8. Modular design: easily expandable for future capacity.

4. PROCESS APPLICATIONS

4.1. Industrial Wastewater Treatment

The various advantages of the MBR process system give it a unique application niche in the treatment of industrial wastewater. Typical wastewater characteristics where MBR becomes a viable technology are as follows:

1. Flow rate: approximately up to 500,000 gpd.
2. COD: greater than approximately 2,000 mg/L.

Industries where this technology can be implemented include chemical, petrochemical, pharmaceutical, fine chemicals, cosmetics, dairy, pulp and paper, automotive, landfill leachate, food, textiles, etc.

An MBR system has been designed for a petrochemical company located in south-east Texas to treat three high-strength industrial wastewater, containing alcohols and sulfur-containing compounds (17). The design was based on a field pilot test conducted by Envirogen, a company in Lawrenceville, NJ. One wastewater stream consisted of approximately 60% isopropanol by weight. The other streams contained light hydrocarbons and organic sulfides. The influent COD to the MBR system was 25,000 mg/L. Removal efficiencies averaged 90–95%, thereby allowing the plant to stay within regulatory limits cost-effectively. The three streams treated accounted for less than 2% of the plant's hydraulic wastewater load, but over 70% of the organic wastewater load.

An industrial plant manager would like to consider possible adoption of an MBR process system for treating the industrial wastewater, usually because of the following reasons:

1. MBR system has smaller plant footprint because it treats low-flow high-strength streams and operates at a much higher MLSS concentration.
2. MBR system has the possibility to recycle 40% of treated effluent to existing RO step (no further pretreatment steps is required).
3. MBR's modularity is suitable to double the capacity in future.
4. MBR system can be installed easily in old unused building.
5. MBR system has much less excess biosolids (sludge) production – its highly concentrated biosolids can be used as supplemental fuel in a boiler.
6. MBR system is the most cost-effective and reliable solution overall.
7. On-site pilot tests have shown simplicity and ease of operation of the MBR system.

4.2. Municipal Wastewater and Leachate Treatments

For treatment of high-flow low-strength municipal wastewater, the MBR process system cannot economically compete with the CAS process system, if

1. the municipality has plenty of land available for wastewater treatment (WWT) facility construction;
2. the treated effluent does not have to meet very stringent effluent standards (including nutrient removal and/or heavy metal removal);

3. the treated effluent does not have to be recycled for reuse;
4. the project does not involve expansion of capacity or treatment by retrofit.

In case one or more of the above factors does/do not apply, the MBR system will have an edge for competition with the CAS process system. The technical as well as economic feasibility of treating municipal wastewater by MBR has been positively demonstrated (17, 18).

For treatment of low-flow high-strength leachate from sanitary landfill sites, MBR process system is superior to the CAS process system in terms of both effluent quality and cost (15, 17, 20, 38).

Two aerobic MBR reactor systems are currently being designed for a municipal WWT district in southern New Jersey, USA. One of these MBR systems will be used to pretreat landfill leachate (17) shipped to the facility from the surrounding area. The effluent from the pretreatment system will then be polished in the existing municipal wastewater treatment plant (WWTP). The design influent flow to the MBR system is 400,000 gpd with a COD of 10,000 mg/L. COD is the measure of the amount of oxygen required to oxidize organic and oxidizable inorganic compounds in wastewater. The COD test is used to determine the degree of pollution in WW. The footprint of the system is approximately 7,000 ft^2 (2,000 ft^2 for reactors and membranes and 5,000 ft^2 for pumps, blowers, and other auxiliary equipment).

The second aerobic MBR system is being designed as a mobile publicly owned treatment works (POTW). It will be capable of treating 80,000 gpd with an influent BOD_5 of 625 mg/L. This system will have phosphorus removal and disinfection capabilities built in. The footprint for this system is approximately 640 ft^2. The system is trailer-mounted (two 40-ft. long by 8 -ft. wide skids) and will be highway transportable (17).

Another case history of an aerobic MBR system for the treatment of a sanitary landfill leachate (20) is presented in Sect. 5.2 of this chapter.

5. PRACTICAL EXAMPLES

5.1. Example 1: Dairy Industry

A dairy plant in central France produces 35,000 mt/year of fruit- and other yogurts from milk. Its wastewater (WW) source includes washing of yogurt vat bottoms, other wash water, and cooling water blowdown. The plant needs a modern WWT system to treat its combined WW properly. The following are the features:

1. The factory had very little floor space.
2. Water recycling (up to 70%) makes the plant less dependent on external water sources, which were not reliable.
3. The receiving water was in a fragile ecosystem, which required as much flow and organic pollution to be removed.

5.1.1. Solution

The modular design of the MBR process system enables the dairy plant to keep up with increased production in a phased manner, without over-investing in initial capital cost.

Comparative data developed as a result of pilot testing are provided in Table 5.1. Both CAS and MBR process systems were piloted at the same time treating the same WW (influent flow = 600 m^3/d, influent COD = 5,000 mg/L, and influent BOD_5 = 3,000 mg/L) for 5 months.

It can be seen from Table 5.1 that the performance of the tested MBR process system was much better than that of the CAS process system.

Based on the pilot tests, the full-scale feasible CAS process system would have required 4,500 m^3 of aeration volume and 1,300 m^2 of total floor area. Although the CAS system did meet the effluent COD standard (over 90% COD reduction), the effluent TSS averaged 30 mg/L, which was too high to recycle the CAS effluent for nonpotable reuse.

The MBR process system, on the contrary, was compact, requiring only 600 m^3 of aeration volume and 260 m^2 of total floor area. The MBR effluent COD, BOD_5, and TSS were 30, 5, and 0 mg/L, respectively. The MBR effluent quality did meet the requirements for recycle as nonpotable water.

A full-scale MBR system was purchased by the dairy plant and started up in May 1998, and is operating successfully.

The reasons for selecting MBR technology for the dairy plant are summarized below:

1. Use of high-quality bacteria-free effluent (suitable for nonpotable reuse).
2. Possibility to recycle/reuse up to 70% of the treated effluent.
3. Small footprint (20% compared to CAS).
4. Ability to expand in the future.
5. 35% less excess biosolids (waste sludge) production.
6. No odor problems.
7. Ease of operation and maintenance (operator friendly).
8. Fits in with equipment in the dairy plant.
9. Lowest cost option overall.

5.2. *Example 2: Landfill Leachate Treatment*

A landfill site in Arnouville, a small town in the suburbs of Paris, France, needs a cost-effective process system for treating its municipal landfill leachate.

Landfill leachates originate mainly from percolation of rainwater and biological decomposition of wastes. Depending on such factors as age of the landfill and waste composition, leachates can contain high levels of organic and inorganic compounds, making treatment mandatory before reuse or discharge into the environment. Although conventional biological or physico-chemical processes (4–11) can efficiently remove SS, organic compounds, and nitrogen, more stringent regulations have been implemented in several countries requiring removal of salts (chlorides and sulfates) and heavy metals.

Reverse osmosis (RO) is a well-known technology with many useful applications, mostly in the desalination of seawater. However, RO treatment of food industry process water (1) requires complete removal of SS and organic matter to avoid rapid fouling and clogging of the membranes (see Sect. 1.2.1).

In view of the respective capabilities of conventional biological processes, conventional physico-chemical processes, MBR, and RO, please recommend a solution to the landfill leachate treatment, which should be technically and economically feasible.

5.2.1. Solution

In view of the respective capabilities of various processes, a combination of MBR processes and RO could provide an integrated system capable of treating highly contaminated leachates and producing high-quality effluent, meeting current and future regulations.

On the basis of results obtained during a 1-year pilot study, a full-scale plant was designed and installed to treat municipal leachates from a sanitary landfill site in the suburbs of Paris, France. The system consisted of an MBR process system followed by an RO unit. Results obtained are provided in Table 5.2.

Table 5.2 shows that the MBR process system (with ultrafiltration and/or microfiltration membrane) was able to achieve the following % removal efficiency while treating the landfill leachate:

1. COD	71.6%
2. TOC	68.9%
3. NH_3–N	98.3%
4. Cl^-	3.3%
5. TSS	100%

While the TSS removal for an MBR process system was 100%, the removals of COD, total organic carbon (TOC), and NH_3–N were moderate-high, and that of chloride was poor.

The MBR system nevertheless was an excellent pretreatment unit for treating the sanitary landfill leachate, prior to the RO process system, due to reduction in silt density index (SDI), which is a very important parameter for satisfactory RO performance.

With the combination of MBR and RO, the overall removal efficiency of COD, TOC, NH_3–N, Cl^-, and TSS was overall 99%, which was very satisfactory.

The definitions of RO are given in Sect. 1.2.1.

Table 5.2
Performance of MBR-RO process system on landfill leachate application in Arnouville (near Paris), France

Parameters	Raw leachate quality	MBR effluent quality	RO effluent quality	Overall removal efficiency, %
COD, mg/L	2,500	710	10	>99%
TOC, mg/L	740	230	1	>99
NH_3–N, mg/L	410	7	3	>99
Cl^-, mg/L	1,500	1,450	50	>95
TSS, mg/L	300	0	0	100

5.3. *Example 3: Coffee Industry*

A coffee processing plant in Belgium produces 625 m^3/d (165,000 gpd) of combined WW, of which 70% is low-strength and 30% is high-strength. The combined WW has the following characteristics:

1. BOD_5	1,150 mg/L
2. S-COD	2,180 mg/L
3. T-COD	2,700 mg/L
4. TSS	280 mg/L
5. Temperature	15°C

The high-strength WW was the plant manager's main concern. Table 5.3 summarizes the high-strength composition.

The government had issued an effluent discharge permit with the following effluent limitations:

1. Total flow	237,600 gpd
2. TSS	500 mg/L, and 220 lb/d
3. S-COD	No limits
4. T-COD	2,000 mg/L and 3,000 lb/d
5. BOD_5	400 mg/L and 600 lb/d

Table 5.3
Coffee factory high-strength streams composition

Influent	Flow		TSS		S-COD		T-COD		BOD_5	
	gpd	gpm	mg/L	lb/d	mg/L	lb/d	mg/L	lb/d	mg/L	lb/d
Stream I	8,500	5.90	1,060	75	6,060	430	8,870	630	3,375	240
Stream II	24,000	16.67	220	44	6,440	1,290	7,000	1,400	3,375	675
Stream III	3,400	2.36	585	17	6,770	190	7,460	210	3,385	96
Stream IV	13,000	9.03	220	24	2,660	290	2,800	305	1,180	124
Total	48,900	33.96	390	160	5,390	2,200	6,240	2,545	2,780	1,135

Contaminant concentrations for the "Total" is an average based on "Total" load and flow.

If you were the plant's environmental engineer responsible for WW compliance at this coffee plant, what would be your recommended engineering solutions to the plant manager?

5.3.1. Solution

The coffee plant's environmental engineer decided to conduct a feasibility or treatability study, and selected an MBR pilot plant with the following specifications:

1. Skid dimensions = 13" × 7" × 8"H
2. Weight = 4,000 lbs. (shipping); 8,000 lbs. (operating)
3. Connections =
   Influent = 1.5" hose clamp
   Discharge = 2" male NPT
   Water supply = 5–10 gpm (3/4" hose clamp)
4. Electrical = three phase, 240 V, 60 Hz, 100 amp, two grounds
5. Flow rate = approximately 1 gpm (depends on influent BOD/COD)
6. Automatic control system.

It was known that the MBR system is technically and economically feasible for treating high-strength and low-flow WW streams. It was then recommended by the plant's environmental engineer that only the low-flow high-strength stream (representing 30% of total combined WW flow) would require treatment in an MBR system. The remaining 70% of untreated low-strength streams could be post-blended with the treated effluent from the MBR, resulting in a combined or blended effluent, which would meet the discharge permit's effluent limitations (see Fig. 5.8).

The 1-gpm pilot plant demonstration was very successful. The above proposed engineering solution was fully proven by the MBR performance. Accordingly, an MBR process system was ordered, installed, started-up, and operated at the coffee plant. Figure 5.8 shows the full-scale WWT flow schematic implemented by the coffee plant. The successful performance data of the installed process system are shown in Table 5.4. It is seen from Table 5.4 that the quality of the MBR effluent was very high. Critical effluent parameters were as follows:

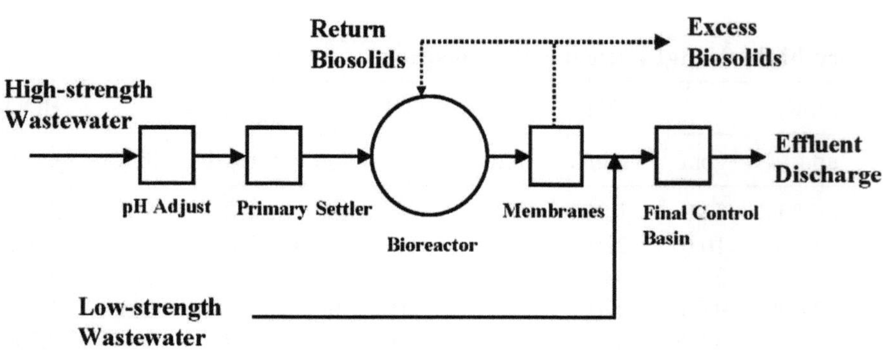

Fig. 5.8. Coffee factory WWT flow schematic.

Table 5.4
Coffee factory effluent characteristics

Effluent		MBR influent	MBR effluent	Untreated	Post-blended	Permitted
Flow	gpd	48,900	–	114,915	163,815	237,600
	gpm	33.96	–	80	114	165
TSS	mg/L	390	0	230	160	500
	lb/d	160	0	220	220	990
S-COD	mg/L	5,390	250	820	645	
	lb/d	2,200	100	785	885	
T-COD	mg/L	6,240	250	1,230	930	2,000
	lb/d	2,545	100	1,180	1,280	3,000
BOD_5	mg/L	2,780	50	460	335	400
	lb/d	1,135	20	440	460	600

TSS = 0 mg/L
S-COD = 250 mg/L
T-COD = 250 mg/L
BOD_5 = 50 mg/L

After blending the MBR treated effluent and the untreated low-strength WW together, the resulting blended final effluent, indeed, met all effluent limitations in the permit.

5.4. *Example 4: Cosmetics Industry*

The WW discharged from a major cosmetics manufacturing plant in northern France was originally treated at the local municipal WWTP. Average flow rate was 160 m^3/d (42,240 gpd). In order to cut down sewer surcharge and fresh water costs, the plant set a goal to remove 90% of the total COD and recycle at least 30% of the treated effluent for non-process uses within the plant.

5.4.1. Solution

A pilot aerobic MBR test program was conducted at the plant using a 1 m^3 (264 US gal) pilot plant to determine treatability and to obtain full-scale design parameters. Results from the 5-month test program demonstrate the excellent overall performance of the MBR process system in terms of efficiency and reliability. Removal efficiencies obtained were 98%+ for COD, 99% for NH_3–N, and 99% for FOG (fats, oils, and greases). Removal of TSS was total; yielding an effluent that could satisfy the recycle criteria within the plant (15, 20).

Following the pilot test program, a full-scale system was designed and installed to handle 150 m^3/d flow and 1,200 kg/d COD. The membrane filtration unit consisted of ceramic microfiltration modules, which were modular and suitable for expansion. The plant has been successfully in operation since the summer of 1994. Despite the variable flow rate and characteristics of the influent WW (COD 2–6 g/L; COD/BOD_5 1.8–2.5), the treated effluent from the innovative MBR process system has been of consistent high quality (COD $<$ 100 mg/L; BOD $<$ 20 mg/L; TSS 0 mg/L; no bacteria). Part of the treated effluent is recycled for

Table 5.5
Aerobic MBR treatment of cosmetic factory wastewater in France

Parameter, mg/L	Influent quality	Effluent quality
COD	6,500	<100
TSS	240	0
NH_3–N	40	0.4
FOG	2,500	<2

reuse in the factory lavatories, and for irrigation. The rest is discharged via a holding pond populated by fish, ducks, and reeds.

The quantity of excess biosolids produced is lower than that in conventional aerobic systems. Approximate net yield is 0.1 kg volatile suspended solids (VSS) per kg COD removed. The plant is easy and economical to operate. One part-time operator is adequate to perform process evaluation duties, which consist of standard analyses of WW parameters. An automatic system facilitates process control at this plant.

Table 5.5 presents the results of aerobic MBR treatment at this facility (20).

6. CONCLUSIONS

6.1. *Industrial Applications*

The MBR process system is a proven, reliable, modular, and compact industrial WWT system that has been successfully implemented in several full-scale installations. Several years of operation has proven its reliability and efficiency for a variety of industrial plant owners, who use it as a water management tool, to conserve precious water resources, and to reduce overall operating costs (39, 40).

6.2. *Municipal Applications*

For treatment of high-flow low-strength WW, the MBR process system will be applicable and cost-effective if one or more of the factors below will apply:

1. The municipality has no space available for expansion.
2. The treated effluent must meet very stringent effluent standards (including nutrient and/or heavy metals removal).
3. The treated effluent has to be recycled for reuse.

In addition to the above municipal WWT applications, leachate from municipal sanitary landfills can also be cost-effectively treated by the newly developed MBR process system.

Although membrane process systems have been widely used for potable water treatment (3, 46), the use of MBR process systems for municipal potable water treatment (mainly aiming at nutrients removal) is still in developmental stage.

6.3. *Acknowledgments*

The MBR process system was originally developed and perfected by Ondeo Degremont (Suez group) and Zenon Environmental Inc. Now the MBR process system has been accepted internationally as one of the mainstream biological WWT systems. More than 200 systems have been installed and are successfully operated.

6.4. *Commercial Availability*

Full-scale MBR process systems are available in USA through companies such as Ondeo Degremont, Zenon Environmental Inc., US Filter, Aqua-Aerobic, Inc., GE Water & Process Technologies, a unit of General Electric Co., etc.

7. RECENT ADVANCES IN MEMBRANE BIOREACTOR TECHNOLOGY

7.1. *Electrodialysis Membrane Bioreactor for Product Separation and pH Control*

The removal of product and pH control are two main issues that limit the technical and commercial viability of such fermentation processes. A bioreactor combining conventional electrodialysis and bipolar membrane electrodialysis has been developed by Li et al. (55) for in situ product removal and pH control in lactic acid fermentation. The electrokinetic process enabled removal of the biocatalytic product (lactic acid) directly from the bioreactor system, in a concentrated form, and also enabled good pH control without generation of troublesome salts. Moreover, end-product inhibition of glucose catabolism was reduced, resulting in a greater generation of the end-product lactic acid.

7.2. *Ethanol Production in Membrane Distillation Bioreactor*

Batch fermentation combined with the removal of ethanol from the broth by means of membrane distillation (MD) process has been investigated by Gryta et al. (56). The porous capillary polypropylene membranes were used for the separation of volatile compounds from the feed (broth), formed as a result of fermentation. The elimination of these compounds allows the increase in the productivity and the rate of conversion of sugar to ethanol, since they act as inhibitors. In the case of fermentation combined with MD, the efficiency of 0.4–0.51 (g EtOH)/(g of sugar) and the production rate of 2.5–4 (g EtOH)/dm^3 h were achieved in relation to 0.35–0.45 (g EtOH)/(g of sugar) and 0.8–2 (g EtOH)/dm^3 h obtained in the classical batch fermentation. The ethanol flux obtained in membrane distillation varied in the range of 1–4 (kg EtOH)/m^2 per day and was dependent on the temperature and the feed composition.

7.3. *Denitrification of Nitrate-Contaminated Drinking Water Using Membrane Bioreactor*

A MBR was investigated by Ergas and Rheinheinmer (57) for denitrification of nitrate-contaminated drinking water. In the MBR, nitrate-contaminated water flows through the lumen of tubular microporous membranes and nitrate diffuses through the membrane pores.

Denitrification takes place on the shell side of the membranes (which are coated with a layer of biofilm), creating a driving force for mass transfer. The microporous membranes provide high nitrate permeability, while separating the treated water from the microbial process, reducing the carryover of organic carbon and sloughed biomass to the product water. The MBR achieved over 99% nitrate removal at an influent concentration of 200 mg/L of nitrate–N. Low effluent turbidity was achieved; however, approximately 8% of the added methanol partitioned into the product water. More research is needed for the process improvement. Recently Pennsylvania State University researchers (58) used MBR for DBP precursor removal from water.

7.4. Treating Contaminated Surface Water for Drinking Water Production Using Membrane Bioreactor

A laboratory MBR using a submerged polyethylene hollow-fiber membrane module with a pore size of 0.4 μm and a total surface area of 0.2 m^2 was used by Li and Chu (59) for treating a raw water supply slightly polluted by domestic sewage. The feeding influent had a TOC of 3–5 mg/L and an ammonia nitrogen (NH_3–N) concentration of 3–4 mg/L. The MBR ran continuously for more than 500 days, with a HRT as short as 1 h or less. Sufficient organic degradation and complete nitrification were achieved in the MBR effluent, which normally had a TOC of less than 2 mg/L and a NH_3–N of lower than 0.2 mg/L. The process was also highly effective for eliminating conventional water impurities, as demonstrated by decreases in turbidity from 4.50 ± 1.11 to 0.08 ± 0.03 NTU, in total coliforms from 10^5/mL to less than 5/mL, and in UV_{254} absorbance from 0.098 ± 0.019 to 0.036 ± 0.007 cm^{-1}. With the MBR treatment, the 3-day trihalomethane formation potential (THMFP) was significantly reduced from 239.5 ± 43.8 to 60.4 ± 23.1 μg/L. The initial chlorine demand for disinfection decreased from 22.3 ± 5.1 to 0.5 ± 0.1 mg/L. The biostability of the effluent improved considerably as the assimilable organic carbon (AOC) decreased from 134.5 ± 52.7 to 25.3 ± 19.9 μg/L. All of these water quality parameters show the superior quality of the MBR treated water. Molecular size distribution analysis and the hydrophobic characterization of the MBR effluent, in comparison to the filtered liquor from the bioreactor, suggest that the MBR had an enhanced filtration mechanism. A sludge layer on the membrane surface could have functioned as an additional barrier to the passage of typical trihalomethane (THM) precursors, such as large organic molecules and hydrophobic compounds. These results indicate that the MBR with a short HRT could be developed as an effective biological water treatment process to address the urgent need of many developing countries that are plagued by the serious contamination of surface water resources.

7.5. Removing Bromate from Drinking Water Using the Ion Exchange Membrane Bioreactor

Bromate is a disinfection byproduct with carcinogenic properties that has to be removed from drinking water to concentrations below 10 or 25 μg/L. The work of Matos et al. (60) evaluates the applicability of the ion exchange membrane bioreactor (IEMB) concept for the removal of bromate from drinking water, in situations where nitrate is also present in concentrations up to three orders of magnitude higher than that of bromate. The experimental results obtained show that the biological reduction of bromate was slow and only occurred after the complete reduction of nitrate. The specific bromate reduction rates varied from 0.027 ± 0.01 mg BrO_3^-/g cell/h to 0.090 mg BrO_3^-/g cell/h for the studied concentrations. The cell weight was gram dry weight. However, transport studies, using anion exchange membranes, showed that Donnan dialysis could efficiently remove bromate from polluted waters. Therefore, the use of a dense, nonporous membrane in the IEMB system isolates the water stream from the biological compartment, allowing for the uncoupling of the water production rate from the biological reduction rate. The IEMB system was used for the treatment of a polluted water stream containing 200 μg/L of BrO_3^- and 60 mg/L of NO_3^-. The concentrations of both ions in the treated water were reduced below the recommended levels. No bromate accumulation was observed in the biocompartment of the IEMB, suggesting its complete reduction in the biofilm formed on the membrane surface contacting the biocompartment. Accordingly, the IEMB has proven to be a technology able to solve specific problems associated with the removal of bromate from water streams, since it efficiently removes bromate from drinking water even in the presence of nitrate, a known competitor of bromate biological reduction, without secondary contamination of the treated water by cells or excess of carbon source.

7.6. New Membrane Bioreactor Design and Applications

Many new MBRs or systems have been developed for various applications (61–81). Banks and Heaven (61) have developed a coarse MBR for two-stage anaerobic digestion of biodegradable municipal wastes.

Energy requirements are of primary interest in MBR systems. Aeration energy is the largest consumer of energy in an MBR system. This comes in two aeration forms: (a) to provide oxygen for biological oxidation and nitrification and (b) to provide scouring of the membrane to keep it clean. The optimization of membrane air scour energy has been done by many universities and MBR manufacturers (73–75). The hydrodynamic conditions and system configuration in an MBR system can greatly affect membrane performance. The membrane filtration performance was improved by 20–60% when a two-phase (air and liquid) crossflow was applied rather than just a single-phase (liquid only) crossflow (74). Air scour energy in an MBR system can provide high levels of turbulence and surface contact to remove solid particles that attach to the surface of the membrane and to guard against irreversible membrane fouling. One application that now uses the slug or plug flow for improving the system performance is the MemPulse MBR system from Siemens Water Technologies (73). The MemPulse MBR system is supplied with a continuous air supply that is accumulated in the base of the module. It periodically releases irregular pulses of air to the MBR module, creating plug flow, which prevents solids accumulation as the mixed liquor on the membrane is continuously refreshed with new liquor, in turn, lowering the air supply up to 60%.

A hybrid MBR system has been developed for biological nitrogen removal through nitrification and denitrification (71). The system includes an aerobic tank and an anoxic tank with an intermediate sludge settler connected to an MBR with a submerged hollow-fiber membrane module. A MLSS concentration of 6 g/L or higher can be maintained in the reactors. COD removal and nitrogen removal can be higher than 95 and 90%, respectively, if the COD loading rate is up to 2,100 mg/L d and the total nitrogen loading rate is up to 170 mg N/L d.

The filterability of membrane bioreactor sludge (MBRS) with or without the CAS and polyelectrolytes was investigated by Yigit et al. (72). It was found that when combining CAS and MBRS, better filterability and higher sludge cake dry solids were achieved with cationic polyelectrolytes compared to anionic and non-ionic polymers for all sludge types investigated.

7.7. *Full-Scale Membrane Bioreactor for Wastewater Treatment: Carnation Wastewater Treatment Plant*

King County, Washington, USA, has replaced its old, failing septic systems with a new, state-of-the-art, US$23-million MBR treatment plant (Carnation Wastewater Treatment Plant) which protects Snoqualmie River and all the creatures that love and depend on it (62). The Carnation plant treats wastewater to such a degree that highly reclaimed water can be used to enhance a wetland at Chinook Bend Natural Area, a mile and a half north of the plant. The membranes are immersed in wastewater. The MBR system sucks wastewater through hollow fibers with microscopic pores small enough to filter out particulate matter and even individual bacteria. The new treatment plant will serve about 2,000 people in Carnation's urban growth area, with the capacity to serve up to 4,000 people in 2030. At design capacity, the plant is expected to treat 480,000 gallons of wastewater a day.

7.8. *Bioremediation Using Membrane Bioreactor–Sequencing Batch Bioreactor*

The aim of Luisa Manigas' experimental work (63) has been the application of a new polluted waters treatment technology for the selection of a bacterial population capable of bioremediating synthetic groundwater polluted by four different chlorinated compounds. The innovative technology applied in Manigas' study was the biological treatment system known as Membrane Bioreactor–Sequencing Batch Bioreactor (MBR–SBR), which consists of a SBR coupled to a membrane module for the filtration of the biological sludge that degrades the chlorinated compounds present in the polluted water fed to the system. The synthetic groundwater used in this experimentation contained four chlorinated compounds: 1,2-dichloroethane, 1,2-dichlorobenzene, 2-chlorophenol, and TCE. During the development of the research, the chlorinated compounds were fed either alone or in different mixtures in order to identify possible effects of each substrate on the degradability of the others. Some operational parameters were varied in order to check the possibility of implementing the system at a larger scale. The MBR–SBR system equipped in this study was effective in the acclimation of a sludge to hardly biodegradable xenobiotic compounds, and the degradation rates achieved were comparable to other similar studies performed, even if in this experimentation, differing from all the others found in the literature, good removal rates were achieved for the degradation of the substrates when they were fed in mixtures. The MBR–SBR is now commercially available (65), and its process descriptions and historical development can be found in the literature (1, 2, 51–54, 66).

7.9. Membrane Bioreactor Design

Copeland et al. (64) have addressed the fundamental design requirements needed for the MBR WWT system's compliance with the regulated effluent limits. They (64) discussed technical issues that were accounted for in the process analysis and biological modeling as a means to help evaluate the design criteria. Their information (64) should help engineers, regulatory agencies, and owners address the minimum requirements for initiating an MBR WWTP. More design considerations of MBR systems can be found in the literature (54, 69, 81–83).

7.10. Using Flotation as a Pretreatment to Membrane Processes

The need for fresh sources of high-quality drinking water or reclaimed water is becoming more and more urgent worldwide. All membrane processes (including MBR), on the contrary, are known to be very sensitive to foulants as colloids, inorganic scale, and biofouling. Pretreatment of their membrane reactor's feed water is often a key step. In membrane plants design, various techniques have been proposed for pretreatment, even other membrane filtrations such as ultrafiltration. However, DAF is highly efficient for removing oil, grease, and suspended solids, such as activated sludge or algae (66, 67).

7.11. Full-Scale Membrane Bioreactor Technology for Water Reclamation

GE Water & Process Technologies, a unit of General Electric Co. (68), is supplying its ZeeWeed MBR ultrafiltration technology as part of a $250 million upgrade and expansion to the Yellow River Water Reclamation Facility (WRF) in Gwinnett County, GA. The selection of the advanced ZeeWeed treatment process contributed to a $50 million savings in the projects capital costs, largely due to the compact size of the ZeeWeed MBR system, which helped to reduce the quantity of concrete, steel, and labor needed to complete the construction dramatically.

GE's ZeeWeed MBR process eliminates the need for large, costly concrete settling tanks that conventional wastewater processes rely on to separate contaminants from treated effluent – instead occupying just a fraction of the space to consistently produce tertiary-quality effluent that can be safely discharged to sensitive receiving bodies or reused for various, nonpotable applications (76–79, 81).

In April 2004, the Couley Creek Water Reuse Facility was commissioned (76). This ZeeWeed MBR plant was the first of its kind in the State of Georgia and was designed to treat an average day wastewater flow of 5 MGD (18,893 m^3/d). The plant incorporated the following upgrades: optimized biological phosphorus removal to minimize chemical consumption, mixed liquor surface wasting to minimize scum and foam in the aerobic zones, and a sludge thickener to minimize the aerobic digester volume. The treated effluent from the reuse facility surpasses tertiary effluent standards and is provided to local golf courses, subdivisions, schools, and churches for irrigation purposes, thereby reducing the amount of water drawn from the Chattahoochee River.

More than a decade of drought in south east Australia, the worst on record, has severely stressed water supplies and raised major concerns about the sustainability of the region's water reserves. For Pennant Hills Golf Club, watering restrictions limited its monthly use to 20,000 m^3 (5.3 MG), an amount that was well short of its requirements during the dry summer months (77). The first of its kind in Australia, the club installed a 650 m^3/d (172,000 gpd) ZeeWeed MBR wastewater reclamation plant in Beecroft, New South Wales, Australia, in March 2008. Figure 5.9 shows the flow diagram of the plant that taps into a municipal sewer line and withdraws wastewater as the source of a new water supply for the golf club's irrigation. By reclaiming high-quality effluent from municipal wastewater, the club now has a drought-proof supply of water for its irrigation needs.

Vancouver Convention & Exhibition Centre (VCEC) in Canada installed a 76 m^3/d (20,000 gpd) wastewater reclamation plant in March 2008 to treat 100% of the VCEC wastewater. The plant effluent is recycled for flush fixtures and irrigation of a 6-acre (24,280 m^2) living roof. The on-site wastewater reclamation plant is located next to the exhibit hall. The influent is first directed to the anoxic chamber for denitrification, then to the bioreactor where active microorganisms consume or digest the biodegradable waste substances, and finally to the ZeeWeed UF chamber. Permeate pumps are used to pull the wastewater gently through thousands of membrane fibers. The MBR effluent flows through an ACF and then the ultraviolet units for further disinfection (78).

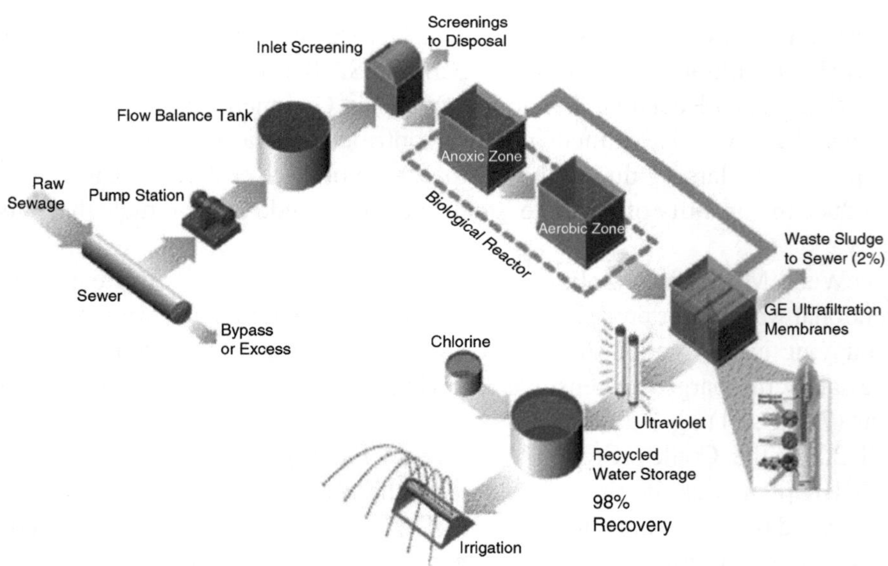

Fig. 5.9. Flow diagram of Pennant Hills golf club water reclamation plant using MBR technology.

Johns Creek Environmental Campus in Fulton County, Georgia, USA, installed a 56,780 m^3/d (15 MGD) MBR wastewater reclamation plant in 2009 (79). This new facility not only houses a water reuse plant, but also features an educational center to promote the environmental and economic benefits of water reuse in Georgia. The small footprint and modular design of the plant allows the entire aerobic biological treatment process and ancillary equipment to be completely contained within the building. With the WWTP being in close proximity to a residential area, this would eliminate any possible noise or odor concerns. The typical treated water quality results are BOD less than 2 mg/L; TSS less than 2 mg/L; and turbidity less than 0.1 NTU.

Corona Wastewater Treatment Plant, in Corona, CA, USA, was built in December 2001 with a 3,785 m^3/d (1 MGD) design capacity. The plant which incorporated reinforced, immersed membranes for biological and physical treatment meets California Title 22 discharge criteria for water reuse (80). The MBR system has been designed to meet the following discharge criteria: BOD below 5 mg/L; total inorganic nitrogen below 10 mg/L; total suspended solids below 5 mg/L; and turbidity below 0.2 NTU. The effluent from the plant is suitable for reuse on a nearby golf course for landscape irrigation, or can be safely discharged to a local creek bed.

The Brighwater Treatment Plant in Woodinville, King County, Washington, USA, will use an advanced MBR technology system instead of the bulky settling tanks used in conventional WWTPs (70). The plant is being designed to meet or exceed stringent water quality standards for effluent discharge or reclaimed water production. Specifically, the plant effluent will be equivalent to Class A Reclaimed Water, meeting the strict standards of the State of Washington Departments of Ecology and Health for use in nondrinking purposes including landscape and agricultural irrigation, heating and cooling, and industrial processing as well as safe discharges into freshwater. The flow diagram of the complete Brighwater Treatment Plant, shown in Fig. 5.10, includes the following liquid treatment processes: preliminary treatment (influent pump station, chemical addition, bar screen, and grit separation), primary treatment (primary sedimentation tank and fine screening), membrane treatment (bioreactor and

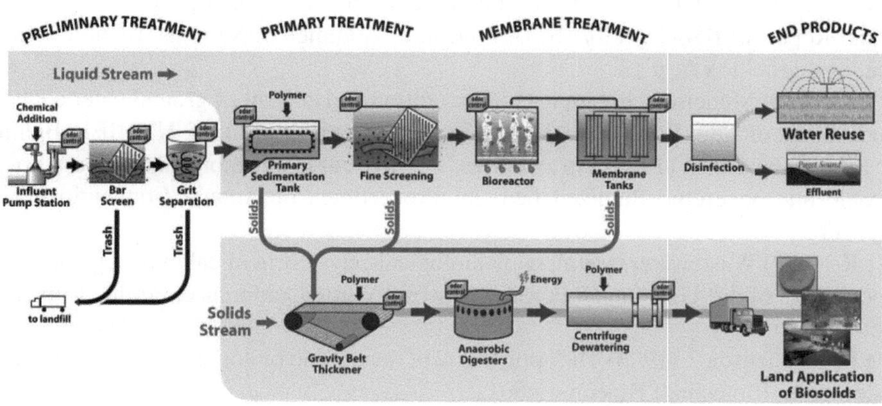

Fig. 5.10. Flow diagram of Pennant Hills brightwater wastewater plant using MBR technology.

membrane tanks), and final treatment (disinfection). The final plant effluent can be reused or discharged into a receiving stream. The solids from the preliminary treatment steps are collected and trucked to a landfill site for disposal. The solids from both the primary treatment and the membrane treatment steps are transported to the sludge treatment facilities (gravity belt thickeners, anaerobic digesters, and centrifuges) for sludge thickening, digestion, and dewatering before final land disposal. More research on water reclamation using MBR can be found form the literature (84–87).

REFERENCES

1. Wang LK, Cheryan M (1995) Application of membrane technology in food industry for clean production. United Nationals Industrial Development Organization (UNIDO) Second International Conference on Waste Minimization and Cleaner Production, Barcelona, Spain. June 7–9, 1995. UNIDO-DTT-8-6-95, Vienna, Austria, p 42.
2. Wang LK, Kropuzek JV, Kounitson U (1995) Case studies of cleaner production and site remediation. Manual No. UNIDO-DTT-5-4-95, United Nationals Industrial Development Organization (UNIDO), Vienna, Austria, p 136.
3. Wang LK, Kopko SP (1997) City of Cape Coral Reverse Osmosis Water Treatment Facility. Technical Report No. PB97-139547, US Department of Commerce, National Technical Information Service, Springfield, VA, p 15.
4. Wang LK, Pereira NC (1986) Handbook of environmental engineering, vol. 3. Humana, Totowa, NJ, p 498.
5. Beer C, Wang LK (1975) Full-scale operations of plug flow activated sludge systems. J New England Water Pollut Control Assoc 9(2):145–173.
6. Wang LK, Wang MHS (1977) Control tests and kinetics of activated sludge process. Water, Air Soil Pollut 8:315–351.
7. Beer C, Wang LK (1978) Activated sludge systems using nitrate respiration – design considerations. J Water Pollut Control Fed 50(9):2120–2131.
8. Krofta M, Wang LK (1983) Improved biological treatment with a secondary flotation clarifier. Civil Eng Practicing Design Eng 2:307–324. (NTIS-PB82-154196).
9. Wang LK, Aulenbach DB (1986) BOD and nutrient removal by biological A/O process systems. Technical Report #PB88-168430/AS. U.S. Dept. of Commerce, National Technical Information Service, Springfield, VA, p 12.
10. Wang LK (1990) Modern technologies for prevention and control of groundwater contamination. In: Proceedings of New York – New Jersey Environmental Exposition, NYNJEE, Belmont, MA.
11. Wang LK, Wang MHS (1990) An overview of environmental biotechnologies with emphasis on aerobic and anaerobic fluidized bed. In: Proceedings of Annual Convention of CAAPS, Albany, NY.
12. Wang LK (1987) Wastewater treatment by biological-physicochemical two-stage process system. In: Proceedings of the 41st Annual Purdue Industrial Waste Conference, Purdue University, West Lafayette, IN, p 67.
13. Brindle K, Stephenson T (1996) The application of membrane biological reactors for the treatment of wastewaters. Biotechnol Bioeng 49:601.
14. Cicek N, Franco JP, Suidan MT, Urbain V (1998) Using a membrane bioreactor to reclaim wastewater. J Am Water Works Assoc 90(11):105.

15. Menon R, Fuchs C (1998) Application of degremont BRM™ membrane bioreactor to the treatment of industrial wastewaters. Membrane Technology Workshop at WEFTEC, Orlando, FL.
16. Chiemchaisri C, Yamamoto K (1993) Biological nutrient removal under low temperatures in a membrane separation bioreactor. Water Sci Technol (G.B.) 28(10):325.
17. Pitre MP, Enegess DN, Unterman R (1999) Bioreactors: the new wave in wastewater treatment. Environ Protect 10(9):30–33.
18. Giese TP (2001) New wastewater treatment technologies. Public Works 132(5):326–332.
19. Zenon Environmental Inc. (2001) Activated sludge without a clarifier. Water Environ Fed 13(9):75.
20. Beaubien A, Trouve E, Urbain V, Amar D, Manem J (1994) Membrane bioreactors offer new solution to old wastewater treatment problems. Environ Solut 34–36.
21. Levy V (1996) The membrane bioreactor and the management of water resources in industry. Presentation at POLLUTEC 96.
22. Urbain V, Benoit R, Manem J (1996) Membrane bioreactor – a new treatment tool. J Am Water Works Assoc 88:75.
23. Kim JS, Lee CH, Chun HD (1998) Comparison of ultrafiltration characteristics between activated sludge and BAC sludge. Water Res (G.B.) 32:3443.
24. Dollorer J, Wilderer PA (1996) Biological treatment of leachates from hazardous waste landfills using SBBR technology. Water Sci Technol 34:437.
25. Livingston AG, Freitas dos Santos LM, Pavasant P, Pistikopoulos EN, Strachen LF (1996) Detoxification of industrial wastewaters in an extractive membrane bioreactor. Water Sci Technol 33(1):1.
26. Strachan LF, Freitas dos Santos LM, Leak DJ, Livingston AG (1996) Minimization of biomass in an extractive membrane bioreactor. Water Sci Technol 34: 273.
27. Cote P, Buisson H, Praderie M (1998) Immersed membranes activated sludge process applied to the treatment of municipal wastewater. Water Sci Technol (G.B.). 38: 437.
28. Chang LS, Lee CH, Ahn KH (1999) Membrane filtration characteristics in membrane-coupled activated sludge system – the effect of floc structure on membrane fouling. Sep Sci Technol 34(9):15.
29. Scott JA, Howell JA, Arnot TC, Smith KL, Brusk M (1996) Enhanced system $k_l a$ and permeate flux with a ceramic membrane bioreactor. Biotechnol Tech 10(4):287.
30. Parratiyar MG, Govind R, Bishop DF (1996) Treatment of trichloroethylene (TCE) in a membrane bioreactor. Biotechnol Bioeng 50:57.
31. Clapp L, Hartono R, Newman M, Park J (1996) Trichloroethylene degradation in a novel membrane bioreactor. The 69th Annual Water Environment Federation Conference, Dallas, TX.
32. Nomura T, Fuji T, Suzuki M (1997) Application of ceramic membrane with hydrophobic skin layer to separation of activated sludge. Water Sci Technol (G.B.) 35(4):137.
33. Choo KH, Lee CH (1998) Hydrodynamic behavior of anaerobic biosolids during crossflow filtration in the membrane anaerobic bioreactor. Water Res (G.B.). 32:3387.
34. van Dijk L, Roncken GCG (1997) Membrane bioreactor for wastewater treatment: the state of the art and new developments. Water Sci Technol (G.B.) 35(10):35.
35. Guender B, Krauth K (1998) Replacement of secondary clarification by membrane separation – results with plate and hollow fiber modules. Water Sci Technol (G.B.). 38:383.
36. Yamamoto K, Hiasa M, Mahmood T, Matsuo T (1989) Direct solid–liquid separation using hollow-fiber membrane in an activated sludge aeration tank. Water Sci Technol (G.B.) 21:43.
37. Choo KH, Stensel HD (1998) Sequencing batch membrane bioreactor treatment: nitrogen removal and membrane fouling evaluation. Water Environ Res 72(4):490–498.

38. Williams D (1997) Membrane treat wastewater and leachate. Pollut Eng 29(1):20–21.
39. Hensel J, Mills T (2002) High and lows across the industry: 2002 executive forecast. Environ Protect 13(1):12–14.
40. Freeman S, Leitner GF, Crook J, Vernon W (2002) A clear advantage. Water Environ Technol 14 (1):16–21.
41. Abbott K, Alpor H (2001) Protection from organic fouling. Environ Protect 12(6):34.
42. Suwa Y, Suzuki T, Toyohara H, Yamagishi T, Urushigawa Y (1992) Single-stage nitrogen removal by an activated-sludge process with crossflow filtration. Water Res (G.B.). 26:1149.
43. van Loosdrecht MCM, Henze M (1999) Maintenance, endogenous respiration, lysis, decay and predation. Water Sci Technol (G.B.) 39(1):107.
44. US Filter (2003) Immersed membrane bioreactor system. Water Eng Manage 150(1):4.
45. Pearson D (2003) Raisin producer's new process cuts wastewater and odors. Water Eng Manage 150(1):26–29.
46. Oreskovich R, Contestables J, Flat K, Watson IC, Rifleman J (2003) Reverse osmosis anion-filtration water plant. Water Eng Manage 150(1): 10–19.
47. Hudkins JM, Schnidt HE (2009) Wastewater Membrane Bioreactors – An Emerging Technology for the Development of a Reclaimed Water Supply. www.consulthai.com/2002.news/papers/waterReuse-2004-MBR
48. Sheindorf GN, Stahl N, Tenenbaum A, Levinsky Y (2003) Membrane bioreactors for final treatment of wastewater. Water Sci Technol 48(8):103–110.
49. Wang LK, Wu Z (2009) Activated sludge processes. In: Wang LK, Pereira NC, Hung YT, Shammas NK (eds) Biological treatment processes. Humana Press, Inc., Totowa, NJ, USA, pp 207–282.
50. Glastra M, Dijk HV, Verberk J (2004) Prospects of Membrane Bioreactors in Water Supply. Berichte aus dem IWW Rheinisch-Westfalisches Institut fur Wasserforschung gemeinnutzige, Gmbtt Band 37b, Mulheim an der Rhur, 2002, ISSN 0941-0961.
51. Wang LK, Li Y (2009) Sequencing batch reactors. In: Wang LK, Pereira NC, Hung YT, Shammas NK (eds) Biological treatment processes. Humana Press, Inc., Totowa, NJ, pp 459–512.
52. Shammas NK, Wang LK (2009) Emerging suspended growth biological processes. In: Wang LK, Shammas NK, Hung YT (eds) Advanced biological treatment processes. Humana Press, Inc., Totowa, NJ, pp. 619–648.
53. Wang LK (2008) Application of Membrane Separation Technologies in Food Processing Industry. Technical paper presented at 2008 National Engineers Week, Practicing Institute of Engineering, Foundation for Engineering Education, and NYS Society of Professional Engineers, Marriott, Albany, NY, February 14–15.
54. Wang LK (2008) Development and Applications of Membrane Bioreactor Technologies. Technical paper presented at 2008 National Engineers Week, Practicing Institute of Engineering, Foundation for Engineering Education, and NYS Society of Professional Engineers, Marriott Albany, NY, February 14–15.
55. Mustacchi R, Knowles CJ, Li H, Skibar W, Sunderland G, Dalrymple I, Jackman SA (2005) The effect of whole Cell immobilisation on the biotransformation of benzonitrile and the use of direct electric current for enhanced product removal. Biotechnology and Bioengineering. 91(4): 436–440. August 2005.
56. Gryta M, Morawski AW, Tiomaszewska M (2000) Ethanol production in membrane distillation bioreactor. Catalysis Today 56: 159–165.
57. Ergas SJ, Rheinheimer DE (2004) Drinking water denitrification using a membrane bioreactor. Water Res 38:3225–3232.

58. Pennsylvania State University (2008) DBP Precursor Removal by Membrane Bioreactor Technology. Pennsylvania State University, PA.
59. Li XY, Chu HP (2003) Membrane bioreactor for the drinking water treatment polluted surface water supplies. Water Res 37:4781–4791.
60. Matos CT, Velizarov S, Reis MAM, Grespo JG (2008) Removal of bromate from drinking water using the ion exchange membrane bioreactor concept. Environ Sci Technol 42(20):7702–7708.
61. Banks CJ, Heaven S (2008) Development of a coarse membrane bioreactor for two-stage anaerobic digestion of biodegradable municipal waste. In: Proceedings of the 5th IWA International Symposium on Anaerobic Digestion of Wastes and Energy Crops, Hammamet, Tunisia, May 25–28.
62. Carnation WWTP, King County, Washington (2008) Carnation Wastewater Treatment – Membrane Bioreactor. Technical paper presented at the 23rd Water Reuse Symposium, Dallas, TX, September 8.
63. Manigas L (2008) Use of Membrane Bioreactors for the Bioremediation of Groundwater Polluted by Chlorinated Compounds. PhD. Thesis, Universita degli Studi di Cagliari, November 6, http://veprints.unica.it/178/
64. Copeland A, Cole K, Barrows R, Pyne JC (2007) The Design Elements of State-of-the-art Treatment Technology: MBR Wastewater Treatment Systems. Technical paper presented at the 2007 Virginia AWWA/WEA Water JAM.
65. Aqua-Aerobic Systems, Inc. (2008) Membrane system coupled with batch reactor. Water Online Newsletter, June 24.
66. Wang LK, Wang MHS (2009) Chemical and Biochemical Technologies for Environmental Infrastructure Sustainability. Technical paper presented at 2009 National Engineers Week, Practicing Institute of Engineering, Foundation for Engineering Education, and NYS Society of Professional Engineers, Marriott, Albany, NY, February 5–6.
67. Wang LK, Shammas NK, Selke WA, Aulenbach DB (eds) (2010) Flotation technology. Humana Press, Totowa, NJ. 680 P.
68. General Electric (2008) GE's ZeeWeed Membrane Bioreactor Technology Selected for One of North America's Largest Wastewater Plant Upgrades and Expansions. General Electric Co., Schenectady, NY. February. www.reuters.com/article/pressRelease/
69. Guglielmi G, Andreottola G (2010) Selection and design of membrane bioreactors in environmental bioengineering. In: Wang LK, Ivanov V, Tay JH, Hung YT (eds) Environmental biotechnology. Humana Press, Totowa NJ, USA. PP 439–516.
70. King County, Washington (2010) Advanced Treatment Technology at Brightwater, Washington. King County, Washington, USA. www.kingtoncounty.gov
71. Chen W, Sun FY, Wang XM, Li XY (2010) A Membrane bioreactor for an innovative biological nitrogen removal process. Water Sci Technol 61(3):671–6.
72. Yigit NO, Civelekogla G, Cinar O, Kitis M (2010) Filterability of membrane bioreactor sludge: impacts of polyelectrolytes and mixing with conventional activated sludge. Water Sci Technol 61 (3):659–669.
73. Kippax V (2010) Importance of air scour in membrane bioreactor system. Water World 2(30). www.waterworld.com.
74. Kang CW, Hua JS, Lou J, Liu W, Jordan E (2008) Bridging the bap between membrane bioreactor pilot and plant studies. J Membr Sci 325:861–871.
75. Cui ZF, Chang S, Fane AG (2003) The use of gas bubbling to enhance membrane processes. J Membr Sci 221:1–35.

76. Zenon Membrane Solutions. (2010) Cauley Creek Water Reclamation Facility. Zenon Membrane Solutions, Oakville, Ontario, Canada. GE Water & Process Technologies, Schenectady, NY, USA. www.gewater.com.
77. Zenon Membrane Solutions (2010) Australian Golf Course Recycles Municipal Wastewater with Onsite ZeeWeed MBR. Zenon Membrane Solutions, Oakville, Ontario, Canada. GE Water & Process Technologies, Schenectady, NY, USA. www.gewater.com.
78. Zenon Membrane Solutions (2010) Vancouver Convention & Exhibition Centre. Zenon Membrane Solutions, Oakville, Ontario, Canada. GE Water & Process Technologies, Schenectady, NY, USA. www.gewater.com.
79. Zenon Membrane Solutions (2010) Johns Creek Environmental Campus. Zenon Membrane Solutions, Oakville, Ontario, Canada. GE Water & Process Technologies, Schenectady, NY, USA. www.gewater.com.
80. Zenon Membrane Solutions (2010) Corona Wastewater Treatment Plant. Zenon Membrane Solutions, Oakville, Ontario, Canada. GE Water & Process Technologies, Schenectady, NY, USA. www.gewater.com.
81. GE Water & Process Technologies (2010) Membrane Bioreactor Design Considerations. http://www.gewater.com/products/equipment/mf_uf_mbr/mbr/design_considerations.jsp GE Water & Process Technologies, Schenectady, NY, USA. www.gewater.com.
82. Fane AG (2002) Membrane bioreactors: design and operational options. Filtration Sep 39 (5):26–29.
83. Van Bentem AGN, Petri CP, Schyns PFT (2007) Membrane Bioreactors: Operation and Results of an MBR Wastewater Treatment Plant. International Water Association. 120.
84. Judd S, Jefferson B (eds) (2003) Membranes for industrial wastewater recovery and re-use. Elsevier Science New York, NY. p 308.
85. Wang LK, Hung YT, Shammas NK (eds) (2010) Handbook of advanced industrial and hazardous waste treatment. CRC Press, New York, NY. 1378 p.
86. Wang LK, Shammas NK, Hung YT (eds) (2009) Advanced biological treatment processes. Humana Press, Totowa, NJ. 737 pages.
87. Wang LK, Tay JH, Tay STL, Hung YT (eds) (2010) Environmental bioengineering. Humana Press, Totowa, NJ. pp. 378–380.

6

Treatment of Food Industry Foods and Wastes by Membrane Filtration

Lawrence K. Wang, Nazih K. Shammas, Munir Cheryan, Yu-Ming Zheng, and Shuai-Wen Zou

CONTENTS

Abstract Membrane separation processes are based on the ability of semipermeable membranes of the appropriate physical and chemical nature to discriminate between molecules primarily on the basis of size and to a certain extent, on shape and chemical composition. A membrane's role is to act as a selective barrier, enriching certain components in a feedstream, and depleting the others. One of the chief attractions of membrane technology is the low energy requirement compared to other food processing technologies. Since membrane processes are nonthermal and do not involve a change of phases, they are energy-efficient and do not change the nature of the foods during their process operation. This chapter presents the membrane process theory and case histories of various production applications in the food industry. Operational problems and recommended engineering solutions for membrane process optimization are presented and discussed.

Key Words Membrane filtration • food industry • food processing • energy conservation • waste minimization • recycle • waste treatment

From: *Handbook of Environmental Engineering, Volume 13: Membrane and Desalination Technologies*
Edited by: L.K. Wang et al., DOI: 10.1007/978-1-59745-278-6_6

1. INTRODUCTION

In the past decade, many developed countries have made major public sector commitments to build awareness of pollution prevention, waste minimization, and energy conservation. These commitments, most notably in the USA, the Netherlands, and Denmark, have led the industrial sectors to investigate and implement pollution prevention measures for existing manufacturing processes and products. As a result, cleaner production is now seen in these developed countries as a potentially cost-effective complement to environmental pollution control.

There are now several efforts of the United Nations Industrial Development Organization (UNIDO) and the United Nations Environment Program (UNEP) to transfer the experience of developed countries in the field of cleaner production for developing countries. All of these efforts are typical examples of technology transfer (that is, the transfer of knowledge, skills, equipment, etc.) to achieve a particular objective: the reduction of pollution intensity and/or energy consumption in the industrial sectors of developing countries through the improvement of manufacturing processes. This chapter is prepared for the transfer of membrane technology – a best available technology (BAT) to the food industry in developing countries.

Since the birth of membrane separation technology in 1960, a vast array of applications unmatched by any other processing technique in its variety has been developed. Desalination and water treatment by reverse osmosis (RO) is probably the earliest and best known application. The decade of the 1970s saw increasing usage in the chemical process industries. In the 1980s, the most significant inroads were made by this technology in the food and biotechnology processing industries, as a gentle and efficient way of fractionating, concentrating, and clarifying a variety of products (1–7). Environment-related cleaner production applications were the focus of the 1990s, such as water reuse and recycling (e.g., in food processing and corn wet-milling plants), the reduction or elimination of waste-generating processes (e.g., filter-aids in processing fruit juices and sweeteners), and the recovery or production of valuable by-products from waste streams (e.g., protein concentrates and biodegradable plastic monomers from cheesy whey).

For membrane process operation, buildup of a layer on the surface of the membrane and deposition of foulants within the membrane pore structure due to high solution concentrations are the major mechanisms responsible for membrane flux decline. Membrane fouling requires frequent chemical cleanings, and in the worst case, membrane replacement. The solute passage of inorganics (measured as sodium chloride) is very dependent upon recovery, while the solute passage of low-molecular weight organics (measured as chemical oxygen demand) remains relatively constant. The potential to efficiently remove both organics and inorganics from food processing effluents makes membrane technology a unique advanced processing method for food manufacturing, water pollution control, water reuse, nutrient separation, and by-product recovery. Engineering experience of the authors and other experts in these membrane separation applications and pretreatment steps can be found in the literature (8–11).

2. MEMBRANE PROCESSES, MODULES, AND EQUIPMENT

2.1. *Membrane Processes*

Membranes are porous filtration mediums, which can be cationic, anionic, or nonionic in nature, and acts as a barrier to prevent mass movement of selected phases, but allows passage of other remaining phases. The phases include solid phase (such as suspended solids, organic solids, inorganic solids, etc.), liquid phase (such as water, ethanol, chloroform, etc.), and gas phase (such as air, nitrogen, oxygen, etc.). Membrane processes include the following five main categories for processing water and wastewater (12, 13):

1. *Microfiltration (MF):* MF is a pressure filtration process for the separation of suspended solids in the particle size range of about 0.08–10 μm. The primary function affecting solids separation from water is the size of the solids. The hydraulic pressure applied in MF is about 1–2 bar, or 15–20 psig, for overcoming the resistance of the "cake."
2. *Ultrafiltration (UF):* UF is another pressure filtration process for the separation of macromolecular solids in the particle size range of about 0.001–0.1 μm. The hydraulic pressure required by UF for overcoming hydraulic resistance of the polarized macromolecular layer on the membrane surface is about 1–7 bar.
3. *Nanofiltration (NF):* NF membranes are multiple-layer thin-film composites of polymer consisting of negatively charged chemical groups, and are used for retaining molecular solids (such as sugar) and certain multivalent salts (such as magnesium sulfate), but passing substantial amounts of most monovalent salts (such as sodium chloride), at an operating pressure of about 14 bar or 200 psig. Both molecular diffusivity and ionic charge play important roles in the separation process. The sizes of molecular solids and multivalent salts to be rejected by NF are normally in the range of 0.0005–0.007 μm.
4. *Reverse osmosis (RO):* RO membranes are mainly made of cellulose acetate (CA) with pore sizes of 5–20 Angstrom units. These membranes are used for rejecting salts (as high as 98%) and organics (as high as 100%), at an operating pressure of 20–50 bar or 300–750 psig. Hydraulic pressure (through a pump) is used to provide the driving force for permeation, or for overcoming the chemical potential difference between the concentrate and the permeate, expressed in terms of osmotic pressure. Sizes of molecular solids and salts (multivalent as well as monovalent) to be rejected by RO are normally in the range of 0.00025–0.003 μm.
5. *Electrodialysis (ED):* ED uses voltage or current as the driving force to separate ionic solutes. The size of ionic solutes to be rejected or separated by ED is normally in the range of 0.00025–0.08 μm, depending on the pore size of ED membranes. EDR is the electrodialysis reversal (or reverse electrodialysis) process – which is similar to ED, but its cathodes and anodes can be reversed for automatic cleaning operation.

Figure 6.1 illustrates the relationships among MF, UF, NF, RO, ED, and conventional separation processes (such as cloth/fiber filters, screens, gel chromatography, dialysis, ion exchange, distillation, freeze concentration, solvent extraction, foam and bubble fractionation, ultracentrifuges centrifuges, liquid cyclones, and gravity sedimentation. Membrane bioreactor (MBR) adopts either MF or UF. Figure 6.2 shows the capabilities of MF, UF, NF, and RO and Fig. 6.3 illustrates the capabilities of ED and EDR processes (14).

The applied pressure for typical membrane filtration process is given in Table 6.1:

Membrane processes are characterized by two basic process parameters:

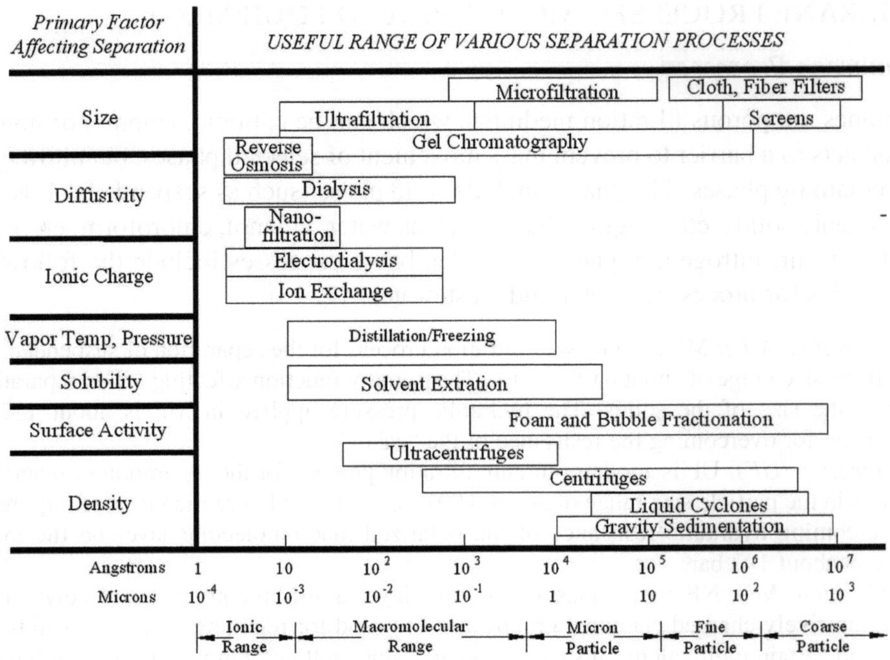

Fig. 6.1. Filtration application guide for membrane separation processes.

1. Flux, which is the rate of transport of solvent or solution through the membrane
2. Rejection, which is the degree of separation of a particular feed component

There are five major variables that affect the two basic process parameters (8):

1. Driving force in terms of applied transmembrane pressure (TMP) and/or electric voltage/current
2. Flow velocity which affects turbulence and mass transfer coefficient
3. Process water temperature which has effects on physical properties, such as density, viscosity, diffusivity, osmotic pressure, surface tension, and others
4. Feed stream characteristics in terms of particle concentration, particle size, viscosity, molecular weight, molecular configuration, ionic charges, and fouling potential
5. Membrane module in terms of materials, pore sizes, membrane configuration membrane ionic charges, and feed compatibility

2.2. *Membrane Modules and Equipment*

There are basically six different designs of membrane modules (12, 13):

1. Tubular modules with channel diameters greater than 3 mm
2. Hollow fiber or capillary modules made of self-supporting tubes, usually 2 mm or less in internal diameters
3. Plate modules
4. Spiral-wound modules
5. Pleated sheet modules
6. Rotary modules

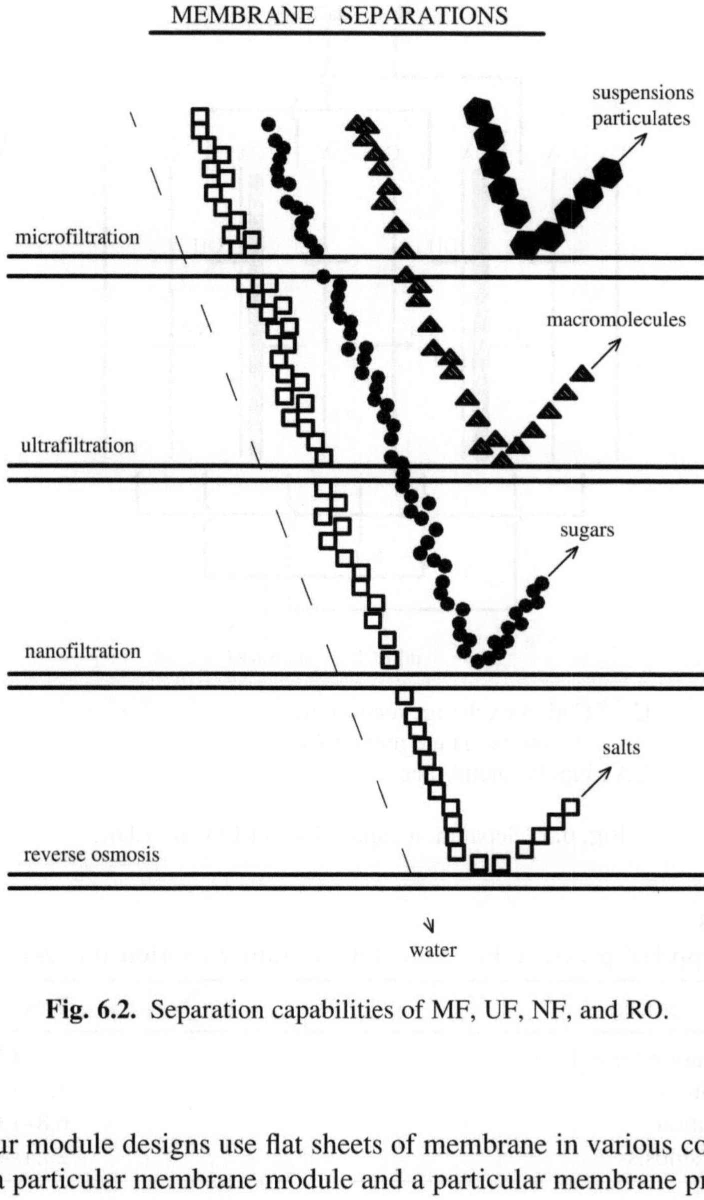

Fig. 6.2. Separation capabilities of MF, UF, NF, and RO.

The latter four module designs use flat sheets of membrane in various configurations.

In selecting a particular membrane module and a particular membrane process, the major criteria are:

1. Feed stream characteristics, which affect the biocompatibility of the membranes.
2. Flux requirements, which are controlled by the volumetric rate of a feed stream.
3. Rejection requirements, which decide the process objectives and treatment efficiencies.
4. Cost requirements, which are affected by the energy consumption, membrane replacement cost and operating and cleaning costs.

Biocompatibility of the membranes relates to the interaction between the membrane module and the feed stream. Major biocompatibility factors include: (a) stability to extremes

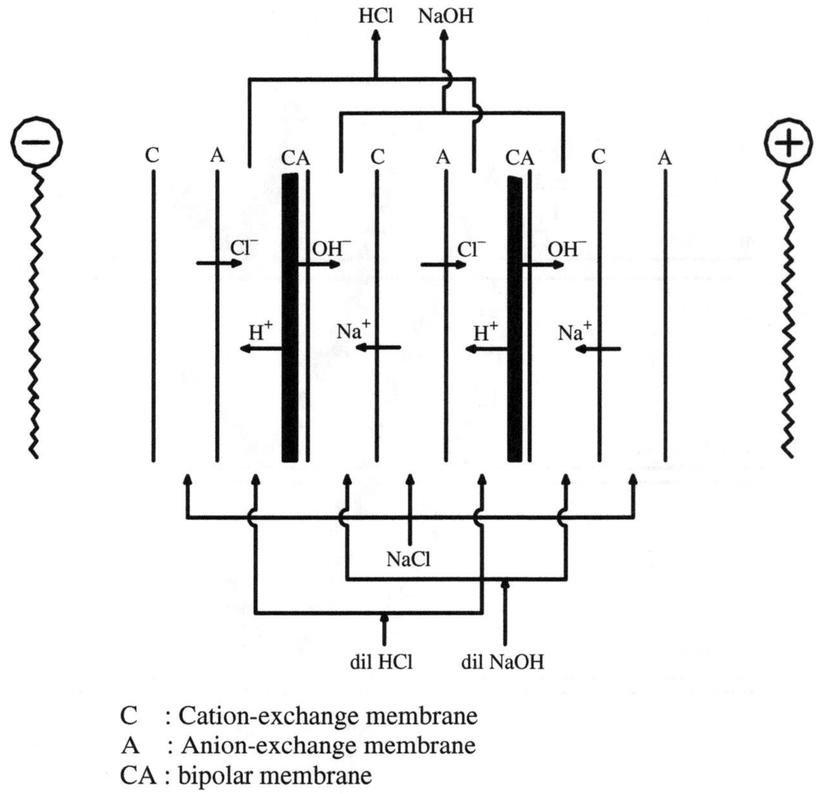

Fig. 6.3. Separation capabilities of ED and EDR.

Table 6.1
List of applied pressure for typical membrane filtration process

Membrane process	Pressure (atm)
Microfiltration (crossflow)	0.7–1.7
Ultrafiltration	1.7–10.2
Nanofiltration	6.8–13.6
Reverse osmosis	20.4–40.8

in temperature, pressure, and pH, especially under cleaning and sanitizing conditions; (b) membrane –solute interactions, which affect the rate of fouling, cleaning, yields, and rejection of individual feed substances, and (c) the acceptability of the membrane as a contact material for the final product, which essentially implies using membrane materials that are inert and do not leach out any toxic substances from the membrane into the final product. In this regard, there are new generations of membranes, made of expensive inorganic inert materials, such as ceramics, stainless steel, carbon–zirconia, etc.

MF membranes are made of a wide range of inorganic materials (such as alumina, zirconia–carbon composites, carbon–carbon composites, ceramics, stainless steel, silica, etc.) and natural and synthetic polymers (such as polypropylene, polycarbonates, polysulfone, polyvinylchloride, PVC, copolymer – cellulose esters, cellulose acetate, etc.) (15–16).

UF membranes are mainly made of polysulfone-type materials (such as polyether sulfone, polyphenylene sulfone, sulfonated polysulfone, etc.). This type of membranes is also available in a wide range of organic materials (such as PVC copolymer, cellulose acetate, etc.) and inorganic materials (such as ceramic composites, stainless steel, etc.).

Most NF membranes are multiple-layer thin-film composites of synthetic polymers. The active NF membrane layer usually consists of negatively charged chemical groups. NF membranes are of porous filter media with an average pore diameter of 2 nm. The nominal molecular weight cutoff ranges from 100 to 200. The active NF membrane layer can be made of polyamide, polyvinyl alcohol, sulfonated polysulfone, and sulfonated polyethersulfone. Salt rejection by NF membranes is mainly due to electrostatic interaction between the ions and the NF membrane. Rejection of neutral substances is by size (17).

CA and derivatives are widely used as the RO membranes, despite their real and perceived limitations. Thin-film composite membranes containing a polyamide separating barrier on a polysulfone or polyethylene supporting layer, generally give better performance for RO applications with regard to temperature and pH stability and cleanability, but have almost zero chlorine resistance. In general, these thin-film composite membranes will be the material of choice for RO applications, unless there is a specific fouling problem with these membranes.

The membranes of UF, NF, and RO have been tried for ED with various results. The use of ionic membranes (18) in ED and EDR processes has been applied as illustrated in Fig. 6.3.

Table 6.2 shows the special features of four types of membrane systems: tubular membrane modules, hollow-fiber membrane modules, plate membrane modules and spiral-wound membrane modules (14). Each design has its own special applications, advantages, and disadvantages.

The large-bore tubular membrane modules (Fig. 6.4) are suitable for food streams with high concentration of suspended solids, such as citrus juices and animal waste streams, even

Table 6.2
Special features of membrane equipment

	Tubular	Hollow fibers	Plates	Spiral
Channel height (mm)	>10	<3	<2.5	<1.5
Spacers	No	No	Yes/no	Yes
Prefiltration	No	Yes	Yes/no	Yes
Flow	Turbulent	Laminar	Laminar	Turbulent
Flux	High	Intermediate	Intermediate	Low
Area/volume	Low	High	Intermediate	High
Energy/flux	High	Intermediate	Intermediate	Low

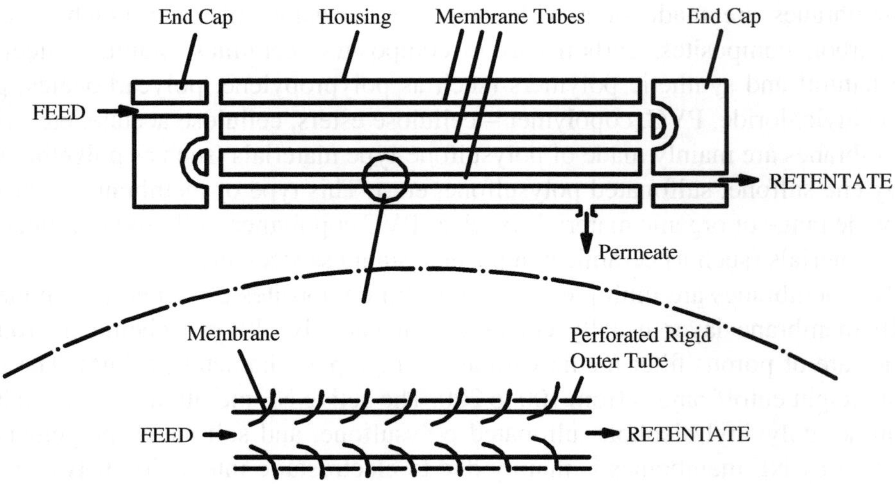

Fig. 6.4. Schematic of tubular membrane module.

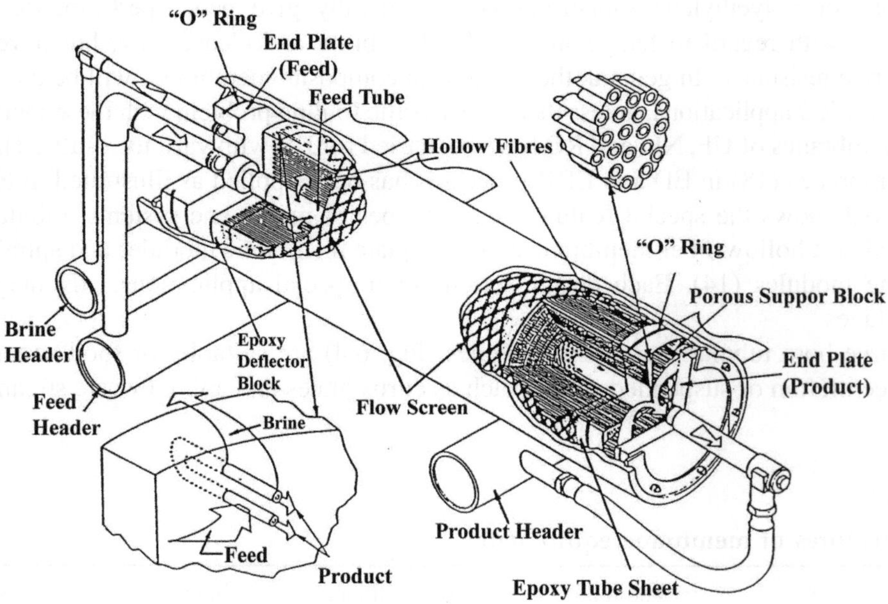

Fig. 6.5. Schematic of hollow fiber membrane module.

though the tubular membrane modules have the lowest packing densities and highest energy consumption among all modules (19). The tubular designs with ceramic inorganic membranes are frequently used in the food processing industries.

The hollow-fiber membrane modules (Fig. 6.5) have extremely high packing density (surface area to volume ratios) and comparatively low energy consumption, and are suitable for comparatively clean feed streams with low concentrations of suspended solids and macromolecules (19). Most food materials in the food processing feed streams, however, contain macromolecules displaying non-Newtonian behavior. Their viscosity will increase dramatically above certain concentrations, making pumping difficult and reducing the mass-transfer rate within the boundary layer. This will eliminate most hollow fiber/capillary modules because they cannot withstand high-pressure drops.

Membrane modules utilizing flat sheets (spiral-wound, plate and pleated sheet modules) usually have a mesh-like spacer between sheets of membrane. This restricts their use to clear feed streams containing only fine suspended solids. Feed streams containing large suspended solids would be poorly treated in spiral-wound modules, owing to the spacers in their feed channels. On the other hand, spiral-wound membrane modules (Fig. 6.7) are the lowest in capital costs and energy consumption. The trend in the food and beverage industries seems to be away from plate modules (Fig. 6.6) and toward spiral-wound modules (Fig. 6.7), with ceramic tubular modules (Fig. 6.4) holding their own (19).

2.3. *Cost of Membrane Filtration*

The cost of a given membrane filtration installation is determined by two parts, the capital cost and the operational cost. The capital cost can be further divided into three parts: membrane modules; cost of piping, pumps, electronics, vessels; pretreatment and posttreatment.

The prices mentioned in here are prices for complete systems, which mean complete systems, typically skid mounted, with membrane elements, internal piping and tubing, valves, pumps, and control equipment. The prices mentioned do not include buildings, installation of the system or auxiliary facilities like water, drain, electric power, and pressurized air. From mid 1970s to 1989, prices for membrane systems in the USA were declining. In 1989, spiral-wound elements were massively introduced into the dairy industry, which reduced the price

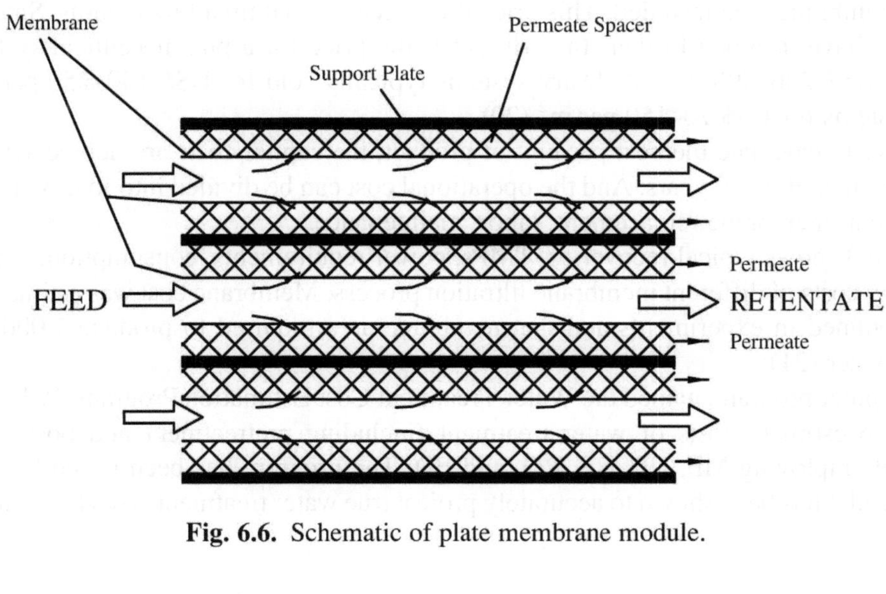

Fig. 6.6. Schematic of plate membrane module.

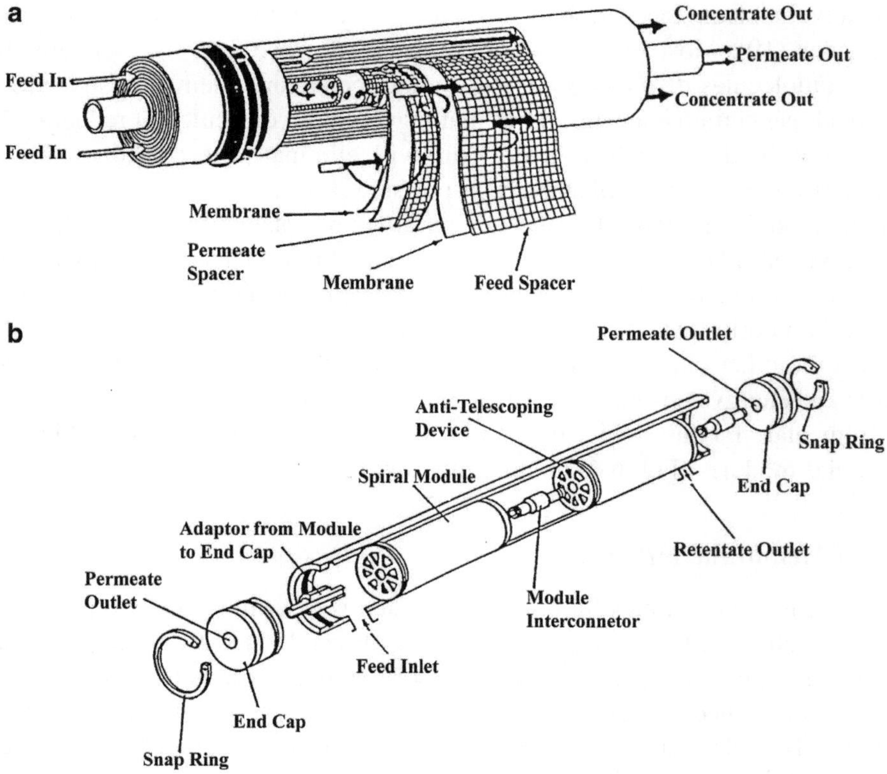

Fig. 6.7. Schematic of spiral-wound membrane module.

for complete systems by a factor of three, from around US$ 1,400/m^2 to as low as US$ 370 per m^2 membrane area installed. This price drop occurred within a few months. Since 1989, the prices have declined further. In early 2001, the price for a pharmaceutical system was typically US$ 200–300 per m^2. Dairy systems typically sold for US$ 150–250 per m^2 and water systems for US$ 70–150 per m^2 (20).

In order to calculate the cost per unit of product, the capital costs are depreciated over a period of time, often 10 years. And the operational cost can be divided into four parts: energy requirement; membrane replacement; labor; maintenance.

Table 6.3 shows typical recoveries, TMPs, power requirements, consumption, and resulting energy costs of different membrane filtration process. Membrane cost was estimated from fluxes obtained in experiments and the membrane area required to produce 1,000 m^3/day product water (21).

A computer program, named the Water Treatment Cost Estimation Program (WTCost) can be used to estimate costs of water treatment (including pretreatment and posttreatment) processed employing MF, UF, NF, RO, and ED. The program has been tested by industry experts, and it has been shown to accurately project true water treatment costs for a number of plants (22).

Table 6.3
Typical recoveries, trans-membrane pressures, power requirements, consumption, and resulting energy costs based on applied pressures with an overall motor/pump efficiency of 40%. Assumed energy cost 0.05 US$/kWh

	Recovery (%)	Pressure (kPa)	Consumption (kWh/day)	Energy cost (US$/m^3)
MF	90–98	100	77.5–70.8	0.004–0.0035
UF	90–98	300	231.5–212.5	0.012–0.011
NF	75–95	500	463.5–365.5	0.023–0.018
RO	50–80	1000	1389–868.3	0.070–0.043

3. OPERATIONAL PROBLEMS AND ENGINEERING SOLUTIONS

All aspects of fouling involve the trapping of some type of material within the membrane process device itself or on the surface of the membrane. Since the causes, symptoms, and cures are different, the following five types of fouling are discussed below (23, 24):

1. Membrane scaling
2. Metal oxides fouling
3. Colloidal fouling
4. Biological fouling
5. Cleaning agent fouling

The problems of membrane scaling and metal oxides fouling are common to all membrane systems, regardless of the type or source of feed water stream (i.e., sea water or food processing water). Colloidal and biological types of fouling are specifically important in the design of a membrane system for treating food processing water and wastewater. To establish the optimal membrane cleaning regime, it behooves the operator to understand clearly the nature and the extent of the interaction of cleaning agents with the membrane.

3.1. *Membrane Scaling*

Precipitation of dissolved salts results from exceeding the solubility limits in a concentrated stream. This may occur because the salts in feed water are normally concentrated from two to four times. This is a function of system recovery; 50% results in a concentration of 2 and 75% results in a concentration of 4. The most common precipitates encountered are those of calcium carbonate, calcium sulfate, and silica. Calcium carbonate precipitation can be prevented by lowering the feed pH by the addition of hydrochloric acid. Carbonates and bicarbonates are then converted to carbon dioxide gas that passes through the membrane without difficulty. The precipitation of calcium sulfate can be retarded by the addition of a sequestering agent, such as sodium hexametaphosphate. The silica precipitation can be prevented by reducing system recovery to keep silica concentrations below saturation.

3.2. *Metal Oxides Fouling*

The oxide or hydroxide forms of iron, manganese, and aluminum are the most common problem causing compounds. Operation of the membrane system at acidic pH levels and possibly, iron and manganese removal by oxidation are recommended pretreatment alternatives.

3.3. *Colloidal Fouling*

Colloidal fouling is caused by the entrapment of colloids on the membrane surface caused by the coagulation of the colloids in the membrane system. Reduction of the concentration of colloids can be accomplished by the following pretreatment schemes (16):

1. Filtration, using sand, carbon or other filter media
2. In-line coagulation with iron, salts, and polyelectrolyte aids, followed by pressure filtration
3. Iron salt coagulation, flocculation, flotation (or sedimentation), and filtration

3.4. *Biological Fouling*

Biological fouling is caused by the growth of microorganisms in membrane modules. Two major factors are considered:

1. Biological fouling or plugging which interferes with the hydraulic flow of the membrane modules.
2. Biological attack of the membrane, which is extremely serious when CA membranes are attacked by the microorganisms' enzyme system.

Periodically feeding a cleaning solution containing free chlorine residuals of 0.5–1.0 mg/L is recommended to prevent detrimental biological growths on the CA membrane surface. Gas chlorine and liquid sodium hypochlorite have been used successfully for this purpose. It should be noted that thin-film composite membranes containing a polyamide separating barrier on a polysulfone or polyethylene supporting layer, generally give better RO performance with regard to temperature stability, pH stability, and cleanability, but have almost zero chlorine resistance. Generally, hydrophobic membranes consisting of polysulfone and polyamides may be more prone to biological fouling than hydrophilic membranes consisting of celluloses.

3.5. *Cleaning Agent Fouling*

Sometimes, membranes "fouling" problems may actually be "cleaning" problems. Ideally, the membrane material, and all other food contact surfaces, should be compatible with normal food and dairy cleaners, such as mild caustic or acid solutions. Severe cleaning problems may require the use of enzyme detergents.

4. MEMBRANE FILTRATION SYSTEM

4.1. *Basic Membrane System*

The major components of any force-driven membrane filtration system (MF, UF, NF, or RO) are the high-pressure booster pump and the membrane module. Figure 6.8 shows a simplified process flow scheme. Feed water enters the suction side of the booster pump, which raises the pressure to a level required for operation, depending upon the membrane process (MF, UF, NF, or RO) and the application (desalination, milk processing, juice processing, or wastewater treatment) (25, 26). The feed water enters the membrane module at pressure P_i, and is split into two streams: (a) permeate at pressure P_p and (b) retentate at pressure P_o.

The permeate that passes through the membrane is the product water, which has drastically reduced suspended and/or dissolved solids concentration, C_P, depending on the type of membrane process applied. The retentate does not actually pass through the membrane, but flows parallel to the surface, exiting on the opposite side. The retentate is the concentrate or reject, retaining high concentration of rejected solids (C_R). A portion of the retentate can be recycled for reprocessing, as shown in Fig. 6.8. The TMP, pressure drop, flux, reject (R), 100% rejection, and 0% rejection are all defined in Fig. 6.8. The flow rate/velocity is a function of pressure drop. The mass balance of flows is as follows:

$$Q_f = Q_p + Q_r. \tag{1}$$

Recovery or system conversion, Y in % is expressed as follows:

$$Y = \frac{Q_p}{Q_f} \times 100, \tag{2}$$

where
Q_f = feed flow
Q_p = permeate flow
Q_r = retentate flow

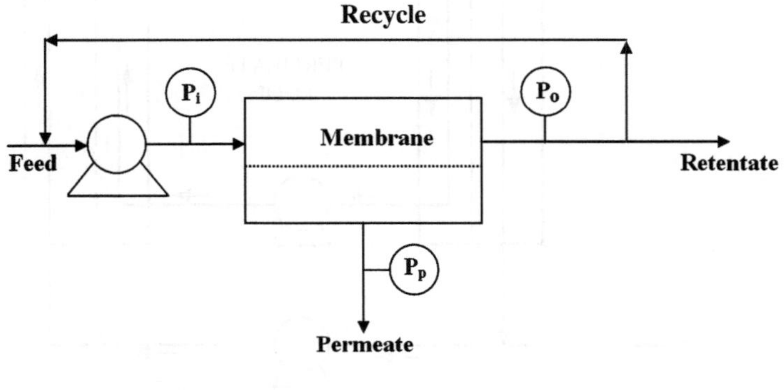

Fig. 6.8. Simplified membrane process system.

4.2. *Uniform Transmembrane Pressure (UTP) System*

Figure 6.9 shows a double loop constant-pressure membrane process system, also known as the Bactocatch process (27). The process requires the simultaneous operation of a retentate pumping loop and a permeate pumping loop, to simulate a backwashing operation, but in a continuous manner rather than the periodic or intermittent traditional practice. With two parallel flows adjusted so that the pressure drop is the same on the permeate and retentate sides of the module, the pressure profile would be more like that shown in Fig. 6.10b instead of the conventional membrane process system's pressure profile shown in Fig. 6.10a. The flux stays high and the rejection remains low (27).

5. APPLICATIONS OF MEMBRANE TECHNOLOGY IN THE FOOD INDUSTRY

5.1. *Production of Dairy Products Using MF, UF, and RO*

Figure 6.11 shows various applications of MF, UF, and RO for processing milk:

1. Production of high-fat cream and fat-free milk using MF
2. Production of cheese, specialty milk products, total protein isolates, lactose, fermented food and nonfood products using UF
3. Production of whole milk powder, bulk milk, and special milk products using RO

Figure 6.12 shows how MF, UF, and RO can be connected in series for the processing of milk by sequential operations.

In each case in Figs. 6.11 and 6.12, the main applications of MF in milk or whey processing are fat separation, bacterial removal, and caseinate recovery. The bacteria and fat in milk products are considerably larger in molecular size than all other components. A membrane

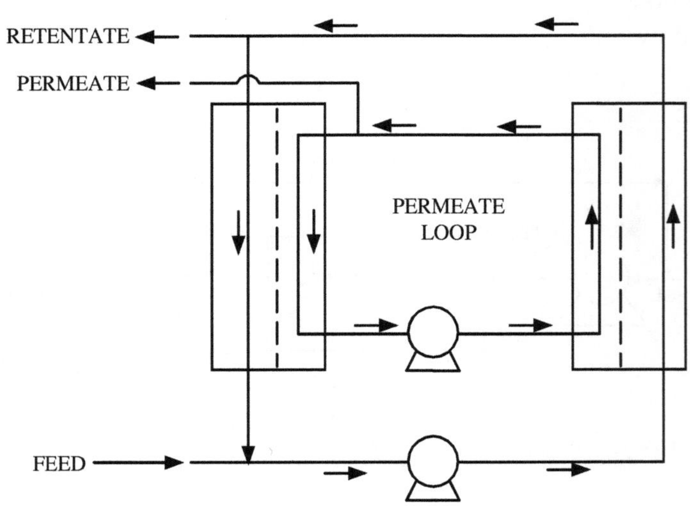

Fig. 6.9. Uniform transmembrane pressure membrane (UTP) process system.

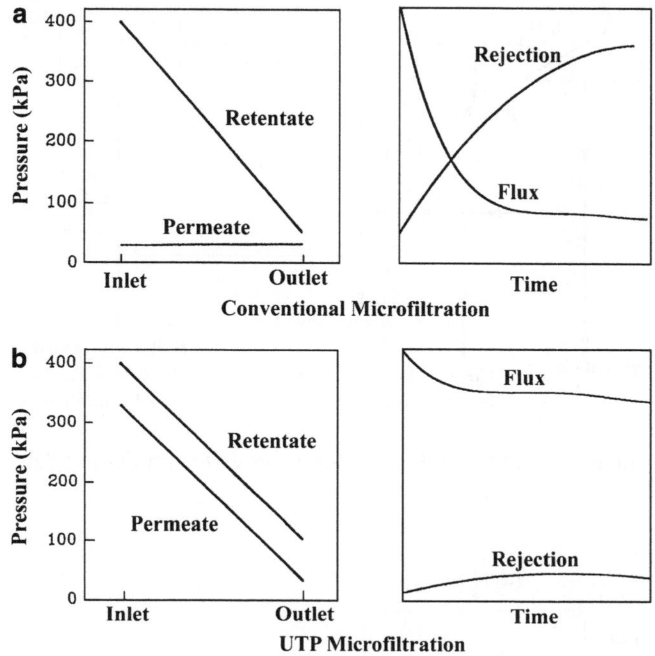

Fig. 6.10. Pressure profiles and performance characteristics of conventional MF and UTP-MF.

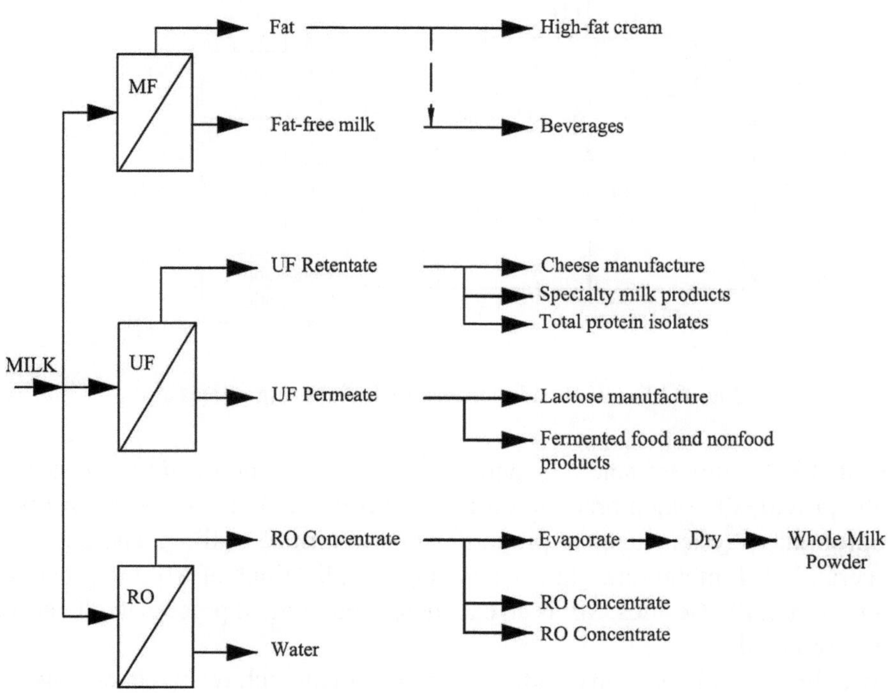

Fig. 6.11. Applications of MF, UF, and RO individually for processing milk.

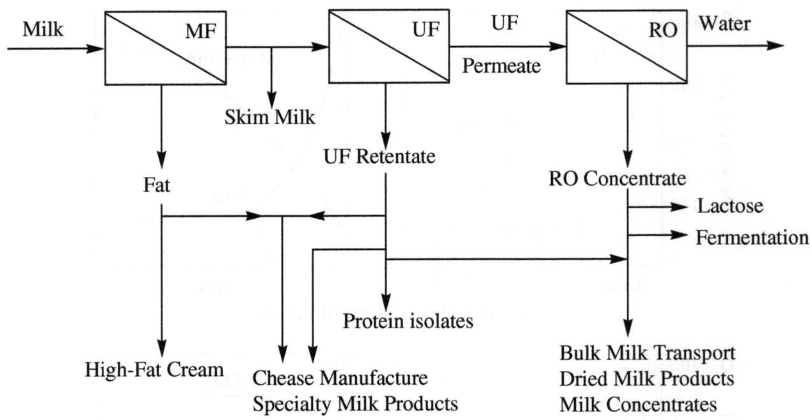

Fig. 6.12. Connection of MF, UF, and RO in series for producing dairy products.

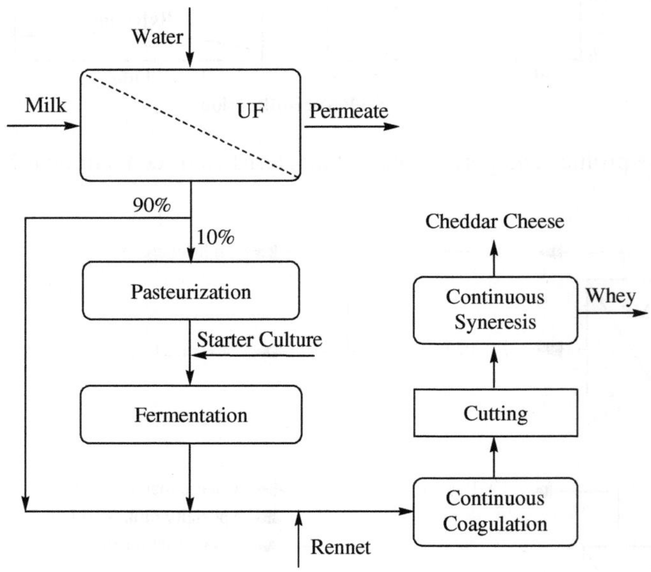

Fig. 6.13. UF cheddar cheese manufacturing process.

with pores of 0.3–1.0 μm are able to separate the fat and bacteria from the rest of the milk components, provided the membrane has a fairly uniform and narrow pore size distribution and the appropriate physicochemical properties that minimize fouling. The development of inorganic ceramic MF membranes has widened the applications of MF for processing milk products and skim milk. Besides, MF is an excellent pretreatment process to UF, aiming at fat and bacterial removals.

UF is widely used in the dairy industry for the manufacture of cheese, as shown in Figs. 6.11–6.13. From a membrane technologist's point of view, cheese manufacturing process can be considered as a fractionation process, whereby casein protein and fat are concentrated in the curd while lactose, soluble proteins, minerals, and other minor components are lost in the whey. Thus, one can readily see the application of UF (Figs. 6.11 and 6.13): concentrating milk to a protein/fat level normally found in cheese, then converting this "pre-cheese" to cheese by conventional or modified cheese-making processes (28, 29). Figure 6.12 shows that the combination of MF and UF membranes forms an appropriate system for manufacturing cheese. A UF cheddar cheese manufacturing plant (Fig. 6.13) with a capacity of 1 million L whole milk per day was installed in 1990 at Land 0' Lakes in Minnesota, USA.

RO treatment of whole milk dry powder and other products is shown in Fig. 6.11. RO treatment of skim milk from MF–UF permeates for the production of skim milk dry powder and other products is shown in Fig. 6.12. In either system, RO is operated primarily for removing water and for reducing energy consumption in milk drying plants. When RO is used as a preconcentrator, it is best to do a 2.5× concentration by RO followed by a conventional evaporation process. A comparison of the energy consumption of RO and thermal evaporation for a 2.5× concentration of milk is shown in Table 6.4. It can be seen that by using RO the potential energy savings are enormous (30).

In Sweden, the Bactocatch process shown in Fig. 6.9 involves the use of both tubular ceramic MF membranes (1.4 μm pore diameter) and the double loop constant pressure operation; it produced stable pasteurized and refrigerated milk products. Fluxes of 500–700 L/m^2/h were achieved for a period of over 6 h. If needed, the TMP can be increased slightly by operating at 0.05–0.1 bar (0.75–1.5 psi) to compensate for fouling. Bacterial retention is 99% with the microbial load usually found in milk.

5.2. *Production of Fruit and Tomato Juices Using MF, UF, and RO*

There are three primary areas where membranes can be applied to the processing of fruit or tomato juices (31):

Table 6.4
Energy consumption of RO vs. thermal evaporation for 2.5 × concentration of milk

Process	Area (m^2)	Energy (Kcal/kg milk)
Thermal concentration		
Open-pan boiling	10.4	455
Evaporator	10.4	455
Double-effect evaporator	25	209
Mechanical vapor recompression	32	136
Membrane processes		
Batch, single-pump	65	80
Batch, dual-pump	65	7
Continuous, one-stage	206	16
Continuous, three-stage	93	7

1. Clarification, e.g., in the production of sparkling clear beverages using MF or UF
2. Concentration, e.g., using RO to produce fruit juice concentrates of greater than 42 degree Brix
3. Deacidification, e.g., using ED to reduce the acidity in fruit juices

Figure 6.14 introduces the Fresh-Note process, also known as Separa-System, which is a multistage process (30): (a) UF is first used with the juice to separate out the pulp (the "bottom solids") from the serum which contains the sugars and flavor components; (b) a pasteurizer is then used to treat the UF retentate, which is about 5–10% of the feed volume for destroying microorganisms and improving the stability of the juice product; (c) a combination of high rejection RO in early stages and low rejection RO in later stages for treating the serum (UF permeate) aiming at the recovery of any traces of sugar or flavor compounds; and (d) a blending unit is finally used to blend the RO concentrated serum and the pasteurized pulp, to produce an orange juice concentrate with 45–60 degree Brix or 45–60% soluble solids.

Concentration of tomato juice presents a different challenge because tomatoes have high pulp content (25% fiber) and high viscosity (behaving in a non-Newtonian manner). A successful RO plant for tomato juice processing has been in operation since 1984 with tubular

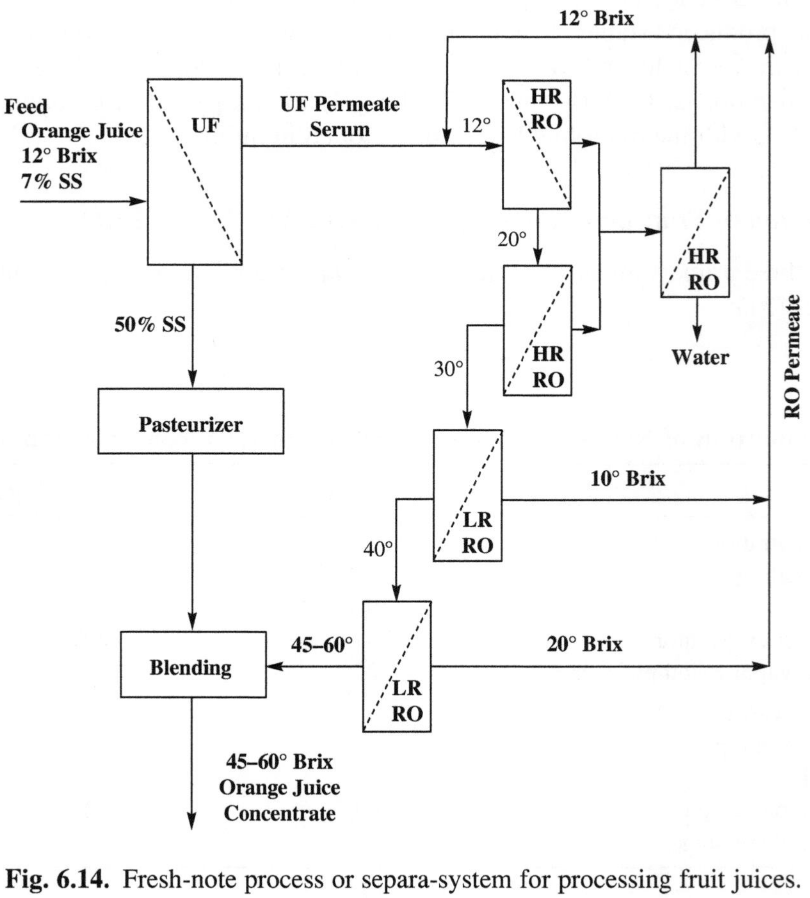

Fig. 6.14. Fresh-note process or separa-system for processing fruit juices.

membranes. The RO plant processes more than 250 tons/h (during the season) of 4.5 degree Brix juice to 8 degree Brix. It should be noted that the usual concentration of natural tomato juice is 4.5–5.0 degree Brix. Commercial tomato sauces are 8–12 degree Brix, and tomato pastes are 28–29 degree Brix. Retention of organic acids (citric and L-malic), sugars (glucose, fructose), mineral ions (K, Mg, Na, F, Cl), and free amino acids is excellent. The color of the tomato juice product is very good and shows none of the browning normally associated with evaporation. This color quality is retained when evaporated further to 28–30 degree Brix paste because of the reduced time in the evaporator.

When the membrane technology is applied to the production of concentrated fruit and tomato juices, the following considerations should be taken into account, according to Dr. A. Sabater de Sabates, Chief of the Food Processing Unit, UNIDO (8):

1. The main advantage of the membrane process system lies in the lower energy consumption. Total liquid volume remains the same. The influent feed stream is equal to the permeate stream plus the retentate stream. The economic advantages depend on energy costs, membrane replacement costs, as well as the duration of the campaign.
2. As the citrus fruit juices are also non-Newtonian, there are flow problems as well as the classical fouling problems. This may increase investment and running costs.
3. For the time being, the membrane process system cannot produce orange juice concentrated up to 65 degree Brix, which is standard. This step, from 55 to 65 degrees, is done by vacuum evaporation.
4. The juice obtained by the membrane process system has fewer microparticles and therefore is more transparent. When consumer preference goes to a more cloudy juice, the juice fibers can be blended.
5. As the rheological behavior of tomato juice is irreversibly shear dependent, the apparent viscosity of the final product is therefore lower than with the conventional process. In any case, tomato juice has to be quickly heated to inactivate pectinase, but it is a fact that the color is better with the membrane technology.
6. As the membrane process system has inherent advantages in energy savings and low temperature processing, some comparative analysis of investment and running vs. the conventional vacuum evaporation methods, would be of great interest.

5.3. *Removal of Salts from Food Processing Water Streams Using NF*

The single most important food manufacturing application of NF membranes is desalting of whey permeate (Fig. 6.15) or other food processing water streams (32). Salty cheese whey, for instance, poses a serious waste disposal problem in the food industry. The salt content, whey solids content and biochemical oxygen demand of the salty cheese whey are about 4–6, 6% and 45,000 mg/L, respectively. Accordingly, the salty cheese whey can neither be discharged directly into a POTW system, nor mixed with normal cheese whey because of its extremely high salt content (33).

Figure 6.15 shows how a major waste disposal problem is solved by desalting of cheese whey permeate using NF membranes. The NF system effectively removes sodium chloride as permeate, which is recycled or discharged. The retentate that is desalted whey is added back to the normal whey. This is a typical example of cleaner production as well as a cost-saving process.

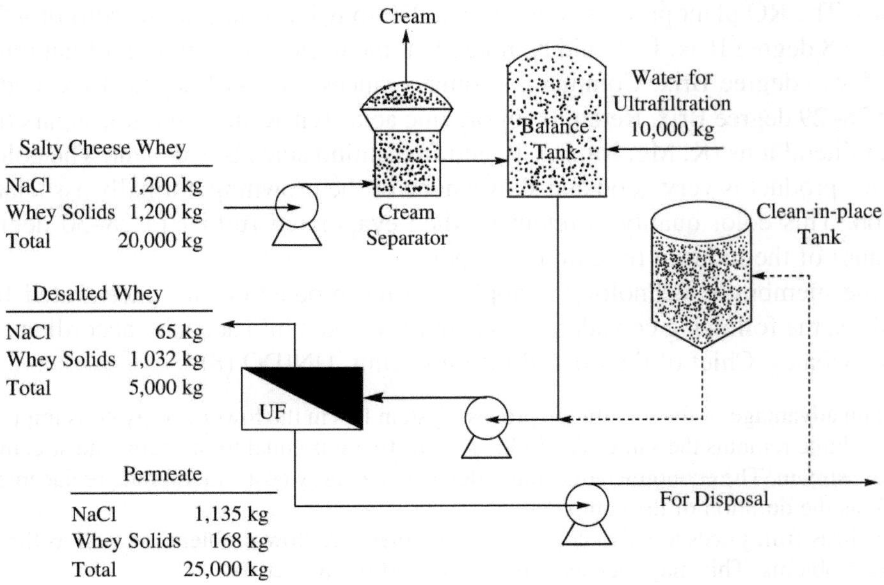

Fig. 6.15. Removal of salts from cheese whey permeate by NF.

Removal of salts from a food processing water stream can also be accomplished by using ED or ion exchange; however, the capital costs of both ED and ion exchange are higher than that of NF.

5.4. Recovery of Potato Starch Effluent Using RO

The Avebe Foxhol plant in the Netherlands produces starch from potatoes (34). As shown in Fig. 6.16, potato production involves washing and grinding potatoes to produce pulpy liquor containing potato water, starch, and fibers. The starch is then extracted, refined, and dried as a product. The starch extraction effluent is treated by centrifugation for the separation of fibers. The residual effluent from the centrifuge contains protein, sugars, and minerals at about 4% TS previously too dilute to recover. The plant originally discharged its 180 m^3/h of effluent into the North Sea and Holland's canals without treatment, causing serious water pollution.

A continuous RO cleaner production system is now operated at Avebe utilizing open-channel tubular RO membranes for handling of high TSS in the centrifuge effluent. With the continuous RO system (34–36), the centrifuge effluent is divided into two streams: (a) the RO permeate amounting to about 90 m^3 of water; and (b) the RO retentate having 8% TS and about 90 m^3 of water. The RO permeate is recycled to the grinding operation for reuse, while the RO retentate is treated by steam heat, coagulation, and mechanical separation, which is then dried as a high grade protein concentrate (protomyl PF) that can be used as animal feed for small animals, such as piglets, poultry, dogs, cats, minks, etc. The residual potato water from mechanical protein separation is evaporated, concentrated to about 60% TS, and

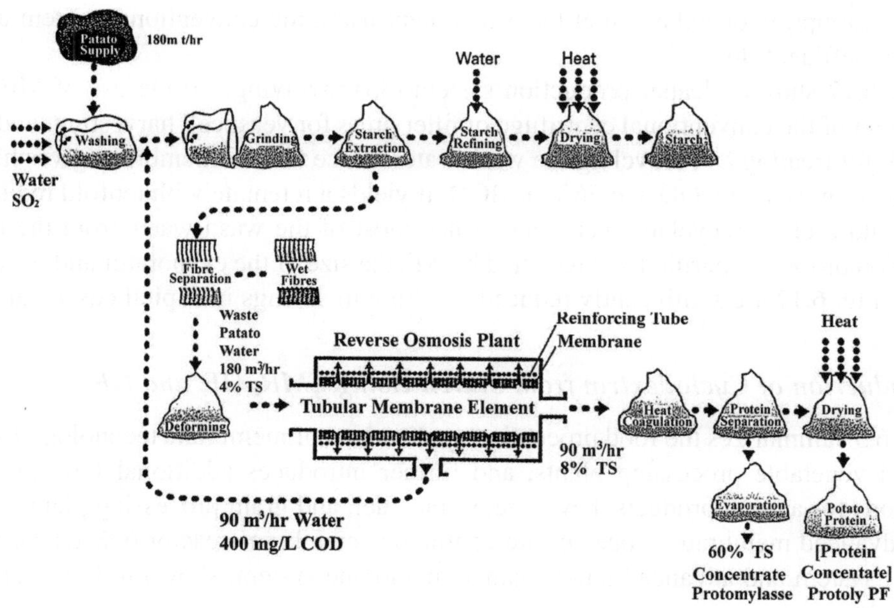

Fig. 6.16. Recovery of protein from potato starch effluent by RO.

marketed as protomylase, which is another saleable product with high potato fiber content that may be used as a cattle feed.

As a result, the Avebe Foxhol plant effluent treatment problem was solved and waste disposal costs were avoided. Water consumption has been reduced from 7 m^3/tonne of potatoes to 0.6 m^3/tonne of potatoes by water recycle, with further cost savings. The two by-products (protomyl PF protein concentration and Protomylase concentration) generated extra revenues. The economic benefits depend upon the market value for the recovered by-products as well as water and wastewater charges.

In summation, the advantages of the continuous RO cleaner production system are (37):

1. Major reduction in the volume of process water consumed through the recycling of RO permeate
2. Total elimination of effluent discharge into the receiving water
3. The reduction in the quantity of water to be handled enabled the steam heat coagulation and evaporation facilities to be reduced to half the size, resulting in capital and energy savings
4. Production of two saleable products from the effluent

5.5. Production of Yeast by Aerobic Fermentation, MF and NF

In a conventional process system for the production of yeast for the bakery industry, aerobic fermentation is used for processing sugar, such as molasses. The yeast produced by fermentation is then harvested by centrifugation or filtration, washed with water, and later dried for sale. The conventional yeast production process generates an industrial effluent which contains high BOD, COD, sugar, dark brown pigments, etc., and must be treated by the evaporation and anaerobic digestion before its discharge into the POTW system. The cost and energy consumption of industrial effluent treatment using the conventional system are high due to high effluent flow.

Figure 6.17 shows a leaner production system (33) involving: (a) the use of MF for the replacement of the conventional centrifuge or filter press for yeast cell harvesting; and (b) the use of NF for treating and recycling the wash water. Since the NF membrane gives high flux while rejecting 97% of COD and 56% of BOD, it yields a retentate with tenfold reduction in volume, and a clean recyclable permeate. Since most of the wash water from the osmotic dehydration process is purified and recycled by NF, the size of the evaporator and the digester shown in Fig. 6.17 are significantly reduced resulting in savings in capital costs and energy.

5.6. Production of Cyclodextrin from Starch Using CMR, UF, and NF

Table 6.5 summarizes the food processing applications of membrane technology for dairy, fruits and vegetable processing plants, and further introduces additional food processing applications for animal products, beverage, sugar, fuel, and grain processing plants (3).

The advanced membrane processes are continuous membrane reactor (CMR), pervaporation (PV) system and advanced ion-exchange membrane system, shown in Figs. 6.18–6.22, respectively.

CMR continuously produces product water (permeate) and recycles the biological cells (retentate) to an upstream fermentation vessel for reuse (Figs. 6.18 and 6.19). CMR has been

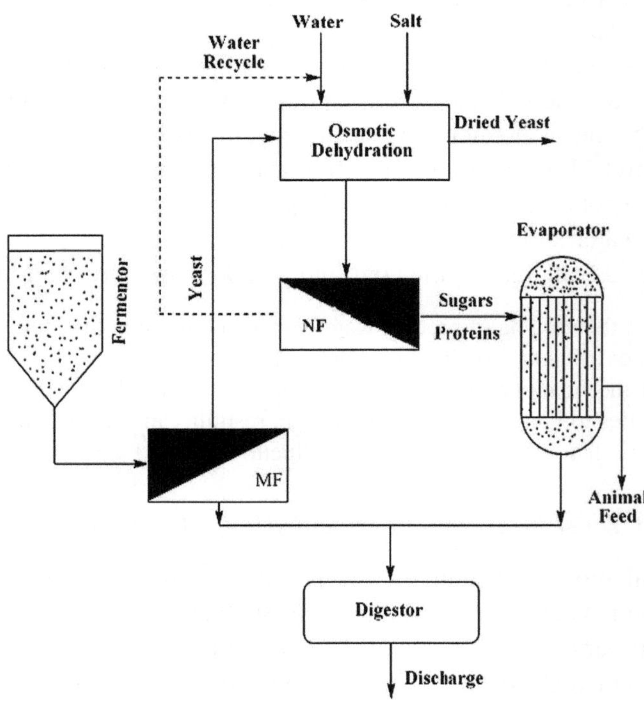

Fig. 6.17. Production of yeast using MF and NF.

Table 6.5
Food processing applications of membrane technology

Dairy	
	(Pre) Concentration of milk and whey prior to evaporation
	Bulk transport (RO)
	Specialty fluid milk products (2–3X/UHT) (RO)
	Partial demineralization and concentration of whey (NF)
	Fractionation of milk for cheese manufacture (UF)
	Fractionation of whey for whey protein concentrates (UF)
	Specialty fluid milk products (UF)
	Clarification of cheese whey (MF)
	Defatting and reducing microbial load of milk (MF)
	Demineralization of milk and whey (ED)
Fruits and vegetables	
Juices	Apple (UF, RO), apricot, citrus (MF/UF, RO, ED), cranberry, grape (UF, RO), kiwi, peach (UF, RO), pear, pineapple (MF/UF, RO), tomato (RO)
Pigments	Anthocyanins, betanins (UF, RO)
Wastewater	Apple, pineapple, potato (UF, RO)
Animal products	
Gelatin	Concentration and deashing (UF)
Eggs	Concentration and reduction of glucose (UF, RO)
By-products	Blood, wastewater treatment (UF)
Beverages	
	Wine, beer, vinegar – clarification (MF/UF)
	Low-alcohol beer (RO)
Sugar refining	
	Beet/cane solutions, maple syrup, candy wastewaters – clarification (MF/UF), preconcentration (RO) Desalting (ED)
Oilseeds, cereals, legumes	
Soybean	Protein concentrates and isolates (UF)
	Protein hydrolyzates (CMR)
	Oil degumming and refining (UF, NF)
	Recovery of soy whey proteins (UF, RO)
	Wastewater treatment (UF, MF, MBR)
Corn	Steepwater concentration (RO)
	Light-middlings treatment: water recycle (RO)
	Saccharification of liquefied starch (CMR)
	Purification of dextrose streams (MF/UF)
	Fermentation of glucose to ethanol (CMR)
	Downstream processing (MF, UF, NF, RO, ED, PV)
	Wastewater treatment (UF, MF, MBR)
Biotechnology	
	Production of high-quality water (MF, UF, RO, ED)
	Downstream processing (MF, UF, RO, ED): cell harvesting, protein fractionation, desalting, concentration
	Bioreactors: enzyme hydrolysis, microbial conversions (MF, UF)
	Tissue culture: plant cells, mammalian cells (MF, UF)

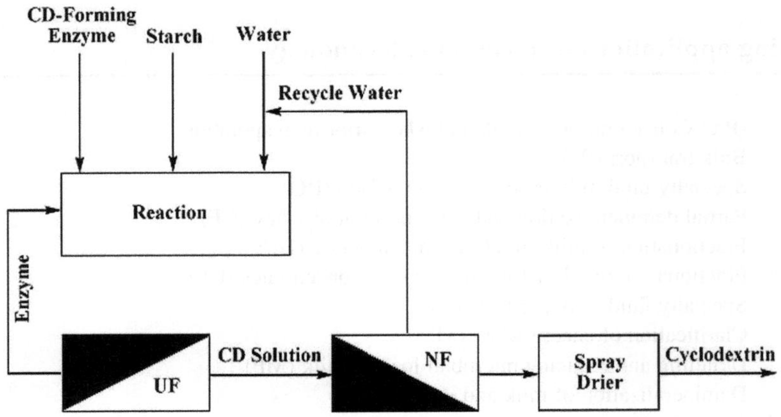

Fig. 6.18. Separation of cyclodextrin from liquefied starch by UF and NF.

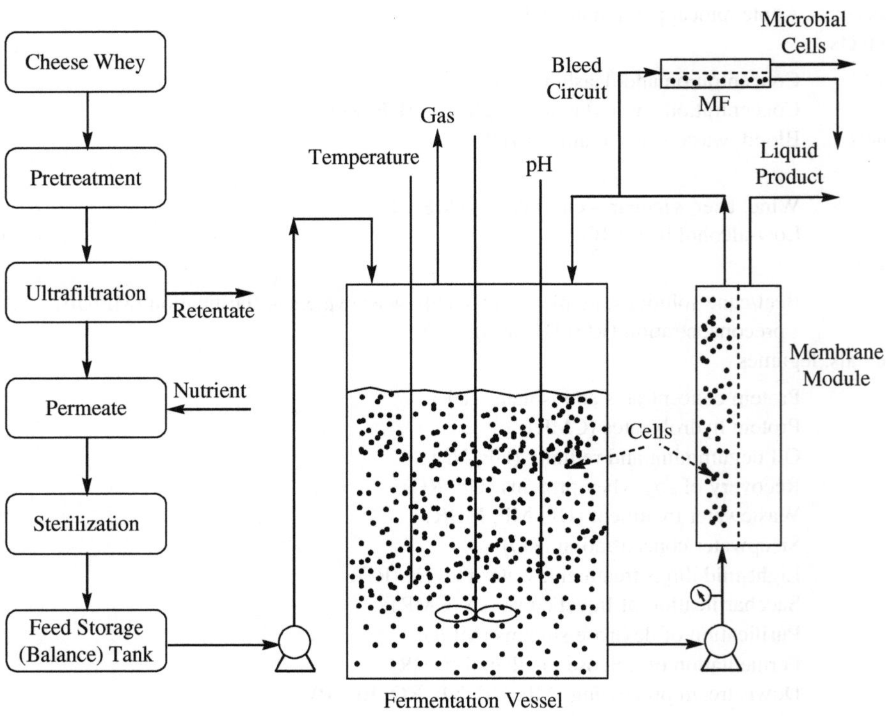

Fig. 6.19. Continuous membrane reactor for fermentation process.

used for fermentation of biomass to ethanol, saccharification of liquefied starch, processing of soybean for producing protein hydrolyzates (Table 6.5) and processing of biomass for the production of biochemicals.

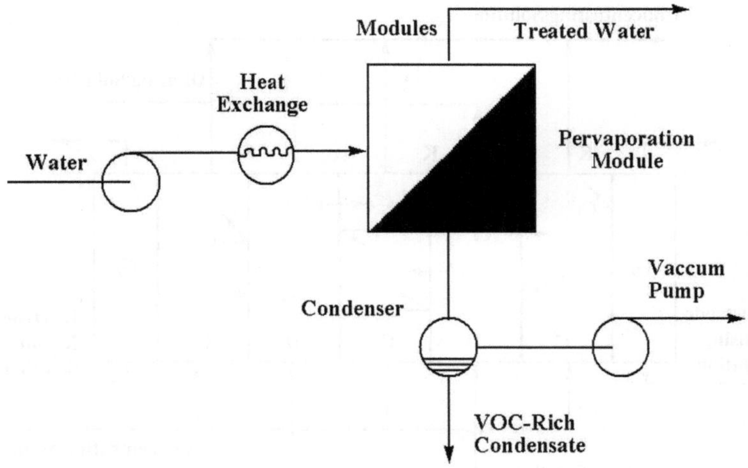

Fig. 6.20. Cross-flow pervaporation system for VOC separation.

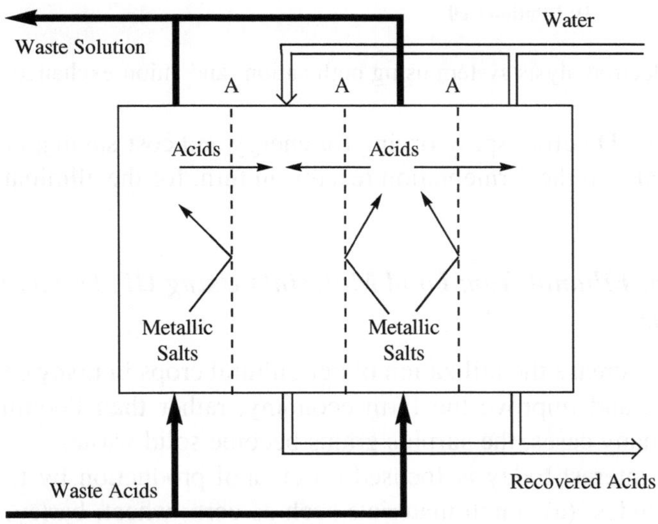

Fig. 6.21. Diffusion dialysis system using anion-exchange membrane. A: Anion Exchange Membrane.

CMR proved to be a feasible process for the treatment of liquefied starch (Table 6.5 and Fig. 6.18). Cyclodextrin (CD) can be produced from starch in the presence of CD-forming enzyme in a fermentation reactor (Fig. 6.18). To enhance the CD productivity, UF membrane is used in conjunction with the fermentation reactor forming a CMR in which an enzyme is continuously separated by UF and recycled to the fermentation reactor for CD production. The fermentation reactor is usually a stirred-tank. The UF membrane separates CD and recycles the enzyme. The functions of the NF membrane shown in Fig. 6.18 are twofold:

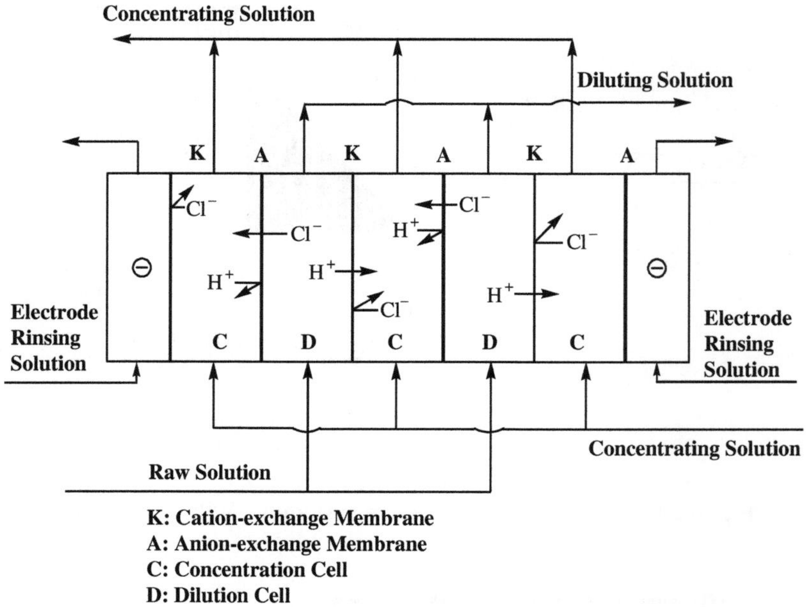

Fig. 6.22. Electrodialysis system using both cation- and anion-exchange membranes.

(a) concentration of CD before spray-drying for energy and cost-saving; and (b) recycling of NF permeate for reuse in the fermentation reactor, in turn, for the elimination of wastewater discharge (37).

5.7. Production of Ethanol from Food Materials Using UF, Fermentation, CMR, and MF

There is need to increase the utilization of agricultural crops in many developed countries to reduce surpluses and improve the farm economy, rather than flooding the world with surplus grains. In many cases, the surplus grains become solid waste.

A great deal of interest today is focused on ethanol production by the fermentation of biomass which includes: (a) starch materials such as corn, wheat, barley, and potatoes; (b) saccharine materials such as sugar cane, sugar beets, molasses, and fruits; and (c) cellulose materials such as wood, corn stalks, straw, and other agricultural biomass.

In the conventional fermentation process shown in Fig. 6.23, the starch materials, such as corn, wheat, and potatoes, must be converted to fermentable sugars by cooking with enzymes. The pretreated fermentable sugars are then fed into a fermentation reactor where yeast, acid, and enzyme are added and ethanol and carbon dioxide gas are produced.

When saccharines, such as sugarcane, beets, and molasses are used as feedstock to an ethanol production system, the saccharine materials can be converted by yeast into ethanol and carbon dioxide gas directly in the fermentation reactor without a preliminary enzymatic cooking treatment. On the other hand, when cellulose materials, such as wood, corn stalks, hay, and straw, are used as feedstock, the cellulose must be broken down to fermentable

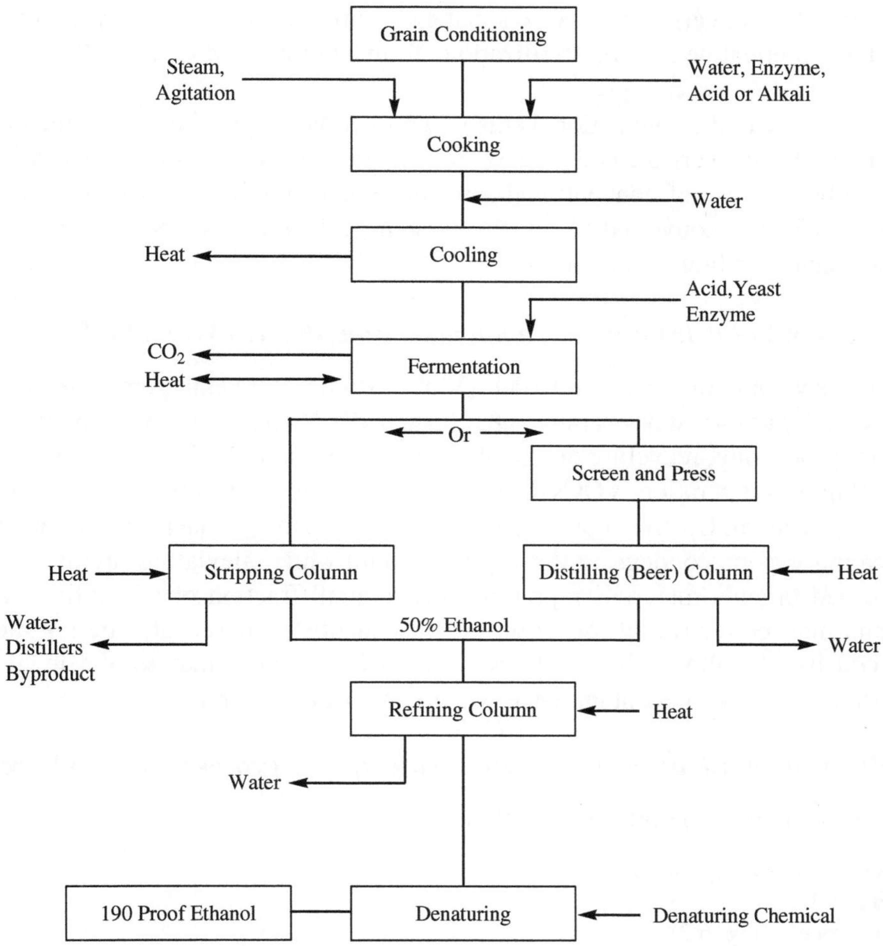

Fig. 6.23. Fermentation process for ethanol production.

sugars using an acid hydrolysis pretreatment process, before it is fed into the fermentation reactor.

The ethanol produced from fermentation (Fig. 6.23) can then be screened, pressed, stripped, distilled, and refined for the production of 190-Proof ethanol. Pure alcohol is 100% or 200-Proof. The 190-Proof product contains 95% alcohol and 5% water. The ethanol can be used as either a beverage or a fuel (8).

In comparison with other fuels, the 200-Proof ethanol's heat value is rated at 84,000 BTU/gallon, which is higher than the 200-Proof methanol (60,000 BTU/gallon), but lower than gasoline (120,000 BTU/gallon) and diesel fuel (140,000 BTU/gallon). It should be noted that methanol can only be produced by synthesis from coal, natural gas, or wood, but not by fermentation (8).

New advances in genetic engineering and membrane technology have vastly improved the productivity of fermentation processes, which could possibly tip the balance in favor of biomass-derived chemicals. A typical schematic of a process system to convert cheese whey to ethanol by pretreatment, UF, sterilization, fermentation, CMR, and MF is shown in Fig. 6.19.

In comparison with the conventional ethanol production system (Fig. 6.23), the innovative system (Fig. 6.19) recovers the cells (yeast and enzymes) for reuse, and produces a product, ethanol, without the use of conventional screening filter pressing, or stripping means. The waste cheese whey is converted to an energy source. This is another example of cleaner production using membrane technology.

5.8. Removal of Volatile Organic Compounds from Process Water by Pervaporation

PV removes volatile organic compounds (VOCs) from water using permeable membranes that preferentially adsorb and separate VOCs (Fig. 6.20). So far PV has been mainly used for downstream processing, recycling of valuable substances, and VOCs separation (38).

In the PV process operation, VOCs diffuse from the membrane-water interface through the membrane by vacuum. Upstream of the PV system's vacuum, a condenser traps and contains the permeating vapors, condensing the vapor to liquid while alleviating fugitive emissions. The condensed organic materials represent only a small fraction of the initial feed water volume, and may be recovered for reuse, or subsequently disposed of. The PV membrane modules consist of hollow fibers with well-defined alignment that results in high mass transfer efficiencies. The decontaminated water is recovered for reuse.

5.9. Application of Advanced Ion Exchange Membrane Processes in Food Processing

Ion-exchange membrane processes include:

1. Diffusion dialysis (Fig. 6.21)
2. ED (Fig. 6.22)
3. Bipolar process (Fig. 6.3)

Diffusion dialysis is the simplest application of ion-exchange membrane technology (23), which has been used for acid recovery. As illustrated in Fig. 6.21, a cell stack of anion-exchange membranes which are much more permeable to acids than to salts, are assembled together using specially designed gasket-spacers. Food processing water containing high concentration of acid can be fed into the bottom of the diffusion dialysis membrane stack while a water source can be fed from the top of the stack. The recovered acid, which is free of contaminated metals, is suitable for recycling and reuse. Diffusion dialysis is an energy-saving process because neither pressure nor electricity is required as a driving force. The recovery of alkali by diffusion dialysis using cation-exchange membranes is a natural extension of its traditional application.

Recent development of anion-exchange membranes with high acid-retention and cation-exchange membranes with hydrogen ions preference has enabled ED to:

1. Concentrate dilute electrolyte solution
2. Demineralize salt solution
3. Separate ions from nonionic materials.

ED utilizes an electrical driving force to transport ions through ion-exchange membranes. The recovery of acid by ED is illustrated in Fig. 6.22. Table 6.5 lists common applications of ED (8):

1. Demineralization of milk and whey
2. Desalting in a sugar refining plant
3. Downstream processing in a corn refining plant
4. Production of a high-quality biochemical product in a biotechnology plant
5. Downstream processing in a biotechnology plant

The bipolar process (Fig. 6.3) is a unique application of ED which can be used to generate equimolar quantities of acid and base from salt. The principle of producing hydrochloric acid and sodium hydroxide from sodium chloride by using the bipolar membrane is illustrated in Fig. 6.3. The economics of using the bipolar process for the generation of acid and base in a food processing plant is to be explored.

6. NONFOOD APPLICATIONS OF MEMBRANE TECHNOLOGY IN THE FOOD INDUSTRY

6.1. Nutrient Removal from Wastewater Streams

Considering the increasingly stringent industrial effluent standards currently being imposed on industrial effluent pretreatment facilities, specifically allowable concentrations of nutrient-containing substances, the ability of RO to remove nutrients should be properly investigated.

CA is still considered the preeminent membrane for wastewater treatment because it is capable of producing the highest flux per unit surface area at specified levels of solute rejection. The rejection performance of RO using a 90% sodium chloride rejection CA membrane has been investigated (23). Phosphorous removal was greater than 95% in all cases (greater than 99% in most cases). Ammonia removals were generally in excess of 90%, and nitrite and nitrate removals generally ranged from 84 to 97%.

In an independent study of nutrient removal by RO conducted by the National Research Council of Canada, Ottawa, Ontario, similar findings were confirmed: 80–90% separation of sodium nitrate, sodium nitrite, and ammonium chloride were realized, with ammonium chloride rejection significantly better than both sodium nitrate and sodium nitrite (39). Rejection of phosphate-containing substances was also in agreement with separation efficiency at greater than 98.5%.

Many food processing plants, including a major dairy plant in Minnesota, USA, use phosphoric acid in their processes, resulting in over 100 mg/L of phosphate in their wastewater streams. RO may be an ideal process for recovery and reuse of nutrients as fertilizers.

6.2. Organics Removal from Wastewater Streams

The rejection of inorganics (such as sodium chloride) generally is well defined. The rejection of organics, which might be found in food industry wastewater, however, remains to be investigated more thoroughly. Some studies employing RO systems using CA membranes have been performed, and their results are documented below.

6.2.1. *Effect of Molecular Size and Shape*

Straight-chain primary monohydric alcohols (methanol, ethanol, 1-propanol, 1-butanol, 1-pentanol, 1-hexanol, l-hepanol, and 1-octanol) were studied. The procedure allowed for the determination of organics rejection as a function of molecular weight (all compounds have identical functional groups). It was found that organics rejection increases with increasing molecular weight (number of carbon atoms) (40).

To further investigate the effect of molecular weight in rejection, a series of alcohols were studied by varying both degrees and positions of branching and positions of the hydroxyl group. In addition to confirming the dependence of solute rejection on molecular weight, these studies also indicated that the degree of branching overrides the effect of molecular weight (40).

The alcohols studied were divided into three distinct groups by the rate of rejection. Each group can be characterized as follows: Group I – straight chain, primary alcohols with no side chains; Group II – primary alcohols with single methyl side group or secondary alcohol with hydroxyl group off to the side; and Group III – tertiary alcohols or double branched primary and secondary alcohols.

It was found that solute rejection increases significantly with increasing molecular branching. Geometrically large, single branched molecules do not diffuse as readily as do linear molecules of the same molecular weight because they present a larger cross-section to the membrane and require larger gaps. Doubly branched molecules have an even larger cross-sectional diameter and require even larger openings. Passage through the membrane is therefore slower, resulting in a higher rejection with increased branching.

6.2.2. *Effect of Hydrogen Bonding*

For the separation of glutamic acid, sodium stearate, ABS, and LAS, the acidity (or the proton-donating characteristics of the molecule), measured as the ease of hydrogen bond formation or the degree of dissociation of the molecule in solution, is probably one of the criteria for RO separation. As hydrogen bonding increases, more solute readily dissolves in the membrane, and consequently, a larger concentration is available for diffusion.

Molecules having the ability to both accept and donate hydrogen atoms to form a covalent bond and form hydrogen bonds more readily than those that can perform only one transaction. This concept is confirmed by experimental data showing that alcohols, amities, amides, and carboxylic acids, which can both donate and accept protons, permeate membranes better (lower rejection) than do esters, aldehydes, ketones, and sulfones, which are proton acceptors only.

In conclusion, studies indicate that greater than 96% separation of organics (measured as TOC and COD) could be obtained by wastewater treatment with RO osmosis. The potential of RO to efficiently remove both organics and inorganics from food-processing industrial wastewaters without depreciation on overall performance makes RO a unique advanced wastewater treatment method for various water pollution control and water reuse applications in the food industry (23, 41). Further details dealing with various types of membranes and food industries can be found in current literature (42–53).

7. NOMENCLATURE

C_P = Suspended and/or dissolved solids concentration in product water, mg/L
C_R = Suspended and/or dissolved solids concentration in retentate, mg/L
Q_f = Feed flow, gpm
Q_p = Permeate flow, gpm
Q_r = Retentate flow, gpm
P_i = Feed water pressure, psi
P_p = Permeate pressure, psi
P_o = Retentate pressure, psi
R = Rejection, decimal fraction

REFERENCES

1. He Y, Xu P, Li C, Zhang B (2005) High-concentration food wastewater treatment by an anaerobic membrane bioreactor. Water Res 39(17):4110
2. Novachis L (2005) Membrane wastewater treatment plants cropping up in food industry. Environmental Science & Engineering. www.esemag.com/0705/zenon.html.July 2005
3. Cheryan M (1980) Membrane separations: mechanisms and models. In: Singh RP, Medina AG (eds) Food properties and computer-aided engineering of food processing systems. Kluwer, Amsterdam, pp 367–391
4. Cheryan M (2003) Membrane concentration systems. In: Heldman DR (ed) Encyclopedia of Agriculture Food, and Biological Engineering, Marcel Dekker, NY, pp 598–601
5. Elia VJ, Clark CS, McGinnis, Cody TE, Kinman RN (1978) Ozonation in a waste water reuse system: examination of products formed. J Water Pollut Control Fed 50:1727–1732
6. Homoda MF (1973) Organics Removal by low pressure reverse osmosis. J Water Pollut Control Fed 45(10):2151–2153
7. Lim MS, Johnston HK (1976) Reverse osmosis as an advanced treatment process. J Water Pollut Control Fed 48:1804–1820
8. Wang LK, Cheryan M (1995) Application of membrane technology in food industry for clean production. United Nations Industrial Development Organization (UNIDO), Technical Paper # DTT-8-6-95
9. Wechsler P (1976) Reverse osmosis on secondary sewage effluent: the effect of recovery. Water Res 11:379–383, Vienna, Austria
10. Yoshiaki N (1994) Waste stream treatment by ionic membrane. In: Proceedings of Sino-Japanese Symposium, Xian, China, pp 175–178
11. Zhou DX, Wang B, Tan Y, Mei H (1994) Fouling and chemical cleansing of UF membranes in separation process of seaweed industry. In: Proceedings of Sino-Japanese Symposium, Xian, China, pp 295–299
12. Chen JP, Mou H, Wang LK, Matsuura T (2006) Membrane filtration. In: Wang LK, Hung YT, Shammas NK (eds) Advanced physicochemical treatment processes. Humana, Totowa, NJ, pp 203–260
13. Water Environment Federation (2006) Membrane systems for wastewater treatment. McGraw-Hill, New York, NY
14. Cheryan M (1987) Membrane technology in food processing. In: The Seventh World Congress of Food Science and Technology, Singapore, pp 3–9

15. Wang LK, Kurylko L, Wang MHS (1993) Contamination removal system employing filtration and plural ultraviolet and chemical treatment steps and treatment mode controller. US Patent 5-90659, 2 Mar 1993
16. Wang LK, Kurylko L, Wang MHS (1993) Method and apparatus for filtration with plural ultraviolet treatment stages. US Patent 5,236,595, 17 Aug 1993
17. AWWA (1999) Reverse osmosis and nanofiltration (M46). American Water Works Association, Denver, CO
18. Xue D, Feng Q (1994) Two decades' progress of ED desalting technique in seaweed industry. In: Proceedings of Sino-Japanese Symposium, Xian, China, pp 170–174
19. Cheryan M (1992) Concentration of liquid foods by reverse osmosis. In: Heidman DR, Lund DB (eds) Handbook of food engineering. Marcel Dekker, New York, NY, pp 393–436
20. Wagner J (2001) Membrane filtration handbook practical tips and hints, 2nd edn. Osmonics, Minnetonka, MN
21. Schafer AI, Fane AG, Waite TD (2001) Factors and chemical pretreatment effects in the membrane filtration of waters containing natural organics matter. Water Res 35(6):1509
22. Moch I Jr, Chapman M, Steward D (2003) Development of a CD_ROM cost program for water treatment projects. Membrane Technology, 2003(6):5
23. Yesselman RJ, Wang LK (1987) Reverse Osmosis, Technical Report No. LIR/02-87/247. Lenox Institute for Research, Lenox, MA
24. US EPA (2003) Membrane filtration guidance manual. US Environmental Protection Agency, Office of Water, Cincinnati, OH, EPA 815-D-03-008
25. US Bureau of Reclamation (1998) The desalting and water treatment membrane manual: a guide to membranes for municipal water treatment, 2nd edn. Technical Service Center, Water Treatment Engineering and Research, Denver, CO
26. BBC Research (2005) Membrane and separations industry review, Report # R2-905. Business Communications Company, Wellesley, MA
27. Cheryan M (1992) Membrane technology in food and bioprocessing. In: Singh RP, Wirakartakusumah MA (eds) Advances in food engineering. CRC, Ann Arbor, MI, pp 165–180
28. Short JL (1994) Membrane technologies in food processing, Report GA-075. Business Communications Company (BCC Research), Wellesley, MA
29. BCC Research (2002) Ultrafiltration membrane industry: developments, markets, Report. Business Communications Company (BCC Research), Wellesley, MA
30. Cheryan M (1991) Membranes in food processing. In: Turner MK (ed) Effective industrial membrane processes, benefits and opportunities. Elsevier Science, London, UK, pp 157–180
31. Mondor M, Brodeur C (2003) Membrane filtration. Food Research and Development Centre, Saint Hyacinthe, QC, Canada. http://sci.agr.ca/crda/pubs/art19_e.htm
32. Schaefer A, Fane A, Waite T (eds) (2004) Nanofiltration: principles and applications. Elsevier, St. Louis, MO
33. Raman LP, Cheryan M, Rajagopalan N (1974) Consider nanofiltration for membrane separations. Chemical Engineering Progress, pp. 68–74
34. UNEP (1993) Cleaner production worldwide. United Nations Environment Program, United Nations Publication, No. 91-lll-D2, 36pp
35. Wang LK (1995) Case studies of cleaner production and site remediation. United Nations Industrial Development Organization (UNIDO), Training Manual No. DTT-5-4-95, Vienna, Austria
36. Casani S, Hansen TB, Christensen J, Knochel FS (2005) Comparison of methods for assessing reverse osmosis membrane treatment of shrimp process water. J Food Prot 68(4):801

37. Yabushita T (1990) Application of membranes to food processing, Technical Report No. 4. Nitto Denko, Osaka, Japan
38. Ditgens B, Laufenberg GH, Kunz B (2002) Clean-production concepts in food industry: treatment of process solutions and recycling of valuable substances using membrane technology. In: 2002 Annual Meeting and Food Expo, Anaheim, CA
39. Takoshi M (1972) Studies on RO for water pollution control. Water Res 6:1077–1085
40. Duvel WA (1975) Removal of wastewater organics by reverse osmosis. J Water Pollut Control Fed 41(1):58–61
41. Bozoglu F, Deak T, Ray B (eds) (2001) Novel processes and control technologies in the food industry. IOS, Amsterdam
42. Wright C (2007) Novel membrane surface science. In: Membrane filtration in the food and drinks industry: problems and solutions. Institution of Chemical Engineers, London, UK
43. Hildebrandt P (2008) Taking another look at membranes. Onsite Water Treat J. http://www.gradingandexcavation.com/ow_0701_taking.html
44. Lindau J (2007) Microfiltration of milk. In: Membrane filtration in the food and drinks industry: problems and solutions. Institution of Chemical Engineers, London, UK
45. Nakhla G, Lugowski A, Patel J, Rivest V (2006) Combined biological and membrane treatment of food-processing wastewater to achieve dry-ditch criteria: pilot and full-scale performance. Bioresource Technol 97(1):1–14
46. Lipnizki F (2007) Membrane optimization in the fruit juice and brewing industries. In: Membrane filtration in the food and drinks industry: problems and solutions. Institution of Chemical Engineers, London, UK
47. Waldron K (ed) (2007) Handbook of waste management and co-product recovery in food processing, vol 1. Woodhead, Cambridge
48. Bird M (2007) Ultrafiltration of tea liquors. In: Membrane filtration in the food and drinks industry: problems and solutions. Institution of Chemical Engineers, London, UK
49. BNET (2008) High organic content industrial wastewater treatment by membrane filtration. CBS Interactive Business Network, (BNET), New York, NY, USA. http://jobfunctions.bnet.com/whitepaper.aspx?docid=125344
50. Wang LK (2008) Application of membrane separation technologies in food processing industry. Technical paper presented at 2008 National Engineers Week, Practicing Institute of Engineering, Foundation for Engineering Education, and NYS Society of Professional Engineers, Marriott, Albany, NY, 14–15, February 2008
51. Wang LK (2008) Development and applications of membrane bioreactor technologies. Technical paper presented at 2008 National Engineers Week, Practicing Institute of Engineering, Foundation for Engineering Education, and NYS Society of Professional Engineers, Marriott, Albany, NY, 14–15, February 2008
52. Wang LK (2009) Chemical and biochemical technologies for environmental infrastructure sustainability. Technical paper presented at 2009 National Engineers Week, Practicing Institute of Engineering, Foundation for Engineering Education, and NYS Society of Professional Engineers, Marriott, Albany, NY, 5–6 February 2009
53. Wang LK, Ivanov V, Tay JH, Hung YT (eds) (2010) Environmental biotechnology, Humana Press, Totowa, NJ, pp 439–516

7
Membrane Separation: Basics and Applications

Jiaping Paul Chen, Honghui Mou, Lawrence K. Wang, Takeshi Matsuura, and Yuting Wei

CONTENTS

Abstract Due to the limited new water resources, the focus of water industry has shifted more towards reclamation, reuse and recycling of raw water/wastewater and seawater desalination. Rising treatment costs and spatial limits also pose a greater pressure on the development of alternative technologies. Compared with traditional water and wastewater treatment technologies, membrane separation has been increasingly received much more considerable interests due to wide applicability, reliable performance, low operating and maintenance costs of membrane systems. The membrane fouling is still a principal obstacle in application of this technology.

This chapter first briefly introduces basic concepts for membrane separation including membrane definition, membrane types, membrane formation and characterization, module configuration, mass transport mechanism in membranes. Key factors such as process design, economic assessment in membrane systems are also covered. Moreover, it describes fouling

From: *Handbook of Environmental Engineering, Volume 13: Membrane and Desalination Technologies*
Edited by: L.K. Wang et al., DOI: 10.1007/978-1-59745-278-6_7 © Springer Science+Business Media, LLC 2011

formation and the strategies to control membrane fouling. Several detailed case studies will be cited to explain membrane separation technology.

Key Words Wastewater treatment • membrane separation • membrane process • filtration • fouling • transport • reverse osmosis • electrodialysis • membrane bioreactor.

1. INTRODUCTION

With the rapid increase of demand in industries and daily life, water has become an increasingly valuable but scarcer resource for human being. Take Singapore as an example. Almost half of its water supply is being imported from Malaysia. The daily water consumption has risen eight times while the population has only grown three times since 1950 (1). This increase has led to a greater pressure on the use of limited raw water resources and the capacity of treatment plants. Potential shortfall between supply and consumption needs to be addressed if domestic and industrial customers are to continue enjoying uninterrupted supply.

Apart from the continuous searching for new water sources from other neighbouring countries via diplomatic routes, the Singapore government has already decided to encourage the use of desalinated seawater as an alternative water source in Singapore. However, comparing with the seawater desalination, reclamation and reuse of raw wastewater and secondary effluent using membrane technology seems more cost effective and is becoming a potential challenge to conventional seawater desalination for water production (2).

Today, membrane separation technologies are widely used in many areas of water and wastewater treatment. Membrane processes can be used to produce potable water from surface water, groundwater, brackish water or seawater, or to treat industrial wastewaters before they are discharged or reused. At the same time, membrane technologies have moved into the area of treating secondary or tertiary municipal wastewater and oil field-related water. Membrane separation systems have many advantages over traditional water or wastewater treatment processes, including (1–12):

- Fewer chemicals are used in the process, which helps minimize the negative impacts of those chemicals on the whole process.
- Formation of absolute barriers to particle and pathogens. Microorganisms such as bacteria and viruses can be removed by size exclusion; ultra-pure water can therefore be produced.
- Lower operating and maintenance costs in comparison to conventional systems consisting of coagulation, clarification, and aerobic and anaerobic treatments.
- Membrane separation systems are easy to operate and the performance is more reliable.
- Membrane systems give a compact and modular construction, which occupies less floor space in comparison to the conventional treatment systems. This becomes extremely abstractive in the land-scarce countries such as Japan and Singapore.
- Membrane systems followed by an evaporator (for low-volume highly concentrated effluent) can enable industries to achieve zero liquid discharge goals.
- One-stop reduction or elimination of most contaminants (impurities) in the wastewater stream, for example, total dissolved solids (TDS), chemical oxygen demand (COD), 5-day biochemical oxygen demand (BOD_5), total organic carbon (TOC), colour, suspended solids, nitrogen, phosphorus and heavy metals.

- Permeate can be suitably reused, resulting in water conservation, which reduces the intake of raw water and provides savings on raw water processing costs.

Membrane fouling can significantly compromise the performance of the membrane process (4–6, 12). Therefore, pre-treatment of feed water and cleaning of fouled membranes are important to the overall achievement of the membrane process.

In this chapter, we introduce fundamental concepts of the membrane and membrane separation processes, such as membrane definition, membrane classification, membrane formation, module configuration, transport mechanism, system design and cost evaluation. Four widely used membrane separation processes in water and wastewater treatment, namely, microfiltration (MF), ultrafiltration (UF), nanofiltration (NF) and reverse osmosis (RO) are discussed in detail. The issue of membrane fouling together with its solutions is addressed. Several examples are given to illustrate the processes.

2. MEMBRANE AND MEMBRANE SEPARATION PROCESSES FOR WATER TREATMENT

The membrane is essentially a barrier, which separates two phases and restricts transport of various chemicals in a selective manner. A membrane can be homogeneous or heterogeneous, symmetric or asymmetric, solid or liquid; it can carry positive or negative charges or can be neutral. Transport through a membrane can be affected by convection or by diffusion of individual molecules, and induced by the chemical gradient or electrical gradient.

2.1. *Basics of Membrane and Membrane Separation System*

According to a definition given by the European Membrane Society, the membrane is an intervening phase separating two phases and/or acting as an active or passive barrier to the transport of matter between phases (3). Basically, it is a thin layer that can separate materials depending on their physical and chemical properties when a driving force, a gradient of either chemical potential (concentration or pressure gradient) or electrical potential, is applied across the membrane.

Some basic requirements for membranes are:

- High flux of the product
- Good mechanical strength for supporting the physical structure
- Good selectivity for the desired substances

Generally, high selectivity is related to membrane properties, such as small pores and high hydraulic resistance or low permeability. It can be compromised by a broad pore size distribution. The permeability increases with increasing density of pores and the overall membrane resistance is directly proportional to its thickness. Therefore, a good membrane must have a narrow range of pore size, a high porosity and a thin layer of material.

Membranes can be either dense or porous. Separation by dense membranes relies on physicochemical interaction between the permeating components and the membrane material. Porous membranes, on the other hand, achieve separation by size exclusion, where the rejected material may be either dissolved or suspended depending on its size relative to that of the pore (3).

Membranes can be organic (polymeric) or inorganic (ceramic or metallic), according to its composition, and their morphology is dependent on the nature of the material, as shown in Fig. 7.1.

A membrane separation system separates an influent stream into two effluent streams: the permeate and the retentate or concentrate, as shown in Fig. 7.2. The permeate is the portion of the fluid, which has passed through the membrane. The retentate (concentrate) contains the constituents that have been rejected by the membrane.

2.2. *Membrane Separation Processes for Water Treatment*

There are various types of membrane separation processes for water treatment, as shown in Fig. 7.3; the details of the four major processes – MF, NF, UF and RO – are discussed here. Feed quality and quantity, requirement of products (e.g. permeate or reject), ease in operation and costs determine selection of process (4–24). The applied pressure for typical membrane filtration processes is given in Table 7.1.

2.2.1. Microfiltration

Microfiltration is a membrane filtration process that remove micron-sized particles from fluids. MF membrane has pore sizes ranging from 0.1 to 10.0 μm, through which microorganisms cannot pass. As a result, microfilters can be used to disinfect water solutions. For

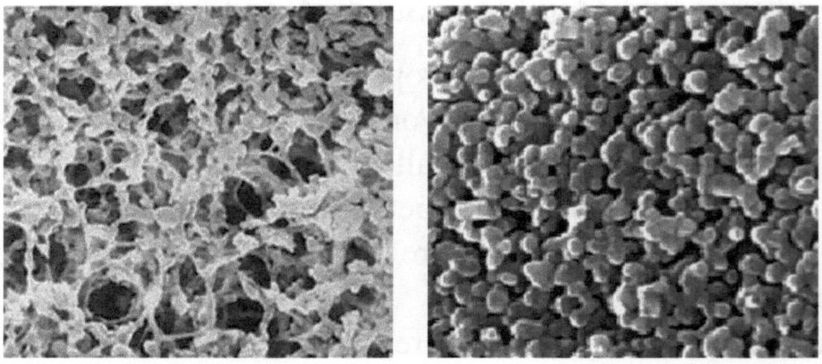

Polymeric membranes   Ceramic membranes

Fig. 7.1. Polymeric and ceramic membranes.

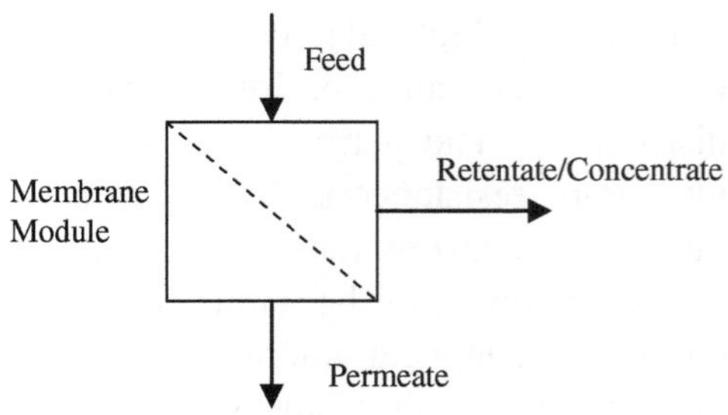

Fig. 7.2. Basic diagram of a membrane separation system.

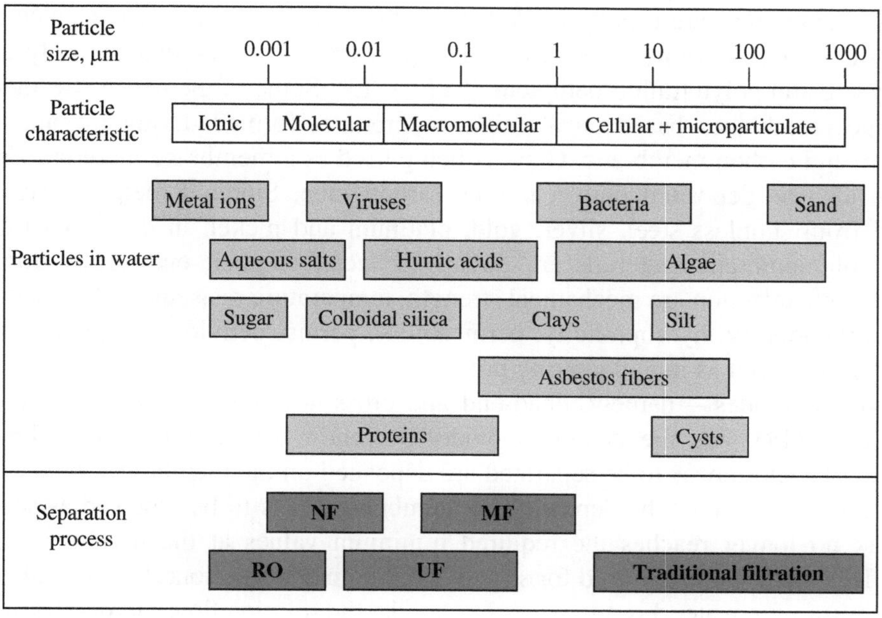

Fig. 7.3. Application range of various membrane processes.

Table 7.1
List of applied pressure for typical membrane filtration processes

Membrane processes	Pressure (atm)
RO seawater	54.4–68.0
RO waste and process	20.4–40.8
RO water purification	13.6–23.8
RO undersink (home)	3.4
NF	6.8–13.6
UF	1.7–10.2
MF (cross-flow)	0.7–1.7

example, the smallest bacterium *Pseudomonas diminuta* with a diameter of 0.3 μm can be disinfected by a MF membrane with smaller pore sizes. On the other hand, both organic and inorganic substances are able to pass through MF membranes.

While the mechanism for conventional depth filtration is mainly adsorption and entrapment, MF membranes use a *sieving mechanism* with distinct pore sizes for retaining particles larger than the pore diameter. Hence, this technology offers membranes with absolute rating, which is highly desirable for critical operations such as sterile filtration of parental fluids, sterile filtration of air, and preparation of particulate, ultra-pure water for the electronics industry.

The MF membranes are usually made from natural or synthetic polymers such as cellulose acetate (CA), polyvinylidene difluoride, polyamides, polysulphone, polycarbonate, polypropylene and polytetrafluoroethylene (PTFE) (13). Some of the newer MF membranes are ceramic membranes based on alumina, membranes formed during the anodizing of aluminium, and carbon membrane. Glass is being used as a membrane material. Zirconium oxide can also be deposited onto a porous carbon tube. Sintered metal membranes are fabricated from stainless steel, silver, gold, platinum and nickel, in discs and tubes. The properties of membrane materials are directly reflected in their end applications. Some criteria for their selection are mechanical strength, temperature resistance, chemical compatibility, hydrophobility, hydrophilicity, permeability, permselectivity and the cost of membrane material as well as manufacturing process.

Two process modes – namely, dead-end and cross-flow modes – are widely used for microfiltration (14). For the dead-end mode, the entire solution is forced through the membrane. The substances to be separated are deposited on the membrane, which increases the hydraulic resistance of the deposit. The membrane needs to be renewed as soon as the filtrate flux no longer reaches the required minimum values at the maximum operation pressure. This mode is mostly used for slightly contaminated solutions, for example, production of the ultra-pure water. For the cross-flow mode, the solution flows across the membrane surface at a rate between 0.5 and 5.0 m/s, which prevents the formation of a cover layer on the membrane surface. A circulation pump produces the cross-flow velocity or the shear force needed to control the thickness of the cover layer. The system is most widely used for periodic back flushing, where part of the filtrate is forced in the opposite direction at certain interval, and breaks up the cover layer. The normal operating pressure for this mode is 1–2 bars.

A MF has a wide range of applications in water and wastewater treatment. The most promising opportunity for MF is the treatment of the municipal water, allowing it to be disinfected without any disinfectants such as ozone and chlorine. A recent study showed that MF membranes can also remove viruses from contaminated surface water. Since viruses are much smaller than the pores in a MF membrane, the finding has been attributed to the viruses being adsorbed on the clay particles, which are large enough to be caught by the microfilter (16). Treatment of municipal sewage is a potential market for MF. Engineers have tried to shift the treatment of sewage to distributed processing, which consists of many small sewage treatment facilities. MF can be used to remove heavy metals from waste streams if pre-treatment chemicals are added to precipitate the metals to particles of filterable size. Although conventional methods of wastewater treatment may use a similar pre-treatment mechanism, the final solid/liquid separation by gravity settling is usually not as effective as membrane filtration. A promising process for the removal of heavy metal ions from aqueous solutions involves bonding the metals firstly to a special bonding agent and then separating the loaded bonding agents from the wastewater stream by separation processes. A new hybrid process of flotation and MF process has been developed to integrate specially designed membrane modules with a flotation reactor, which can treat heavy metal cationic wastewater up to portable-grade level (17). This made it possible to combine the advantages of both flotation and membrane separation while overcoming the limitations. MF is applicable for treatment of organic contaminants in groundwater and process water. Cross-flow membrane filtration may be applicable to wastewater containing high molecular weight or non-polar organic contaminants, such as polychlorinated biphenyls. A study by Futamura and coworkers (18) showed that an organic wastewater was treated by an activated sludge process, which featured the integrated type membrane separation (hollow-fibre microfiltration membrane) modules, and the treated water quality was favourable.

2.2.2. Ultrafiltration

Today, ultrafiltration technology is being used worldwide for treating various waters. It is reported that 50% of the UF membrane plants have been applied on surface waters, including river, reservoir and lake waters. This technology has been used in municipal drinking water application for more than 10 years (21). UF is most commonly used to separate a solution that has a mixture of some desirable components and some that are not desirable. It acts as a selective barrier. UF membranes are capable of retaining species in the range of 300–500,000 Da of molecular weight, with pore sizes ranging from 10 to 1,000 Å (0.001–0.1 μm). Typical rejected species include sugars, biomolecules, polymers and colloidal particles. They are typically rated by molecular weight cut-off (MWCO), a convenient but fictitious value giving the molecular weight of a hypothetical macrosolute that the membrane will just retain (15). The driving force for transport across the membrane is a pressure difference. UF processes operate at 2–10 bars, although in some cases up to 25–30 bars can be used. The separation is achieved by concentrating the large molecules present in the feed on one side of the membrane, while the solvent and microsolutes are depleted as they pass through the membrane.

To distinguish UF from two related processes – MF and RO – one can use the following example. Consider a solution of protein, water and salt. The UF process will separate the protein (macrosolute) from the solution. As the water and salts pass through the membrane, the protein is held back. The protein concentration increases and the salts, whose concentration relative to the solvent is unchanged, are depleted relative to the protein. The result is that the protein is both concentrated and purified by the ultrafiltration. For the RO membrane, it will pass only the water, retaining both salt and protein. The protein is concentrated, but not purified, as the rate of salt to protein remains unchanged. For the MF membrane, it will pass protein, salt and water all together. Therefore, the protein will be neither concentrated nor purified.

UF usually implies separation of macromolecules such as protein from low molecular weight solvents. Pores in the support layer of the membrane are relatively larger than those of the surface layer. Material passing into fine pores can readily be transported through the open-celled, sponge-like structure of the support layer. For example, in electrodeposition paint recovery, the paint – composed of resin, a pigment and water – is separated into two streams that can be reused. The first stream includes the water and a small amount of paint resin, which can be used to rinse the parts later in the process. The paint pigment is separated from that stream and can be reused in the paint bath, allowing the bath to be concentrated to a useable level.

Most UF membranes are made from polymeric materials, such as polysulphone, polypropylene, nylon 6, PTFE, polyvinyl chloride and acrylic copolymer. Inorganic materials such as ceramics, carbon-based membranes and zirconia have been commercialized by several vendors. The important characteristics for membrane materials are porosity, morphology, surface properties, mechanical strength and chemical resistance. The membrane is tested with dilute solutions of well-characterized macromolecules, such as proteins, polysaccharides and surfactants of known molecular weight and size, to determine the MWCO.

Most UF processes are operated in cross-flow mode. When the solvent of a mixture flows through the membrane, retained species are locally concentrated at the membrane surface and resist the flow. In the case of processing solution, this localized concentration of solute normally results in precipitation of a solute gel over the membrane. When processing a suspension, the solids collect as a porous layer over the membrane surface. In view of the above, it is clear that the permeate rate can be effectively controlled by the rate of transport through the polarization layer rather than by membrane properties. Hence, UF throughput depends on physical properties of the membrane, such as permeability, thickness, process and system variables like feed consumption, feed concentration, system pressure, velocity and temperature.

There are three primary UF module configurations: tubular, spiral wound and hollow fibre (15). The tubular module is often used for small-scale, high-value applications. As mechanical cleaning is easy for tubular structure, the tubular module is the choice for cases involving severe fouling. The spiral-wound module is often used for high-volume applications. Its structure is more vulnerable to fouling and mechanical cleaning is difficult to perform. The hollow-fibre module consists of a membrane wound into a hollow cylinder with the inside diameters ranging from 500 to 1,000 μm. The choice of diameter value depends on whether the application is high-fouling process or low-fouling process.

UF tests showed good results using the PAN 40-kDa membrane reaching an average copper concentration in the permeate of 2 mg/L and a flux of 135.4 and 156.5 L/h m^2 at 2 and 4 bars, respectively (22). Metal rejection, permeate flow rate and the possibility of regenerating and recycling the polymer make the polymer-assisted ultrafiltration process (PAUF) very interesting for metal ion removal from waters. Satisfactory results were also obtained by Wu and coworkers. A poly(vinyl alcohol) (PVA) membrane was used for ultrafiltration of oily water and demonstrated high oil retention and permeation flux (23). Some other applications of UF process for water and wastewater treatment include:

- Production of ultra-pure water for electronics industry
- Reduction of high COD levels in corn starch plants
- Selective removal of dissolved toxic metals from groundwater in combination with chemical treatment
- Treatment of whey in dairy industries
- Wine or fruit juice clarification

2.2.3. Nanofiltration

Nanofiltration is a form of filtration that uses membranes to separate different fluids or ions. NF is typically referred to as “loose” RO due to its larger membrane pore structure as compared to the membranes used in RO, which allows more salt passage through the membrane. As it can operate at much lower pressures, normally 7–14 bars, and passes some of the inorganic salts, NF is used in applications where high organic removal and moderate inorganic removals are desired. NF is capable of concentrating sugars, divalent salts, bacteria, proteins, particles, dyes and other constituents that have a molecular weight greater than 1,000.

Membranes used for NF are made of cellulose acetate and aromatic polyamide with characteristics such as salt rejections from 95% for divalent salts to 40% for monovalent salts and an approximate MWCO of 300 for organics. An advantage of NF over RO is that NF can typically operate at higher recoveries, thereby conserving total water usage due to a lower concentrate stream flow rate. NF is not effective on small molecular weight organics, such as methanol.

NF membranes can be used to produce drinking water, because they have good molecular rejection characteristics for divalent cations, for example, calcium and magnesium, and they can effectively remove hardness in place of conventional chemical softening. NF membranes can also reject higher molecular weight organics that contribute to taste and odour or that can react with chlorine to form trihalomethanes or other particles, which improves the efficiency of downstream disinfection processes. In Flagler Beach, FL, a study was conducted to evaluate the rejection of six synthetic organic compounds from a potable water sources by a NF membrane process (15). In another study, a NF process was used for recovering tannins and water from exhausted baths and reusing them as tanning agents and washings, and the advantages are particularly significant in terms of reduction of environmental impact (up to 75% of COD reduction of the global effluent), simplification of cleaning-up processes of wastewaters, saving of chemicals and water, easy reuse of sludges and decrease of disposal costs (19). Tertiary treatment of distillery wastewater by NF was also reported by Rai and coworkers (20) who found that NF membrane achieved very high separation efficiency for both organic and inorganic compounds (around 96–99.5%, 85–95% and 98–99.5% removal of COD, TDS and cooler, respectively).

2.2.4. *Reverse Osmosis*

Unlike conventional filtration, which can only remove suspended solids with sizes >1 μm, the reverse osmosis process is able to eliminate the dissolved solids, bacteria, viruses and other germs contained in the water. An RO is essentially a pressure-driven membrane diffusion process for separating dissolved solutes. During the process, a pressurized feed solution is passed over one surface of the membrane. As long as the operation pressure, ranging from 10 to 70 bars depending on its application, is greater than the osmotic pressure of the feed solution, water will flow from the more concentrated solution to the more dilute solution through the membrane. An RO is generally used for desalination of seawater for its conversion into potable water. The salient features of the process are that it involves no phase change and it is a relatively low-energy process.

In normal process, solvent (e.g. water) can flow across the semi-permeable membrane from a dilute concentration to a more concentrated solution until equilibrium is reached (15).

The application of high pressure to the concentrated side will cause this process to reverse. This results in solvent flow away from the concentrated solution, leaving a higher concentration of solute. In application, the waste stream flows past the membrane; while the solvent (e.g. water) is driven through the membrane, the remaining solutes (e.g. organic or inorganic components) do not pass through, and become increasingly concentrated on the feed side of the membrane.

Almost all RO membranes are made of polymers, cellulosic acetate and matic polyamide types, and are rated at 96–99% NaCl rejection. Generally, there are two types of RO membranes: asymmetric or skinned membranes and thin-film composite membranes. The support material is commonly polysulphones, while the thin film is made from various types of polyamines and polyureas.

RO membranes have the smallest pore structure with pore diameter ranging from approximately 5 to 15 Å (0.5–1.5 nm). The extremely small size of RO pores allows only the smallest organic molecules and unchanged solutes to pass through the semi-permeable membrane along with the water. Greater than 95–99% of inorganic salts and charged organics will also be rejected by the membrane due to charge repulsion established at the membrane surface.

The major advantage of RO for handling process effluents is its ability to concentrate dilute solutions for recovery of salts and chemicals with low power requirements. No latent heat of vaporization or fusion is required for effecting separations; the main energy requirement is for a high-pressure pump. It also requires relatively limited floor space for compact, high capacity units, and it exhibits good recovery and rejection rates for a number of typical process solutions (15).

The major problem of RO is its higher potential of fouling. This is caused by particulate and colloidal matters that become concentrated at the feed side of the membrane surface. Fouling of membranes by slightly soluble components in solution or colloids has caused failures, and fouling of membranes by feed water with high levels of suspended solids can be a problem. Pre-treatment is used to remove particulate matters from the feed water. A system designed to operate at a high permeate flux is likely to experience high fouling rates and will require frequently chemical cleaning. Limited operational temperature range is another limitation for RO process. For cellulose acetate membranes, the preferred range is 18–30°C, lower temperatures will cause decreased fluxes and higher temperatures will increase the rate of membrane hydrolysis, hence reduce system life. These membranes are also chemically sensitive. Strongly acidic or basic solutions, strong oxidizing agents, solvents and other organic compounds can cause dissolution of the membrane. Poor rejection of some compounds such as borates and low molecular weight organics is another problem. Finally, there are some concentrated solutions having initial osmotic pressures, which are so high that they may exceed the available operating pressures and make it uneconomical (15).

The product water from an RO unit will have a low pH and most probably a high concentration of carbon dioxide (15). The carbon dioxide can be removed and the pH of the product increased by use of a decarbonator. A decarbonator is a packed column in which product water is introduced at the top while either forced or induced air is introduced at the bottom. The air and water flow counter-currently over and around the column packing. The carbon dioxide is stripped from the water and exits from the decarbonator at the top in the air stream.

In the last 6–8 years, the technology has gained industry acceptance as a viable water treatment option for many different fluid separation applications. Low operating costs and the ability to remove organic contaminants and 95–99% of inorganic salts with minimal chemical requirements make RO an attractive technology for many industrial applications.

A study by Subhi and Anne (24) shows that RO was capable of separating oil from oily wastewater with a rejection of 99% when oil content was up to 30%. For industrial applications, RO was generally used to treat groundwater for production of pharmaceutical-grade water; the retention reported by Belkacem and coauthors was more than 95% for the totality of ions and conductivity was decreased successfully (25). RO also has extensive applications in the following water treatments:

- Boiler feed water and cooling tower blow down recycle for utilities and power generation
- Cleaning of contaminated surface water and groundwater
- Potable water from sea or brackish water
- Ultra-pure water for food processing and electronic industries
- Water for chemical, pulp and paper industry

2.2.5. Other Membrane Technologies

Apart from the above four major separation processes, there are other membrane technology applications in water and wastewater treatment:

1. *Dialysis*. Dialysis is a process where solutes travel from one side of the membrane to the other side according to their concentration gradients (13, 26, 27). It uses a semi-permeable membrane capable of passing small solute molecules, such as salts and small organic species, while retaining colloids and solutes of higher molecular weight. The transfer through the membrane is by diffusion, rather than by the hydrodynamic flow that would occur in a porous medium. The current applications of dialysis include:
   - Hemodialysis
   - Purification techniques in pharmaceutical and biochemical laboratories
   - Separating nickel sulphate in the electrolytic copper refining industry
2. *Electrodialysis*. Electrodialysis (ED) is a process in which ions are transported through ion-permeable membranes from one solution to another under the influence of a potential gradient (26, 27). The electrical charges on the ions allow them to be driven through the membranes fabricated from ion-exchange polymers. Applying a voltage between two end electrodes generates the potential field required for this. As the membranes used in ED have the ability to selectively transport ions having positive or negative charge and reject ions of the opposite charge, useful concentration, removal or separation of electrolytes can be achieved by ED. Sadrzadeh and Mohammadi (28) reported that ED was also very efficient for desalination of seawater especially at lower concentrations. The current applications of ED also include:
   - Electroplating rinse water
   - Etch bath rinse water
   - Removal of organic acids from wine and fruit juices
   - Radioactive wastewater treatment
   - Regeneration of ion-exchange resins
   - Ultra-pure water production
3. *Donnan dialysis*. Donnan dialysis uses an anion- or cation-selective membrane, which functions similarly to ion-exchange resins (15, 26). For an anion-exchange membrane, cations in both

solutions are prevented from diffusing through the membrane, but anions will get through until equilibrium is reached. For a cation-selective membrane, cations will diffuse through the membrane while anions will be retained. Potential difference is the driving force behind these separation processes. It is created by the displacement of the system from the equilibrium ratios, which can be controlled by adjusting solution concentrations. The application for ion exchange includes:
- pH control without adding acid or base
- Recovery of acids and bases from salts
- Water softening

4. *Membrane bioreactor*. Membrane bioreactor (MBR) can be defined as integration of conventional biological degradation and membrane separation into a single process where microorganisms responsible for biodegradation and suspended solids are separated from biologically treated water by a membrane filtration unit. The entire biomass is confined within the system, providing both perfect control of the sludge retention time for the microorganism in the bioreactor and physical disinfection of the effluent (12, 29, 30).

As the MBR process does not need the secondary clarifier for the solid–liquid separation, it can prevent these "inherited" problems that occur in the conventional processes for the wastewater reclamation (30). The resulting high-quality and completely disinfected effluent means that the MBR process can be used for many purposes, that is, industrial and municipal wastewater treatment and reuse. More important, the land requirement of MBR is much less than that of conventional wastewater reclamation processes.

2.3. *Case Studies on Membrane Applications in Water Treatment*

2.3.1. *Case 1: Desalination of Seawater by RO*

Desalination of seawater is one of the important applications of membrane processes. There are various ways to produce freshwater such as distillation, electrodialysismembrane distillation, freezing, membrane bioreactor and reverse osmosis. Among them, distillation is the most used technique, but RO is becoming more popular in the desalination industry. A flow diagram of a single-stage RO system is shown in Fig. 7.4.

To reduce fouling and protect membrane, pre-treatment must be performed for seawater. A pre-treatment system removes suspended solids and other potential fouling materials. Flocculation agents such as iron chloride or polyelectrolytes are added in order to remove suspended solids. Chlorine is added to remove bacteria and algae.

The water quality is determined by membrane materials and configurations. Cellulose acetate membranes can be degraded by the formation of biological slimes on the membrane surface, while polyamide membranes cannot tolerate chlorine. Thus, sodium hydrogen sulphite is normally needed to remove chlorine in the water. Operation at higher temperatures and pressures can cause the compaction of the membrane layer and thus reduce output. A stagnant boundary layer at the membrane surface can cause concentration polarization of salts, increasing the chance of precipitation and reducing output (26, 27). The operating brine concentration, pressure and water conversion factor need to be selected to limit both compaction and concentration polarization effects.

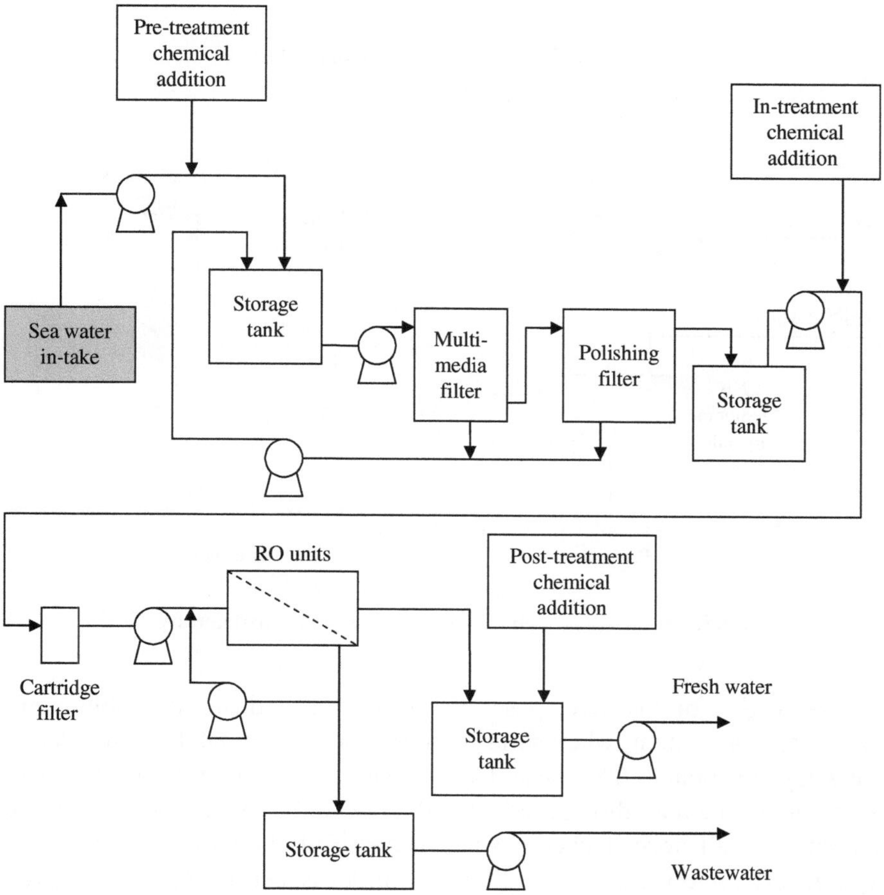

Fig. 7.4. RO system for seawater desalination.

When a membrane-based desalination process is used, seawater is first collected and pumped to the water treatment plant. Sodium hypochlorite is injected periodically after the intake pumps to prevent biological growth within the water treatment system. Suspended solids are retained by sand filters or MF. The filtered water is then acidified, treated with a sequestering agent and dechlorinated. A security filter precedes the high-pressure pumps which transfer the water to the RO membrane modules. The concentrate leaves the membrane modules via energy-recovery turbines directly coupled to the high-pressure pumps. The product water with a quantity of carbon dioxide and a residual salinity of less than 690 mg/L is led to the degasifier, treated for pH stabilization and finally collected in a reservoir.

2.3.2. Case 2: Removal of Salts from Food Processing Water Stream by NF

The single most important food manufacturing application of NF membranes is desalting of whey permeate, as shown in Fig. 7.5, or other food processing water streams.

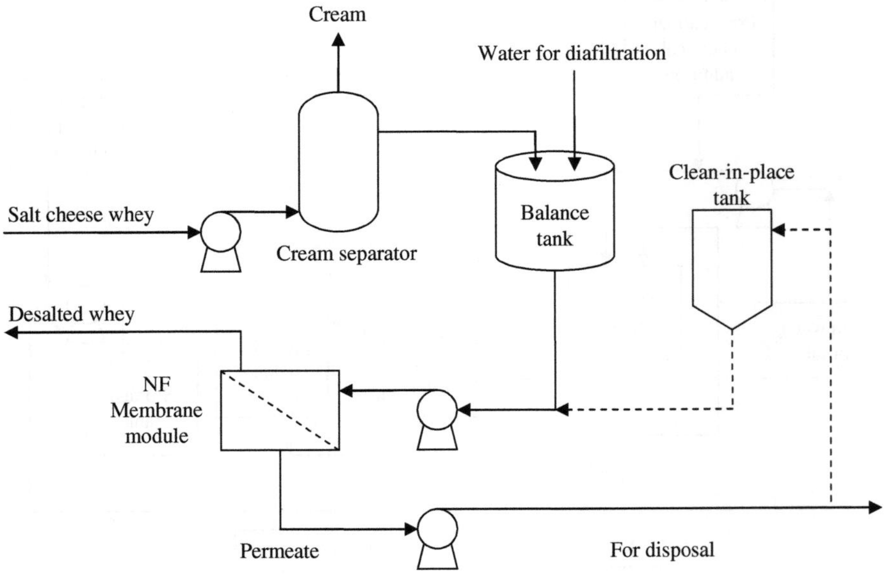

Fig. 7.5. Removal of salts from cheese whey permeate by NF.

Salty cheese whey, for instance, poses a serious waste disposal problem in the food industry (31). The salt content, whey solids content and biochemical oxygen demand of the salty cheese whey are about 4–6%, 6% and 45,000 mg/L, respectively. Accordingly, the salty cheese whey can be neither discharged directly into a Public Owned Treatment Works system, nor mixed with normal cheese whey because of its extremely high salt content. The NF system effectively removes sodium chloride as permeate, which is recycled or discharged. The retentate, which is desalted whey, is added back to the normal whey. This is a typical example of cleaner production and cost-saving measurement.

The desalting process system has double benefits. It not only increases profitability, but also solves a tough waste disposal problem. Removal of salts from a food processing water stream can also be accomplished by using electrodialysis or ion exchange. The capital costs of both ED and ion exchange are higher than that of NF.

2.3.3. *Case 3: Nutrient Removal from Wastewater Streams by RO*

Considering the increasingly stringent industrial effluent standards currently being imposed on industrial effluent pre-treatment facilities, specifically allowable concentrations of nutrient-containing substances, the capacity of RO to remove nutrients should be properly documented.

CA is still considered the preeminent membrane for wastewater treatment because it is capable of producing the highest flux per unit surface area at specified levels of solute rejection. The rejection performance of RO using a 90% sodium chloride rejection CA membrane was studied. Phosphorous removal was greater than 95% in all cases. Ammonia removals were generally in excess of 90%, and nitrite and nitrate removals generally ranged from 84 to 97% (32).

In another study of nutrient removal by RO, similar findings were confirmed: 80–90% separation of sodium nitrate, sodium nitrite and ammonium chloride were realized, with ammonium chloride rejection significantly better than both sodium nitrate and sodium nitrite. Rejection of phosphate-containing substances was also in agreement with separation efficiency at greater than 98.5% (11).

Meanwhile, many food processing plants, including a major dairy plant in Minnesota, USA, use phosphoric acid in their processes, resulting in over 100 mg/L of phosphate in their wastewater streams. RO is an ideal process unit for recovery and reuse of nutrients as fertilizers.

3. MEMBRANE MATERIALS: PREPARATION AND MODIFICATION

It is essential to understand the basic characteristics of the membrane, such as membrane materials and their formation, so that proper selection of membrane separation processes can be made.

3.1. *Membrane Materials*

The range of available membrane materials used in water and wastewater treatment is quite broad, but most of them are synthetic membranes. Synthetic membranes can be organic and inorganic; however, the most important class of membrane materials is organic or polymer membrane. The choice of a given polymer as a membrane material is not arbitrary (13). Inorganic materials generally possess superior chemical and thermal stability relative to polymeric materials. However, both types of membranes have different applications. A list of common membranes is shown in Table 7.2.

3.2. *Types of Membrane and Their Formation*

3.2.1. *Introduction*

Symmetric membranes and *asymmetric membranes* are two basic types of membrane based on their structure. *Symmetric membranes* include non-porous (dense) symmetric membranes and porous symmetric membranes, while *asymmetric membranes* include integrally skinned asymmetric membranes, coated asymmetric membranes and composite membranes. A number of different methods are used to prepare these membranes. The most important techniques are sintering, stretching, track-etching, template leaching, phase inversion and coating (13, 33).

3.2.2. *Phase Inversion Membranes*

Generally, in phase inversion technique, a polymer transforms from its liquid phase to solid phase in a well-controlled manner. Usually, the process is carried out by starting the transition from one liquid state to two liquid states. One of them has high polymer content. At some point during this transition, the polymer-rich liquid phase begins to solidify, forming a solid matrix. The morphology of the membrane can be controlled by adjustment of the initial stage of the phase transition.

Table 7.2
List of commercial membranes: compositions and applications

Membranes	Membrane materials	Applications
Organic membrane	Cellulose regenerated	D, UF, MF
	Cellulose nitrate	MF
	Cellulose acetate	GS, RO, D, UF, MF
	Polyamide	RO, NF, D, UF, MF
	Polysulphone	GS, UF, MF
	Poly(ether sulphone)	UF, MF
	Polycarbonate	GS, D, UF, MF
	Poly(ether imide)	UF, MF
	Poly(2,6-dimethyl-1,4-phenylene oxide)	GS
	Polyimide	GS
	Poly(vinylidene fluoride)	UF, MF
	Polytetrafluoroethylene	MF
	Polypropylene	MF
	Polyacrylonitrile	D, UF, MF
	Poly(methyl methacrylate)	D, UF
	Poly(vinyl alcohol)	PV
	Polydimethylsiloxane	PV, GS
Inorganic membrane		
Ceramic	Metal (Al, Zn, Ti, Si, etc.) oxide	PV, MF
	Metal (Al, Zn, Ti, Si, etc.) nitride	
	Metal (Al, Zn, Ti, Si, etc.) carbide	
Metallic	Metal (Al, Zn, Ti, Si, etc.) oxide	PV, MF
Zeolite	SiO_2	PV, GS

MF microfiltration, *UF* ultrafiltration, *NF* nanofiltration, *D* dialysis, *PV* pervaporation, *GS* gas separation, *RO* reverse osmosis.

There are a number of different techniques belonging to the category of phase inversion: solvent evaporation, precipitation by controlled evaporation, precipitation from the vapour phase, thermal precipitation and immersion precipitation (13, 34–36). The most commercially available membranes are prepared by the last method.

3.2.3. *Spinning of Hollow Fibres*

The spinning of hollow fibres is very often applied as a preparation technique for UF membranes. During the spinning process, nascent hollow fibres must be mechanically self-supporting, which means that the spinning solution must have high enough viscosity to support the nascent fibre (37–40). The spinning solution passes through a spinneret with a precision orifice containing a centrally positioned inlet tube for the delivery of the internal coagulation medium. The internal coagulant is required to form the hollow fibre inside bore (37–39). A schematic diagram of the wet spinning apparatus is shown in Fig. 7.6.

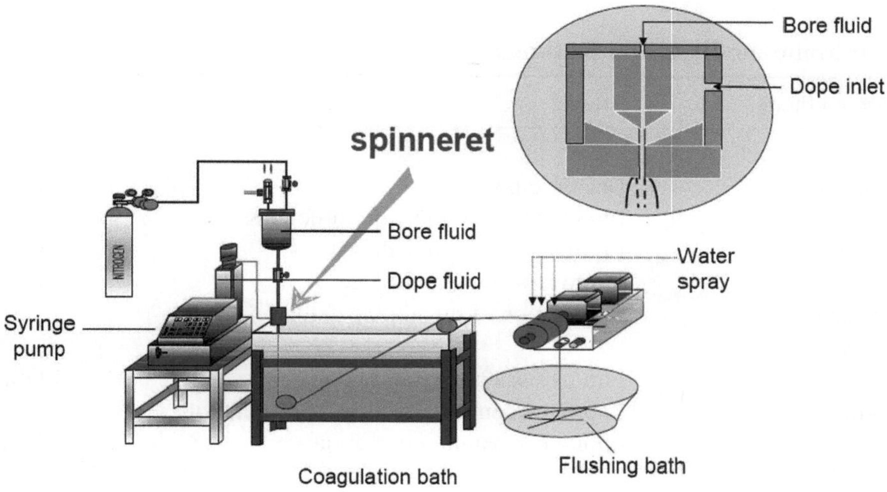

Fig. 7.6. Wet spinning apparatus.

In the spinning of UF hollow-fibre membranes, three options exist for spinneret placement: wet spinning, dry–wet spinning and dry spinning. The shape and morphological structure of the membranes obtained is related to the rheological characteristics and phase behaviour of the polymer solution during spinning. At the bore side, pore structure is determined by the solvent/non-solvent ratio of the internal bore liquid. For the outer surface, pore structure is mainly dependent on the conditions in the air gap.

Spinning conditions affect membrane physical dimensions and morphology, which in turn influence membrane performance such as permeability. A satisfactory spinning process would produce fibres having the requisite surface pore size and mechanical strength. Parameters involved in a spinning process are:

- Polymer concentration and composition
- Additives
- Velocity of spinning solution through spinneret
- Type and velocity of internal coagulant through the inner tube
- External coagulant
- Spinning temperature
- Distance of the air gap

3.2.4. Membrane Modification

To enhance the overall performance of the membrane, it is necessary to modify the membrane material or the structure (41). The objectives for modification of the existing membranes are to increase flux, selectivity and chemical resistance (solvent resistance, swelling resistance and fouling resistance). Some of the most commonly practiced membrane modification methods are listed in Table 7.3.

Table 7.3
List of membrane modification methods

Modification methods	Effects	Applications
Annealing		
• Heat treatment	Decrease pore size	RO, GS, UF
• Solvent treatment	Eliminate membrane micro-defects	
Surface coating	Eliminate membrane defects	GS
	Improve fouling resistance	RO, NF, UF
Solvent exchange	Elimination of membrane defects	GS, UF
Chemical treatment		
• Fluorination	Improve selectivity and reduce flux	GS, RO
• Cross-linking	Improve chemical stability and selectivity	UF, PV
• Pyrolysis	Improve separation performance	RO, GS, PV

The first reported membrane modification method involved annealing of porous membranes by heat treatment. Zsigmondy and Bachmann (42) in 1922 demonstrated that the pore size of a preformed nitrocellulose membrane could be decreased with a hot water or stream treatment. An alternative way of annealing is solvent treatment. Micro-defects in a membrane can be eliminated by swelling the thin separating layer with a vapour or a liquid. As a result of swelling, the modulus of the polymer decreases sharply. It has been suggested that capillary forces can pull the swollen polymer layer together, and eliminate the membrane defects (43).

One setback for the production of thin-film composite membranes and integrally skinned asymmetric membranes with separating layer thickness of less than 0.2 μm is the defects. A thin coating of a highly permeable polymer can help eliminate the defects. Surface coatings are also applicable in improving the fouling resistance of membranes for UF or NF applications (44).

Another way to minimize the membrane defects is *solvent exchange*. Typical solvent exchange methods involve replacing water first with iso-propanol and then with *n*-hexane (41). This solvent exchange is essential as collapse of water-wet membranes occurs upon drying. This collapse is due to the strong capillary forces within the finely porous structure during the drying process. According to the Young–Laplace relationship, the capillary pressure is proportional to the surface tension of a liquid and inversely proportional to the pore size. If this capillary force is higher than the modulus of the membrane material, the pores will collapse. As water usually has a very high surface tension, it is often difficult to dry water-wet membranes without destroying the membrane structure. One possible solution is substituting water with liquids having lower surface tension, such as alcohols or aliphatic hydrocarbons, which will maintain the original membrane structure upon drying.

Chemical surface modification methods of gas separation membranes include treatment with fluorine, chlorine, bromine or ozone. These treatments result in an increase in membrane selectivity with a decrease in flux. Cross-linking of polymers is often applied to improve the chemical stability and selectivity of membranes for reverse osmosis, pervaporation and gas separation applications (41). Mosqueda-Jimenez and coworkers (45) studied the addition of surface-modifying macromolecules and the use of the additive polyvinylpyrrolidone (PVP). Their results showed that membranes prepared with PVP had pure water permeation rates significantly higher than membranes prepared without the PVP additive. Pyrolysis method can lead to significantly improved separation performance of synthetic membranes, especially molecular sieve membranes made from pyrolyzed polyacrylonitrile and polyimide (46).

4. MEMBRANE CHARACTERIZATION

Membranes need to be characterized to ascertain which may be used for a certain separation or class of separations (13). Membrane characterization is to measure structural membrane properties, such as pore size, pore size distribution, free volume and crystallinity to membrane separation properties. It helps gather information for predicting membrane performance for a given application.

Membrane characterization means the determination of structural and morphological properties of a given membrane. As membranes range from porous to non-porous depending on the type of separation problem involved, different characterization techniques are required in each case. For example, in MF or UF membranes, fixed pores are present. MF membranes have macropores (pore diameter > 50 nm) while UF membranes have mesopores (2 nm < pore diameter < 50 nm). The pore size (and size distribution) mainly determines which particles or molecules are retained or pass through. On the other hand, for dense or non-porous membranes, no fixed pores are present and the material chemistry itself mainly determines the performance.

4.1. Porous Membrane

Two different types of characterization method for porous membranes can be distinguished (13):

1. *Structure-related parameters*: determination of pore size, pore size distribution, top layer thickness and surface porosity
2. *Permeation-related parameters*: determination of the actual separation parameters using solutes that are more or less retained by the membrane

There are a number of characterization techniques available for porous membranes, the following methods are usually used:

- Electron microscopy
- Atomic force microscopy (AFM)
- Bubble point method
- Permeation measurement

The first three methods involve the measurement of structural-related parameters while the last one is a typical permeation-related technique. Both electron microscopy and AFM can

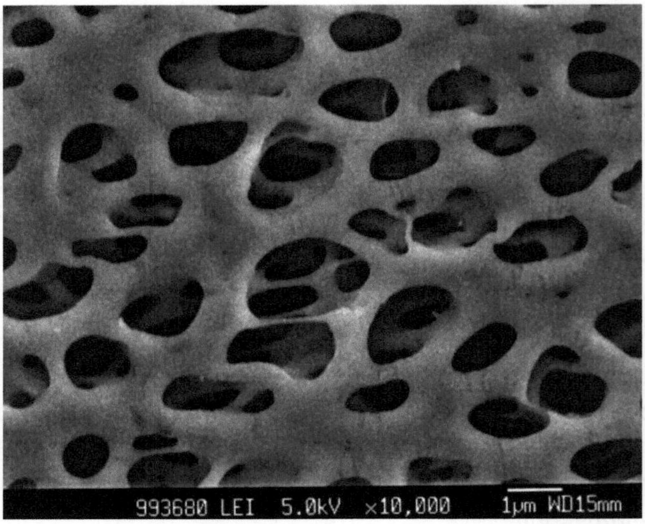

Fig. 7.7. Visualization of top surface of a porous polymeric membrane by scanning electron microscopy (10,000×).

provide qualitative measurement of membrane materials. Figure 7.7 shows the top surface of porous polymeric membrane observed by scanning electron microscopy (SEM). The bubble point method and permeation measurement, on the other hand, provide quantitative information of membrane materials.

4.2. *Non-porous Membrane*

Transport through non-porous membranes follows the solution–diffusion mechanism and separation is achieved by differences in either solubility or diffusivity. Therefore, the characterization method is different from those in the previous section. There are several methods used for non-porous membrane:

- Permeability methods
- Differential scanning calorimetry/differential thermal analysis methods
- Surface analysis (such as X-ray photoelectron spectroscopy)

Details of these techniques will not be discussed in this book and readers can find relative information in manuals of the respective analytical instruments.

5. MASS TRANSPORT IN MEMBRANES

A membrane can be defined as a barrier between two phases. A molecule or particle is transported from one phase to another through the membrane because a driving force acts on that molecule or particle. This driving force can be chemical potential gradient, for example, concentration gradient, pressure difference, or electrical potential, or combination of these (10). There is a proportional relationship between the flux and the driving force:

$$\begin{aligned} \text{Flux} &= \text{proportionality factor} \times \text{driving force} \\ J &= A \times X, \end{aligned} \tag{1}$$

where J is the flux, A is the proportionality factor and X is the driving force. The proportionality factor (A) determines how fast the component is transported through the membrane, that is, A is a measurement of the resistance exerted by the membrane as a diffusion medium, when a given force is acting on this component (13). Various models have been established to study both the proportionality factor and the driving force. Among those models, two of the most useful models will be discussed next.

5.1. *The Solution–Diffusion Model*

The solution–diffusion model is currently being used by the majority of the membrane community. The most general description of the mass transport across a membrane is based on irreversible thermodynamics (10):

$$J_i = L_{ii}X_i + \sum L_{ij}X_j, \tag{2}$$

where J_i is the mass flux of component i (mol/m^2 s), L_{ii}, L_{ij} are phenomenological coefficients and X_i, X_j are driving forces. The first term of Eq. (2) shows the flux of the ith component. The second term shows the relation between the flux of the ith component and the forces acting on components other than ith component. It should be noted that L_{ii} is always positive and $L_{ij} = L_{ji}$.

The most appropriate choice of the force X for the molecular diffusion through the membrane under isothermal conditions without external forces being applied to the mass transfer of the ith component is the chemical potential gradient:

$$X_i = -\nabla \mu_i, \tag{3}$$

where μ_i is the chemical potential of the ith component (J/mol).

The flux is given by

$$J_i = -L_{ii}\nabla \mu_i. \tag{4}$$

In Eq. (4), the second term of Eq. (2) is eliminated, which means that the flow of the ith component is totally decoupled from the flow of other components.

The chemical potential gradient can be described as

$$\nabla \mu_i = RT\nabla \ln a_{im} + v_i \nabla p, \tag{5}$$

where a_{im} is the activity in the membrane (mol/m^3); v_i is the partial molar volume of the ith component (m^3/mol); p is the pressure (Pa); R is the gas constant, 8.314 J/mol K or 0.08206 atm L/K mol and T is the temperature (K).

Considering the membrane process as a binary system, the transport of solvent (e.g. water) and solute is involved. Designating solute and solvent by subscripts A and B, Eq. (5) can be written for solvent B as

$$\nabla \mu_B = RT \nabla \ln a_{Bm} + v_B \nabla p. \tag{6}$$

Assuming v_B is constant, integration over the membrane thickness yields

$$\Delta \mu_B = v_B \left(\frac{RT}{v_B} \Delta \ln a_{Bm} + \Delta p \right). \tag{7}$$

(Δ is defined here as "permeate side–feed side".)

Assuming further that thermodynamic equilibrium is established at the membrane–solution boundaries, $a_{Bm} = a_B$, where a_B is the activity of solvent (mol/m^3) outside of the membrane. This relationship should be valid on both sides of the membrane. Since the osmotic pressure, Π (Pa), is defined as

$$\Pi = -\frac{RT}{v_B} \ln a_B, \tag{8}$$

Eq. (7) can be written as

$$\Delta \mu_B = v_B (\Delta p - \Delta \Pi). \tag{9}$$

For practical purposes, the osmotic pressure of a feed may be estimated by van't Hoff's equation

$$\Pi = R(T + 273) \sum M_i \tag{10a}$$

or for NaCl,

$$\Pi = 7{,}720(T + 273) \sum M_i, \tag{10b}$$

where T is the temperature (°C) and $\sum M_i$ is the sum of molalities or molarities of ions and non-ionic compounds (mol/kg or mol/L).

As for solute A, we shall further assume that the activity coefficient remains constant, then Eq. (5) becomes

$$\begin{aligned} \nabla \mu_A &= RT \nabla \ln a_{Am} + v_A \nabla p \\ &= RT \nabla \ln c_{Am} + v_A \nabla p, \end{aligned} \tag{11}$$

where a_{Am} is the activity of solute in the membrane (mol/m^3), c_{Am} is the concentration of solute in the membrane (mol/m^3) and v_A is the molar volume of solute (m^3/mol).

Integration over the membrane thickness yields

$$\Delta \mu_A = RT \Delta \ln c_{Am} + v_A \Delta p. \tag{12}$$

The second term of this equation can be ignored when the solute concentration at the low-pressure side of the membrane is less than 90% of that at the high-pressure side of the membrane, which is usually the case for RO process. Therefore, we have

$$\Delta\mu_{\mathrm{A}} = RT\Delta \ln c_{\mathrm{Am}}. \tag{13}$$

As for L_{ii}, the following physical meaning is usually given to the phenomenological coefficient L_{BB}:

$$L_{\mathrm{BB}} = \frac{c_{\mathrm{Bm}}}{f_{\mathrm{Bm}}}, \tag{14}$$

where c_{Bm} is the concentration of the solvent in the membrane (mol/m^3) and f_{Bm} is the friction between a unit mole of the solvent and the membrane material (J s/m^2).

Combining Eq. (4) and Eq. (14) and integrating over the membrane thickness, we obtain

$$J_{\mathrm{B}} = -\frac{c_{\mathrm{Bm}}}{f_{\mathrm{Bm}}}\nabla\mu_{\mathrm{B}} = -\frac{c_{\mathrm{Bm}}}{f_{\mathrm{Bm}}}\frac{\Delta\mu_{\mathrm{B}}}{l}, \tag{15}$$

where l is the membrane thickness (m). Combining further Eq. (9) and Eq. (15), we obtain

$$J_{\mathrm{B}} = -\frac{c_{\mathrm{Bm}}}{f_{\mathrm{Bm}}} v_{\mathrm{B}} \frac{\Delta p - \Delta\Pi}{l}. \tag{16}$$

As the friction f_{Bm} can be given as

$$f_{\mathrm{Bm}} = \frac{RT}{D_{\mathrm{Bm}}}, \tag{17}$$

where D_{Bm} is the diffusion coefficient (m^2/s) of solvent in the membrane, Eq. (16) becomes

$$J_{\mathrm{B}} = -\frac{c_{\mathrm{Bm}} D_{\mathrm{Bm}} v_{\mathrm{B}}}{RTl}(\Delta p - \Delta\Pi). \tag{18}$$

As for the solute, the phenomenological coefficient L_{AA} can be given as

$$L_{\mathrm{AA}} = \frac{c_{\mathrm{Am}}}{f_{\mathrm{Am}}}. \tag{19}$$

Then the solute flux can be written as

$$J_{\mathrm{A}} = -\frac{c_{\mathrm{Am}}}{f_{\mathrm{Am}}}\nabla\mu_{\mathrm{A}} = -\frac{c_{\mathrm{Am}}}{f_{\mathrm{Am}}}\frac{\Delta\mu_{\mathrm{A}}}{l}. \tag{20}$$

Combining Eq. (13) and Eq. (20), we obtain

$$J_{\mathrm{A}} = -\frac{c_{\mathrm{Am}}}{f_{\mathrm{Am}}}\frac{RT\Delta \ln c_{\mathrm{Am}}}{l}. \tag{21}$$

Approximating $\ln c_{\mathrm{Am}}$ by $(\Delta c_{\mathrm{Am}})/c_{\mathrm{Am}}$ and using the relation $f_{\mathrm{Am}} = RT/D_{\mathrm{Am}}$, we obtain

$$J_{\mathrm{A}} = -D_{\mathrm{Am}} \frac{\Delta c_{\mathrm{Am}}}{l}. \tag{22}$$

Assuming further thermodynamic equilibrium at both sides of the membrane and a proportional relationship between the concentration inside the membrane, c_{Am}, and outside the membrane, c_{A},

$$c_{\mathrm{Am}} = K_{\mathrm{A}} c_{\mathrm{A}}, \tag{23}$$

where K_{A} is the distribution constant. Hence, Eq. (22) can be written as

$$J_{\mathrm{A}} = -D_{\mathrm{Am}} K_{\mathrm{A}} \frac{\Delta c_{\mathrm{A}}}{l}. \tag{24}$$

Therefore, the fluxes of the solvent and solute can be described by Eq. (18) and Eq. (24), or

$$J_{\mathrm{B}} = -\frac{c_{\mathrm{Bm}} D_{\mathrm{Bm}} v_{\mathrm{B}}}{RTl}(\Delta p - \Delta \Pi) \quad \text{and} \quad J_{\mathrm{A}} = -D_{\mathrm{Am}} K_{\mathrm{A}} \frac{\Delta c_{\mathrm{A}}}{l}.$$

Other than the flux of solvent, the process is also evaluated in terms of solute rejection, R, defined as

$$R = 1 - \frac{c_{\mathrm{AP}}}{c_{\mathrm{AF}}}, \tag{25}$$

where c_{AP} is the concentration of solute in the permeate (mol/m^3) and c_{AF} is the concentration of solute in the feed (mol/m^3).

Because

$$\frac{J_{\mathrm{A}}}{J_{\mathrm{B}}} = \frac{c_{\mathrm{AP}}}{c_{\mathrm{BP}}}, \tag{26}$$

$$R = 1 - \frac{J_{\mathrm{A}} c_{\mathrm{BP}}}{J_{\mathrm{B}} c_{\mathrm{AF}}}, \tag{27}$$

$$\begin{aligned} R &= 1 - \frac{D_{\mathrm{Am}} K_{\mathrm{A}} RT \Delta c_{\mathrm{A}} c_{\mathrm{BP}}}{c_{\mathrm{Bm}} D_{\mathrm{Bm}} v_{\mathrm{B}} (\Delta p - \Delta \Pi) c_{\mathrm{AF}}} \\ &= 1 - \frac{D_{\mathrm{Am}} K_{\mathrm{A}} RT (c_{\mathrm{AP}} - c_{\mathrm{AF}}) c_{\mathrm{BP}}}{c_{\mathrm{Bm}} D_{\mathrm{Bm}} v_{\mathrm{B}} (\Delta p - \Delta \Pi) c_{\mathrm{AF}}} \\ &= 1 + \frac{D_{\mathrm{Am}} K_{\mathrm{A}} RT c_{\mathrm{BP}}}{c_{\mathrm{Bm}} D_{\mathrm{Bm}} v_{\mathrm{B}} (\Delta p - \Delta \Pi)} \frac{c_{\mathrm{AF}} - c_{\mathrm{AP}}}{c_{\mathrm{AF}}} \\ &= 1 + \frac{D_{\mathrm{Am}} K_{\mathrm{A}} RT c_{\mathrm{BP}}}{c_{\mathrm{Bm}} D_{\mathrm{Bm}} v_{\mathrm{B}} (\Delta p - \Delta \Pi)} R. \end{aligned} \tag{28}$$

Rearranging,

$$R = \left[1 - \frac{D_{\text{Am}} K_{\text{A}} RT c_{\text{BP}}}{c_{\text{Bm}} D_{\text{Bm}} v_{\text{B}} (\Delta p - \Delta \Pi)}\right]^{-1}. \tag{29}$$

Example 7.1

The following numerical values were given for reverse osmosis of aqueous sodium chloride solution at 25°C (47):

$$c_{\text{Bm}} D_{\text{Bm}} = 2.7 \times 10^{-8}\,\text{kg/m s},$$

$$D_{\text{Am}} K_{\text{A}} = 4.2 \times 10^{-14}\,\text{m}^2/\text{s}.$$

Calculate the solute separation and water flux when the feed sodium chloride molality is 0.1 mol/kg and the operating pressure is 4.134×10^6 Pa (gauge). The thickness of the selective layer is 10^{-7} m.

Use the following numerical values:

$$RT = 2.479 \times 10^3\,\text{J/mol for } 25°\text{C},$$

$$c_{\text{BP}} = 10^3\,\text{kg/m}^3 \quad \text{and} \quad v_{\text{B}} = 18.02 \times 10^{-6}\,\text{m}^3/\text{mol}.$$

From Eq. (10),

$$\Pi = 7{,}720 \times (25 + 273) \times 0.2 = 0.460 \times 10^6\,\text{Pa}.$$

As $\Pi = 0$ for pure water, firstly, we can assume pure water on the permeate side. Approximating $\Delta\Pi = (0 - 0.460 \times 10^6) = -0.460 \times 10^6$ Pa, from Eq. (29)

$$R = \left[1 - \frac{(4.2 \times 10^{-14})(2.479 \times 10^3)(10^3)}{(2.7 \times 10^{-8})(18.02 \times 10^{-6})[(-4.134 \times 10^6) - (-0.460 \times 10^6)]}\right]^{-1} = 0.945.$$

Then, the sodium chloride molality in the permeate will be

$$0.1 \times (1 - 0.945) = 0.0055\,\text{mol/kg}.$$

From Eq. (10), the osmotic pressure of the permeate will be

$$\Pi = 7{,}720 \times (25 + 273) \times 2 \times 0.0055 = 0.0253 \times 10^6\,\text{Pa}.$$

Then, the osmotic pressure difference, $\Delta\Pi$, between the permeate and feed will be

$$\Delta\Pi = (0.0253 - 0.460) \times 10^6 = -0.4347 \times 10^6\,\text{Pa},$$

$$R = \left[1 - \frac{(4.2 \times 10^{-14})(2.479 \times 10^3)(10^3)}{(2.7 \times 10^{-8})(18.02 \times 10^{-6})[(-4.134 \times 10^6) - (-0.4347 \times 10^6)]}\right] = 0.945.$$

Hence, $R = 0.945$ is an accurate answer.

The flux (kg/m^2 s) is calculated from Eq. (18) as

$$J_{\text{B}} = -\frac{(2.7 \times 10^{-8})(18.02 \times 10^{-6})[(-4.314 \times 10^6) - (-0.4347 \times 10^6)]}{(2.479 \times 10^3)(10^{-7})}$$

$$= 76.14 \times 10^{-4}\,\text{kg/m}^2\,\text{s}.$$

5.2. The Pore Model

The fine-pore model was developed assuming the presence of open micropores on the active surface layer of the membrane through which the mass transport occurs (10). The existence of these different pore geometries also means that different models have been developed to describe transport adequately. The simplest representation is one in which the membrane is considered as a number of parallel cylindrical pores perpendicular to the membrane surface. The length of each of the cylindrical pores is equal or almost equal to the membrane thickness. The volume flux through these pores can be described by the Hagen–Poiseuille equation. Assuming all the pores have the same radius, then we have

$$J = \frac{\varepsilon r^2}{8\mu\tau} \frac{\Delta p}{\Delta x}, \quad (30)$$

where J is the flux (m^3/m^2 s); r is the pore radius (m); ε is the surface porosity, $\varepsilon = n_p \pi r^2 / A_m$ (A_m is the membrane area, n_p is the number of pores); μ is the fluid viscosity (Pa s); τ is the pore tortuosity; Δp is the pressure difference (Pa) (Δ is defined here as "feed side–permeate side") and Δx is the thickness of the membrane (m).

This equation clearly shows the effect of membrane structure on transport and indicates that the solvent flux is proportional to pressure difference as the driving force.

Eq. (30) gives a good description of transport through membranes consisting of a number of parallel pores. However, very few membranes possess such a structure in reality. Membranes consist of a system of closed spheres, which can be found in organic and inorganic sintered membranes or in phase inversion membranes with a nodular top layer structure. Such membranes can best be described by the Kozeny–Carman relationship:

$$J = \frac{\varepsilon^3}{K\mu\, S^2(1-\varepsilon)^2} \frac{\Delta p}{\Delta x}, \quad (31)$$

where J is the flux (m^2/m s); ε is the volume fraction of the pores; μ is the fluid viscosity (Pa s); S is the specific surface area (m^2/m^3); K is the Kozeny–Carman constant, which is dependent on the pore shape and tortuosity; Δp is the pressure difference (Pa) and Δx is the thickness of the membrane (m).

Phase inversion membranes frequently shows a nodule-like structure. The volume flux through these membranes is described by the Hagen–Poiseuille or the Kozeny–Carman relation, although the morphology is completely different.

Example 7.2

Schulz and Asunmaa (48), based on their SEM observation, assumed that the selective layer of an asymmetric cellulose acetate membrane for reverse osmosis consists of closely packed spherical nodules with a diameter of 18.8 nm. Water flows through the void spaces between the nodules. Calculate the water flux by Eq. (30) assuming circular pores, the cross-sectional area of which is equal to the area of the triangular void surrounded by three circles with a diameter of 18.8 nm (as shown in Fig. 7.8):

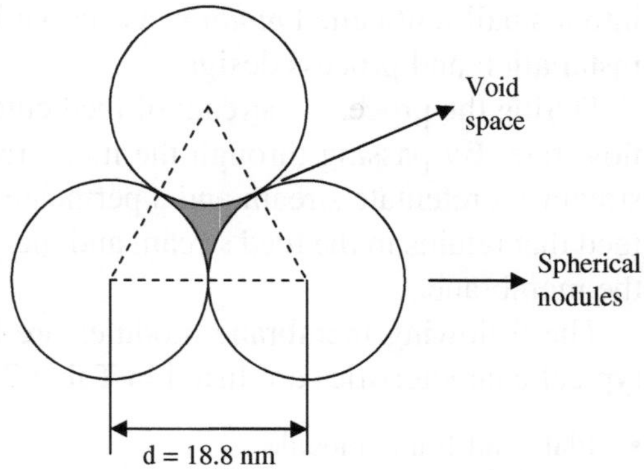

Fig. 7.8. Calculating void space between spherical nodules.

1. Calculate the area of the triangular void space.
2. Calculate the radius of the circular pore whose cross-sectional area is equal to the triangular area calculated above.
3. Subtract the thickness of the monolayer of water, 0.28 nm, which is adsorbed at the pore wall and assumed to be immobile.
4. Calculate the water flux using the relevant data in Example 7.1 and the following numerical values:

$$n_p/A_m = 6.5 \times 10^{15}\,\text{m}^{-2},$$
$$\Delta p = 4.134 \times 10^6\,\text{Pa},$$
$$\Delta x = 10^{-7}\,\text{m},$$
$$\tau = 2.5,$$
$$\mu = 0.035\,\text{Pa s (viscosity of water that fills nano-sized pores)}.$$

Answer

1. Area of the triangular void space is $9.4^2 \times (\sqrt{3} - \pi/2) = 14.25\,\text{nm}^2$.
2. The radius is $\sqrt{14.30/\pi} = 2.13\,\text{nm}$.
3. The effective radius of the mobile water channel is $r = 2.13 - 0.28 = 1.85\,\text{nm}$.
4. Water flux is

$$J = \frac{(6.5 \times 10^{15})(3.1416)(1.85 \times 10^{-9})^4(4.134 \times 10^6)}{(8)(3.5 \times 10^{-2})(2.5)(10^{-7})} = 141.26 \times 10^{-7}\,\text{m}^3/\text{m}^2\,\text{s}.$$

Assuming water density of 10^3 kg/m^3, the water flux is 141.26×10^{-4} kg/m^2 s.

6. MEMBRANE MODULE AND PROCESS DESIGN

6.1. Introduction

As we know, large surface areas are required for industrial applications of membrane processes. A practical solution for providing this large surface area is packing the membranes into a small unit called *module*, as shown in Fig. 7.2. The module is the base for membrane installation and process design.

During the process, a stream of feed enters the module with a specific content at a specific flow rate. By passing through the membrane module, the feed stream is separated into two streams: a retentate stream and a permeate stream. The retentate stream is the fraction of the feed that retains in the feed stream and the permeate stream is the fraction that passes through the membrane.

The following membrane modules are largely used for industrial applications, of which typical characteristics are listed in Table 7.4:

- Plate-and-frame module
- Spiral-wound module
- Tubular module
- Hollow-fibre module

The choice of module configuration, as well as the arrangement of the modules in a system, is based on economic considerations with correct engineering parameters being employed to achieve this, which include the type of separation problem, ease of cleaning, ease of maintenance, ease of operation, compactness of the system, scale and the possibility of membrane replacement (13). Next, we will discuss these typical membrane modules.

6.2. *Typical Membrane Modules*

6.2.1. Plate-and-Frame Module

A schematic drawing of a plate-and-frame module is given in Fig. 7.9.

This type of module appeared in the earliest stage of industrial membrane applications. The structure is simple and the membrane replacement is easy. As illustrated, sets of two membranes are placed in a sandwich-like fashion with their feed sides facing each other. In each feed and permeate compartment, a suitable spacer is placed. The number of sets needed for a given membrane area furnished with sealing rings and two end plates then builds up to a plate-and-frame stack. The membrane permeate is collected from each support plate. The spacer surface is made uneven in order to promote turbulence of the feed fluid and minimize concentration polarization. The module diameter is 20–30 cm. The total membrane area in one module is up to 19 m^2, depending on the height of the module (10).

6.2.2. Spiral-Wound Module

The spiral-wound module is in fact a plate-and-frame system wrapped around a central collection pipe, similar to a sandwich roll. The basic structure of this module is illustrated in Fig. 7.10. Membrane and permeate-side spacer material are then glued along three edges to build a membrane envelope. The feed-side spacer separating the top layer of the two flat membranes also acts as a turbulence promoter. The feed flows axial through the cylindrical module parallel along the central pipe and the permeate flows radially towards the central pipe. To make the membrane length shorter, several membrane envelopes are wound simultaneously. The spiral-wound module is featured by:

Table 7.4
Typical characteristics of membrane modules

	Plate and frame	Spiral wound	Tubular	Hollow fibre
Packing density (m^2/m^3)	30–500	200–800	30–200	500–9,000
Resistance to fouling	Good	Moderate	Very good	Poor
Ease of cleaning	Good	Fair	Excellent	Poor
Membrane material choices	Many	Many	Few	Few
Relative cost	High	Low	High	Low
Applications	D, RO, PV, UF, MF	D, RO, GP, UF, MF	RO, UF	D, RO, GP, UF

MF microfiltration, *UF* ultrafiltration, *D* dialysis, *PV* pervaporation, *RO* reverse osmosis.

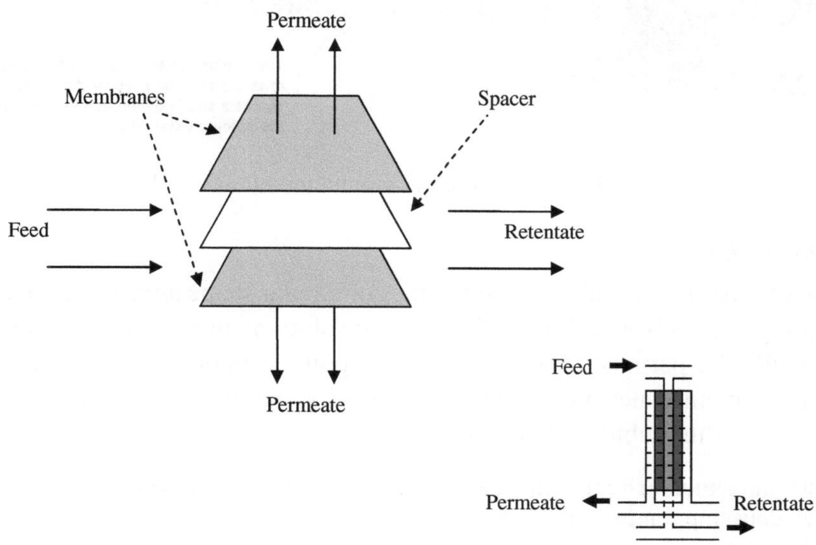

Fig. 7.9. A plate-and-frame module.

- A compact structure
- High-pressure durability
- Less contamination
- Less pressure drop at the permeate channel
- Minimum concentration polarization

Usually, a number of spiral-wound modules are assembled in one pressure vessel and are connected in series via the central permeate tubes. Some industrial-scale spiral-wound modules are shown in Fig. 7.11.

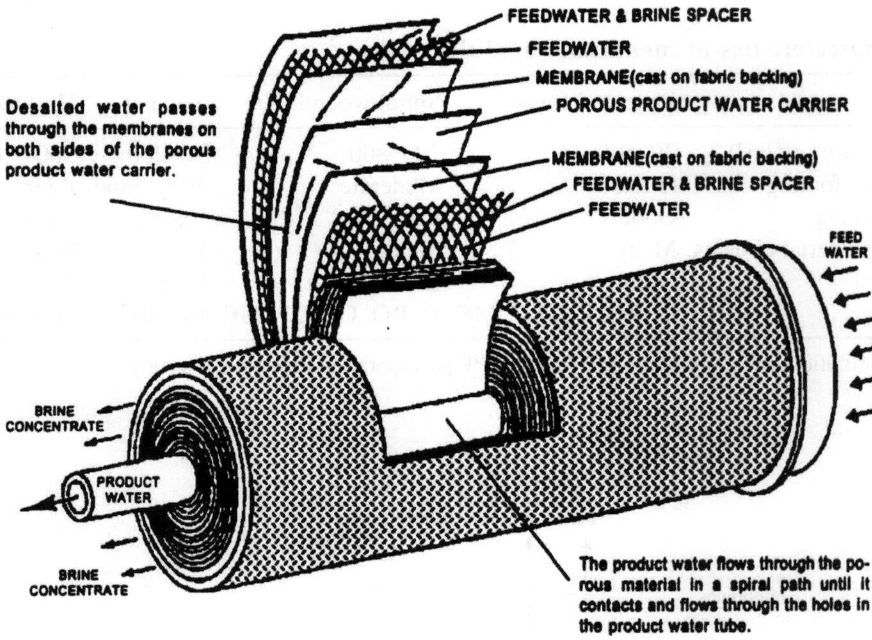

Fig. 7.10. A spiral-wound module.

6.2.3. *Tubular Module*

In this type of module, a number of membranes of tubular shape are encased in a container. A schematic diagram is given in Fig. 7.12. The feed solution always flows through the centre of the tubes while the permeate flows through the porous supporting tube into the module housing. Ceramic membranes are mostly assembled in such tubular module configurations. The main features of the tubular module are:

- Convenience in membrane replacement and easy cleaning of surface contamination
- High-energy consumption per unit amount of liquid treated
- High feed flow rate help reduce the membrane contamination
- Low packing density of the module
- Simple pre-treatment of feed liquid

6.2.4. *Hollow-Fibre Module*

Hollow-fibre module consists of a large number of fibres assembled together in a module, as shown in Fig. 7.13. The free ends of the fibres are often potted with agents such as epoxy resins, polyurethanes or silicon rubber. The membranes are self-supporting for this module. There are two basic types of arrangement for this module:

1. *Inside-out*, as shown in Fig. 7.14a, where the feed solution passes through the bore of the fibre and the permeate is collected on the outside of the fibre
2. *Outside-in*, as shown in Fig. 7.14b, where the feed solution enters the module on the shell side of the fibres and the permeate passes into the fibre bore

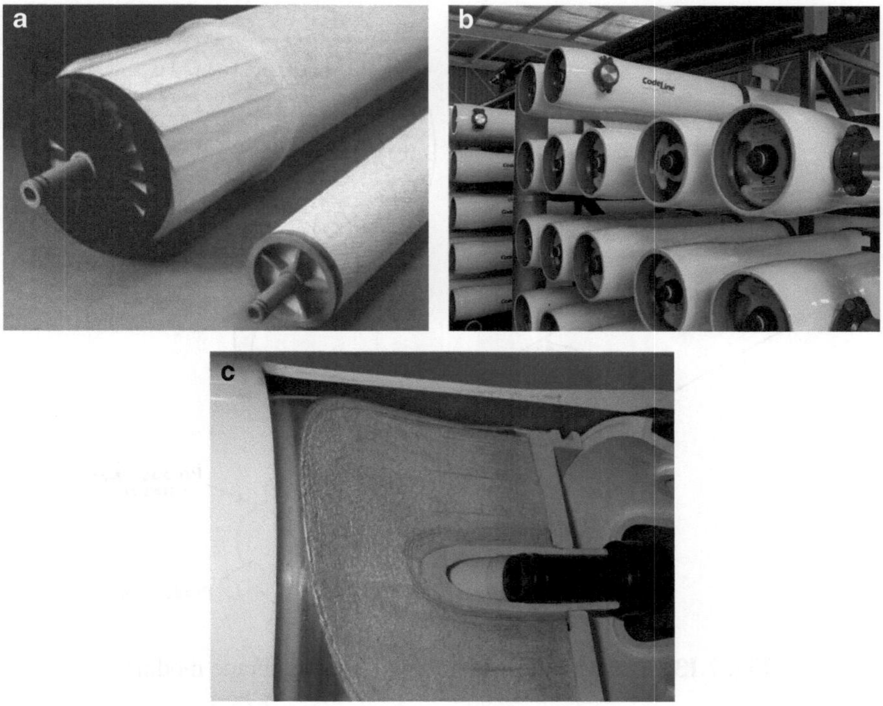

Fig. 7.11. Industrial-scale spiral-wound modules.

Fig. 7.12. A tubular module.

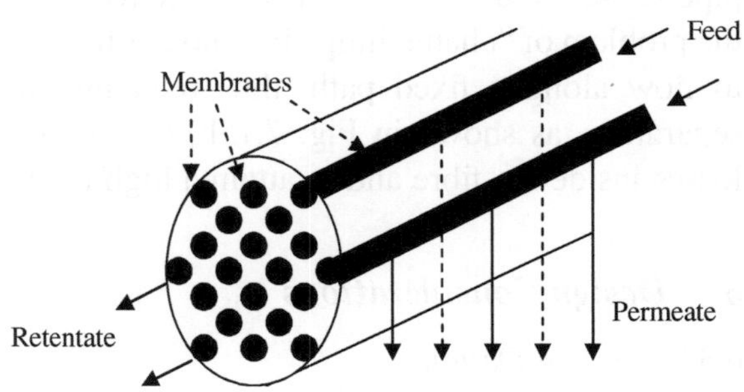

The choice between the two concepts is mainly based on some parameters such as operation pressure, pressure drop, type of membrane available, etc. The fibre wall has a structure of the asymmetric membrane, and the active skin layer is placed to the feed side. The hollow-fibre module is featured by a very high packing density, which can reach values of 30,000 m^2/m^3.

The hollow-fibre module is often used when the feed stream is relatively clean, such as in gas separation. It has also been used in the case of seawater desalination, but pre-treatment is needed. The module construction given in Fig. 7.15a is a typical RO module, where a central

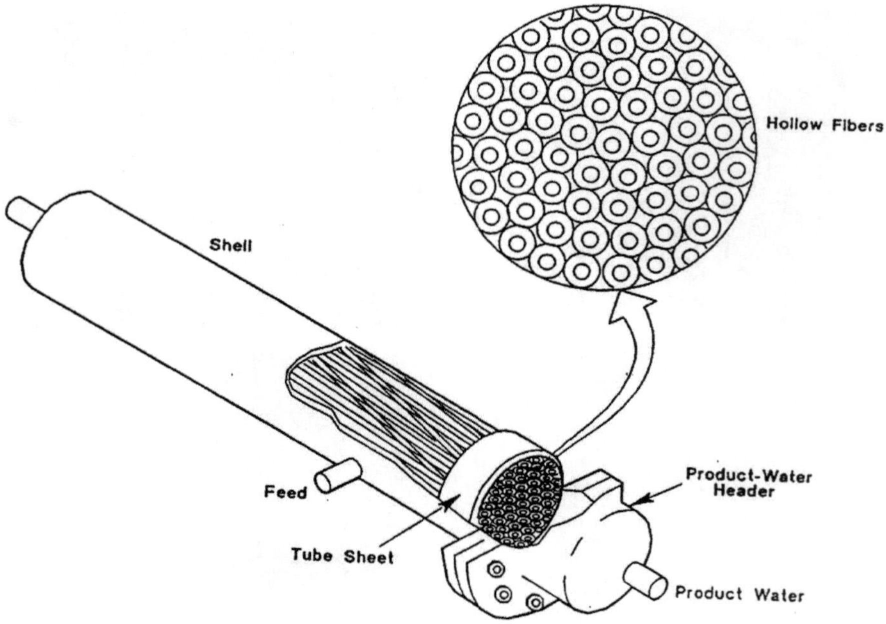

Fig. 7.13. Cutaway view of hollow-fibre membrane module.

pipe is used to uniformly distribute the feed solution throughout the module. This is to avoid the problem of "channelling" in "outside-in" model, which means that the feed has a tendency to flow along a fixed path, thus reducing the effective membrane surface area. In gas separation, as shown in Fig. 7.15b, the "outside-in" model is used to avoid high-pressure losses inside the fibre and to attain a high membrane area (13).

6.3. Design Considerations

6.3.1. System Design

The design of the membrane operation system can be very different for different applications and module configurations (13). A number of modules are connected together to form a stage. It is desirable to optimize the whole separation system in order to achieve low capital and operational costs and good quantity and quality of treated water. Normally, MF and UF do not require pre-treatment units; however, the feed must be pretreated before entering NF or RO unit.

The simplest design is a dead-end operation, as shown in Fig. 7.16a. As the feed is forced through the membrane, the concentration of rejected components in the feed increases and accumulates at the membrane interface; hence, the permeate quality decreases with time. Therefore, for industrial applications, a cross-flow operation, as shown in Fig. 7.16b, is preferred for its lower fouling tendency comparing with the dead-end mode.

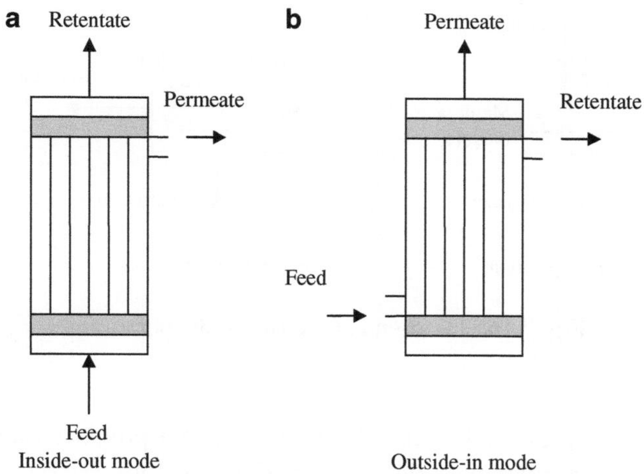

Fig. 7.14. Hollow-fibre module operations.

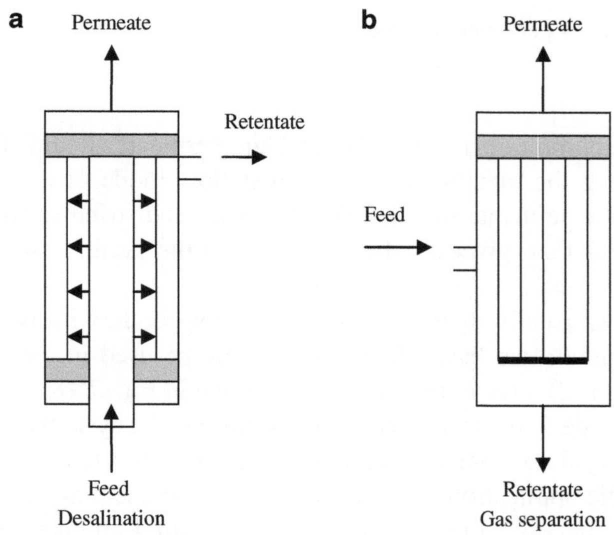

Fig. 7.15. Hollow-fibre module applications.

In the cross-flow operation, the inlet feed stream enters the module at a certain composition and it flows parallel to the membrane surface. The composition of the stream changes along the module and the stream is separated into two parts: a permeate stream and a retentate stream. Flux decline is relatively smaller with cross-flow and can be controlled and adjusted by proper module configuration and cross-flow velocities.

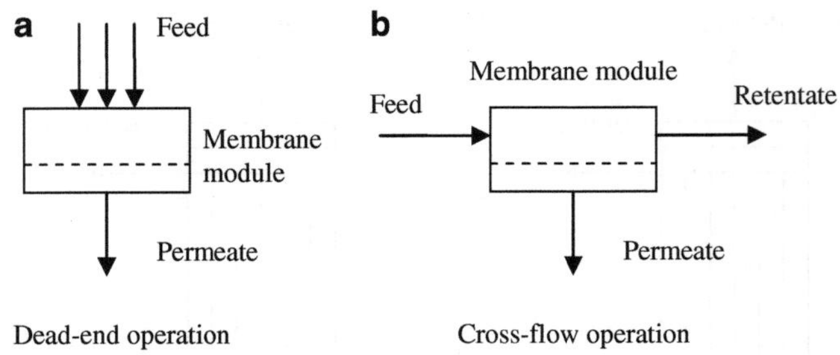

Fig. 7.16. Dead-end and cross-flow operation.

As cross-flow mode is preferred over a dead-end one, the proper choice of the module is the next crucial step. For a given module design and feed solution, the cross-flow velocity is the main parameter that determines mass transfer in the module. Four cross-flow operations shown in Fig. 7.17 are used:

1. Co-current
2. Counter-current
3. Cross-flow with perfect permeate mixing
4. Perfect mixing

In co- and counter-current operations, the feed and permeate streams flow co-currently or counter-currently along the membrane. In the cross-flow mode, it is assumed that mixing occurs so rapidly on the permeate side that the composition distributes equally. Among these modes, counter-current flow gives the best results and the perfect mixing gives the worst result.

In practice, systems usually operate in the cross-flow mode with perfect mixing on the permeate side. In general, two basic flow schemes can be used in the system design (1) a single-pass system and (2) a recycling system, as shown in Fig. 7.18.

In the single-pass system, the feed stream passes through the system only once, and there is no recycling. In a recycling system, a recirculation pump is used to recycle the retentate stream. For small-scale applications, a batch system can be used, as shown in Fig. 7.19.

Usually, a single-stage operation does not give the desired product quality, and a second stage is required. A combination of stages is called a *cascade*. In a cascade operation, a large number of units are used, and the permeate of the first stage is the feed for the second stage and so on. An example of a two-stage operation process is shown in Fig. 7.20. When more stages are required, the optimization of the process becomes very complex and difficult.

6.3.2. Nanofiltration and Reverse Osmosis

A typical NF or RO unit includes raw water pumps, pre-treatment, membrane modules, disinfection units, storage, and distribution elements (3). The principal design considerations for NF and RO units are:

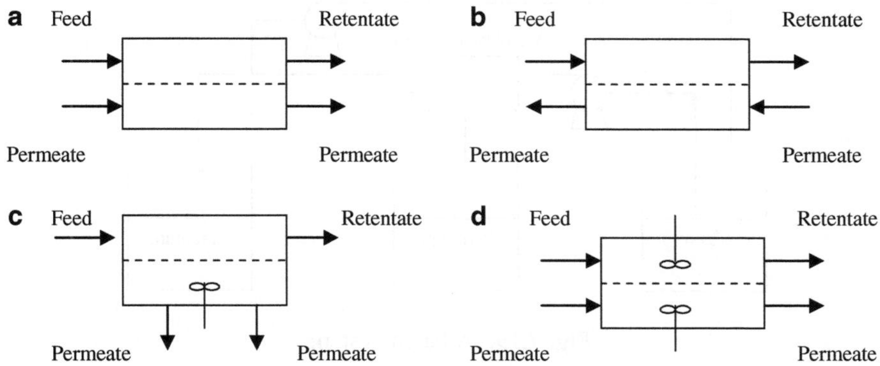

Fig. 7.17. Some cross-flow operations: (**a**) co-current, (**b**) counter-current, (**c**) cross-flow with perfect permeate mixing, and (**d**) perfect mixing.

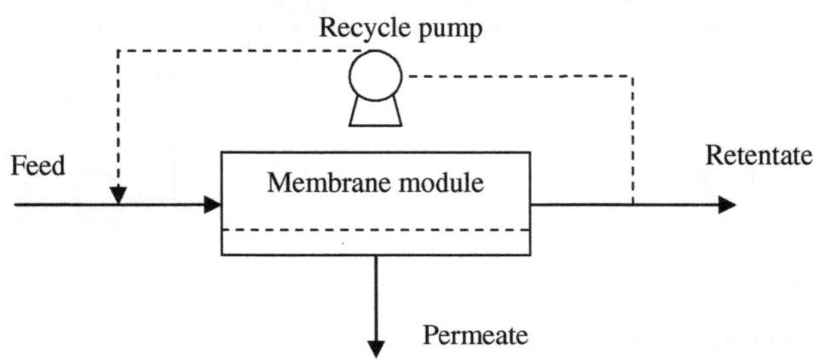

Fig. 7.18. Single-pass and recycling system.

- *Membrane material.* During the early days, membranes were usually made of cellulose acetate. At present, membranes can also be made from aromatic polyamide and thin-filmed polymer composites. Different membrane materials have their own distinctive characteristics, such as hydraulic resistance, pH range, temperature range, chlorine tolerance and biodegradation tolerance.
- *Module type.* Spiral wound and hollow fibre are the two membrane modules usually selected for RO systems. The spiral-wound unit clogs less frequently, while the hollow-fibre membrane has much greater surface area per unit volume.
- *Operating pressure.* Applied pressure of NF and RO can be found in Table 7.1. It is essential to the quantity and quality of water. RO systems operating at pressure less than 17 bars are classified as low-pressure units, while those operating above 24 bars are classified as high-pressure units. High operating pressure can trigger noise, vibration and corrosion problems, but they are more effective. Recently, developments have been made on new models that perform well at low pressures.
- *Pre-treatment requirements.* Pre-treatment requirements include (a) contaminant concentration, (b) ionic size of the contaminants, (c) membrane type, (d) presence of competing ions, (e) suspended solid concentration and (f) water temperature. Pre-treatment is commonly used to prevent fouling of the membrane. Typical pre-treatment for NF and RO includes particle removal by filtration, sequestering hardness ions by precipitates and pH control to prevent clogging.

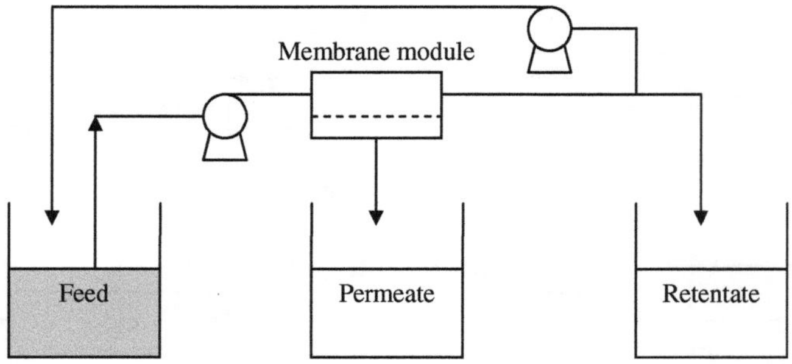

Fig. 7.19. A batch system.

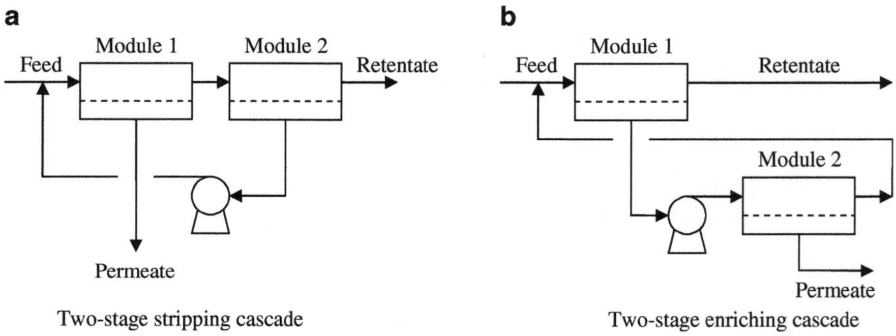

Fig. 7.20. Two-stage membrane process: (**a**) two-stage stripping cascade and (**b**) two-stage enriching cascade.

- *Product conversion rate*. It depends on several factors, mainly ionic charge and ionic size. The higher the ionic charge and the larger the ionic size of the contaminant, the more easily the ion is removed and the more finished water is recovered relative to the amount of reject water.

6.4. Engineering Design

A membrane system can be designed by empirical or semiempirical approaches and computer-simulation models. The former can be found in the literature such as Geankoplis' manuscript (27); the later is illustrated by Judd and Jefferson (3). A frequently used design tool based on a semiempirical approach is given below.

6.4.1. Reverse Osmosis

Mass transfer of substances in a membrane unit can be considered as steady-state diffusion, which can be described by the following equations (6, 12, 40, 49). A conceptual illustration of RO is given in Fig. 7.21. Noted that 1 and 2 represent feed and permeate (product) sides, respectively:

Table 7.5
Physical properties of pure water

Temperature (°C)	Density, ρ (kg/m^3)	Specific weight, γ (kN/m^3)	Dynamic viscosity, μ (mPa s)	Kinematic viscosity, v (μm^2/s)
0	999.842	9.805	1.787	1.787
3.98	1,000.000	9.807	1.567	1.567
5	999.967	9.807	1.519	1.519
10	999.703	9.804	1.307	1.307
12	999.500	9.802	1.235	1.236
15	999.103	9.798	1.139	1.140
17	998.778	9.795	1.081	1.082
18	998.599	9.793	1.053	1.054
19	998.408	9.791	1.027	1.029
20	998.207	9.789	1.002	1.004
21	997.996	9.787	0.998	1.000
22	997.774	9.785	0.955	0.957
23	997.542	9.783	0.932	0.934
24	997.300	9.781	0.911	0.913
25	997.048	9.778	0.890	0.893
26	996.787	9.775	0.870	0.873
27	996.516	9.773	0.851	0.854
28	996.236	9.770	0.833	0.836
29	995.948	9.767	0.815	0.818
30	995.650	9.764	0.798	0.801
35	994.035	9.749	0.719	0.723
40	992.219	9.731	0.653	0.658
45	990.216	9.711	0.596	0.602
50	988.039	9.690	0.547	0.554
60	983.202	9.642	0.466	0.474
70	977.773	9.589	0.404	0.413
80	971.801	9.530	0.355	0.365
90	965.323	9.467	0.315	0.326
100	958.366	9.399	0.282	0.294

Pa s = (mPa s) $\times 10^{-3}$.
m^2/s = (μm^2/s) $\times 10^{-12}$.

$$N_{\mathrm{w}} = A_{\mathrm{w}}(\Delta p - \Delta \Pi), \tag{32}$$

$$N_{\mathrm{s}} = A_{\mathrm{s}}(c_1 - c_2), \tag{33}$$

$$\Delta P = P_1 - P_2, \tag{34}$$

$$\Delta \Pi = \Pi_1 - \Pi_2, \tag{35}$$

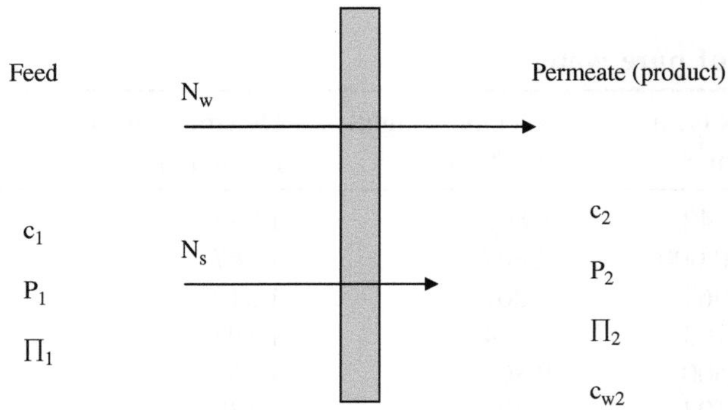

Fig. 7.21. Mass transfer in RO membrane.

where N_w is the water flux (kg water/s m^2), N_s is the salt flux (kg salt/s m^2), A_w is the water permeability constant (kg water/s m^2 atm), A_s is the salt permeability constant (m/s), P_1 is the pressure applied on feed solution (atm), P_2 is the pressure of permeate (product) solution (atm), Π_1 is the osmotic pressure of feed solution (atm), Π_2 is the osmotic pressure of permeate (product) solution (atm), c_1 is the salt concentration in feed solution (kg salt/m^3) and c_2 is the salt concentration in permeate (product) solution (kg salt/m^3).

Water permeability constant (A_w) is a function of the membrane material such as water membrane permeability and membrane thickness. The presence of salt, however, is less important for the constant. For CA membranes, A_w typically ranges from 1×10^{-4} to 5×10^{-4} kg water/s m^2 atm. Salt permeability constant depends on the types of salts in the solution. For CA membranes, A_s values range from 1×10^{-7} to 6×10^{-7} m/s. Some of the typical values are 4×10^{-7} m/s for NaCl, 2.2×10^{-7} m/s for $MgCl_2$, 2.4×10^{-7} m/s for $CaCl_2$, 6×10^{-7} m/s for KCl and 4×10^{-7} m/s for Na_2SO_4, respectively.

The solution volumetric flow rate in the feed side is approximately equal to that in the permeate side. Therefore, we have

$$A_m \frac{N_s}{c_2} = A_m \frac{N_w}{c_{w2}},$$

where c_{w2} is the concentration of water in permeate (product) (kg water/m^3).

Thus, we can have

$$N_s = \frac{N_w}{c_{w2}} c_2. \tag{36}$$

If the salt concentration in permeate (product) (c_2) is low, its osmotic pressure (Π_2) is normally ignored and the concentration of water in permeate (product) (c_{w2}) is approximately the pure water density that is a function of temperature (Table 7.5).

Define the salt rejection (R) as

$$R = \frac{c_1 - c_2}{c_1}. \tag{37}$$

Combining Eq. (32), Eq. (33) and Eq. (37) yields

$$R = \frac{B(\Delta P - \Delta\Pi)}{B(\Delta P - \Delta\Pi) + 1}, \tag{38}$$

$$B = \frac{A_{\mathrm{w}}}{A_{\mathrm{s}} c_{\mathrm{w2}}}. \tag{39}$$

Noted that the newly combined parameter B has a unit of atm^{-1}.

During membrane operations, localized concentration of salt builds up at the boundary layer of membrane. This is called the *concentration polarization* (40, 49). The concentration polarization can be reduced by increasing turbulence.

Owing to the presence of concentration polarization, Eq. (33) and Eq. (35) will have to be revised as

$$N_{\mathrm{s}} = A_{\mathrm{s}}(\beta c_1 - c_2), \tag{40}$$

$$\Delta\Pi = \beta\Pi_1 - \Pi_2, \tag{41}$$

where β is the concentration polarization, which normally ranges from 1.2 to 2.0. When the feed solution is well mixed and the salt concentration is about 1%, the RO membrane filtration unit can be treated as a continuously stirred tank reactor (27).

Example 3

An RO membrane system is used for the desalination of a feed water that contains 2.5 g NaCl/L. The temperature is 25°C and the density of the feed is 999 kg/m^3. The applied pressure (ΔP) is 27.6 atm. The water permeability constant (A_{w}) and the NaCl permeability constant (A_{s}) are 4.8×10^{-4} kg/s m^2 atm and 4.4×10^{-7} m/s, respectively:

(a) Assume that the concentration polarization can be neglected. Calculate the salt rejection, the water flux and the salt flux through the membrane, and the salt concentration in the permeate.
(b) Owing to the incomplete mixing in the RO module, the concentration polarization can be assumed to be 2.0. What are the salt rejection, the water flux, the salt flux and the salt concentration in the permeate? Also comment on the applied pressure (ΔP) if the quality of the product water remains the same as in (a).

Solution

(a) According to Eq. (10a), the osmotic pressure in the feed can be determined by

$$\sum M_i = 2 \times 2.5/58.45 = 0.0856\,\mathrm{M},$$

$$\Pi_1 = RT\sum M_i = 0.08206 \times (273 + 25) \times 0.0856 = 2.09\,\mathrm{atm}.$$

Assume that the salt concentration in the permeate is very low. Therefore, we have

$$\Pi_2 = 0,$$

$$N_{\mathrm{w}} = A_{\mathrm{w}}(\Delta p - \Delta\Pi) = 4.8 \times 10^{-4} \times (27.6 - 2.09) = 0.01224\,\mathrm{kg\,water/s\,m^2}.$$

From Table 7.5, at a temperature of 25°C, the water density is 997.05 kg/m^3. Thus, we have

$$c_{w2} = 997.05\,\text{kg/m}^3,$$

$$B = \frac{A_w}{A_s c_{w2}} = \frac{4.8 \times 10^{-4}}{4.4 \times 10^{-7} \times 997.05} = 1.094\,\text{atm}^{-1},$$

$$R = \frac{B(\Delta P - \Delta\Pi)}{B(\Delta P - \Delta\Pi) + 1} = \frac{1.094 \times (27.6 - 2.09)}{1.094 \times (27.6 - 2.09) + 1} = 0.965.$$

From $R = (c_1 - c_2)/c_1$, we have

$$c_2 = c_1(1 - R),$$

$$c_2 = 2.5 \times (1 - 0.965) = 0.0865\,\text{g NaCl/L},$$

$$N_s = A_s(c_1 - c_2) = 4.4 \times 10^{-7} \times (2.5 - 0.0865) = 1.062 \times 10^{-6}\,\text{kg NaCl/s m}^2.$$

(b) The concentration polarization is 2.0. According to Eq. (41), we have

$$\Delta\Pi = \beta\Pi_1 - \Pi_2 = 2.0 \times 2.09 = 4.18\,\text{atm},$$

$$N_w = A_w(\Delta p - \Delta\Pi) = 4.8 \times 10^{-4} \times (27.6 - 4.18) = 0.01124\,\text{kg water/s m}^2,$$

$$R = \frac{B(\Delta P - \Delta\Pi)}{B(\Delta P - \Delta\Pi) + 1} = \frac{1.094 \times (27.6 - 4.18)}{1.094 \times (27.6 - 4.18) + 1} = 0.962,$$

$$c_2 = c_1(1 - R) = 2.5 \times (1 - 0.962) = 0.095\,\text{g NaCl/L}.$$

From Eq. (40), we have

$$N_s = A_s(\beta c_1 - c_2) = 4.4 \times 10^{-7} \times (2 \times 2.5 - 0.095) = 2.158 \times 10^{-6}\,\text{kg NaCl/s m}^2.$$

From the above calculation, we can find that, owing to the concentration polarization, the salt concentration in the permeate (c_2) is increased by about 10%, the water flux (N_w) is reduced by 8% and the salt flux (N_s) is increased by 103%. If the applied pressure is increased to overcome the effect of concentration polarization (is increased by 2.09 atm), the same quality of water as in part (a) can be obtained.

6.4.2. *Microfiltration and Ultrafiltration*

The water flux of MF and UF can be calculated by Eq. (32). As the membrane is more porous than RO membrane, it is only good for removal of "large" molecules with a molecular weight ranging from 500 to 1,000,000. Thus, it is reasonable to consider that the osmotic pressure ($\Delta\Pi$) is very low and can be ignored. Eq. (32) is therefore modified as

$$N_w = A_w \Delta P. \tag{42}$$

The concentration polarization is much more severe than RO. Therefore, the pressure drop must be considered in the calculation. The reader can refer to Geankoplis' manuscript (27) for more detailed information.

6.5. Membrane Testing

Before installing a membrane operation system, membrane testing is required to determine the capability of the particular membrane polymer for the separation, optimum membrane configuration in the application and optimum processing conditions (e.g. pressure, flow rates and temperature).

In general, every stream must be tested to determine design factors such as the specific membrane polymer, membrane element design, total membrane area, applied pressure, system recovery, flow conditions, membrane element array and pre-treatment requirements.

For an ideal system, all contaminants to be removed are separated by the membrane and exit in the concentrate stream. In reality, no membrane is perfect. The actual quantity of solute that passes through the membrane depends on the chemistry of solute on the feed, nature of membrane material and the operational conditions. Cell test, applications test and pilot test can be used to evaluate membrane technology with a particular stream.

Cell test uses small, approximately 100 cm^2 cut pieces, of sheet membrane mounted in a "cell" that exposes the membrane to the test solution by cross-flow mode. This test is effective for quick evaluation of a number of different membrane polymers to determine the degree of separation. However, it cannot determine the long-term chemical effect of a solution on the polymer and does not provide engineering scale-up data.

An *applications* test is quite similar to the cell test. It typically involves the evaluation of treatment efficiency of a sample with a slightly higher volume, such as 100–150 L. The test can be completed within 1–2 h. The test is fast and provides scale-up data, such as flow, element efficiency and pressure requirements. It can also provide an indication of membrane stability. However, it provides neither the long-term chemical effects nor sufficient data on the fouling effects of the test solution. Figure 7.22 illustrates a typical applications test process.

Unlike the above two tests, the *pilot* test can provide more detailed information. The test has a test machine in the process (similar to that in the applications test), which is operated for a longer period (normally a minimum of 30 days). It provides long-term membrane operation data such as fouling and chemical stability data. However, it can be expensive to perform, in terms of cost of equipment, and daily monitoring.

6.6. Economics of Membrane Processes

Membrane processes are mainly used in processes where very high purity of the component is required. Despite the technical advantages, such as being gentle to the products, the large-scale applications of membrane processes are often affected by their cost. The cost of a given installation is determined by two parts: the capital cost and the operational cost. The capital cost, or the installation investment, can be further divided into three parts:

1. Membrane modules
2. Costs of piping, pumps, electronics and vessels
3. Pre-treatment and post-treatment

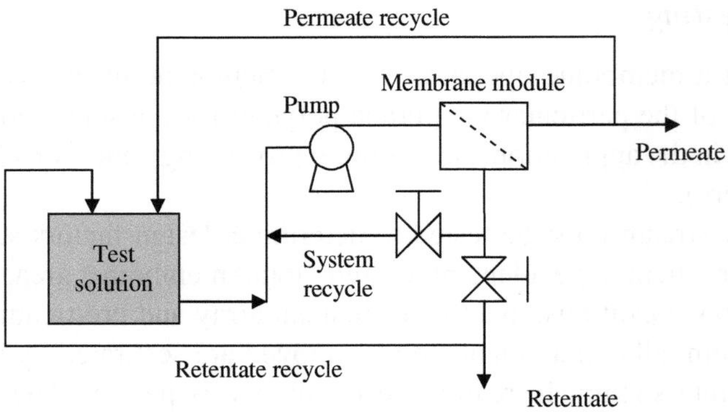

Fig. 7.22. A typical membrane application test.

To calculate the cost per unit (volume or mass) of product, the capital costs are depreciated over a period of time, often 10 years. Interest has to be paid over this time on this amount of money. And, the operational cost can be divided into four parts:

1. Energy requirement
2. Membrane replacement
3. Labour
4. Maintenance

A more detailed calculation of the cost of membrane operation can be found in some articles and books (50, 51). A computer program, named the Water Treatment Cost Estimation Programme (WTCost), can be used to estimate costs of water treatment processes employing MF, UF, NF, RO and ED. The costs of pre-treatment and post-treatment unit operations can also be estimated. The WTCost has been tested by industry experts, and it has been shown to accurately project true water treatment costs for a number of plants of different brackish and seawater capacities (51).

7. MEMBRANE FOULING AND PREVENTION

7.1. *Mechanisms*

The major drawback hindering the use of membranes is the reversible and irreversible fouling that impedes the flux of clean water through the membranes (52). The consequence of membrane fouling is a reduction of permeate production rate and/or an increase in solute passage across the membrane with time. Fouling also causes increase in energy consumption as transmembrane pressure can increase substantially due to fouling. In addition, fouling increases downtime and may shorten membrane lifespan. Figure 7.23 demonstrates membrane fouling due to the presence of organic compounds, inorganic salts and microorganisms.

Fouling can be defined as irreversible deposition of materials onto or into the membrane, causing loss of flux and altered rejection (53–55). Membrane fouling can be caused by:

Fig. 7.23. Visualization of membrane fouling by scanning electron microscopy: (**a**) inorganic fouling due to calcium carbonate, calcium sulphate, silica, iron, barium and strontium sulphate (35,000×); (**b**) organic fouling due to humic acid (35,000×) and (**c**) flow channels in membrane fouled with biological growth.

- Particulate deposition (colloidal fouling)
- Adsorption of organic molecules (organic fouling)
- Inorganic deposits (scaling)
- Microbial adhesion and growth (biofilm formation)

Colloidal fouling is caused by accumulation of particles and macromolecule on, in and near a membrane. Materials accumulated at the membrane surface create an additional layer of resistance to permeation. Early work on colloidal fouling of RO membrane used to treat secondary effluents indicated that particles smaller than 5 μm contribute more substantially to fouling than larger particles (56). It was postulated that as particle size increases, it is subjected to higher velocity and shear force at the membrane surface. Therefore, larger particles tend to be swept away in bulk flow rather than deposits on the membrane surface. In addition to surface deposition, some particles may be small enough to penetrate and remain within the pores of the membrane. The tendency of pore blockage is likely to be higher in UF and MF as their pore size is bigger and hence allows more particles to penetrate and deposit in the membrane pores. In another study, factors influencing the flux decline of NF membranes for the treatment of dye bath wastewaters were investigated. It was found that cake layer formation of dye molecules on membrane surface, especially at low salt concentrations, was the principal cause of flux decline (57).

The adsorption of organic matter on membrane surfaces is detrimental to permeate flux and affects the salt rejection of membranes (Fig. 7.23b). The negatively charged functional groups on organic foulants have an affinity for charged membrane surfaces, thereby forming a permeate-resistant layer. Organic foulants also interact with colloidal foulants. Polyphenolic compounds, proteins and polysaccharides bind colloids and particles and increase their cohesion to the membrane surface (58). It was also found that membrane fouling by natural organic matter (NOM) adsorption is a major factor limiting the use of MF for drinking water treatment (59). UF is commonly used for isolation and fractionation of humic substances. In addition, the biochemical interaction between organics and microorganisms promotes biofilm formation and growth. Insufficient knowledge on the composition of dissolved organics in water makes the control of organic fouling difficult.

Inorganic fouling is caused by deposit of iron, silica, aluminium, calcium, phosphorus and sulphate (Fig. 7.23a). The fouling mechanism at the membrane surface can be caused by the concentration polarization effect. A concentrated boundary layer is created on the separation surface as product water passes through the membrane. Within this boundary layer, the high concentration causes the salts to precipitate and suspended solids can start to deposit on the membrane surface leading to scaling and fouling (58). Scale deposition onto and into RO membranes impairs the hydrodynamic conditions of the feed flow. When fouling conditions are not controlled properly, scaling becomes a self-sustaining phenomenon (60). Under severe concentration polarization, channelling, failure of RO performance and damage to membrane surface occur.

Biofouling is the term given to the adhesion of microorganisms and growth of biofilm on the membrane surface (Fig. 7.23c). Sewage carries a very high load of potentially pathogenic and nonpathogenic microorganisms in addition to chemical pollutants and nutrients (61). Besides the detrimental effects of increasing transmembrane pressure and decreased permeate flux, biofouling may cause chemical degradation of the membrane material. This could result from direct enzymatic biodegradation of the membrane surface or by generation of extreme local pH that may hydrolyse the membrane polymer (61–63). Such fouling can significantly reduce the membrane lifetime.

The formation of biofilm involves three steps:

1. Formation of a conditioning film composed of macromolecules, proteins, etc.
2. Primary bioadhesion by microorganisms
3. Biofilm development

A recent study has shown that some RO membranes are more prone to bioadhesion than others. The study involved a bioadhesion assay, which utilizes a model bacterium, SW8, known to adhere to membranes. Examination of bacteria adhered to the membranes using optical microscopy revealed that membranes that are less susceptible to bioadhesion are hydrophilic in nature (62, 63). Further investigation with municipal wastewater carried out on RO simulators that consisted of flat sheet membranes to simulate spiral-wound module revealed biofilm characteristics under field emission scanning electron microscope. Microorganisms cover the surface of all types of RO membranes used in the experiment to a density of about 2.25×10^8 cells/cm^2 (62). Bacteria of different shapes (mostly rod-shaped) with dimensions between 1 and 3 μm were observed. The organisms appeared to excrete extracellular polymeric substances (EPSs).

The ratio of carbon to nitrogen to phosphorus (C:N:P) has important influences on the rate of biofilm development. It has been reported that membranes which suffered severe biofouling were found to contain a high percentage (typically >60%) of organics. Laboratory characterization of membrane biofilms found that a typical biofilm contains:

- 90% moisture
- 50% total organic matter

- Up to 40% humic substances
- Low inorganic content
- High microbiological counts ($>10^6$ cfu/cm^2) including bacteria and fungi

Membrane autopsy of desalination RO membranes that had been in service for 2.5 years in Saudi Arabia revealed bacterial deposits that are slimy and very adherent (64). This is primarily due to accumulation of extracellular polysaccharides excreted by microorganisms, thus resulting in biofilm formation (54). Bacteria embedded in a biofilm are found to be more resistant to biocides than freely suspended ones (65).

Understanding membrane fouling creates the foundation for researchers to devise an approach to counter or minimize fouling in order to maintain high RO performance. One strategy to alleviate membrane fouling is feed pre-treatment to reduce or remove fouling constituents in the feed. A proper membrane cleaning and regeneration protocol to periodically remove foulants from the membrane surface is also essential in maximizing RO efficiency.

7.2. Feed Pre-treatment

Applicable pre-treatment prior to membrane separation are screening, coagulation, flotation and, in extreme cases, activated carbon or resin adsorption and even ultrafiltration (66). Economics has a major role to play in deciding the pre-treatment sequence. For example, membrane life estimates may be 1 and 3 years for minimal and extensive pre-treatment, respectively. A study shows that prefiltration of the humic acid solution significantly reduced the rate and extent of fouling by removing the large humic acid aggregates/particles from the solution (67).

RO membranes are generally not robust enough to operate directly on feed such as secondary wastewater. The success of RO process is highly dependent on appropriate feed pre-treatment. Pre-treatment must be effective in reducing RO fouling potential in a reliable and consistent manner. Feed pre-treatment continues to be one of the fields that are extensively studied. Water Factory 21, which has been operating for more than two decades, continues to evaluate feed pre-treatment to maximize RO efficiency (68). In the development of a pre-treatment programme, the focus is on removing as much fouling constituents in the feed water as possible before RO processing.

7.2.1. Conventional Pre-treatment

Wastewater reclamation was pioneered using advanced conventional treatment processes to upgrade the water quality of wastewater to reusable standards. When RO was first introduced to produce water closer to drinking water quality from wastewater, a conventional treatment process was employed as pre-treatment to the RO. A typical conventional pre-treatment configuration would include flocculation, lime or alum clarification, recarbonation, settling, filtration and activated carbon adsorption. Biological activity is controlled by chlorination.

Water Factory 21, which employs such a train of pre-treatment, reported that 26% of TOC was removed by lime clarification. Concentration of inorganic constituents such as calcium, magnesium, iron, fluoride and silica was significantly reduced. Over 99% of coliform bacteria were removed and viruses were effectively inactivated by the high pH. Multimedia filtration produced an average effluent turbidity of 0.14–0.16 NTU. Granular activated carbon (GAC) removed 30–50% of the organics. The average TOC concentration of GAC effluent was 5.5 mg/L (69).

Although clarification following lime or alum coagulation is a very effective pre-treatment for removing colloidal and suspended matter, the process is expensive, as chemical dosage is required. The process is also difficult to operate since optimum dosage depends on influent quality. Furthermore, the coagulation–clarification process generates solid waste that requires additional handling and disposal. In cases where over-dosage occurs, high metal salt content of the pre-treatment water may result in metal hydroxide precipitation on the subsequent RO membranes (70). Another disadvantage of the clarification process is a large requirement in land space.

A simplified process used in smaller systems is in-line flocculation followed by pressure filtration. The simplified process produces water of lower quality than the lime clarification process but the equipment is smaller and simpler to operate (71). Experience with in-line filtration showed that optimal dosage of alum was rarely achieved due to fluctuating influent turbidity (72, 73).

Alum, alum–polyelectrolyte and polyelectrolyte in-line flocculation filtration systems were compared by Cikurel et al. (74). Their results show that high molecular weight (2×10^6) branched-chain polyacrylamide at dosage as low as 0.5 mg/L is more effective at turbidity removal than alum dosed at 10–20 mg/L. The results are consistent at both low (6–12 NTU) and high (20–27 NTU) feed turbidity. The performance of polyacrylamide as primary flocculant was comparable to the alum–polyelectrolyte system. Various polyelectrolytes of different molecular weights (MW) were evaluated. High MW polyacrylamide was more effective than medium and low MW polyamides at same dosage of 0.5 mg/L. Effectiveness of medium MW polyamide increased when dosage was increased above 5 mg/L. Low MW polyamide was not effective in reducing turbidity even at dosage of 7 mg/L. The charge density of the polymers did not play a significant role in particle removal. It was postulated that the main mechanism of high MW polymer is bridging in contrast to adsorption and charge neutralization action of alum.

Studies on in-line filtration showed that effluent turbidity of less than 2 NTU is achievable with alum dosage of 0–8 mg/L and cationic polymer dosage of 0–0.5 mg/L (73). However, the performance of this system is dependent on feed water quality, hydraulic loading rate of the granular media column, and good control over the dosage of alum and the cationic polymer. High feed water turbidity and high polymer dosage have adverse effect on the duration of filtration cycle.

7.2.2. Membrane Pre-treatment

MF and UF membrane processes are increasingly being used in the water and wastewater treatment. The outcome of rapid developments in membrane industry is low-cost high-productivity membranes, making membrane processes economically feasible. Numerous studies and site experience have led to better understanding of process parameters, allowing process optimization making membrane processes more technically feasible.

Water Factory 21 has demonstrated that MF pre-treatment can provide good pre-treatment for the RO processes with multiple benefits over conventional pre-treatment (75). The benefits of MF pre-treatment include increased RO flux and overall efficiency, prolonged operation time between cleans, and reduction in operating and chemical costs. Following this successful demonstration at Water Factory 21, the industry is moving from lime clarification towards MF and UF as a pre-treatment. There are multiple pilot projects evaluating MF as a pre-treatment to the RO process. Pilot study performed in San Francisco utilizes MF as pre-treatment to RO for desalination of municipal wastewater for horticultural reuse. The average turbidity removal was $99.4 \pm 0.4\%$. The average silt density index (SDI) was 1.15 ± 0.53.

Pilot plant studies conducted at Canary Islands (Spain) showed that microfiltered secondary effluent from Tías WWTP contained below 1.0 mg/L of suspended solids and turbidity below 1.0 NTU. Total and fecal coliforms were also absent from the microfiltered water. The SDI of the microfiltered water was below 3.0. Average removal achieved for BOD_5, COD and TOC were 81, 40 and 27%, respectively. The MF achieved water recovery of about 85% (76).

A study was performed to investigate MF pre-treatment performance in treating a broad range of water. Results showed that the performance of MF remains satisfactory when subjected to very cold water (0.2°C), water with high iron content and water with high organic load and biofouling potential (77). Acid wash has to be included in the operation procedures when treating high iron content water to prolong run time between chemical cleaning. The addition of acid keeps the iron content in the water in a reduced and dissolved form, preventing precipitation and scaling on the membrane.

A lab-scale evaluation of pre-treatment for RO recycling of secondary effluent from refinery demonstrated that UF is able to provide good pre-treatment for subsequent RO process. UF was capable of removing over 98% of suspended solids and colloids. Partial removal (30%) of COD was also achieved. The removal efficiency was consistent and was independent of influent water quality and operating conditions (78). Similar results reported by Qin and coworkers (79) demonstrate that an appropriate UF pre-treatment could reduce fouling of RO membrane and increase the flux of RO membrane by 30–50%.

Another pilot study on UF–RO membrane treatment of industrial effluent (pulp and paper mill effluent) showed that UF permeate flux increased with temperature of feed water. The fluxes were 1.44 and 1.84 times higher at 30 and 40°C, respectively, compared to flux at 20°C. The improvement in flux was attributed to decreased wastewater viscosity with increased temperature. However, at higher temperatures, more organic matter was able to diffuse through the membrane. Comparison of UF pre-treatment at feed pH 2.4, 5.3 and 7.0 showed that higher flux is achieved at pH 7.0 (80).

MF and UF can reduce biofouling tendency in RO membrane as they pose a physical barrier to these microorganisms. MF is capable of removal of protozoa (~10 μm), coliform (~1 μm) and cysts (~0.1 μm). The pore size of UF is smaller and thus can further remove viruses (~0.01–0.1 μm) (81). It has been reported that UF showed more than five-log reduction of microbial content (82). Membranes are attractive as disinfection process because it reduces dosage of aggressive chemicals such as chlorine and ozone. In addition, undesirable disinfection by-products can be minimized or avoided. The interest in using membrane as part of the disinfection process has intensified with the emergence of chlorine-resistant pathogens. Chlorine-resistant *Cryptosporidium parvum* has been reported to cause outbreak of diarrhoea epidemic in US and UK. Water supply authorities are looking to UF and MF application to act as an absolute barrier to *Cryptosporidium* oocysts, which range from 4 to 6 μm (83).

Chlorination of secondary effluent prior to membrane pre-treatment may extend membrane run times between cleans. Over 90 h of MF operation were achieved with prechlorinated secondary effluent compared to 42-h operation reported when secondary effluent was not chlorinated (72, 73). Similar observations were reported with dosage of chloramine prior to microfiltration pre-treatment (77). It was speculated that preoxidation due to chlorination altered the chemistry of EPS produced by the microorganisms in the secondary effluent. This could weaken the attachment of the EPS on the membrane and thus offset the detrimental effect on the membrane flux. However, care must be taken to verify compatibility of membrane with chlorination as some membranes are not tolerant to the aggressive action of chlorine.

Although MF and UF membranes have been shown to be a viable option as feed pre-treatment for RO, long-term operating and cost data are required to verify that membrane pre-treatment is more cost effective than conventional pre-treatment.

From a fouling perspective, use of membrane pre-treatment has made it possible to segregate flux loss due to colloidal fouling and biofouling from flux loss due to scaling (84). Anti-fouling strategy can be developed with more focus on relevant type of fouling. Although microbes can escape defective portions of pre-treatment membrane and cause biofouling, the potential and degree of biofouling have been greatly reduced. This can reduce the application of disinfection, increase operation interval between cleans and prolong membrane life.

Overall, studies on membrane pre-treatment demonstrate the following advantages:

- Addition of chemicals is not required
- Effluent quality is independent of feed quality
- Operation at ambient temperature
- Forms an absolute barrier to pathogens
- Space efficient

The simplicity of membrane operation makes it an attractive option in the field of wastewater reclamation.

8. MEMBRANE CLEANING AND FLUX RESTORATION

Fouling is almost an inevitable consequence of the nature of the RO process itself even when good pre-treatment is employed. The challenge is therefore to reduce and control fouling sufficiently to minimize the rate of RO flux decline and prolong membrane lifetime.

This can be accomplished by a combination of good feed pre-treatment and well-developed membrane cleaning programs.

It is essential to have a good knowledge of fouling to successfully clean and regenerate fouled membranes. Types of foulants that could occur on RO membrane surfaces include suspended solids, colloids, metal oxides, scales, biological slime, organics, oil and grease. The type of fouling that could occur on the membrane surface is a function of the make-up of the feed water, the pre-treatment applied prior to the RO membrane and the interaction between the different foulants. The effects of cleaning parameters (e.g. pH and temperature) can be obtained through factorial design approach as described by Chen et al. (4).

The most dominant of fouling that can occur in wastewater reclamation is biological fouling and organic fouling (70). The major factor in controlling fouling on membranes used in secondary effluent treatment is the dissolved organic content of the feed water (56). Dissolved organics such as humic acids, proteins, carbohydrates and tannins in addition to biological growth were the major fouling constituents of RO membranes in wastewater reclamation (85).

8.1. Chemical Cleaning Methods

Most membrane manufacturers recommend chemical methods for membrane cleaning and regeneration (70). Chemical cleaning is the most common method to clean NF membranes (86). Chemical cleaning methods depend on chemical reactions to weaken the cohesion forces between the foulants and the adhesion forces between the foulants and the membrane surface. Chemical reactions involved in cleaning include hydrolysis, peptization, saponification, solubilization, dispersion, chelation, sequestering and suspending (87). It was found that cleaning with strong chelating agents, such as EDTA, most effectively removes the fouling layer and restores permeate flux for NF membranes (88). Chemical used for cleaning membranes should ideally serve the following functions and posses the following desirable properties:

- Loosen and dissolve foulants from membrane surface
- Keep foulants in dispersed and soluble form
- Avoid fresh fouling
- Does not cause damage to membrane material
- Easily rinsed away after cleaning
- Chemically stable before, during and after use
- Cost effective

Table 7.6 gives a brief description of cleaning action of some chemicals commonly used in membrane cleaning.

Important cleaning parameters that vary with foulant and membrane material are type of cleaning agent, pH, concentration, temperature and time. However, as a rule, mineral deposits are removed by acidic solutions and organic compounds by alkaline solutions (89). The choice of chemical cleaning agents not only depends on type of foulants present in the membrane system, but also depends on the chemical resistance of the membrane material

Table 7.6
Chemicals commonly used in membrane cleaning and their cleaning actions

Type of chemicals	Common chemicals used	Typical concentration (%)	Cleaning actions (87)
Acid	HCl	0.1–0.2	• Dissolve precipitation of inorganic salts
	HNO_3	0.3–0.5	
	H_3PO_4	0.3–0.5	
	Citric acid	1–2	
Carbonate	Na_2CO_3	0.5–1.0	• Buffer pH of cleaning solution
	$NaHCO_3$		
Disinfectant	H_2O_2	0.1	• Inactivate microorganisms
	NaOCl	0.002–0.02	
	Na_2SO_3	0.25	
Enzymes	Lipase		• Break down high molecular weight organic compounds
	Protease		
Hydroxide	NaOH	0.5–1.0	• Remove organic fouling
	KOH		
Polyphosphate	Polymeric phosphonate		• Act as dispersant • Solubilize carbonates • Bind ion salts • Regulate pH • Emulsify fats • Peptize proteins
Chelating agent	EDTA-Na_4	0.5–1.0	• Increase ease of rinsing • Improve contact between cleaning chemicals and the foulants
	EDTA + NaOH		• Minimize amount of water required
	EDTA-Na_4 + NaOH		• Shorten rinsing time

and the whole system. Caution must be taken in applying these chemicals as the aggressive nature of these chemicals may adversely affect system or membrane integrity when not applied properly. This is especially true for aromatic polyamide membranes, which are less robust than cellulose acetate and polysulphone membranes. For example, some cationic and non-ionic surfactant may be adsorbed onto polyamide membranes and cause a flux decrease.

Polyamide membranes are also not resistant to the strong oxidizing actions of disinfectants such as hydrogen peroxide and hypochlorite. Acids used to clean membrane system must be rinsed out thoroughly before application of hypochlorite for disinfection as hypochlorite at low pH can cause corrosion in stainless steel. The pH of the chemical cleaning solution must fall within the tolerable range of pH 1–13 for polysulphone membranes and pH 3–8 for CA membranes (87).

In studies to remove organic fouling on RO membranes, it has been reported that several commercial cleaners developed by Pfizer gave excellent results. Differential pressure of the membrane system could be reduced by 42% using a neutral pH liquid formulation (Floclean 107) designed to remove organics, silt and other particulates from cellulose acetate RO membranes. For polyamide, polysulphone and thin-film composite membranes, a high pH commercial cleaner (Floclean 411) was able to recover permeate flow by 23.8% (90). In a comparison study, commercial cleaners proved to perform better than cleaning recipes recommended by membrane manufacturers. Ebrahim and El-Dessouky (91) compared commercial cleaner (Floclean 403 and 411) from Pfizer with the membrane manufacturer's recommended recipe of 2% citric acid at pH 4.0 for cleaning seawater RO membranes. An improvement of 11.6–30.8% was achieved.

High biological activity can be expected from secondary effluent due to biological treatment. Bacterial concentrations between 10^3 and 10^4 cfu/mL are common in secondary effluent (92). Biofouled membranes cleaned with a combination of a high pH cleaner designed for biofouled membranes and a low pH cleaner designed for bioslimes reportedly resulted in 76% restoration in permeate flow. This was accompanied by a reduction in differential pressure of 41% (90). Alternating between caustic detergent at pH 12 and peracetic acid and hydrogen peroxide at pH 2 was effective in eliminating sulphate-reducing bacteria on seawater RO membranes (93).

Biocides have commonly been used in the water treatment industry to counter biofouling problems. Unfortunately, the application of biocide to a membrane system has its limitations. Application of biocide such as hypochlorite can remove about 80% of the biofouling layer on an RO membrane. This can result in considerable process improvement in the RO system. However, the remaining 20% of biofilm provide nutrients for rapid regrowth of the biofilm (94).

Systems with chlorination prior to the membrane processes were easier to clean due to the presence of chloramines, which act as a disinfectant and reduce the tendency of biofouling. However, membranes in such systems are more susceptible to structural damage owing to prolong exposure to the aggressive action of the combined chlorine. Cellulose acetate membranes are reported to become more brittle when chlorine dosage was in the range 15–20 mg/L (92). It was also suggested that certain organochlorine derivatives might modify the molecular structure of the membranes, thus resulting in flux decline and decreased salt rejecting efficiency.

Results of a study on regeneration of RO membranes treating brackish water showed that dosing of a high molecular weight polymer colloidal solution following high pH alkaline chemical cleaning could reduce salt passage through the membranes. Tannic acid was chosen as the polymer for this study. Average salt passage across the RO membrane was reduced from 14 to 6–10% using this cleaning method (95). However, the system suffered an average of 10% flux loss, which was claimed to be tolerable. It was postulated that a dynamic layer capable of reducing salt passage was formed on the membrane surface during such a polymer post-treatment following chemical cleaning.

It was reported that a combination of anionic cleaning agent and enzymes could effectively remove biofilm (92). The disadvantage in using enzymes is that they are very costly and formulating enzymes into effective cleaners is very expensive (87). In addition, enzymes are very specific and may not work well in all applications. Enzyme cleaning is dependent on many factors such as temperature and pH. Enzymes act slowly, and consequently longer cleaning times are required.

Inorganic fouling in the form of scale formation is generally removed by acidic cleaning (54, 96, 97). Membranes treating water with high inorganic potential, such as a high TDS or hardness level, often include low pH cleaning in their chemical cleaning procedures. Table 7.7 shows a survey of chemical methods used to counter scaling on membrane.

Although chemical cleaning agents are categorized with respect to their action on different type of foulant, their combined effects are much more complex. These cleaning agents may interfere with the cleaning effects of each other. Some may provide efficient control over particular foulants while adversely affecting fouling control of another foulant. For example, some cationic polymers while effective in silica-scale inhibition reduce calcium carbonate-scale inhibition by 30–40% (98). Humic and fulvic acids act as good calcium-scale inhibitor, but they promote biofouling in the membrane system.

8.2. *Physical Cleaning Methods*

Physical cleaning methods depend on mechanical forces to dislodge and remove foulants from the membrane surface. Physical methods used include forward flushing, reverse flushing, backwashing, vibrations, air sparge and CO_2 back permeation.

MF and UF used in pre-treatment to RO are more frequently cleaned by physical cleaning and less frequently by chemical cleaning. Cleaning frequencies reported in literature varied widely. Physical cleaning frequency is approximately every 40 min with a chemical clean scheduled every 6 months (98). An air backwash frequency of 15–20 min is sufficient for

Table 7.7
Chemicals used in removing scale from RO membranes

Target foulants	Recipe for chemical cleaning	Results	References
$CaSO_4$ $CaCO_3$	2% EDTA and citric acid stabilized by ammonia to pH 7	• Remove calcium scale effectively	(96)
Metal hydroxide deposits	Floclean 103A by Pfizer (low pH formulation)	• Removed stubborn metal hydroxide deposits • 30% improvement in permeate flow • 33% reduction in differential pressure	(90)
Ca and Mg deposits	1% EDTA at pH 10	• Remove Ca and Mg deposits effectively • Removal is faster in dynamic cleaner than in static cleaning	(97)
Fe deposits	1–2% sodium hydrosulphite at pH 3.2–3.8	• Remove Fe deposits effectively	(97)

hollow-fibre MF membranes (77). In a UF evaluation study, backwashing was able to achieve an average flux recovery of 86.5% (99). It was observed in the same study that flux restoration could be achieved even when backwash was reduced from 10 to 1 min.

There has not been adequate investigation into physical and physicochemical cleaning methods for RO membranes (70). It was commented that there is little published information on membrane cleaning and regeneration. Most cleaning studies reported are based on trial and error (4, 87). A more systematic approach is required to study the various aspects of fouling control (98). A better understanding of cleaning mechanisms and the effect of different cleaning agents on different foulants and different types of membranes is much needed. Further efforts are needed to develop more feasible and cost-effective cleaning and restoration procedures for different types of membranes.

A new generation of low fouling composite membranes has been introduced. The salt rejection layer on the membrane has been modified to make it more hydrophilic, thereby reducing its affinity to organics (98). The novel process is the incorporation of an electromagnetic device, which produces an electromagnetic field surrounding the membrane elements. This electromagnetic field reduces the precipitation of sparingly soluble salts. Organic fouling can also be reduced by this method as the electromagnetic field can neutralize organics and affect their ability to interact with the membrane surface.

9. RECENT ADVANCES IN MEMBRANE SEPARATION

9.1. Membrane Bioreactors for Wastewater Treatment

Different types of MBRs are increasingly being applied in water treatment due to good-quality effluent, capability of retaining high content of biomass, ensured biomass-water separation, low footprint requirement, well-controlled sludge concentration, etc. MBR process can be considered as a modification of the activated sludge process: membrane module is submerged in the bioreactor and combined with aerobic or anaerobic systems. However, membrane fouling is still the major issue during the large-scale application of the MBR processes (100–102).

A novel submerged UF membrane coagulation bioreactor (MCBR) has been developed by Tian and coauthors for drinking water treatment at a short HRT (0.5 s). Such reactor can not only separate particulates and microorganisms, but also achieve 96% of nitrification and 97.8% of dissolved phosphate removal. Due to the addition of coagulant in the bioreactor, MCBR can efficiently remove organic matter in terms of TOC, COD, DOC, etc. Therefore, MCBR technology is considered to be one promising technology in drinking water treatment (103). Sridang and coworkers used a submerged membrane bioreactor to treat seafood processing wastewater with 943–1,923 mg/L and 560–1,127 mg/L of COD and BOD_5, respectively. It was found that during 1,000 h of filtration, removal efficiencies for the COD and BOD_5 were 85 and 99%, respectively (104). A modified membrane bioreactor has also reported for the treatment of synthetic high strength water. Removal of organic matter and phosphate achieved 95 and 90.5%, respectively. Moreover, simultaneous nitrification and denitrification processes occurred and total nitrogen removal efficiency was around 90.6% (105).

9.2. *Gas Separation*

Because of the rising energy costs, membrane technology for gas separation plays an important role in minimizing the environmental impact and costs of industrial processes. In comparison with traditional technologies, gas separation membranes do not require a gas-to-liquid phase change in gas mixture, thus significantly decreasing the energy costs. Recently, gas separation membranes have been most widely applied, such as hydrogen separation, CO_2 capture, organic vapour removal, etc. (http://pubs.acs.org).

Yoshino and coworkers (106) fabricated a heat-resistant hydrogen-selective membrane unit – an all-ceramic module with silica membrane tubes for hydrogen separation at high temperature; and results showed that the H_2/N_2 separation factor was as high as 1,300 and the H_2 permeance was around 1.9×10^{-7} mol/m^2 s Pa at 873 K, respectively. A novel CO_2-selective polymeric membrane with the facilitated transport mechanism has been developed by Huang and coauthors in order to capture CO_2 from industrial gas mixtures (e.g. flue gas). The membrane demonstrated desirable CO_2 permeability of 6196 Barrers and CO_2/N_2 selectivity of 493 at 110°C. Moreover, under a feed gas flow rate of 60 mL/min, CO_2 in the retentate can be reduced from a concentration of 17% to around 100 ppm (107). Liu and coworkers also reported a new carbon/zeolite nanocomposite membrane by pyrolysis of polyamic acid/zeolite nanocomposites. Small molecules (H_2, CO_2, O_2 and N_2) were used for gas separation. It was found that the composite membrane had both good permeability and high selectivity, making it an attractive membrane material in gas separation fields (108).

10. SUMMARY

In conclusion, the use of membrane technology in water and wastewater treatment has begun to emerge as the most significant advancement in water treatment in the past 20 years, and its presence is expected to become widely spread in the future (109–117).

The potential to use membranes exists whenever they provide the ability to remove contaminants that cannot be removed by other technologies, remove contaminants at less cost than other alternatives, or require less land area than competing technologies. The biggest single technical challenge with the use of membranes for water treatment is the high fouling that occurs universally, which results in an increase in feed pressure and requires frequent cleaning of membranes. This leads to a reduction in overall facility efficiency and a shorter membrane life. Pre-treatment prior to membrane filtration is generally required for a full-scale plant (109, 110).

With extensive research and development efforts on tackling the membrane fouling and emerging of new cleaning technology, membrane technology has become a more promising option for both water treatment (109, 110, 114) and wastewater treatment (111, 115–117) when considering urban reuse, agricultural reuse, industrial recycle, groundwater recharge, salinity barriers, and augmentation of potable water or ultra-pure water supplies. MBR is the latest environmental process development which combines both biotechnology and membrane technology together for wastewater treatment (111, 115–117). Another recent development involves the use of membrane technology for gas separation (37, 112). The future of membrane technology is unlimited.

11. ABBREVIATIONS

AFM = Atomic force microscopy
BOD_5 = 5-day Biochemical oxygen demand
CA = Cellulose acetate
COD = Chemical oxygen demand
ED = Electrodialysis
EPS = Extracellular polymeric substance
GAC = Granular activated carbon
MBR = Membrane bioreactor
MF = Microfiltration
MWCO = Molecular weight cut-off
NF = Nanofiltration
PTFE = Polytetrafluoroethylene
PVP = Polyvinylpyrrolidone
RO = Reverse osmosis
SDI = Silt density index
TDS = Total dissolved solids
TOC = Total organic carbon
UF = Ultrafiltration
WTCost = Water treatment cost estimation programme

12. NOMENCLATURE

A = Proportionality factor
A_m = Membrane area (m^2)
A_s = Salt permeability constant (m/s)
A_w = Water permeability constant (kg water/s m^2 atm)
a_{Am} = Activity of solute in the membrane (mol/m^3)
a_B = Activity of solvent outside of the membrane (mol/m^3)
a_{Bm} = Activity of solvent in the membrane (mol/m^3)
a_{im} = Activity in the membrane (mol/m^3)
B = Combined parameter (atm^{-1})
β = Concentration polarization, 1.2–2.0
c_1 = Salt concentration in feed solution (kg salt/m^3)
c_2 = Salt concentration in permeate (product) solution (kg salt/m^3)
c_A = Concentration of solute outside the membrane (mol/m^3)
c_{AF} = Concentration of solute in the feed (mol/m^3)
c_{Am} = Concentration of solute in the membrane (mol/m^3)
c_{AP} = Concentration of solute in the permeate (mol/m^3)
c_{Bm} = Concentration of the solvent in the membrane (mol/m^3)
c_{BP} = Concentration of the solvent in the permeate (kg/m^3)

c_{w2} = Concentration of water in permeate (product) (kg water/m^3)
D_{Am} = Diffusion coefficient of solute in the membrane (m^2/s)
D_{Bm} = Diffusion coefficient of solvent in the membrane (m^2/s)
Δ = Quantity difference
ε = Surface porosity
f_{Am} = Friction between a unit mole of the solute and the membrane material (J s/m^2)
f_{Bm} = Friction between a unit mole of the solvent and the membrane material (J s/m^2)
∇ = Gradient
J = Flux (mol/m^2 s (by quantity) or kg/m^2 s (by mass) or m^3/m^2 s (by volume))
J_A = Flux of solute (mol/m^2 s)
J_B = Flux of solvent (mol/m^2 s)
J_i = Flux of component i (mol/m^2 s)
K = Kozeny–Carman constant
K_A = Distribution constant of solute
L = Phenomenological coefficients
l = Membrane thickness (m)
M_i = Molalities or molarities of ions and non-ionic compounds (mol/kg or mol/L)
μ = Fluid viscosity (Pa s) or chemical potential (J/mol)
μ_A = Chemical potential of solute (J/mol)
μ_B = Chemical potential of solvent (J/mol)
μ_I = Chemical potential of the ith component (J/mol)
n_p = Number of pores
N_s = Salt flux (kg salt/s m^2)
N_w = Water flux (kg water/s m^2)
p = Pressure (Pa)
P_1 = Pressure applied on feed solution (atm)
P_2 = Pressure of permeate (product) (atm)
Π = Osmotic pressure (Pa)
Π_1 = Osmotic pressure of feed solution (atm)
Π_2 = Osmotic pressure of permeate (product) (atm)
r = Pore radius (m)
R = Gas constant (8.314 J/mol K) or solute rejection rate
S = Specific surface area (m^2/m^3)
T = Temperature (°C or K)
τ = Pore tortuosity
v = Molar volume (m^3/mol)
v_A = Molar volume of solute (m^3/mol)
v_B = Molar volume of solvent (m^3/mol)
v_i = Partial molar volume of the ith component (m^3/mol)
x = Axial coordinate in the pore (m)
X = Driving force

13. SUBSCRIPTS

1 = Quantities concerning feed solution
2 = Quantities concerning permeate (product) solution
A = Quantities concerning solute
B = Quantities concerning solvent
F = Quantities concerning the feed
I = Quantities concerning *i*th component
j = Quantities concerning *j*th component
m = Quantities concerning inside the membrane or membrane
P = Quantities concerning the permeate, or pore
s = Quantities concerning salt
w = Quantities concerning water

ACKNOWLEDGEMENT

The authors wish to thank Professor T.S. Neal Chung of Department of Chemical and Biomolecular Engineering, National University of Singapore for providing Fig. 7.6 and input for this manuscript.

REFERENCES

1. Ong HS (1997) Challenges ahead for Singapore's water supply. Seminar on *Ensuring Singapore's Water Supply: Options and Issues*, Shangri-la Hotel, 10–11 November
2. Joel M, Peter EO, Mark RW (1996) Water treatment membrane process. McGraw-Hill, New York, pp 17.1–17.31
3. Judd S, Jefferson B (eds) (2003) Membrane for industrial wastewater recovery and re-use. Elsevier Advanced Technology, Oxford
4. Chen JP, Kim SL, Ting YP (2003) Optimization of feed pretreatment for membrane filtration of secondary effluent. J Memb Sci 219:27–45
5. Kim SL, Chen JP, Ting YP (2002) Study on feed pretreatment for membrane filtration of secondary effluent. Sep Purif Technol 29:171–179
6. Letterman RD (ed) (1999) Water quality and treatment: a handbook of community water supplies, 5th edn. McGraw-Hill, New York
7. Ho WSW, Sirkar KK (eds) (1992) Membrane handbook. Chapman & Hall, New York
8. Singapore Public Utilities Board (2002) Singapore water reclamation study, expert panel review and findings. Singapore Government, Singapore
9. Matsuura T (2001) Progress in membrane science and technology for seawater desalination – a review. Desalination 134:47–54
10. Matsuura T (1994) Synthetic membranes and membrane separation processes. CRC, Boca Raton, FL
11. Matsuura T, Sourirajan S (1972) Studies on reverse osmosis for water pollution control. Water Res 6:1073–1086
12. Metcalf and Eddy, Inc. (ed) (2002) Wastewater engineering: treatment disposal and reuse, 4th edn. McGraw-Hill, New York
13. Mulder M (1996) Basic principles of membrane technology. Kluwer, Dordrecht

14. Economic Commission for Europe (1990) Membrane technology in the chemical industry. United Nations Publication, New York
15. Noyes R (ed) (1994) Unit operations in environmental engineering. Noyes, Park Ridge, NJ, pp 239–264
16. Madaeni SS, Fane AG, Grohmann GS (1995) Virus removal from water and wastewater using membranes. J Memb Sci 102:65–75
17. Sudilovskiy PS, Kagramanov GG, Trushin AM, Kolesnikov VA (2007) Use of membranes for heavy metal cationic wastewater treatment: flotation and membrane filtration. Clean Technol Environ Policy 9:189–198
18. Futamura O, Katoh M, Takeuchi K (1994) Organic waste water treatment by activated sludge process using integrated type membrane separation. Desalination 98:17–25
19. Cassano A, Adzet J, Molinari R, Buonomenna MG, Roig J, Drioli E (2003) Membrane treatment by nanofiltration of exhausted vegetable tanning liquors from the leather industry. Water Res 37:2426–2434
20. Rai UK, Muthukrishnan M, Guha BK (2008) Tertiary treatment of distillery wastewater by nanofiltration. Desalination 230:70–78
21. Laine JM, Vial D, Moulart P (2000) Status after 10 years of operation – overview of UF technology today. Desalination 131:17–25
22. Molinari R, Gallo S, Argurio P (2004) Metal ions removal from wastewater or washing water contaminated soil by ultrafiltration-complexation. Water Res 38:593–600
23. Wu CJ, Li A, Li L, Zhang L, Wang H, Qi XH, Zhang QX (2008) Treatment of oily water by a poly (vinyl alcohol) ultrafiltration membrane. Desalination 225:312–321
24. Subhi AJ, Anne N (2008) An experimental evaluation of reverse osmosis membrane performance in oily water. Desalination 228:287–294
25. Belkacem M, Bensadok K, Refes A, Charvier PM, Nezzal G (2008) Water produce for pharmaceutical industry role of reverse osmosis stage. Desalination 221:298–302
26. Seader JD, Henley EJ (1998) Separation process principles. Wiley, New York
27. Geankoplis CJ (2003) Transport processes and separation process principles, 4th edn. Prentice Hall, Upper Saddle River, NJ
28. Sadrzadeh M, Mohammadi T (2008) Sea water desalination using electrodialysis. Desalination 221:440–447
29. Reddad Z, Gerente C, Andres Y, Thibault JF (2003) Cadmium and lead adsorption by a natural polysaccharide in MF membrane reactor: experimental analysis and modelling. Water Res 37:3983–3991
30. Van Dijk L, Roncken GCG (1997) Membrane bioreactor for wastewater treatment: the state of art and new developments. Water Sci Technol 10:35–41
31. Raman LP, Cheryan M, Rajagopalan N (1994) Consider nanofiltration for membrane separations. Chem Eng Progr 90:68–74
32. Yesselman RJ, Wang LK (1987) Reverse osmosis. Technical Report No. LIR/02-87/2.47. Lenox Institute of Water Technology, Lenox, MA, 115 pp
33. Sa-nguanruksa J, Rujiravanit R, Supaphol P, Tokura S (2004) Porous polyethylene membranes by template-leaching technique: preparation and characterization. Polym Test 23:91–99
34. Young TH, Huang YH, Chen LY (2000) Effect of solvent evaporation on the formation of asymmetric and symmetric membranes with crystallizable EVAL polymer. J Memb Sci 164:111–120
35. Wijmans JG, Baaij JPB, Smolders CA (1983) The mechanism of formation of microporous or skinned membranes produced by immersion precipitation. J Memb Sci 14:263–274

36. Reuvers AJ, Van den Berg JWA, Smolders CA (1987) Formation of membranes by means of immersion precipitation: part 1. A model to describe mass transfer during immersion precipitation. J Memb Sci 34:45–65
37. Chung TS, Shieh JJ, Qin J, Lin WH, Wang R (2001) Polymeric membranes for reverse osmosis, ultrafiltration, microfiltration, gas separation, pervaporation, and reactor applications. In: Nalwa HS (ed) Advanced functional molecules and polymers, Chap. 7. Gordon & Breach, New York, pp 219–264
38. Wang KY, Chung TS (2005) The characterization of flat composite nanofiltration membranes and their applications in the separation of Cephalexin. J Memb Sci 247:37–50
39. Zhu G, Chung TS, Loh KC (2000) Activated carbon-filled cellulose acetate hollow fibre membrane for cell immobilisation and phenol degradation. J Appl Polym Sci 76:695–707
40. Porter MC (1997) Membrane filtration. In: Schweitzer PA (ed) Handbook of separation technologies for chemical engineers, 3rd edn, Sect. 21. McGraw-Hill, New York
41. Pinnau I, Freeman BD (2000) Formation and modification of polymeric membranes: overview. In: Pinnau I, Freeman BD (eds) Membrane formation and modification. American Chemical Society, Washington, DC, pp 1–22
42. Zsigmondy R, Bachmann W (1922) Filter and method of producing same. U.S. Patent No. 1,421,341. U.S. Patent and Trademarks Office, Washington, DC
43. Rezac ME, Le Roux JD, Chen H, Paul DR, Koros WJ (1994) Effect of mild solvent post-treatments on the gas transport properties of glassy polymer membranes. J Memb Sci 90:213–229
44. Li RH, Barbari TA (1995) Performance of poly (vinyl alcohol) thin-gel composite ultrafiltration membranes. J Memb Sci 105:71–78
45. Mosqueda-Jimenez DB, Narbaitz RM, Matsuura T, Chowdhury G, Pleizier G, Santerre JP (2004) Influence of processing conditions on the properties of ultrafiltration membranes. J Memb Sci 231:209–224
46. Jones CW, Koros WJ (1994) Carbon molecular sieve gas separation membranes-I. Preparation and characterization based on polyimide precursors. Carbon 32:1419–1425
47. Lonsdale HK (1996) Properties of cellulose acetate membranes. In: Merten U (ed) Desalination by reverse osmosis, Chap. 4. MIT, Cambridge
48. Schulz RD, Asunmaa SK (1970) Ordered water and the ultrastructure of the cellular plasma membrane. In: Danielli JF, Riddiford AC, Rosenberg M (eds) Recent progress in surface science, vol. 3. Academic, New York, pp 291–332
49. Geankoplis J (2003) Transport processes and separation processes principles, 4th edn. Pearson Education, Upper Saddle River, NJ
50. Duranceau SJ (ed) (2001) Membrane practices for water treatment. American Water Works Association, Denver, CO, pp 3–42
51. Moch I Jr, Chapman M, Steward D (2003) Development of a CD-ROM cost program for water treatment projects. Membr Technol 6:5–8
52. Wetterau GE, Clark MM, Anselme C (1996) A dynamic model for predicting fouling effects during the ultrafiltration of a groundwater. J Memb Sci 109:185–204
53. Cabassud C, Anselme C, Bersillon JL, Aptel P (1991) Ultrafiltration as a nonpolluting alternative to traditional clarification in water treatment. Filtr Sep 28:194–198
54. Dudley LY (1998) Membrane autopsies for reversing fouling in reverse osmosis. Membr Technol 95:9–12

55. Basu OD, Huck PM (2004) Integrated biofilter-immersed membrane system for the treatment of humic waters. Water Res 38:655–662
56. Winfield BA (1979) A study of the factors affecting the rate of fouling of reverse osmosis membranes treating secondary sewage effluent. Water Res 13:565–569
57. Koyuncu I, Topacik D, Wiesner MR (2004) Factors influencing flux decline during nanofiltration of solutions containing dyes and salts. Water Res 38:432–440
58. Weisner MR, Aptel P (1996) Mass transport and permeate flux and fouling in pressure-driven processes. In: Odendaal PE, Wiesner MR, Mallevialle J (eds) Water treatment membrane processes. McGraw-Hill, New York, pp 4.1–4.30
59. Carroll T, Booker NA, Meier-Haack J (2002) Polyelectrolyte-grafted microfiltration membranes to control fouling by natural organic matter in drinking water. J Memb Sci 203:3–13
60. Khedr MG (1998) A case study of RO plant failure due to membrane fouling, analysis and diagnosis. Desalination 120:107–113
61. Sadr Ghayeni SB, Madaeni SS, Fane AG, Schneider RP (1996) Aspects of microfiltration and reverse osmosis in municipal wastewater reuse. Desalination 106:25–29
62. Sadr Ghayeni SB, Beatson PJ, Schneider RP, Fane AG (1998) Water reclamation from municipal wastewater using combined microfiltration-reverse osmosis (ME-RO): preliminary performance data and microbiological aspects of system operation. Desalination 116:65–80
63. Sadr Ghayeni SB, Beatson PJ, Schneider RP, Fane AJ (1999) Bacterial passage through microfiltration membranes in wastewater applications. J Memb Sci 153:71–82
64. Butt FH, Rahman F, Baduruthamal U (1997) Characterisation of foulants by autopsy of RO desalination membranes. Desalination 114:51–64
65. Baker JS, Dudley LY (1998) Biofouling in membrane systems – a review. Desalination 118: 81–90
66. Groves GR (1983) Application of membrane separation processes to the treatment of industrial effluents for water reuse. Desalination 47:277–284
67. Yuan W, Zydney AL (1999) Humic acid fouling during microfiltration. J Memb Sci 157:1–12
68. Kim SL (2002) M. Eng Thesis, National University of Singapore
69. Mills WR Jr, Bradford SM, Rigby M, Wehner MP (1998) Groundwater recharge at the orange county water district. In: Takashi A (ed) Wastewater reclamation and reuse. Technomic Publishing, Lancaster, PA, pp 1105–1142
70. Ebrahim S (1994) Cleaning and regeneration of membranes in desalination and wastewater applications: state-of-the-art. Desalination 96:225–238
71. Wilf M (1998) Reverse osmosis membranes for wastewater reclamation. In: Takashi A (ed) Wastewater reclamation and reuse. Technomic Publishing, Lancaster, PA, pp 236–344
72. Jolis D, Campana R, Hirano RA, Pitt P, Mariñas B (1995) Desalination of municipal wastewater for horticultural reuse: process description and evaluation. Desalination 103:1–10
73. Jolis D, Hirano RA, Pitt PA, Müller A, Mamais D (1996) Assessment of tertiary treatment technology for water reclamation in San Francisco, California. Water Sci Technol 33(10–11): 181–192
74. Cikurel H, Rebhun M, Amirtharajah A, Adin A (1996) Wastewater effluent reuse by in-line flocculation filtration process. Water Sci Technol 33:203–211
75. Reith C, Birkenhead B (1998) Membranes enabling the affordable and cost effective reuse of wastewater as an alternative water source. Desalination 117:203–210
76. Del Pino MP, Durham B (1999) Wastewater reuse through dual-membrane processes: opportunities for sustainable water resources. Desalination 124:271–277

77. Van Houtte E, Verbauwhede J, Vanlerberghe F, Demunter S, Cabooter J (1998) Treating different types of raw water with micro- and ultrafiltration for further desalination using reverse osmosis. Desalination 117:49–60
78. Teodosiu CC, Kennedy MD, Van Straten HA, Schippers JC (1999) Evaluation of secondary refinery effluent treatment using ultrafiltration membranes. Water Res 33:2172–2180
79. Qin JJ, Wai MN, Oo MH, Wong FS (2002) A feasibility study on the treatment and recycling of a wastewater from metal plating. J Memb Sci 208:213–221
80. Sierka RA, Cooper SP, Pagoria PS (1997) Ultrafiltration and reverse osmosis treatment of an acid stage wastewater. Water Sci Technol 35:155–161
81. Fane AG (1996) Membranes for water production and wastewater reuse. Desalination 106:1–9
82. Kruithof JC, Schippers JC, Kamp PC, Folmer HC, Hofman JAMH (1998) Integrated multi-objective membrane systems for surface water treatment: pretreatment of reverse osmosis by conventional treatment and ultrafiltration. Desalination 117:37–48
83. Hills P, Padley MB, Powell NI, Gallegher PM (1998) Effects of backwash conditions on out-to-in membrane microfiltration. Desalination 118:197–204
84. Freeman SDN, Morin OJ (1995) Recent developments in membrane water reuse projects. Desalination 103:19–30
85. Belfort G (1977) Pretreatment and cleaning of hyperfiltration (reverse osmosis) membranes in municipal wastewater renovation. Desalination 21:285–300
86. Liikanen R, Yli-Kuivila J, Laukkanen R (2002) Efficiency of various chemical cleanings for nanofiltration membrane fouled by conventionally-treated surface water. J Memb Sci 195: 265–276
87. Trägårdh G (1989) Membrane cleaning. Desalination 71:325–335
88. Hong S, Elimelech M (1997) Chemical and physical aspects of natural organic matter (NOM) fouling of nanofiltration membranes. J Memb Sci 132:159–181
89. Lindau J, Jonsson AS (1994) Cleaning of ultrafiltration membranes after treatment of oily waste water. J Memb Sci 87:71–78
90. Graham SI, Reitz RL, Hickman CE (1989) Improving reverse osmosis performance by periodic cleaning. Desalination 74:113–124
91. Ebrahim S, El-Dessouky H (1994) Evaluation of commercial cleaning agents for seawater reverse osmosis membranes. Desalination 99:169–188
92. Ridgway HF, Justice CA, Whittaker C, Argo DG, Olson BH (1984) Biofilm fouling of RO membranes – its nature and effect on treatment of water reuse. J Am Water Works Assoc 76: 94–102
93. Darton EG, Turner AG (1991) Operating experiences in a sea water reverse osmosis plant in Gibraltar (1987–1990). Desalination 82:51–69
94. Flemming HC, Schaule G, McDonogh R, Ridgway HF (1994) Effects and extent of biofilm accumulation in membrane systems. In: Geesey GG, Lewandowski Z, Flemming HC (eds) Biofouling and biocorrosion in industrial water systems. Lewis, Chelsea, MI, pp 63–89
95. Wilf M, Glueckstern P (1985) Restoration of commercial reverse osmosis membranes under field conditions. Desalination 54:343–350
96. Pervov AG (1991) Scale formation prognosis and cleaning procedure schedules in reverse osmosis systems operation. Desalination 83:77–118
97. Farinas M, Granda JM, Gurtubi L, Villagra MJ (1987) Pilot experiences on recovery of polluted reverse osmosis membranes. Desalination 66:385–402
98. Sheikholeslami R (1999) Fouling mitigation in membrane processes. Desalination 123:45–53

99. Teodosiu CC, Kennedy MD, Van Straten HA, Schippers JC (1999) Evaluation of secondary refinery effluent treatment using ultrafiltration membranes. Water Res 33:2172–2180
100. Wang Y, Huang X, Yuan Q (2005) Nitrogen and carbon removals from food processing wastewater by an anoxic/aerobic membrane bioreactor. Process Biochem 40:1733–1739
101. Nakhla G, Lugowski A, Patel J, Rivest V (2006) Combined biological and membrane treatment of food-processing wastewater to achieve dry-ditch criteria: pilot and full-scale performance. Bioresour Technol 97:1–14
102. Chapman S, Leslie G, Law I (2004) Membrane bioreactors (MBR) for municipal wastewater treatment – an Australian perspective. Available online from http://www.membrane.unsw.edu.au/staff/papers/gleslie/mbr_for_reuse_awa.pdf
103. Fu ZM, Yang FL, Zhou FF, Xue Y (2009) Control of COD/N ratio for nutrient removal in a modified membrane bioreactor (MBR) treating high strength wastewater. Bioresour Technol 100:136–141
104. Sridang PC, Pottier A, Wisniewski C, Grasmick A (2008) Performance and microbial surveying in submerged membrane bioreactor for seafood processing wastewater treatment. J Memb Sci 317:43–49
105. Tian JY, Liang H, Li X, You SJ, Tian S, Li GB (2008) Membrane coagulation bioreactor (MCBR) for drinking water treatment. Water Res 42:3910–3920
106. Yoshino Y, Suzuki T, Taguchi H, Nomura M, Nakao S, Itoh N (2008) Silica membrane tubes for high temperature hydrogen separation. Sep Sci Technol 43:3432–3447
107. Huang J, Zou J, Winston Ho WS (2008) Carbon dioxide capture using a CO_2-selective facilitated transport membrane. Ind Eng Chem Res 47:1261–1267
108. Liu Q, Wang TH, Liang CH, Zhang B, Liu SL, Cao YM, Qiu JS (2008) Zeolite married to carbon: a new family of membrane materials with excellent gas separation performance. Chem Mater 18:6283–6288
109. Bartlett M, Bird MR, Howell JA (1995) An experimental study for the development of a qualitative membrane cleaning model. J Memb Sci 105:147–157
110. Wang LK, Hung YT, Shammas NK (eds) (2005) Physicochemical treatment processes. Humana Press, Totowa, NJ, 723 pp
111. Wang LK, Kopko S (1997) City of cape coral reverse osmosis water treatment facility. Technical Report No. PB97-139547. U.S. Department of Commerce, National Technical Information Service, Springfield, VA
112. Wang LK, Shammas NK, Hung YT (eds) (2009) Advanced biological treatment processes. Humana Press, Totowa, NJ, 738 pp
113. Wang LK, Pereira NC, Hung YT (eds) (2005) Advanced air and noise pollution control. Humana Press, Totowa, NJ, 526 pp
114. California Coastal Commission (2010) Seawater desalination in California. California Coastal Commission, Sacramento, CA. http://www.coastal.ca.gov/desalrpt/dchap1.html
115. Levesque S, Wallis-Large C, Hemken B, Bontrager S, Kreuzwiesner S (2009) Plan ahead with MBRs. Water Environ Fed 21(1):34–37
116. Wang LK, Hung YT, Shammas NK (eds) (2010) Handbook of advanced industrial and hazardous wastes treatment. CRC, Boca Raton, FL, 1346 pp
117. Wang LK, Ivanov V, Tay JH, Hung YT (eds) (2010) Environmental biotechnology. Humana Press, Totowa, NJ

8
Membrane Systems Planning and Design

Nazih K. Shammas and Lawrence K. Wang

CONTENTS

Abstract The purpose of this chapter is to provide the basic information on the use of membrane filtration and application of the technology in the design of potable water facilities. The main issues involved in the planning and design of membrane systems are covered: pilot testing; the considerations that influence system design and operation including operational unit processes, system design considerations, and residuals treatment and disposal; and the initial start-up phase which must be completed prior to placing the system into service and actual water production. The initial start-up phase is a critical step in the successful installation of a full-scale membrane filtration system and thus is an essential consideration in the facility planning and design process. This phase includes such tasks as initial system flushing and disinfection, system diagnostic checks, membrane module installation, integrity testing new equipment, and operator training, all of which must be completed prior to placing the system into service.

Key Words Membrane filtration • Pilot testing • System planning • Design • Operation • Start-up phase

From: *Handbook of Environmental Engineering, Volume 13: Membrane and Desalination Technologies*
Edited by: L.K. Wang et al., DOI: 10.1007/978-1-59745-278-6_8 © Springer Science+Business Media, LLC 2011

1. INTRODUCTION

The purpose of this chapter is to provide the basic information on the use of membrane filtration and application of the technology in the design of potable water facilities. The main issues involved in the planning and design of membrane systems are covered including pilot testing; the considerations that influence system design and operation; and the initial start-up phase, which must be completed prior to placing the system into service and actual water production.

Pilot testing a membrane filtration process is often conducted as part of the design effort for a membrane treatment facility. This chapter is intended to provide general guidance in terms of widely recognized industry practices. However, any particular state requirements would supersede the information provided in here. This chapter is intended to highlight some of the important benefits and considerations associated with pilot testing.

The successful implementation of a membrane filtration system requires a general understanding of the major considerations that influence system design and operation. Although there are many such considerations, this chapter focuses on the most significant of these issues, categorized as operational unit processes, system design considerations, and residuals treatment and disposal. The purpose is to present a general overview of these critical considerations, highlighting their respective roles in the implementation of the technology.

The initial start-up phase is a critical step in the successful installation of a full-scale membrane filtration system and thus is an essential consideration in the facility planning and design process. This phase includes tasks such as initial system flushing and disinfection, system diagnostic checks, membrane module installation, integrity testing new equipment, and operator training, all of which must be completed prior to placing the system into service.

2. PILOT TESTING

Pilot testing should not be confused with challenge testing. Challenge testing, as shown in a previous chapter, is typically conducted on a product-specific basis and characterizes a membrane filtration module in terms of removal efficiency. Pilot testing, by contrast, is conducted on a site- and system-specific basis and is used to collect information for full-scale facility design. Pilot testing need not include challenge testing unless specifically required by the state. In cases in which a particular membrane product that has not undergone challenge testing is proposed for compliance with United States Environmental Protection Agency (US EPA) requirements, it may be convenient to include a challenge test in the scope of pilot testing (1).

The primary goal of a membrane pilot study is to obtain information such as treated water quality (e.g., turbidity) and operating parameters (e.g., flux) that are necessary for the design of a membrane filtration facility. The treated water quality data provides assurance that the treatment objectives can be achieved, while the pre-determination of operating parameters allows for proper sizing of the membrane facility and minimizes the uncertainties regarding footprint and utility requirements. In general, the use of pilot data helps account for unforeseen conditions that may otherwise have gone undetected. Pilot testing also helps to familiarize operators with the membrane treatment equipment.

Most of the information contained in this section is generally applicable to both microfiltration (MF)/ultrafiltration (UF) and nanofiltration (NF)/reverse osmosis (RO) systems (2–7). Distinctions between the two systems are noted where differences occur. Pilot testing for membrane cartridge filtration (MCF) systems is typically simpler relative to testing other types of membrane filtration and is often performed simply to verify the replacement frequency and productivity for the filters. Consequently, pilot testing MCF systems is not specifically addressed in this chapter. However, MCF piloting guidelines are generally similar to those for MF/UF with the exception of references to cleaning intervals, since membrane cartridges are not typically cleaned and instead disposed off when fouled.

2.1. *Planning*

The planning phase prior to implementing pilot testing is an important component of an overall pilot test program. Careful planning helps ensure that all the pilot test objectives are achieved both efficiently and economically without unexpected delays. The most important element of the planning process is the development of a comprehensive pilot test protocol specifying how the testing should be conducted. This protocol should include not only instructions for carrying out the testing, but also specific testing objectives and strategies for optimizing performance in terms of flux, backwashing, and chemical cleaning. A plan for collecting water quality, integrity test, and microbial (where applicable) data should also be incorporated in the test protocol. In addition, it is recommended that the pilot protocol be developed in conjunction with the state to ensure that any particular state requirements are satisfied.

In addition to the test protocol, a number of other planning considerations should be taken into account. For example, it should be determined whether the pilot test units can be provided by the membrane manufacturers at no cost, rented or custom-fabricated, as budget constraints allow. Also, once the particular membrane filtration systems to be tested are selected, the utility requirements (i.e., water, electricity, and/or air) for the pilot units must be accommodated at the test site. Appropriately sized plumbing connections should be provided for the feed, filtrate, and concentrate streams, and provisions must be made for the disposal of backwash (for MF/UF) and chemical cleaning residuals. Labor requirements for operating the pilot units should be estimated, and it is also important to ensure that the designated testing site has sufficient area to accommodate all the units to be tested. One consideration that is sometimes overlooked is shipping the pilot units; the proper equipment for loading and unloading the pilot units must be available at the testing site or be provided with the unit (s). Special attention in the planning phase should be given to process considerations such as scalability, screening appropriate membrane filtration systems to test, and test scheduling.

2.1.1. *Process Considerations*

It is important that the pilot process be representative of the full-scale system. For example, if pretreatment is part of the planned full-scale process, then the pilot process should incorporate similar pretreatment. Likewise, piloting should also accommodate production (e.g., filtration) and intermittent (e.g., backwashing) design considerations to the extent that these parameters have been defined prior to the pilot testing phase. Thus, if it has been determined that the membrane filtration system(s) that will be used at full-scale (e.g., if an MF/UF system has been preselected) requires air at a specified flow rate and duration during a backwash process, then the pilot system should be designed and operated in a similar manner. The design of the membrane system should also mimic the hydraulic configuration of the full-scale system. However, it should be recognized that it might not be possible to design a pilot system using the same hydraulic configuration used at full-scale for all membrane processes. For example, it may not be economically feasible to design a NF/RO pilot unit with hydraulic characteristics identical to the full-scale system. In this case, the methodology of conversion from pilot-scale data to full-scale design should be included in the development of the pilot test protocol.

2.1.2. *Screening and System Selection*

Spiral-wound NF and RO membrane modules are standardized such that the membranes from different manufacturers are interchangeable and system design is somewhat uniform. Thus, in terms of screening for appropriate membranes, the primary consideration involves the selection of a membrane material that provides desirable productivity, resistance to fouling, and removal characteristics.

Screening is somewhat more complex for MF/UF systems, which are largely proprietary in design. The various commercially available systems may use either pressure or vacuum as the driving force and can be designed to filter from the inside-out or outside-in direction relative to the fiber lumen. In addition, the various membrane materials have differences that may be important, including removal efficiency and pH and oxidant tolerance. Membranes of different materials also have varying degrees of compatibility with water treatment chemicals such as coagulants (8) and powdered activated carbon (PAC) (9) that may affect performance and cost. It is important to consider such differences in MF/UF membranes and membrane systems and how these may impact system selection. For example, if the source water were periodically pre-chlorinated (10), a membrane that is not compatible with chlorine would be undesirable. A list of some questions to consider when screening membranes for a pilot study is provided below (1):

1. What are the treatment objectives of the application?
2. What operational constraints/goals are to be considered in membrane selection?
3. Has the membrane been used on similar waters at other sites?
4. What are the pH and oxidant tolerances of the membrane? Are these compatible with the application?
5. Is the membrane compatible with any pretreatment chemicals aside from oxidants that may be in use, such as alum, ferric chloride, PAC, and polymers?
6. Is this membrane compatible with solids and total organic carbon (TOC) levels in the raw water?
7. Does the membrane have prior applicable state regulatory approval and any required certifications?
8. Does the supplier have experience with full-scale operation for a facility of this size and for treating similar water with the same configuration to be used in the pilot?
9. Does the system require proprietary items such as spare parts or cleaning chemicals?
10. Are there any unusual operational considerations associated with the membrane filtration system, such as significant power requirements, frequent membrane replacement, or substantial or undesirable chemical use?

These types of questions should be considered as a part of the screening process before selecting any membrane filtration systems to pilot. An additional consideration for MF/UF systems is that because these systems are proprietary, the membrane modules vary in design and are not presently interchangeable. Thus, a utility is limited to obtaining replacement membranes from only one supplier after a system has been installed.

In some cases, screening tools for MF/UF membranes may be limited to experience at other sites or chemical incompatibility. Therefore, pilot testing is an important consideration. For NF/RO, proprietary software programs that are available from the various membrane manufacturers can predict membrane system removal performance with a fairly good degree of accuracy for a particular membrane product. Small-scale module testing may be an option for screening some MF/UF membranes. Similarly, NF/RO membranes could be screened for some parameters using flat-sheet studies or single element tests.

2.1.3. Scheduling

Scheduling is another important aspect of pilot test planning. One factor to consider when scheduling a pilot test for surface waters is seasonal variations in water quality, since turbidity, temperature, algae content, taste and odor, and other parameters can potentially vary significantly throughout the year. A description of typical seasonal variation in surface water quality is summarized below (1). Note that not all of the seasonal variations described will be applicable for every site, or necessarily pertinent for all types of membrane filtration systems.

Summer: Because user demand is typically highest in the summer, the filtration system may also have to produce more water during this season. However, greater production is facilitated by warmer water temperature. The degree to which a membrane process is affected by a change in water temperature is related to the viscosity of the water. Since the viscosity of the water is lower at higher temperatures, the membrane flux will likely be at its peak in this season. Warmer water also facilitates enhanced algal growth that may be problematic for membrane system operation. Taste and odor events may also manifest during the summer months.

Autumn: In areas with hardwood cover in the watershed, autumn months typically bring an increase in the organic content of surface water resulting from the decay of fallen leaves. Turbidity may sharply increase, as well. Furthermore, cooler air temperatures and wind cause surface waters in reservoirs to "turnover," bringing deeper, more anaerobic water to the surface and thus creating the potential for both taste and odor, and iron and manganese problems.

Winter: Although winter months yield the coldest water temperatures and thus typically the lowest membrane fluxes, the demand for water is typically at its lowest, as well. Both the cold temperature and decrease in demand are likely to minimize required membrane flux in this season. Most membrane manufacturers have membrane-specific temperature compensation factors for cold-water operations.

Spring: Spring months usually yield increases in water temperature, as well as in the potential for turbidity spikes related to run-off caused from snow melt. There is also potential for springtime turnover of a reservoir, resulting in subsequent degradation in feed water quality.

Ideally, the pilot study should be conducted during the time of year yielding the most difficult water quality to treat, so that design parameters resulting from the study, such as flux and chemical cleaning frequency, would be conservative for year round operation. Some water quality effects can be accurately modeled, such as that of temperature on productivity. Multi-season piloting is advantageous if scheduling and cost constraints will allow. In fact, some states require multi-season piloting for membrane treatment facilities.

The duration of the pilot study is also an important scheduling consideration. In general, piloting through at least three cleaning cycles is recommended practice. The target cleaning frequency for hollow-fiber MF/UF systems is typically at least 30 days of continuous operation. Through the first cleaning cycle, the membrane flux and backwash frequency are usually targeted to provide for 30 days of operation before cleaning is required. The second cycle provides an opportunity for optimization and operational improvement. The third cycle establishes repeatability if the operating conditions remain the same. Note that because new membranes typically perform better than membranes that have been previously fouled and subsequently cleaned, it may be beneficial to add an extra cycle; to study the effects of repeated fouling and cleaning on membrane performance. Because this strategy represents three operational cycles of at least 30 days, pilot test duration of 90 days (or 3 months) is recommended, if possible. For spiral-wound NF/RO membrane systems, longer cleaning intervals are desirable, which results in fewer operational cycles than for a MF/UF pilot of similar duration.

For more thorough MF/UF pilot studies, approximately 4–7 months (3,000–5,000 h) of cumulative operational time is usually recommended. A longer pilot study may be appropriate for newer, less proven membrane filtration systems, or for applications in which the water quality is extremely variable. NF/RO pilot studies generally range from about 2–7 months (1,500–5,000 h) of cumulative operational time, with longer studies used for waters of variable quality. The state may also have minimum requirements for the duration of pilot studies.

2.2. *Testing Objectives*

Membrane flux and system productivity are typically the most important design parameters to optimize, as these dictate the number of membranes (and hence a large portion of the capital cost) required for the full-scale plant. Because these two parameters are inversely related to a certain extent, the pilot testing process may help to establish the optimum balance. For example, typically operating at higher fluxes increases the rate of fouling, in turn requires more frequent backwashing and chemical cleaning. However, the system productivity is limited by the backwashing (where applicable) and chemical cleaning frequency. Backwashing and chemical cleaning not only use filtrate as process water (thus affecting the productivity), but also represents time during which filtrate cannot be produced (thus affecting overall system productivity). The effect of more frequent backwashes and chemical cleanings on the system productivity may be an important consideration if the source water is limited, or alternatively if waste disposal is problematic. In addition, chemical cleaning may be a labor-intensive operation that requires the handling of harsh chemicals and produces a waste stream that may be difficult to dispose. However, chemical cleaning is a necessary process associated with membrane filtration, and establishing operating practices that extend the time between cleanings is one objective of a membrane pilot study.

This section provides guidelines for balancing flux, productivity, backwash frequency (where applicable), and chemical cleaning intervals during pilot testing. It is important to note that in order for cause and effect to be analyzed properly only one process variable should be changed at a time during the pilot study. Economic and time constraints often dictate the duration of the pilot study and may not allow complete optimization of each of these parameters. Therefore, it is important to understand which of these parameters is most important for a particular application of membrane filtration and to structure the pilot test protocol accordingly. Thus, the result of a pilot study is not necessarily the "optimum" design, but rather a set of operational conditions that will result in feasible and economic water treatment over the anticipated range of operating and source water conditions.

2.2.1. Membrane Flux Optimization

Membrane manufacturers can recommend fluxes for particular applications and a given water quality. Table 8.1 lists some of the important water quality data that should be provided to membrane suppliers to facilitate a fairly accurate initial estimation of anticipated membrane productivity.

If scheduling permits, it is typically advantageous to begin piloting MF/UF systems at a conservative flux and then increase it based upon the rate of fouling observed. The flux may be increased either after a chemical cleaning or during a filter run if the pilot unit has undergone sustained operation with only a nominal increase in fouling. In subsequent filter runs (i.e., between chemical cleanings) the pilot unit flux may be either increased if the fouling rate of the previous run was still within acceptable tolerances, or decreased if the fouling occurred at an unacceptably high rate. Note that the backwash frequency may also be adjusted either during a filter run or between filter runs to minimize fouling at a particular flux. This fine-tuning process of adjusting the flux and/or backwashing interval (if possible, only one parameter should be adjusted at a time) between filter runs may be repeated to the extent that budget and scheduling constraints allow, or to the point at which such adjustments reach the point of diminishing returns. NF/RO performance can typically be accurately gauged by the manufacturers based on water quality models; although it is often beneficial to adjust the flux over a series of filter runs to optimize productivity based on pilot test results. (Note, however, that manufacturers' models do not predict fouling.) The pilot test protocol should include the expected range of membrane fluxes and guidelines on how these fluxes should be adjusted based upon the results of early testing. Models may also indicate chemical pretreatment requirements necessary to control scaling on NF/RO membranes and thus to help maintain acceptable fluxes.

Table 8.1
Water quality parameters to measure prior to piloting

Parameter	MF/UF	NF/RO
Cations		
Aluminum		X
Ammonia		X
Barium		X
Calcium	X	X
Iron	X	X
Magnesium	X	X
Manganese	X	X
Potassium		X
Sodium		X
Strontium		X
Anions		
Chloride		X
Fluoride		X
Nitrate		X
Silica	X	X
Sulfate		X
Other chemical/physical parameters		
Algae	X	
Alkalinity		X
pH	X	X
SDI		X
TDS		X
TOC	X	X
TSS	X	
Turbidity	X	X
UV-254	X	X

2.2.2. Backwash Optimization

Backwashing is a periodic reverse flow process used with many hollow-fiber MF/UF systems to remove accumulated contaminants from the membrane surface, thus maintaining sufficient flux and minimizing the rate of long-term, irreversible fouling. A MF/UF membrane manufacturer can provide information on the backwash protocol required for its system, as well as a recommended backwash frequency. It is important to understand the following issues regarding the backwash process, which are applicable to both piloting and full-scale operation:

1. Large water and/or air flows over short periods of time may be included as a part of the backwash process. The flows of air and water vary depending on the system manufacturer.
2. Backwash effluent water contains approximately 10–20 times the concentration of the feed contaminants.

3. Many manufacturers add free chlorine to the backwash process water to reduce membrane fouling via disinfection (a process often referred to as a "chemically enhanced backwash"). Other manufacturers may include acid or caustic solution to remove inorganic or organic foulants. As a result, the backwash effluent may contain both solids and/or chemical residuals.
4. In locations where the disposal of backwash water may be problematic, consideration should be given to the volume of backwash water generated. Backwash water treatment should also be considered in order to maximize system productivity and minimize residuals.
5. Increased backwash efficiency can significantly enhance system performance, allowing higher fluxes and recoveries.

Variations in the backwash frequency will influence the waste flow, as well as the productivity of the system. If higher residual flows can be easily accommodated and productivity is not a critical design consideration, increasing the backwashing frequency may be a viable strategy for achieving longer filter run times or higher membrane fluxes.

Based upon the water quality data provided, MF/UF manufacturers can estimate the initial membrane flux and backwash frequency for the pilot test. The backwash frequency can be optimized during pilot testing to provide the most economical process. The potential for increased flux and useful membrane life that may be commensurate with increased backwash frequency should be balanced against decreased productivity, increased waste volume, and higher chemical usage (if applicable).

2.2.3. *Chemical Cleaning Optimization*

The chemical cleaning of a membrane filtration system is a necessary process that results in lost production time, produces chemical waste, and requires operator attention. Consequently, pilot testing should be used to minimize chemical cleaning frequency for both MF/UF and NF/RO systems, while maintaining acceptable productivity and controlling long-term fouling.

The typical membrane-cleaning cycle consists of recirculating a heated cleaning solution for a period of several hours. Some manufacturers require that the cleaning solution be prepared with softened or demineralized water, which should be accounted for during the pilot test. The specific cleaning regime will depend upon the feed water quality and the particular membrane system. Some membranes may be degraded by excessive cleaning frequencies, shortening membrane life. For MF/UF systems, the typical chemical cleaning frequency assumed for design purposes is approximately once every 30–60 days. NF/RO systems are generally designed with significantly longer cleaning intervals, typically in the range of 3 months to 1 year. It is important for a utility to understand the impacts of chemical cleaning on the system and to determine whether to increase the flux at the expense of the consequent increase in chemical cleaning frequency. One of the objectives of a pilot test should be to determine a chemical cleaning strategy that restores membrane permeability without damaging the membrane integrity. Thus, the pilot test should be designed to include at least one chemical cleaning, even if the pilot schedule is very restrictive. If necessary, the flux can be significantly increased to induce membrane fouling for the purposes of conducting chemical cleaning.

It is important to demonstrate the effectiveness of the chemical cleaning during the pilot study. Because membrane cleaning is a time-, labor-, and chemical-intensive process, it is important to avoid experimentation on the full-scale plant. The best way to evaluate the success of a particular chemical cleaning scheme is to conduct the cleaning in a methodical manner and evaluate the performance of each step in the process. Before and after each step, a plot of the trans-membrane pressure (TMP) as a function of flux (or filtrate flow) should be developed. This method of evaluation is normally described as a "clean water flux test." This test requires that the membrane unit be placed into filtration mode in between each of the various steps in the chemical cleaning process in order to observe the effect on the flux (or filtrate flow). (Note that filtrate produced during this process should be considered chemical waste.) For example, if a chemical cleaning process consists of recirculating citric acid and caustic in succession, three plots of TMP vs. flux (or filtrate flow) should be generated. Figure 8.1 illustrates the clean water flux test technique for this example. The three applicable plots shown in Fig. 8.1 are as follows:

1. TMP vs. flux (or filtrate flow) for the fouled membrane prior to cleaning
2. TMP vs. flux (or filtrate flow) after citric acid recirculation/cleaning
3. TMP vs. flux (or filtrate flow) after caustic recirculation/cleaning (i.e., after all cleaning steps are completed)

In an ideal cleaning, the final plot (i.e., after caustic) of TMP vs. flux would be similar to that after the previous clean such that the plots generated after each successive chemical cleaning (i.e., the final plot of each cleaning) would overlap. Thus, the plots established during a clean water flux test facilitate this comparison and provide an indication as to the effectiveness of the cleaning regimen. This is illustrated in Fig. 8.1 with plots showing the increase in TMP per incremental change in flux. Smaller slopes indicate lower pressure requirements for operation at a given flux, and thus a more effective step in the chemical cleaning process.

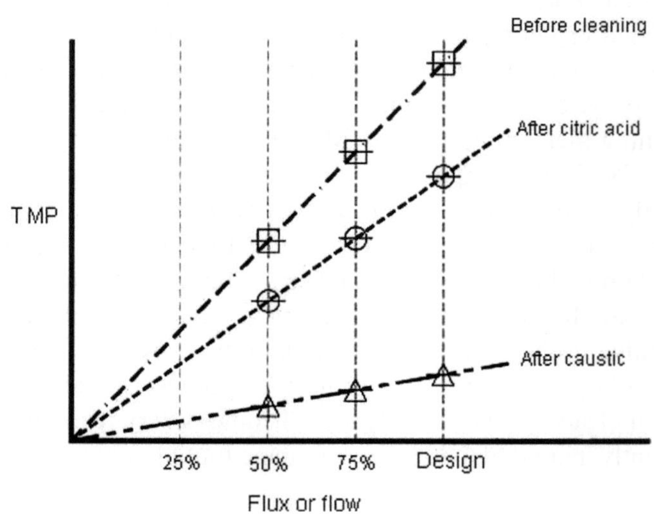

Fig. 8.1. Sample chemical cleaning test profile.

After cleaning, a direct integrity test should be conducted before the pilot unit is returned to service. If the membrane is integral and an acceptable flux has been restored, then the cleaning is deemed successful. If not, alternate cleaning strategies should be considered. A flow chart outlining a general pilot study sequence of events is shown in Fig. 8.2, illustrating a typical series of test runs and subsequent chemical cleaning events.

2.3. *Testing and Monitoring*

A thorough and carefully developed testing and monitoring plan is a critical component of a pilot test program, as this is the means by which system performance is assessed. There are several categories of testing and monitoring that should be addressed in a pilot test protocol, including operational parameters, water quality, and microbial monitoring (where applicable), and integrity testing.

2.3.1. Operational Parameter Monitoring

Monitoring pilot unit operational parameters is an important means of assessing system performance and tracking the rate of membrane fouling. The following operational parameters should be monitored continuously, if possible:

1. Elapsed run time
2. Pressure (feed, filtrate, and concentrate)
3. Flow (feed, filtrate, and concentrate)
4. Temperature (feed or filtrate)

Another important consideration in the evaluation of a membrane process is operational data collection for the intermittent process sequences, such as backwashing. For these intermittent processes, the design parameters (e.g., flows, times, volumes) associated with air, water, or chemical usage during the sequence should be established and verified during pilot testing. It is not uncommon for the pilot unit to operate under different parameters than those that would be considered appropriate for a full-scale unit. For example, a pilot system generally has faster pneumatic valve actuation times and a shorter overall backwash sequence than a full-scale unit. Some of these discrepancies are unavoidable; however, these scale-up issues should be noted for consideration prior to initiating pilot testing.

2.3.2. Water Quality Monitoring

The particular water quality data collected during a pilot study will depend upon the type of membrane filtration system, site-specific treatment objectives, and foulants of concern. In general, Table 8.2 can be used as a guideline.

In addition to those flows listed in Table 8.2, it is recommended that the MF/UF backwash flow be checked weekly for total suspended solids (TSS) and turbidity. Note that the recommended parameters and sampling frequencies should be modified to meet the requirements and objectives of each particular site-specific pilot study. The sampling frequencies may also be modified during the course of a pilot study if conditions warrant.

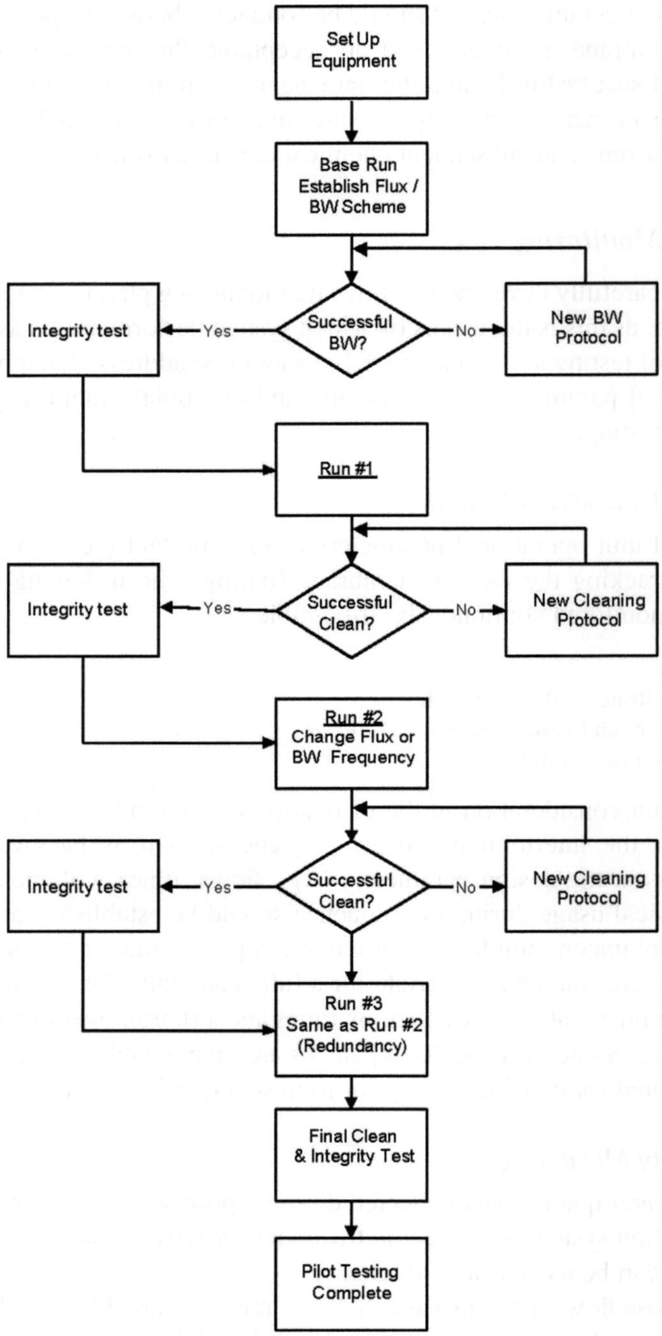

Fig. 8.2. Sample pilot study sequence overview.

Table 8.2
Suggested water quality sampling schedule for membrane piloting[a]

Parameter	MF/UF			NF/RO		
	Feed	Concentrate[b]	Filtrate	Feed	Concentrate	Filtrate
General water quality parameters						
Algae	W		W	W		
Color	W	M	W	W	M	W
HPC	M		M	M		M
Particle counts	C		C			
pH	D	W	W	C	W	W
SDI				W		
Taste and odor	M	M	M	M	M	M
Temperature	C			C		
TOC	W	M	W	W	M	W
Total coliform	W		W	W		
TSS	M	M	M			
Turbidity	C	M	C	C		W
UV-254	W	M	W	W		W
Dissolved solids						
Alkalinity	W	M	W	W	W	W
Barium				W	W	W
Hardness			M	W	W	W
Iron	W			W	W	W
Manganese	W			W	W	W
Silica				W	W	W
Sulfate	W			W	W	W
TDS			M	C	W	C
Simulated distribution system disinfection byproducts						
HAAs			M			M
TTHMs			M			M

C continuous; *D* daily; *W* weekly; *M* monthly.
[a]Generic recommendations only; the specific applicable parameters will vary with each application.
[b]Where applicable, based on particular hydraulic configuration used.

2.3.3. Microbial Monitoring

One advantage of membrane filtration is that it provides a physical barrier through which water must pass, virtually eliminating pathogens larger than the membrane exclusion characteristic. As with the full-scale facility, it is important that the pilot test demonstrate that this barrier is intact and is rejecting pathogens. A site-specific microbial challenge test may not be required by the state during the pilot study for a membrane filtration system that has received prior regulatory approval. For the purposes of regulatory compliance, product-specific demonstration of *Cryptosporidium* removal efficiency is accomplished during challenge testing, although microbial monitoring may be conducted during a pilot study for additional verification of membrane performance at the discretion of the utility or if required by the state. Although additional microbial testing can be conducted for any pathogen of concern during piloting, typically, coliform bacteria, and sometimes heterotrophic plate count (HPC) are used as an indicator for microbial removal efficiency.

2.3.4. Integrity Testing

The importance of integrity testing necessitates that greater emphasis be placed upon integrity testing during piloting for applications of membrane filtration intended for regulatory compliance. For direct integrity testing, the membrane manufacturer should use a method approved or mandated by the state. The standard direct integrity test sequences intended for the full-scale system should be incorporated into the pilot unit. Direct integrity testing should be conducted at least as frequently as required by the state for the full-scale facility. In the absence of a particular state requirement, it is generally recommended that direct integrity testing be conducted on a daily basis. If an integrity breach occurs during the piloting testing, then diagnostic testing and membrane repair techniques may also be practiced. Continuous indirect integrity monitoring can usually be accomplished using data that are collected during a typical pilot test program (e.g., turbidity or particle count data). However, any state requirements regarding data that must be collected for the purposes of continuous indirect integrity monitoring for the full-scale facility should also be implemented during the pilot test, and may be required by the state in any case.

Note that because the number of membrane modules to which an integrity test is applied affects the sensitivity of the test (for both direct integrity testing and continuous indirect integrity monitoring), any control limits established for a full-scale system will generally not be applicable to a pilot unit, which typically utilizes only a small fraction of the number of modules in a full-scale membrane unit. Guidance on establishing full-scale system control limits for both direct integrity testing and continuous indirect integrity monitoring for the purposes of regulatory compliance are provided in Chap. 7.

2.4. Report Development

After the pilot test is complete, a report should be prepared to summarize the procedures and the test results. The pilot study report should contain sufficient detail to establish design parameters for the full-scale plant to the extent for which the testing was intended. Flux, chemical cleaning, backwashing, and integrity testing, should be addressed, where applicable. Collected water quality data should also be included with emphasis given to unanticipated test results. Details on the operational parameters for each filter run (i.e., between chemical cleanings) should be summarized. The state may also have specific requirements for the report if the piloting is required.

3. OPERATIONAL UNIT PROCESSES

All membrane filtration systems have associated operational unit processes that are essential for maintaining and optimizing system performance and therefore critical to the successful implementation of the technology. These operational processes include:

1. Pretreatment
2. Backwashing
3. Chemical cleaning
4. Integrity testing
5. Post-treatment

3.1. Pretreatment

Pretreatment is typically applied to the feed water prior to entering the membrane system in order to minimize membrane fouling, but in some cases may be used to address other water quality concerns or treatment objectives. Pretreatment is most often utilized to remove foulants, optimize recovery and system productivity, and extend membrane life. Pretreatment may also be used to prevent physical damage to the membranes. Different types of pretreatment can be used in conjunction with any given membrane filtration system, as determined by site-specific conditions and treatment objectives. Pilot testing can be used to compare various pretreatment options, optimize pretreatment, and/or demonstrate pretreatment performance. Several different types of commonly used pretreatment for membrane filtration systems are discussed in the following subsections.

3.1.1. Prefiltration

Prefiltration, including screening or coarse filtration, is a common means of pretreatment for membrane filtration systems that is designed to remove large particles and debris. Prefiltration can either be applied to the membrane filtration system as a whole or to each membrane unit separately. The particular pore size associated with the prefiltration process varies depending on the type of membrane filtration system and the feed water quality. For example, although hollow-fiber MF and UF systems are designed specifically to remove suspended solids, large particulate matter can damage or plug the membranes fibers. For these types of systems the pore size of the selected prefiltration process may range from as small as 100 μm to as large as 3,000 μm or higher (11, 12), depending on the influent water quality and manufacturer specifications. Generally, hollow-fiber MF/UF systems that are operated in an inside-out mode are more susceptible to fiber plugging and thus may require finer prefiltration.

Because NF and RO utilize non-porous semipermeable membranes that cannot be backwashed and are almost exclusively designed in a spiral-wound configuration for municipal water treatment, these systems must utilize much finer prefiltration in order to minimize exposure of the membranes to particulate matter of any size. Spiral-wound modules are highly susceptible to particulate fouling, which can reduce system productivity, create operational problems, reduce membrane life, or in some cases damage or destroy the membranes. If the feed water has turbidity less than approximately 1 NTU or a silt density index (SDI) less than approximately 5, cartridge filters with a pore size range from about

5–20 μm are commonly used for NF/RO prefiltration. However, if the feed water turbidity or SDI exceeds these values, a more rigorous method of particulate removal, such as conventional treatment (including media filtration) (11) or MF/UF membranes, is recommended as pretreatment for NF/RO.

In conventional applications of cartridge filtration technology, high quality source waters are treated, and thus additional prefiltration is typically not required. However, MCF systems may warrant some level of prefiltration, possibly using more conventional cartridge or bag filters, to protect and extend the life of the more effective and more expensive MCF, particularly if the MCF system is applied to surface waters. Since MCF is a new class of technology, pretreatment practices for MCF will likely evolve as this technology is applied. Prefiltration requirements will likely be similar to those for MF/UF. A summary of the typical prefiltration requirements associated with the various types of membrane filtration is presented in Table 8.3.

In some cases, one type of membrane filtration may be used as prefiltration for another. This type of treatment scheme is commonly known as an integrated membrane system (IMS). Typically, this involves the use of MF/UF as pretreatment for NF/RO in applications that require the removal of particulate matter and microorganisms as well as some dissolved contaminants such as hardness, iron and manganese, or disinfection byproduct (DBP) precursors. One of the most significant advantages of an IMS treatment scheme is that the MF/UF filtrate is of consistently high quality and allows the NF/RO system to maintain stable operation by reducing the rate of membrane fouling.

3.1.2. *Chemical Conditioning*

Chemical conditioning may be used for a number of pretreatment purposes, including pH adjustment, disinfection, biofouling control, scale inhibition, or coagulation. Some type of chemical conditioning is almost always used with NF/RO systems, most often the addition of an acid (to reduce the pH) or a proprietary scale inhibitor recommended by the membrane manufacturer to prevent the precipitation of sparingly soluble salts such as calcium carbonate ($CaCO_3$), barium sulfate ($BaSO_4$), strontium sulfate ($SrSO_4$), or silica species (e.g., SiO_2).

Software programs that simulate NF/RO scaling potential based on feed water quality are available from the various membrane manufactures. In some cases, such as for those NF and RO membranes manufactured from cellulose acetate, the feed water pH must also

Table 8.3
Typical membrane system prefiltration requirements

Membrane system		Prefiltration requirements	
Classification	Configuration	Size (μm)	Type(s)
Membrane cartridge filtration (MCF)	Cartridge	300–3,000	Strainers; bag filters
Microfiltration (MF)/ultrafiltration (UF)	Hollow-fiber, inside-out	100–300	Strainers; bag filters
	Hollow-fiber, outside-in	300–3,000	Strainers; bag filters
Nanofiltration (NF)/reverse osmosis (RO)	Spiral-wound	5–20	Cartridge filters

be adjusted to maintain the pH within an acceptable operating range to minimize the hydrolysis (i.e., chemical deterioration) of the membrane. The addition of chlorine or other disinfectants may also be used as pretreatment for primary disinfection or to control biofouling. However, because some NF/RO membrane materials are readily damaged by oxidants, it is important that any disinfectants added upstream be quenched prior to contact with such membranes.

A number of different chemicals may be added as pretreatment for MF or UF, depending on the treatment objectives for the system. For example, lime may be added for softening applications; coagulants may be added to enhance removal of TOC with the intent of minimizing the formation of DBPs or increasing particulate removal; disinfectants may be applied for either primary disinfection or biofouling control; and various oxidants can be used to precipitate metals such iron and manganese for subsequent filtration. However, as with NF and RO membranes, it is important to ensure that the applied pretreatment chemicals are compatible with the particular membrane material used. As with conventional media filters, pre-settling may be used in conjunction with pretreatment processes such as coagulation and lime softening. While a MF/UF system may be able to operate efficiently with the in-line addition of lime or coagulants, pre-settling in association with these pretreatment processes can enhance membrane flux and increase system productivity by minimizing backwashing and chemical cleaning frequency.

Although chemical pretreatment is generally not associated with cartridge filters, if the filters are compatible, disinfectants may be added upstream of MCF systems to maximize the time for primary disinfection (i.e., CT). Because cartridge filters are designed to be disposable and are generally not backwashed, pretreatment chemicals such as lime or coagulants could rapidly foul the cartridges and thus are not applied upstream these systems.

With any form of chemical pretreatment, it is very important to understand whether any chemical under consideration for use is compatible with the membrane material. In addition to irreversible fouling and/or physical damage to the membranes, the use of an incompatible chemical may void the manufacturer's warranty. Some chemicals such as oxidants can be quenched upstream, while others such as coagulants and lime cannot be counteracted prior to membrane exposure. In general, most NF/RO membranes and some MF/UF membranes are not compatible with disinfectants and other oxidants. However, some MF/UF membranes that are incompatible with stronger oxidants such as chlorine may have a greater tolerance for weaker disinfectants such as chloramine, which may allow for a measure of biofouling control without damaging the membranes. Certain types of both MF/UF and NF/RO membranes require operation within a certain pH range. Coagulants and lime are incompatible with many NF/RO membranes but are typically compatible with most types of MF/UF membranes. Polymers are incompatible with NF/RO membranes, and generally not compatible with MF/UF membranes, as well, although this depends to some degree on the charge of the polymer relative to the charge associated with the membrane. A polymer with a charge opposite to that of the membrane is likely to cause rapid and potentially irreversible fouling. It is critical to consult with the membrane manufacturer prior to implementing any form of chemical pretreatment.

3.2. Backwashing

The backwash process for membrane filtration systems is similar in principle to that for conventional media filters and is designed to remove contaminants accumulated on the membrane surface. Each membrane unit is backwashed separately and in a staggered pattern so as to minimize the number of units in simultaneous backwash at any given time. During a backwash cycle, the direction of flow is reversed for a period ranging from about 30 s to 3 min. The force and direction of the flow dislodge the contaminants at the membrane surface and wash accumulated solids out through the discharge line. Membrane filtration systems are generally backwashed more frequently than conventional media filters, with intervals of approximately 15–60 min between backwash events. Typically the membrane backwash process reduces system productivity in the range of 5–10% due to the volume of filtrate used during the backwash operation.

Of the various types of membrane filtration systems, backwashing is almost exclusively associated with hollow-fiber MF and UF processes. Some MCF systems may also be backwashed, depending on how the system is designed and configured; however, it is more common to replace the cartridge filters, which are manufactured to be disposable, when they become fouled.

Backwashing is conducted periodically according to manufacturer specifications and site-specific considerations. Although more frequent backwashing maintains higher fluxes, this benefit is counterbalanced by the decrease in system productivity. In general, a backwash cycle is triggered when a performance-based benchmark is exceeded, such as a threshold for operating time, volumetric throughput, increase in TMP, and/or flux decline. Ideally, the backwash process restores the TMP to its baseline (i.e., clean) level; however, most membranes exhibit a gradual increase in the TMP that is observed after each backwash, indicating that the accumulation of foulants cannot be removed by the backwash process alone. These foulants are addressed through chemical cleaning. Some systems also utilize pressurized air and/or chlorine (if a compatible membrane material is used) in combination with filtered water to remove solids, provide a measure of pathogen inactivation and biofouling control, and improve overall backwash effectiveness. A disinfectant such as chlorine may be added at a frequency ranging from every backwash to about once per day. If chloramines are added upstream of the membrane process, the benefits of adding chlorine during the backwash process should be weighed against the cost of breakpoint chlorination to achieve a free chlorine residual and possible subsequent ammonia addition to regenerate a chloramine residual for secondary disinfection in the distribution system.

Some manufacturers have implemented backwash strategies for their proprietary MF/UF systems that involve the use of chemicals other than chlorine (such as acids, bases, surfactants, or other proprietary chemicals). These strategies are used as a method to enhance membrane flux and extend intervals between chemical cleanings, thus lowering the cost of operation. In these cases state regulators may require enhanced cross-connection control measures for the backwash piping similar to those used for chemical cleaning systems, depending on the backwash frequency and the type(s), and concentration(s) of chemical(s) used. Special provisions for rinsing the membranes at the end of the backwash process may also be required. Because the design of spiral-wound membranes generally does not permit reverse flow, NF and RO membrane systems are not backwashed. For these systems, membrane fouling is controlled primarily with chemical cleaning, as well as through flux control and crossflow velocity. The inability of spiral-wound membranes to be backwashed is one reason that NF and RO membranes are seldom applied to directly treat water with high turbidity and/or suspended solids.

3.3. *Chemical Cleaning*

Chemical cleaning is another means of controlling membrane fouling, particularly those foulants such as inorganic scaling and some forms of organic and biofouling that are not removed via the backwash process. As with backwashing, chemical cleaning is conducted for each membrane unit separately and is typically staggered to minimize the number of units undergoing cleaning at any time. While chemical cleaning is conducted on both MF/UF and NF/RO systems, because non-porous, semi-permeable membranes cannot be backwashed, chemical cleaning represents the primary means of removing foulants in NF/RO systems.

Although cleaning intervals may vary widely on a system-by-system basis, the gradual accumulation of foulants makes eventual chemical cleaning virtually inevitable. Membrane cartridge filters are an exception; however, in that, cartridge filters are usually designed to be disposable and thus are typically not subject to chemical cleaning.

As with backwashing, the goal of chemical cleaning is to restore the TMP of the system to its baseline (i.e., clean) level. Any foulant that is removed by either the backwash or chemical cleaning process is known as reversible fouling. Over time, membrane processes will also typically experience some degree of irreversible fouling, which cannot be removed through either chemical cleaning or backwashing. Irreversible fouling occurs in all membrane systems, albeit over a wide range of rates, and eventually necessitates membrane replacement.

There are a variety of different chemicals that may be used for membrane cleaning, and each is generally targeted to remove a specific form of fouling. For example, citric acid is commonly used to dissolve inorganic scaling, and other acids may be used for this purpose as well. Strong bases such as caustic are typically employed to dissolve organic material. Detergents and surfactants may also be used to remove organic and particulate foulants, particularly those that are difficult to dissolve. Chemical cleaning may also utilize concentrated disinfectants such as a strong chlorine solution to control biofouling. Due to the variety of foulants that are present in many source waters, it is often necessary to use a combination of different chemicals in series to address multiple types of fouling. The various types of chemical cleaning agents used are summarized in Table 8.4.

Numerous proprietary cleaning chemicals are also available, and these specialty cleaning agents may be useful in cases in which more conventional chemicals are ineffective. For example, under some circumstances, enzymatic cleaners have been found to be effective at dissolving organic contaminants. Chemical cleaning options are more limited for membranes that cannot tolerate oxidants and/or extreme pH levels. A chemical cleaning regimen may be specified by the manufacturer or identified based on site-specific pilot testing and source

Table 8.4
Chemical cleaning agents

Category	Chemicals commonly used	Typical target contaminant(s)
Acid	Citric acid Hydrochloric acid (HCl)	Inorganic scale
Base	Caustic (NaOH)	Organics
Oxidants/disinfectants	Sodium hypochlorite (NaOCl) Chlorine (Cl_2) gas Hydrogen peroxide	Organics; biofilms
Surfactants	Various	Organics; inert particles

water quality analyses to determine the prevalent form(s) of fouling experienced at a particular facility.

The term clean-in-place (CIP) is often used to describe the chemical cleaning process, since it is typically conducted while the membrane modules remain in the membrane unit (i.e., in situ). The cleaning process generally involves recirculating a cleaning solution through the membrane system at high velocities (to generate scouring action) and elevated temperature (to enhance the solubility of the foulants). A soak cycle follows the recirculation phase. After the soak cycle is completed, the membrane system is flushed to remove residual traces of the cleaning solution(s). The process may be repeated using a different cleaning solution to target different types of foulants until the membranes have been successfully cleaned. Under some circumstances, the use of softened or demineralized water may be required for the cleaning solution or as rinse water.

While backwashing may be conducted at more regular intervals or on a routine basis, chemical cleaning is typically conducted only when necessary. Chemical cleaning is generally necessary for MF and UF systems when the ability of periodic backwashing to restore system productivity (i.e., increase flux and reduce TMP) reaches a point of diminishing returns. For NF and RO systems, a 10–15% decline in temperature- and pressure-normalized flux or about a 50% increase in differential pressure may indicate the need for chemical cleaning. Delaying necessary chemical cleaning can accelerate irreversible fouling, reduce production capacity, and shorten membrane life.

Isolating the cleaning chemicals from the treated (i.e., filtered) water is an important consideration for membrane filtration systems. It is important to properly flush the membrane unit after the cleaning process and before restarting the filtration cycle. The flushed water should be diverted to waste until filtrate water quality parameters (such as turbidity for MF/UF systems and pH for NF/RO systems) return to normal production mode levels. Note that the volume of flushed water can be significant in cases in which surfactants are used.

For MF/UF systems, it is common to recycle as much as 90% of the cleaning chemicals for reuse, thus reducing the volume of chemical waste as well as the cost associated with cleaning. Recycling cleaning solutions is less common with NF/RO systems, since the used cleaning solutions accumulate dissolved constituents with repeated use, diminishing the effectiveness of the cleaning agents.

3.4. *Integrity Testing*

Integrity testing is a means of determining whether or not a membrane system is "integral" or free of any breaches, leaks, or defects that might allow unfiltered water to bypass the membrane barrier, passing contaminants that are normally removed. As with both backwashing and chemical cleaning, integrity testing is conducted on each membrane unit (or smaller division in some cases) separately and is typically staggered to minimize the number of units simultaneously undergoing testing. The use of periodic or continuous integrity testing and monitoring methods allows ongoing operational verification that the membranes are performing as expected based on their established exclusion characteristics. This verification is an essential component of any membrane filtration system, particularly when the constituents of concern are pathogenic microorganisms.

There are a number of potential modes of failure associated with membrane filtration systems that would result in an integrity breach. For example, membranes may become damaged via exposure to oxidants, pH levels outside the recommended range, or other chemicals or operating conditions to which the membranes are sensitive. In addition, membranes may break or puncture as a result of extreme pressure, scratches or abrasions, or operational fatigue over time. Spiral-wound membranes can be damaged at glue lines if the pressure on the filtrate side of the membrane exceeds that on the feed side. Factory imperfections such as glue line gaps or potting defects may cause integrity breaches as well. Improper installation of membrane modules can also create integrity problems at o-rings or interconnections.

The various types of integrity tests are generally divided into two categories: direct and indirect test methods. Direct methods are physical tests that are applied specifically to the membrane unit to detect integrity breaches and/or determine their sources. Indirect methods are surrogate measures of integrity that involve filtrate water quality monitoring such that a significant decline in filtrate quality may indicate an integrity problem. Both direct integrity testing and continuous indirect integrity monitoring are required for compliance with US EPA regulations (13–17).

Direct test methods are non-destructive techniques that can be used to identify and/or isolate leaks. While these methods yield direct information about membrane integrity, they cannot be conducted continuously while the membrane filtration system is in operation. Thus, the longer and more often that a direct test is conducted, the greater the impact on the overall system productivity. The minimum frequency requirements for direct testing vary among different states, although daily testing is commonly recommended as good practice and is required for those utilities using membrane filtration for US EPA compliance.

Although there are a number of types of direct integrity tests, the most common method is the pressure decay test, which measures the rate of pressure loss across the membrane relative to a maximum acceptable threshold. Almost all currently available proprietary MF and UF systems are designed with the capability to conduct this test automatically at regular intervals. MCF systems may also be equipped to conduct automatic pressure decay tests. A similar test using a vacuum can also be conducted on NF and RO membranes, although currently spiral-wound systems are generally not designed to conduct this testing automatically during the course of normal operation. Other on-line direct test methods measure the flow of air or water from integrity breaches, or the concentration of either spiked inert particles or a molecular marker.

The various indirect methods consist primarily of water quality monitoring practices that are common throughout the water treatment industry, such as turbidity monitoring (18) and particle counting (19–22). Particulate-based indirect monitoring techniques are applicable to all membrane classes used for filtration. For NF/RO membranes capable of removing dissolved contaminants, other parameters such as conductivity or sulfide may also potentially be used as surrogate measures of membrane integrity. The effectiveness of indirect methods is a function of the ability of membrane filtration to produce very consistent, high quality filtrate, such that a marked decline in filtrate quality is likely to indicate an integrity problem. Although indirect methods are not as sensitive as the direct methods for detecting integrity breaches, the primary advantage of indirect methods is that they allow continuous monitoring, a capability not possible with currently available direct testing methods. Typically, if an indirect method indicates a potential integrity problem, a direct test will be conducted to determine more conclusively whether or not a breach has occurred.

3.5. *Post-Treatment*

Post-treatment for membrane filtration systems typically consists of chemical conditioning and/or disinfection and is typically applied to combined filtrate. Most chemical conditioning is associated with NF and RO systems, because the removal of dissolved constituents that is achieved by these processes has a more significant impact on water chemistry than the filtering of suspended solids alone. For example, because NF and RO pretreatment often includes acid addition to lower the pH and, consequently, increase the solubility of potential inorganic foulants, a portion of the carbonate and bicarbonate alkalinity in the water is converted to aqueous carbon dioxide, which is not rejected by the membranes. The resulting filtrate can thus be corrosive given the combination of a low pH, elevated carbon dioxide levels, and minimal buffering capacity of the filtrate. Other dissolved gases such as hydrogen sulfide, will also readily pass through the semi-permeable membranes, further augmenting the corrosivity of the filtrate and potentially causing turbidity and taste and odor problems. As a result, the primary goal of chemical conditioning is the stabilization of NF/RO filtrate with respect to pH, buffering capacity, and dissolved gases.

Degasification is commonly achieved via packed tower aeration (i.e., air stripping) (23). Air stripping also increases dissolved oxygen levels, which may be very low in the case of an anaerobic groundwater source. The pH of the water may subsequently be readjusted to typical finished water levels (i.e., approximately 6.5–8.5) by adding a base such as lime or caustic. Alkalinity (e.g., in the form of sodium bicarbonate) may also be added, if necessary, to increase the buffering capacity of the water. Alternatively, if the pH is raised prior to degasification (thus converting the dissolved carbon dioxide to bicarbonate), much of the alkalinity may be recovered. However, this post-treatment strategy also converts any dissolved hydrogen sulfide gas into dissociated sulfide, which may readily react with other dissolved species to produce sparingly soluble sulfide compounds that may precipitate.

Because MF, UF, and MCF systems do not directly affect the pH or remove alkalinity, these processes do not generally require chemical conditioning to stabilize the filtrate.

While the use of membrane filtration does not specifically necessitate disinfection post-treatment as a result of process considerations, the need for post-disinfection is generally required by regulation for primary and/or secondary disinfection. However, in some states, the use of membrane filtration may reduce primary disinfection (i.e., CT) requirements, thus helping to control DBP formation. Because membrane filtration is often the last major process in the treatment scheme, it is common to apply a disinfectant to the filtrate prior to entry into a clearwell and/or the distribution system (24). This application is particularly important if disinfectants were either neutralized or not added at all prior to the membrane filtration process to avoid damaging oxidant-intolerant membranes.

For NF and RO membrane processes, if the disinfectant is applied prior to filtrate pH adjustment, post-disinfection may have the additional benefit of oxidizing sulfide to sulfate, thus reducing the potential for both sulfide precipitation and taste and odor concerns. Corrosion inhibitors may also be added prior to distribution, particularly for NF and RO systems that produce more corrosive water.

4. BASIC PRINCIPLES OF SYSTEM DESIGN AND OPERATION

The material presented in this section is intended to provide an overview of the basic principles underlying membrane filtration system design and operation. Although all of the types of membrane filtration (i.e., MF, UF, NF, RO, and MCF) utilize pressure (or vacuum) as a driving force, there are fundamental differences in the models used to describe systems using porous membranes (MF, UF, and MCF) and semi-permeable membranes (NF and RO). The basic principles of these respective models, along with some general concepts that are applicable to all membrane filtration systems, are discussed in this section.

4.1. General Concepts

There are a number of general concepts that are applicable to all types of pressure-driven membrane filtration systems and which serve as the underlying basic principles for system design and operation. These concepts include

1. Flux
2. Recovery
3. Flow balance

Membrane filtration system throughput or productivity is typically characterized by the system flux, which is defined as the filtrate flow per unit of membrane filtration area, as shown in Eq. (1):

$$J = \frac{Q_{\mathrm{p}}}{A_{\mathrm{m}}}, \tag{1}$$

where J is the flux; gal/ft^2/d, Q_{p} is the filtrate flow, gpd; and A_{m} is the membrane surface area, ft^2.

The recovery of a membrane unit is defined as the amount of feed flow that is converted to filtrate flow, expressed as a decimal percent, as shown in Eq. (2):

$$R = \frac{Q_p}{Q_f}, \tag{2}$$

where R is the recovery of the membrane unit, decimal percent, Q_p is the filtrate flow produced by the membrane unit, gpd, and Q_f is the feed flow to the membrane unit, gpd.

The recovery does not account for the use of filtrate for routine maintenance purposes (such as chemical cleaning or backwashing) or lost production during these maintenance operations. Because the definition of recovery is not necessarily consistent throughout the water treatment industry, it is important to identify how recovery is defined in any particular discussion. However, the use of the term recovery as defined in this discussion by Eq. (2) is consistent with the applicable US EPA regulations and the *Membrane Filtration Guidance Manual* (1). Note that for some types of membrane systems, particularly those that operate in suspension mode that can be modeled as plug flow reactors (PFRs), recovery can also be defined as a function of position within a membrane unit. This is simply a variation of Eq. (2) for systems in which the filtrate flow increases in the direction of feed flow through the membrane unit, thus increasing the recovery in direct proportion. The limit of this recovery in the direction of flow (i.e., the recovery at the furthest position in the unit) is equivalent to the overall membrane unit recovery, as defined in Eq. (2).

A general flow balance that can be applied to all membrane filtration systems is shown in Eq. (3):

$$Q_f = Q_c + Q_p, \tag{3}$$

where Q_f is the feed flow to the membrane unit, gpd, Q_c is the concentrate flow from the membrane unit, gpd, and Q_p is the filtrate flow from the membrane unit, gpd.

Note that the concentrate (i.e., bleed or reject) flow, Q_c, is zero for systems operating in deposition (i.e., dead-end) mode. For the purpose of sizing a membrane filtration system, it may be desirable to account for the additional filtered water used for both backwashing and chemical cleaning in the determination of the filtrate flow, Q_p. Similarly, an estimate of the total required feed flow Q_f to the system should incorporate any raw water that may be used in these processes.

4.2. MF, UF, and MCF Processes

The driving force for the transport of water across a microporous membrane – that utilized by MF, UF, and MCF processes – is a pressure gradient across the membrane, or the TMP. The TMP is defined by the pressure on the feed side of the membrane minus the filtrate pressure, commonly called the backpressure, as shown in Eq. (4):

$$\mathrm{TMP} = P_f - P_p, \tag{4}$$

where TMP is the trans-membrane pressure, psi, P_f is the feed pressure, psi, and P_p is the filtrate pressure (i.e., backpressure), psi.

For systems that operate in suspension mode and thus utilize a concentrate stream that is wasted or recirculated, the pressure on the feed side of the membrane is not constant, but can instead be approximated by a linear pressure gradient from the feed inlet to the concentrate outlet. In this case, the pressure on the feed side of the membrane may be represented by the average of the feed and concentrate pressures, as shown in Eq. (5):

$$\mathrm{TMP} = \frac{P_{\mathrm{f}} + P_{\mathrm{c}}}{2} - P_{\mathrm{c}}, \tag{5}$$

where

TMP = trans-membrane pressure, psi
P_{f} = feed pressure, psi
P_{c} = concentrate pressure, psi
P_{p} = filtrate pressure (i.e., backpressure), psi

The resistance to flow acting in opposition to the driving force and inhibiting the transport of water across the membrane can also be quantified. This resistance has two components: the intrinsic resistance of the membrane and the resistance attributable to the accumulated foulant layer at the membrane surface at any point during operation. The total resistance is represented by the sum of these two components, as shown in Eq. (6):

$$R_{\mathrm{t}} = R_{\mathrm{m}} + R_{\mathrm{f}}, \tag{6}$$

where R_{t} is the total membrane resistance, psi/gal/ft^2/d/cp, R_{m} is the intrinsic membrane resistance, psi/gal/ft^2/d/cp, and R_{f} is the resistance of the foulant layer, psi/gal/ft^2/d/cp.

While the intrinsic resistance of the membrane should remain constant for all practical purposes and can generally be obtained from the membrane manufacturer (if necessary), the increase in fouling during normal operation and the decrease in fouling as a result of backwashing and chemical cleaning causes the fouling resistance to fluctuate. If the total membrane resistance is known, the flux can be calculated as a function of the TMP and water viscosity, as shown in Eq. (7):

$$J_T = \frac{\mathrm{TMP}}{(R_{\mathrm{t}})(\mu_T)}, \tag{7}$$

where

J_T = flux, gal/ft^2/d
TMP = trans-membrane pressure, psi
R_{t} = total membrane resistance, psi/gal/ft^2/d/cp
μ_T = viscosity of water, cp

Because the viscosity of water increases with decreasing temperature, larger TMPs (by application of increased pressure or vacuum) are required to maintain constant flux. Values for water viscosity can be found in the literature or approximated using the empirical relationship expressed in Eq. (8):

$$\mu_T = 1.784 - 0.0575\,T + 0.0011\,T^2 - 10^{-5}\,T^3, \tag{8}$$

where μ_T is the viscosity of water at temperature T, cp, T is the water temperature, °C.

Since water temperature can have a significant impact on flux, it is common practice to "normalize" the flux to a reference temperature during operation for the purposes of monitoring system productivity independent of changes in water temperature. For MF/UF and MCF processes, a reference temperature of 20°C is typically used for convenience, since the viscosity of water is approximately 1 cp at 20°C. For constant TMP and total membrane resistance, Eq. (9) can be used to illustrate the relationship between the normalized flux and viscosity at 20°C and the actual flux and viscosity at a given temperature of interest T:

$$J_{20}(\mu_{20}) = J_T(\mu_T), \tag{9}$$

where

J_{20} = normalized flux at 20°C, gal/ft^2/d
μ_{20} = viscosity of water at 20°C, cp
J_T = actual flux at temperature T, gal/ft^2/d
μ_T = viscosity of water at temperature T, cp

Substituting the value of 1 cp for the viscosity at 20°C (μ_{20}) and Eq. (8) for the viscosity of water at temperature T (μ_T), yields an expression for normalized flux at 20°C as a function of the actual flux and the temperature, as shown in Eq. (10):

$$J_{20} = J_T(1.784 - 0.0575\ T + 0.0011\ T^2 - 10^{-5}\ T^3), \tag{10}$$

where J_{20} is the normalized flux at 20°C, gal/ft^2/d, J_T is the actual flux at temperature T, gal/ft^2/d, and T is the water temperature, °C.

It is important to note that the normalized flux (J_{20}) does not represent an actual operating condition. This term simply represents what the flux would be at 20°C for a constant TMP and total membrane resistance. Thus, changes in the value of J_{20} during the course of normal operation are indicative of changes in pressure and/or membrane resistance due to fouling. If values for viscosity are known, the polynomial expression for viscosity as a function of temperature in Eq. (10) may be simplified to a temperature correction factor (TCF). For a MF, UF, or MCF process, the TCF is defined as the ratio of the viscosity at temperature T to the viscosity at 20°C, as shown in Eq. (11):

$$\text{TCF} = \frac{\mu_T}{\mu_{20}}, \tag{11}$$

where TCF is the temperature correction factor, dimensionless, μ_T is the viscosity of water at temperature T, cp, and μ_{20} is the viscosity of water at 20°C, cp.

Note that the term TCF is often used generically to refer to any type correction factor used to adjust a parameter for temperature. Thus, the specific equation for the TCF may vary depending on the parameter to which it is applied. For example, in the context of membrane filtration, the TCF applied to reference MF, UF, and MCF flux to a standard temperature, as defined in Eq. (11), is different than that applied to NF and RO flux to a standard temperature, as shall be shown in Eq. (15). Thus, it is important to always consider the context in which the term TCF is used.

Because the TCF is a dimensionless ratio, the values for viscosity can be expressed in any convenient and consistent units. Thus, the temperature-normalized flux can be expressed in simplified terms, as shown in Eq. (12):

$$J_{20} = J_T(\text{TCF}), \tag{12}$$

where J_{20} is the normalized flux at 20°C, gal/ft^2/d, J_T is the actual flux at temperature T, gal/ft^2/d, and TCF is the temperature correction factor, dimensionless.

Generally, in order to identify changes in productivity (as measured by flux) that are specifically attributable to membrane fouling, it is desirable to normalize the flux for pressure as well as temperature, as shown in Eq. (13). Note that the temperature- and pressure-normalized flux is often referred to as either the specific flux or permeability.

$$M = \frac{J_{20}}{\text{TMP}}, \tag{13}$$

where M is the temperature- and pressure-normalized flux, gal/ft^2/d/psi, J_{20} is the normalized flux at 20°C, gal/ft^2/d, and TMP is the trans-membrane pressure, psi.

4.3. *NF and RO Processes*

As with the microporous MF, UF, and MCF membranes, the driving force for the transport of water across a semi-permeable membrane – such as that utilized by NF and RO processes – is a pressure gradient across the membrane. However, because NF and RO processes reject dissolved salts, the resulting osmotic pressure gradient, which acts against the transport of water from the feed to the filtrate side of the membrane, must also be taken into account. Typically, the osmotic pressure gradient is approximated from the concentration of total dissolved solids (TDS) on the feed and filtrate sides of the membrane. The corrected driving force across semi-permeable membrane is termed the net driving pressure (NDP) and can be calculated using Eq. (14) (5):

$$\text{NDP} = \frac{P_f + P_c}{2} - 0.01\left[\frac{\text{TDS} + \text{TDS}_c}{2} - \text{TDS}_p\right] - P_p, \tag{14}$$

where:

NDP = net driving pressure, psi
P_f = feed pressure, psi
P_c = concentrate pressure, psi
TDS_f = feed TDS concentration, mg/L
TDS_c = concentrate TDS concentration, mg/L
TDS_p = filtrate TDS concentration, mg/L
P_p = filtrate pressure (i.e., backpressure), psi

Equation (14) can be considered as three distinct components. The first term represents the average pressure on the feed side of the membrane; the second term represents the average osmotic backpressure; and the third term represents the filtrate backpressure. The conversion factor of 0.01 in the osmotic pressure term comes from a widely used rule of thumb for fresh and brackish waters indicating that there is approximately 1 psi of osmotic pressure for every 100 mg/L of TDS. In many cases the filtrate TDS concentration (TDS_p) is small and can be neglected.

While the flux associated with MF, UF, and MCF systems is typically referenced to a temperature of 20°C for the purposes of assessing operational performance, it is common to reference the flux associated with NF and RO systems to 25°C (298 K). Accordingly, the appropriate TCF for NF and RO systems is shown in Eq. (15):

$$\mathrm{TCF} = \exp\ U\left[\frac{1}{T+273} - \frac{1}{298}\right], \tag{15}$$

where TCF is the temperature correction factor, dimensionless, T is the water temperature, °C, U is the membrane-specific manufacturer-supplied constant, 1/°K.

Once the TCF has been determined, the flux normalized to 25°C can be calculated according to Eq. (16):

$$J_{25} = J_T(\mathrm{TCF}), \tag{16}$$

where J_{25} is the normalized flux at 25°C, gal/ft^2/d, J_T is the actual flux at temperature T, gal/ft^2/d, TCF is the temperature correction factor, dimensionless.

As with MF, UF, and MCF systems, it is important to note that the normalized flux (J_{25}) for NF and RO systems does not represent an actual operating condition. This term simply represents what the flux would be at 25°C for the purposes of comparing membrane performance independent of temperature-related affects. Similarly, it is also common to normalize the flux for pressure in order to identify changes in productivity that are attributable to fouling, as shown in Eq. (17):

$$M = \frac{J_{25}}{\mathrm{NDP}}, \tag{17}$$

where M is the temperature- and pressure-normalized flux, gal/ft^2/d/psi, J_{25} is the normalized flux at 25°C, gal/ft^2/d, and NDP is the net driving pressure, psi.

4.4. Hydraulic Configurations

The term hydraulic configuration is used to describe the manner in which the feed water and associated suspended solids are processed by a membrane filtration system. Although there are a number of different hydraulic configurations in which the various membrane filtration systems can operate, each of these configurations can be categorized into one of two basic modes of operation:

1. Deposition mode
2. Suspension mode

The hydraulic configuration of a system is determined from operational conditions such as backwash, concentrate flow, and recycle flow, where applicable

One of the most important implications of a system's hydraulic configuration is its impact on the degree to which suspended solids are concentrated on the feed side of the membrane.

This concentration effect is characterized by the volumetric concentration factor (VCF), a dimensionless parameter representing the ratio of the concentration of suspended solids on the feed side of the membrane relative to that of the influent feed to the membrane filtration process, as shown in Eq. (18):

$$\mathrm{VCF} = \frac{C_{\mathrm{m}}}{C_{\mathrm{f}}}, \tag{18}$$

where VCF is the volumetric concentration factor, dimensionless, C_{m} is the concentration of suspended solids maintained on the feed side of the membrane, mg/L, and C_{f} is the concentration of suspended solids in the influent feed water to the membrane system, mg/L.

By definition, the VCF is equal to 1 for a system that does not concentrate suspended solids on the feed side of the membrane (i.e., $C_{\mathrm{m}} = C_{\mathrm{f}}$); these are defined as deposition mode systems. However, some hydraulic configurations concentrate suspended solids on the feed side of the membrane to degrees much greater than the influent feed concentration, with a corresponding VCF greater than 1; these are the suspension mode systems. Consequently, the VCF can significantly affect the quantity of particulate matter that can flow through an integrity breach, and thus must be considered in the determination of direct integrity test method sensitivity. For example, if a system concentrates particulate matter by a factor of 10 (i.e., VCF = 10), then the direct integrity test used to monitor the system must be ten times more sensitive than a test applied to a system that does not concentrate particulate matter (i.e., VCF = 1), in order to demonstrate the same log removal value (LRV).

The purpose of this subsection is to indicate the various hydraulic configurations in which membrane filtration systems can operate and present the associated equations used to determine the respective VCFs. The various hydraulic configurations are based on three assumptions:

1. In the absence of a hydraulic force tangential to the membrane surface, particulate matter in the feed stream is deposited on the membrane and held in place by the TMP.
2. In the presence of a hydraulic force tangential to the membrane surface, all of the particulate matter remains in suspension, resulting in elevated concentrations of suspended particulate matter on the feed side of the membrane. This increase in concentration is characterized by the VCF, which can vary as a function of position and/or time for various hydraulic configurations.
3. The membrane is a complete barrier to the passage of the particulate contaminants.

Although the majority of systems utilize one of the hydraulic configurations given below, there may be unusual cases in which a system-specific analysis is necessary to characterize the VCF. For example, some systems that are designed to operate in deposition mode may still exhibit some degree of particle suspension, and conversely, some degree of particle deposition may occur in a system operating primarily in suspension mode. In such cases in which characteristics of both types of hydraulic configurations may be observed, the VCF should be calculated using the most conservative applicable assumptions that result in the highest anticipated VCF values. (Note that the most conservative condition for a particular system is that which results in the highest estimated concentration of suspended particulate matter). As an alternative to the theoretical calculations, the VCF may be measured experimentally under realistic operating conditions. This approach may be advantageous for systems that may not necessarily be well described by mathematical modeling. In addition to both the modeling and experimental approaches, there may be some cases in which a manufacturer of a proprietary membrane filtration system has developed a system-specific method for determining the VCF.

4.4.1. Deposition Mode

Membrane filtration systems operating in deposition mode utilize no concentrate stream such that there is only one influent (i.e., the feed) and one effluent (i.e., the filtrate) stream. These systems are also commonly call "dead-end" or "direct" filtration systems and are analogous to conventional granular media filters in terms of hydraulic configuration. In the deposition mode of operation, contaminants suspended in the feed stream accumulate on the membrane surface and are held in place by hydraulic forces acting perpendicular to the membrane, forming a cake layer, as illustrated in Fig. 8.3.

In a deposition mode hydraulic configuration, the concentration of suspended material on the feed side of the membrane (C_m) is assumed to be equivalent to the concentration of influent feed stream (C_f), independent of time or position in the membrane system, as the suspended contaminants are removed from the process stream and deposited in the accumulated cake layer. Therefore, all systems operating in deposition mode have a VCF equal to one. MCF and most hollow-fiber MF and UF systems operate in deposition mode. Typically, the accumulated solids are removed from MF/UF systems by backwashing, while most MCF systems simply operate until the accumulated solids reduce the flow and/or TMP to an unacceptable level, at which point the membrane cartridge is replaced.

Some MF/UF systems utilize a periodic "back-pulse" – a short interval of reverse flow (which may include air and/or the addition of small doses of oxidants) designed to dislodge particles from the membrane surface without removing these solids from the system. This process re-suspends particles, effectively concentrating the suspended solids in the feed near the membrane surface and increasing the potential for pathogens or other particulate to pass through an integrity breach and contaminate the filtrate. Consequently, systems that do not utilize a concentrate stream, but still practice back-pulsing may be more appropriately and conservatively modeled as operating in suspension mode.

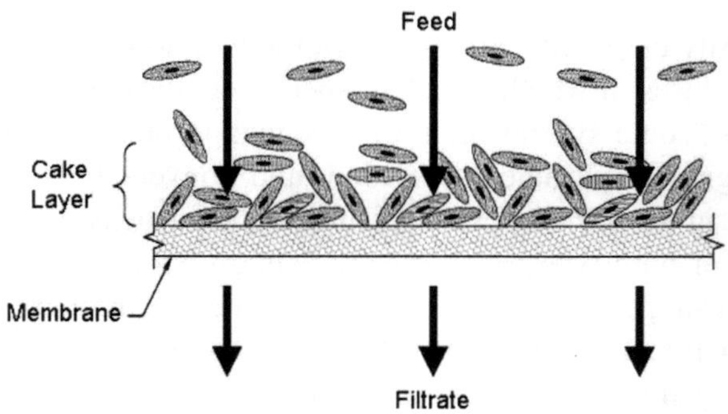

Fig. 8.3. Conceptual illustration of deposition mode operation.

4.4.2. Suspension Mode

In membrane filtration systems that operate in suspension mode, a scouring force using water and/or air is applied parallel (i.e., tangential) to the membrane surface during the production of filtrate in a continuous or intermittent manner, as illustrated in Fig. 8.4. The objective of operating in this mode is to minimize the accumulation of contaminants at the membrane surface or boundary layer, thus reducing fouling. As shown in Eq. (18), the VCF quantifies the increase in the feed side concentration of suspended solids relative to that of the influent feed stream that occurs in a suspension mode of operation.

The three most common suspension mode hydraulic configurations are the PFR model, the crossflow model, and the continuous stirred tank reactor (CSTR) model. Systems operating under a crossflow hydraulic configuration may be further categorized as either small volume or large volume systems, since the volume affects the manner in which suspended solids are concentrated in such a system. Depending on the particular hydraulic configuration, the VCF may vary temporally or spatially. In a PFR, the VCF increases in the direction of feed flow as a function of position in the system. Examples of such systems include spiral-wound NF/RO systems and vacuum-driven MF/UF systems submerged in tanks with large length-to-width ratios. In a CSTR or crossflow reactor, the VCF increases with time over the course of a filtration cycle.

The objective of crossflow filtration is to maintain a high scour velocity across the membrane surface to minimize particle deposition and membrane fouling. Crossflow membrane processes operate in an unsteady-state manner, in which suspended solids accumulate on the feed side of the membrane over the course of a filtration cycle. Thus, in crossflow systems, the VCF varies as a function of time. At the end of each filtration cycle, the membrane unit is backwashed to remove the accumulated solids. Crossflow filtration has traditionally been used in conjunction with inside-out hollow-fiber membrane processes to increase the scouring velocity in the fiber lumen in order to minimize fouling.

The CSTR (also known as feed-and-bleed in some applications) hydraulic configuration is similar to that of a crossflow system in that the particulate matter is held in suspension and increases in concentration on the feed side of the membrane as a function of time. However, the CSTR incorporates a continuous concentrate waste stream (also referred to as the reject or

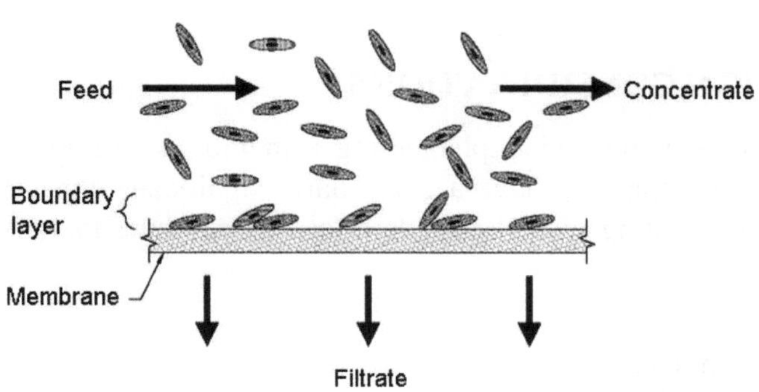

Fig. 8.4. Conceptual illustration of suspension mode operation.

bleed stream) that removes suspended solids from the system. Since the solids are continuously removed from the system, steady-state operation is achieved when the rate at which solids are removed with the concentrate stream is equal to the rate at which solids enter the system with the feed water. Although a PFR also utilizes a continuous concentrate stream and operates at steady state, the CSTR model assumes complete mixing and thus a uniform concentration of suspended solids on the feed side of the membrane. A PFR, by contrast, has a feed side concentration profile that increases in the direction of feed flow through the membrane unit. Thus, the CSTR model is more appropriate than the PFR model for systems that are expected to have a relatively uniform suspended solids concentration on the feed side of the membrane.

CSTR theory generally describes systems that operate continuously without backwashing. For example, some MF and UF systems utilize a periodic short duration reverse flow operation (i.e., a back-pulse) to remove solids from the membrane surface, but collect the dislodged solids in the feed tank. Since this operation does not constitute backwashing (i.e., removal of solids from the system), these MF and UF systems can be modeled as CSTRs without backwashing. In addition, a small-scale NF or RO membrane unit that is too small to be considered a PFR (e.g., a single stage system using a high recirculation flow or a single module system) can also be modeled as CSTR that does not utilize backwashing. However, there may be types of membrane filtration systems that can be effectively modeled as a CSTR with backwashing, such as a small, immersed vacuum-driven membrane system in which the backwash water is removed from the tank after the backwash operation.

A summary of the equations used to calculate the average and maximum VCFs for each of the various hydraulic configurations is presented in Table 8.5. The table also indicates the types of membrane filtration systems that are generally associated with the respective hydraulic configurations. Note that it is also possible to determine the VCF experimentally for a particular system (either full-scale or representative pilot-scale) by measuring the influent feed concentration (C_m) and the concentration on the feed side of the membrane (C_f). Systems that operate in deposition mode have a VCF equal to one, while the various suspension mode hydraulic configurations have VCFs greater than one, as summarized in Table 8.6.

5. SYSTEM DESIGN CONSIDERATIONS

In the process of planning and implementing a membrane filtration system, there are several important issues that can have a particularly significant impact on system design and operation, thus warranting special consideration. These issues include:

1. Membrane flux
2. Water quality
3. Temperature compensation
4. cross-connection control
5. System reliability

Table 8.5
Summary of VCF equations for various hydraulic configurations

Hydraulic configuration	Typical membrane process(es)	VCF equation			
		VCF_{avg}	Equation No.	VCF_{max}	Equation No.
Deposition mode	MCF MF/UF	1		1	
Small volume crossflow $[(t_b \cdot Q_b) < V_f]$	MF/UF (w/o feed tank)	$\frac{t_f}{2\tau}$	(19)	$\frac{t_f}{\tau}$	(20)
Large volume crossflow $[(t_b \cdot Q_b) > V_f]$	MF/UF (w/ feed tank)	$\frac{t_f \cdot Q_f}{t_b \cdot Q_b}\left[1 - \frac{t_f}{2 \cdot \tau \cdot \frac{t_f \cdot Q_f}{t_b \cdot Q_b}}\right]$	(21)	$\frac{t_f \cdot Q_f}{t_b \cdot Q_b}$	(22)
CSTR w/o backwashing	NF/RO (small-scale) MF/UF (w/ bleed)	$\frac{1}{1-R}$	(23)	$\frac{1}{1-R}$	(24)
CSTR w/ backwashing	MF/UF (w/ bleed)	$\frac{1}{1-R} \cdot \left[1 - \left(\frac{\tau}{t_f} \cdot \left[1 - \exp\left(\frac{-t_f}{\tau}\right)\right]\right)\right]$	(25)	$\left[\frac{1}{1-R}\right] \cdot \left[1 - \exp\left(\frac{-t_f}{\tau}\right)\right]$	(26)
PFR	NF/RO	$\frac{\sum VCF(x) \cdot Q_p(x)}{Q_p}$	(27)	$\frac{1}{1-R}$	(28)

Q_b = backwash flow, gpm.
Q_f = feed flow, gpm.
Q_p = filtrate flow, gpd.
$Q_p(X)$ = filtrate flow at position "X" within the membrane unit, gpd.
R = membrane unit recovery, decimal percent.
t_b = backwash duration, min.
t_f = filtration cycle duration, min.
τ = solids retention time, min.
VCF_{avg} = average VCF, dimensionless.
VCF_{max} = maximum VCF, dimensionless.
$VCF(X)$ = VCF at discrete position "X" within the membrane unit, dimensionless.
V_f = volume of feed, gal.
X = position in the membrane unit in the direction of tangential flow (i.e., $X = 0$ at the entrance to the first module), ft.

Table 8.6
Typical range of VCF values for various hydraulic configurations

Hydraulic configuration		VCF
Deposition mode	Dead-end	1
Suspension mode	PFR	3–20
	Crossflow	4–20
	CSTR	4–20

5.1. *Membrane Flux*

The flux – the flow per unit of membrane area is one of the most fundamental considerations in the design of a membrane filtration system, since this parameter dictates the amount of membrane area necessary to achieve the desired system capacity and thus the number of membrane modules required. Because the membrane modules represent a substantial component of the capital cost of a membrane filtration system, considerable attention is given to maximizing the membrane flux without inducing excessive reversible fouling, thereby minimizing the number of modules required.

Typically, the maximum flux associated with a particular membrane filtration system is determined during pilot testing, mandated by the state, or established via a combination of these two in cases in which the state specifies a maximum operating flux based on the pilot results. Independent of maximum flux, pilot testing is also commonly used to determine a reasonable operating range that balances flux with backwash and chemical cleaning frequencies. Because higher fluxes accelerate fouling, backwashing and chemical cleaning must usually be conducted more frequently at higher fluxes. The upper bound of the range of acceptable operating fluxes (provided this bound does not exceed the state-mandate maximum) is sometimes called the "critical" flux, or the point at which a small increase in flux results in a significant decrease in the run time between chemical cleanings. A membrane filtration system should operate below this critical flux to avoid excessive downtime for cleaning and the consequent wear on the membranes over time due to increased chemical exposure.

The flux through a membrane is influenced by a number of factors, including:

1. Pore size (for MF, UF, and MCF membranes)
2. Module type (i.e., cartridge, hollow-fiber, spiral-wound, etc.)
3. Membrane material
4. Water quality

However, it is important to note that higher fluxes do not necessarily indicate that one membrane is better than another for a particular application. Factors such as:

1. Estimated membrane life
2. Fouling potential
3. Frequency and effectiveness of chemical cleaning

4. Chemical use, and
5. Energy requirements to maintain a given flux

5.2. *Water Quality*

Because water quality can have a significant impact on membrane flux, feed water quality is also a primary design consideration for membrane filtration systems. Poorer water quality will require lower fluxes, which in turn increase the necessary membrane area and required number of modules, augmenting both the cost and the size of the system. However, pretreatment can often improve feed water quality at a lower cost than additional membrane area. Conversely, better water quality will allow higher fluxes, reducing the required membrane area, the size of the system, and the capital cost. Typically, membrane flux is determined through pilot testing. In the absence of pilot test data, it is important to have some understanding of how critical the following water quality parameters affect the flux:

1. SDI
2. Turbidity
3. Organic carbon
4. Dissolved solids

The influence of each of these parameters on flux is briefly described in this section. Temperature also has a significant impact on membrane flux, and this relationship is discussed separately in the following section.

5.2.1. Silt Density Index

The SDI is an empirical dimensionless measure of particulate matter in water and is generally useful as a rough gauge of the suitability of source water for efficient treatment using NF/RO processes. The American Society for Testing and Materials (ASTM) *Standard Test Method for SDI of Water* (25) details the procedure for determining SDI. In general, SDI measurements are taken by filtering a water sample through a 0.45-μm flat sheet filter with a 47-mm diameter at a pressure of 30 psi. The time required to collect two separate 500 mL volumes of filtrate is measured, and the resulting data become the inputs to a formula used to calculate SDI. Water samples that contain greater quantities of particulate matter require longer filtering and thus having higher SDI values.

As a general rule of thumb, spiral-wound NF and RO modules are not effective for treating water with a SDI of 5 or greater, as this quality of water contains too much particulate matter for the non-porous, semi-permeable membranes, which would foul at an unacceptably high rate. Thus, some form of pretreatment to remove particulate matter is generally required for SDI values exceeding five in NF/RO applications. NF/RO membrane module manufacturers can usually provide a rough estimate of the range of anticipated operating fluxes based on the type of source water, which is roughly associated with a corresponding range of SDI values. A summary of these estimates is included in Table 8.7.

However, SDI is only one measure of water quality, and there are a number of site- and system-specific water quality and operational factors that combine to dictate the flux for a

Table 8.7
Estimated NF/RO membrane fluxes as a function of SDI

Source	SDI (dimensionless)	Estimated NF/RO flux (gal/ft^2/d)
Surface water	2–4	8–14
Ground water	<2	14–18

given system. Thus, the ranges cited in Table 8.7 should only be used as a rough guideline. Caution should also be exercised when interpreting SDI results, as measurements can vary from test to test and with the analyst, as well as with both temperature and the specific type of membrane used. Consequently, it is important that the results are given for comparable conditions when evaluating SDI data. Note that because SDI is a batch process, it is not conducted continuously on line and thus is not typically utilized as gauge of water quality, or system performance during daily operation in the way turbidity or conductivity monitoring are often employed.

SDI is typically not used as a tool for estimating flux for MF, UF, and MCF systems. Since these systems are designed to filter particulate matter, parameters such as turbidity, more commonly associated with conventional drinking water filtration, are used to assess system performance.

5.2.2. Turbidity

Turbidity is a measure of the scatter of incident light caused by particulate matter in water. Because turbidity is widely used as a performance gauge for conventional media filters (26), among the various types of membrane filtration systems turbidity is most often used as an assessment tool (27, 28) for MF, UF, and MCF, since these systems are specifically designed to remove particulate matter. Higher turbidity measurements are indicative of greater quantities of suspended solids, and thus the potential to cause more rapid membrane fouling. Therefore, water with higher turbidity is usually filtered at lower fluxes to minimize fouling and the consequent backwash and chemical cleaning frequency. In some cases when turbidity levels are extremely elevated, it may be more economical to provide pretreatment for a MF/UF system to reduce the solids loading to the membranes. In general, if the turbidity of the water normally exceeds 10 NTU on a sustained basis, some type of pretreatment (e.g., prefiltration or pre-settling) should be considered. MCF systems may be applied to untreated source waters; however, as turbidity levels increase, disposable MCF cartridges will foul more rapidly, requiring more frequent replacement. In some applications with higher turbidity source water, coarse bag or cartridge filters may be used as pretreatment for MCF systems.

Because spiral-wound NF and RO membrane modules are not designed to handle significant solids loading, these systems are typically not applied to treat water with turbidity levels exceeding approximately 1 NTU. For water turbidity levels greater than 1 NTU, pretreatment would be necessary to reduce solids loading upstream of NF or RO.

5.2.3. Organic Carbon

Another water quality constituent that influences membrane flux is the organic carbon content, which is typically expressed in terms of either total (TOC) or dissolved organic carbon (DOC). Organic carbon in the feed water can contribute to membrane fouling, either by adsorption of the dissolved fraction onto the membrane material or obstruction by the particulate fraction. Thus, lower fluxes may be necessary if membrane filtration is applied to treat water with significant organic carbon content. The tendency for a membrane to be affected by TOC is partially influenced by the nature of the organic matter in the water. TOC can be characterized as either hydrophilic or hydrophobic in composition, and studies suggest that the hydrophobic fraction contributes more significantly to membrane fouling. The character of the organic carbon content can be roughly quantified by measuring the specific ultraviolet absorbance (SUVA) of the water, as calculated using the following equation:

$$\text{SUVA} = \frac{\text{UV}_{254}}{\text{DOC}}, \tag{29}$$

where SUVA is the specific UV absorbance, L/mg m, UV_{254} is the UV absorbance at 254 nm, 1/m, and DOC is the dissolved organic carbon, mg/L.

Because TOC is more commonly measured than DOC in drinking water treatment, SUVA is sometimes estimated using values for TOC in place of those for DOC.

Higher SUVA values tend to indicate a greater fraction of hydrophobic organic material, thus suggesting a greater potential for membrane fouling. Generally SUVA values exceeding 4 L/mg m are considered somewhat more difficult to treat. However, organic carbon (as well as turbidity) can often be removed effectively via coagulation and pre-settling, particularly if more hydrophobic in character, thus minimizing the potential for membrane fouling and facilitating operation at higher fluxes. Coagulation can also be conducted in-line (i.e., without pre-settling) with MF/UF systems. Pretreatment using the injection of PAC may also reduce DOC in the membrane feed; however, because spiral-wound membrane modules cannot be backwashed, PAC should not be used in conjunction with NF/RO systems unless provisions are made to remove the particles upstream.

5.2.4. Dissolved Solids

The total dissolved solids (TDS) and the particular species of dissolved solids present in the membrane feed are both critical considerations for NF/RO systems. Species such as silica, calcium, barium, and strontium, which can precipitate as sparingly soluble salts, can cause scaling and a consequent rapid decline in flux under certain conditions. Scaling is typically controlled using pretreatment chemicals such as an acid to lower the pH and/or a proprietary scale inhibitor. However, the total quantity of dissolved solids of any species also influences system operation, as the NDP required to achieve a target flux is related to the osmotic pressure of the system, which is directly proportional to the TDS. Thus, as the TDS increases, so does the required feed pressure.

TDS is generally not a significant consideration for MF, UF, and MCF systems, since these processes do not remove dissolved solids. In some cases, however, the use of upstream oxidants may cause the precipitation of iron or manganese salts (either unintentionally or by design as a pretreatment process), which could accelerate membrane fouling.

5.3. Temperature Compensation

Like other water quality parameters such as turbidity and TDS (for NF/RO systems), the temperature of the feed water also affects the flux of a membrane filtration system. At lower temperatures water becomes increasingly viscous; thus, lower temperatures reduce the flux across the membrane at constant TMP or require an increase in pressure to maintain constant flux. The means of compensating for this phenomenon varies with the type of membrane filtration system used. General viscosity-based means of compensating for temperature fluctuations for both MF/UF and NF/RO systems are described below, although membrane manufacturers may have a preferred product-specific approach.

MF/UF membrane systems usually operate within a relatively narrow range of TMPs, which may limit increasing the TMP in order to maintain constant flux as the water temperature decreases. Because the membrane modules can be damaged if the TMP exceeds an upper limit, as specified by the manufacturer, it may not be possible to operate the system at a TMP that is sufficient to meet the required treated water production during colder months if demand remains high. As a result, additional treatment capacity (i.e., membrane area or number of membrane modules) is incorporated into the design of the system such that the water treatment production requirements can be satisfied throughout the year.

For the microporous MF/UF membranes, the relationship between flux, TMP, and water viscosity is given by Eq. (7):

$$J_T = \frac{\mathrm{TMP}}{R_\mathrm{t}(\mu_T)}, \tag{7}$$

where

J_T = flux, gal/ft^2/d
TMP = trans-membrane pressure, psi
R_t = total membrane resistance, psi/gal/ft^2/d/cp
μ_T = viscosity of water, cp

If the system is operated at constant flux, then increases in viscosity require proportional increases in operating TMP (assuming constant membrane resistance). However, once the TMP approaches the rated maximum for the membranes, further increases in viscosity necessitate a reduction in flux. Thus, in order to maintain the required filtered water production flow (so as to satisfy customer demand), the membrane area must increase in proportion to the flux decrease, as was shown in Eq. (1):

$$J = \frac{Q_\mathrm{p}}{A_\mathrm{m}}, \tag{1}$$

where J is the flux, gal/ft^2/d, Q_p is the filtrate flow, gpd, and A_m is the membrane surface area, ft^2.

Combining Eqs. (1) and (7) demonstrates that the additional membrane area required is directly proportional to the increase in water viscosity (for constant flow, TMP, and membrane resistance), as shown in Eq. (30):

$$\frac{Q_{\mathrm{p}}}{A_{\mathrm{m}}} = \frac{\mathrm{TMP}}{R_{\mathrm{t}}(\mu_T)}, \tag{30}$$

where

Q_{p} = filtrate flow, gpd
A_{m} = membrane surface area, ft^2
TMP = trans-membrane pressure, psi
R_{t} = total membrane resistance, psi/gal/ft^2/d/cp
μ_T = viscosity of water, cp

Membrane filtration systems are commonly designed to operate at a particular flux (e.g., as determined via pilot testing or mandated by the state) to produce a specific flow (i.e., rated system capacity) at a reference temperature of 20°C. Thus, the required membrane area at 20°C can be calculated using Eq. (1). The increased membrane area required to compensate for cold weather flow can be determined by multiplying this area by the ratio of the viscosity at the coldest anticipated temperature (e.g., the coldest average monthly temperature) to that at the reference temperature of 20°C. Values for water viscosity can be found in the literature or approximated using Eq. (8):

$$\mu_T = 1.784 - 0.0575\,T + 0.0011\,T^2 - 10^{-5}\,T^3, \tag{8}$$

where μ_T is the viscosity of water at temperature T, cp and T is the water temperature, °C.

After the appropriate values for water viscosity have been determined for both the reference temperature (commonly 20°C for MF/UF systems) and the coldest anticipated temperature, then the design membrane area, as compensated for seasonal temperature variation, can be calculated as shown in Eq. (31):

$$A_{\mathrm{d}} = A_{20}\left(\frac{\mu_T}{\mu_{20}}\right), \tag{31}$$

where

A_{d} = design membrane area (as adjusted for temperature), ft^2
A_{20} = membrane area required at 20°C reference temperature, ft^2
μ_T = viscosity of water at temperature T°C, cp
μ_{20} = viscosity of water at temperature 20°C, cp

Equation (30) can also be utilized to calculate the required membrane area using a less conservative approach that accounts for seasonal fluctuations in demand. First, if the information is available, the average daily flow and temperature over each calendar month can be tabulated, and the temperature data converted into associated values for water viscosity using Eq. (8). Then, the 12 sets of paired flow (Q_{p}) and viscosity (μ) data can be applied to Eq. (30) to generate 12 values of membrane area. The largest of these values for the required membrane area over each of the 12 calendar months, along with its corresponding flow, is applied to Eq. (1) to generate an associated flux. If this flux is less than the maximum permitted value, then this largest of the 12 calculated values for membrane area represents the design value for the system. However, if the resulting flux exceeds this threshold, the design area must be increased to lower the flux to the maximum permitted value.

Note that use of Eq. (30) in this approach requires values for both the TMP and the total membrane resistance, R_t, both of which should be considered constant for the purposes of calculating the various values for membrane area. While a reasonable TMP can be easily identified, appropriate values of R_t are more difficult to determine. The total membrane resistance represents the sum of the intrinsic resistance of the membrane (which may be considered a constant and can generally be obtained from the manufacturer) and the resistance attributable to fouling at any given point during operation, as shown in Eq. (6):

$$R_t = R_m + R_f, \tag{6}$$

where R_t is the total membrane resistance, psi/gal/ft^2/d/cp, R_m is the intrinsic membrane resistance, psi/gal/ft^2/d/cp, and R_f is the resistance of the foulant layer, psi/gal/ft^2/d/cp.

Because it is difficult to identify and justify a single specific value for the fouling resistance for use with this approach, the contribution attributable to fouling may be ignored for practical purposes. Thus, for the purpose of calculating the membrane area, the total resistance used in Eq. (30) may be approximated by the membrane's intrinsic resistance. This approximation may be reasonable for the membrane system at the start of a filtration cycle when the fouling resistance is minimal. Since the minimum value for membrane resistance expected over a filtration cycle is used, the minimum TMP anticipated over a filtration cycle should also be used. This minimum TMP occurs at the beginning of a filtration cycle before gradually increasing until the system must be backwashed. Use of the intrinsic membrane resistance and minimum TMP should result in a reasonable and not excessively conservative estimate of the membrane area requirements. In rare cases in which the membrane is experiencing significant fouling under conditions of peak demand such that the flux and/or TMP are approaching their maximum values, the backwash and/or chemical cleaning frequencies can be increased temporarily to compensate and keep fouling to a minimum. Alternatively, standby membrane units may be used when necessary.

A MF/UF membrane manufacturer may have an alternate preferred method of determining the design membrane area for a particular application based on temperature, membrane material, or other site- or system-specific factors. It is recommended that the utility collaborate with the state, the membrane manufacturer, and its engineer (if applicable) to select the most appropriate method for determining the required area. Note that the addition of membrane area to compensate for low-temperature flow will also help the system to meet higher flow demands during warm weather without operating at an exceedingly high membrane flux. Temperature compensation for MCF systems, if necessary, can be determined using the methodology for MF/UF systems.

Because spiral-wound NF/RO membrane modules are designed to operate over a larger range of TMPs than MF/UF modules, in NF/RO systems the TMP is simply increased to maintain constant flux as the temperature of the feed water decreases. The required increase in TMP at the temperature of interest relative to that at a given reference temperature (typically 25°C for NF/RO systems) is dependent on the specific proprietary membrane used and can be calculated by means of a TCF, as shown in Eq. (32):

$$\mathrm{TCF} = \exp\ U\left(\frac{1}{T+273} + \frac{1}{298}\right), \tag{32}$$

where TCF is the temperature correction factor, dimensionless, T is the water temperature, °C, and U is the membrane-specific manufacturer-supplied constant, 1/°K.

Alternatively, many NF/RO membrane manufacturers may supply tables specifying TCF values over a range of temperatures for a given membrane. Once the appropriate TCF is known, the required TMP at temperature T of interest can be calculated by dividing the TMP at the reference (i.e., design) temperature (commonly 25°C for NF/RO systems) by the TCF value, as shown in Eq. (33):

$$\mathrm{TMP}_T = \frac{\mathrm{TMP}_{25}}{\mathrm{TCF}}, \tag{33}$$

where TMP_T is the trans-membrane pressure (temperature-corrected), psi, TMP_{25} is the trans-membrane pressure at 25°C reference temperature, psi, and TCF is the temperature correction factor, dimensionless.

Note that Eqs. (6) and (32) also allow the calculation of the TMP reduction resulting from decreased water viscosity at higher feed water temperatures during warmer months of the year.

5.4. Cross-Connection Control

In the context of membrane filtration systems, cross-connection control measures are implemented to prevent chemicals from the cleaning process from contaminating the feed or filtrate streams. States may have particular requirements for cross-connection control, although in general there are two strategies that are commonly used: a double block and bleed valving arrangement or a removable spool. These strategies are applicable to both MF/UF and NF/RO systems; however, because MCF systems typically utilize modules that are disposable cartridges, these systems are not usually subject to chemical cleaning and thus cross-connection control measure are generally unnecessary for these systems.

The double block and bleed valve arrangement is the most common method of cross-connection control for large membrane filtration systems. A schematic illustrating this method is shown in Fig. 8.5. In summary, two isolation valves (V-1A and V-2A) are placed in the feed line to isolate it from the cleaning chemicals. These are the block and bleed valves, respectively. During the cleaning process, valve V-4A and V-6A are opened to bring cleaning solution(s) into membrane rack (i.e., unit) A. Although valve V-3A is kept closed during this operation, if it were to leak it would allow chemicals to pass into the common feed manifold, causing contamination. In order to prevent this potential contamination, valve V-1A is also closed and valve V-2A is opened. With this configuration, if valve V-3A leaks, the cleaning solution(s) simply flows through valve V-2A to waste. A similar block and bleed valve

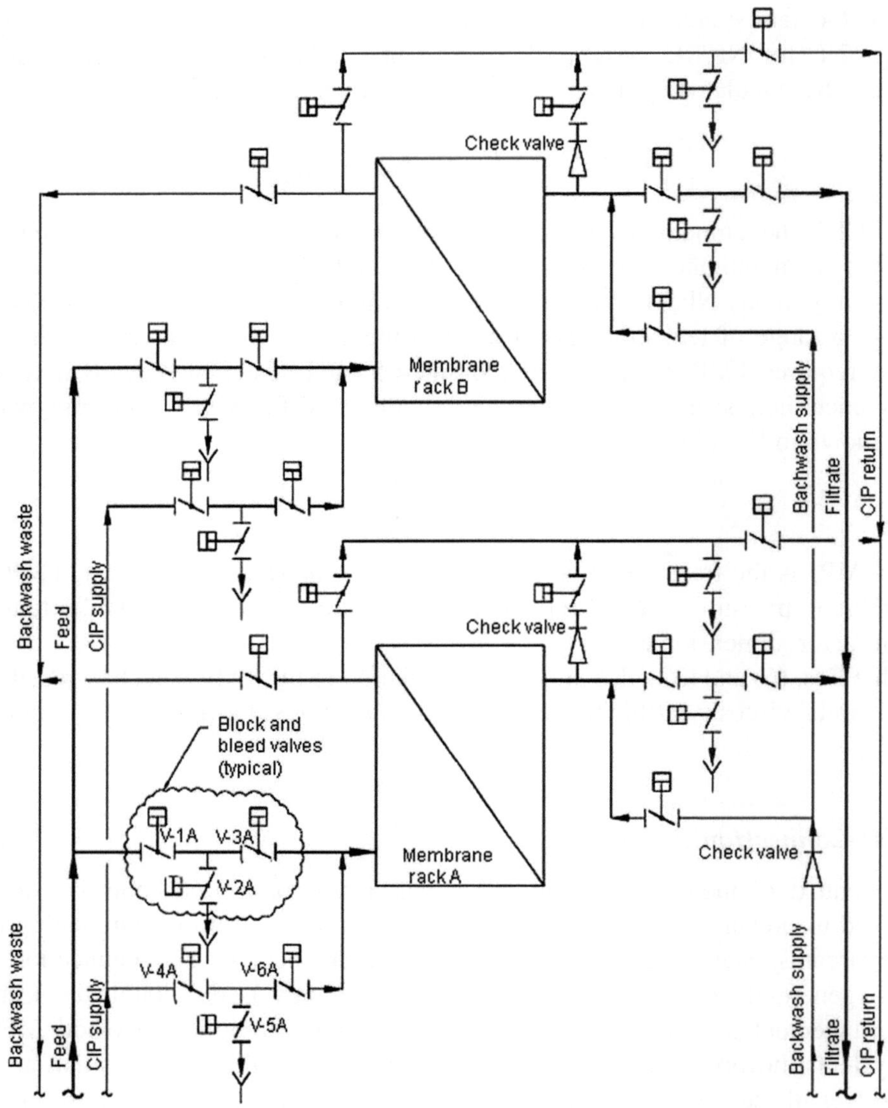

Fig. 8.5. Double block and bleed valving arrangement.

arrangement is utilized to prevent the cleaning chemicals from contaminating the filtrate manifold as they are recirculated, which gives rise to the term "double" block and bleed. If the cleaning process starts automatically, valves V-1A and V-2A (i.e., the block and bleed valves) must be actuated automatically.

As an alternative to the double block and bleed valving arrangement, a removable spool (i.e., a short section of pipe) can be placed between valves V-1A and V-3A in lieu of utilizing valve V-2A. The spool is then removed during the chemical cleaning. A removal spool should not be used for cross-connection control if chemical cleaning is automated, since the process could potentially be initiated with the spool still in place.

5.5. *System Reliability*

System reliability is an important consideration in the design of a membrane filtration system. Design capacities are typically maximum values and may not necessarily account for the possibility of one or more units being taken out of service for repair or routine maintenance. Even standard operational unit processes such as backwashing, chemical cleaning, and integrity testing that are normally accounted for in accurately sizing the facility and determining system capacity may be problematic if it becomes necessary to conduct these processes more frequently than was planned. For example, most membrane filtration systems are designed with sufficient storage and equalization capacity to meet average demand even when a unit is taken out of service for chemical cleaning. However, this storage would generally not allow the system to operate at capacity for extended periods (e.g., longer than about 24 h) with a unit of out service for repair. Reliability issues such as this may be particularly pronounced for smaller systems with fewer membrane units, since one out-of-service unit may significantly impact overall system capacity. Note that any state sizing and redundancy requirements should be considered in the design of the facility.

In order to ensure system reliability, it is common for filtration systems to incorporate some measure of redundancy. One method of providing this redundancy is by oversizing the membrane units at the design flux. Oversizing the membrane units generally allows for operation at average flow as a minimum with one unit out of service and may allow for operation at or near system capacity in some cases. For example, a water treatment plant with rated capacity of 7 MGD may be designed with four membrane units rated at 2 MGD each rather than 1.75 MGD. Thus, with one unit out of service and the remaining three operating at capacity, the system can produce as much as 6 MGD (at the design flux) consistently. Although the design in this example would not allow the system to operate at maximum capacity, it is likely that the average flow could be met or exceeded.

A system could also be designed to operate at the maximum rated capacity with one unit out of service (in this example, using four units at 2.33 MGD each), although this ability may be cost prohibitive, particularly for smaller systems, since the percentage that each unit must be oversized increases as the total number of units is reduced. Note that because this method of providing system redundancy is based on the assumption that the rated capacities are all specified with respect to the same design flux, the rate of fouling would not be increased by operating three of the four oversized units at capacity. In addition, continuing this example of a treatment plant with 7 MGD of permitted capacity consisting of four 2.33-MGD membrane filtration units, with all four units in operation the system could operate at reduced flux and still produce 7 MGD in accordance with its permit. This operation at reduced flux would decrease the rate of fouling and lower operating costs to partially offset the increase in capital cost associated with the extra unit capacity.

Another method of adding system redundancy is by providing an additional membrane unit. Thus, using the same example of a water treatment plant permitted for a maximum rated capacity of 7 MGD, the membrane filtration system would consist of five 1.75-MGD units, rather than four. This design would always allow for operation at capacity with one unit either out of service or in standby mode. In this case, the standby unit can be rotated such that any time a unit is taken offline for chemical cleaning the standby unit is activated. The newly cleaned unit then reverts to standby mode until the next unit is due for chemical cleaning. This method of adding redundancy is more cost effective for larger facilities with a greater number of membrane units, such that the addition of an extra unit does not represent a large percent increase in capital cost. This approach is recommended when the treatment facility utilizing membrane filtration is the primary source of drinking water for the public water system, such that operating the membrane system at capacity may be critical to the ability of the utility to satisfy customer demand. Alternatively, under this scenario all the units may be operated at lower flux to extend the interval between chemical cleanings.

Independent of the particular strategy for providing some redundant capacity for the membrane filtration system, utilities should also provide some redundancy for any major ancillary mechanical equipment that may be utilized, such as pumps, compressors, and blowers. Utilities must comply with any state-mandated redundancy requirements for either membrane capacity or ancillary equipment.

Membrane filtration systems may also have some inherent redundancy if the system is normally operated significantly below the maximum flux permitted by the state. In this case, if one membrane unit is taken out of the service, the utility can increase the temperature- and pressure-normalized flux through the remaining units to partially or fully compensate for the loss of production attributable to the out-of-service unit. However, if a unit is out of service for a prolonged period, the increase in flux may accelerate fouling in the remaining in-service units to an unacceptable rate.

Timely membrane replacement is another consideration for maintaining system performance and reliability. Membrane replacement is usually conducted on an as-needed basis, typically either in cases in which the membranes have been damaged or the flux has declined to an unacceptable level as a result of irreversible fouling. Although the useful life of membranes is commonly cited in the range of 5–10 years (a period generally consistent with manufacturer warranties), the use of membrane technology – particularly MF and UF – has increased almost exponentially in the last decade, and thus most membrane filtration systems have not been in continuous operation for more than 5–10 years. Consequently, there is limited field data available to document the typical useful life of membrane filtration modules. Nevertheless, it is recommended that utilities keep a small number of surplus membrane modules on site in the event that emergency replacement becomes necessary.

6. RESIDUALS TREATMENT AND DISPOSAL

As with many water treatment processes, membrane filtration systems may generate several different types of residuals that must be treated and/or disposed of, including backwash and chemical cleaning residuals and concentrate. The types of residuals that are generated vary with both the type of membrane filtration system and the hydraulic configuration in which the system in operated. For example, NF/RO systems produce a continuous concentrate stream and periodic chemical cleaning waste, but because these systems are not backwashed, no backwash residuals are generated. MF/UF systems are regularly backwashed and undergo periodic chemical cleanings and thus produce residuals from both these operational unit processes.

However, only MF/UF systems that are operated in a crossflow hydraulic configuration and waste (rather than recirculate or recycle to the plant influent) the unfiltered flow – the so-called "feed and bleed" mode – will generate a concentrate stream. Because cartridge filters operate in a deposition mode hydraulic configuration and are designed to be disposable, there are typically no residual streams associated with MCF systems. Nevertheless, the spent filter cartridges (as well as all membrane filtration residuals) must be properly disposed off in accordance with any applicable state and local regulations, particularly if the cartridges have been used to filter any potentially hazardous materials. The various potential residuals streams from membrane filtration systems include:

1. Backwash residuals
2. Chemical cleaning residuals
3. Concentrate

Note that the following discussions are not meant to represent a comprehensive review of residuals treatment and disposal, but rather a general overview of some of the primary considerations that should be taken into account when planning and designing a membrane filtration system. For additional information, the reader is referred to Refs. (3, 5, 29–34).

6.1. *Backwash Residuals*

Among the various types of membrane filtration, only MF/UF systems employ backwashing, thus generating backwash residuals. Although the frequency varies on a site- and system-specific basis, backwashing is typically conducted every 15–60 min. Under normal operating conditions the backwash frequency should remain relatively consistent, allowing for the quantity of residuals generated to be estimated fairly accurately. As a general rule of thumb, the residual stream produced from backwashing MF/UF membranes has a concentration of suspended solids that is approximately 10–20 times greater than that of the feed water. Although MF/UF systems remove approximately the same types of feed water constituents as conventional media filters, the volume and characteristics of the residuals may be significantly different. In many current applications of MF/UF for municipal water treatment, filter aids such as coagulants and polymers are not necessary. In these cases, the amount of solids removed in the backwash process may be significantly less than that for a comparable conventional filtration plant. In addition, disposal of these coagulant- and polymer-free MF/UF backwash residuals may be less problematic. However, in some applications of MF/UF, coagulants may be added in-line (i.e., without pre-settling) to help facilitate the removal of TOC, which is not normally removed to a significant degree by MF or UF. In such applications the MF/UF backwash residuals characteristics will generally be similar to those for conventional media filtration.

Disposal options for MF/UF backwash residuals are similar to those for conventional water treatment plants, and typically include the following:

1. Discharge to a suitable surface water body
2. Discharge to the sanitary sewer
3. Treatment with supernatant recycle and solids disposal

The discharge of backwash residuals to surface water bodies or the sanitary sewer is likely to be subject to state and/or local regulations and, in the case of surface water discharge, to require a permit. Moreover, the potential to utilize one of these options may be complicated if the residuals include chemical wastes. In addition to the use of coagulants added to the feed, some backwash procedures utilize chlorine or other chemicals. Small amounts of chlorine may be quenched and acids or bases can be neutralized prior to discharge, although larger amounts or other types of chemicals added to the backwashing processes may require additional treatment or preclude discharge to a surface water body or sanitary sewer.

On-site treatment options for MF/UF backwash residuals are also similar to those that might be used with conventional media filtration, and include clarification, sedimentation lagoons, gravity thickening, centrifuging, belt filter presses, or a combination of these processes (26). A second stage of MF or UF may also be utilized to further concentrate residuals and increase process recovery. If a sedimentation process is used to treat MF/UF backwash residuals, the addition of a coagulant may be necessary to improve the settling characteristics of the solids if coagulant is not already applied in the MF/UF pretreatment process. With on-site treatment, the supernatant is generally recycled to the treatment plant influent while the concentrated solids are transported off site for landfilling or other means of disposal. As with discharging, the addition of chlorine or other chemicals to the backwash process may complicate residuals treatment. It is important to note that backwash residuals concentrate any pathogenic organisms that are present in the feed water, as well as other suspended solids. Although the control of such concentrated pathogens in filter backwash residuals is largely unregulated, the potential treatment of this stream should be taken into consideration if these residuals are to be discharged into a surface water receiving body.

6.2. Chemical Cleaning Residuals

Both MF/UF and NF/RO membranes undergo periodic chemical cleaning, and thus both types of systems generate spent chemical waste as a byproduct of these processes. As with the backwashing process for MF/UF, the frequency of chemical cleaning varies on both a site- and system-specific basis. Although chemical cleaning is conducted much less frequently than backwashing, the frequency is also more difficult to predict. Generally, MF/UF systems are cleaned no more frequently than once every month for efficient operation and the minimization of system downtime, although it is not uncommon for these systems to operate for much longer without requiring chemical cleaning. The cleaning frequency for NF/RO may vary anywhere from 3 months to 1 year, depending on the feed water quality and the effectiveness of feed water pretreatment for minimizing fouling. However, because chemical cleaning is a relatively infrequent batch process, estimating the quantity of residuals generated is not as critical for day-to-day operation, as is the case with backwashing.

Chemical cleaning residuals are generally treated on-site and discharged to either a suitable surface water body or sanitary sewer, subject to state and/or local regulations. Oxidants such as chlorine used in the chemical cleaning process can be quenched prior to discharge, and acids and bases can be neutralized. The use of other chemicals, such as surfactants or proprietary cleaning agents, may complicate the process of obtaining regulatory approval for discharge and could require additional treatment.

Note that the rinse water applied to the membranes after the cleaning process also represents a chemical waste and must be treated prior to discharge. Although the rinse water increases the volume of the chemical cleaning residuals, this increase can be balanced somewhat by the recovery and reuse of a significant portion of the cleaning solutions. In some cases, as much as 90% of the applied cleaning solutions can be reused, reducing residuals treatment and disposal costs, as well as chemical usage.

6.3. Concentrate

The term "concentrate" is usually associated with the continuous waste stream of concentrated dissolved solids produced by NF and RO processes. This waste stream is typically 4–10 times more concentrated than the feed with respect to suspended and dissolved constituents and represents about 15–25% of the total feed flow, although it can exceed 50% or more in some cases. As a result, concentrate disposal is a significant logistical and regulatory concern for utilities and is often a critical factor in the planning and design of an NF or RO facility. There are a number of methods for concentrate treatment and disposal, including the following options:

1. Surface water discharge
2. Sanitary sewer discharge
3. Land application/irrigation
4. Deep well injection
5. Evaporation

Because there are complicating factors associated with each of these options, no single option is ideal or most appropriate for every application. In many cases surface water discharge is the least expensive option, although the permitting process may be difficult, since there are potential environmental impacts if the salinity of the concentrate is significantly higher than that of the receiving body. Discharge to the sanitary sewer may have similar issues, since the wastewater treatment process does not typically affect dissolved solids concentrations, and the treatment plant effluent may ultimately be discharged to surface water receiving body. High salinity or major ion toxicity may also preclude land application of the concentrate if levels exceed threshold levels tolerated by the irrigated crops. Bioaccumulation of metals has also been cited as a potential concern for land application of NF/RO concentrate. Deep well injection is an effective and commonly used technique for concentrate disposal, although this method risks the escape of brackish water into less saline or freshwater aquifers and may have unknown long-term environmental affects. The use of evaporation ponds is generally limited to areas with low precipitation and high evaporation rates, as well as an abundance of inexpensive and available land.

Another option for dealing with NF/RO concentrate is the concept of "zero liquid discharge," which involves sufficiently concentrating the residual stream through the use of such technologies as crystallizers and evaporators to allow remaining solids to be landfilled. While zero liquid discharge is typically an expensive option, it does offer a number of advantages, including avoiding the discharge permitting process and the ability to be utilized at any location independent of factors such as the proximity to a suitable surface water body or available land for evaporation ponds. In addition, zero liquid discharge maximizes facility recovery and has minimal environmental impact. Although still fairly uncommon, this option has become increasingly feasible relative to other methods as environmental and discharge regulations have become more stringent.

Some MF/UF systems that are operated in a suspension mode and waste the unfiltered flow generate a continuous concentrate stream. This stream differs from that associated with NF/RO systems in two significant ways: MF/UF systems concentrate suspended rather than dissolved solids; and the concentrate stream represents only a small fraction of total feed flow. Note that NF/RO membranes represent a barrier to particulate matter, and thus these systems will also concentrate suspended solids; however, because suspended solids rapidly foul the semipermeable membranes (which cannot be backwashed), most particulate matter is typically removed with prefilters. An MF/UF concentrate stream has characteristics similar to those of backwash residuals, and therefore can be considered comparable for the purposes of treatment and disposal. Alternatively, these two residuals streams may be blended together.

7. INITIAL START-UP

The primary objectives of the initial phase are to ensure that the installation is successfully completed and that the equipment is in proper working order and ready to produce potable water that achieves all target quality standards. A well-planned initial start-up phase can thus help facilitate the proper execution of the membrane filtration system start-up and create a smooth transition from testing to drinking water production. This section discusses important general start-up considerations, pointing out any significant differences between NF/RO and MF/UF systems. For the purposes of this discussion, MCF systems are considered to be similar to MF/UF systems, except as otherwise noted.

7.1. Temporary System Interconnections

During the start-up process, the filtered water produced from the membrane units may not be acceptable for distribution. Therefore, facility design should include provisions for the recycle and/or temporary disposal of feed and filtrate water. Provisions for disposal usually consist of a removable pipe spool or a tee with a "dump" valve placed in the common feed or filtrate line. This water can usually be diverted to the sanitary sewer. If the feed and filtrate piping are interconnected, the connection is removed and replaced with blind flanges after startup has been completed. Some facilities simply divert and recycle the treated water to the inlet structure until the system is fully commissioned. Any state requirements governing the disposal and potential recycle of the water produced during the start-up phase should be incorporated into the planning process.

7.2. *Flushing and Testing Without Membranes*

Prior to installing the membrane modules, any debris introduced during construction should be flushed from the system. This flushing is usually conducted by running the appropriate pump(s) at high velocity and low pressure through the piping and discharging the water to a suitable drain. Typically, the flush water can be discharged to the sanitary sewer, although the utility should comply with any state requirements for disposal. Because the intent of the flushing process is to remove debris from the system, this water should not be recycled unless pretreatment processes would be expected to remove contaminants flushed from the system.

After the piping system has been flushed, the operational sequences and chemical addition systems should be tested to ensure that they are operating properly before installing the membrane modules. Because of the complexity of the equipment involved, it is suggested that a plan for conducting these diagnostic checks be developed in advance. In general, the recommended testing can be divided into the following categories (1):

1. *Mechanical equipment* – Inspect automated mechanical equipment to ensure that it is properly installed and that there are no leaks in the piping system. Operate mechanical equipment in manual mode and then in automatic mode to verify that it is working as designed.
2. *Instrumentation* – Verify that the instrumentation is properly installed and calibrated. Confirm that the instruments are operating correctly and are responsive to the control system.
3. *Control system* – Verify control inputs and outputs (both digital and analog), instrumentation alarm limits, programing logic, instrumentation loops, and operational sequences.
4. *Membrane units and related systems* – Verify that the membrane filtration system and each respective membrane unit can be both started and shutdown smoothly. Operate the chemical feed systems (if any) to verify that they are each in proper working condition and that all chemicals are delivered at the proper dosages.

For NF/RO systems, the procedure for testing the operating sequences is commonly known as the "48-h test," as it may take two or more days to complete. Unlike the initial flushing, the 48-h test should be conducted at actual operating parameters, including both typical pressures and flows. Therefore, it is common to insert flow restriction devices in the system in place of the membrane modules during this operation to simulate the anticipated system backpressure. These devices typically consist of orifice plugs placed in the permeate (i.e., filtrate) port inside the pressure vessel. The orifice plugs are sized to simulate the flow and pressure parameters of the operating membrane filtration system. Alternatively, an isolation valve on the permeate piping could be throttled to the appropriate backpressure. Although the term "48-h test" is less commonly used in association with MF and UF, these systems are also generally tested either with "dummy" modules that have similarly designed orifices or by throttling filtrate effluent valves to create sufficient backpressure during the testing sequences.

7.3. Membrane Installation

Membrane module installation should be conducted according to the instructions obtained from the manufacturer. Care should be taken not to damage the membranes during the installation process. It is recommended that the location of each individual membrane in the system be recorded according to its serial number. Factory test data are often shipped with the membranes, and this information should be collected and filed.

Membrane modules are typically shipped "wet" with a liquid preservative solution. The particular preservative depends upon the type of membrane. Most membranes are preserved with a 1% solution of sodium bisulfite, a reducing agent that acts as a biocide to control microbial growth. Prior to installation, the membrane modules should be stored in an appropriate manner such that they are not subjected to freezing conditions that could damage the membranes. Regardless of the membrane type, the preservative should be flushed to waste after the membranes are installed. Note that the disposal of any preservative solutions should be conducted in accordance with any applicable regulations.

Some membrane modules are shipped with more problematic storage solutions. For example, glycerin solutions may pose a waste treatment issue because of the biochemical oxygen demand (BOD) that may result from the solution being flushed down the drain. The use of formaldehyde, once a common membrane preservative is generally no longer acceptable and it may pose a significant disposal problem. Both state and local regulations regarding chemical disposal may apply to membrane preservative solutions, and these may dictate whether it is permissible to discharge these preservatives to the sanitary sewer, or if collection and alternate disposal and/or treatment is required.

7.4. System Disinfection

Initial system disinfection typically involves the application of a disinfectant, such as a chlorine solution, throughout the entire system, including both the feed and filtrate piping. Although both soaking and recirculating procedures are used, recirculating the solution through the system will generally provide more effective disinfection. This disinfection step may be required by the state, and in any case is recommended to inactivate bacteria or other pathogenic organisms that may contaminate the membrane filtration system and associated piping. Because some membranes have limited compatibility with disinfecting chemicals, the manufacturer may have specific requirements for the disinfection process, and approval will be necessary if manufacturer recommendations differ from state requirements. General procedures for the disinfection of both chlorine-tolerant and chlorine-intolerant membranes are described in the following sections. In either case, once the system disinfection is complete, the entire system (including the membranes) should be flushed prior to placing the membrane unit(s) into continuous service. This final flush can be conducted according to guidelines and should continue until the desired filtered water quality is attained as measured by site- or membrane system-specific water quality parameters.

7.4.1. Chlorine-Tolerant Membranes

The American Water Works Association (AWWA) and the American National Standards Institute (ANSI) have developed standards pertaining to the disinfection of water treatment facilities. They provide guidance for the disinfection of water treatment plants and associated piping systems. The procedure involves surface contact with a high strength chlorine solution for a specific time period. The disinfection is complete when bacteriological sampling and testing indicate the absence of coliform organisms. For systems that use chlorine-tolerant membranes, the membrane modules should be installed prior to initiating the disinfection procedure. If the disinfection process is conducted while the membranes are in place, sufficient pressure should be applied to ensure adequate flow across the membranes such that the filtrate side piping is thoroughly disinfected. Some MF/UF membranes and most MCF membranes are chlorine-tolerant.

7.4.2. Chlorine-Intolerant Membranes

Most NF and RO membranes, as well as some MF and UF membranes are not chlorine tolerant. For systems that use these types of membranes, there are two options that are generally available for system disinfection. One option consists of conducting the system disinfection with chlorine, but prior to membrane installation. Subsequently, the membranes are installed and flushed, and then another disinfection specific to the chlorine-intolerant membrane is conducted. This second disinfection typically utilizes a high pH (i.e., ~10 or greater) solution of caustic or other alkaline chemical that is very effective for bacteriological inactivation. The membrane manufacturer should be consulted to ensure that this second disinfection is conducted within the pH tolerance limits of the membrane. A second option, if permissible under state regulatory requirements, involves the elimination of chlorine or other oxidants and using caustic (or other alkaline chemical) to disinfect the entire system, including the membranes. The state should be consulted prior to performing this type of alternative disinfection procedure. If an alternate disinfection procedure is used, additional microbiological monitoring (35) may be warranted to ensure the efficacy of the disinfection process.

7.5. Initial Direct Integrity Testing

Once the system is thoroughly flushed and disinfected, a direct integrity test should be conducted on each membrane unit. If the direct integrity test utilized requires that the membrane be fully wetted, the system may need to be operated for a period of time prior to conducting the test. After the direct integrity test has been completed (36–40), any defective membrane modules or leaks in o-rings, pipe connections, or valve seals should be repaired or replaced, as necessary. Until the direct integrity test has been successfully completed, the filtrate should be discharged to waste or recycled, as appropriate.

For most membrane systems it is advisable to conduct relatively frequent direct integrity testing during the initial start-up phase, as there may be a higher incidence of integrity failure observed during the initial stages of facility operation. As a general rule, most membrane manufacturing related defects will manifest within 72 h of continuous operation. However, the propensity to exhibit such defects varies significantly among different membranes. In order to ensure that all initial manufacturing defects are detected, it is recommended that direct integrity testing be conducted 2–6 times per day during start-up until the test results are stable.

7.6. *Acceptance Testing*

The purpose of acceptance testing is to demonstrate equipment performance as a condition for transferring responsibility over the membrane filtration equipment from the manufacturer or contractor to the utility. Acceptance testing is conducted after other phases of the start-up and commissioning process (i.e., flushing disinfection and integrity testing) have been completed. Typically, acceptance testing is the final phase of the commissioning process and consists of the following two criteria:

1. Operation: The entire system, including all membrane units, is continuously operated for a predetermined period of time, usually from 3 to 30 days. Any interruptions in continuous operation as a result of faulty equipment or controls during this period of time may result in restarting or temporarily suspending the test. Design parameters (e.g., flux and backwash frequency (if applicable)) are also typically verified as a component of the operational criterion.
2. Water quality testing: Periodic sampling and analysis is conducted to ensure that the treated water quality objectives are continuously satisfied. The required sampling can vary widely in terms of the number of constituents of interest and is very site-specific. The sampling frequency for each parameter is typically once per day. If the system is unable to produce the required water quality,

Table 8.8
Schedule of training events

Time period	Topic	Duration (hours)
During construction and commissioning	System overview/review of unit operations	1–4
	Facility walk through	1–2
	Principles of membrane operation	1–4
	Control system	4–24
	Pretreatment unit operations	1–4
	Operation/pumps/instrumentation	4–16
	Post-treatment	1–2
	Chemical cleaning (including demonstration)	4–16
	Integrity testing/module isolation and repair	4–8
	Monitoring/troubleshooting/data normalization	2–4
	Details of control system/remote monitoring	1–2
	Total	24–86
After commissioning	Open discussion with membrane manufacturer	1–3 days
After 3 months	Open discussion with membrane manufacturer	1–3 days

the manufacturer or contractor should be obligated to correct the problem before the acceptance testing is determined to be successfully completed. Utilities that apply membrane filtration for compliance with US EPA regulations may use the acceptance testing period to establish the required control limits for direct integrity testing and continuous indirect integrity monitoring, subject to any particular state requirements.

7.7. Operator Training

Because operators may not be familiar with the various types of membrane filtration systems, operator training is an important program element to include in the initial start-up phase. Even if some experience was gained during pilot testing, this experience should serve as a supplement to training on the completed full-scale facility, rather than a substitute. This training can also help facilitate a smooth transition of responsibility from the equipment supplier to the utility. A sample operator training schedule is shown in Table 8.8.

8. ACRONYMS

ANSI	American National Standards Institute
AWWA	American Water Works Association
CIP	Clean-in-place
CSTR	Continuous stirred tank reactor
DBP	Disinfection byproduct
DOC	Dissolved organic carbon
HPC	Heterotrophic plate count
IMS	Integrated membrane system
LRV	Log removal value
MCF	Membrane cartridge filtration
MF	Microfiltration
NDP	Net driving pressure
NF	Nanofiltration
PAC	Powdered activated carbon
PFR	Plug flow reactor
RO	Reverse osmosis
SDI	Silt density index
SUVA	Specific UV absorbance
TCF	Temperature correction factor
TMP	Trans-membrane pressure
TOC	Total organic carbon
TSS	Total suspended solids
UF	Ultrafiltration
US EPA	United States Environmental Protection Agency
VCF	Volumetric concentration factor

9. NOMENCLATURE

A_d	Design membrane area (as adjusted for temperature), ft^2
A_m	Membrane surface area, ft^2
A_{20}	Membrane area required at 20°C reference temperature, ft^2
C_f	Concentration of suspended solids in influent feed to membrane system, mg/L
C_m	Concentration of suspended solids on feed side of membrane, mg/L
DOC	Dissolved organic carbon, mg/L
J	Flux, gal/ft^2/d
J_T	Flux at temperature T, gal/ft^2/d
J_{20}	Normalized flux at 20°C, gal/ft^2/d
J_{25}	Normalized flux at 25°C, gal/ft^2/d
M	Temperature- and pressure-normalized flux, gal/ft^2/d/psi
NDP	Net driving pressure, psi
P_c	Concentrate pressure, psi
P_f	Feed pressure, psi
P_p	Filtrate pressure (i.e., backpressure), psi
Q_b	Backwash flow, gpd (gpm)
Q_c	Concentrate flow from the membrane unit, gpd (gpm)
Q_f	Feed flow to the membrane unit, gpd (gpm)
Q_p	Filtrate flow, gpd (gpm)
$Q_p(x)$	Filtrate flow at position "x" within the membrane unit, gpd (gpm)
R	Recovery of the membrane unit, decimal percent
R_f	Resistance of the foulant layer, psi/gal/ft^2/d/cp
R_m	Intrinsic membrane resistance, psi/gal/ft^2/d/cp
R_t	Total membrane resistance, psi/gal/ft^2/d/cp
SUVA	Specific UV absorbance, L/mg m
t_b	Backwash duration, min
t_f	Filtration cycle duration, min
T	Water temperature, °C
TCF	Temperature correction factor, dimensionless
TDS_c	Concentrate TDS concentration, mg/L
TDS_f	Feed TDS concentration, mg/L
TDS_p	Filtrate TDS concentration, mg/L
TMP	Trans-membrane pressure, psi
TMP_T	Trans-membrane pressure (temperature-corrected), psi
TMP_{25}	Trans-membrane pressure at 25°C reference temperature, psi
U	Membrane-specific manufacturer-supplied constant, 1/°K
UV_{254}	UV absorbance at 254 nm, 1/m
V_f	Volume of feed, gal
VCF	Volumetric concentration factor, dimensionless
VCF_{avg}	Average VCF, dimensionless
VCF_{max}	Maximum VCF, dimensionless
VCF(x)	VCF at discrete position "x" within the membrane unit, dimensionless

(Continued)

x	Position in the membrane unit in the direction of tangential flow (i.e., $x = 0$ at the entrance to the first module), ft
μ_T	Viscosity of water at temperature T, cp
μ_{20}	Viscosity of water at 20°C, cp
τ	Solids retention time, min

REFERENCES

1. US EPA (2003) Membrane filtration guidance manual. US Environmental Protection Agency, Office of Water, EPA 815-D-03-008, Cincinnati, OH, June 2003
2. Chen JP, Yang L, Ng W-J, Wang LK, Song SL (2006) Membrane filtration. In: Wang LK, Hung YT, Shammas NK (eds) Advanced physicochemical treatment processes. Humana Press, Inc., Totowa, NJ
3. US Bureau of Reclamation (1998) The desalting and water treatment membrane manual: a guide to membranes for municipal water treatment, 2nd edn. Technical Service Center, Water Treatment Engineering and Research, Denver, CO
4. WEF (2006) Membrane systems for wastewater treatment. Water Environment Federation, New York, NY
5. AMTA (2008) Technology transfer workshop – membrane applications for all your thirsty needs. American Membrane Technology Association, Stuart, FL. http://www.amtaorg.com/publications.html, March 12–13 2008
6. Pearce G (2007) Introduction to membranes: filtration for water and wastewater treatment. Filtr Sep 44(2):24–27
7. Degremont Technologies (2010) Membranes. http://www.degremont-technologies.com/dgtech.php?rubrique56
8. Hung YT, Taricska JR, Lo HH, Wang LK, Li KH (2006) Powdered activated carbon adsorption. In: Wang LK, Hung YT, Shammas NK (eds) Advanced physicochemical treatment processes. Humana Press, Inc., Totowa, NJ
9. Wang LK (2006) Water chlorination and chloramination. In: Wang LK, Hung YT, Shammas NK (eds) Advanced physicochemical treatment processes. Humana Press, Inc., Totowa, NJ
10. Asano T, Burton FL, Leverenz HL, Tsuchihashi R, Tchobanoglous G (2007) Water reuse, issues, technologies and applications. McGraw-Hill, New York, NY, p 1570
11. Shammas NK, Chang CC, Wang LK (2006) Tertiary microscreening. In: Wang LK, Hung YT, Shammas NK (eds) Advanced physicochemical treatment processes. Humana Press, Inc., Totowa, NJ
12. Côté P, Sutherland I, Adams N, Cadera J (2003) Validation of membrane integrity methods in a challenge test with Bacillus subtillis. In: AWWA membrane technology conference proceedings, Atlanta, GA, 2–5 March 2003
13. Johnson WT (1997) Predicting log removal performance of membrane systems using in-situ integrity testing. In: AWWA annual conference proceedings, Atlanta, GA, 15–19 June 1997
14. Adham SS, Jacangelo JG, Laine J-M (1995) Low-pressure membranes: assessing integrity. J Am Water Works Assoc 87(3):62
15. Farahbakhsh K, Smith D (2003) Estimating air diffusion contribution to pressure decay during membrane integrity tests. In: AWWA membrane technology conference proceedings, Atlanta, GA, 2–5 March 2003

16. Anerjee A, Carlson K, Lozier J (2000) Monitoring membrane integrity using ultra high sensitive laser light. In: AWWA water quality technology conference proceedings, Salt Lake City, UT, 5–9 November 2000
17. US EPA (1999) Guidance manual for compliance with the interim enhanced surface water treatment rule: turbidity provisions. US Environmental Protection Agency, EPA 815-R-99-010, Cincinnati, OH, April 1999
18. American Water Works Association Research Foundation (2000) Fundamentals of drinking water particle counting. McGraw-Hill, New York
19. American Water Works Association Research Foundation (2000) Particle count method development for concentration standards and sample stabilization. McGraw-Hill, New York
20. International Organization for Standardization (1999) Hydraulic fluid power – on-line automatic particle counting systems for liquids – methods of calibration and validation, 11943, Geneva, Switzerland
21. ASTM (2000) Standard practice for calibration of a liquid-borne particle counter using an optical system based upon light extinction. American Society for Testing and Materials, F 658-00a, West Conshohocken, PA
22. Huang J-C, Shang C (2006) Air stripping. In: Wang LK, Hung YT, Shammas NK (eds) Advanced physicochemical treatment processes. Humana Press, Inc., Totowa, NJ
23. Neemann J, Freeman S, Cooke T, Delphos P (2010) Filtration membranes for drinking water applications. Water World. http://ww.pennnet.com/display_article/322060/141/ARTCL/none/none/1/Filtration-Membranes-for-Drinking-Water-Applications
24. ASTM (1995) Standard testing method for silt density index (SDI) of water. American Society for Testing and Materials, D 4189-95, West Conshohocken, PA
25. Energy Center of Wisconsin (2010) Membrane filtration fact sheet. Madison, WI. http://www.ecw.org/prod/425-1
26. Banerjee A, Hanson F, Paoli E, Korbe C, Kolman R, Nelson D, Smith K, Lambertson M (1999) Ultra low range instrument increases turbidimetric sensitivity by two orders of magnitude. In: AWWA water quality technology conference proceedings, Tampa, FL, October 31–November 4 1999
27. Colvin C, Brauer R, DiNatale N, Scribner T (2001) Comparing laser turbidimetry with conventional methods for monitoring MF and UF membrane integrity. In: AWWA membrane technology conference proceedings, San Antonio, TX, 4–7 March 2001
28. Cornwell D, Bishop M, Gould R, Vandermeyden C (1987) Water treatment plant waste management. American Water Works Association Research Foundation, Denver, CO
29. MacPhee MJ, LeGouellec JY, Cornwell D (2002) Treatment of MF residuals for contaminant removal prior to recycle. American Water Works Association Research Foundation, Denver, CO
30. Kenna EN, Zander AK (2000) Current management of membrane plant concentrate. American Water Works Association Research Foundation, Denver, CO
31. Mickley MM, Hamilton R, Gallegos L, Truesdall J (1993) Membrane concentrate disposal. American Water Works Association Research Foundation, Denver, CO
32. Mickley MM (2000) Major ion toxicity in membrane concentrate. American Water Works Association Research Foundation, Denver, CO
33. Mickley MM (2001) Membrane concentrate disposal: practices and regulation. United States Bureau of Reclamation Technical Service Center, Water Treatment Engineering and Research, Denver, CO
34. NSF (2002) ETV protocol for equipment verification testing for physical removal of microbiological and particulate contaminants. National Sanitation Foundation, Ann Arbor, MI

35. Vickers J (1993) Aspects of integrity testing and module construction for microporous membrane filters. Technical Paper, Memtec America Corporation
36. Glucina K, Laine J-M, Anselme C, Chamant M, Simonie P (1999) Acoustic sensor: a novel technique for low pressure membrane integrity monitoring. In: AWWA membrane technology conference proceedings, Long Beach, CA, February 28–March 3 1999
37. Lozier J, Colvin C, Kim J-H, Kitis M, Mariñas B, Mi B (2003) Microbial removal and integrity monitoring of high-pressure membranes. American Water Works Association Research Foundation, Denver, CO
38. ASTM (1998) Standard practices for detecting leaks in reverse osmosis devices. American Society for Testing And Materials, D 3923–94, West Conshohocken, PA
39. Banerjee A, Lozier J, Carlson K (2001) An on-line, multi-sensor, membrane filtration permeate water quality monitoring system. AWWA membrane technology conference proceedings, San Antonio, TX, 4–7 March 2001
40. GEA Process Engineering Inc. (2010) Membrane filtration. http://www.niroinc.com/membrane_filtration.asp

9
Adsorption Desalination: A Novel Method

Kim Choon Ng, Ibrahim I. El-Sharkawy, Bidyut Baran Saha, and Anutosh Chakraborty

CONTENTS

Abstract The search for potable water for quenching global thirst remains a pressing concern throughout many regions of the world. The demand for new and sustainable sources and the associated technologies for producing fresh water are intrinsically linked to the solving of potable water availability and hitherto, innovative and energy efficient desalination methods seems to be the practical solutions. Quenching global thirst by adsorption desalination is a practical and inexpensive method of desalinating the saline and brackish water to produce fresh water for agriculture irrigation, industrial, and building applications. This chapter provides a general overview of the adsorption fundamentals in terms of adsorption isotherms, kinetics, and heat of adsorption. It is then being more focused on the principles of thermally driven adsorption desalination methods. The recent developments of adsorption desalination plants and the effect of operating conditions on the system performance in terms of specific daily water production and performance ratio are presented. Design of a large commercial adsorption desalination plant is also discussed herein.

Key Words Adsorption desalination • adsorption kinetics • adsorption isotherms • heat of adsorption • heat recovery • mass recovery • performance ratio • specific daily water production

From: *Handbook of Environmental Engineering, Volume 13: Membrane and Desalination Technologies*
Edited by: L.K. Wang et al., DOI: 10.1007/978-1-59745-278-6_9

1. INTRODUCTION

With increasing world's population and the geographical mal-distribution of water resources, the scarcity of potable water is a serious problem that calls for practical solutions. The total amount of global water resources reserves is estimated to be about 1.4 billion cubic kilometers, but fresh water constitutes only about 2.5% of the total reserves. Furthermore, the majority of fresh water in the world is trapped in the atmosphere, polar ice, underground, and other inaccessible forms, and only about 0.014% of the total fresh water is directly available to human activities (1).

Desalination is a process by which the salinity in sea or brackish water is removed and converted to potable or fresh water and it is a practical solution for solving water shortage problems. Such processes can be divided into two groups: The first group includes heat-driven processes (distillation) such as multistage flashing (MSF), multiple effect distillation (MED), and solar distillation. The second method employs electric power-driven processes that include freezing, mechanical vapor compression, electrodialysis, and reverse osmosis (RO).

All of the aforementioned types of desalination are found to be either highly energy intensive or are prone to serious corrosion and fouling problems when operated at elevated evaporating temperatures (2). However, adsorption desalination (AD) plant differs in concept from the conventional desalination methods and the attractive features are as follows:

- It employs low-temperature waste heat from industrial processes which are available in refineries, the co-generation plants, marine engines of ships, and the power plants. The temperature of waste heat needed to power the AD cycle is in the range of 65–85°C which is essentially "free" if unused, and the heat would have been purged into the atmosphere.
- It is environment friendly as only silica gel ($SiO_2 \cdot nH_2O$) and water are employed as adsorbent–adsorbate pair.
- The AD cycle has almost no major moving parts except the valves for the vapor and coolant flows.
- It has low electricity usage as only water pumps are used.

The advantages of an AD cycle over other conventional methods would become apparent with the analysis of the specific energy consumption (kWh/m^3 of water). Table 9.1 compares the energy cost of water production between the AD cycle and the conventional methods, such as the MSF, MED, and RO. In spite of the laboratory scale experiments, the AD cycle is found to give the lowest energy consumption at about 5.63 kWh/m^3 or equivalent to US$0.25/m^3, while the highest cost of production is derived by MSF at US$0.647/m^3. With economy of scale of a full-scale plant, the energy consumption rate of AD method could be reduced to below 1.5 kWh/m^3 as the parasitic pumping consumption is decreased.

To understand the principles of adsorption desalination operation, one should have a background about the theory of adsorption, including adsorption isotherms, adsorption kinetics, heat of adsorption, and the operation of adsorption systems. These backgrounds will be discussed in the following subsections.

Table 9.1
Energy costs comparison for various methods of desalination

Method of desalination	Thermal energy consumed kWh/m^3 (A)	Electric energy kWh/m^3 consumed (B)	Primary fuel input kWh/m^3 $C = (A/\eta_b + B/\eta_c)$	Energy cost of water US\$/m^3 = 5(C × 3.6)/1,055
Multistage flash (MSF)	19.4	5.2	37.9	0.647
Multieffect distillation (MED)	16.4	3.8	30.5	0.520
Vapour compression (VC)	–	11.1	29.2	0.497
Reverse osmosis (RO) – single pass	–	8.2	21.5	0.366
Reverse osmosis (RO) – double pass	–	9.0	23.7	0.403
Adsorption desalination (*high-grade* water)	Energy from waste heat	5.63[a] Author's data (3)	14.8	0.25

All data are extracted from Seawater Desalination in California, California coastal commission Chapter 1: Energy Use section, http://www.coastal.ca.gov/index.html. The conversion units of 1 AF = 1,345 m^3, one million BTU (1.055 GJ) of Natural gas costs US$5 (adopted from Singapore's natural gas prices in 2005). The electricity conversion efficiency, η_c, of power plants is 38% and the efficiency of boiler is 80%.
[a]The authors believe that the parasitic pumping energy could be further reduced (about 1.2 kWh/m^3) in a full-scale plant where the pumping operation is much more efficient.

1.1. Description of Sorption Processes

The rubric "sorption" represents both adsorption and absorption processes. Adsorption (solid–vapor) phenomenon concerns the separation of a substance (adsorbate) from one phase, accompanied by its accumulation or concentration of another (adsorbent). The adsorbing substance is the adsorbent and the material adsorbed on that substance is the adsorbate. On the contrary, absorption process is a process in which material transferred from one phase to another (e.g., liquid) interpenetrates the second phase to form a solution (4).

Adsorption processes can be classified into physical adsorption (physisorption) or chemical adsorption (chemisorption). Physical adsorption involves only relatively weak intermolecular forces which are caused by Van der Waals forces, while chemical adsorption involves essentially the formation of a chemical bond between the sorbate molecule and the surface of adsorbent (5, 6). It should be mentioned that the adsorptive actions is physical for almost all of solid adsorbents that are commonly used in adsorption cooling systems.

1.2. Adsorption Equilibrium

When an adsorbent is in direct contact with a surrounding fluid (usually vapor), adsorption phenomenon takes place and after a sufficient long time, both the adsorbent and the surrounding fluid reach the equilibrium. In this state, the amount of adsorbate uptake per kg of adsorbent, W (kg/kg), is called adsorption equilibrium capacity or equilibrium uptake. Both the definitions are commonly used in the adsorption field.

Adsorption equilibrium capacity, W, depends on the equilibrium pressure, P; the adsorbent temperature, T; and the nature of gas–solid system, thus it may be written as (7)

$$W = f(P, T, \text{physio-properties of adsorbent}). \tag{1}$$

The functional form of the equation, f, may be complex and it is usually determined experimentally. If the refrigerant vapor pressure is changed while the temperature is kept constant, the plot of the amount adsorbed against the pressure is called the adsorption isotherm. When the refrigerant is a vapor, the results can be expressed in terms of relative vapor pressure. This implies that W is plotted against P/P_s rather than against P itself, where P_s denotes saturated vapor pressure. There is no simple quantitative theory for predicting adsorption isotherms in detail from known parameters. However, most of the gas and/or vapor adsorption isotherms can be categorized by means of Brunauer classification (8). According to the Brunauer classification, adsorption isotherms can be classified into five types, as shown in Fig. 9.1. Isotherms of type I are generally true for microporous adsorbents. Adsorption of ethanol on activated carbon fiber is an example of this type of isotherms (9). Type II is observed in adsorbents having a wide range of pore sizes, with either mono- or multimolecular adsorption layers. Adsorption of benzene vapor on graphitized carbon represents an example of this type of isotherms. Type III is uncommon and they include capillary condensation in addition to the multimolecular adsorption layer. Adsorption of bromine on silica gel is an example of this type of isotherms. Isotherms of type IV involve the formation of two surface layers on the adsorbents having pore size much larger than the molecule diameter of the adsorbate. Isotherms of type V are observed on the adsorbents involving large forces produced by intermolecular attraction, such as adsorption of phosphorus on NaX.

1.3. *Adsorption Kinetics*

In practical adsorption applications, especially in adsorption cooling/heat pump systems, the maximum adsorption capacity of the adsorbent cannot be fully utilized. This is because it takes a long time to reach the equilibrium state. In order to estimate practical or dynamic adsorption capacity, adsorption kinetics should be conducted (6). LDF approximation, which is driven by Glueckauf (10), as shown by Eq. (2), is the most widely used and powerful method to estimate the adsorption/desorption rates.

$$\frac{\mathrm{d}w}{\mathrm{d}t} = k_s a_p (W - w). \tag{2}$$

In this equation, W is the equilibrium adsorption capacity of adsorbate per kg of adsorbent, w is the instantaneous adsorption capacity (uptake), and the left hand side of the equation represents the adsorption/desorption rates. The overall mass transfer coefficient ($k_s a_p$) for adsorption is given by

$$k_s a_p = F_0 \frac{D_s}{R_p^2}. \tag{3}$$

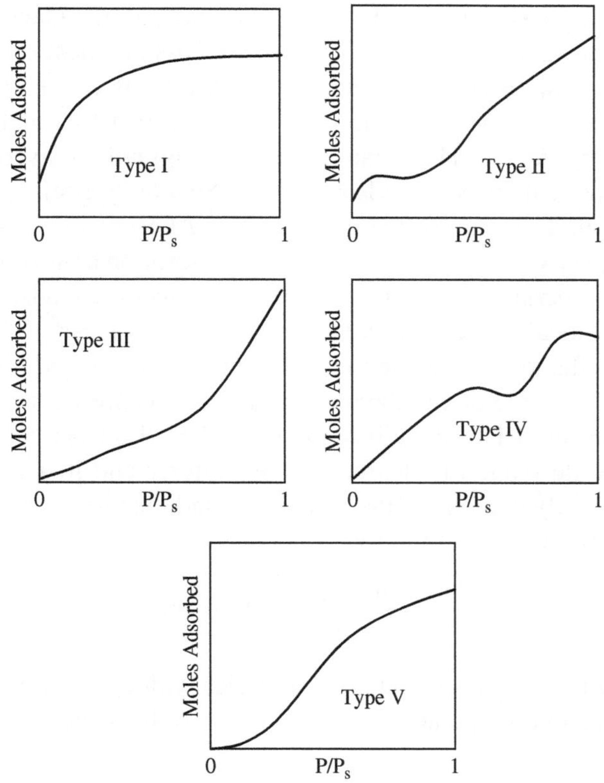

Fig. 9.1. Five different Brunauer adsorption isotherms (8).

Here, R_p denotes the average radius of an adsorbent particle and F_0 is a constant depending on the adsorbent shape. The surface diffusivity D_s is expressed as

$$D_s = D_{so} \exp\left(-\frac{E_a}{RT}\right), \tag{4}$$

where D_{so} is a pre-exponential term, E_a denotes activation energy, R denotes the universal gas constant, and T stands for temperature.

In the linear-driving force (LDF) method, the key parameter is the determination of the overall particle mass transfer coefficient. Ruthven (5) reported that both the diffusion time constant and the overall mass transfer coefficient can be estimated by tracking the experimental vapor-uptake behavior using the solution of a diffusion equation where the particle mass transfer coefficient ($k_s a_p$) could be determined. There exits a constant relationship between the overall mass transfer coefficient to the particle shape (e.g., described by the radius) and local concentration profile within the adsorbent. Liaw et al. (11) show that a parabolic concentration of the form, $w(t,r) = a_0 + a_2 r^2$, as prescribed within the spherical particle, led to the well-known LDF model. Li and Yang (12) have demonstrated a general profile of the LDF model but modified with an exponent index n, where n is an integer ≥2. They commented that the parabolic profile of Liaw et al. (11) is one of the many possible local solutions. An article of Sircar and Hufton (13) highlighted that as long as the local concentration profile of the adsorbent is continuous, a constant relationship exits between $k_s a_v$ and the diffusion time, D_s/R^2, which is given by Eq. (3). This relationship was originally derived by Glueckauf (10), where F_0 is 15 for spherical shapes of adsorbents. El-Sharkawy et al. (14, 15) measured the adsorption rate of ethanol onto activated carbon fiber (ACF) of type A-20 using a thermo-gravimetric analyzer (TGA) unit over a wide range of adsorption temperatures, typically from 27 to 60°C, a region of operation which is useful for the processes of adsorption chillers. They have proposed a novel local concentration profile, $w(t,r) = a_0 + a_2 r^{2+k}$, which is effective in attributing the higher ethanol uptake due to the presence of meso- and micropores within the ACFs (see Fig. 9.2). The value of k in the local concentration profile has been independently validated using the measured data and it is found to be about 1.5 when F_0 is 11. The value of k is also valid for the wide range of adsorption temperatures and it is much higher than the conventional value of 8. By combining the values of k and the diffusion time constant, D_s/R^2, the LDF model is proposed to be of the form

$$\frac{\mathrm{d}w}{\mathrm{d}t} = \frac{(8+3k)D_\mathrm{s}}{R_\mathrm{p}^2}(W - w), \tag{5}$$

which is valid for a cylindrical adsorbent. The numerical values of k and D_s/R^2 are valid over a wide range of adsorption temperatures (27–60°C) as well as the adsorbent surfaces with

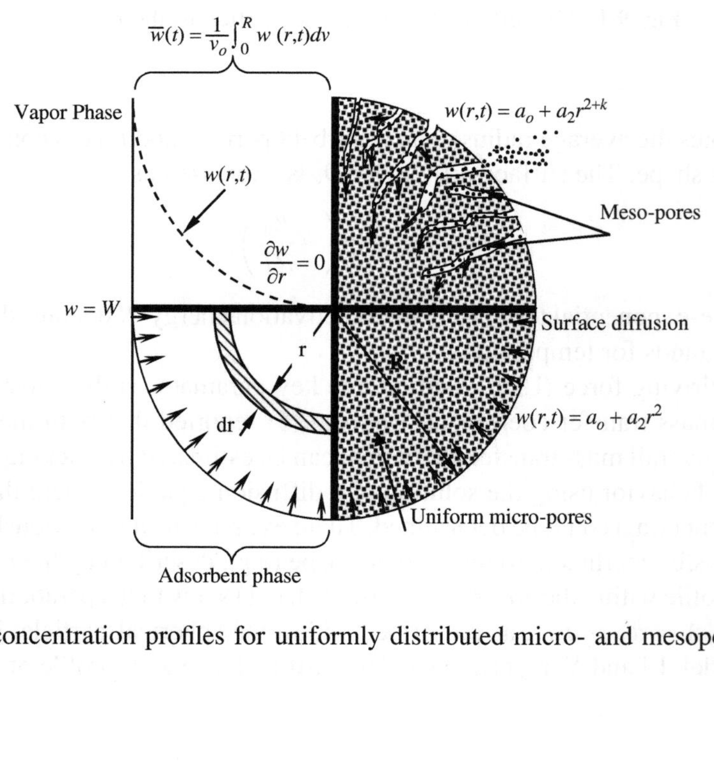

Fig. 9.2. Local concentration profiles for uniformly distributed micro- and mesopores structures in ACF (14).

meso- and micropores. Similar LDF models for one-dimensional slab and spherical shapes where the parentheses terms are $(3 + k)$ and $(15 + 3k)$, respectively, are reported.

1.4. Heat of Adsorption

The isosteric heat of adsorption is a combined property of an adsorbent–adsorbate combination. It is a major contributor to the heating inventories of adsorption refrigeration and gas storage systems and also cooling requirements of adsorption heat pumps. Isosteric heat of adsorption is traditionally expressed as a function of concentration as its dependency on temperature is relatively weak (16, 17). Clausius–Clapeyron equation is commonly used to estimate the heat of adsorption at a constant concentration, as given by Eq. (6):

$$Q_{\mathrm{st}(W)} = \frac{-R\,\partial \ln P}{\partial(1/T)}. \tag{6}$$

Here, Q_{st} denotes the isosteric heat of adsorption in kJ/kg, R is the gas constant in kJ/kg K, P is the equilibrium pressure in kPa, and T is the adsorbent temperature in K. Critoph (18) proposed a relation (Eq. 7) that is derived by direct integration of Eq. (6) to estimate the isosteric heat of adsorption for practical applications in solar adsorption cooling systems.

$$Q_{\mathrm{st}(W)} = \frac{R\,\ln(P_2/P_1)}{(1/T_1) - (1/T_2)}. \tag{7}$$

This assumption is validated by the linearity of Clausius–Clapeyron diagram for a short temperature range. However, as the adsorbate concentration changes during adsorption and desorption processes, the isosteric heat of adsorption is usually calculated at the mean concentration. In adsorption cooling heat pump applications, an average value of heat of adsorption within a certain concentration range is usually used for system analysis and design. In 1995, Cacciola and Restuccia (19) presented a correlation to estimate heat of adsorption as a function of concentration (Eq. 8) for activated carbon methanol and zeolite water pairs.

$$\frac{Q_{\mathrm{st}(W)}}{R} = a_0 + a_1 W + a_2 W^2 + a_3 W^3. \tag{8}$$

They determined the values of the constants a_0, a_1, a_2, and a_3 based on their experimental data of Zeolite 13X-water and AC-methanol, and these are furnished in Table 9.2.

It can be noticed that the correlation derived by Critoph (18) is only valid for a relatively narrow temperature range. Moreover, his correlation is valid only for a constant concentration.

El-Sharkawy et al. (20) proposed a non-dimensional empirical correlation which partitions the contributions of the concentration and temperature dependence (Eq. 9). The correlation is tested out against data obtained from experimental isotherms of ethanol adsorption on ACFs [ACF (A-20) and ACF (A-15) (9) and HFC 134a] on two specimens of activated carbon powders and one specimen of carbon granules (16).

$$\frac{Q_{\mathrm{st}} - h_{\mathrm{fg}}}{E} = \left[\ln\left(\frac{W_0}{W}\right)^{1/n}\right] + A\left(\frac{T}{T_{\mathrm{c}}}\right)^{B}. \tag{9}$$

Table 9.2
Numerical coefficient of Eq. (8) (19)

Adsorbent–adsorbate pair			
Constants	Zeolite 4A-water	Zeolite 13X-water	Activated carbon-methanol
a_0	−7,698.85	−7,373.78	−6,003.58
a_1	214.981	67.2292	63.1516
a_2	−18.4589	0.562447	−2.60587
a_3	0.152605	-3.4867×10^{-3}	40.5379×10^{-3}

Table 9.3
Thermophysical and experimental constants of the assorted refrigerants (20)

Adsorbent	Ethanol	R134a
T_c (K)	489.15	374.21
(A × E) (kJ/kg)	927	156
B	9.75	6.25

Table 9.4
Adsorption parameters of the assorted adsorbent/adsorbate pairs (20)

Adsorbent	Adsorbate	n	A	W_0 (kg/kg)	E (kJ/kg)	r^2
ACF (A-20)	Ethanol	2	6.717	0.797	138	0.999
ACF (A-15)	Ethanol	2	5.297	0.570	175	0.973
Maxsorb II	HFC 134a	1.3	2.136	2.058	73	0.994
Fluka	HFC 134a	0.9	2.079	0.592	75	0.987
Chemviron	HFC 134a	1.45	1.173	0.354	133	0.96

Here Q_{st} is the isosteric heat of adsorption, h_{fg} is the latent heat, and E is the adsorption characteristic parameter that can be evaluated experimentally. W is the mass of adsorbate per kg of adsorbent and W_0 is the maximum adsorption capacity, kg/kg, T stands for the temperature in K, and T_c is the critical temperature. n is an exponential constant that gives the best fitting of Dubinin–Astakhov (DA) equation. A and B are experimental constants that depend on the adsorbent/adsorbate pair. The relevant constants and the parameters are shown in Tables 9.3 and 9.4.

Chakraborty et al. (21, 22) proposed a theoretical framework for the estimation of the isosteric heat of adsorption between an adsorbate (vapor) and an adsorbent (solid) based on the thermodynamic requirements of chemical equilibrium, Maxwell relations, and the entropy of the adsorbed phase. Using these relations and the chemical potential gradients of pressure, *P*, and temperature, *T*, with respect to entropy and specific volumes, respectively, a working expression for heat of adsorption is formulated (Eq. 10). The adsorbate properties and the vapor uptake isotherms are also employed to predict the Q_{st} behaviors.

$$Q_{st} = T v_g \left(\frac{\partial p}{\partial m_a} \right)_T \frac{dm_a}{dT} + Q_{st}|_{m=\text{const}}, \tag{10}$$

where v_g is the specific volume of the gas phase, m_a is the mass of adsorbed phase, and *P* stands for pressure. In this equation, the first term includes the behavior of adsorbed mass with respect to both the pressure and the temperature changes during an adsorbate uptake, while the second term is the conventional form of the heat of adsorption. The predictions are verified against a previously published experimental adsorption data. Figure 9.3a–c shows three sets of published data, and they are used to validate Eq. (10). The first set of data pertains to the Q_{st} of methane [CH_4] on silicate at 23°C (23). Modeling these measured data

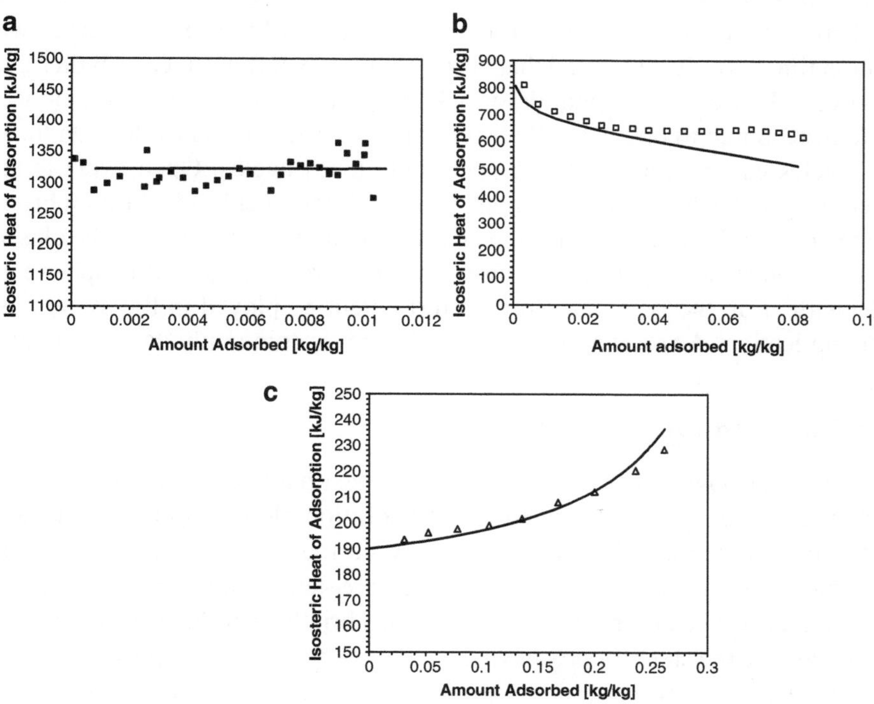

Fig. 9.3. (**a**) The isosteric heat of adsorption of methane (CH_4) on silicate at 23°C; *filled diamond* denotes the experimental data (23) and *solid line* denotes the predictions by Eq. (10) where the adsorption isotherm is modeled with the Langmuir equation (21). (**b**) The isosteric heat of adsorption of carbon dioxide (CO_2) on zeolites (H-ZMS-5) at 22.5°C. *Open square* denotes the experiments and *solid line* denotes the predictions of Eq. (10) where the adsorption isotherm is modeled with the Dubinin–Astakhov equation (21). (**c**) The isosteric heat of adsorption of sulfur hexafluoride (SF_6) on zeolites (NaX) at 31.6°C. *Open triangle* denotes experiments and *solid line* denotes predictions of Eq. (10). The adsorption isotherm is modeled with the Langmuir equation (21).

with the Langmuir isotherms, the right-hand side of Eq. (10) gives the full predictions of Q_{st}. A good agreement between experiments and prediction is observed for vapor uptake up to 10.5% of the adsorbent mass. The second set of experimental data relates to the heat of adsorption of CO_2 on zeolites [H-ZMS-5] at 22.5°C (24), as shown in Fig. 9.3b. Good agreement is observed for an initial uptake up to 3% of the adsorbent mass. At higher vapor loading, however, the experimental values are observed to be higher. The isotherm model employed here is the DA equation that has been applied to obtain predictions of Eq. (10). Lastly, the heat of adsorption data of sulfur hexafluoride [SF_6] on zeolites [NaX] at 31.6°C (24) show again a good agreement for an adsorbate uptake up to 27% of the mass of adsorbent, as shown in Fig. 9.3c, and the Langmuir isotherm model has been assumed.

1.5. *Classification of Thermally Driven Sorption Systems*

Heat-driven sorption systems can be simply classified into absorption (liquid–gas) and adsorption (solid–gas) systems, as shown in Fig. 9.4 (25). Generally speaking, the main difference between adsorption and absorption is the nature of the sorbent pair as well as the cycle duration time (26). It is reported that absorption cycle performance is higher than that of adsorption cycle. However, the main drawback of absorption cycles is that they need a high temperature heat source to operate. All absorption cycles are closed cycles. On the contrary, adsorption cycles can be classified into open and closed cycles. Open cycle is applied in desiccant cooling and dehumidification applications. Calcium chloride, lithium bromide, and silica gel are commonly used as desiccants. Closed adsorption systems can be classified into high-pressure (above atmospheric pressure) and low-pressure (sub-atmospheric pressure) systems depending on the type of adsorbent-refrigerant employed in the system as well as the system application. As shown in Fig. 9.4, silica gel is commonly used as an adsorbent.

1.6. *Basic Closed Adsorption Cycles*

The simplest heat-powered adsorption cooling or heat pump cycle composes of three heat exchangers, namely, evaporator, condenser, and sorption element (adsorber/desorber reactor), as shown in Fig. 9.5a. The cycle consists of four batch-operated thermodynamic processes which are shown in Fig.9.5b, the pressure–temperature–concentration diagram (Clapeyron diagram). The processes can be explained briefly as follows, where the subscripts *a*, *b*, *c*, and *d* indicate the states of processes *a–b*, *b–c*, *c–d*, and *d–a*, respectively, of Fig. 9.5a.

Adsorption and evaporation process (a–b): the sorption element is connected to the evaporator, which makes it possible to adsorb the refrigerant vapor from the evaporator. During the adsorption process, the adsorber is cooled from T_a to T_b. Adsorption heat is removed by a coolant that flows inside the sorption element. Refrigerant concentration in the bed increases until reaches its maximum value, W_b.

Preheating process (b–c): the sorption element is isolated from both the evaporator and the condenser. Heat is added by an external heat source to the sorption element at constant refrigerant concentration, which results in increase in temperature of refrigerant from T_b to T_c. Sorption element pressure increases from that of the evaporator to the condenser pressure.

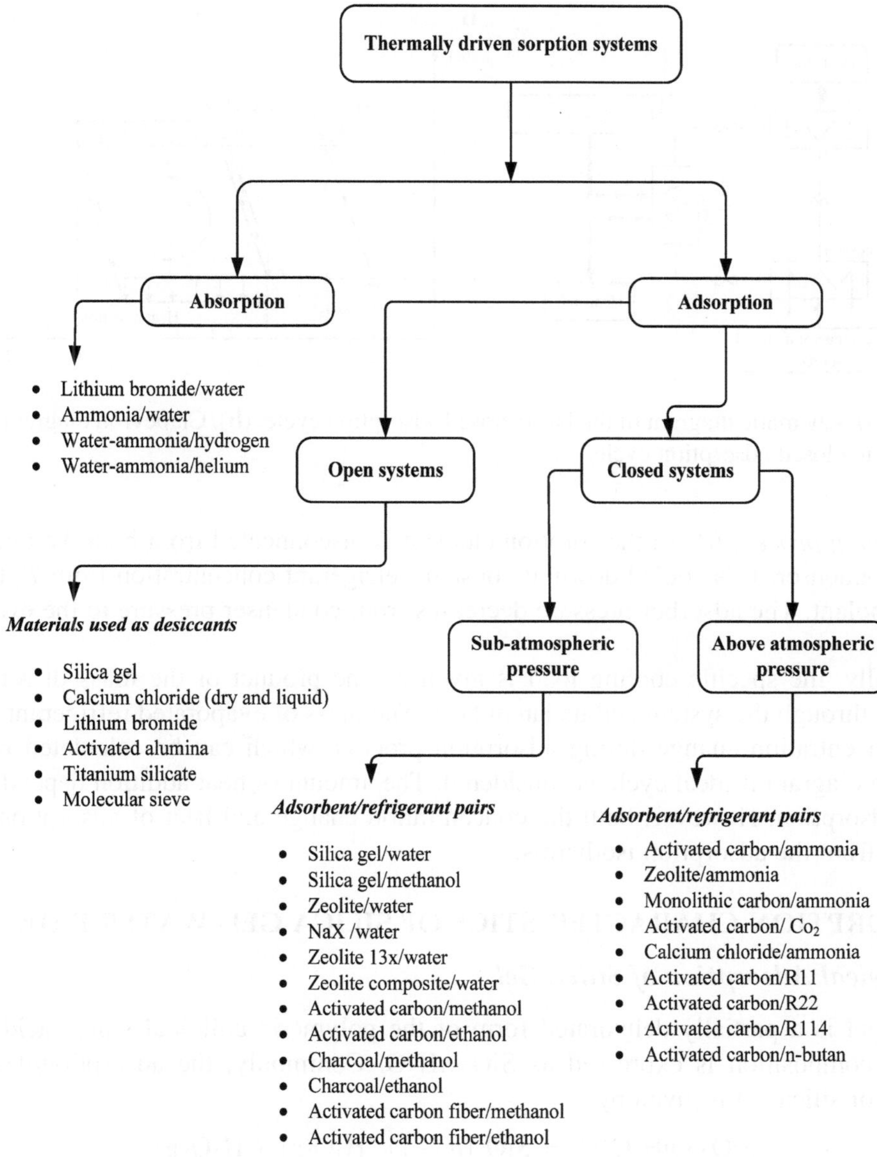

Fig. 9.4. Schematic diagram of the classifications of sorption systems (25).

Desorption and condensation process (c–d): the sorption element is connected to the condenser and the heat input from the external source continues in this process. The refrigerant vapor is desorbed at constant pressure, P_{cond}, while the temperature increases until it reaches the regeneration temperature, T_d. The desorbed refrigerant is continually provided to the condenser, where condensation takes place at T_{cond}. Condensation heat is removed by the coolant that flows inside the condenser heat transfer tubes. In this process, the concentration decreases until it reaches as equilibrium at T_{d}.

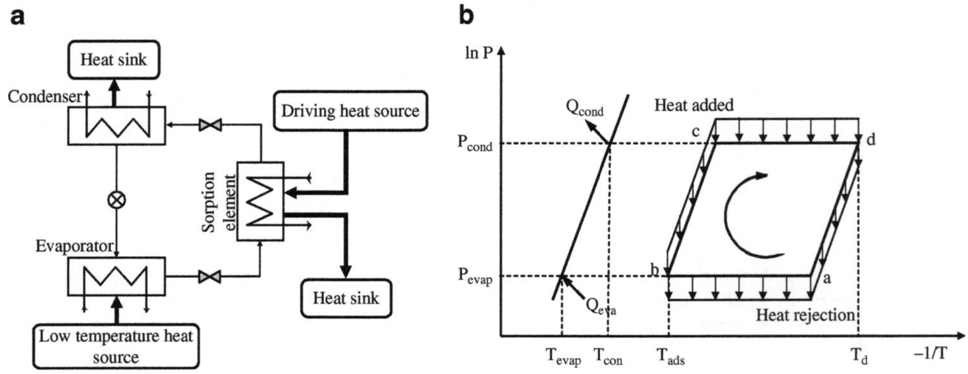

Fig. 9.5. (**a**) Schematic diagram of the basic closed adsorption cycle. (**b**). Clapeyron diagram (ln *P* vs. 1/T) of basic closed adsorption cycle.

Precooling process (d–a): the sorption element is disconnected from both the evaporator and the condenser. It is cooled down at constant refrigerant concentration from T_d to T_a by using a coolant. The adsorber pressure decreases from condenser pressure to the evaporator pressure.

Generally, the specific cooling load is given by the product of the mass of refrigerant circulated through the system and its latent heat. The mass of evaporated refrigerant is equal to the concentration change during adsorption process, which can be calculated by using Clapeyron diagram if ideal cycle is considered. The amount of heat addition depends on the heat of adsorption. However, both the concentration change and heat of adsorption can be extracted from the adsorption isotherms.

2. ADSORPTION CHARACTERISTICS OF SILICA GEL–WATER PAIR

2.1. Physical Adsorption of Silica Gel

Silica gel is a partially dehydrated form of the polymeric colloidal silicic acid and its chemical composition is expressed as $SiO_2{\cdot}nH_2O$. Commonly, the adsorption/desorption equation for silica gel is given by

$$SiO_2 \cdot nH_2O(s) \Leftrightarrow SiO_2(n-1) \cdot H_2O(s) + H_2O(g), \tag{11}$$

where s and g denote, respectively, solid phase and vapor phase. Commercially available silica gels are produced from silicate (20 wt% of SiO_2), the basic material in a sol–gel process that employs aluminum ions as a growth inhibitor for the primary silica gel. The production processes of silica gel, as described by Ito et al. (27), comprise the following:

(i) Reaction: Sodium silicate (SiO_2 – 20 wt%) + H_2SO_4, (ii) monomer, (iii) dimmer, (iv) particle, (v) silica solution (aqueous aluminum sulfate), (vi) three-dimensional gel network, (vii) pH_2 washing of aluminum sulfate, and (viii) drying.

Silica gel is also described as a dehydrated colloidal gel produced under controlled conditions from sulfuric acid and sodium silicate solution (28). Being chemically oxidized to silicon dioxide (SiO_2), it is made of an extremely hard and inert material which does not dust in service and it is quite inert when it comes into contact with most chemicals.

The adsorptive action of silica gel for vapors is a physical effect and not a chemical reaction: The particles suffer no change in size or shape even when they are saturated with the adsorbate, and the particles appear perfectly dry. The adsorptive property of the gel emanates from its high porosity, where the dimensions of the pores are sub-microscopic. The internal structure of silica gel has been proposed for the sorption behavior, that is, silica gel is described as a body with sub-microscopic pores piled together in a homogeneous mass. The pores are formed by the juxtaposition of silica particles with a distribution of sizes.

Normal commercial silica gel will adsorb approximately 25–50% of its weight of adsorbate (water vapor) in a single component or mixture of saturated air. It has been estimated that 1 m^3 of adsorbent contains surface pores having a surface area of about 2.8×10^7 m^2. This enormous internal surface and infinite number of small diameter capillaries attract vapors and hold them as "adsorbed phase" in the pores of capillaries. The weight of any compound adsorbed on silica gel depends on its partial pressure in the original gas mixture, the temperature of the silica gel and of the gas mixture, the compressibility and surface tension of the condensed liquid, its ability to wet the pore surface, and the volume and shape of the pores. Silica gel can adsorb vapor from a gas mixture until the pores of the gel are filled to a point where the internal vapor pressure of the adsorbed phase in the pores at a given temperature approaches as a limit the partial pressure of the vapor in the surroundings at the same temperature.

The amount of an adsorbate that is adsorbed in silica gel at any gel temperature increases as the partial pressure of the vapor increases, and they would exist if the gas were "saturated" at the gel temperature. For example, silica gel at 28°C in contact with air saturated at this temperature will adsorb an equilibrium uptake from 0.4 to 0.45 kg of water per kg of gel. Increasing the temperature of the gel could result in a decrease in its adsorption capacity. Thus, it is clear that higher adsorption occurs with decreasing temperature. The maximum concentration of the vapor that is adsorbed occurs when the pores are saturated with vapor.

In a reactivation process, silica gel would arrive at a residual moisture content, and the useful concentration in the gel is the difference between the equilibrium concentration and the residual concentration under the conditions of use. Adsorption of an adsorbate liberates an equivalent amount of heat that is slightly greater than its latent heat of evaporation. The additional amount of heat is known as the "heat of wetting," and the sum of the latent heat plus the heat of wetting is the heat of adsorption. This heat of adsorption is released during the vapor uptake, causing a rise in temperature of the adsorbent and the associated bed. On the contrary, if heat is added to the silica gel, it could be regenerated or activated and the process of adsorption and desorption is almost "reversible" and could be repeated practically unlimited number of times without loss in efficiency or capacity. Properly activated silica gel will contain 2–6% of residual water, which is also the water content of the material as commercially supplied. The length of regeneration period depends upon the rate and temperature at which the heat is supplied and, it may vary from a few minutes to several hours.

The amount of heat required to regenerate silica gel varies with the design of the equipment, in addition to the supply of the heat necessary to release the adsorbed refrigerant from the gel, which is equivalent to the heat of adsorption. Heat must be added for rising the temperature of the adsorbent bed and the adsorber, and to overcome heat leaks in the system. The action of silica gel is practically instantaneous under dynamic adsorption conditions, and the length of the adsorption period may be arbitrarily established.

2.2. *Porous Properties of Various Silica Gels*

2.2.1. *Nitrogen Adsorption Experiment*

The suitability of an adsorbent for adsorption desalination application is determined by both the maximum adsorption capacity and the rate of adsorption and desorption of water vapor. These factors are inherent in the pore structure and total pore volume of the adsorbent, as well as the purity and hardness of the adsorbent. Three types of Fuji Davison silica gels (type A^{++}, type 3A, and type RD) have been selected for nitrogen adsorption isotherm and adsorption characteristic measurements. The porosity, total pore volume, average pore diameter, and surface area of the three different samples were evaluated from the adsorption isotherm of nitrogen (N_2), and the conditions of preparation of the samples are (a) supplying nitrogen at 77.4 K after out-gassing at temperature 300°C for 3 h, (b) a residual pressure at 10^{-3} Pa by using an Autosorb 1 – MP machine, and (c) the apparatus must be equipped with windows based built-in data reduction software, as shown in Fig. 9.6 (29).

The Autosorb 1 – MP machine operates by measuring the quantity of adsorbate in gas phase adsorbed onto or desorbed from an adsorbent surface (solid surface) at some

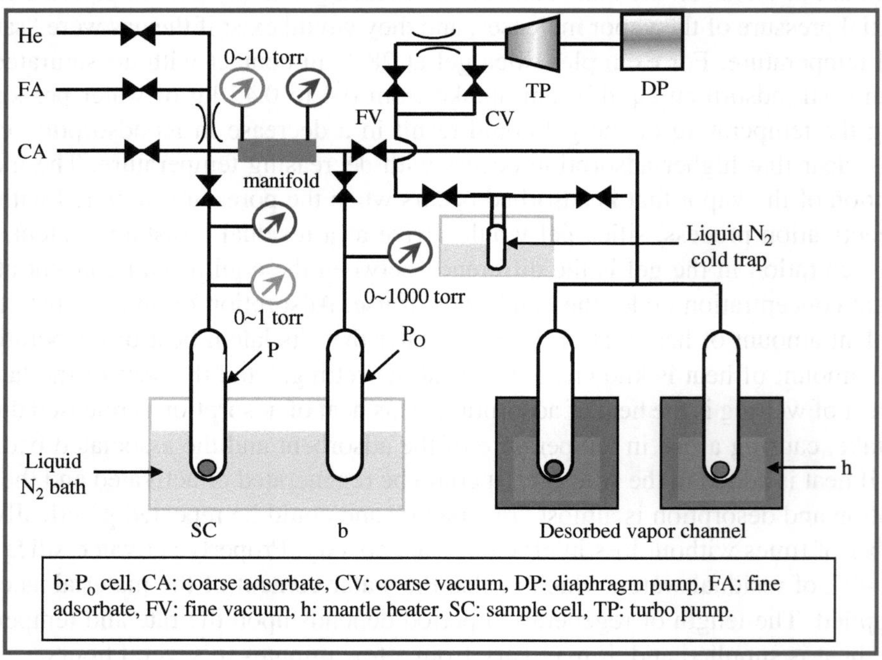

Fig. 9.6. Schematic of the Autosorb 1-MP machine (29).

equilibrium vapor pressure by the static volumetric method. The data are obtained by admitting or removing a known quantity of adsorbate gas into or out of a sample cell (SC) containing the solid adsorbent maintained at a constant temperature below the critical temperature of the adsorbate. As adsorption or desorption occurs, the pressure in the SC changes until an equilibrium is established. The quantity of adsorbate gas adsorbed or desorbed at the equilibrium pressure is the difference between the amount of adsorbate gas admitted or removed and the amount required to fill the space around the adsorbent (void space).

2.2.2. Porous Properties of Silica Gels

Figure 9.7 shows the N_2 adsorption/desorption isotherm data of three types of assorted silica gel samples. As the adsorption experiments are run at a temperature lower than the critical capillary condensation temperature, the adsorption theory of infinite cylinder without any external surface predicts type IV adsorption isotherm for the silica gel curves, which are characterized by solid lines for adsorption and dotted lines for desorption. Characteristic features of type IV isotherm are its hysteresis loop that is associated with capillary condensation and the limiting uptake over a range of high P/P_0 (30). It can be seen from Fig. 9.7 that the RD type silica gel has the highest adsorption capacity, followed by the type A^{++} gel. The breakthrough curves for all the three silica gels show that the adsorption and desorption branches for each gel do not pile on one another when the value of P/P_0 is greater than 0.4, which implies the presence of adsorption–desorption hysteresis. This implies that the silica gel isotherms are not completely reversible. Surface area, pore volume, and pore size data for all the three silica gel samples are presented in Table 9.5. As can be seen from Table 9.5, the A^{++} and RD silica gels possess similar BET surface area and total pore volume measured by DR and HK methods, which is followed by 3A type silica gel. However, the N_2 adsorption

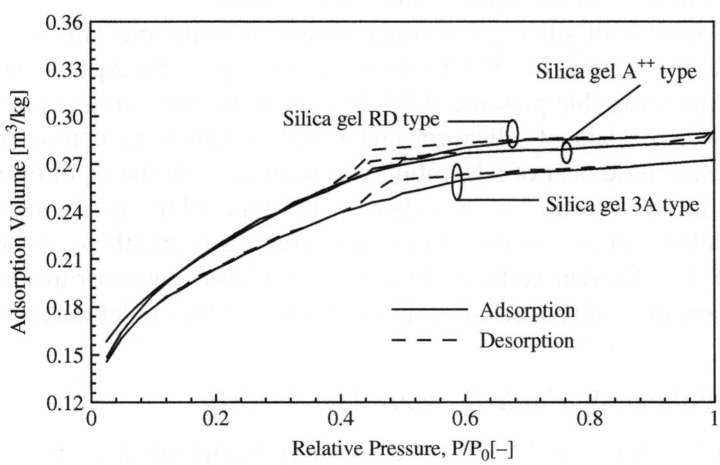

Fig. 9.7. Adsorption/desorption isotherm of various silica gels (29).

Table 9.5
Surface area, micropore volume, and pore size data of various parent silica gels (29)

Sample name	S_{BET} (m^2/kg)	$V_{0.99}$ (m^3/kg)	Daver (nm)	MPD (Å)	Dubinin–Radushkevich		Horvath–Kawazoe	
					S_{mic} (m^2/kg)	V_{mic} (m^3/kg)	V_{mic} (m^3/kg)	MPW (Å)
SiO_2 (3A)	0.69×10^6	4.22×10^{-4}	2.44	24.48	0.90×10^6	3.20×10^{-4}	3.00×10^{-4}	12.23
SiO_2 (RD)	0.74×10^6	4.48×10^{-4}	2.42	24.27	0.93×10^6	3.29×10^{-4}	3.17×10^{-4}	12.52
SiO_2 (A^{++})	0.75×10^6	4.53×10^{-4}	2.42	24.08	0.96×10^6	3.41×10^{-4}	3.17×10^{-4}	12.27

S_{BET}: specific surface area measured by BET method; $V_{0.99}$: the total pore volume estimated at relative pressure 0.99; Daver = (4.0 × $V_{0.99}$)/S_{BET}, according to the cylindrical pore model; MPD: mean pore diameter; S_{mic}: micropore surface area; V_{mic}: volume of micropore; MPW: mean pore width.

capacity of RD type silica gel is remarkably higher than that of type A^{++} silica gel when P/P_0 exceeds 0.4. So RD silica gel is recommended for adsorption desalination application.

3. ISOTHERMAL ADSORPTION OF WATER VAPOR ONTO TWO TYPES OF SILICA GELS

The common types of silica gel found in the commercial adsorption chillers are the Fuji Davison type “A,” and “RD.” Thermo-physical properties of both types of silica gels are reported by Ng et al. (31); the data are provided by Fiji Silysia Chemical Ltd., Japan, and summarized in Table 9.6. One can observe that the thermo-physical properties of these two types of silica gels are similar. Although the type “RD” silica gel has a slightly higher thermal conductivity than type “A” silica gel, the most significant difference between these two types of silica gels lies in their water vapor uptake characteristics.

Adsorption isotherms of silica gel + water have were experimentally investigated by Ng et al. (31) and later by Wang et al. (32). They used two independent experimental methods, viz. the constant volume–variable pressure (CVVP) system and the variable pressure TGA. The experimental isotherm data of silica gel and water system were compared by these two methods and the isosteric heat of adsorption was derived from the equilibrium information. Two types of silica gels (Fuji Davison type A and type RD) were investigated under an assortment of chillers’ operation conditions (temperatures from 303 to 358 K and pressures from 500 to 7,000 Pa). The data collected from the two independent experiments were compared with each other and their trends are consistent with those of the chiller manufacturer.

3.1. *Constant Volume–Variable Pressure Test Facility*

As shown in Fig. 9.8, the CVVP system mainly comprises a charging tank (material SS304) with volume of 572.64 ± 27 cm^3 inclusive of rated piping and valves, a dosing tank (material SS304) with volume of 698.47 ± 32 cm^3 inclusive of rated piping and valves,

Table 9.6
Thermo-physical properties of silica gels (31)

Property	Value	
	Type "A"	Type "RD"
Specific surface area (m^2/g)	650	650
Porous volume (ml/g)	0.36	0.35
Average pore diameter (Å)	22	21
Apparent density (kg/m^3)	730	800
pH value	5.0	4.0
Water content (wt%)	<2.0	–
Specific heat capacity (kJ/kg K)	0.921	0.921
Thermal conductivity (W/m K)	0.174	0.198
Mesh size	10–40	10–20

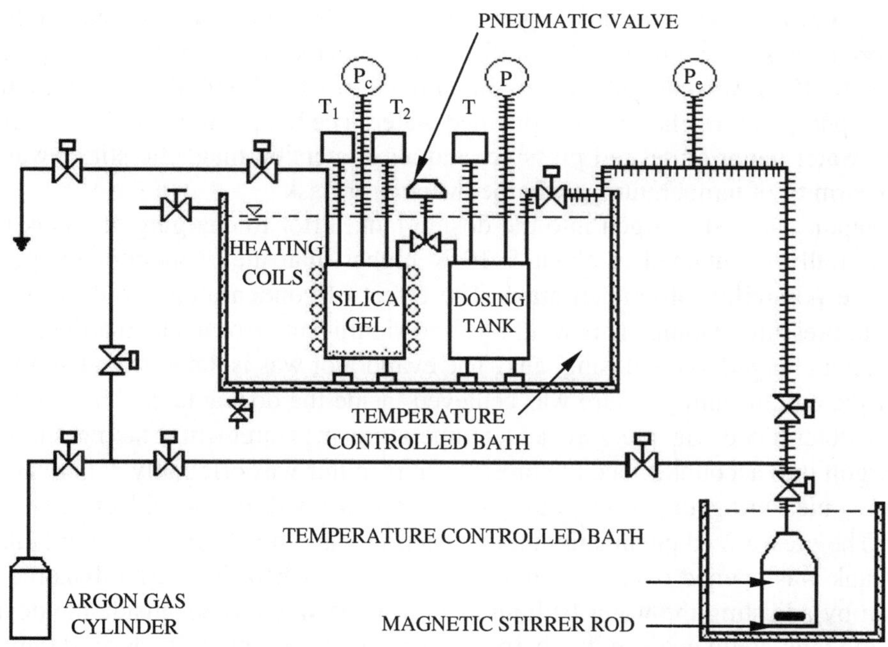

Fig. 9.8. Schematic diagram of the constant volume–variable pressure test facility (32).

an evaporator flask with a volume of 2,000 cm^3, and a vacuum system (a two-stage Edwards bubbler pump). The charging tank had been designed to have a high aspect ratio so that the adsorbent could be spread on the large flat base. Its external wall was covered by the heating tape to enable in situ regeneration of the silica gel. The silica gel is placed in the big bottom plate and could be assumed to be in contact with the pure water vapor only, which closely resembled the actual condition in an adsorption chiller. The dry mass of silica gel was determined by the calibrated moisture balance (Satorious MA40 moisture analyzer, uncertainty: 0.05% traceable to DKD standard) at 413 K. For the two types of silica gel, type "A" and type "RD" used in our experiments, mass of about 0.1–1.25 g was introduced into the charging tank for different pressure and temperature. Prior to the experiment, the dosing tank and charging tank plus related piping system, so-called test system, must be purged by the purified and dried argon, and evacuated and subsequently isolated. Based on the measurements involving only argon and silica gel, it was concluded that there was no measurable interaction between the inert gas and the adsorbent. The effect of the partial pressure of argon in the tanks was found to be very small and we also adjusted the vapor partial pressure to avoid the additional system error.

In order to maintain the system isothermal, the test system was immersed in a temperature-controlled bath (precision of control: $\pm$0.01 K). A two-stage rotary vane vacuum pump (Edwards bubbler pump) with a water vapor pumping rate of 315×10^{-6} m^3/s was equipped with this system for purging and vacuuming purposes. Argon with a purity of 99.9995% was set through a column of packed calcium sulfate before being used to purge the vacuum system. Upon evacuation, a residual argon pressure of 100 Pa can be achieved in the set-up. The evaporator flask was immersed in a different temperature-controlled bath (precision of control: $\pm$0.01 K). It was purged with argon with a purity of 99.9995% and evacuated using the vacuum pump before charging the distilled water. The temperature could be regulated to supply the water vapor at desired pressure, and an immersion magnetic stirrer was used to ensure uniformity of temperature inside the evaporator flask.

Water vapor was first charged into the dosing tank. Prior to charging, the tank temperature was initially maintained at about 5–10 K higher than the evaporator temperature to eliminate the possibility of condensation. The effect of condensation would cause significant error in pressure reading. This would lead to significant error in calculating actual mass of water vapor charged to the dosing tank. The evaporator was isolated from the dosing tank soon after the equilibrium pressure was achieved inside the dosing tank. The mass of water vapor was determined via pressure and temperature measurements, taking the effect of residual argon into account. Since the pressure transmitter was originally calibrated at room temperature, measuring the water vapor mass at this condition would ensure maximum accuracy. The pressurized pneumatic interconnecting valve between the dosing tank and the charging tank was opened once the whole test system reached the thermodynamic equilibrium. Then by adjusting the water bath temperature, the water vapor inside the dosing tank and charging tank could achieve the thermodynamic equilibrium at the desired temperature. The temperatures and pressures inside the tanks could be monitored and recorded by the temperature and pressure sensors. Having the recorded pressures and temperatures data, the water vapor could be calculated by using the real gas equation before and after being adsorbed. The differences will be counted as uptake of the adsorbent. Since the above procedure can be repeated for all the desired temperature and pressure, the isotherm of the adsorbent can be determined.

3.2. *TGA System: Experimental Set-Up and Procedure*

The TGA, as shown in Fig. 9.9, comprises the Cahn TGA unit, an evaporator, a vacuum pumping subsystem, and the helium gas supply subsystem. Water vapor is generated separately in a heater bath fitted with a PID controller. The pressure of the vapor is measured with a capacitance monometer, the Edwards barocel type-622 with 10 kPa span. The steam generator generates a flow of 50–300 ml/min and the flow rate is within the free delivery of the vacuum pump (about 315 × 10^{-5} m^3/s). The silica gel sample was put in the reaction chamber of TGA that has a radiant heater for maintaining constant temperature of the sample. Sufficient thermal insulation is incorporated to reduce heat loss. The changes in the sample mass were measured using a microbalance. A barocel pressure sensor mounted at the bottom of the reaction chamber is used to read partial vacuum in the system. A RTD is inserted from the bottom of a "tee-joint" and the whole glass tube is insulated. A vacuum pump is used for vacuuming and purging the system before the test and for maintaining the vapor testing condition in terms of pressure and temperature during the experiment. The argon gas, for protecting weighing bridge, is introduced from the top (from the inside) of a double-walled tubing, while the pure adsorbate is introduced upward to the silica gel sample that is contained within a quartz holder, and the whole assembly is subjected to a preset pressure (P) and temperature (T). Isotherm environment of silica gel is maintained by direct radiant heating where the heaters are placed outside (non-vacuum) the reaction chamber. Cooling to the outside chamber (non-vacuum) is effected by passing cool air from a vortex chiller and external fan. By balancing the heating and the cooling effects (external to reaction chamber), a steady temperature environment for the silica gel can then be achieved readily.

In this experiment, efficiency of the Fuji Davison type A and type RD silica gel has been tested. The silica gel, about 100–500 mg is placed inside the quartz holder and the reaction

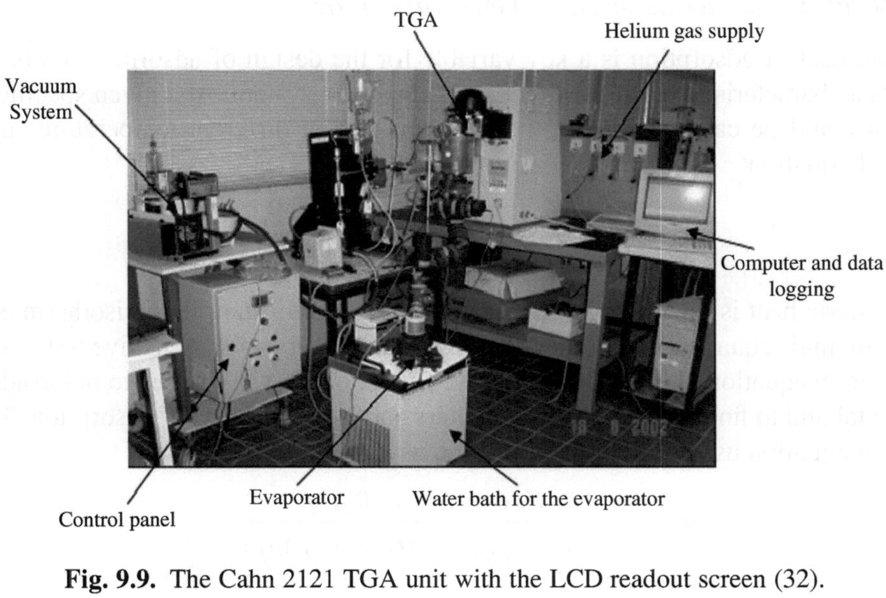

Fig. 9.9. The Cahn 2121 TGA unit with the LCD readout screen (32).

chamber is then vacuum-sealed and suitably insulated to prevent any possibility of condensation. The dry mass of silica gel is first measured by using a calibrated moisture balance (MA40), same as that used in the CVVP system. Thereafter, it is further confirmed within the reaction chamber of TGA, and the temperature of silica gel is held at 413 K for about 8 h. The dry mass is read using the TGA balance, which is accurate to within ±0.2 μg. At the end of the regeneration process, the test system is purged by helium gas and evacuated. Thereafter, the water vapor is continuously charged into the reaction cell and the vacuum pump keeps pumping the vapor–gas mixture to maintain the desired vacuum and temperature conditions. Prior to the experiment, the Platinum standard mass was used to verify the TGA system under the moisture environment.

To avoid condensation, the temperature of the reaction cell and piping system is always kept at least 5–10 K higher than the saturated temperature of the vapor, which is achieved by wrapping heater tapes around the piping system and they are also well insulated to minimize heat loss. The system is deemed to have reached thermodynamic equilibrium at desired pressure and the temperature when the mass of the adsorbed adsorbate remains unchanged for a period of half hour. The temperature is measured by two RTD sensors, one is used for temperature control and other is connected to the TGA data recorder for data collection. The pressure and the adsorbed adsorbate are measured by the capacitance monometer and TGA balance, respectively. By repeating the above procedures, assorted range of tests could be carried out under different settings of temperature and pressure. For every new test, system checks should be performed diligently, namely, checking whether (a) the system is evacuated, (b) the silica gel is regenerated, and (c) the chamber is purged with the purified inert gas. Uncertainty of the measurements is reduced by repeating the tests (at least three times) and a fresh batch of the silica gel is always recommended.

3.3. *Adsorption Isotherms of Silica Gel–Water Pair*

Isosteric heat of adsorption is a key variable for the design of adsorption chiller, which represents a characteristic of the adsorbent and adsorbate system. At a given specific surface loading, it could be calculated from the isotherm data at different temperatures using the Von't Hoff equation:

$$Q_{\mathrm{st}} = RT^2 \left[\frac{\mathrm{d}(\ln P)}{\mathrm{d}(T)}\right]_q. \tag{12}$$

The isosteric heat is related to the thermal equilibrium constant in the isotherm equation, such as Langmuir equation and Tóth equation. Therefore, it is very sensitive to the selection of the isotherm equation (33, 34). In this study, Tóth equation (35) is used to fit the adsorption experimental and to find the isotherm parameters and isosteric heats of adsorption. The form of the Tóth equation used in this study is given as

$$q* = \frac{K_0 \exp(\Delta H_{\mathrm{ads}}/R_{\mathrm{u}}T)p}{[1 + (K_0/q_{\mathrm{m}} \exp(\Delta H_{\mathrm{ads}}/R_{\mathrm{u}}T)p)^t]^{1/t}}, \tag{13}$$

where $q*$ is the quantity of adsorbate adsorbed by the adsorbent under equilibrium conditions, q_m denotes the monolayer capacity, P, T is the equilibrium pressure and temperature of the adsorbate in gas phase, respectively, R_u is the universal constant, ΔH_{ads} is the isosteric enthalpies of adsorption, K_0 the pre-exponential constant, and t is the dimensionless Tóth's constant. Tóth observed in his experiments that the empirical constant t was between zero and one, but Valenzuela and Myers (33) reported that t could actually be greater than one.

Adsorption equilibrium data of water vapor in the above-mentioned silica gels were obtained from these two independent methods. Figure 9.10 shows the comparison between the isothermal data from the two independent experimental methods and the manufacturer's data (36). It could be found that the equilibrium data from the two methods were well consistent with each other and the trends are quite compatible with the manufacturer's data.

The isothermal maps covering entire chiller operation range for the two types of silica gel are depicted in Figs. 9.11 and 9.12, respectively. One could observe that the equilibrium data from the two methods under different temperature and pressure were in very good agreement with each other. This implies that both methods are accurate of isothermal measurements and they can be used to verify the experimental results for each test procedure. In Figs. 9.11 and 9.12, the solid lines are experimental curve fits obtained by using Tóth equation to fit the experimental data from CVVP method. At the temperature of 348 and 358 K, the exploited data from Tóth equation was well consistent with the experimental data from TGA.

The isotherm parameters and isosteric heats of adsorption are shown in Table 9.7. It can be seen that ΔH_{ads} and q_m are significantly different from those reported by Chihara and Suzuki (37). Cremer and Davis (38) first reported a value of 2.89×10^3 kJ/kg for ΔH_{ads}, and Sakoda and Suzuki (39) later reported a value of 2.8×10^3 kJ/kg for ΔH_{ads}. It could be easily explained according to the effects of the selected isothermal equation on the calculation of

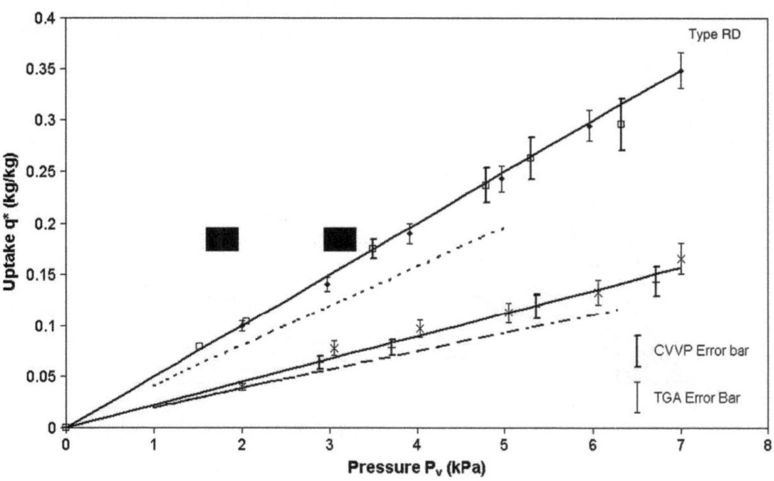

Fig. 9.10. Comparison of isotherm data for type RD silica gel + water system. CVVP experimental data points: *open square* $T = 323$ K, *plus* $T = 338$ K; variable pressure TGA experimental data points: *filled circle* $T = 323$ K, *asterisk* $T = 338$ K; *solid lines* are experimental curve fits using the Tóth equation; a chiller manufacturer's data: *dotted line* T = 323 K, *dashed line* $T = 338$ K (32).

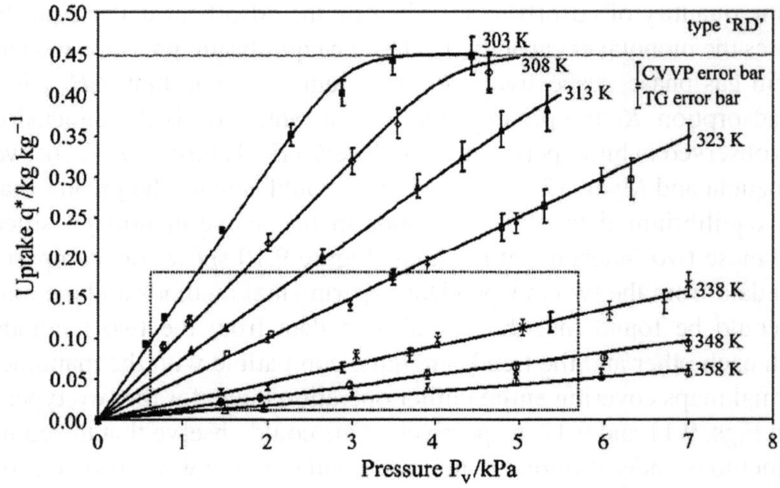

Fig. 9.11. Isotherm data for type RD silica gel + water system. CVVP experimental data points: *filled square* T = 303 K, *open diamond* T = 308 K, *filled triangle* T = 313 K, *open square* T = 323 K, *plus* T = 338 K; variable pressure TGA experimental data points: *filled circle* T = 323 K, *asterisk* T = 338 K; *open circle* T = 348 K; *open triangle* T = 358 K; *solid lines* are experimental curve fits using the Tóth equation (32).

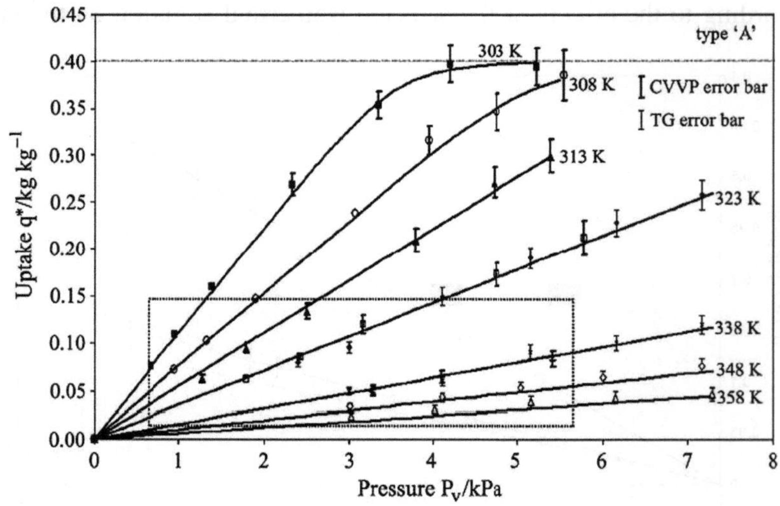

Fig. 9.12. Isotherm data for type A silica gel + water system. CVVP experimental data points: *filled square* T = 303 K, *open circle* T = 308 K, *filled triangle* T = 313 K, *open square* T = 323 K, *plus* T = 338 K; variable pressure TGA experimental data points: *filled circle* T = 323 K, *asterisk* T = 338 K; *open diamond* T = 348 K; *open triangle* T = 358 K; *solid lines* are experimental curve fits using the Tóth equation (32).

Table 9.7
Correlation coefficients for the two grades of Fuji Davison silica gel + water systems (32)

Type	K_0 (kg/kg kPa)	Q_{st} (kJ/kg)	q_m (kg/kg)	t	Remarks
"A"	(4.65 ± 0.9) × 10−10	(2.71 ± 0.1) × 103	0.4	10	By Tóth equation
"RD"	(7.30 ± 2) × 10−10	(2.693 ± 0.1) × 103	0.45	12	By Tóth equation
"RD" (manufacturer)	2.0 × 10−9	2.51 × 103	–	–	Henry Law

isosteric heat. In the literature, most of them use the Henry equation or the Langmuir equation to fit the isothermal data and calculate isosteric heat. As shown in the figures, both Henry equation and Langmuir equation could not fit the experiment data well, especially in the saturation region. In addition, if only the chiller operation range was considered, both Henry equation and Langmuir equation can fit the experimental data. Therefore, isosteric heat can only be calculated based on the experimental data within this region.

4. THERMALLY DRIVEN ADSORPTION DESALINATION

4.1. Laboratory Scale Prototype Testing

A recent patented adsorption desalination (AD) plant, by Ng et al. (3), employs a low temperature waste-heat source that are available in abundance from industrial processes. A pilot plant has been constructed and operated in the Air Conditioning Laboratory at National University of Singapore (NUS). Figure 9.13a–c show the schematic diagram and pictorial views of the four-bed adsorption desalination plant, respectively. The AD plant would comprise six heat exchangers, namely, four units of adsorption/desorption reactors or beds, the condenser, and evaporator. The desalt operation of the AD cycle is similar to that of the four-bed adsorption chiller (41), the detailed operation of which was recently reported by Wang and Ng (40). The main difference between the operations of the AD and the chiller cycles is that in AD, saline solution is pretreated before supplying to the evaporator and the condensate water that emanates from the condenser is removed as pure water, using either a vacuum pump or a 10-m high liquid-filled U-tube. The overall system operation can be summarized as follows:

1. The saline or brackish water is first pretreated (e.g., coagulation, filtering, and de-aeration) and fed to the evaporator. Purging of solution from the evaporator is conducted periodically for salt concentration control.
2. The seawater is dosed into the stainless steel evaporator, where an external water circuit provides the heating load for maintaining the evaporator's temperature and pressure. Evaporation inside the evaporator is enhanced by using the spray water method instead of pool boiling technique.
3. The water level inside the evaporator is regulated for both the water level and concentration and this prevents the corrosion of the evaporation unit.

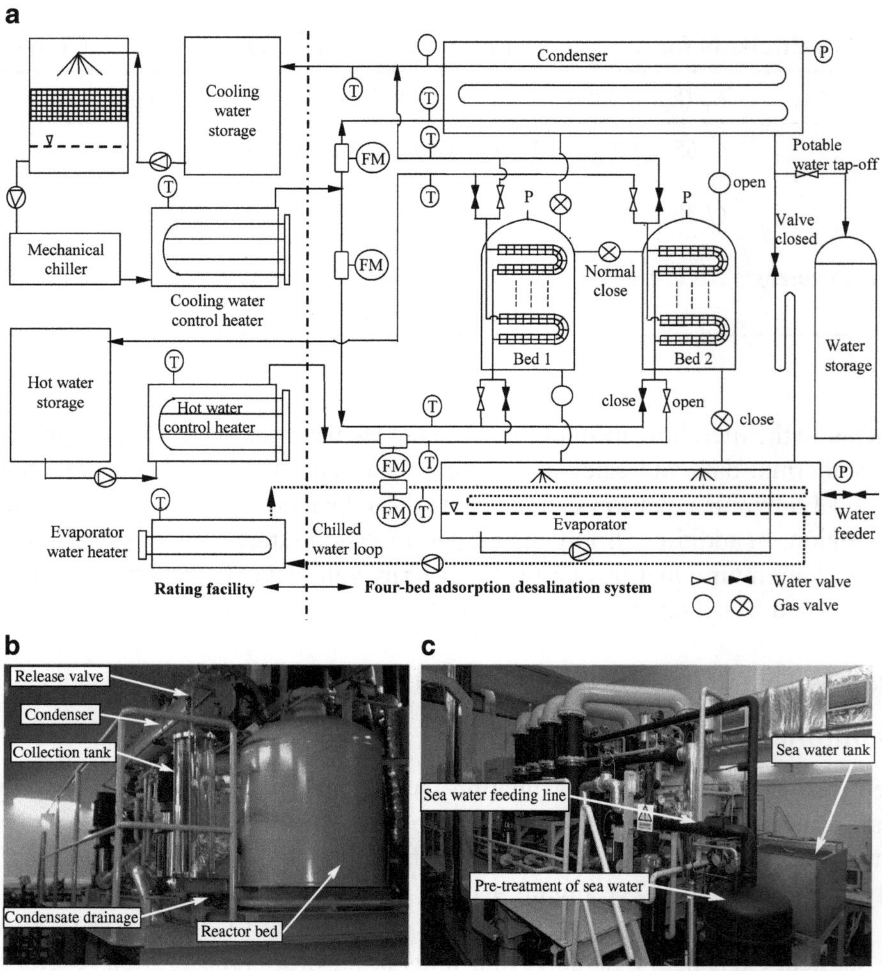

Fig. 9.13. (**a**) A schematic of a four-bed adsorption desalination plant (40). (**b**) Pictorial view of a four-bed adsorption desalination plant in NUS (condenser side view). (**c**) Pictorial view of a four-bed adsorption desalination plant (evaporator side view).

4. The evaporator is connected with the adsorbers where the evaporated water vapor is adsorbed by the silica gel. The heat of adsorption is removed by a coolant from cooling tower during the adsorption cycle.
5. The desorption–condensation process takes place at the condenser pressure where the desorbers are heated up to the desorption temperature with an external heat input. The desorbed vapor is condensed in the condenser, where the heat of condensation is removed by the cooling water from the cooling tower.
6. The condensate is collected in a collection tank, which is either intermittently pumped out to the atmospheric pressure or extracted via a 10-m high U-tube.

Tests are conducted under steady state conditions of the coolant (hot water source and cooling water source temperatures) which are accurately conditioned by a rating facility and thus it is weather independent. The cooling water that returns from the adsorber beds and the condenser is first mixed in a cooling water storage tank to minimize its temperature fluctuations. It is then pumped serially through the cooling tower and an evaporator coil of a mechanical chiller, dropping the coolant temperature to a level below that of set-point temperature of the heater controller. Within the heater tank, the cooling water temperature is further fine-tuned by an electrical heat input using a cascaded two-loop PID controller. Excellent control of the outlet water temperatures is thus achieved by this control arrangement, capitalizing on the faster time constant of heaters vs. those of the processes of cooling tower and mechanical chiller. The control strategy used in the hot water loop is similar to that of the cold water loop. A hot water storage tank is used to damp temperature fluctuations of hot water returning from the desorption processes. Hot water from the mixed tank is fed into the hot water heater tank where its temperature can be fine-tuned electrically with a similar PID controller. Similar to the hot water circuit, the evaporator in the adsorption plant is maintained by a constant chilled water return, which is also electrically controlled with heaters, that simulates the cooling load. All the above-mentioned circuits have been tested to have an accuracy of $\pm 0.3°C$.

4.2. Definitions and Modeling

The energy balance equations, the isotherm equation, and the kinetics that are used for the evaluation of the performance of the adsorption desalination plant are presented below. Tóth isotherm equation (Eq. 13) is used to estimate the amount of water vapor that the silica gel can adsorb at the equilibrium conditions of temperature and pressure. The adsorption kinetics is estimated by using the well-known LDF correlation (10) as follows:

$$\frac{\mathrm{d}q}{\mathrm{d}t} = \frac{15D_{\mathrm{so}}\exp(-E_{\mathrm{a}}/RT)}{R_{\mathrm{p}}^2}(q^* - q). \tag{14}$$

Here, D_{sO} is the pre-exponential constant, E_{a} is the activation energy, R_{p} is the particle radius, and q denotes the instantaneous uptake.

Figure 9.14 shows the energy balance of adsorption desalination system. As can be seen from the control volume CV1 in Fig. 9.14, the sensible cooling heat (Q_{ads}) is necessary to reject the adsorption latent heat $\left(Q_{\mathrm{ads}}^{\mathrm{latent}}\right)$ that is developed due to adsorption of water vapor by silica gel during adsorption process, while extracted heat by chilled water (Q_{chill}) drives the heat of evaporation $\left(Q_{\mathrm{des}}^{\mathrm{latent}}\right)$ inside the evaporator. The control volume CV2 in the same figure shows that the heating source (Q_{des}) is needed to reject the latent heat $\left(Q_{\mathrm{des}}^{\mathrm{latent}}\right)$ that is generated due to desorption. Heat of condensation $\left(Q_{\mathrm{ads}}^{\mathrm{latent}}\right)$ is rejected through the condenser by a sensible cooling (Q_{cond}). The sensible and latent heats of the AD system can be defined as follows:

1. For the adsorbent bed in adsorption mode

$$Q_{\mathrm{ads}} = \left(\dot{m}c_{\mathrm{p}}\right)_{\mathrm{ads}}\left(T_{\mathrm{cw,o}} - T_{\mathrm{cw,in}}\right), \tag{15}$$

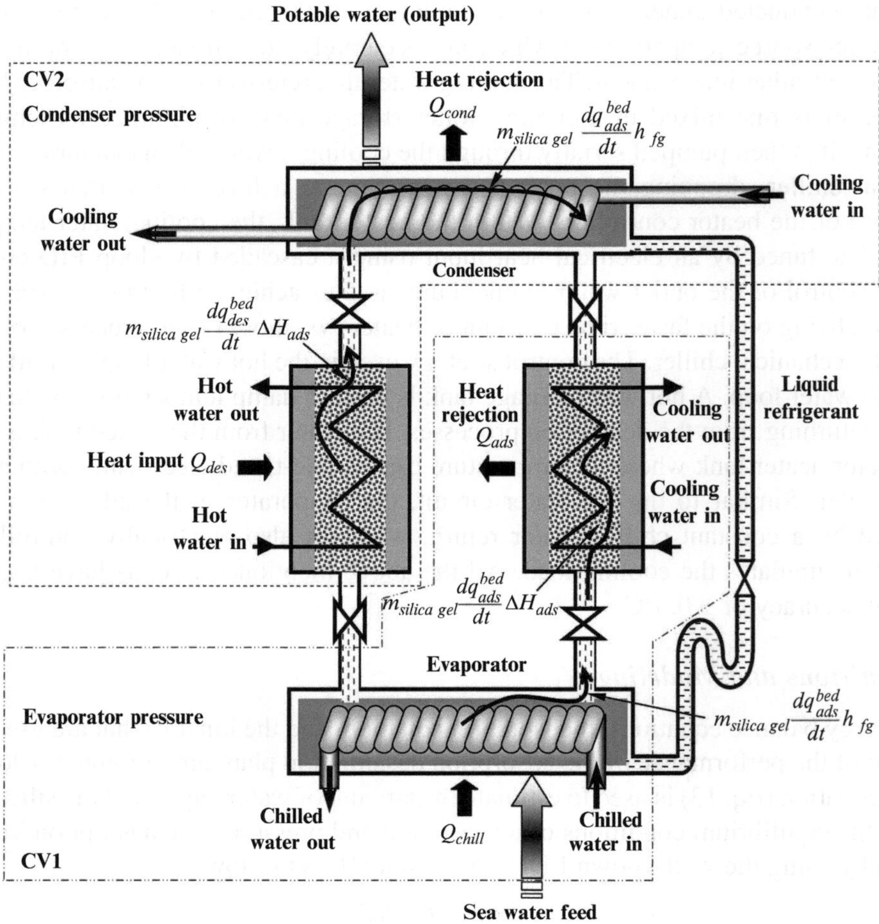

Fig. 9.14. Energy balance of AD system (Adapted from Saha et al. (42) and El-Sharkawy (25)).

$$Q_{\text{ads}}^{\text{latent}} = m_{\text{silica gel}} \frac{\mathrm{d}q_{\text{ads}}^{\text{bed}}}{\mathrm{d}t} \Delta H_{\text{ads}}. \tag{16}$$

2. For the adsorbent bed in desorption operation

$$Q_{\text{des}} = \left(\dot{m}c_{\text{p}}\right)_{\text{des}}\left(T_{\text{hw,in}} - T_{\text{hw,o}}\right), \tag{17}$$

$$Q_{\text{des}}^{\text{latent}} = m_{\text{silica gel}} \frac{\mathrm{d}q_{\text{des}}^{\text{bed}}}{\mathrm{d}t} \Delta H_{\text{ads}}. \tag{18}$$

3. For the condenser

$$Q_{\text{cond}} = \left(\dot{m}c_{\text{p}}\right)_{\text{cond}}\left(T_{\text{cw,o}} - T_{\text{cw,in}}\right), \tag{19}$$

$$Q_{\mathrm{cond}}^{\mathrm{latent}} = m_{\mathrm{silica\ gel}} \frac{\mathrm{d}q_{\mathrm{des}}^{\mathrm{bed}}}{\mathrm{d}t} h_{\mathrm{fg}}. \tag{20}$$

4. For the evaporator

$$Q_{\mathrm{chill}} = \left(\dot{m}c_{\mathrm{p}}\right)_{\mathrm{chill}} \left(T_{\mathrm{chill,in}} - T_{\mathrm{chill,o}}\right), \tag{21}$$

$$Q_{\mathrm{evap}}^{\mathrm{latent}} = m_{\mathrm{silica\ gel}} \frac{\mathrm{d}q_{\mathrm{ads}}^{\mathrm{bed}}}{\mathrm{d}t} h_{\mathrm{fg}}. \tag{22}$$

The sensible and latent heat balances of the adsorption cooling system are

$$Q_{\mathrm{ads}} + Q_{\mathrm{cond}} = Q_{\mathrm{des}} + Q_{\mathrm{chill}} \tag{23}$$

and

$$Q_{\mathrm{ads}}^{\mathrm{latent}} + Q_{\mathrm{cond}}^{\mathrm{latent}} = Q_{\mathrm{des}}^{\mathrm{latent}} + Q_{\mathrm{evap}}^{\mathrm{latent}}, \tag{24}$$

respectively.

In the above-mentioned equations, $\dot{m}_{\mathrm{des}}$, $\dot{m}_{\mathrm{cooling}}$, $\dot{m}_{\mathrm{chilled}}$, and $\dot{m}_{\mathrm{cond}}$ are the mass flow rates of hot water, adsorber cooling water, chilled water, and condenser cooling water, respectively. $T_{\mathrm{hw,in}}$, $T_{\mathrm{cw,in}}$, and $T_{\mathrm{chll,in}}$ stand for a hot, cooling, and chilled water inlet temperatures. $T_{\mathrm{hw,o}}$, $T_{\mathrm{cw,o}}$, and $T_{\mathrm{chll,o}}$ are hot, cooling, and chilled water outlet temperatures, respectively.

As AD is aimed to produce potable water, the end user may be more concerned with the specific daily water production and the system performance ratio. In this context, the SDWP and PR are, respectively, defined as given in the following relations:

$$\mathrm{SDWP} = \int \frac{3,600\, dQ_{\mathrm{cond}}}{h_{\mathrm{fg}(T_{\mathrm{cond}})} m_{\mathrm{silica\ gel}}} \tau\, \mathrm{d}t, \tag{25}$$

where τ is the number of running hours of the plant. For more accurate and realistic results, Q_{cond} is used to evaluate SDWP. However, in real calculation of the water production, the readings from the flow meter are used as follows:

$$\mathrm{SDWP} = \frac{1,440}{N} \frac{\sum_{i=1}^{N} (F)_i}{m_{\mathrm{sg}}}, \tag{26}$$

where F is condensate water flow rate measured in LPM and m_{sg} is the mass of silica gel used. The performance ratio of the adsorption desalination process is expressed as

$$\mathrm{PR} = \int \frac{\dot{m}_{\mathrm{water}} h_{\mathrm{fg}(T_{\mathrm{cond}})}}{\mathrm{d}Q_{\mathrm{des}}} \mathrm{d}t, \tag{27}$$

where $\mathrm{d}Q_{\mathrm{cond}}$ is the heat rejected at condenser and $\mathrm{d}Q_{\mathrm{des}}$ is the heat of desorption during the desorption process.

4.3. Specific Daily Production Vs. Cycle Time

Presently, there is no available standard rating for the testing of an adsorption desalination plant. The work of Wang and Ng (40) is the first report where they have tested the plant at the following rating conditions:

1. The inlet temperatures of hot, cooling, and chilled water supplied to the beds and heat exchangers of the plant are set at 85, 29.4, and 12.2°C, respectively.
2. The coolant flow rates for the evaporator is 0.0455 l/(s kW), where the kW is based on the maximum cooling capacity across the evaporator.
3. The total cooling water flow rate is set to about 2.5 times of the evaporator flow rate, i.e., 0.114 l/(s kW), where the ratio of water supplied to the designated adsorption beds and the condenser is 1–1.5, respectively.
4. The hot water flow rate supplied to the beds (in series flow configuration) is set to have the same flow rate as that of the evaporator, i.e., 0.0455 l/(s kW).

Based on the above-mentioned rating conditions, the switching (preheating or precooling cycle) time is held constant at 40 s. Too long or too short a switching time may deprive the performance of the chiller. The performance of the experimental desalination plant is evaluated at different half-cycle times, typically from 120 to 600 s, while the switching time is held constant at 40 s. As can be seen from Fig. 9.15, the maximum specified water production for a standard AD cycle is 4.7 kg/kg silica gel and this occurs at an operating cycle time of 180 s, while increasing the half-cycle time of the adsorption cycle leads to its reduction. It is noted that the COP (defined here as the ratio of cooling to heat input) is generally low at 0.24–0.38 due mainly to the temperature of chilled water being produced, typically less than 7°C. A practical half-cycle time for the desalination cycle is about 250 s without sacrificing much of the SDWP yield, i.e., 4.6 kg/kg of silica gel. It should be pointed

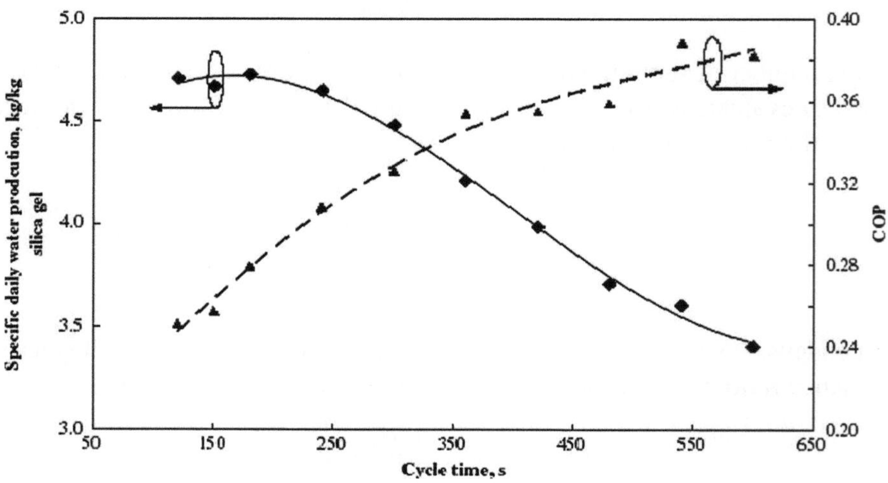

Fig. 9.15. Performance of the four-bed adsorption desalination system under standard operation condition (40).

out that as the inlet temperature of evaporator is raised, near to the ambient temperature, the SDWP could be increased to almost two- or threefold.

4.4. Effect of Heat Source Temperature on the Cycle Performance

Figure 9.16 shows the effect of the hot water inlet temperature on the cycle COP and SDWP for two half-cycle times of 180 and 300 s. It is noted that the SDWP increases linearly with the increase of hot water inlet temperature at both cycle times. The improvement is attributed to the increase in refrigerant flow within the AD cycle at higher driving source temperatures. The rate of increase in COP is higher when regeneration temperature is below 80°C and beyond this temperature, the increase in COP is slower. This happens due to the fact that the requirement of heat input becomes significantly large when the temperature difference between heat source and heat sink becomes high.

4.5. Effect of Cooling and Chilled Water Temperature on the Cycle Performance

Effects of chilled and cooling water inlet temperatures on the cycle performance are shown in Fig. 9.17. The SDWP can be improved by about 10% when the cooling water inlet temperature decreases by about 1.6°C. This is because of the increase in adsorption capacity of silica gel with the decrease in adsorption temperature. It is noticed that the SDWP also increases with the increase in chilled water inlet temperature, which implies that if the chilled water temperature is raised, further enhancement of the SDWP is possible.

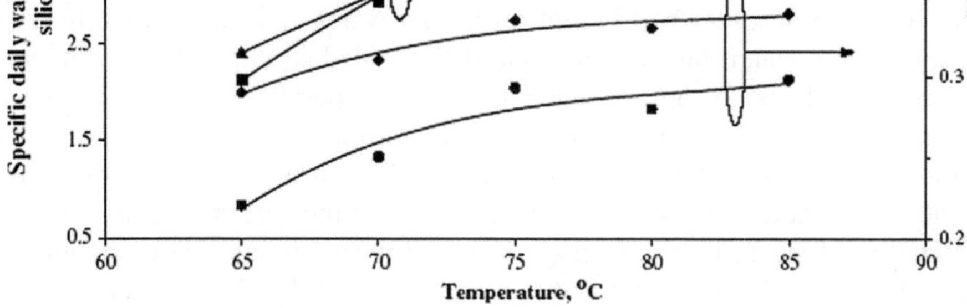

Fig. 9.16. Effects of the hot water temperature on the system performance (40).

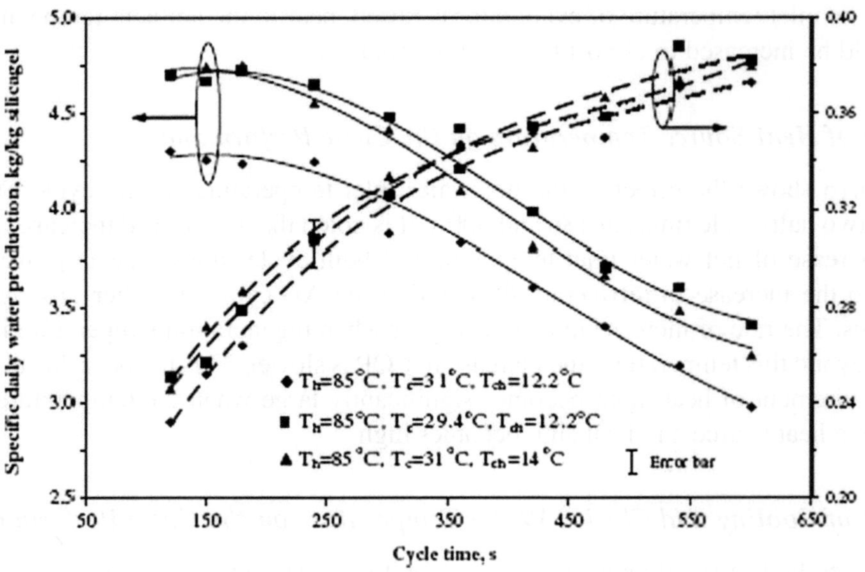

Fig. 9.17. Effects of the cooling water temperature and chilled water temperature on the system performance (40).

4.6. Adsorption Desalination Plant with Heat Recovery

The COP of an AD plant can be significantly improved by using the regenerative schemes. Passive heat recovery is deemed one of the most practical schemes as it requires minimal hardware alterations of the plant. Ng et al. (41) and Wang et al. (43) clarify in detail the principles of passive heat scheme in a four-bed adsorption chiller. Passive recovery can be achieved by using a valve delay scheme where the residual heat left in the flow channels in the beds could be recovered without having to purge the cooling tower. In the same manner, cooling water from the master (cooler) adsorber is used to cool. Valve delay scheme is an effective method for heat recovery and it involves no piping alteration of the existing flow circuits of the AD plant.

During switching period, the master desorber needs to be cooled down to adsorption temperature by the cooling water. The cooling water that pushes the residual hot water and gets the heat from the desorber bed is sent to the hot water tank rather than to the cooling water tank, and the hot water that becomes relatively cooled is sent to the cooling water tank instead of sending it back to the hot water tank. This operation can be achieved by just delaying the opening of the valves that control the cooling water and hot water. The valve delay operation can be utilized by just modifying the PLC programming without any major changes in the system hardware. The performance ratio of the system increases for the better with the utilization of isosteric heat energy. The energy saving and the reduction of energy can be seen in Fig. 9.18. It can be noticed that heat recovery by valve delay results in more significant improvement in the performance ratio of the system only when master and slave configuration is operated, as the cooling water can get more heat energy as well as the hot

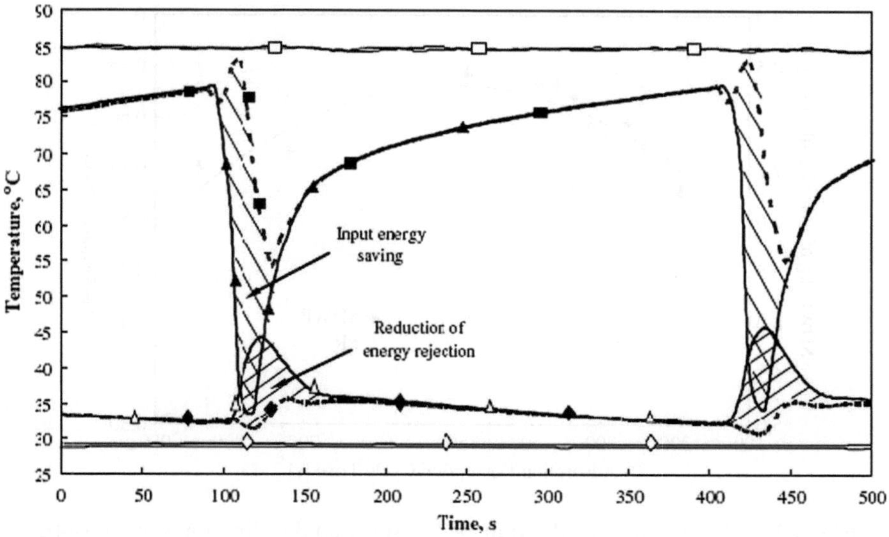

Fig. 9.18. Temperature profile of the system hot and cooling water under standard operation scheme and passive heat recovery scheme. *Open square* Hot water inlet temperature using two different operation schemes. *Filled triangle* Hot water outlet temperature using standard operation scheme. *Filled square* Hot water outlet temperature using passive heat recovery scheme. *Open diamond* Cooling water inlet temperature using two different operation schemes. *Filled diamond* Cooling water outlet temperature using passive heat recovery scheme. *Open triangle* Cooling water outlet temperature using standard operation scheme (40).

water can be cooled down nearly to the cooling water temperature. The hot water flow rate and cooling water flow rate to the bed must be equal to maintain the water storage capacity when the valve delay scheme is operated.

5. RECENT IMPROVEMENTS OF ADSORPTION DESALINATION PLANT

5.1. *Adsorption Desalination Operation at a Relatively Higher Evaporation Temperature*

At a constant adsorption temperature, the adsorption capacity of silica gel + water pair increases with the increase in adsorption pressure. El-Sharkawy et al. (44) reported that the SDWP can be significantly increased with an increase in the evaporation temperature of the evaporator. Evaporator pressure can be increased by raising the chilled water inlet temperature. It is noted that the SDWP can also be increased by using (a) the recovery of heat by cooling the cooling water supplied to the adsorption process using a plate heat exchanger (3) and (b) the tapping of heat to the evaporator by injecting part of the flow from cooling tower circuit. In the former method, the advantage lies in the cooling of silica gel during the adsorption process only to a temperature lower than the ambient and thus enhancing the adsorption process of silica gel. The latter method lowers the supply temperature of the coolant from the cooling tower, benefiting both the adsorber and the condenser. Figure 9.19

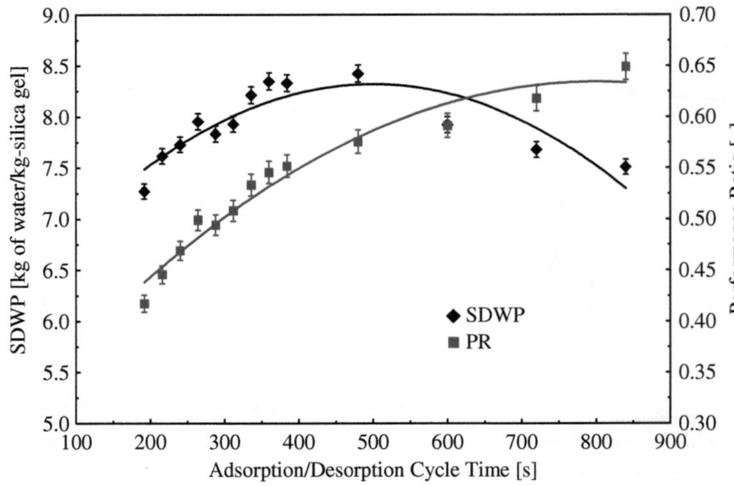

Fig. 9.19. Performance of the four-bed adsorption desalination system (44).

Table 9.8
Rated conditions (44)

Hot water inlet (desorber)		Cooling water inlet (adsorber)		Cooling water inlet (condenser)	
Temperature (°C)	Flow rate (LPM)	Temperature (°C)	Flow rate (LPM)	Temperature (°C)	Flow rate (LPM)
85	48	29.5	48	30	120
Mass of silica gel (kg)			36		
Switching time (s)			40		

shows the improvements in SDWP and PR with the rage of operation time at the rated operation condition shown in Table 9.8. The chilled water inlet temperature to the evaporator is kept constant at 30°C. Experimental data show that the SDWP of the plant is increased by twofold, to about 8.2 kg of water/kg of adsorbent per day.

Figure 9.20 shows the effect of the hot water inlet temperature on the cycle SDWP and PR using a half-cycle time of 360 s. It is noted that the SDWP increases linearly with increasing hot water inlet temperatures. The improvement is attributed to the increased in vapor flow within the AD cycle at higher driving source temperatures. However, the increase in PR is only marginal. The increase in SDWP occurs due to the larger temperature difference between heat source and heat sink. Figure 9.21 shows the effects of chilled water inlet temperatures on the cycle performance, with and without heat exchange between the chilled water coming out from the evaporator and the cooling water supplied to the adsorber. Experimental results show that the SDWP increases linearly with the increase in chilled

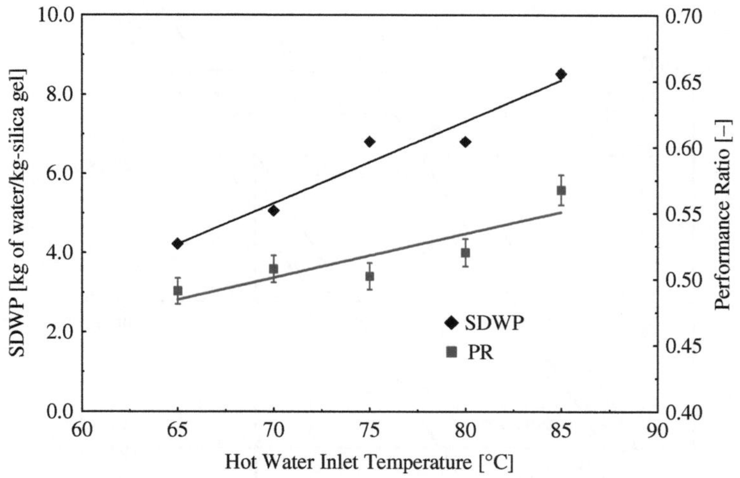

Fig. 9.20. Effects of the hot water inlet temperatures on the system performance (44).

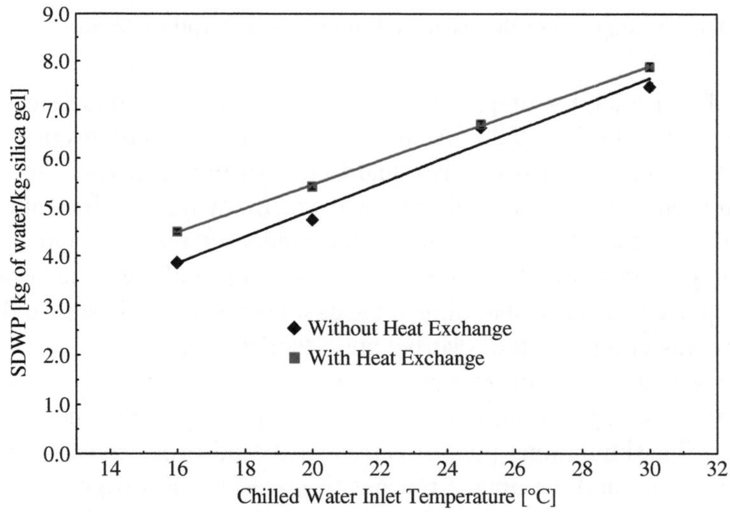

Fig. 9.21. Effects of the chilled water inlet temperatures on the system performance (44).

water inlet temperature. Due to the limitation of the used heat exchanger capacity, the improvement in the SDWP is slightly low.

5.2. *Adsorption Desalination Plant with Mass Recovery*

A practical and yet simple scheme that has been demonstrated in the prototype of a four-bed adsorption desalination plant is the mass recovery by using a pressure equalization (PE)

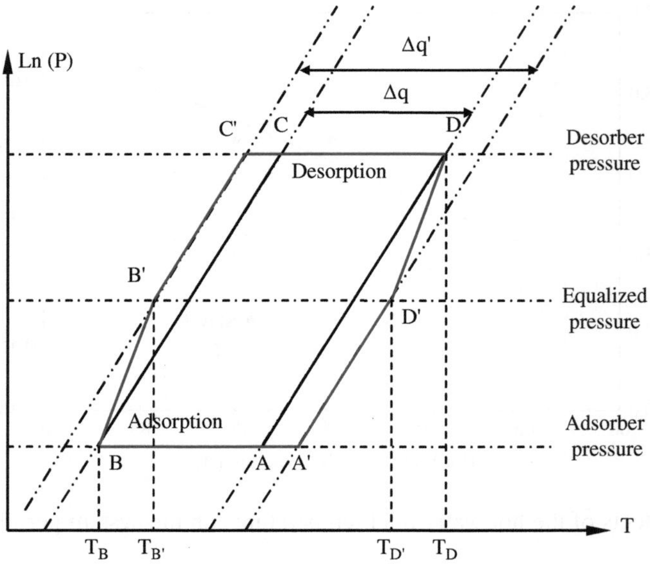

Fig. 9.22. Dühring diagram of the basic and improved adsorption desalination plant (44).

technique (45). The principles of pressure equalization are best understood by using the *P–T–q* diagram shown in Fig. 9.22. At the end of adsorption/desorption processes, the pressure of the desorber reactor is nearly equal to that of the condenser, and the refrigerant (water) concentration reaches its minimum value (q_D), which is the equilibrium uptake corresponding to the desorption temperature and condenser pressure. On the contrary, the pressure of adsorber is controlled by that of the evaporator and the concentration reaches it the equilibrium uptake (q_B), corresponding to the adsorber reactor pressure and the adsorption temperature. By connecting both the adsorber and desorber reactors for a short time at the end of adsorption/desorption processes, refrigerant vapor in the desorber is forced to move to the adsorber by pressure swing action, making the silica gel dryer and the refrigerant uptake decreases to $q_{D'}$. The simulations pressurization in the adsorber causes silica gel state to adsorb more refrigerant, and the uptake reaches to $q_{B'}$. Mass of refrigerant circulated in the system is then increased to $(q_{B'} - q_{D'})$, thus improving the SDWP. While maintaining the optimal cycle time, Fig. 9.23 shows the performance of adsorption desalination plant at different pressure equalization time intervals, namely, 10, 20, and 30 s. It can be seen that the SDWP increases by as much as 10 kg of water/kg of adsorbent per day at a PE of 10 s. However, an increase in the PE time beyond the optimal value reduces the SDWP due to the decrease in the mass transfer rates.

6. DESIGN FOR LARGE COMMERCIAL ADSORPTION DESALINATION PLANT

Most of the commercial scale desalination plants are designed to the order of million gallons of water per day (MGD). Ng et al. (3) proposed a modular design for reactor towers of

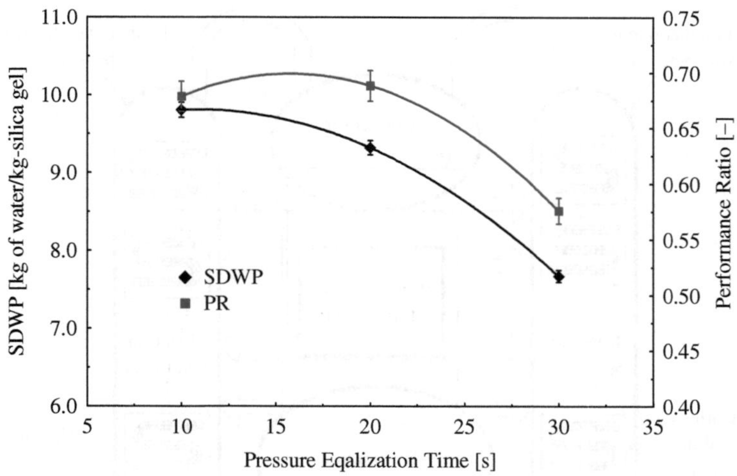

Fig. 9.23. Performance of four-bed adsorption desalination plant with different pressure equalization times (44).

20-tonne capacity of adsorbent per "silo"-type tower with a diameter of about 3 m and a height of 4 m. The towers would be constructed by using high strength foam concrete as the structural cover, lined internally with a thin inner steel sheet for maintaining the necessary vacuum conditions. Moreover, foam concrete towers provide good thermal insulation and reduce heat leaks to/from ambient. For such a plant capacity and operating at the standard conditions, the predicted yield of potable water is about 0.2 MGD.

Figure 9.24 shows a schematic view of the proposed system, which is a two reaction bed towers AD plant. The process involves the evaporator having a supply of saline water subjected to a relatively low temperature environment, for instance in the range 15–40°C. During adsorption process, water vapor is directed to an array of adsorbent beds arranged in reaction bed towers. The adsorbent, which is silica gel, then adsorbs the water vapor. To reject heat of adsorption, the adsorbent beds include heat exchangers subject to the circulation of a coolant. During the desorption mode, the supply of water vapor is disconnected from the adsorbent beds and a conduit to the condenser is opened. In order to achieve equilibrium, the saturation point of the adsorbent material is such that water vapor is then directed to the condenser. However, the heat exchangers are subjected to supply of heat in order to extract the adsorbed water vapor. The condenser captures and condenses the desorbed water vapor, and subsequently directs this to a water storage tank containing the desalinated water.

To maintain the desired temperature within the array of adsorbent beds during each of the phases (adsorption and desorption), the heating and cooling supply may further include a re-circulation system, whereby the coolant/heating supply may be maintained at the appropriate temperature within the array through constant replenishment.

For efficient extraction of waste heat, plant with four reaction bed towers (3) is usually recommended; each tower contains adsorbent bed arrays. Saline water is fed into the evaporator with the water vapor being directed to each of the reaction bed towers.

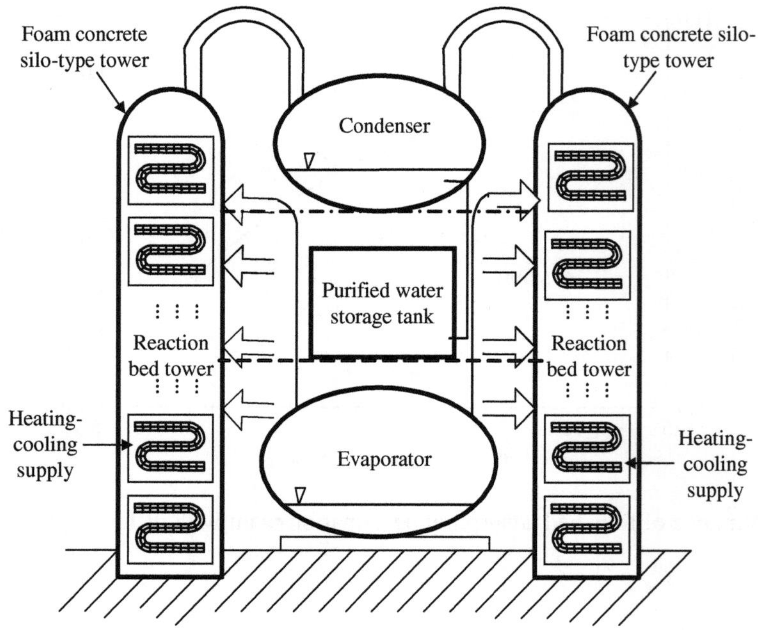

Fig. 9.24. Schematic diagram of a two reaction bed towers AD plant (extracted from patent document of Ng et al. (3)).

The reaction bed towers are each connected to a supply of hot water, cooling water, and chilled water. In this schematic view, each of the aforementioned supplies are made available through common supply lines, with the communication of these sources being subject to selectively operable valves, which are opened or closed subject to which phase the respective tower is operating under. The subsequent waste water is fed into corresponding common lines for re-circulation to the various supplies. More detailed description of these systems is found in Ng et al. (3).

Heat and mass recovery schemes are also used to improve the system performance. They reported that among heat and mass recovery methods, pressure equalization scheme is one of the best and applicable techniques. Figure 9.25 shows the regenerative heat flows from the adsorber towers and the condenser into the evaporator of the adsorption desalination plant. From first Law analysis, the heat input is given simply by

$$Q_{\text{in}} = (1-n)(Q_{\text{cond}} + Q_{\text{ads}}), \tag{28}$$

where n is the fraction of the amount of regenerated heat to the total heat in the process (assuming equal amount in both the adsorber towers and the condenser). The COP_{AD} is given by the ratio of the energy used to generate the condensate to the heat input, i.e.,

$$\text{COP}_{\text{AD}} = \frac{\int_0^t \dot{m}_{\text{w}} h_{\text{fg}}}{\int_0^t Q_{\text{input}}} \approx \frac{n}{1-n}. \tag{29}$$

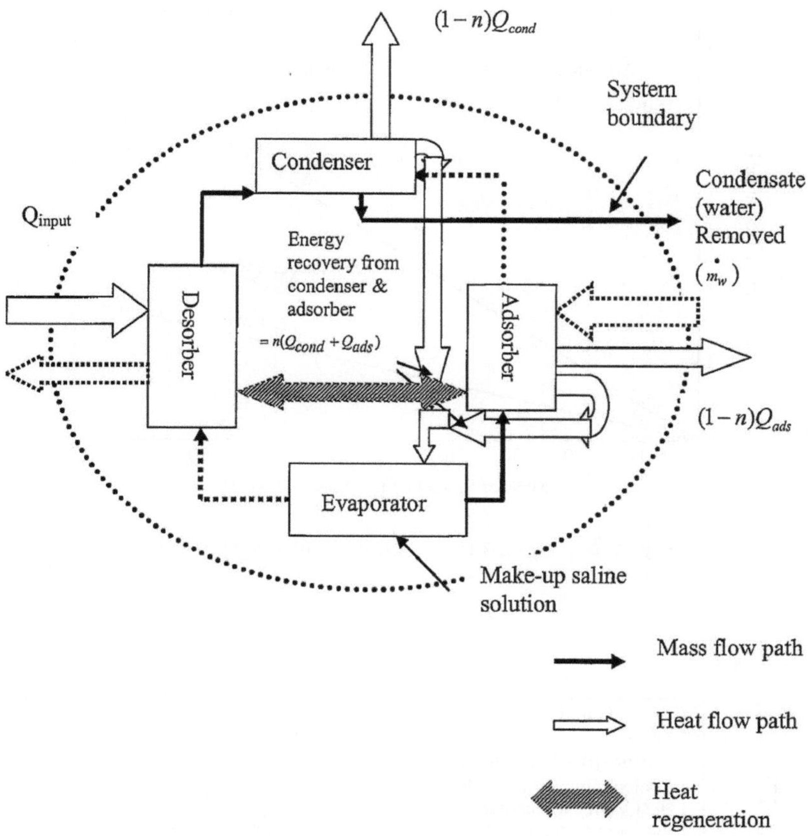

Fig. 9.25. Regenerative heat flows from the adsorber towers (extracted from patent document of Ng et al. (3)).

The range of COP_{AD} would be comparatively better than that of the "kettle-type" boiler or desalinator when the heat regenerative index, *n*, varies from 0.3 to 0.55. Between these values, the regeneration index "*n*" would have an optima where the water production is maximum and thus, a realistic operating point for the adsorption plant. For the water temperature of evaporator (that could be utilized by the regenerative processes), it should be equal or lower than the cooling tower temperature. However, the COP of the proposed heat-driven adsorption desalination plant is comparably better than that of the conventional "kettle" boiler, and it has the added advantage of much lower fouling at the evaporator.

For a heat source temperature less than 85°C and to assist the designer in sizing the large-scale AD plants, Figs. 9.26 and 9.27 provide the dimensions of the silo-type tower where the rate of heat input could be estimated with respect to the amount of silica gel used. For a packing density of silica gel of 700 kg/m^3, the required tower height is estimated at different masses of silica gel and different tower diameters, namely, 3, 4, and 5 m, respectively. Similarly, the heat input required to operate such AD plants can be calculated by using the following relation:

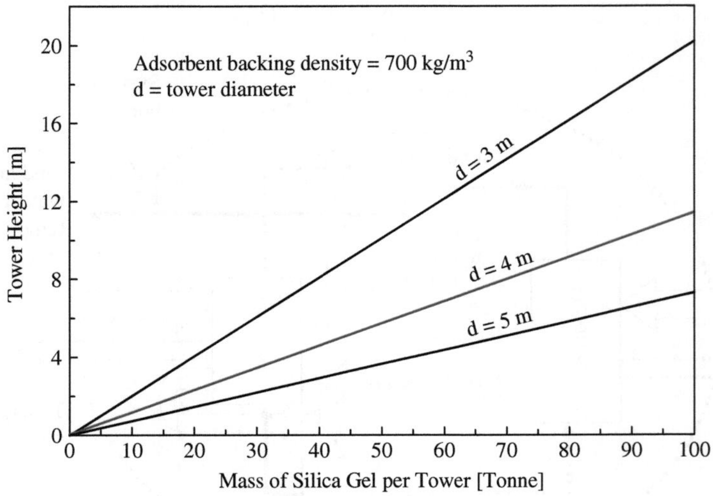

Fig. 9.26. Design dimensions of silo-type tower.

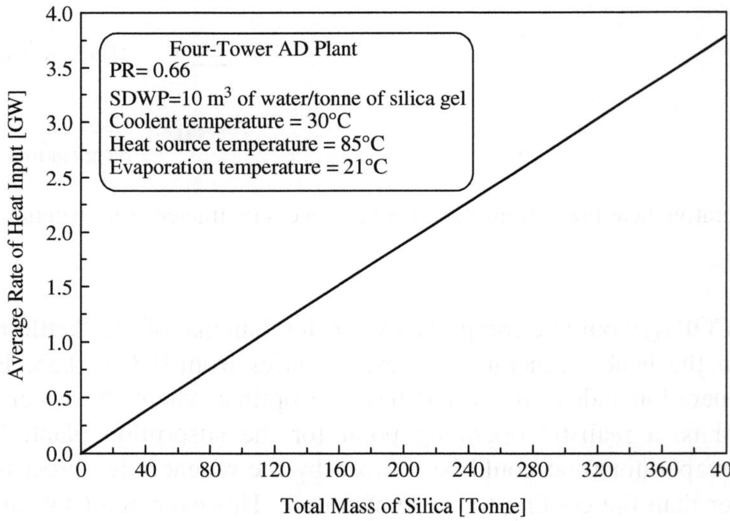

Fig. 9.27. Average heat input to the AD plant for assorted mass of silica gel.

$$\text{Average heat input (GW)} = \frac{\text{SDWP} \times m_{\text{silica gel}} \times h_{\text{fg(water)}}}{\text{PR} \times 3,600 \times 10^9}. \quad (30)$$

7. CLOSURE

The chapter provides fundamental and design information that are necessary for engineers and scientists to make a foray into the adsorption desalination plant design. These data are deemed to be novel at this junction of writing the monographs, but the authors believe that further improvements to the adsorption desalination plants design could be forthcoming as more research will be performed in the future. With further improvements to the AD cycle, it is deemed possible for the AD cycle to be most energy efficient amongst all the known desalination methods. The specific daily water production from the AD cycle varies from 8 to 11 m^3/tonne of silica gel. The parasitic power consumption of the cycle could be improved from 1.2 to 1.5 kWh/m^3 of high-grade water, where the TDS of such water is usually less than 7 ppm. The other major bonus obtained from the AD cycle is the vast amount of cooling effect, estimated to be about 145 kW/tonne of silica gel when the cycle is operated at the rated conditions.

ACKNOWLEDGMENTS

The authors express sincere thanks to King Abdullah University of Science and Technology (KAUST) for the financial support through the project (WBS R265-000-286-597). The authors would also like to thank Mr. Kyaw Thu, a NUS Ph.D. student in the ME Department, for his help in experimental investigations of the AD plant (46–47).

REFERENCES

1. Al-kharabsheh S, Goswami DY (2004) Theoretical analysis of a water desalination system using low grade solar heat. J Solar Energy Eng Trans ASME 126(2):774–780
2. Ehrenman G (2004) From sea to sink. Mech Eng 126(10):38–43
3. Ng KC, Wang XL, Gao LZ, Chakraborty A, Saha BB, Koyama S, Akisawa A, Kashiwagi T (2006) Apparatus and method for desalination. Patent, Publication Number: WO/2006/121414
4. Papadopoulos AM, Oxizidis S, Kyriakis N (2003) Perspectives of solar cooling in view of the developments in the air-conditioning sector. Renew Sustain Energy Rev 7(5):419–438
5. Ruthven DM (1984) Principles of adsorption and adsorption process. Wiley, Boston, MA
6. Suzuki M (1990) Adsorption engineering. Elsevier, Amsterdam
7. Rouquerol F, Rouquerol J, Sing K (1999) Adsorption by powders and porous solids. Academic, San Diego
8. Srivastava NC, Eames IW (1998) A review of adsorbents and adsorbates in solid–vapour adsorption heat pump systems. Appl Therm Eng 18(9–10):707–714
9. El-Sharkawy II, Kuwahara K, Saha BB, Koyama S, Ng KC (2006) Experimental investigation of activated carbon fibers/ethanol pairs for adsorption cooling system application. Appl Therm Eng 26:859–865
10. Glueckauf E (1955) Formula for diffusion into sphere and their application to chromatography. Trans Faraday Soc 51:1540–1551
11. Liaw CH, Wang JSP, Greenkorn RH, Chao KC (1979) Kinetics of fixed-bed adsorption: a new solution. AIChE J 25:376–381
12. Li Z, Yang RT (1999) Concentration profile for linear driving force model for diffusion in a particle. AIChE J 45(1):196–200

13. Sircar S, Hufton JR (2000) Interparticle adsorbate concentration profile for linear driving force model. AIChE J 46(3):659–660
14. El-Sharkawy II, Saha BB, Koyama S, Ng KC (2006) A study on the kinetics of ethanol-activated carbon fiber: theory and experiments. Int J Heat Mass Transf 49:3104–3110
15. El-Sharkawy II, Saha BB, Chakraborty A, Kuwahara K, Koyama S, Ng KC (2006) Determination of an improved linear driving force equation for adsorption of ethanol on activated carbon fiber. In: Proceedings of the 3rd Asian conference on refrigeration and air-conditioning, Gyeongju, Korea, vol 1, pp 129–132
16. Akkimaradi BS, Prasad M, Dutta P, Srinivasan K (2001) Adsorption of 1,1,1,2-tetrafluoroethane (HFC 134a) on activated carbon. J Chem Eng Data 46:417–422
17. Prasad M, Akkimaradi BS, Rastogi SC, Rao RR, Srinivasan K (1999) Heats of adsorption for charcoal-nitrogen systems. Carbon 37:1641–1642
18. Critoph RE (1988) Performance limitations of adsorption cycles for solar cooling. Solar Energy 41(1):21–31
19. Cacciola G, Restuccia G (1995) Reversible adsorption heat pump: a thermodynamic model. Int J Refrig 18(2):100–106
20. El-Sharkawy II, Saha BB, Koyama S, Srinivasan K (2007) Isosteric heats of adsorption extracted from experiments of ethanol and HFC 134a on carbon based adsorbents. Int J Heat Mass Transf 50 (5–6):902–907
21. Chakraborty A, Saha BB, Koyama S, Ng KC (2006) On the thermodynamic modeling of the isosteric heat of adsorption and comparison with experiments. Appl Phys Lett 89:171901
22. Chakraborty A, Saha BB, Ng KC, Koyama S, Srinivasan K (2009) Theoretical insight of physical adsorption for a single component adsorbent–adsorbate system: I. Thermodynamic property surfaces. Langmuir 25(4):2204–2211
23. Dunne AJ, Mariwala R, Rao M, Sircar S, Gorte RJ, Myers AL (1996) Calorimetric heats of adsorption and adsorption isotherms. 1. O_2, N_2, Ar, CO_2, CH_4, C_2H_6, and SF_6 on silicalite. Langmuir 12(24):5888–5895
24. Dunne AJ, Rao M, Sircar S, Gorte RJ, Myers AL (1996) Calorimetric heats of adsorption and adsorption isotherms. 2. O_2, N_2, Ar, CO_2, CH_4, C_2H_6, and SF_6 on NaX, H-ZSM-5, and Na-ZSM-5 zeolites. Langmuir 12(24):5896–5904
25. El-Sharkawy II (2006) Development of adsorption systems powered by renewable energy or waste heat sources. Ph.D. Thesis, Kyushu University, Japan
26. Fan Y, Luo L, Souyri B (2007) Review of solar sorption refrigeration technologies: development and applications. Renew Sustain Energy Rev 11(8):1758–1775
27. Ito M, Watanabe F, Hasatani M (1996) Improvement of both adsorption performances of silica gel and heat transfer characteristics by means of heat exchange modulation for a heat pump. Heat Transf Jpn Res 25(7):420–431
28. Dehler FC (1940) Silica gel adsorption. Chem Metall Eng 37:307–310
29. Saha BB, Koyama S, Alam KCA, Hamamoto Y, Akisawa A Kashiwagi T, Ng KC, Chua HT (2003) Isothermal adsorption measurement for the development of high performance solid sorption cooling system. Trans JSRAE 20(3):421–427
30. Ryu Z, Zheng J, Wang M, Zhang B (1999) Characterization of pore size distributions on carbonaceous adsorbents by DFT. Carbon 37(8):1257–1264
31. Ng KC, Chua HT, Chung CY, Loke CH, Kashiwagi T, Akisawa A, Saha BB (2001) Experimental investigation of the silica gel–water adsorption isotherm characteristics. Appl Therm Eng 21 (16):1631–1642

32. Wang X, Zimmermann W, Ng KC, Chakraborty A, Keller JU (2004) Investigation on the isotherm of silica gel + water systems. J Therm Anal Calorim 76(2):659–669
33. Valenzuela DP, Myers DP (1989) Adsorption equilibrium data handbook. Prentice Hall, Englewood Cliffs, NJ
34. Sircar S (1991) Isosteric heats of multicomponent gas adsorption on heterogeneous adsorbents. Langmuir 7(12):3065–3069
35. Tóth J (1971) State equations of the solid–gas interface layers. Acta Chim Acad Sci Hung 69:311–328
36. NACC (1992) PTX data for silica gel/water pair, Manufacturer's proprietary data. Nishiyodo Air Conditioning Co Ltd., Tokyo, Japan
37. Chihara K, Suzuki M (1983) Air drying by pressure swing adsorption. J Chem Eng Jpn 16:293–298
38. Cremer HW, Davis T (1958) Chemical engineering practice, vol 6. Butterworths, London, pp 286–287
39. Sakoda A, Suzuki M (1984) Fundamental study on solar powered adsorption cooling system. J Chem Eng Jpn 17:52–57
40. Wang XL, Ng KC (2005) Experimental investigation of an adsorption desalination plant using low-temperature waste heat. Appl Therm Eng 25:2780–2789
41. Ng KC, Chua HT, Wang XL, Kashiwagi T, Saha BB (2003) Prototype testing of a novel four-bed regenerative silica gel–water adsorption chiller. In: ICR 0042, Washington, DC
42. Saha BB, El-Sharkawy II, Chakraborty A, Koyama S (2007) Study on an activated carbon fiber–ethanol adsorption chiller: Part II – performance evaluation. Int J Refrig 30(1):96–102
43. Wang XL, Chua HT, Ng KC (2006) Experimental investigation of silica gel adsorption chillers with and without a passive heat recovery scheme. Int J Refrig 28:756–765
44. El-Sharkawy II, Thu K, Ng KC, Saha BB, Chakraborty A, Koyama S (2007) Performance improvement of adsorption desalination plant: experimental investigation. Int Rev Mech Eng 1 (1):25–31
45. Ng KC, Wang X, Lim YS, Saha BB, Chakraborty A, Koyama S, Akisawa A, Kashiwagi T (2006) Experimental study on performance improvement of a four-bed adsorption chiller by using heat and mass recovery. Int J Heat Mass Transf 49(19–20):3343–3348
46. NUS (2008) Adsorption desalination to quench global thirst. Engineering Research, 23(3), Oct. 2008. National University of Singapore (NUS)
47. Thu K, Ng KC, Saha BB, Chakraborty A, Koyama S (2009) Operational strategy of adsorption desalination systems. Int J Heat Mass Transf 52(7–8):1811–1816, March

10

Membrane Processes for Reclamation of Municipal Wastewater

Jianjun Qin and Kiran A. Kekre

Contents

Abstract In this chapter, advanced membrane technology for the reclamation of municipal wastewater has been introduced. The design of various membrane processes for this application is briefed. The investigations on the most common membrane (integrated) processes are emphasized, including (a) UF for the tertiary treatment of municipal wastewater; (b) MF-RO for the reclamation of the secondary domestic effluent; (c) Total organic carbon (TOC) removal in the reclamation of municipal wastewater by reverse osmosis (RO); (d) New option of MBR-RO for the reclamation of municipal wastewater; (e) the reclamation of a mixed sewage effluent using UF-RO. The recent R&D is highlighted.

Key Words Microfiltration • ultrafiltration • reverse osmosis • membrane bioreactor • water reclamation • municipal wastewater.

From: *Handbook of Environmental Engineering, Volume 13: Membrane and Desalination Technologies*
Edited by: L.K. Wang et al., DOI: 10.1007/978-1-59745-278-6_10

1. INTRODUCTION

The shortage of fresh water supplies has become a global issue because of the rapid development of population and industry. A lot of efforts in many countries have been put into developing nonconventional water resources, such as desalination of seawater and the reclamation of municipal wastewater. Advanced membrane technology as one of the most promising technologies has been increasingly attractive because of its high efficiency, ease and economic of operation in those two applications mentioned above (1–29). Since membrane technology for seawater desalination has been introduced in detail in Chap. 13, membrane processes for the reclamation of municipal wastewater are focused in this chapter.

The reclamation of municipal wastewater by membrane processes is not new, and it has been practiced for 30 years. For example, at Water Factory 21, Southern California, high-quality water reclaimed from treated used water by reverse osmosis (RO) has been injected into ground water as a source of water supply (so-called indirect potable use) since 1976 (7). Similarly, high-quality water reclaimed at Upper Occoquan Sewage Authority, North Virginia as indirect potable use has been discharged into Occuquan Reservoir which is a source of water for the vicinity of Washington, DC since 1978. Nowadays, water reclamation for different applications has been fast growing around the world. In the USA, today there are about 40 full-scale membrane facilities for municipal wastewater treatment and reuse of various purposes. An example of them for indirect potable use is the membrane plant with the treatment capacity of 50 MGD (189,000 m^3/d) at Gwinnett near Atlanta, Georgia.

Research on membrane bioreactor (MBR) technology for municipal wastewater treatment started more than 30 years ago (9) while commercialization of MBR for large-scale municipal wastewater applications was in the mid 1990s (10). In Europe, MBR technology has been widely applied to treat municipal wastewater for irrigation. Examples of commercial facility are hollow fiber MBR plants with the capacity of 37,850 m^3/day at Brescia in Italy, 45,144 m^3/day in Germany, 96,000 m^3/day in the Netherlands as well as flat sheet MBR plant of 13,000 m^3/day in Swanage, UK.

In Singapore, the reclamation of treated effluent as a secondary source of water for nondomestic use is a national strategy as water is a scarce resource. Since a demonstration plant with the capacity of 10,000 m^3/day for the production of potable grade water (so-called NEWater) from the secondary treated domestic sewage effluent was successfully operated in May 2000 (19), four factories with the total capacity of 234,000 m^3/day, which is more than 15% of the total water demand, have been commenced to supply NEWater to the wafer fab industry, cooling towers, and indirect potable use. The fifth NEWater factory with the capacity of 225,000 m^3/day is also brought forward. Hence by 2011, NEWater supply will meet 30% of Singapore's water needs (20).

This chapter focuses on membrane technology for the reclamation of municipal wastewater as follows. The design of various membrane processes is introduced, the investigations on the most commonly used membrane (integrated) processes are emphasized and the recent R&D are highlighted.

2. PROCESS DESIGN

2.1. Typical Flow Schematics of Membrane Processes

Design of membrane (integrated) processes in the reclamation of municipal wastewater varies and depends on the requirement of the reclaimed water quality or the applications of the reclaimed water. Figure 10.1 provides a summary of five typical flow schematics of membrane processes that are commonly used in the reclamation of municipal wastewater. Basically, membrane (integrated) processes are applied after the municipal wastewater treatment with the conventional activated sludge process (ASP). In case that an MBR which is a combination of ASP and membrane separation is implemented, the overall treatment process is much more simplified when Process (b) is compared to Process (a) or Process (e) is compared to Process (d). As a consequence, the following benefits can be obtained after the MBR is implemented.

- A small footprint (savings in space by over 30%)
- High efficiency to remove ammonium and organism content
- Better consistency in effluent quality
- Less sludge disposal
- Cheaper in total cost

In general, Processes (a) and (b) are applied to produce reclaimed water with low levels of colloidal materials, suspended solids, and microorganisms while Processes (c)–(e) are

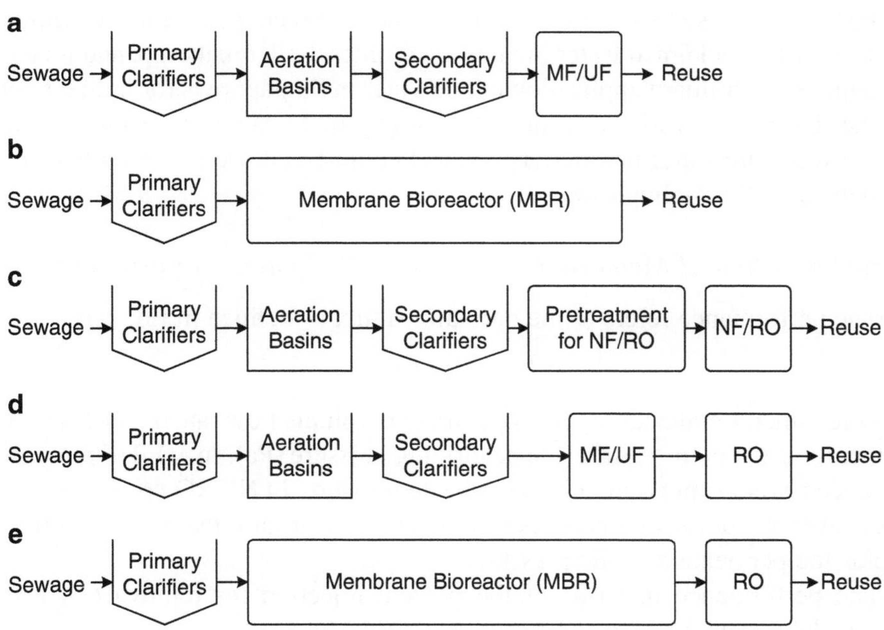

Fig. 10.1. Flow schematics of membrane processes in reclamation of municipal wastewater.

utilized to achieve high grade water with low levels of both inorganic ions and organic compounds. An MF/UF is used as a pretreatment step prior to RO to remove suspended solids and colloidal materials in Process (d) while the pretreatment of NF/RO may include coagulation, multimedia filtration, and cartridge filtration, etc., in Process (c). However, in Process (e), an MBR is a combination of the biological treatment of wastewater and the pretreatment prior to RO.

2.2. *Applications of Reclaimed Water*

The reclamation of municipal wastewater by membrane processes has been attractive in various applications. Some important applications are as follows.

- Indirect potable uses
- Agricultural irrigation
- Industrial uses
- Commercial use (cooling tower)
- Fire-fighting, gardening, car-washing
- Toilet flushing, construction use, and cleaning, etc.

Reuse of the reclaimed water can be classified as indirect potable use and direct nonpotable use. Examples of indirect potable uses of the reclaimed water are groundwater recharge and blending water-supply reservoirs. Groundwater recharge is used to prevent coastal aquifers from saltwater intrusion, protect groundwater level decline, and store underground freshwater in the aquifer system for future use. The high grade reclaimed water can also be mixed and blended with a natural groundwater source (river or reservoir) and then undergo conventional water treatment, such as sedimentation, filtration, and disinfection, to produce drinking water. Industrial uses of the reclaimed water include electronics, boiler make-up, and process needs, etc. For example, the primary application of NEWater in Singapore is for indirect nonpotable use. It is the first in the world to achieve the supply of NEWater to wafer manufacturing processes for ultrapure water production. Typical commercial use of the reclaimed water to replace tap water is for cooling towers.

2.3. *Characterization of Membrane and Membrane System Performance*

Membrane performance refers to the normalized flux as defined in Eq. (1):

$$J = Q/(A\Delta P), \tag{1}$$

where J is pressure normalized flux (L/m^2 h Pa), Q is volume flow rate of permeate (L/h), A is effective membrane area (m^2), ΔP is the net driving pressure (Pa). In MF/UF processes, ΔP is equal to the feed average pressure minus permeate pressure. In NF/RO processes, ΔP is equal to the feed average pressure minus permeate pressure minus the feed average osmotic pressure plus the permeate osmotic pressure.

Membrane performance in terms of the percent rejection (or removal) of a particular component is defined as Eq. (2):

$$R = (1 - C_\mathrm{p}/C_\mathrm{f}) \times 100\%, \tag{2}$$

where R is rejection or removal (%), C_f and C_p are the component concentration in the feed and permeate, respectively.

The water percent recovery of the membrane system is defined as Eq. (3):

$$\text{Recovery} = Q_p/Q_f \times 100\%, \tag{3}$$

where Q_p is permeate flow rate (L/h) and Q_f is feed flow rate (L/h).

3. UF FOR TERTIARY TREATMENT OF MUNICIPAL WASTEWATER

3.1. Background

Water reclamation from the secondary treated effluent by way of producing NEWater is one of the National Taps identified in Singapore. The fourth NEWater factory with the capacity of 25 MGD (112,500 m^3/day) was commenced at Ulu Pandan Water Reclamation Plant (Ulu Pandan) in 2006. Figure 10.2 shows a typical flow chart for the reclamation of municipal wastewater using a dual UF/RO membrane process. In the process, UF as a pretreatment prior to RO is to remove suspended solids and colloidal materials for reducing the fouling of RO.

Nowadays, the hollow fiber configuration in UF has become a favorite choice because the hollow fiber membrane has three major advantages (21): (1) hollow fiber modules have much larger ratio of membrane area to unit volume (several thousand m^2/m^3) compared to flat and spiral-wound modules (a few hundred m^2/m^3), and hence higher productivity per unit volume of membrane module; (2) they are self-supporting which can be backwashed to recover the flux; and (3) they have good flexibility in the mode of operation. Dead-end operation has advantages of low energy consumption (low operation pressure and flow) and high water recovery over cross-flow operation. The former, however, has a major problem of more serious membrane fouling than the latter. Therefore, it is a challenge to solve the fouling problem and achieve a stable operation of the UF system in the dead-end mode.

The objective of the study was to establish the process of alum injection to achieve a stable operation of a UF system in the dead-end mode to treat the secondary treated effluent at Ulu Pandan. The XIGA hollow fiber UF membrane technology was used in the study. The plant was operated from March to June 2003. The study has established the optimal concentration of alum dosing to achieve a stable operation of the UF system in the dead-end mode.

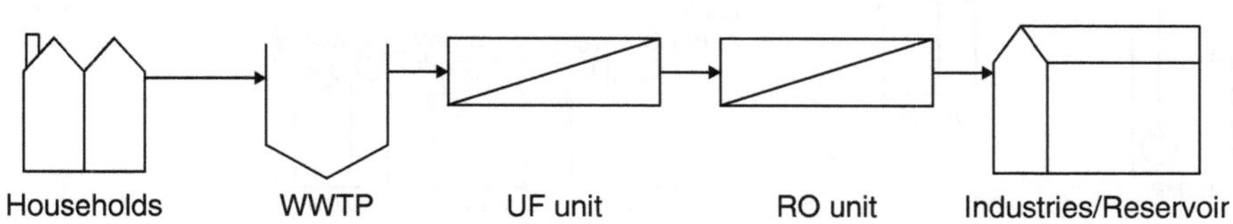

Fig. 10.2. Typical flow chart using UF/RO process for reclamation of municipal wastewater (13).

3.2. Description of Overall Process

3.2.1. Plant Process

The schematic diagram of the plant process is shown in Fig. 10.3. The secondary treated effluent was transferred by a submersible pump from the outfall sump through a duplex basket strainer (370 μm) to the feed tank. The large suspended solids in the raw feed water were removed by the strainer to prevent membrane fibers from being blocked or abraded. The filtrate water was then transferred to the UF unit by the UF feed pump. Alum was injected into the suction pipe before the feed pump by a dosing pump. The UF unit further removed suspended solids and colloidal materials from the filtrate water. The UF product was collected in the permeate tank and then overflowed into the drain.

The UF unit could be operated in the filtration mode, backwash mode, and chemically enhanced backwash (CEB) mode (22). In the filtration mode, the UF unit was operated in dead-end and in-out flow at a fixed water flow rate by the use of a speed variable feed pump, and no retentate was discharged. In the backwash mode, the clean water from the permeate tank was transferred by the backwash pump into the shell side of the UF module, then went out from the lumen side of the fibers, and then carried the floccules on the fiber inner surface, finally was discharged into the drain from the retentate line. The CEB mode was similar to the backwash mode; in addition, chemicals were dosed into the suction pipe before the backwash pump as shown in Fig. 10.3. The cleaning chemicals used were NaOCl, NaOH, and H_2SO_4 to recover flux from the biofouling and scaling.

3.2.2. Hollow Fiber Unit and Trial Runs

The plant had two S-225 FSFC UF modules with total membrane area of 70 m^2 and capacity of 2–7 m^3/h. The UF unit was operated in a dead-end and in-out flow mode. Newly

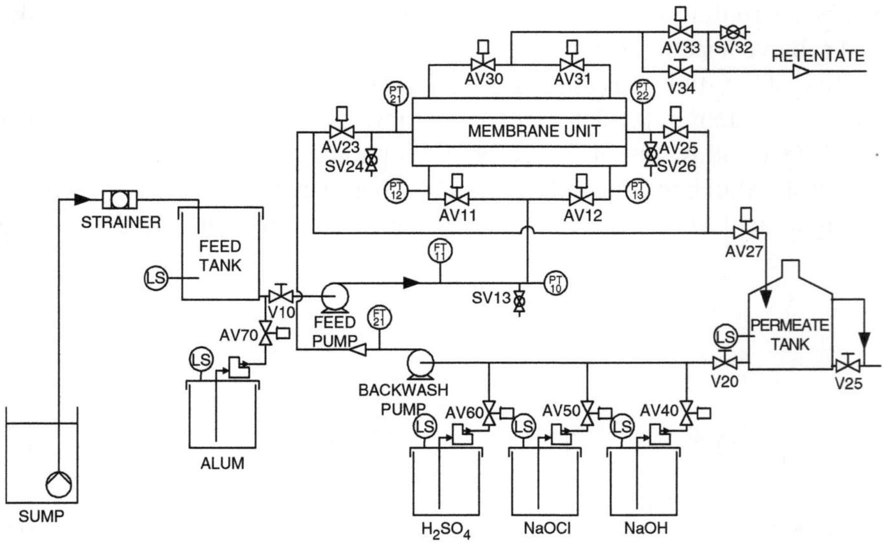

Fig. 10.3. Schematic diagram of the plant process (13).

loaded UF elements were flushed with town water. Then, the initial performance of UF membranes was tested with town water and the normalized flux was 330 L m^2 h at 0.1 MPa at 20°C. Before and after the trial, the membrane integrity test was performed using the pressure decay method. The results showed the membrane integrity was good.

Trial runs on various concentrations of alum dosage at different water recoveries were conducted. The common and variable operating conditions of the plant are given in Tables 10.1 and 10.2, respectively. The plant was later continuously operated for 4 weeks under the conditions showed in Test 5 of Table 10.2. Transmembrane pressure (TMP) and normalized flux of UF membranes as a function of time were observed. Product quality and chemical consumption were determined. An energy meter was purposely installed to monitor the energy consumption.

3.3. *Plant Performance*

3.3.1. Permeate Quality of UF Membrane

Table 10.3 shows typical characteristics of the raw feed water in the study. The raw feed water contained a wide range of suspended solids and colloidal substances. The micro-strainer was required to remove large suspended solids for protecting the UF membranes. The UF membranes further removed smaller suspended solids and colloidal substances.

Table 10.4 shows the permeate quality of UF membrane during the study and the target quality for RO feed (19). It can be seen that the average silt density index (SDI) and turbidity were 1.6 and 0.13 NTU, respectively. Particle (1–15 μm) was 2 counts/mL in average. Both fecal coliform and total coliform in the permeate were not detectable. Total bacteria were 69 cfu/mL in average. Total suspended solid (TSS) was lower than 2.5 mg/L (detectable

Table 10.1
Common operating conditions of the plant (13)

Parameter	Value
Feed flow (m^3/h)	4.9–5.0
Backwash flux (L/m^2 h)	250–300
Backwash duration (s)	30–40
Backwash frequency (time/h)	2–4
CEB interval (day)	1–3

Table 10.2
Variable operating conditions of the plant (13)

Parameter	Test 1	Test 2	Test 3	Test 4	Test 5
Alum dosing concentration (mg/L as Al)	4.8	1.3	0.6	2.5	2.5
Water recovery (%)	87.5	87.5	87.5	87.5	90.1

Table 10.3
Typical characteristics of the raw feed water (13)

Parameter	Range
Total suspended solid (TSS) (mg/L)	3–26
Aluminum (mg/L)	0.05–0.56
Ammonia (as N) (mg/L)	2–25
Chloride (mg/L)	100–300
Chemical oxygen demand (COD) (mg/L)	30–200
Fecal coliform (cfu/100 mL)	$3–15 \times 10^4$
Total coliform (cfu/100 mL)	$3.8–5.0 \times 10^5$
Total bacteria (cfu/mL)	$1.1–1.6 \times 10^5$
Total alkalinity as $CaCO_3$ (mg/L)	40–160
Total nitrogen (mg/L)	10–50
Total dissolved solids (TDS) (mg/L)	400–1,000
pH	6.5–7.5

Table 10.4
Comparison of the UF product and the target quality (13)

Parameter	Quality of UF product		Target quality
	Range	Average	
Silt density index (SDI)	1.1–2.1	1.6	<3
Turbidity (NTU)	0.12–0.14	0.13	0.5
Particle (1–15 μm, counts/mL)	0–4	2	<5
Fecal coliform (cfu/100 mL)	<1	<1	<1
Total coliform (cfu/100 mL)	<1	<1	<1
Total bacteria (cfu/mL)	56–80	69	<300
TSS (mg/L)	<2.5	<2.5	<2.5
Aluminum (mg/L)	0.019–0.049	0.026	–

limit) throughout the study. It needs to be pointed out that the residual aluminum in the UF permeate was 0.026 mg/L, which was even much lower than that in the raw feed (0.05–0.56 mg/L as shown in Table 10.4) although Alum was dosed into the feed. In addition, from the data in Tables 10.4 and 10.5, it can be seen that the total coliform rejection was high at log reduction value (LRV) >5, whereas, the total bacteria rejection was low at LRV <4. The former indicated the membrane had excellent performance for removal of water borne bacteria and the latter was probably due to contamination from the environment which caused total bacteria of 56–80 cfu/mL in the UF permeate. In summary, the results showed that the UF membrane product met the target quality and was well suitable for RO feed for producing NEWater from the secondary treated effluent.

Table 10.5
Consumption of chemical and power and cost per cubic meter product for the pilot trials (13)

Item	Consumption	Cost (S$)
Chemicals		
Flocculants (as Al)	2.77 g	0.0174
NaOCl	0.41 g	0.0027
H_2SO_4	0.56 g	0.0019
Subtotal		0.022
Power	0.194 kW-h	0.0248
Total		0.047

Note: *S$* Singapore dollar; 1 US dollar = 1.75 Singapore dollar.

3.3.2. *Transmembrane Pressure as a Function of Time*

The effect of alum dosing concentration on stabilization of the plant operation was intensively investigated. Figure 10.4 illustrates the TMP of the UF membranes as a function of time throughout the study under different concentrations of alum dosage. It should be pointed out that a TMP band appeared because TMP increased with time within one cycle of the operation mode (20–30 min). The high level and low level corresponded to the TMP at the time immediately before and after a backwash operation, respectively. Figure 10.4a indicated that TMP was 160–210 mbar when alum concentration was 4.8 mg/L and the water recovery was 87.5% at the beginning, then TMP increased to 200–270 mbar as alum concentration decreased to 1.3 mg/L. TMP tended to slightly increase with time during these two tests. However, TMP increased sharply and was unstable when alum concentration further decreased to 0.6 mg/L. Figure 10.4b illustrated that TMP was stable in the range of 200–240 mbar as alum concentration was enhanced to 2.5 mg/L during 29 April–6 May. High TMP on 7 May appeared because the air lock in the feed line caused the feed pump shutdown; further, the plant shutdown when water level in the feed tank became gradually low due to the clogging of the strainer. After that, the water recovery was increased to 90.1% on 10 May while the alum concentration remained the same. TMP had been stable in the range of 220–260 mbar since then. The results showed that alum concentration was critical for maintaining stable TMP and achieving a smooth operation of the plant, which could be explained as follows from two aspects.

On the one hand, within a backwash interval, a colloidal fouling of the membrane was formed. In the dead-end UF process for this particular application, alum was dosed into the feed line just prior to feed pump, then well mixed with the feed inside the pump. It took a few seconds for the flocculating feed to reach the membranes. Theoretically, flocs formed should be much larger than the membrane pores before they reached the membrane surface with an appropriate dosing concentration of alum. During the filtration operation, the flocs gradually cumulated on the membrane surface and formed a temporary cake layer of suspended solids. After the backwash, the back washable suspended solids could be easily removed and the

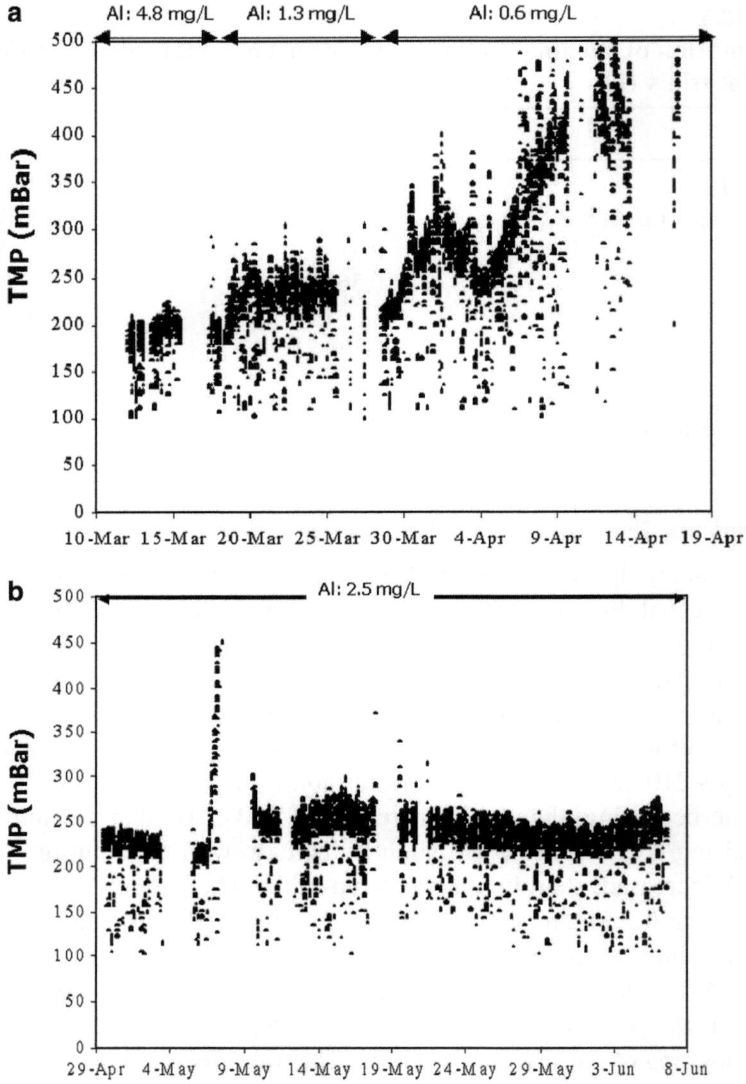

Fig. 10.4. Transmembrane pressure (TMP) as a function of time (13). (a) Concentration of alum dosage: 4.8, 1.3 and 0.6 mg/L (b) Concentration of alum dosage: 2.5 mg/L

membrane flux was mostly recovered. However, the flocs formed should not be too large to precipitate on the membrane surface and to block the lumen of the fibers. For a fixed duration which allowed the flocs growing, the size of flocs reaching the membrane surface should increase with an increase in alum dosing concentration. An optimal alum concentration to the membrane filtration meant the lowest alum dosing concentration that could make the size of flocs formed meet the requirements mentioned above, which resulted in a low colloidal fouling of the membrane and a stable plant operation. If the alum concentration was higher, the chemical cost would be more while no significant benefit to the flux recovery was obtained. The results obtained from the pilot trial study showed that 2.5 mg/L Al was an appropriate dosing concentration in this particular application. Hagmeyer et al. (23) worked on drinking water treatment of reservoir water using a UF process with alum dosage and concluded that the optimal alum dosing concentration was depending on the feed water quality.

On the other hand, within an NaOCl CEB interval, biofouling of the UF membrane was formed due to the substances of microbial origin that clung to the membrane surface and could not be removed through the backwash. It was increased with time. The biofouling could be eliminated after the NaOCl CEB at pH 11.5. Immediately after that, H_2SO_4 CEB at pH 2 was followed to remove the scaling. Finally, the membrane flux was recovered again.

3.3.3. Normalized Flux as a Function of Time

Normalized flux (corrected to 20°C) of the UF membranes as a function of time throughout the study under different concentrations of alum dosage is shown in Fig. 10.5. Corresponding to TMP, a low TMP resulted in a high normalized flux since the water flow rate was fixed, and vice versa. A stable TMP also provided a stable flux. Figure 10.5a indicated that the normalized flux dropped from 250–300 L/m^2 h at 0.1 MPa to 200–250 L/m^2 h at 0.1 MPa as alum concentration was reduced from 4.8 to 1.3 mg/L, which could be explained as follows. Since the flocculation capacity of alum reduced with a decrease in alum concentration, the size of flocculated particles decreased for the fixed flocculation duration. Subsequently, small colloidal might easily enter the surface pores of the UF membrane, and they were more difficult to be removed through the normal backwash than the large flocs. As a consequence, the membrane resistance was enhanced and high membrane fouling was resulted. Figure 10.5b illustrated that the normalized flux had been stable when alum concentration was 2.5 mg/L since 29 April, and it was between 200 and 250 L/m^2 h at 0.1 Map from 9 May–6 June. The results showed that the optimum alum concentration was 2.5 mg/L for maintaining a stable flux of the UF membranes.

3.3.4. Overall Process Consideration

It is a proper combination of the dead-end flow operation of hollow fiber membranes with alum dosage for treating the secondary treated effluent because the flocs formed are much larger than the membrane pores and can be easily removed with a normal backwash, which benefits flux recovery. Alum dosing concentration was found critical for achieving stable plant operation in dead-end flow. The containerized pilot plant is equipped with an advanced programmable logic controller (PLC) control system and is able to operate automatically with well-designed programs. The successful pilot trial runs have demonstrated that the employment of a UF membrane process in dead-end operation mode is capable of recovering 90.1% of the secondary treated sewage effluent and the reclaimed water met target quality as RO feed for NEWater production.

Based on the results of the pilot trials under the operating conditions of feed flow 5.0 m^3/h and backwash interval 25 min during the stable operation of the plant, an economic evaluation in terms of consumption of chemical and power is summarized in Table 10.5. It can be seen that for 1 m^3 of product water, the cost on chemical consumption is S$0.02 or

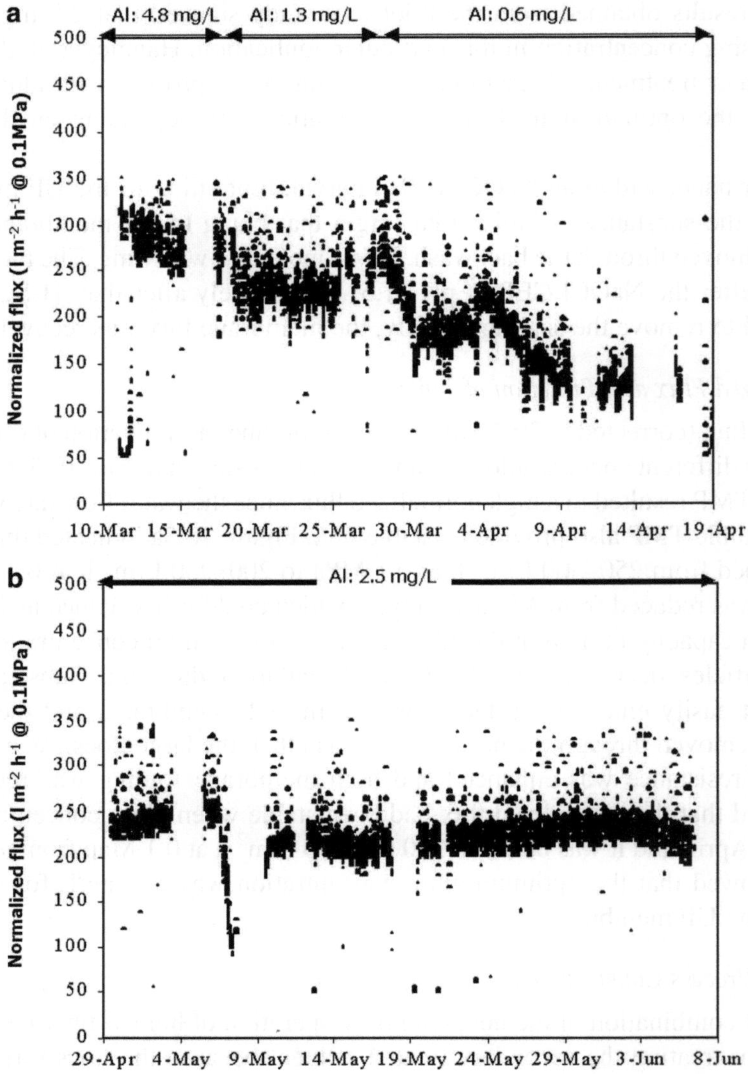

Fig. 10.5. Normalized flux as a function of time (13). (a) Concentration of alum dosage: 4.8, 1.3 and 0.6 mg/L (b) Concentration of alum dosage: 2.5 mg/L

1.3 cents (US) and the total power consumption is 0.194 kW-h/m^3 which is quite low compared to previous studies (24–26). The total cost, including chemical and energy consumption for producing 1 m^3 of product water, is around S$0.05 or 2.7 cents (US). The results of the economic evaluation indicate that the UF membrane process is attractive for the reclamation of the secondary treated sewage effluent. It should be noted that the inclusion of this would have further improved the economics when the chemicals are purchased as a bulk amount, and the pump efficiency is enhanced for a full-scale plant.

3.4. Conclusions

The conclusions from the study may be summarized as follows:

1. A stable operation of the plant in dead-end flow has been achieved by optimizing the operating conditions.
2. Alum dosing concentration was critical and the optimum concentration was determined at 2.5 mg/L as Al.
3. The pilot plant was operated stably for 4 weeks under the optimum conditions that TMP was between 0.22 and 0.26 bar and normalized flux of UF membrane was $200-250 \times 10^{-5}$ L/m^2 h Pa at 20°C.
4. Product water with average SDI of 1.6 and turbidity of 0.13 NTU was produced from the plant at a water recovery of over 90%. Particle (1–15 μm) in the UF permeate was 2 counts/mL. Total coliform was not detectable. Total bacteria were 69 cfu/mL.
5. The total power consumption was 0.194 kW-h/m^3 (of which, 0.100 kW-h/m^3 was for the operation of UF unit and 0.094 kW-h/m^3 was for the submersible pump and the air compressor), and the chemical cost was 1.3 cents (US)/m^3 product water at the capacity of 5 m^3/h.
6. The UF membrane used could be attractive as a pretreatment prior to RO for the reclamation of the secondary treated sewage effluent.

4. MF-RO FOR RECLAMATION OF THE SECONDARY DOMESTIC EFFLUENT

4.1. Background

A dual membrane MF (or UF)-RO process has become increasingly attractive as a promising technology for the reclamation of municipal wastewater (8, 11–16, 19–20, 27–29) since it was demonstrated at Water Factory 21 (8, 27). In this process, an MF or an UF is used as a pretreatment step prior to RO to remove suspended solids and colloidal materials. RO membranes remove dissolved solids and organics.

Pino and Durham introduced pilot studies of treating an effluent from municipal wastewater treatment plant for irrigation using an MF/RO process. They reported that rejections of the RO membrane in terms of conductivity, total organic carbon (TOC), ammonium and nitrate were 95, 91, 91, and 82% in average, respectively (27). Abdel-Jawad et al. treated a tertiary-level effluent of municipal wastewater using an RO unit at operating pressure of 9 bars and showed that rejections of the RO membrane in terms of conductivity and ammonium were 97.5 and 92%, respectively (29). A demonstration plant of a dual membrane MF/RO process with the capacity of 10,000 m^3/day for the production of NEWater from the secondary treated domestic sewage effluent has been successfully operated since May 2000 in Singapore (19). It was reported that rejections of the RO unit in terms of conductivity, TOC, ammonium and nitrate were >97, 97, 90, and 80%, respectively (19). However, attempts are still being made to optimize the RO process in this application because the process effectiveness depends on many factors, including the type of the RO membrane, feed concentration, and operating conditions, etc. (30–31).

This study presents an MF-RO dual membrane process for the reclamation of the secondary effluent in Singapore. The study focused on the performance of the RO membrane under different operating conditions to produce the reclaimed water with NEWater quality for use in the electronics industry. The study also established the optimal operating flux rate of the RO membrane in the application.

4.2. Description of Overall Process

4.2.1. Plant Process

During the study, the plant was continuously operated at Bedok Wastewater Treatment Works, Singapore. The schematic diagram of the process is shown in Fig. 10.6. The secondary treated sewage effluent as the raw feed was transferred by a submersible pump from outfall sump through an automatic strainer (200 μm) to the MF feed tank. The large suspended solids in the raw feed water were removed by the strainer to prevent membrane fibers from being blocked or abraded. The filtrate water was then transferred to the MF unit by the MF feed pump. The MF unit further removed suspended solids and colloidal materials. NaOCl solution was dosed into an MF feed to reduce biofouling (16). The MF permeate was collected in the backwash tank and then transferred to RO feed tank. RO feed was transferred by feed pump and high-pressure pump to RO unit. Sodium metabisulfite was dosed into RO feed to protect the RO membranes from the attack of free chlorine (31). An antiscalant PermaTreat 191 was also dosed into RO feed to minimize the fouling of RO membranes (32). Concentrate water was discharged to drain and permeate was stored in RO permeate tank.

The MF unit was composed of polyvinylidene fluoride (PVDF) hollow fiber membranes with pore size of 0.1 mm and had a pretreatment capacity of 2.7 m^3/h. It was equipped with advanced PLC auto-control system. The programs for filtration and backwash had been programed. The RO unit contained six elements of RE4040-FL membranes installed in two pressure vessels in series as single stage configuration, and it had a capacity of 20 m^3/day. There was also a Clean-In-Place (CIP) system attached with the RO unit. Two percent citric acid was used for cleaning at pH 2.0 over 3 h. Then, 1% EDTA was used at pH 12.0 over 1 h operation followed by 16 h soaking and 0.5 h operation (31).

4.2.2. Trial Runs

Pilot trials on various membrane flux rates in a range of 10–26 gallon/ft^2 day at operating pressure of 52–141 psi and product recovery of 50–66% for the RO unit were conducted from 12 May–20 August 2003 to determine the optimum flux rate of the RO membrane for this

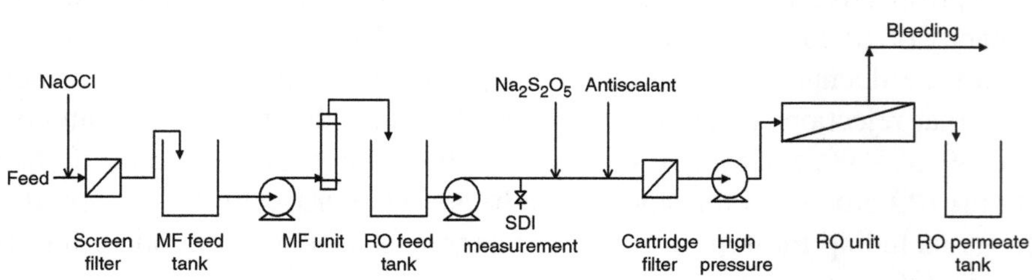

Fig. 10.6. Schematic diagram of the plant process (14).

particular application. Variable operating conditions of the RO unit are shown in Table 10.6. CIP on the RO membrane was conducted for twice to examine the recovery capability of the RO membrane flux after fouled. The RO membrane unit was operated under the conditions of Test 4 shown in Table 10.6 during the last 5-week study.

Operating pressure of RO unit as a function of time was observed and the corresponding normalized flux of the RO membrane was calculated. Qualities of RO feed and permeate were measured.

4.3. *Plant Performance*

Table 10.7 shows typical characteristics of the raw feed (secondary treated sewage effluent) in the study. The raw feed water contained a wide range of suspended solids and colloidal substances. A microstrainer was required to remove large suspended solids for

Table 10.6
Variable operating conditions of the RO unit (14)

Parameter	Test 1	Test 2	Test 3	Test 4
Flux rate (gallon/ft^2 day)	26	20	15	10
Feed flow (m^3/h)	3.2	2.8	2.1	1.6
Recovery (%)	66	58	56	50
Operating pressure (psi)	136–141	117–136	75–83	52–62

Table 10.7
Typical characteristics of the raw feed water (14)

Parameters	Value
pH	6.9
Conductivity (μS/cm)	1,100
Suspended solid (mg/L)	5.8
True color (HU)	51.9
Ammonium as N (mg/L)	5.7
Total alkalinity as $CaCO_3$ (mg/L)	66.7
COD (mg/L)	28.3
B (mg/L)	0.07
Ca (mg/L)	36.0
Fe (mg/L)	0.10
Mg (mg/L)	21.2
Sr (mg/L)	0.16
F (mg/L)	0.84
SO_4 (mg/L)	78.1
Total silica (mg/L)	9.15

protecting the MF membranes. The MF membranes further removed smaller suspended solids and colloidal substances. Table 10.8 shows the quality of the feed (MF permeate) to the RO unit.

Free chlorine and total chlorine of an MF feed after dosing NaOCl of 6 mg/L were found to be 0.36 and 0.82 mg/L, respectively. The MF permeate in the RO feed tank contained free chlorine of 0.10 mg/L and total chlorine of 0.52 mg/L, respectively. Sodium metabisulfite of 2 mg/L was dosed into RO feed to protect the RO membrane from the attack of free chlorine.

4.3.1. *Normalized Flux and Rejection of the RO Membranes as a Function of Time*

Normalized flux of the RE4040-FL RO membrane under different flux rates as a function of time throughout the study is shown in Fig. 10.7. The results are summarized in Table 10.9.

Table 10.8
Quality of the RO feed (14)

Parameters	Value
pH	6.6–7.0
Conductivity (μS/cm)	840–1,160
Ammonium as N (mg/L)	0.4–8.4
Nitrate (mg/L)	18.6–53.0
Hardness as $CaCO_3$ (mg/L)	81–130
TOC (mg/L)	5.2–11.9
Turbidity (NTU)	0.1
SDI	3.1

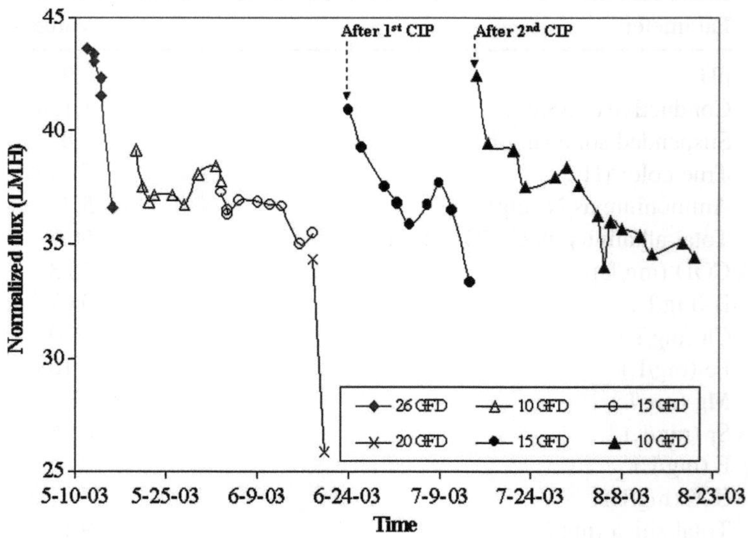

Fig. 10.7. Normalized flux at different flux rates as a function of time (14).

Table 10.9
Normalized flux and salt rejection of the RO membrane (14)

Parameter	Test 1	Test 2	Test 3	Test 4
Flux rate (gallon/ft^2 day)	26	20	15	10
Operating pressure (psi)	136–141	117–136	75–83	52–62
Normalized flux (L/m^2 h MPa at 25°C)	43.6–36.6	34.3–25.9	40.9–33.4	42.4–34.0
Rejection (%)	98.4–98.6	98.1–97.8	96.9–97.5	96.1–96.5

Figure 10.7 and Table 10.9 show that the normalized flux of the RO membrane dropped obviously by 3.2% per day from 43.6 to 36.6 L/m^2 h MPa at 25°C within 5 days from 12 to 16 May when the RO unit was operated at the flux rate of 26 gallon/ft^2 day. Then, it was operated at 10 gallon/ft^2 day for 2 weeks from 20 May to 3 June; the normalized flux dropped slightly by 0.3% per day from 39.1 to 37.7 L/m^2 h. And then, the flux rate was increased to 15 gallon/ ft^2 day during 3–18 June, the normalized flux dropped slightly by 0.3% per day from 37.2 to 35.5 L/m^2 h. After that, the flux rate was further increased to 20 gallon/ft^2 day; the normalized flux declined significantly 12.2% per day from 34.3 to 25.9 L/m^2 h within 2 days, 18–20 June. Then, the first CIP was conducted. When RO unit was operated again at 15 gallon/ft^2 day during 24 June–14 July, the normalized flux declined by 1% per day from 40.9 to 33.4 L/m^2 h over 3 weeks. CIP was done for the second time. Finally, RO unit was operated at 10 gallon/ft^2 day during 15 July–20 August, and the normalized flux declined by 0.6% per day from 42.4 to 34.0 L/m^2 h over 5 weeks. The average normalized flux was 38 L/m^2 h MPa at 25°C. Both of the normalized flux decline rates for the first operating flux rate of 10 and 15 gallon/ft^2 day were 0.3% per day, whereas they increased to 0.6 and 1% per day for the second 10 and 15 gallon/ft^2 day, respectively. The increases in the normalized flux decline rates could be attributed to the facts that they were calculated based on the respective initial normalized fluxes after the system operation was stable for the first run but immediately after a CIP was conducted for the second run. However, the normalized flux decline rates of 0.3–0.6% per day at 10 gallon/ft^2 day and of 0.3–1% per day at 15 gallon/ft^2 day were much lower than those of 3.2–12.2% per day at higher operation flux rates of 20–26 gallon/ft^2 day. The above results indicate that the optimal operation flux rate of the RO membrane ranged from 10 to 15 gallon/ft^2 day for this application. Figure 10.6 also shows that the normalized flux recovery after the first CIP and second CIP comparing to the initial flux was 94 and 97%, respectively.

Table 10.9 also indicates that salt rejections of the RO membrane ranged of 96–97% at 10 gallon/ft^2 day, 97–98% at 15 gallon/ft^2 day, and over 98% at 20 gallon/ft^2 day. Higher flux rate was the RO membrane operated at, higher rejection was obtained. It appeared that salt rejection increased with time by 0.8% of the maximum at the beginning of each test at a fixed operation flux rate and then fluctuated within 0.5%.

Table 10.10 shows the permeate quality of the RO membrane during the study and the target quality as NEWater (20). It can be seen that the quality of the RO permeate meets the target quality as NEWater in terms of the parameters measured except pH which could be easily raised by adding NaOH.

Table 10.10
Comparison of the RO permeate quality and the target quality (14)

Parameter	RO permeate quality	Target quality
pH	5.7–6.1	7–8.5
Conductivity (μS/cm)	18.4–46.4	<200
Ammonium–N (mg/L)	0.01–0.47	<0.5
Nitrate (mg/L)	1.8–7.1	<15
Hardness as $CaCO_3$ (mg/L)	0.03–0.38	<20
TOC (mg/L)	0.06–0.37	<0.5
Total bacteria HPC (cfu/mL)	<3	<300
Total coliform (cfu/100 mL)	<1	<1
Turbidity (NTU)	<0.1	<5

The analytical results also indicated that the average rejections of the RO membrane for TOC, ammonium and nitrate were >97, 90, and 85%, respectively, which matched the RO membrane performance used at the demonstration plant (19).

4.4. Conclusions

The conclusions from the pilot study are summarized as follows.

1. The optimal operation flux rate of the RO membrane ranged of 10–15 gallon/ft^2 day for this application. The normalized flux after CIP was 97% of the initial one.
2. At a flux rate of 10 gallon/ft^2 day and water recovery of 50% the average operating pressure of 57 psi was noted corresponding to a high normalized flux of 38 L/m^2 h MPa at 25°C.
3. Rejections of the RO membrane in terms of conductivity, TOC, ammonium, and nitrate were higher than 96, 97, 90, and 85%, respectively.
4. RO permeate quality in terms of conductivity, turbidity, TOC, ammonium, nitrate, hardness, total bacteria, and total coliform matched the quality of NEWater.

5. TOC REMOVAL IN RECLAMATION OF MUNICIPAL WASTEWATER BY RO

5.1. Background

NEWater as an alternative water source has been supplied to the wafer fab industry, cooling towers and indirect potable use in Singapore. High quality of NEWater with a low-level TOC of <0.5 mg/L has been achieved (20), which is an important criteria in NEWater production to supply for the electronics industry to further produce ultrapure water. Therefore, it is a challenge for RO membranes to produce NEWater with low TOC concentration. This study focused on TOC removal in the reclamation of the secondary treated sewage effluent in Singapore using RO membranes to achieve the target with TOC level less than 40 ppb (μg/L).

5.2. *Description of Overall Process*

An MF/RO pilot plant with the capacity of 20 m^3/day was used in this study. The overall pilot plant process and pilot operation had been described previously in detail (14). The RO unit contained six elements of membranes installed in series as single stage configuration. Operating conditions of the RO unit are shown in Table 10.11. For each RO membrane, the pilot plant was continuously operated over 2 weeks at Bedok Water Reclamation Plant, Singapore.

During the study, online TOC of RO permeate was monitored using a model 820 Sievers portable TOC analyzer and online measurements of conductivity were conducted by a model Ecomet C75.

5.3. *Plant Performance*

Figure 10.8 shows TOC of RE4040-FL RO permeate as a function of time over 360 h (or 15 days) in Test 1. It should be pointed out that the plant operation had been stabilized over 3 months before the online TOC measurements were conducted (14). It can be seen that the permeate TOC fluctuated regularly between 36.2 and 65.7 ppb. Figure 10.8 also illustrates that the peak valley of TOC appeared daily. It might be due to high and low loading of influent to the wastewater treatment works, which subsequently produced the secondary treated effluent with high and low TOC, respectively. Also it is well understood that the pattern of the TOC peak valley is consistent with the daily life of human beings as the raw sewage is from domestic sources.

Figure 10.9 shows TOC of RE4040-FE RO permeate as a function of time over 340 h (or 14 days). It can be seen that the TOC level dropped significantly over the first 48 h at operating pressure of 70 psi. After the stabilization of plant operation, the permeate TOC over 26 h at operating pressure of 107 psi in Test 2 was in the range of 30–36 ppb. The low TOC level of the reclaimed water does not appear to be reported previously for the application of RO in the reclamation of municipal wastewater. In Test 3 at 70 psi, the permeate TOC over 260 h (or 11 days) fluctuated between 32 and 44 ppb, which was lower than that of RE4040-FL membrane as shown in Fig. 10.8. The RE4040-FE RO permeate with the low TOC level greatly benefits the special application where the reclaimed water from municipal wastewater is to supply for the electronics industry to further produce ultrapure water.

Table 10.11
Operating conditions of the RO unit for different membranes (15)

Parameter	RE4040-FL	RE4040-FE	
	Test 1	Test 2	Test 3
Flux rate (gallon/ft^2 day)	10	15	10
Feed flow (m^3/h)	1.7	2.3	1.7
Recovery (%)	50	55	50
Operating pressure (psi)	60	107	70

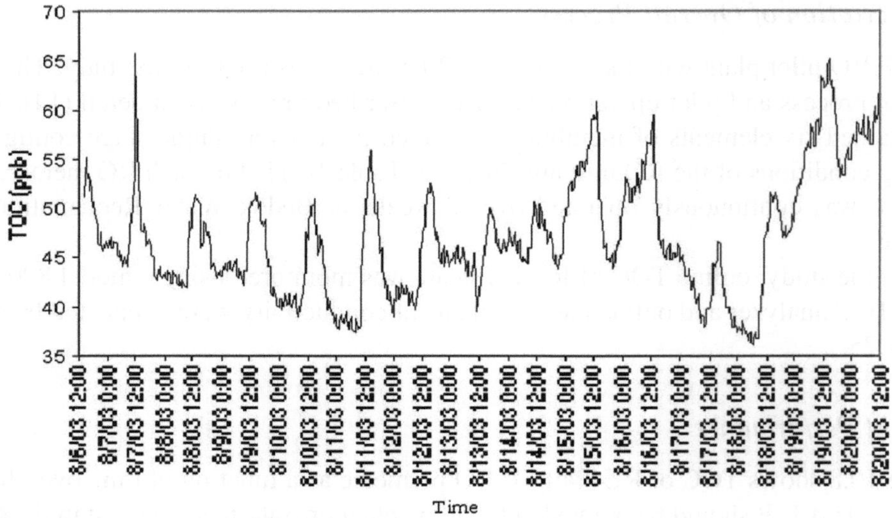

Fig. 10.8. Total organic carbon (TOC) (on-line) of the RE4040-FL RO permeate as a function of time (15).

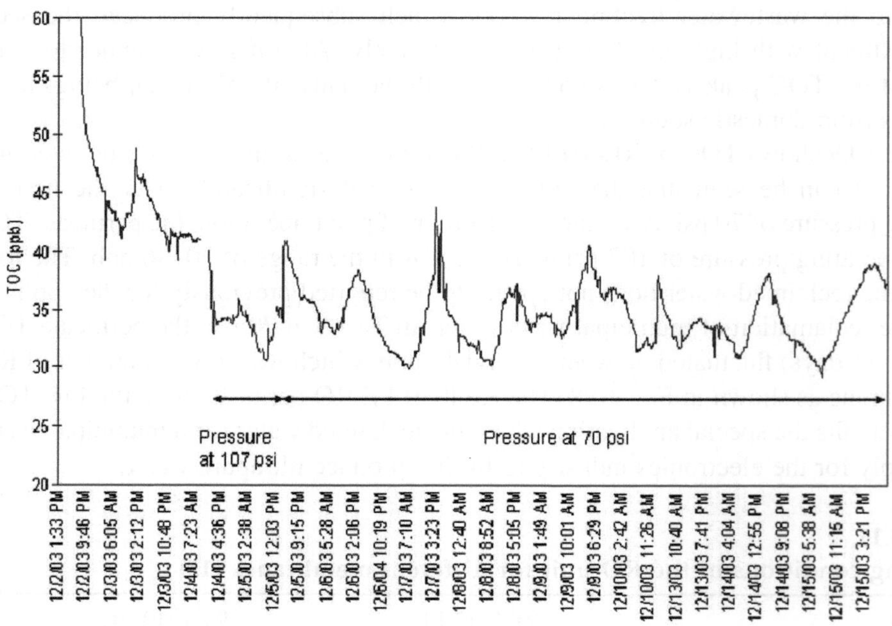

Fig. 10.9. TOC (on-line) of the RE4040-FE RO permeate as a function of time (15).

Table 10.12 shows the performance of RE4040-FE and RE4040-FL membranes tested at the flux rate of 10 gallon/ft^2 day. It can be seen that the permeate quality and rejections of RE4040-FE were better than that of RE4040-FL which was comparable to NEWater quality

Table 10.12
Comparison of the performance of RE4040-FE and RE4040-FL membranes (15)

Parameter	Feed	Permeate		Rejection (%)	
		RE4040-FE	RE4040-FL	RE4040-FE	RE4040-FL
pH	6.9	6.1	5.8	–	–
Conductivity (μS/cm)	978–1,500	20.4–32.6	35.3–45.3	97.9–98.0	96.1–96.5
TOC (μg/L)	7,900–9,000	32–44	36–66	99.5–99.6	99.2–99.6
Normalized flux (L/m^2 h MPa at 25°C)	–	30	35	–	–

(14, 19). Salt and TOC rejections of the RE4040-FE membrane were higher than 97.8 and 99.5%, respectively. Moreover, a high normalized flux of 30 L/m^2 h MPa at 25°C was noted for RE4040-FE membrane at operating pressure of 70 psi and water recovery of 50%.

5.4. Conclusions

The conclusions from the study are summarized as follows.

1. Online TOC measurements of the RO permeate were in the range of 36–66 ppb at operating pressure of 60 psi for RE4040-FL membrane. For RE4040-FE membrane, they were in the range of 32–44 and 30–36 ppb at operating pressure of 70 and 107 psi, respectively. The low TOC level of the reclaimed water does not appear to be reported previously for the application of RO in the reclamation of municipal wastewater.
2. TOC and salt rejections of the RE4040-FE membrane were higher than 99.5 and 97.8%, respectively. The permeate quality and rejections of RE4040-FE membrane were better than that of RE4040-FL which was comparable to NEWater quality.
3. A high normalized flux of 30 L/m^2 h MPa at 25°C was noted for RE4040-FE RO membrane at operating pressure of 70 psi and water recovery of 50%.

6. NEW OPTION OF MBR-RO FOR RECLAMATION OF MUNICIPAL WASTEWATER

6.1. Background

There has been significant research and development in recent years on MBR technology (10, 33–45). An MBR is a combination of the conventional ASP and MF/UF membrane separation, which has advantages of small foot print, reduced sludge production and consistently high quality of effluent in the treatment of domestic sewage over ASP. Commercial MBR systems with different capacities have proven both reliable and simple to operate. These encourage introducing a new concept of an MBR-RO process in the production of NEWater. If the MBR is applied in NEWater production, MBR will compress the conventional ASP composed of aeration unit and final clarification, and MF/UF pretreatment into

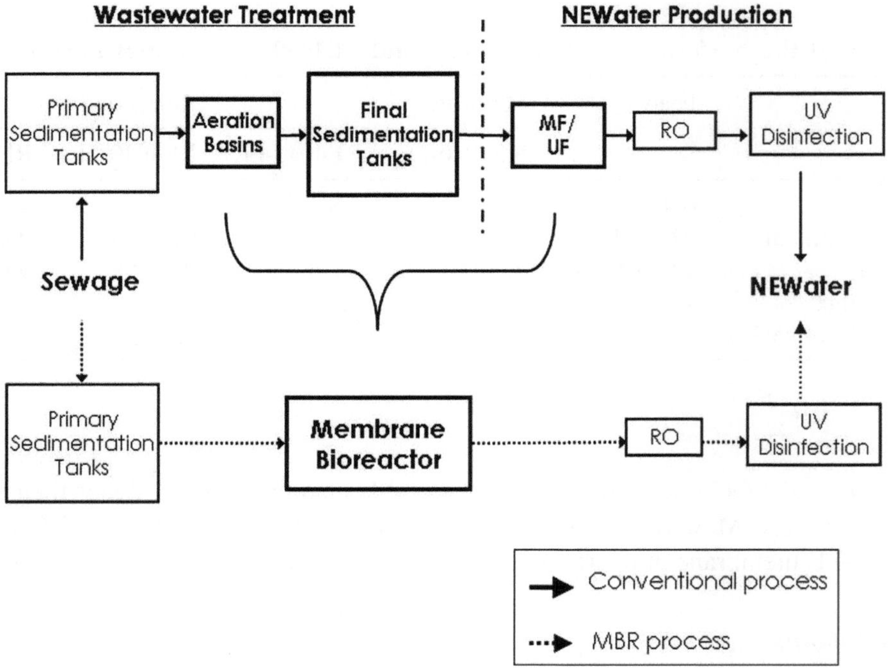

Fig. 10.10. Schematics of NEWater production with the conventional and membrane bioreactor (MBR) processes (33).

one as shown in Fig. 10.10. Compared to the conventional process, the new MBR process is significantly simplify the overall process for NEWater production. The objective of this study was to investigate the new option of MBR-RO process to produce the high grade water with lower level of TOC directly from domestic sewage under the tropical conditions.

6.2. Description of Overall Process

An MBR-RO system with the capacity of 20 m^3/day was used to conduct the study for the reclamation of the domestic sewage at Bedok Water Reclamation Plant in Singapore. The settled sewage was the raw feed of an MBR and the filtrate of MBR was used as the feed of RO. A branch stream of the filtrate from an MBR pilot plant (45) was fed to the RO plant. An MBR with hollow fiber membrane of pore size 0.4 μm which was defined as MBR-A was used in this study unless specified elsewhere. Another MBR with hollow fiber membrane of pore size 0.035 μm (defined as MBR-B) was used just as a comparison. The details about the MBRs were given in the previous study (45). Operation conditions of ASP and MBR plants are summarized in Table 10.13. Typical analysis of the settled sewage is given in Table 10.14. Soluble TOC concentration in the supernatant of mixed liquors in the membrane tank was 7.6–10.5 and 7.8–13.3 mg/L for MBR-A and MBR-B, respectively.

The RO pilot consisted of six spiral-wound RE4040-FE elements configured in single stage. RO pilot trials started on 14 May 2004 and ended on 6 October 2004. The pilot system

Table 10.13
Operation conditions of activated sludge process (ASP) and MBR plants (45)

Parameter	ASP	MBR-A	MBR-B
Surface area (m^2)	–	1,120	1,008
Mixed liquor suspended solids (MLSS) (mg/L)	2,500–3,000	10,000	10,000
HRT (h)	6–8	6	6
SRT (d)	10	21	21
Recirculation flow ratio	1:1	3:1	3:1
Anoxic volume ratio (%)	25	50	36
Net flux ($m^3/m^2/d$)	–	0.27	0.30
Net flux ($L/m^2/h$)	–	11.2	12.4
Aeration demand (Nm^3 air/m^3 permeate)	DO at 1.5 mg/L	24	30
Initial permeability ($L/m^2/h/bar$)	–	66	124
Frequency of clean-in-place	–	Once per 4 months	2–3 MC/week
Operation mode	Continuously	On: Idle = 13 min:2 min	On: back pulse = 12 min:30 s 1 idle for every 5 back pulses
Initial transmembrane pressure (kPa)	–	17	10

Table 10.14
Typical analysis of the settled sewage (45)

Parameter	Average value
COD (mg/L)	265
SCOD (mg/L)	109
BOD_5 (mg/L)	99
TSS (mg/L)	77
TN (mg/L)	45
TKN (mg/L)	33
NH_4–N (mg/L)	26
TP (mg/L)	9.2
TDS (mg/L)	817
Alkalinity as $CaCO_3$ (mg/L)	173
pH	7.05

ran continuously (24-h) during the study. Trial runs on various fluxes of the RO membrane were conducted over 5 months. Comparison of the new MBR-RO process to the ASP-MF-RO process was also made. The operating conditions of the RO pilot during the study if not specified elsewhere are shown in Table 10.15.

Table 10.15
Operating conditions of the RO pilot plant (33)

Parameter	Range of values
Operating pressure (psi)	84–92
Flux rate (L/m^2 h)	22
Feed flow (m^3/h)	2.1
Permeate flow (m^3/h)	1.06
Recovery (%)	50
Dosage of NaOCl in MBR product (mg/L) (free Cl_2 of the RO feed was 0.06 mg/L)	2
Dosage of PT-191 as an antiscalant in RO feed (mg/L)	2

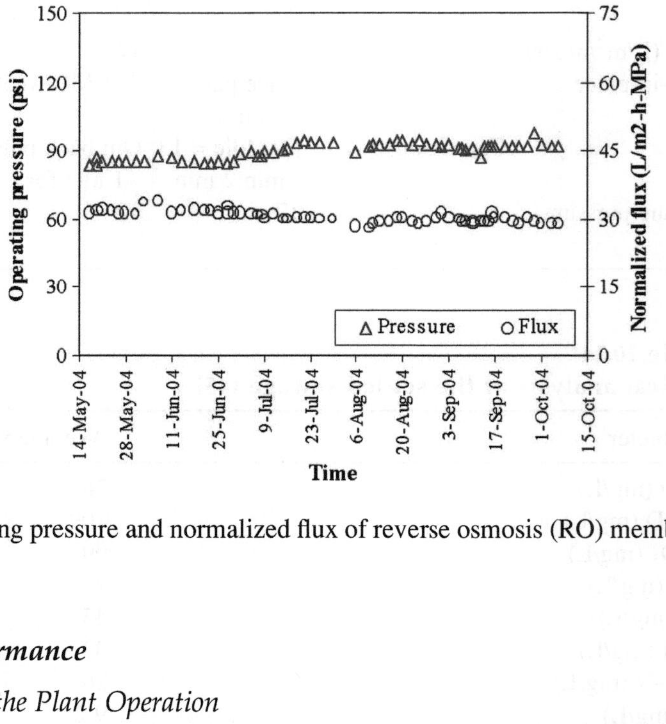

Fig. 10.11. Operating pressure and normalized flux of reverse osmosis (RO) membrane vs. time (33).

6.3. *Plant Performance*

6.3.1. *Stability of the Plant Operation*

Figure 10.11 shows operating pressure and normalized flux of the RO membrane at the MBR-RO pilot plant as a function of time. The results in Fig. 10.11 indicated that the operating pressure increased by 11% over 5 months and the normalized flux of RO membranes decreased from 32.0 to 29.0 L/m^2 h MPa at 25°C by 9%. There was no CIP conducted throughout the study when the RO pilot was operated at a membrane flux of 22 L/m^2 h, which was 30% higher than that at the NEWater Plant (currently operated at 17 L/m^2 h which was recommended by the RO membrane supplier for this particular application). These are the indications of low fouling on the RO membrane when an MBR product was used as the RO feed.

Figure 10.12 shows conductivities of RO feed and permeate at the plant throughout the study. The conductivity of RO feed was in the range of 780–2,310 μS/cm while the corresponding conductivity of RO permeate was 12.1–49.3 μS/cm. The results indicated that the RO membrane performance in terms of salt rejection was stable in the range of 98.1–98.8% throughout the 5-month study.

6.3.2. Online Monitoring of RO Permeate TOC

Figure 10.13 shows the online TOC of RO permeate during 16 May–26 June 2004 when the permeate from MBR-A was used as the RO feed. It can be seen that TOC was in the range of 23–32 ppb during 16 May–10 June. A high peak of TOC over 60 ppb appeared on 27 May, which was probably because the wastewater influent received a spike of nonbiodegradable organic that was capable of passing through the RO membrane. Then, TOC level slightly increased at 26–37 ppb during 11–23 June, which might be due to change of the feed TOC. Figure 10.13 also illustrates that the peak valley of TOC appeared daily. It might be due to high and low loading of influent to the wastewater treatment works. Also, it is well understood that the pattern of the TOC peak valley is consistent with the daily life of human beings as the raw sewage is from domestic sources.

6.3.3. RO Permeate Quality at Different Fluxes

Table 10.16 shows analytical result of RO permeates at different fluxes of the RO membrane, where the operating pressure was adjusted while the RO system recovery was remained as 50%. At the flux of 17 L/m^2 h, RO permeate quality in terms of total dissolved solids (TDS), B, Ca, Na, Cl, F, NO_3, and SiO_2 were relatively high. With an increase in the membrane flux, RO permeate quality became better than that at 17 L/m^2 h. Higher operating

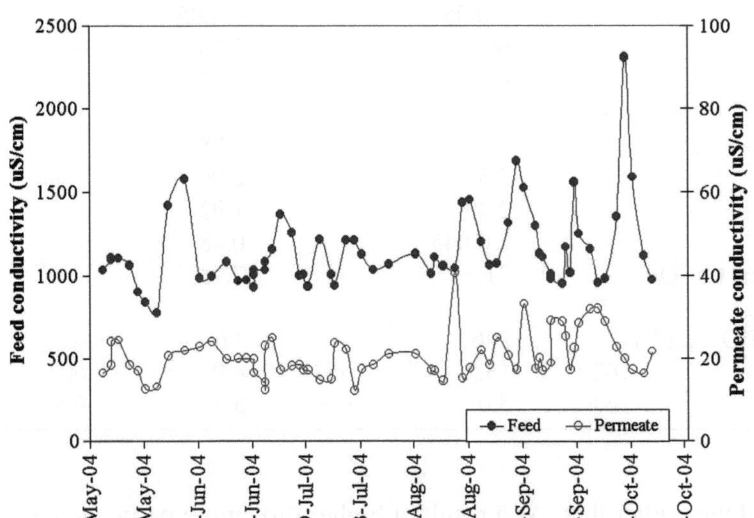

Fig. 10.12. Conductivities of RO feed and permeate vs. time (33).

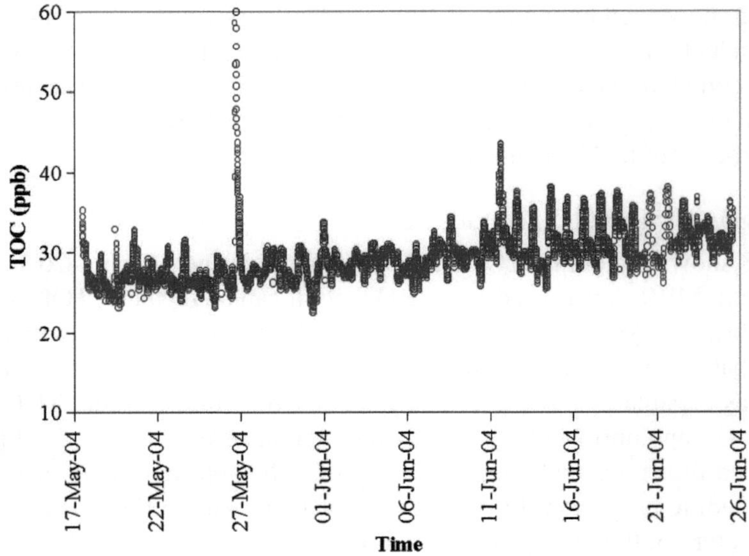

Fig. 10.13. Online TOC of RO permeates vs. time (17 May–26 June 2004) (33).

Table 10.16
Product quality at different fluxes of RO membrane (33)

Parameter	17 L/m^2 h		22 L/m^2 h		25.4 L/m^2 h	
	RO feed	RO permeate	RO feed	RO permeate	RO feed	RO permeate
NH_4–N (mg/L)	–	<0.05	–	<0.05	–	<0.05
B (μg/L)	–	60.3	–	50.5	–	49.6
Ca (μg/L)	–	4.13	–	1.53	–	1.45
Na (mg/L)	–	3.66	–	2.85	–	2.45
F (μg/L)	–	35.1	–	32.8	–	32.7
Cl (mg/L)	–	2.57	–	2.04	–	2.02
NO_3 (mg/L)	–	2.49	–	1.92	–	2.04
SO_4 (mg/L)	–	<0.045	–	0.48	–	0.47
Total hardness as $CaCO_3$ (μg/L)	–	21.5	–	7.93	–	8.36
Total silica as SiO_2 (μg/L)	–	246	–	176	–	146
TDS (mg/L)	620	13.2	557	8.9	603	9.0
pH	6.8	5.0	6.8	5.1	6.8	5.0

pressure caused the higher flux. As a result of higher flux, more permeate passed through the RO membrane, and the better quality of RO permeate was produced.

A further analysis to compare the reduction of TDS in RO permeate at different fluxes is shown in Table 10.16. On the one hand, assuming the net salt passage remained constant, expected TDS concentration in permeate at 22 and 25.4 L/m^2 h would be 1/1.3 and 1/1.5 times that at 17 L/m^2 h, respectively. As a result, TDS concentration in the permeate at 22 and 25.4 L/m^2 h compared to that at 17 L/m^2 h would be reduced by 23 and 33%, respectively. On the other hand, based on the measured TDS results as shown in Table 10.5, measured the reduction of TDS in the permeate at 22 and 25.4 L/m^2 h compared to that at 17 L/m^2 h was calculated as 25 and 30%, respectively, which matched the expectation. TDS rejections of the RO system at 17, 22, and 25.4 L/m^2 h can be calculated as 98.6, 98.7, and 98.5%, respectively.

6.3.4. Comparison of the MBR-RO Process to the ASP-MF-RO Process

In order to compare the new MBR-RO process with the ASP-MF-RO process, an MF-RO pilot was operated with the secondary treated effluent that was from the conventional ASP during 10–15 September. 0.1 mm MF membrane (PVDF hollow fiber) and the same RO membrane were used. The operating conditions of the RO pilot remained the same as that in the MBR-RO process. Analytical results of MBR product and MF permeate are given in Table 10.17. The results show that NH_4–N, NO_3^-, TOC, and pH of the MBR product were lower than that of the MF permeate, which was in agreement with previous studies (38, 45). It could be attributed to better performance of MBR than ASP on the reduction of the biodegradable components. The lower concentration of NO_3^- in the MBR effluent suggested that the MBR process included a greater degree of denitrification than the ASP. It was due to the much higher mixed liquor suspended solids (MLSS) concentration and anoxic volume ratio in the MBR than in the ASP. It should be pointed out that the better performance of the MBR product than the MF permeate was due to differences in the biological treatment efficiency of the MBR process as opposed to the physical separation process.

Figure 10.14 shows the online TOC of both RO permeates at the RO pilot. TOC level of the RO permeate from the ASP-MF-RO process fluctuated in the range of 33–53 ppb while the TOC level from the MBR-RO process during 16–22 September was in the range of 24–33 ppb, which was similar to the TOC level in May–June. The results showed that TOC level of RO permeate from the new MBR-RO process was not only lower, but also less fluctuated than that from the ASP-MF-RO process. The lower TOC level for the MBR-RO permeate was due to the lower TOC of RO feed (MBR product) in the MBR-RO process than that in the ASP-MF-RO process as shown in Table 10.17, i.e., because of the higher TOC removal efficiency of the MBR process.

Three samples for each stream of RO feeds and permeates in the ASP-MF-RO and the MBR-RO processes were analyzed. The ranges of RO feeds, permeates, and rejections are

Table 10.17
Comparison of quality of MBR product and MF permeate (33)

Parameter	MBR product	MF permeate
NH_4–N (mg/L)	0.05–0.62	0.97–2.57
NO_3 (mg/L)	17.6–22.8	25.2–42.2
TOC (mg/L)	4.9–5.1	6.8–6.9
pH	6.2–6.4	6.7–6.8

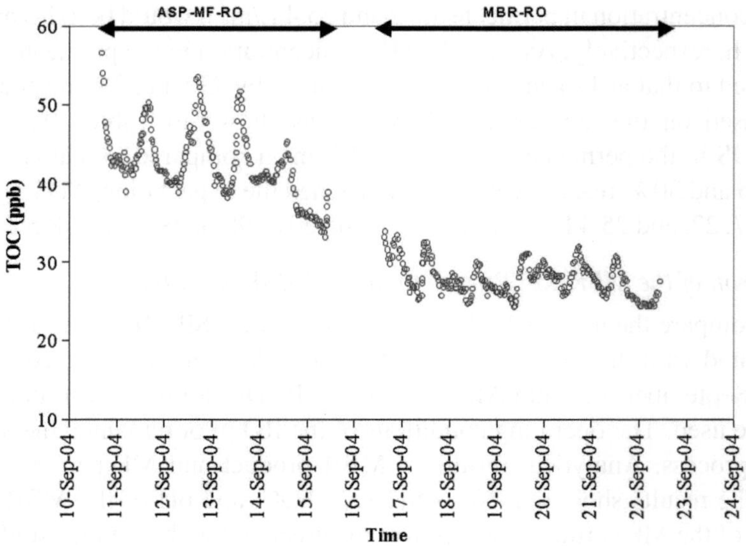

Fig. 10.14. Online TOC of RO permeate from ASP-MF-RO and MBR-RO processes vs. time (33).

compared in Table 10.18. General speaking, rejections of the RO membrane for all components were similar in both the ASP-MF-RO and the MBR-RO processes although the quality of both RO feeds fluctuated. For the parameters involved in biological process, NH_4–N and NO_3 in RO permeate from the MBR-RO process were 0.05–0.22 and 0.53–1.92 mg/L, respectively, which were lower than those (NH_4–N 0.07–0.42 mg/L and NO_3 1.62–3.86 mg/L) from the ASP-MF-RO process. It could be attributed to better performance of MBR than ASP on the reduction of the biodegradable components, including TOC mentioned above. pH (5.0 in average) of the MBR-RO permeate appeared lower than that (5.5 in average) of ASP-MF-RO permeate although pHs of their RO feeds were similar. However, for other inorganic ions, the quality in both permeates were similar since the same RO membranes were used.

In summary, the new MBR-RO process could produce same or more consistent quality (in terms of TOC, NH_4, and NO_3) of RO permeate compared to the conventional ASP-MF-RO process. The RO permeates with the lower TOC level from the new MBR-RO process than the ASP-MF-RO process more greatly benefits the special application where the reclaimed water from municipal wastewater is to supply for the electronics industry to further produce ultrapure water since ultrapure water with low TOC contamination is required.

6.4. *Conclusions*

The conclusions from the study can be summarized as follows.

1. The new MBR-RO process demonstrated the capability of producing the same or better (in terms of TOC, NH_4, and NO_3) product quality compared to the conventional ASP-MF-RO process in the reclamation of domestic sewage.

Table 10.18
Comparison of RO performance for the removal of inorganic between ASP-MF-RO and MBR-RO processes (33)

Parameter	RO feed (mg/L)*		RO permeate (mg/L)*		RO rejection (%)	
	ASP-MF-RO	MBR-RO	ASP-MF-RO	MBR-RO	ASP-MF-RO	MBR-RO
NH_4–N	1.96–5.54	1.0–1.7	0.07–0.42	0.05–0.22	87.7–97.4	93.6–97.4
B	0.080–0.100	0.071–0.104	0.059–0.074	0.051–0.078	24.1–34.7	27.4–33.1
Ca	24.8–30.0	30.0–33.5	0.001–0.004	0.002–0.006	>99.9	>99.9
Na	123–169	149–185	2.83–3.86	2.85–3.84	97.6–98.2	98.3–98.4
F	1.29–1.42	1.51–1.55	0.012–0.029	0.015–0.033	98.2–99.3	99.1–99.2
Cl	146–193	164–214	1.51–2.51	1.75–2.69	98.9–99.1	98.9–99.1
NO_3	26.2–60.5	6.37–28.4	1.62–3.86	0.53–1.92	94.6–95.3	93.0–94.9
SO_4	66.2–71.7	71.9–82.1	<0.010–0.034	0.029–0.48	>99.9	>99.9
Hardness as $CaCO_3$	104–129	125–136	0.011–0.018	0.008–0.042	>99.9	>99.9
Total silica as SiO_2	7.43–8.34	9.22–9.82	0.140–0.160	0.151–0.201	98.3–98.4	98.2–98.7
TDS	573–703	557–749	9.8–13.3	8.9–13.1	98.1–98.6	98.5–98.7
pH, unit	6.6–6.8	6.7–6.9	5.4–5.5	4.7–5.2	–	–

*except as noted

2. RO membranes in the MBR-RO process could be operated at 22 L/m^2 h without CIP during the whole study period of 5 months, which was 30% higher than that (17 L/m^2 h) in the ASP-MF/UF-RO process for NEWater production.
3. The study provided a new option of the MBR-RO process for NEWater production.

7. RECLAMATION OF A MIXED SEWAGE EFFLUENT USING UF-RO

7.1. Background

Water reclamation from the secondary treated effluent by way of producing NEWater is one of the National Taps identified in Singapore. Jurong Water Reclamation Plant (JWRP) is treating a mixed industrial (60%) and domestic (40%) sewage stream. Additional source of water to serve the needs of industries would be introduced potentially if NEWater quality water from the secondary treated mixed sewage effluent could be produced. There were, however, considerable uncertainties about the long-term performance of UF-RO membrane process in this application because of the high content of industrial waste. The objective of the study was to determine the feasibility of reclaiming such a mixed sewage effluent to NEWater quality for industrial use using the dual membrane process. This study was conducted from September 2003 to February 2004, over 6 months.

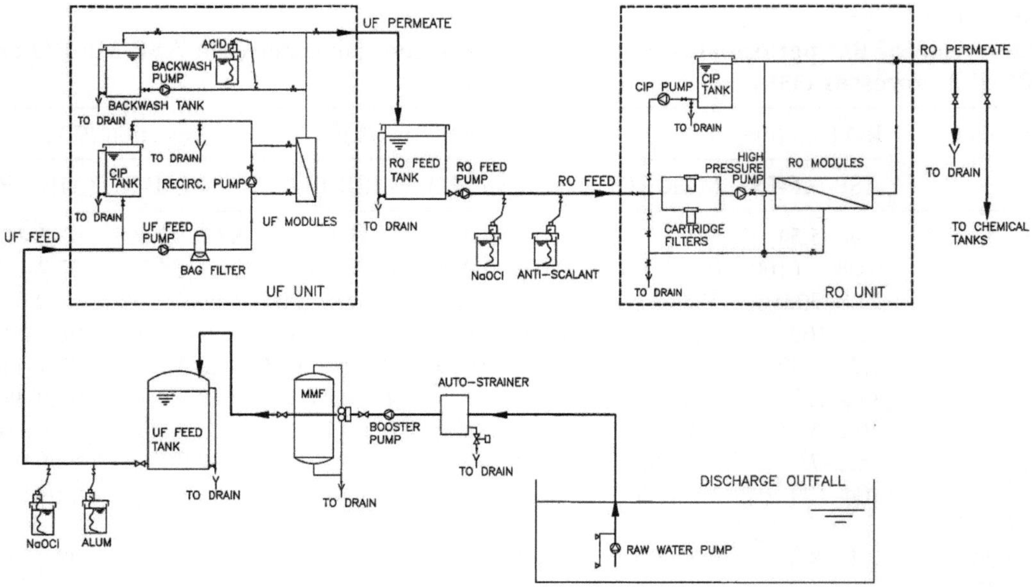

Fig. 10.15. Schematic flow diagram of the pilot plant process (46).

7.2. Description of Overall Process

7.2.1. Plant Process

A schematic diagram of the pilot process is shown in Fig. 10.15.

The raw feed was pumped at a flow of 3 m^3/h to the pilot plant by a submersible pump, installed in the discharge outfall sump. Debris and large particles in the raw feed were removed by a fully automated self-cleaning strainer. The auto-strainer was fitted with 300-μm pore-size rotating stainless steel screen filter and a cleaning blade, assembled in a stainless steel vessel and coupled with electrically actuated valve at the drain line. It was equipped with programmable timer controller controlling screen rotation and drain valve opening. After passing through the auto-strainer, the raw water was then sent to Multimedia filter (MMF) by an online booster pump for the removal of large suspended solids. The MMF had three layers of different media, 3–6 mm size gravel, 0.4–0.6 mm size filtered sand, and 0.8–1.6 mm size anthracite, packed inside an FRP pressure vessel. It was equipped with programmable timer controlled valve assembly and programed to activate 30-min automatic backwash cycle everyday. The filtrate stream from MMF was collected in the feed tank of UF unit.

The filtrate was then pumped to the UF unit by the UF feed pump to remove the fine suspended solids and coagulated colloidal materials. Sodium hypochlorite as a disinfectant was dosed to the UF feed to reduce biofouling (16) and alum was dosed to the UF feed as a coagulant (13). A 200-μm pore-size bag filter was installed prior to UF membranes to protect them from foreign materials. The permeate water from the UF unit was collected in RO feed tank. The reject stream was bled off to drain and sent back to the outfall sump. The UF permeate water was then pumped to RO unit by RO feed pump to remove dissolved solids and organics. Sodium hypochlorite and antiscalant PermaTreat 191 were dosed to the RO feed line to prevent biofouling and scaling to RO membranes (47). A 5-μm pore-size cartridge filter was installed prior to RO membranes to protect them from foreign materials. Both the product water from the RO unit and the concentrate stream were sent back to the outfall sump during the study.

7.2.2. UF and RO Units

The UF unit contained four HF66–45-XM50P modules installed in parallel from Koch Membrane Systems, Inc. The membranes were polyacrylonitrile hollow fibers with a molecular weight cut-off of 50,000 Da. The UF unit was designed to operate at high velocity of feed water by recirculation pump, periodic reversal of the feed flow direction and periodic backwashing sequences to reduce fouling of UF membranes. A Clean-In-Place (CIP) system was provided for manual chemical cleaning to UF membranes. Kochkleen® Liquid Detergent was used for cleaning at pH 10.0 and 2% citric acid was used for cleaning at pH 2.0.

The RO unit was constructed with three pressure vessels in 2:1 configuration. Each pressure vessel held two LFC1-4040 spiral-wound modules that were made of thin film polyamide composite membranes. A solenoid valve was installed in parallel with the throttled SS glove valve on the concentrate line and was automatically controlled by PLC to periodically flush the RO membrane surface under high feed flow and low pressure. Pressure, flow, temperature, pH, and conductivity indicators and online instruments were installed at the RO unit. There was also a CIP system attached with the RO unit. Two percent citric acid was used for cleaning at pH 2.0 over 3 h. Then, 1% EDTA was used at pH 12.0 over 1 h operation followed by 16 h soaking and 0.5 h operation (31).

7.2.3. Trial Runs

Preliminary investigation was conducted to monitor the raw feed quality and understand the characteristics of the raw feed from March to August 2003 prior to the plant operation. Samples were collected on hourly basis with an automatic sampler from the discharge outfall sump for 24 h. pH, conductivity, turbidity, and TOC of hourly samples were measured and analyses of daily composite samples were conducted.

The study was conducted from September 2003 to February 2004, over 6 months. The conductivity probing was conducted after RO membranes were loaded into the pressure vessels to ensure no leakage between the interconnections of membrane elements. The plant was disinfected thoroughly.

In order to increase RO operating pressure while the RO feed flow remained the same, the number of RO elements was cut down from previously running with 6–4. One pressure vessel in RO train was isolated, and the RO unit was operated with four elements in series from November 2003. The operating pressure of RO unit was increased to 100 psi. The operating conditions of the pilot plant during the study are summarized in Table 10.19.

Table 10.19
Operating conditions of the plant (46)

Parameter	Sep 03	Oct 03	Nov 03	Dec 03	Jan 04	Feb 04
UF unit						
Dosage of NaOCl in UF feed (mg/L)	6	6	6	6	6	6
Dosage of alum in UF feed (mg/L)	3.5	3.5	3.5	3.5	3.5	3.5
UF feed flow (m^3/h)	2.25	2.0	1.9	1.9	1.7	1.7
UF permeate flow (m^3/h)	1.8	1.5	1.4	1.4	1.2	1.2
UF recovery (%)	80	75	74	74	71	71
RO unit						
Dosage of NaOCl in RO feed (mg/L)	2	2	2	2	2	2
Dosage of PT-191 in RO feed (mg/L)	2	2	2	2	2	2
RO feed flow (m^3/h)	1.3	1.1	1.1	1.0	1.0	1.0
RO permeate flow (m^3/h)	0.52	0.44	0.44	0.39	0.38	0.38
RO water recovery (%)	40	40	40	39	38	38
RO operating pressure (psi)	55	60	100	100	100	100
Normalized flux ($\times 10^{-6}$ L m^2 h Pa)	29.4	22.8	20.5	18.0	17.5	17.5

7.3. *Plant Performance*

7.3.1. *Characteristics of the Raw Feed Water*

Raw feeds for hourly samples on different days and for composite sample of 24 hourly samples on each day were characterized during March–August 2003. Overall ranges of the analytical results are summarized in Table 10.20. Comparing the analytical results for hourly samples with that of daily composite samples, the overall range of the composite samples for each parameter was narrower than that of hourly samples. It indicated that the fluctuation of raw water quality was wider with the time of the day than the daily average of different days. Overall range of the analytical results for daily composite samples shows that levels of ammonia, TOC, suspended solids, and conductivity as well as heavy metals, such as iron and strontium of the raw feed were high.

7.3.2. *Monitoring of Conductivity and pH Values in the Process*

Figures 10.16 and 10.17 show conductivity of RO feed and permeate as a function of time during the study. It can be seen that conductivity of the RO feed was in the range of 1,300–2,100 μS/cm and most of the time between 1,400 and 1,700 μS/cm. The RO permeate conductivity ranged of 60–80 μS/cm when operating pressure was 55–60 psi during September–October 2003 as shown in Fig. 10.16. In Fig. 10.17, after the operating pressure was increased to 100 psi, the permeate conductivity was better and ranged of 30–50 μS/cm in November 2003, and then 35–75 μS/cm in December 2003. However, the permeate conductivity suddenly increased to 120 μS/cm since 24 December, which was probably due to serious fouling of the RO membranes. Later, integrity check on UF modules was conducted, and a few broken fibers from one of the modules were found. After the damaged UF module

Table 10.20
Characteristics of the raw feed water (46)

Parameter	Hourly samples	Daily composite samples
Ammonia–N (mg/L)	–	37.8–49.6
Alkalinity as $CaCO_3$ (mg/L)	–	100–283
Barium (mg/L)	–	0.025–0.088
Boron (mg/L)	–	0.952–1.14
Calcium (mg/L)	–	29.3–44.7
COD (mg/L)	–	42–83
Fluoride (mg/L)	–	4.36–9.92
Iron (mg/L)	–	0.329–2.06
Magnesium (mg/L)	–	9.39–15.2
Nitrate–N (mg/L)	–	<0.01–0.14
Oil and grease (mg/L)		<1
Total silica (mg/L)	–	9.02–10.7
Strontium (mg/L)	–	1.09–2.17
Sulfate (mg/L)	–	135–230
DOC (filtered TOC) (mg/L)	3.2–49.0	4.34–29.1
Total PO_4 as P (mg/L)	–	3.52–7.05
True color (HU)	–	50–70
TSS (mg/L)		6.0–67.5
Turbidity (NTU)	1.0–27.1	1.4–24.3
Conductivity (μS/cm)	1,220–2,180	1,517–2,044
pH, unit	6.7–8.7	6.7–7.7

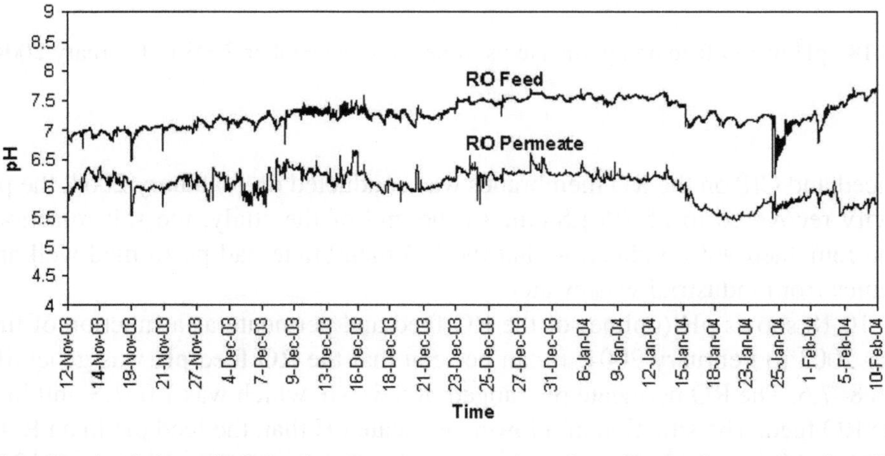

Fig. 10.16. Conductivity of RO feed and permeate vs. time from September to October 2003 (46).

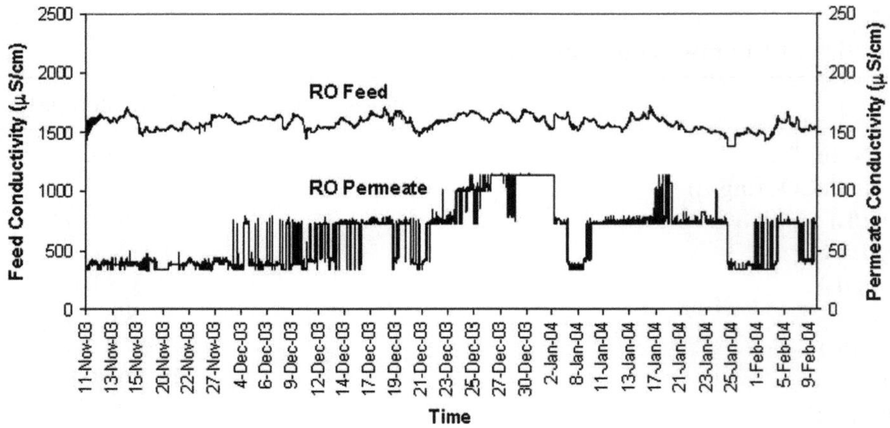

Fig. 10.17. Conductivity of RO feed and permeate vs. time from November 2003 to February 2004 (46).

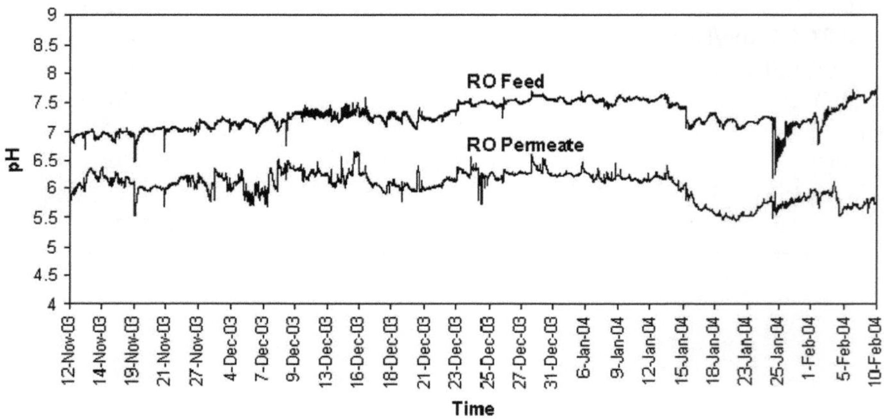

Fig. 10.18. pH of RO feed and permeate vs. time from November 2003 to February 2004 (46).

was replaced and CIP on the RO membranes was conducted on 6 January 2004, the permeate conductivity recovered to 35–75 μS/cm. At the end of the study, the salt rejection of RO membrane remained 96%, indicating that the RO membrane had performed well and tolerated organics from industrial wastewater.

Figure 10.18 shows pH (online) of the RO feed and permeate as a function of time from November 2003 to February 2004. It can be seen that the RO feed pH was generally in the range of 6.8–7.6. The RO permeate pH ranged of 5.5–6.6, which was 1.0–1.3 unit lower than that of the RO feed. The situation of a lower permeate pH than the feed pH in an RO process appears to be quite a common occurrence in water and wastewater treatment (18, 48–51), which has been explained by the influence of HCO_3^- elsewhere (50–51).

7.3.3. Monitoring of TOC in the Process

Figure 10.19 shows TOC (online) of the RO permeate as a function of time during September 2003 to January 2004. It can be seen that the online TOC fluctuated significantly. The lowest TOC was 75 ppb in October 2003, and the highest TOC was higher than 500 ppb. High TOC was found during weekdays, probably due to high loading of organics and upset operation of sewage treatment. However, TOC of the RO permeate was well below 500 ppb for most of the time. It was below 200 ppb at most of the time in December 2003 before 25th and in January 2004.

Table 10.21 shows the results of TOC retention in UF and RO processes during the study. DOC reduction in the UF process was observed. It is because natural organic matters (NOMs) are negatively charged and the dosage of alum into the raw feed water favors the formation of insoluble charge-neutral products due to charge neutralization between cationic aluminum

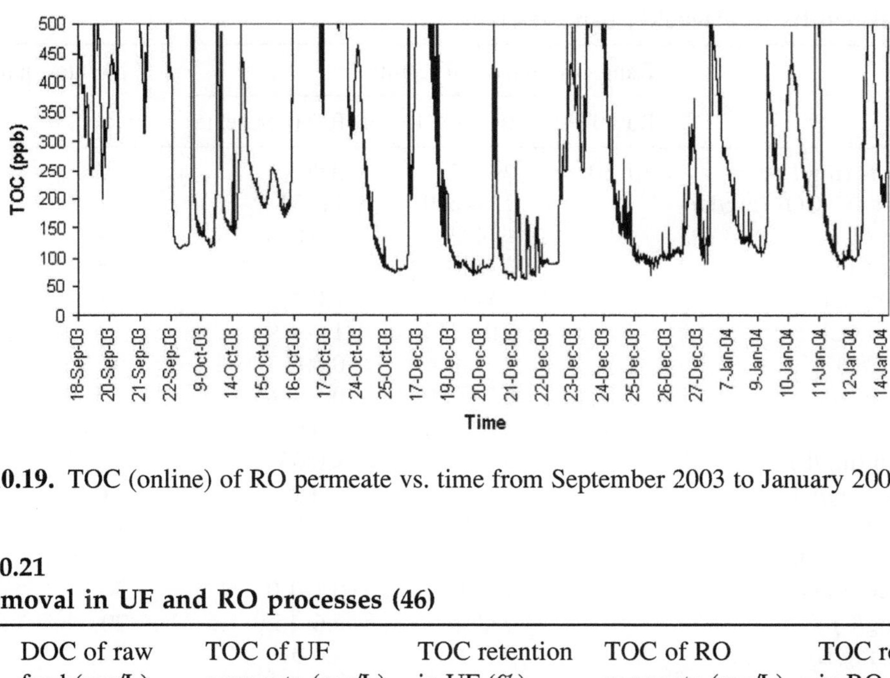

Fig. 10.19. TOC (online) of RO permeate vs. time from September 2003 to January 2004 (46).

Table 10.21
TOC removal in UF and RO processes (46)

Date	DOC of raw feed (mg/L)	TOC of UF permeate (mg/L)	TOC retention in UF (%)	TOC of RO permeate (mg/L)	TOC retention in RO (%)
19/09/03	16.8	14.6	13	0.51	96.5
07/10/03	27.8	22.8	18	1.84	90
16/10/03	34.4	24.9	28	3.54	76
17/11/03	Not measured	21.4	–	3.83	82
18/11/03	Not measured	38.6	–	3.21	92
17/12/03	56.7	37.6	34	3.11	92
24/12/03	19.6	18.1	8	0.52	97
30/12/03	33.4	29.1	13	8.50	71
08/01/04	42.0	17.8	58	1.12	94

and anionic NOMs (52–53), resulting in the removal of NOMs after UF. However, the retention of DOC significantly varied in the range of 8–58%. It was probably due to fluctuation of the percentage of NOMs among total organic contents. TOC retention in RO process was 71–97% dependent on TOC level and compositions in the UF permeate (RO feed).

7.3.4. *Analytical Results of Product Quality*

Range of analytical results of biweekly samples of raw, RO feed and RO permeate during the study are summarized in Table 10.22. For comparison, NEWater quality (6) is also given in Table 10.22. Table 10.22 shows that most parameters (18 out of 22) of the treated water (RO permeate) from the UF-RO membrane process met the quality of NEWater. However, ammonia–N in the RO permeate was 0.9–3.4 mg/L, which was high due to consistent high

Table 10.22
Analytical results of biweekly samples (46)

Parameter	Range of analytical results			NEWater quality (20)
	Raw feed	RO (feed)	RO (permeate)	
Ammonia–N (mg/L)	30–50	29–42	0.9–3.4	<0.5
Alkalinity as $CaCO_3$ (mg/L)		152–239	11–34	<20
Barium (mg/L)		<0.02	<0.004	<0.1
Boron (mg/L)		0.49–1.07	0.3–0.7	<0.5
Calcium (mg/L)		30–39	<0.2	<20
Chloride (mg/L)		219–344	1.9–6.9	<20
Copper (mg/L)			0.002–0.011	<0.05
Fluoride (mg/L)		6.3–15.0	0.14–0.54	<0.5
Iron (mg/L)	0.3–6.4	0.03–0.48	<0.01	<0.04
Magnesium (mg/L)		9.1–16.5	<0.03	–
Manganese (mg/L)			<0.003	<0.05
Nitrate (mg/L)		<0.2	<0.01	<15
Oil and grease (mg/L)	<1			–
Silica as SiO_2 (mg/L)		7.1–12.5	0.1–1.0	<3
Sodium (mg/L)		158–277	6.6–15.0	<20
Strontium (mg/L)		0.7–1.3	<0.01	<0.1
Sulfate (mg/L)		199–286	<0.8	<5
Zinc (mg/L)			<0.01	<0.1
SDI (mg/L)		1.7–2.2		–
Trihalomethane (mg/L)			<0.02	<0.08
TOC (filtered) (mg/L)	3.2–56.7	14.6–38.6	0.075–8.5	<0.5
Total PO_4 as P (mg/L)	4.8–11.7	0.32–7.65	<0.03	–
True color (HU)		45–75	<5–10	<5
Turbidity (NTU)	1–27.1	0.1–0.6	<0.1	<5
Conductivity (μS/cm)		1,251–2,475	32–100	<200
pH, unit		7.0–7.6	5.9–6.4	7–8.5

Notes: All units are in mg/L unless specified.

ammonia in the feed (29–42 mg/L). Alkalinity in the RO permeate ranged of 11–34 mg/L as $CaCO_3$ and boron in the RO permeate was 0.3–0.7 mg/L, which sometimes did not meet the requirements. TOC in the RO permeate fluctuated in a wide range of 0.075–8.5 mg/L due to the fluctuation of TOC contamination in the RO feed (3.2–56.7 mg/L).

The very high TOC level in the feed could be caused by high contamination of organic solvent in industrial waste which was mainly due to illegal discharges from the industries, resulting in the upset operation of sewage treatment. A strong smell of solvent was occasionally noted near the discharge outfall of the effluent. The RO permeate with TOC 8.5 mg/L on December 20, 2004 was analyzed by GC and methanol of 2.4 mg/L was determined.

Nitrate in the RO feed was low at <0.2 mg/L, which indicated that there was almost no nitrification at upstream sewage treatment plant. Table 10.22 also shows that in the raw feed iron ranged from 0.3 to 3.7 mg/L and turbidity fluctuated in the range of 1–27.1 NTU. The high levels of iron and turbidity resulted in fouling of the UF membranes, consequently frequent acid cleaning of the UF unit required once per 2–3 days (the UF membranes were soaked at pH 2 for 10 min and then flushed with UF permeate for 2 min). High concentration of TOC caused fast flux reduction of the RO membranes because of organic fouling although the antiscalant was used.

It should be noted that when comparing data of raw feed to RO feed (UF permeate) for parameters of iron, turbidity, TOC, and PO_4 as shown in Table 10.22, UF in this study as a pretreatment prior to RO not only effectively removed particle foulants as SDI of UF permeate was low at 1.7–2.2 (iron from 0.3–6.4 mg/L down to 0.03–0.48 mg/L and turbidity from 1–27.1 NTU down to 0.1–0.6 NTU), but also reduced certain amount of organic foulants (TOC removal of 8–58%) and phosphate scalants (PO_4 from 4.8–11.7 mg/L down to 0.32–7.65 mg/L) by the dosage of alum.

7.4. Conclusions

The conclusions from the pilot study are summarized as follows:

1. Wide variations in the effluent quality of sewage treatment plant were observed. Typical values were, pH in the range of 6.7–7.8 and conductivity in the range of 1,220–2,180 μS/cm, Ammonia–N consistently high at 30–50 mg/L. Very high fluctuations in iron (0.3–3.7 mg/L), turbidity (1–27.1 NTU), and TOC (3.2–56.7 mg/L) were observed. Nitrate was low at <0.2 mg/L.
2. For the UF-RO membrane process, it was found that UF required frequent cleaning. The polymeric RO membrane could tolerate organics from industrial wastewater and had salt rejection of more than 96% at the end of the study after 6 months.
3. The quality of the treated water from the UF-RO membrane process is comparable to that of NEWater for 18 out of 22 specified parameters and can be used for industrial purposes.

8. RECENT R&D

As global population and industry are further rapidly developed, membrane technology advancements for the treatment and reclamation of municipal wastewater have been driven by the needs to protect the environment and to provide an alternative reliable water resource. Fouling is a common issue in membrane processes. Fouling control and reduction have become one of the most active research fields in the development of membrane and membrane processes because they reduce operating cost, cleaning frequency, membrane area required, and energy consumption. Current R&D on membranes and membrane processes for converting municipal sewage to water resources is focused on the follow aspects (54–81).

8.1. Membrane Development

It is first consideration to use a hydrophilic membrane for the reduction of membrane fouling. Development of hydrophilic PVDF hollow fiber MF membranes has been very active due to its high tolerance to chlorine and oxidant. Research on the applications of the PVDF hollow fiber membranes are intensively focused on MBRs for the treatment of municipal wastewater (54–56). Investigation on fabrication of hydrophilic membranes from cellulose is also a trend as cellulose is the most abundant natural organic resource in the world and has excellent hydrophilicity as well as resistance to acid, alkali, and organic solvents (57–59).

Research on temperature sensitive membranes with the unique characteristic to favorite the membrane cleaning is increasing (60–61). The membranes are temperature sensitive as the temperature is higher than the normal operating temperature range but not temperature sensitive as the temperature is within the normal operating temperature range. Therefore, the membrane pores can be enlarged for the ease of foulant removal by controlling the temperature. In addition, the membrane can be cleaned with pure water alone, without chemical agents.

Development of membranes with long life has been paid much attention due to the requirement of end-users for low cost of membrane replacement. For example, the period of warrantee on hollow fiber membranes tends to increase from 1–2 to 3–5 years even up to 5–7 years. The challenge is to develop hollow fiber membranes with high mechanical strength of >25 kg/fiber. New membrane modules with lower packing density and sludge thickening membranes are very attractive for the MBR application for municipal wastewater treatment.

A high flux RO membrane with high salt rejection and low fouling tendency for municipal wastewater treatment is another challenge. The RO membrane can be operated under a low pressure of 5 bar with the normal designed flux as an operating pressure of 10 bar is required normally for current RO membrane system. It has potential on energy saving of 50%.

8.2. Membrane Processes

Recent development of new membrane (integrated) processes for the reclamation of municipal wastewater has been focused on both the reduction of membrane fouling and the enhancement of effluent quality. It includes (a) hybrid coagulation-UF process; (b) gas scouring in UF and MF systems; (c) submerged MF/UF membrane systems; (d) direct nanofiltration; (e) osmotic pressure cleaning technique; (f) ultrasonic technology; and (g) integrated membrane technology.

Hybrid coagulation-UF process can not only efficiently reduce phosphate contaminant in the effluent, but also control the membrane fouling. Gas scouring or sparging in UF and MF systems can significantly reduce fouling and enhance the membrane flux (62, 63). Submerged MF/UF membrane systems have demonstrated low power consumption and higher water recovery.

Direct nanofiltration (64, 65) is used in the treatment of municipal effluent to remove microcontaminants and ensure a high-quality effluent suitable for reuse because the microcontaminants in municipal effluent cause eco-environmental problems (e.g., estrogenic substances result in only female fish).

Research on the application of ultrasonic technology in membrane systems has been paid attention recently (66). Ultrasound irradiation can break the cake layer on the membrane surface and reduce the concentration polarization. As a consequence, the fouling of membranes is prevented, and the membrane flux is enhanced although the intrinsic permeability of the membrane is not affected by ultrasound irradiation. Investigations focus on the effects of frequency, power intensity, irradiation time, feed properties, membrane properties, crossflow velocity, temperature, and pressure on the effectiveness of ultrasound irradiation. It is reported that the fouling reduction could be up to 80% by ultrasonic technology (67).

In an RO process, an osmotic pressure exists and causes extra energy consumption. However, the osmotic pressure has been recently utilized to drive the permeate to flow back the feed in spiral-wound RO elements to remove the foulants on the membrane surface just like the backwash of hollow fiber membranes (68, 69). The osmotic pressure cleaning method offers a novel approach for the RO membrane operation by preventive, online, short membrane backwash without stoppage of high pressure pump or interruption of the operational process. It is an environment and membrane friendly technique. The direct osmosis treatment by the introduction of a high salinity solution (68) or by an immediate low operating pressure (69) allows keeping RO membrane continuously clean even in heavy biofouling conditions and operating RO membranes at high flux.

Development and application of MBR technology have been the most rapidly increased (56). Research on integrated membrane technology is focused on NF-MBR (70, 71), photo catalyst oxidation-MBR (72), coagulation-MBR (73–74), and MBR-RO (33, 75). For example, it is reported that the flux enhancement in the coagulation-MBR process could be over 50% for hollow fiber membranes compared to the MBR without the addition of coagulant (73). The MBR-RO process can produce better quality of the reclaimed water and are more economic compared to ASP-UF-RO process.

Several other challenges have to be faced in the development of better membrane processes (76). They can be (a) the economic factors, including capital cost, such as membranes, modules and systems, and operating cost, such as chemical-free and energy reduction; (b) the operability, including good fouling control and usable hardware for nonexperts; (c) reliability, including the development of membranes with good integrity, the development of effective, low cost online monitors; and (d) module standardization, including hollow fiber modules and interchange from different sources (77–81).

9. ACRONYMS

ASP = Activated sludge process
BOD = Biochemical oxygen demand
CEB = Chemically enhanced backwash
CIP = Clean-in-place

COD = Chemical oxygen demand
LRV = Log reduction value
MBR = Membrane bioreactor
MF = Microfiltration
MLSS = Mixed liquor-suspended solids
MMF = Multimedia filter
NF = Nanofiltration
NOM = Natural organic matter
NTU = Nephelometric turbidity unit
PLC = Programmable logic controller
RO = Reverse osmosis
SDI = Silt density index
TDS = Total dissolved solids
TMP = Transmembrane pressure
TSS = Total suspended solid
UF = Ultrafiltration

10. NOMENCLATURE

A = Effective membrane area (m^2)
C_f = Component concentration in the feed (ppm or mg/L)
C_p = Component concentration in the permeate (ppm or mg/L)
J = Pressure normalized flux (L/m^2 h Pa)
ΔP = The net driving pressure (Pa)
Q = Volume flow rate of permeate (L/h)
Q_p = Permeate flow rate (L/h)
Q_f = Feed flow rate (L/h)
R = Rejection or removal (%)
T = Temperature (K)

REFERENCES

1. Arasu S, Kiang FH, Ratnayaka DD, Yong WL (2005) Supply of desalinated water by the private sector Singapore's first public-private-partnership initiative. In: Proceedings of international desalination association world congress, SPO5–142, Singapore, 11–16 September 2005, pp 1–16
2. Fethi K (2005) RO 22500m3/d Gabès plant: 10 years of operation without membrane replacement. In: Proceedings of international desalination association world congress, SPO5–032, Singapore, 11–16 September 2005, pp 1–11
3. Mubeen FM (2005) Desalination and water reuse in Pakistan – an overview. In: Proceedings of international desalination association world congress, SPO5–003, Singapore, 11–16 September 2005, pp 1–12
4. Vlaski A, Ragab A, Kleinendorst T (2005) Desalination prospects in Egypt. In: Proceedings of international desalination association world congress, Singapore, SPO5–075, 11–16 September 2005, pp 1–15

5. Jordan BL, Veerapaneni V (2005) Desalination in the United States: Panacea or Pandora's Box. In: Proceedings of international desalination association world congress, SPO5–128, Singapore, 11–16 September 2005, pp 1–7
6. Castle B, Frenkel V, Pankratz T, Reynolds T (2005) Seawater desalination in the San Francisco Bay. In: Proceedings of international desalination association world congress, SPO5–037, Singapore, 11–16 September 2005, pp 1–15
7. Nusbaum I, Argo DG (1984) Design, operation and maintenance of a 5-MGD wastewater reclamation reverse osmosis plant. In: Belfort G (ed) Synthetic membrane processes, fundamental and water applications. Academic, New York, pp 377–436
8. Reith C, Birkenhead B (1998) Membranes enabling the affordable and cost effective reuse of wastewater as an alternative water source. Desalination 117:203–209
9. Hardt FW, Clesceri LS, Nemerow NL, Washington DR (1970) Solids separation by ultrafiltration for concentrated activated sludge. J Water Pollut Control Fed 42:2135–2148
10. Stephenson T, Judd S, Jefferson B, Brindle K (2001) Membrane bioreactors for wastewater treatment. IWA, Padstow, Cornwall
11. Lopez-Ramirez JA, Sahuquillo S, Sales D, Quiroga JM (2003) Pre-treatment optimization studies for secondary effluent reclamation with reverse osmosis. Water Res 37:1177–1184
12. Kim SL, Chen JP, Ting YP (2002) Study on feed pretreatment for membrane filtration of secondary effluent. Sep Purif Technol 29:171–179
13. Qin JJ, Oo MH, Lee H, Kolkman R (2004) Dead-end ultrafiltration for pretreatment of RO in reclamation of municipal wastewater effluent. J Membr Sci 243:107–113
14. Qin JJ, Oo MH, Wai MN, Lee H, Xing YJ, Zhang MC (2005) Pilot study for reclamation of the secondary treated sewage effluent. Desalination 171:299–305
15. Qin JJ, Oo MH, Wai MN, Kekre KA (2005) TOC removal in reclamation of municipal wastewater by RO. Sep Purif Technol 46:125–128
16. Wilf M, Alt S (2000) Application of low fouling RO membrane elements for reclamation of municipal wastewater. Desalination 132:11–19
17. Abdessemed D, Nezzal G, Ben Aim R (1999) Treatment of wastewater by ultrafiltration. Desalination 126:1
18. Sourirajan S, Matsuura T (eds) (1985) Reverse osmosis/ultrafiltration process principles. NRCC, Ottawa
19. Tan JY, Lee MF (2001) NEWater – an alternative source of water for the wafer fab industry in Singapore. In: Proceedings of recycling and alternate sources for the wafer/PCB industries conference, Singapore, pp 1–4
20. Public Utilities Board of Singapore (2007) NEWater quality. http://www.pub.gov.sg Accessed 27 July 2007
21. Qin JJ, Li Y, Lee LS, Lee H (2003) Cellulose acetate hollow fiber ultrafiltration membranes made from CA/PVP(360K)/NMP/water. J Membr Sci 218:173–183
22. Futselaar H, van Hoof S, Slotman R, Jansan Yang J (2002) Koning, dead-end ultrafiltration for effluent polishing and pre-treatment for reverse osmosis. In: Proceedings of the 2nd international conference on application of membrane technology, Beijing, China, pp 212–216
23. Hagmeyer G, Gimbel R, Kiepke O, Dautzenberg W (2001) Flocclation/ultrafiltration for drinking water treatment of reservoir water. In: Proceedings of American water works association (AWWA) membrane technology conference, San Antonio, TX
24. Pall Corporation (1998) High quality water and low cost of ownership at San Patricio municipal water district Ingleside, Texas. Water Desalination Report, Vol. 34, No. 41, October 1998

25. Parsons Brinckerhoff Quade & Douglas Inc. and Milian Swain & Associates Inc (2001) Project cost estimate peer review of microfiltration supplemental technology demonstration project, 7 May 2001, New York, NY
26. Wilf M, Allam J (2001) Reclamation of sand filter backwash effluent using HYDRAcap LD capillary UF membrane technology. Hydranautics Technical Report, December 2001
27. del Pino MP, Durham B (1999) Wastewater reuse through dual-membrane processes: opportunities for sustainable water resources. Desalination 124:271–277
28. Ahn K-H, Song K-G (1999) Treatment of domestic wastewater using microfiltration for reuse of wastewater. Desalination 126:7–14
29. Abdel-Jawad M, Al-Shammari S, Al-Sulaimi J (2002) Non-conventional treatment of treated municipal wastewater for reverse osmosis. Desalination 142:11–18
30. Porter MC (ed) (1992) Handbook of industrial membrane technology. Noyes, Park Ridge, NJ
31. Byrne W (ed) (1995) Reverse osmosis. Tall Oaks, Littleton, CO
32. Bonne PAC, Hofman JAMH, van der Hoek JP (2000) Scaling control of RO membranes and direct treatment of surface water. Desalination 132:109–119
33. Qin J-J, Kekre KA, Tao G, Oo MH, Wai MN, Ting CL, Viswanath B, Seah H (2006) New option of MBR-RO process for production of NEWater from domestic sewage. J Membr Sci 272:70–77
34. Gander M, Jefferson B, Judd S (2000) Aerobic MBRs for domestic wastewater treatment: a review with cost considerations. Sep Purif Technol 18:119–130
35. Kishino H, Ishida H, Iwabi H, Nakano I (1996) Domestic wastewater reuse using a submerged membrane bioreactor. Desalination 106:115–119
36. Lozier J, Fernandez A (2001) Using a membrane bioreactor/reverse osmosis system for indirect potable reuse. Water Supply 1:303–313
37. Poitel D, Courant P, Primi C, Mandin JM (1999) Various leachate treatment plants in France. In: Proceedings of Sardinia 99 7th international waste management and landfill symposium, Cagliari, Italy, 4–8 October 1999
38. Lawrence P, Adham S, Barrott L (2002) Ensuring water re-use projects succeed-institutional and technical issues for treated water re-use. Desalination 152:291–298
39. Trouve E, Urbain V, Manem J (1994) Treatment of municipal waste-water by a membrane bioreactor-results of a semi industrial pilot-scale study. Water Sci Technol 30:151–157
40. Ghyoot W, Verstraete W (1999) Reduced sludge production in a two stage membrane assisted bioreactor. Water Res 34:205–215
41. Fan XJ, Urbain V, Qian Y, Manem J (1996) Nitrification and mass balance with a membrane bioreactor for municipal wastewater treatment. Water Sci Technol 34:129–136
42. Yoon TI, Lee HS, Kim CG (2004) Comparison of pilot scale performances between membrane bioreactor and hybrid conventional wastewater treatment systems. J Membr Sci 242:5–12
43. Holler S, Trosch W (2001) Treatment of urban wastewater in a membrane bioreactor at high organic loading rates. J Biotechnol 92:95–103
44. van der Roest HF, van Bentem AGN, Lawrence DP (2002) MBR-technology in municipal wastewater treatment: challenging the traditional treatment technologies. Water Sci Technol 46:273–280
45. Tao G, Kiran KA, Zhao W, Ting CL, Bala V, Seah H (2005) Membrane bioreactors for water reclamation. Water Sci Technol 51:431–440
46. Qin J-J, Wai MN, Oo MH, Lee H, Kekre KA, Seah H (2006) Feasibility study for reclamation of a secondary treated sewage effluent mainly from industrial sources using a dual membrane process. Sep Purif Technol 50:380–387

47. Bonne PAC, Hofman JAMH, van der Hoek JP (2000) Scaling control of RO membranes and direct treatment of surface water. Desalination 132:109–119
48. Prats D, Chillon-Arias MF, Rodriguez-Pastor M (2000) Analysis of the influence of pH and pressure on the elimination of boron in reverse osmosis. Desalination 128:269–273
49. Kang M, Kawasaki M, Tamada S, Kamei T, Magara Y (2000) Effect of pH on the removal of arsenic and antimony using reverse osmosis membranes. Desalination 131:293–298
50. Dey A, Thomas G, Kekre KA, Tao GH (2001) Membranes. Part 2: impact of caustic dosing on contaminant removal using double-pass RO. Ultrapure Water 9:43–48
51. Qin JJ, Oo MH, Wai MN, Wong FS (2003) Effect of feed pH on an integrated membrane process for the reclamation of a combined rinse water from electroless nickel-plating. J Membr Sci 217:261–268
52. Edwards GA, Amirtharajah A (1985) Removing colour caused by humic acids. J AWWA 77:50–57
53. Greger JE, Nokes CJ, Penton E (1997) Optimising natural organic matter removal from low turbidity waters by control pH adjustment of aluminium coagulation. Water Res 12:2949–2958
54. Schaefer J (2000) Advance water reuse saves money. Industrial Water World. Pall Corporation, Port Washington, NY
55. Business Communications Company Inc (2005) Membrane and separations industry review. Business Communications Company Inc, Lubbock, Texas, pp 1–363
56. Yang W, Cicek N, Ilg J (2006) State-of-the-art of membrane bioreactors: worldwide research and commercial applications in North America. J Membr Sci 270:201–211
57. Zhou JP, Zhang LN, Cai J, Shu H (2002) Cellulose microporous membranes prepared from NaOH/urea aqueous solution. J Membr Sci 210:77–90
58. Chen Y, Zhang LN, Gu JM, Liu J (2004) Physical properties of microporous membranes prepared by hydrolyzing cellulose/soy protein blends. J Membr Sci 241:393–402
59. Li H-J, Cao Y-M, Qin J-J, Jie X-M, Wang T-H, Liu J-H, Yuan Q (2006) Development and characterization of anti-fouling cellulose hollow fiber UF membranes for oil-water separation. J Membr Sci 279:328–335
60. Ichimura S, Asakawa K, Morihara N, Tsujikawa H, Akamatsu K, Yamaguchi T (2005) Development of thermo-sensitive porous hollow fiber membranes for novel filtering systems. In: Proceedings of the international congress on membranes and membrane processes, Seoul, Korea, 21–26 August 2005, p 115
61. Petrov S, Ivanova T, Christova D, Ivanova S (2005) Modification of polyacrylonitrile membranes with temperature sensitive poly(vinylalcohol-co-vinylacetal). J Membr Sci 261:1–6
62. Bellara SR, Cui ZF, Pepper DS (1996) Gas sparging to enhance permeate in ultrafiltration using hollow fiber membranes. J Membr Sci 121:175–184
63. Smith SR, Cui ZF (2004) Gas-slug enhanced hollow fibre ultrafiltration – an experimental study. J Membr Sci 242:117–128
64. Ventresque C, Gisclon V, Chagneau V, Chagneau G, Bablon G (2000) An outstanding feat of modern technology: the Mery Sar Oise nanofiltration treatment plant. Desalination 131:1–16
65. Chang IS, Lee EW, Oh S, Kim Y (2005) Comparison of SAR (sodium adsorption ratio) between RO and NF processes for the reclamation of secondary effluent. Water Sci Technol 51:313–318
66. Kyllönen HM, Pirkonen P, Nyström M (2005) Membrane filtration enhanced by ultrasound: a review. Desalination 181:319–335
67. Sanderson RD, Hallbauer DK, Reinecke FJ, Swapan SK, Mbanjwa MK (2005) The challenge of detecting the very initial onset of biofouling on membranes by novel ultrasonic wavelet

techniques. In: Proceedings of the international congress on membranes and membrane processes, Seoul, Korea, 21–26 August 2005, p 128
68. Liberman B, Liberman I (2005) Replacing membrane CIP by direct osmosis cleaning. Desalination Water Reuse 15:28–31
69. Sagiv A, Semiat R (2005) Backwash of RO spiral wound membranes. Desalination 179:1–9
70. Choi JH, Cockko S, Fukushi K, Yamamoto K (2002) A novel application of a submerged nanofiltration membrane bioreactor (NF MBR) for wastewater treatment. Desalination 146:413–420
71. Choi JH, Fukushi K, Yamamoto K (2005) Comparison of treatment efficiency of submerged nanofiltration membrane bioreactors using cellulose triacetate and polyamide membrane. Water Sci Technol 51:305–312
72. Ryu J, Choi W, Choo KH (2005) A pilot-scale photocatalyst-membrane hybrid reactor: performance and characterization. Water Sci Technol 51:491–497
73. Yoon SH, Collins JH, Musale D, Dave B, Kong JF, Hallsby GA, Yoo KH, Tsai SP, Koppes J, Cachia P (2005) A novel flux enhancing method for membrane bioreactor process using polymer. In: Proceedings of the international congress on membranes and membrane processes, Seoul, Korea, 21–26 August 2005, p 100
74. Yoon SH, Collins JH, Musale D, Sundararajan S, Tsai SP, Hallsby GA, Kong JF, Koppes J, Cachia P (2005) Effects of flux enhancing polymer on the characteristics of sludge in membrane bioreactor process. Water Sci Technol 51:151–157
75. Lew CH, Hu JY, Song LF, Lee LY, Ong SL, Ng WJ, Seah H (2005) Development of an integrated membrane process for water reclamation. Water Sci Technol 51:455–463
76. Fane AG (2005) Towards sustainability in membrane processes for water and wastewater processing. In: Proceedings of the international congress on membranes and membrane processes, Seoul, Korea, 21–26 August 2005, p 104
77. Vigneswaran S, Ngo HH, Chaudhary DS, Hung YT (2005) Physicochemical treatment processes for water reuse. In: Wang LK, Hung YT, Shammas NK (eds) Physicochemical treatment processes. Humana, Totowa, NJ, pp 635–676
78. Wang LK, Wang MHS (2008) Development and applications of membrane bioreactor technologies. 2008 national engineers week conference, Albany Marriott, Albany, NY, 14–15 February 2008
79. Wang LK, Shammas NK, Hung YT (eds) (2009) Advanced biological treatment processes. Humana, Totowa, NJ, p 738
80. Wang LK, Wang MHS, Suozzo T, Dixon RA, Wright TL, Sarraino S (2009) Chemical and biochemical technologies for environmental infrastructure sustainability. 2009 national engineers week conference, Albany Marriott, Albany, NY, 5–6 February 2009
81. Guglielmi G, Andreottola G (2010) Selection and design of membrane bioreactors in environmental bioengineering. In: Wang LK, Ivano V, Tay JH, Hung YT (eds). Environmental biotechnology. Humana, Totowa, NJ, pp 439–516.

11

Potable Water Biotechnology, Membrane Filtration and Biofiltration

Puangrat Kajitvichyanukul, Nazih K. Shammas, Yung-Tse Hung, Lawrence K. Wang, and Jirapat Ananpattarachai

Contents

Abstract Membrane filtration is considered as a simplified drinking water treatment process, which can remove organic impurities, as well as metal ions and other ions. Nowadays, membrane processes are increasingly employed for removal of bacteria and other microorganisms, particulate material and natural organic matter, which can impart color, tastes, and odors to the water and react with disinfectants to form disinfection by-products (DBPs). Recently, there have been several advanced technologies derived from the combination of biotechnology and filtration with application for potable water treatment. This chapter describes these techniques which includes biofiltration, membrane bioreactor, ion-exchange membrane bioreactor, and biological activated carbon adsorption-filtration. Several case studies in applying biofiltration for DBP control in bench- and pilot-scale are also demonstrated.

Key Words Potable water • biotechnology • biofiltration • membrane filtration • membrane bioreactor • ion-exchange membrane bioreactor filtration • biological activated carbon adsorption-filtration • biofilm

From: *Handbook of Environmental Engineering, Volume 13: Membrane and Desalination Technologies*
Edited by: L.K. Wang et al., DOI: 10.1007/978-1-59745-278-6_11

1. INTRODUCTION

Drinking water treatment provides good quality potable water that is chemically and microbiologically safe for human consumption. The water treatment plant generally consists of physical, chemical and biological treatment processes. Membranes have a major role in liquid–solid separation especially in the removal of bacteria and other microorganisms, particulates, and natural organic matter. Membrane application is considered a simplified drinking water treatment process with several advantages such as avoiding the need for chemical pretreatment, reducing chemical usage, minimizing recirculation and repumping of concentrate and eliminating the need for booster pumps. Membrane filtration can remove organic impurities, as well as metal ions and other ions. Nowadays, membrane processes are increasingly employed for removal of bacteria and other microorganisms, particulate material, and natural organic matter, which can impart color, tastes, and odors to the water and react with disinfectants to form disinfection by-products (DBPs). As advancements are made in membrane production and module design, capital and operating costs continue to decline (1).

A filtration unit is a thin layer of filter media capable of separating substances when a driving force is applied across the filter media. Typical filter media include membranes, sand, granular activated carbon (GAC), cloth, etc (1). This filtration technology removes particles from the water source. Though this removal depends mainly on the pore size of the filter media, application of these either in single or in combination provides complete solution for various requirements, based on their sizes and configuration. This chapter introduces mainly slow sand filtration, membrane filtration, and biological GAC filtration in conjunction with biotechnology for potable water treatment plants. Both slow sand filtration and biological GAC filtration have been applied to drinking water treatment in full-scale. The four main types of membrane filtration used in drinking water treatment plants are reverse osmosis, microfiltration, ultrafiltration, and nanofiltration membranes. Comparison of filtration application of these membranes is shown in Fig. 11.1. Details and role of each type in drinking water treatment will be discussed later in the chapter.

2. TREATMENT OF DRINKING WATER USING FILTRATION AND BIOTECHNOLOGY

2.1. *Turbidity Removal*

Low turbidity is one requirement of drinking water treatment plant. As a result, membrane systems especially microfiltration and ultrafiltration become the alternative method in drinking water treatment. Generally, most surface water used for the production of drinking water requires filtration to remove turbidity and suspended solids before it can be used safely. Conventional treatment systems employ a wide range of techniques to separate suspended solids, such as coagulation and flocculation, settling, or flotation. These techniques require constant monitoring, evaluation, and adjustment by operators. However, with the membrane process, minimal operator monitoring is required. Moreover, membrane processes provide the advantage in turbidity removal under varying raw water conditions. Several studies have evaluated the removal of turbidity and particles by microfiltration and ultrafiltration.

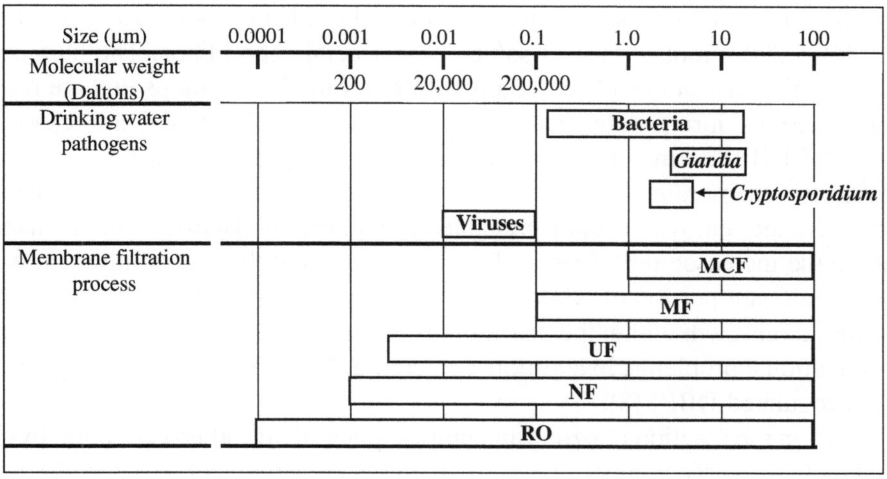

Fig. 11.1. Comparison of membrane filtration applications in drinking water treatment process (adapted from (2)).

2.2. *Desalination*

Desalination is the area of growth for membrane technology. Various types of membranes used for desalination include reverse osmosis, nanofiltration, and electrodialysis. Other techniques such as the distillation process, ion exchange, freezing, and other miscellaneous processes are also widely employed. Approximately 60% of the worldwide desalting capacity is attributable to distillation technologies (3). However, the percent of worldwide desalting capacity by membrane processes has been steadily increasing (4).

To obtain drinking water from brackish water or seawater, the membrane system is a good alternative treatment method. A combination of ultrafiltration followed by reverse osmosis can be very cost effective. Ultrafiltration cartridges used as pretreatment can considerably extend the useful life of reverse osmosis elements. When treating brackish water with relatively low total dissolved solids (TDS), ultrafiltration can be used as pretreatment to reverse osmosis and to produce permeate for blending with reverse osmosis product water. The result is water for distribution that not only tastes good, but is less expensive to produce.

2.3. *Control of Disinfection By-Products*

Recently there has been a growing interest in employing membrane processes for removal of precursors to DBPs. The major precursor of DBPs is natural organic matter (NOM). NOM is normally derived from plant or microbial residues, or is produced in situ in water by life cycles and by a variety of decomposition pathways (5). NOM can be categorized into humic, microbial by-products, and colloidal natural organic matter, depending on their source and characteristics. Microbial by-products are composed of acids, with a relatively high charge density, polysaccharides, amino sugars, and proteins. Humic material, normally found in surface water, consists of fulvic acid and humic acid. Fulvic acid has an average molar mass of 500–2,000 g/mol. It is soluble in water at all pH values. Humic acid has an average molar mass of 2,000–5,000 g/mol. It is well soluble under acidic conditions. The humic material consists of organic macromolecules of many different functional groups that are responsible for many interactions during water treatment processes. In addition, they are responsible for the formation of DBPs during chlorination or ozonation of potable water (6).

The advance in membrane systems is mainly focused on removal of DBPs from drinking water, especially the removal of biodegradable organic matter (BOM) from ozonated water. This is due to the increased use of ozone for disinfection and the recognition that ozone can transform a significant portion of NOM into BOM. The presence of BOM in drinking water leads to bacterial regrowth within the distribution system (7), the loss of residual disinfectant, and other operational problems. In addition, decrease in NOM enhances the diminution of the formation of regulated DBPs (8).

The DBPs from the reaction of ozone and humus material are called ozonation by-products (OBPs) that include a large number of new, low molecular weight compounds like aldehydes, ketones, and carboxylic acids (9–12). OBPs are polar and present at low concentrations. From toxicological information, mutagenicity studies with ozonated water have shown no mutagenicity or significantly lower mutagenicity than in chlorinated water indicating that OBPs are not a similar public health concern as chlorination by-products (13). However, some OBPs such as formaldehyde, glyoxal, and glyoxylic acid have been reported to be mutagenic (14).

In water chlorination, the chlorinated disinfection by-products (CBPs) result from the reaction of chlorine with dissolved organic matter (DOM) especially humic organic substances from natural sources. The principal chlorinated disinfection by-products are volatile hydrophobic compounds referred to as trihalomethanes (THM) with chloroform and bromodichloromethane being the most common chemical forms. In addition, a variety of chlorinated and unchlorinated aromatic and aliphatic compounds are also produced. The THM chloroform, dichlorobromomethane, and bromoform have been classified by the US Environmental Protection Agency (US EPA) as probable human carcinogens (cancer group B2), while dibromochloromethane, chloral hydrate, dichloroacetonitrile, and dibromoacetonitrile as possible human carcinogens (cancer group C). In addition, THM are regulated to have maximum contaminant levels (MCL) of 0.08 mg/L in drinking water.

Membrane is one of the feasible technologies in removal of NOMs from drinking water. Biologically active filters seem to be the most effective process to reduce the biodegradable fraction in DBPs (15–17). The application of biofilters in removal of both BOM and NOM concentrations can therefore lead to the decrease in DBPs formation potential. Such control of DBPs by biological filters will be discussed in detail in a later section.

2.4. *Inactivation and Removal of Targeted Microorganisms*

Generally, the primary requirement of microfiltration and ultrafiltration is to remove the microorganisms such as viruses, bacteria, and protozoa. Viruses are the smallest organisms of concern to the water quality. Their size ranges from 0.02 to 0.08 μm. Size of bacteria is in the range from 0.5 to 10 μm. For protozoa and oocysts, size ranges from 3 to 15 μm. The protozoa

Giardia and *Cryptosporidium* have been the principal organisms controlling disinfection regulations in the US over the last decade.

Giardia is a protozoan parasite that occurs in a trophozoite and an oval-shaped cyst form. Cysts excreted in the feces of an infected host move passively through the environment (18). If cysts are ingested, infection may be transmitted to another vertebrate host. *Giardia* is distributed worldwide in lakes, ponds, rivers, and streams. Reported *Giardia* levels have ranged from 10,000 to 100,000 cysts/L in untreated sewage, 10 to 100 cysts/L in treated sewage, and ten or fewer cysts/L in surface water sources and tap water (18). According to the US EPA's Surface Water Treatment Rule (SWTR), the public water systems filter, except in rare circumstances, and disinfect surface water and groundwater that is directly impacted by surface water; 99.9% of *Giardia* should be removed or killed. Commonly used water disinfectants can effectively inactivate *Giardia* cysts depending on the disinfectant concentration and contact time. Cysts are relatively more resistant to disinfectants than bacteria and viruses, and high doses and lengthy contact times may be needed. This may result in high levels of DBPs that are regulated by the US EPA (18). The appropriate technology for removal of cysts is filtration, especially by using membranes. Microfiltration and ultrafiltration can also effectively remove *Giardia* cysts from water.

Cryptosporidium is a parasite that can live inside the intestines of humans, farm and wild animals, and pets. The source of *Cryptosporidium* is animal or human fecal waste. *Cryptosporidium* can form a protective shell, allowing it to survive under harsh conditions. As a result, *Cryptosporidium* has been found in water, soil, and food, after contamination with fecal waste. It is reported that the conventional filtration process used by treatment plants may not remove all *Cryptosporidium* from the drinking water supply. These unfiltered organisms may also survive chlorine treatment. *Cryptosporidium* can be effectively removed from drinking water by reverse osmosis, or filtration using 1 μm or smaller filter.

The SWTR and Groundwater Disinfection Rule have led to the investigation of ultrafiltration and microfiltration for microbial removal. Many membrane filtration methods have been developed and used to detect the presence of both *Cryptosporidium* and *Giardia* in surface and treated water samples. Through examination of the sizes of the target organisms and the range of membrane pore sizes, it is apparent that removal of these organisms is specific to the particular membrane and its pore size distribution (4). The selection of membrane type with the appropriate pore size is a crucial consideration for the removal of all microorganisms especially *Cryptosporidium* and *Giardia* from drinking water.

2.5. Nitrate Removal from Drinking Water

Nitrate (NO_3^-) removal in drinking water is one promising area in application of membrane biotechnology. Contamination of drinking water sources with NO_3^- originates from over fertilization (both artificial fertilizer and animal manure) in agriculture, human and animal waste disposal, disposal of wastewater from food processing operations, and anthropogenic sources (19). Nowadays, NO_3^- contamination of groundwater is becoming a worldwide problem. A significant fraction of groundwater used for drinking water exceeds the MCL for NO_3^- in many regions of the world (20). In the United States, 10–25% of the water from wells exceeds the MCL for NO_3^- and long-term trends indicate increasing concentrations of NO_3^- in groundwater (21).

NO_3^- are water-soluble and do not bind to soil thus causing their migration to groundwater sources (22). Once consumed, NO_3^- are converted into nitrites (NO_2^-) in the body. Methemoglobinemia or blue baby syndrome in infants is related to the ingestion of water with high NO_3^- concentration (23). Moreover, there exists an increasing concern that consumption of water containing high NO_3^- may lead to some other diseases including carcinoma, malformation, and mutation when transformed into nitrosoamines (24–26).

Removing NO_3^- (NO_3^-) and NO_2^- from drinking water has gained tremendous attention in recent years. It is reported by the World Health Organization that NO_3^- concentration in drinking water should be lower than 10 mg-N/L (27). In the USA, US EPA set MCL of 10 mg NO_3^--N/L and 1 mg nitrite-N/L. The European standard for NO_2^- is stricter (0.03 mg-N/L) to account for the direct toxic effects from NO_2^- (28). A recommended level for NO_3^- is 5.6 mg-N/L for the European Community (29). However, these standards are difficult to meet by merely controlling the contamination at the source; denitrification processes constitute the major technology for the removal of NO_3^- from potable water.

2.5.1. *Biological Nitrate Removal*

Biological nitrate removal is conducted by a wide variety of organisms by either assimilatory or dissimilatory pathways (30, 31). In assimilatory pathway, organisms use nitrate, rather than ammonia, as a biosynthetic nitrogen source for nitrate reduction. This process can occur in either aerobic or anaerobic condition. Assimilatory nitrate reduction occurs in the absence of more reduced inorganic nitrogen species (e.g., ammonia). In this process, the inorganic nitrogen is converted to organic nitrogen, thus, no net removal of inorganic nitrogen is accomplished.

Dissimilatory pathway consists of two groups of prokaryotic organisms. In the first pathway, dissimilatory nitrate reduction to ammonia, NO_2^- is reduced to either nitrite or ammonia by anaerobic and facultative anaerobic bacteria. In denitrification pathway, nitrate is reduced to NO_2^- and finally to gaseous nitrogen (N_2) as the end products by the second group of bacteria, facultative anaerobic bacteria.

2.5.2. *Biological Denitrification*

Biological denitrification is generally used in treating municipal and industrial wastewaters. Recently, research has lead to more knowledge and experience in using biological denitrification as a viable technology in nitrate removal from drinking water. Denitrification is a biological process that reduces NO_3^- to NO_2^-, nitric oxide (NO), nitrous oxide (N_2O), and finally to N_2 as shown in Eq. (1) (31):

$$NO_3^- \rightarrow NO_2^- \rightarrow NO \rightarrow N_2O \rightarrow N_2. \tag{1}$$

Denitrification takes place under anoxic conditions. Oxygen concentrations in the range of 0.1–0.2 mg/L can exert inhibitory effects on these processes (32). Both heterotrophic and autotrophic bacteria can be involved in denitrification (29). Heterotrophic bacteria under anoxic conditions utilize organic substrates such as methanol, ethanol, and acetic acid for the conversion of nitrates to nitrogen. Gaseous organic substrates such as methane and carbon monoxide can also serve as substrates for denitrification in water. Heterotrophic denitrification is very efficient in nitrate removal if adequate amounts of organic carbon are available.

Autotrophic bacteria such as *Thiobacillus denitrificans* are capable of denitrification (33). In autotrophic denitrification, hydrogen or reduced sulfur compounds serve as substrates, and carbon dioxide or bicarbonate serve as the carbon source for cell synthesis (29). Autotrophic denitrification has been booming recently due to two major advantages compared with heterotrophic denitrification (34, 35): (a) no external carbon is needed, which reduces the cost and risk of operation; and (b) less sludge is produced, which minimizes the bacteria in the effluent.

Both hydrogen gas and elemental sulfur can be used as ideal electron donor for autotrophic denitrification because they are completely harmless to human health, and no further steps are needed to remove either excess substrate or its derivatives (35–37).

However, in drinking water, the concentration of biodegradable organic materials is insufficient for denitrification. The shortage of organic carbon may limit the application of heterotrophic denitrification. Thus, the reduction of NO_3^- (or NO_2^-) requires addition of an electron donor substrate, and many organic (heterotrophic denitrification) and a few inorganic electron donors (autotrophic denitrification) are possible.

Hydrogen gas (H_2) is an excellent electron donor for autotrophic choice because of its clean nature, low biomass yield, and relatively low cost, as well as because hydrogen gas does not persist in the treated water to create biological instability (32, 38). The hydrogen gas can form flammable and explosive mixtures with air. It is poorly soluble in water (1.6 mg/L at 20°C), and thus utilization of this gas in water treatment is limited (39).

Elemental sulfur is also used as electron donor for autotrophic denitrification (34). This application is available as sulfur limestone autotrophic denitrification (SLAD) systems, which have been studied widely in Europe and USA (35, 40–42). The denitrification efficiency can be compared with that of the heterotrophic denitrification in the nitrate-contaminated water treatment. In such processes, limestone is used to adjust the pH and 7.54 mg/L sulfate will be produced stoichiometrically when 1 mg-N/L NO_3^- is removed. However, hardness may increase in treated water because of Ca^{2+} produced by the limestone for pH adjustment (34).

It is reported that there is a wider application of heterotrophic denitrification in comparison with autotrophic denitrification, especially on a full-scale level. Heterotrophic denitrification processes possess higher specific volumetric denitrification rates (0.4–24 kg NO_3–N/m^3/day) than autotrophic denitrification (0.5–1.3 kg NO_3–N/m^3/day). The autotrophic reaction rate is low; thus, a large volume of reactor is required to achieve sufficient residence time for denitrification that increases the capital cost. The disadvantage of biological denitrification is that it requires additional carbon sources for the activity of microorganisms and consequently some post-treatment operations for the elimination of the process contaminants (22). The post-treatment processes include filtration through sand and active carbon beds, aeration, and chlorination. In conventional denitrification processes these post-treatment operations are usually carried out in sequence (22).

The effective membrane technology that can be successfully applied in biological denitrification is membrane bioreactor (MBR) such as Zenon Zee weed® membrane bioreactor (43). The pilot plant using this membrane system for biological denitrification is built in Italy to remove NO_3^- from drinking water. The plant comprises two units: an anoxic tank where NO_3^- are biologically reduced to nitrogen, followed by an anaerobic tank that contains the Zenon ZW-10 membranes for excess carbon removal. Ethanol is used as the carbon source. The optimum C/N ratio was found to be at 2.2 g C/g N. A maximum NO_3^- removal rate close to 20 mg NO_3^-–N/g VSS/h was reported.

3. TYPES OF FILTRATION PROCESSES FOR WATER PURIFICATION

3.1. *Biological Slow Sand Filtration*

Slow sand filters, first used in the USA in 1872, have been widely used in Europe since the early 1800s and are the oldest type of municipal water filtration units. Slow sand filtration is a simple and reliable process, which is suitable for small systems with low turbidity or algae containing water sources.

A slow sand filtration unit for drinking water treatment consists of a tank, a bed of fine sand, a layer of gravel to support the sand, a system of underdrains to collect the filtered water and a flow regulator to control the filtration rate. In this process, no chemicals are required to aid the filtration process. The process requires a very low filtration rate at 0.015 gal/min/ft^2 of bed area (44), which depends on the gradation of the filter medium and the quality of the raw water. Untreated water is spread over the filter sand bed and is drained out from the bottom. Slow sand filtration does not require any pretreatment or extensive operator control.

The removal mechanism in slow sand filters includes a biological process in addition to the physical and chemical processes. The biological process occurs by a sticky mat of biological matter, called a "schmutzdecke", forming on the sand surface, where particles are trapped and organic matter is biologically degraded. To avoid disturbing the schmutzdecke, flows near the upper filter surface must be very gentle. A sufficient water depth to drive water through the schmutzdecke is required. The depth of filter bed is about 1 m. The bed depth has a strong influence on the effectiveness of filtration. The filter media effective size is typically 0.15–0.35 mm and its uniformity coefficient is between 2 and 3 (45).

Slow sand filtration can reduce bacteria, cloudiness, and organic levels. This process reduces the need for disinfection and consequently, the presence of DBPs in the finished water. For turbidity removal, the filtration process is effective in removing suspended particles without any pretreatment when the turbidity levels are lower than 10 NTU (44). Turbidity can be reduced to below 1.0 NTU; however, water with a very low nutrient content may impair turbidity removal since some nutrients must be present to promote biological growth in the filter bed (44). Slow sand filters can also achieve 90–99% (or higher) reductions in bacteria and viruses, and provide virtually complete *Giardia lamblia* cyst and *Cryptosporidium* oocyst removal (44). For organic and inorganic removal, slow sand filters do not completely remove ions such as heavy metals or THM precursors from water. Slow sand filters are less effective at removing microorganisms from cold water because as temperatures decrease, the biological activity within the filter bed declines (46).

As the purification mechanism in a slow sand filtration is based on a balanced community in the schmutzdecke, the filter should operate at a constant rate. When operation is stopped, the microorganisms causing bacteriological degradation of trapped impurities lose their effectiveness. To solve this problem, filter operation using declining rate filtration for additional water is recommended (44).

3.2. *Microfiltration*

Microfiltration membrane is defined as a membrane separation process that has the ability to remove suspended or colloidal particles via a sieving mechanism based on the membrane pore size (0.1–0.2 μm), which is related to the sizes of particulate matter (2). However, the microfiltration membranes with pore sizes of up to 10 μm are also available. The microfiltration typically has a molecular weight cut off of greater than 100,000 Da with a relatively low feedwater operating pressure of approximately 100–400 kPa (15–60 psi).

For drinking water applications, commercially available configurations for microfiltration include spiral wound, tubular, and hollow capillary fiber. The hollow capillary fiber is the most commonly used. In this configuration, the membranes are cast into small diameter tubes or straws, which are bundled together longitudinally, potted in a resin on both ends, and encased in a pressure vessel that is included as a part of the hollow-fiber module. The housing of these bundles is usually made from PVC or stainless steel. The microfiltration membrane may be constructed from a wide variety of materials including cellulose acetate (CA), polyvinylidene fluoride (PVDF), polyacrylonitrile (PAN), polypropylene (PP), polysulfone (PS), polyethersulfone (PES), or other polymers. Types of material are selected based on different properties with respect to surface charge, degree of hydrophobicity, pH and oxidant tolerance, strength and flexibility (2).

In potable water treatment, microfiltration can remove various types of suspended particles including sand, silt, and clays, *Giardia lamblia* and *Cryptosporidium* cysts, algae, and some bacterial species. In addition, this technology is feasible in removal of natural or synthetic organic matter when pretreatment is applied, as well as retardation of membrane fouling. As a pretreatment step, microfiltration can be used to reduce fouling that might occur for the following reverse osmosis and nanofiltration processes.

Generally, microfiltration provides the particulate removal from a feed stream by a separation process based on retention of contaminants on a membrane surface. This system can be operated under ultralow pressure conditions. Typically, the microfiltration process involves prescreening raw water and pumping it under pressure onto a membrane. Pretreatment process, such as filtration, is necessary to remove large particles that may block the inlet to the fibers within the membrane module. To enhance the removal of viruses and DOM, the complex pretreatments, such as addition of coagulants or adsorption by powdered activated carbon can be employed. Potable water plants treating 1 MGD normally utilize membrane units as their major component. Other units of the plant may include prescreens, feed pump, cleaning tank, automatic gas backwash system, air compressor, membrane integrity monitor, backwash water transfer tank, pressure break reservoir, air filter for the gas backwash, controls for the programmable logic controller, and a coalescer.

3.3. Ultrafiltration

Ultrafiltration membranes have pore sizes in the range of 0.01–0.05 μm (nominally 0.01 μm), which are relative to the sizes of the particulate matter. Ultrafiltration can remove particulate contaminants via a size exclusion mechanism according to pore size with respect to microbial and particulate removal capabilities. This type of membrane is often characterized by a molecular weight cut off, which is a measure of the removal characteristic of a membrane in terms of atomic weight (or mass) rather than a particular pore size. The molecular weight cut off levels for ultrafiltration range from 10,000 to 500,000 Da, with most membranes used for water treatment at approximately 100,000 Da (2). Molecules larger than the cutoff size are completely rejected. Typically, a pressure differential of at least 100 kPa (15 psi) across the ultrafiltration membrane is required. The operating pressure of this membrane is approximately 200–700 kPa (30–100 psi) (1).

The commercially available configuration for ultrafiltration membranes includes fibers and tubes with the active membranes sealed in each module. The flat sheets stacked in frames or rolled into a spiral configuration are also available. Hollow-fiber ultrafiltration membranes are the most widely used for this type of membrane. Diameters of membranes are usually in the range of 0.2–2.0 mm. This geometry does not require extensive pretreatment because the fibers can be periodically backwashed. Flow in these hollow fiber systems can be either from the inside of the membrane fiber to the outside or from the outside to the inside of the fibers.

Ultrafiltration membranes provide an absolute barrier to particles, bacteria, high molecular weight organic molecules, emulsified oils and colloids. Moreover, removal of chlorine-resistant pathogens and their spores, such as *Cryptosporidium* and *Giardia* from water supply is also possible with the ultrafiltration system. Ultrafiltration can be used independently or in association with other treatments such as coagulation or precipitation. The required pretreatment for ultrafiltration membranes include the process to remove suspended solids such as in-line screen or cartridge filter, and oil and grease removal. A feed pump is required to feed raw water onto a membrane surface. The systems can be designed to be fully automatic with on-line analysis or backwashing. Ultrafiltration can be used effectively as a method of pretreatment for osmosis. Its benefits in producing a high quality water stream for use are widely applied for drinking water treatment plants.

3.4. Nanofiltration

Nanofiltration is classified as a pressure-driven membrane separation process that employs the principles of reverse osmosis to remove dissolved contaminants from water. They are typically applied in softening, or the removal of dissolved organic contaminants. Nanofiltration is usually referred to as a membrane process that rejects solutes approximately 1 nm (10 Å) in size with molecular weights above 200 Da. Nanofiltration membranes have a nominal pore size of approximately 0.001 μm. The typical range of molecular weight cut off levels is between 200 and 1,000 Da. However, in some cases, the molecular weight cut off can be increased up to 100,000 Da. Operating pressures of this membrane are from 600 kPa (90 psi) to as high as 1,000 kPa (150 psi) (1). Nanofiltration membranes represent the class of membrane processes that is most often used in drinking water treatment applications that require the removal of dissolved contaminants, as in the case of softening or desalination.

Nanofiltration membrane modules are commonly fabricated in a spiral configuration for drinking water treatment. An important consideration of spiral elements is the design of the feed spacer, which promotes turbulence to reduce fouling.

Nanofiltration membranes are capable of removing bacteria and viruses as well as organic-related color without generating undesirable chlorinated hydrocarbons and THMs. Nanofiltration is used to remove pesticides and other organic contaminants from surface and ground waters to help insure the safety of public drinking water supplies. It is an attractive alternative to lime softening or zeolite softening technologies for groundwater treatment. For many groundwater applications, the pretreatment required for nanofiltration consists in adding acid or antiscalant and then passing the feed through a cartridge filter. However, for surface waters, more extensive pretreatment such as conventional treatment, microfiltration, ultrafiltration, slow sand filtration, and/or GAC adsorption is necessary.

3.5. *Reverse Osmosis*

Reverse osmosis is the pressure-driven membrane separation process that employs the principles of reverse osmosis to remove dissolved contaminants from water. Process mechanism of this membrane is to employ the reverse of the natural osmosis process – i.e., the passage of a solvent (e.g., water) through a semipermeable membrane from a solution of higher concentration to a solution of lower concentration against the concentration gradient, achieved by applying pressure greater than the osmotic pressure to the more concentrated solution (2). An approximate osmotic pressure of fresh or brackish water is approximately 1 psi for every 100 mg/L difference in TDS concentration on opposite sides of the membrane.

Reverse osmosis can remove contaminants from water using a semipermeable membrane that permits only water, and not dissolved ions (such as sodium and chloride), to pass through its pores. Contaminated water is subject to a high pressure that forces pure water through the membrane, leaving contaminants behind in a brine solution. Reverse osmosis is effective in rejecting organic solutes with molecular weights, such as fulvic acids, lignins, humic acids, and detergents. Low molecular weight, nonpolar, water soluble solutes (for example, methanol, ethanol, and ethylene glycol) are poorly rejected. In addition, undissociated organic acids and amines are also poorly rejected, while their salts are readily rejected.

Generally, reverse osmosis units for potable water treatment plants include raw water pumps, pretreatment, membranes, disinfection, storage, and distribution elements. These units are able to process virtually any desired quantity or quality of water by configuring units sequentially to reprocess waste brine from the earlier stages of the process.

Reverse osmosis membranes have the ability to screen microorganisms and particulate matter in the feed water. In addition, it can remove nearly all contaminant ions and dissolved nonions. It is relatively insensitive to flow and TDS level and simple to operate due to its automation, which makes it suitable for small systems with a high degree of seasonal fluctuation in water demand. However, membrane units require high capital and operating costs.

4. COMBINATION OF BIOTECHNOLOGY AND FILTRATION TECHNOLOGY

4.1. Biofiltration

4.1.1. General Description

Biotechnology is normally important in industrial and commercial wastewater renovation. However, specific problems in potable water treatment such as NO_3^- and NOM have opened the door for application of biotechnology to this field. Biofiltration is one of biological technology applications in potable water treatment. In general, biofilters in drinking water treatment are typically operated for the dual purpose of particle removal and removal of dissolved BOM (47). Biofiltration is most often used after flocculation and clarification (sedimentation or flotation), either in the filtration unit itself or as a separate subsequent unit. Other uses of biological technology in drinking water are:

(a) conversion of ammonia to NO_2^- and then to NO_3^- and finally to nitrogen gas
(b) BOD removal
(c) oxygen addition
(d) carbon dioxide removal
(e) removal of excess nitrogen and other inert gasses
(f) turbidity removal and water clarification
(g) removal of various organic contaminants

Biofiltration is distinguished from other biological waste treatments by the fact that there is a separation between the microorganisms and the treated waste. In biofiltration, the microbial biomass is static, immobilized to the bedding material, while the treated water is mobile and it flows through the filter (48).

A biofilter consists of one or more beds of solid biologically active materials, such as peat, compost, soil, leaves, or woodbark (49). The packing may be mixed with inert support materials such as porous clay, polystyrene spheres, GAC, diatomaceous earth, perlite, or vermiculite to increase its reactive surface and durability, to reduce back pressure, and to prevent compaction. In biofilters, bacteria are attached not only to the filter media, but also form a biofilm on the surface of the filter grains. A large available surface area for biofilm growth is one of the most essential properties of media used in this process. The available surface area is dependent on the grain size and density (for porous materials) of the media.

The feed water with biodegradable organics, suspended cells and other particulates, dissolved oxygen (DO), and other dissolved nutrients are pumped into the top of the filter, and flow down through the filter bed. On the filter media, biomass accumulates and grows as feed water flows through the filter. Biodegradation of substrates and growth of cellular mass also occurs in the bulk liquid. Biomass on the filter media is removed by decay, fluid shear, and backwashing. The major mechanism of pollutant removal in biofiltration is by biological degradation of the contaminants. The contaminants are incorporated into the microbial biomass or used as energy sources (electron donors or electron acceptors) (48). In addition, the pollutants may also be removed from the fluid by adsorbing to the microbial film or to the bedding material.

In Biofiltration, the immobilization of microorganisms is due to two main processes (48):

(a) the self-attachment of microorganisms to the filter bedding material, which is defined as "attached growth".
(b) the artificial immobilization of microorganisms to the bedding material.

In attached growth systems, several forces are involved in microbial attachment to the surface (48):

1. *Electrostatic interactions*. The electrostatic forces are highly significant in the initial stages of adsorption. As reported by Kolot (50), all microorganisms are negatively charged and if the bedding material is also negatively charged, electrostatic repulsive forces can prevent attachment. In other cases, where the attachment surface is positively charged, the electrostatic forces can dominate the attachment process (51).
2. *Covalent bond formation*. Covalent bonds are usually formed between organic groups on the bedding material and the microbial ligands (52).
3. *Hydrophobic interactions*. Hydrophobic groups on the microbial surface can interact with hydrophobic groups on the bedding material, while removing the water molecules, which separate the microorganisms from the surface.
4. *Partial covalent bond formation* between the microorganisms and the hydroxyl groups on surfaces.

The efficiency of the biofilm is also influenced by the film thickness (48). Various factors can affect the film thickness including the rate of flow through the biofilter, the bedding material construction, and the different treatment system designs. The rate of flow through the biofilter controls the growth of bacterial films. High flow rates limit the bacterial film growth and small film thickness will prevail (48). Kosaric and Blaszczyk (53) found that microorganisms form a thinner layer upon smooth surfaces in comparison to microbial films formed upon porous material. The biofilm thickness usually varies from tens of micrometers to more than 1 cm with an average of 1 mm or less usually observed (54). However, it is necessary to determine the active thickness. Previous work has shown that there is a considerable nonuniformity within the biofilm, even though it seems to consist of a homogeneous layer. The DO diffusivity is not constant throughout the biofilm, but decreases with depth (55).

There are various methods for artificial immobilization of microorganisms (48):

1. *Microencapsulation*. In this method, microorganisms will be wrapped as a droplet with a thin membrane. The microorganisms can freely move with their own capsule. They consume the substrates that penetrate through the membrane. The diameters of these microcapsules vary from 10 to 100 μm (56). The main advantage of this method is the low diffusion restriction of the thin membrane. However, this method can bring several problems including the toxicity of the membranes and the mechanical rupture of the membrane from the growth of cells or from the production of gas in the capsule (57).
2. *Membrane separation*. The major concept of this method is the separation of the microorganism from water by the use of sheets of membranes. The membranes allow the substrates to penetrate to the microorganism's zone, while preventing the microorganisms from mixing with the treated water (58, 59). Either porous or nonporous membranes are used for this purpose. However the main problem of this technique is fouling. The pores in a porous membrane may clog and a

nonporous membrane can be covered by a biofilm that alters its performance (60). The conventional method of membrane cleaning may damage the microbial cells.

3. *Covalent bonding and covalent crosslinking*. The covalent bonding method involves the creation of covalent bonds between the reactive groups on the outer surface of microbial cells and different ligands on the bedding material. To activate the ligands on the microbial cells, coupling agents are applied. The most commonly used coupling agents are glutaraldehyde, carbodiimine, isocyanate, and amino silane (57). The covalent crosslinking method is support free and involves the joining of the microorganisms to each other to form a large, three-dimensional complex structure, which is used as a bedding material. The major problems of these two methods are the toxicity effects associated with the covalent bonding method (48).
4. *Entrapment within polymers*. This method consists of trapping microorganisms within a three-dimensional polymer matrix. The pores in the matrix are smaller than the microbial cells, keeping them trapped within the material. But the pores still allow the penetration of substrates through the polymer matrix towards the trapped microorganisms (48).

4.1.2. Application of Biofiltration in Natural Organic Matter Removal

The factors affecting biofilter efficiency in removal of NOM include the use of preozonation, NOM characteristics, water temperature, and biofilter backwashing (61).

In using preozonation prior to the biofilter, it was found that (61) the removal of organic carbon by biodegradation is inversely proportional to the UV absorbance (254 nm)-to-TOC ratio and directly proportional to the percentage of low molecular weight material (as determined by ultrafiltration). The extent and rate of total organic carbon (TOC) removal typically increased as ozone dose increased, but the effects were highly dependent on NOM characteristics. NOM with a higher percentage of high molecular weight material provided the greatest enhancement in biodegradability by ozonation.

In biofilters operating for OBPs removal, the efficiency of the biofilter is a function of empty bed contact time (EBCT), or filter velocity, and influent concentration (62). In biological potable water treatment, substrate concentrations in influent are relatively low and biofilter performance can be modeled as a first-order process (62). Huck et al. (63) found a first-order relationship for the influent concentration and the removal rate of assimilable organic carbon (AOC), biodegradable dissolved organic carbon (BDOC) and THM formation potential in biological filters treating ozonated water. Gagnon et al. (64) reported a first-order relationship for the removal of carboxylic acids. For the relation between EBCT and rates, it was reported that in most cases higher specific rates were obtained with shorter EBCT (63).

To evaluate the efficiency of biofilters, a simple respirometric method was developed and applied for the measurement of biomass activity in bench-scale potable water biofilters (65). The term "Biomass respiration potential" (BRP) was introduced based on the consumption of DO resulting from the biodegradation of BOM by aerobic respiration in a water sample containing a given amount of biomass-laden filter media. Such a method represents a useful tool for water utilities that are operating their filters biologically. Recently, a numerical model was developed to simulate the nonsteady state behavior of biologically active filters used for drinking water treatment (49, 66). The biofilter simulation model is called "BIOFILT" that simulates the substrate (normally BOM) and biomass (both attached and suspended) profiles in a biofilter as a function of time through incorporation of the fundamental processes of biofiltration as follows (48):

(a) deposition of suspended biomass onto filter media
(b) transport of substrate through the filter by advection and dispersion
(c) diffusion of soluble substrate across a liquid thin film layer
(d) diffusion of substrate in the biofilm
(e) biodegradation of substrate by immobilized (in biofilm matrix) bacteria
(f) biomass growth
(g) biomass decay
(h) biomass loss from the biofilm due to fluid shear
(i) biomass loss during a backwashing event

Hozalski and Bouwer (66) reported that BIOFILT appears to be an effective model for simulation of the biofiltration process and is the first model to effectively simulate the highly nonsteady-state behavior encountered in a periodically backwashed drinking water biofilter. The information for biofiltration design based on the "BIOFILT" model is reported as follows (66):

(a) BOM with a greater diffusivity or with faster degradation kinetics experienced greater pseudo steady state BOM removals and also contributed to shorter biofilter start-up times.
(b) The presence of readily degradable substrate can significantly enhance the removal of slowly degradable material primarily due to the ability to maintain greater biomass levels in the biofilters.
(c) A temperature decrease from 22.5 to 3°C resulted in declines in pseudo steady state BOM removal and increases in biofilter start-up time.
(d) Periodic backwashing should not significantly impair the BOM removal performance of biofiltration.

Recently, biofiltration technologies in NOM removal are progressively advanced. Various methods were developed to evaluate the performance of biofiltration in removal of NOM from water. GAC and expanded clay (EC) were used to examine biofiltration of surface water (67). Particle removal was measured by flow cytometry, which enabled discrimination between total- and autofluorescent particles (microalgae) in size ranges of 0.4–1 and 1–15 μm, and measured by on-line particle counting. It was found that biofilters were also challenged with 1 μm fluorescent microspheres with hydrophobic and hydrophilic surface characteristics and bacteriophages (*Salmonella typhimurium* 28B) (67). Yavich et al. (68) developed a simple procedure in order to describe the kinetics of biodegradation of NOM in drinking water and also used this procedure to evaluate changes in the concentration of BOM during ozonation and biotreatment. The results showed that ozonation of NOM depends on source of water that might result in either minimal formation of biodegradable organic carbon, or the formation of predominantly rapidly biodegradable NOM, or in the formation of both rapidly and slowly biodegradable NOM (68). The distribution, composition, and activity of microbial communities developing in biofilters treating nonozonated and ozonated drinking water were evaluated by Fonseca et al. (69) using tetrazolium reduction assays, phospholipid analysis, and 16S rRNA (rDNA) sequence analysis. The response of media-attached biomass to operating temperature (3 vs. >12°C) and ozone application point was also evaluated. It was found that phospholipid and rDNA were a common structure response to ozone application and operating temperature, but no response to different ozone application points (69).

Biofiltration with GAC as a media to support biological growth successfully removed both geosmin and 2-methylisoborneol (MIB), and taste and odor causing compounds in drinking water (70). In biofiltration, while biomass is breaking down naturally occurring organic matter, cometabolism of taste and odor compounds, such as geosmin and MIB, also occurs. MIB removal can be accomplished with biofiltration at varying empty-bed contact times, or bed depths. An Arizona utility increased removal (up to about 67%) by increasing empty-bed times and the addition of PAC at 10–20 mg/L. The combination of ozonation and biofiltration can provide an excellent and low-cost solution for removing high percentages of taste and odor compounds, while the GAC serves as a backstop against any taste and odor escaping the plant (70).

The respirometric method was developed and applied for the measurement of biomass activity in bench-scale drinking water biofilters (65). The results obtained with BRP, indicated a high sensitivity allowing the quantification of the activity of low amounts of biomass. The ratio between biomass activity and the amount of viable biomass (phospholipid) at different filter depths indicated a substantial increase of this ratio with filter depth, where biofilm thickness is low. The comparison of the filter profiles of biomass activity and dissolved BOM, expressed as theoretical oxygen demand, showed a high correlation between these profiles. Consequently, BRP results appear to be good indicators of the BOM removal capacity of the filter biomass and can potentially be used in certain cases instead of BOM measurements for the assessment of the BOM removal capacity of drinking water biofilters, operated under different conditions (65).

4.1.3. Application of Fixed-Film Bioreactor in Nitrate Removal

Fixed film bioreactor is one type of biofilm reactors that is widely used in NO_3^- removal. In this reactor, bacteria grow on inert packing. In the aerobic process, oxygen is provided by air sparging. In some cases, GAC is utilized as a media providing benefits for microbial growth in sorption of contaminants on activated carbon surface.

Soares et al. (71) utilized sand as a media in denitrification of groundwater containing NO_3^- at 22.6 mg NO_3–N/L using a downflow sand column. In this process sucrose is used as a substrate. It was reported that complete NO_3^- removal was achieved at a C:N ratio of two. Kappelhof et al. (72) described the application of upflow fixed bed denitrification reactors using three kinds of support media including expanded schist, anthracite, and sand. The support media were compared based on their ability to attach biomass, NO_3^- removal capacity, ease of stripping the biomass, and abrasivity. Sand was found to be the most preferable support media.

Harremoes et al. (73) and Hoek et al. (74) addressed the issue of clogging of fixed bed bioreactors. The presence of dissolved nitrogen and oxygen can affect the operation of fixed bed denitrification reactors. Release of nitrogen bubbles requires special care to avoid gas entrapment and filter blockage. The presence of oxygen can lead to production of biomass and can cause clogging of the reactor. To solve the problem, vacuum degassing preceding biological denitrification is an effective method to control the clogging problem.

Woodbury et al. (75) investigated the performance of fixed-film two-stage reversible biodenitrification reactors operated at very short detention times for the removal of NO_3^- from contaminated groundwater. The results demonstrate that these systems imparted lower concentrations of organics and suspended solids into the treated effluent than traditional single-stage systems, while maintaining higher levels of NO_3^- removal rates at hydraulic residence time (HRT) values as low as 30 min. The lower detention times translate into substantial reductions in capital cost on scaled-up systems. In 2001, Woodbury and Dahab (76) also compared the operation of a traditional single-stage system with a two-stage fixed bed bioreactor, reversible flow biodenitrification system for removing NO_3^- from drinking water. They investigated the ability of these two-stage systems to remove NO_3^- and residual organics from treated water as compared to single-stage units. In the reversible flow system, the second-stage reactor is operated in series with the first-stage reactor. After a given period of operation, the flow regime is reversed. The active solids remaining in the follow reactor (previously the lead one) are capable of removing residual soluble organics and NO_3^- to levels below the concentrations provided by single-stage units. Nitrate-nitrogen removal efficiency improved slightly from 98 to 99.5% for the single- and two-stage systems, respectively. Most notably, reversible flow reactors were found to reduce long-term effluent residual organics concentrations with an average of approximately 1/3 that of the single-stage system. Also the reversible flow system, with its design redundancy, demonstrated the ability to receive extreme shock loads with no sustained loss of treatment efficiency.

Min and et al. (77) examined the treatment of perchlorate and nitrate from contaminated groundwater using two side-by-side pilot-scale fixed-bed bioreactors packed with sand or plastic media, and bioaugmented with the perchlorate-degrading bacterium *Dechlorosoma* sp. KJ. Groundwater containing perchlorate (77 μg/L), NO_3^- (4 mg-NO_3/L), and DO (7.5 mg/L) was amended with acetic acid as a carbon source and ammonium phosphate as nutrients. Perchlorate was completely removed (<4 μg/L) in the sand medium bioreactor at low rates of 0.063–0.126 L/s (hydraulic loading rate of 0.34–0.68 L/m^2/s) and in the plastic medium reactor at flow rates of <0.063 L/s. Acetate in the sand reactor was removed from 43 ± 8 to 13 ± 8 mg/L (after day 100), and NO_3^- was completely removed in the reactor. A regular backwashing cycle was necessary to achieve consistent reactor performance and avoid short-circuiting in the reactors. This study demonstrated for the first time in a pilot-scale system, that with regular backwashing cycles, fixed-bed bioreactors could be used to remove perchlorate and NO_3^- in groundwater to a suitable level for drinking water.

Recently, Rittmann et al. (78) used the transient-state multiple-species biofilm model (TSMSBM) to obtain more in-depth biofilm information in contaminant removal from drinking water. Multiple-species biofilms that experience time-varying conditions, particularly including periodic detachment by backwashing were described by the TSMSBM. The model includes six features that are essential for describing multiple-species biofilms that undergo changes over time:

1. Four biomass types: heterotrophs, ammonia oxidizers, NO_2^- oxidizers, and inert biomass
2. Seven chemical species: input biodegradable organic material (BOM), NH_4^+–N, NO_2^-–N, NO_3^-–N, utilization-associated products, biomass-associated products, and DO

3. Eight reactions that describe the rates of consumption or production of the different species, as well as the stoichiometric linkages among the rates
4. Reaction with diffusion of all the soluble species in the biofilm
5. Growth, decay, detachment, and flux of each biomass type by location in the biofilm
6. Constant or periodic detachment of biofilm, both of which allow for protection of biomass deep inside the biofilm.

A series of examples illustrates insights that the TSMSBM can provide the transient development of multiple-species biofilms; the roles of soluble microbial products, and detachment in controlling the distribution of biomass types and process performance; and how backwashing affects the biofilm in drinking-water biofiltration.

4.2. Membrane Bioreactor

The MBR is usually referred to a bioreactor integrated to a membrane module system. This technology has been developed over 30 years ago (79) and it is mostly applied to treat industrial wastewater, domestic wastewater, and specific small municipal wastewater treatment plants (80). In drinking water treatment, a promising area of MBR application is NO_3^- removal. Advantages of the MBR include better control of biological activity, effluent that is free of bacteria and pathogens, smaller plant size, and higher organic loading rates (81).

In general, membrane bioreactors consist of two compartments, the biological unit and the membrane module (82). The biodegradation of the compounds occur in the biological unit, while the physical separation of the treated water from the mixed liquor takes place in the membrane module. MBR systems can be classified into two groups: integrated or submerged membrane bioreactors and external or recirculated membrane bioreactors (82). The first group, integrated membrane bioreactors, involves outer skin membranes that are internal to the bioreactor. The driving force across the membrane is achieved by pressurizing the bioreactor or creating negative pressure on the permeate side (82–85). A diffuser is provided for mixing and facilitating the filtration surface scouring. The second type, external or recirculated membrane bioreactors, involves the recirculation of the mixed liquor through a membrane module that is outside the bioreactor (85). The driving force is the pressure created by high cross-flow velocity along the membrane surface (86, 87). For drinking water and groundwater treatment, external membrane bioreactors are widely used for pesticide removal and denitrification.

Types and configurations of membranes used in MBR applications include tubular, plate and frame, rotary disk, hollow fiber, organic (polyethylene, polyethersulfone, polysulfone, polyolefin, etc.), metallic and inorganic (ceramic) microfiltration, and ultra-filtration membranes (88). The pore size of membranes ranges from 0.01 to 0.4 μm (82). The fluxes range from 0.05 to 10 m/day (m^3/m^2/day), strongly depending on the configuration and membrane material (82).

In drinking water treatment, major application of membrane bioreactors is denitrification. This process has many advantages over traditional biological denitrification (22, 82). In a typical biological denitrification system, post treatment processes are required to separate microorganisms and DOM. By using a membrane bioreactor, the number of post treatment processes can be reduced. In addition, long denitrifying culture retention times and short hydraulic retention times can be maintained in the process (22).

The performance of a pilot-scale MBR for denitrification of groundwater was tested as a function of hydraulic and biological parameters by Buttiglieri et al. (43). Synthetic groundwater was prepared by taking lake water and adding known amounts of ethanol and sodium nitrate to study the NO_3^- removal capacity of the sludge, to search for an optimum C/N ratio, and to measure filtering ability for microorganisms through the membrane. The optimum C/N ratio was 2.2 g C/g N, resulting in an effluent NO_3^- concentration within the limits for drinking water use. The membrane module, Zenon ZW-10, was monitored and performed well except for a short stress episode due to low airflowthat was rapidly corrected afterwards and thus putting the membrane back to its previous stable behavior. Total bacterial count for the treated effluent was lower than influent water, and 100% removal was observed for both total coliforms and *Escherichia coli*.

Hollow fiber membrane bioreactor is one of the major configurations widely used in denitrification from drinking water. Ergas and Rheinheimer (20) have described the working process of this reactor and developed a mathematical model of NO_3^- mass transfer. In their study, NO_3^- contaminated water flows through the lumen of the tubular membranes and NO_3^- diffuses through the membrane pores. A denitrifying population is circulated on the shell side of the MBR, creating a driving force for NO_3^- mass transfer. Nutrients and electron donors are added to the shell side of the membrane bioreactor to support the microorganisms. The membranes provide high NO_3^- permeability, while separating the microbial process from the water being treated (20). The advantages of this system is described by Ergas and Rheinheimer as: (a) the contamination of the product water with sloughed biomass is eliminated by the separation of water from biomass using a microporous membrane; (b) a number of low cost electron donors such as ethanol, methanol, and acetic acid can be used to drive the high rate denitrification process; and (c) reactors can be constructed of low cost tubular and hollow-fiber membranes. It was also reported that pressure drops across these modules are low.

A MBR that combines the advantages of biological conversion of NO_3^- to nitrogen with that of hollow fiber ultrafiltration technology to produce high quality drinking water was studied by Chang et al. (89). The influence of several parameters, both biological and hydraulic, on the overall performance of the process has been investigated. With adequate membrane backwashing frequency, a crossflow velocity was maintained for more than two months with a net permeate flow rate of over 100 L/h/m^2. A high permeate flow rate was not detrimental to the overall denitrification process, since permeate NO_3^- and NO_2^- concentration remained below 20 and 0.1 mg/L, respectively, for a NO_3^- volumetric loading rate of 2.8 kg/m^3/day at a hydraulic retention time of either 60 or 30 min.

Li and Chu (90) investigated the performance of a hollow-fiber membrane reactor in treating raw water supply. A submerged polyethylene hollow-fiber membrane module with a pore size of 0.4 μm and a total surface area of 0.2 m^2 was used for treating a raw water supply slightly polluted by domestic sewage with a TOC level of 3–5 mg/L and ammonia nitrogen (NH_3–N) concentration of 3–4 mg/L. The process was highly effective in eliminating conventional water impurities, as demonstrated by decreases in turbidity, total coliforms, and UV_{254} absorbance. With the MBR treatment, the 3-day trihalomethane

Formation potential (THMFP) was significantly reduced. The initial chlorine demand for disinfection, the biostability of the effluent improved considerably and the AOC also decreased. Such significant effects on the water quality demonstrate the superior advantage of the MBR in producing an effluent water quality, which is comparable to or even better than of local tap water.

Flat-sheet membrane bioreactors were investigated by McCleaf and Schroeder (91), and Mansell and Schroeder (92). A denitrification rate of 4–5.8 g NO_3–N/m^2/day was demonstrated. However, significant methanol and microbial contamination with product water was reported.

The inclined-plate membrane bioreactor (IPMBR) was investigated on a pilot scale for 123 days in municipal wastewater treatment at zero excess sludge discharge (93). Results indicated that at a hydraulic retention time of 6 h, average removals of chemical oxygen demand (COD), ammonia nitrogen, and turbidity were 92.1%, 93% and 99.9%, resulting in daily averages in the effluent at 12.6 mg COD/L, 1.3 mg NH_3–N/L and 0.03 NTU, respectively. The inclined-plates in the presence of intermittent air blowing were much effective to confine concentrated sludge within the anoxic tank. The sludge concentration difference between anoxic and aerobic tanks was found to be from 0.1 to 13.1 g SS/L of mixed liquor.

4.3. *Ion-Exchange Membrane Bioreactor*

In applying biotechnology to drinking water, one major problem is the contamination of treated water by microbial cells, nutrients, and metabolic by-products. IEMB is an integrated process for treating water contaminated with inorganically charged pollutants (94–96). Typically, ion-exchange membrane bioreactor employs Donnan dialysis principles, which uses an ion-exchange membrane as a barrier between the polluted water stream and a biocompartment in which a suitable mixed microbial culture degrades or transforms the pollutants to harmless products (96).

Donnan dialysis is simple and one of the most economic membrane processes (97). During Donnan dialysis there is a stoichiometric exchange of ions of the same charge (so called counter-ions) through an ion-exchange membrane until equilibrium is reached. A new direction for the process is a bioreactor with an ion-exchange membrane (IEMB), where NO_3^- removed from water are subject to biological denitrification (97).

The removal of low molecular mass charged compounds from water streams in an IEMB was studied by Velizarov et al (98). This integrated membrane process combines continuous ion exchange transport (Donnan dialysis) of the pollutant through a nonporous membrane, selective for the transport of mono-anions, and its simultaneous bioconversion to harmless products in a biocompartment. The mechanism of the process, governed by the Donnan equilibrium principles, allows for regulation of the direction and magnitude of flux of the individual anions present by adjusting the ratio of the impermeable co-ions between the biocompartment and polluted water compartment. Experimental results, obtained for NO_3^-

removal from drinking water showed that the IEMB process allows for the most selective removal of the target pollutant simultaneously avoiding microbial and secondary contamination of the treated water stream. Velizarov et al. (99) investigation on the IEMB uses a mono-anion permselective membrane as a barrier between a water stream, containing a target polluting anion, and a biocompartment, containing a suitable driving counter-ion for coupled counter-diffusion of the target pollutant, and a microbial culture capable of its bioreduction to harmless product(s). Resistances-in-series to trace counter-ion approximation model based on the Fick formalism and the Donnan equilibrium principle was proposed for analysis of the transport rate of dilute inorganic anionic pollutants with relevance to drinking water treatment. Transport of co-ions (cations) is negligible due to their electrostatic repulsion (Donnan exclusion) from the positively charged membrane. The model was tested in a system, containing NO_3^- as the target pollutant and chlorides as the driving bulk counter-ions. The effect of the most important process variables on the target pollutant flux were assessed using the model, making possible the comparison between different IEMB process options, which can be quantified and used for process design. Donnan dialysis was also evaluated for the possibility of using various membranes in IEMB by Velizarov et al. in 2004 (96). The anion-exchange membrane, Neosepta ACS, proved to be a good choice for perchlorate and NO_3^-, but was not suitable for arsenate, for which the Ionac MA-3475 allowed more than a three-order magnitude increase in flux. The cation-exchange membrane, Nation-112, showed promising results for the transport of mercuric ions, while the Neosepta CMX was almost impermeable to this micropollutant. Tap water from the Lisbon public distribution network was supplemented with 100 μg/L of perchlorate and 60 mg/L of NO_3^- and was continuously treated in the IEMB incorporating a Neosepta ACS membrane for a period of about 2 months. Highly selective simultaneous removal of both pollutants was observed, in spite of their very different concentrations in the contaminated water. There was no secondary pollution of the treated water by microbial cells and residual organics (96).

The results of the experiments on selected anions removal from water investigated by Wiśniewski et al. (97) indicated the removal of some ions, especially 'troublesome' anions (sulfates and bicarbonates), and a high degree of water desalination by means of electrodialysis. Selemion (AMV, DMV) and Neosepta (AFN, AMX) anion-exchange membranes (AEMs) were used in an ion-exchange study. For each of the four examined membranes, the following high efficiencies of anions removal were obtained: 87–98% removal of NO_3^-, 94–100% removal of sulfates, whereas for bicarbonates the removal efficiency was 77–99%. It was found out that with higher concentrations of NaCl (300 and 500 mM), the equilibrium concentration of the exchanged ions in the feed was lower. However, with the higher ratio of feed and receiver volumes (4:1), the concentration at the point of Donnan equilibrium was set at an elevated level. The best transport properties for the examined anions were shown for the Neosepta AFN membrane. The average flux of NO_3^- and sulfates was up to two times higher than with other membranes and bicarbonate flux was even three times higher (97).

4.4. Biological Activated Carbon Adsorption: Biofilm

Biological activated carbon (BAC) has been attracting great attention and is being applied to various treatment processes to remove pollutants from drinking water. BAC processes were discovered and developed during the 1970s in drinking water treatment plants, and now are in full-scale operation in some 60 plants or more in Europe, USA, and Japan (100, 101).

BAC was first defined by Rice and Robson (102) as a water or wastewater treatment system in which aerobic microbial activity is deliberately promoted in a GAC system. This process provides simultaneous adsorption of nonbiodegradable matter and oxidation of biodegradable contaminants in a single reactor (103). GAC is widely used in drinking water after ozonation to eliminate by-products such as aldehydes and ketones, or chlorination to eliminate chlorinated disinfection by-products such as toxic halogenated organic compounds and unpleasant odor. These pollutants are removed by adsorption, chemical reduction, and mechanical filtration. This new combination process has an advantage over conventional activated carbon treatment for the removal of organic substances. The use of BAC treatment makes the whole facilities compact and the life of carbon longer (104). With the assistance of the bioactivity on BAC, the reduction of TOC and other organic substances is considered to maintain a longer service time before GAC regeneration (105). In addition, BAC is more efficient than simple biological oxidation systems without an adsorption surface. Thermal regeneration of activated carbon is not often required because of the influence of the biofilm. Furthermore, partial biological regeneration of GAC owing to sorption, and later, metabolism of slowly biodegradable carbon can be observed in practice (106, 107). This also leads to an extended life of the carbon column.

Two noncontradictory hypotheses about the mechanism of bioregeneration were purposed. The degradation mechanism of adsorbed substances due to exoenzymes from microorganisms was proposed (107). A biofilm containing bacteria in the macropores and their "exoenzymes" in the micropores was established. The biodegradation in pores takes place by means of exocellular enzymes, which are capable of diving into micropores and interact with adsorbed substrate and promote its hydrolytic decay (107, 108).

In another hypothesis (108, 109), the development of biological processes on the activated carbon surface in combination with adsorption processes, desorption, and diffusion in pores leads to partial regeneration of activated carbon. The reverse of an adsorption concentration gradient can be created by reaching a critical rate of biodegradation in the biofilm and in the liquid phase. The adsorbed substances are desorbed back into the biofilm, and also through it into the liquid phase, and become accessible to the microbial degradation (109). It was reported that biofilm develops on BAC even with very high shear stress from backwashing (110). The negative influence of toxic substances on bacterial biofilm is also protected by activated carbon (110). The biological regeneration (bioregeneration) of activated carbon leads to higher efficiency of BAC than the consecutive adsorption and biological treatment.

Many researches have focused on application of BAC in drinking water treatment. Amsterdam Water Supply (AWS) used biological activated carbon filtration (BACF) for the removal of NOM in general and the removal of organic micro-pollutants in particular. Six years of operation of BACF in the River-Lake Waterworks (31 Mm^3/year) have shown that successive reactivations do not affect the dissolved organic carbon (DOC) removal capacity of the carbon. Three years of operation of BACF in the River-Dune Waterworks (70 Mm^3/year) have shown that carbon retains its pesticide removal capacity. The use of BACF in an integrated membrane system shows important perspectives in minimizing the fouling of reverse osmosis membranes, and in minimizing the organic carbon content in the membrane concentrate (111). BAC pilot plant was operated using eutrophic lake water as raw water (112). The particle separation performance of ozone-biological activated carbon was evaluated based on the analyses of the fate of organic substances in the process. It was found that the ozonation not only converted refractory organic matter into biodegradable matter, but also converted particulate organic carbon (POC) into DOC. Total decrease in adsorbable and nonbiodegradable dissolved organic carbon fraction (ADOC) after ozonation was only 16% of the influent into the biofiltration process followed by ozonation. The ozone-biological activated carbon process before membrane separation could reduce organic loading to the membrane system. The smaller loading on microfiltration resulted in long intervals of back washing and less frequent membrane fouling (112).

The effects of preozonation on BAC performance for the removal of DOC and extension of activated carbon service time were evaluated by Kim et al. (113). It was reported that although the increased BDOC had possibilities to be removed both by adsorption and biodegradation, most of it was removed by biodegradation on BAC, which resulted in longer BAC service time. Biodegradation of the increased BDOC was responsible for the higher removal of DOC after saturation with DOC in the ozone-biological activated carbon process.

Changes of adsorption capacity and pore distribution of BAC on advanced water treatment were investigated by Kameya et al. (114). In BAC, the volume of smaller pores under 2 nm in diameter mainly decreased. Accumulations of minerals such as aluminum and calcium were small, and the pore volume decreases were mainly caused by the accumulation of organic substances. Almost all of the organic substances that accumulated in activated carbon could be extracted by sodium hydroxide solution. The mean density of the organic substances that accumulated in the activated carbon was estimated to be 0.91 g/mL. Since the pore volume decrease of the activated carbon was small compared with the removal amounts by the adsorption effect, a large amount of organic substances that had adsorbed once disappeared and the pore volume of the activated carbon was regenerated. The readers are referred to the literature for recent developments in the area of filtration and biological technologies for drinking water treatment (115–119).

5. US EPA STUDIES ON DBP CONTROL THROUGH BIOFILTRATION

DBP control through biofiltration is defined as the removal of DBP precursor material (PM) by bacteria attached to the filter media. The PM consists of DOM and is utilized by the filter bacteria as a substrate for cell maintenance, growth, and replication. The PM utilized by bacteria is no longer available to react with chlorine to form DBPs. All other things being equal, a water with lower PM concentration will yield lower DBP concentrations, at a given chlorine dose, after a given time period. The biological filtration process is cost effective, since the bacteria are naturally present in the water supply, can colonize existing filter media, do not produce a residual that needs disposal, and require almost no modification of ambient conditions. The only prerequisite for maximizing bacterial substrate utilization in filters is the absence of disinfectant in the filter influent or backwash water. The filter media colonized by bacteria can be sand, anthracite, or GAC. Anthracite and sand are considered inert because neither interacts chemically with PM. Thus, any removal of PM would be solely to biological activity. GAC that has been colonized by bacteria will initially remove DOM through adsorption and biological substrate utilization. After the GAC's adsorptive capacity has been exhausted, PM removal is achieved only through substrate utilization, and the GAC is defined as BAC. All drinking water filters will become biologically active in the absence of applied disinfectant residuals. The process of biological colonization and substrate utilization is enhanced by the ozonation of filter influent water.

In the US, preozonation is practiced to remove color, taste, and odor, to inactivate *Giardia* and *Cryptosporidium*, and to serve as an alternative to chlorine disinfection. Ozonation decreases the average molecular size and weight of the PM, allowing indigenous bacteria to utilize more of it as substrate in a given amount of contact time. Some fraction of microbes will always survive ozonation. As long as no liquid phase ozone residual is present in the filter influent, the surviving microbes will eventually colonize the filter media. A preozonated biological filter will achieve greater PM removals than one with influent that has not been preozonated. PM is measured as TOC, DOC, THMFP, or haloacetic acid (HAA) formation potential (HAAFP). All of the studies discussed in this section measure DBP control using one or some combination of the parameters TOC, DOC, THMFP, and HAAFP. Because of its potential to control DBP precursors and its economic advantages, the US EPA has performed or funded a number of research studies to characterize the impact of biological filtration on the control of DBPs (120).

5.1. Pilot-Scale Study, Shreveport, LA

A pilot-scale study was executed to investigate the combination of ozone and GAC for THM precursor removal (121). Following conventional treatment (coagulation, flocculation, and sedimentation), the process water was split to a GAC filter, and to an ozone contactor followed by a GAC filter (O_3–GAC). The results are summarized in Table 11.1. Analysis of

Table 11.1
TOC and THMFP removal in pilot-scale GAC columns, Shreveport, LA (121)

	% TOC removal		% THMFP removal	
Week	GAC	O_3–GAC	Week	GAC
0–21	82	83	0–21	82
22–52	55	55	22–52	55
53–62	10	19	53–62	10
65–83	24	27	65–83	24

Note: Average filter influent TOC = 30 mg/L.

the data showed microbial activity to be a significant contributor to the removal process for TOC and THMFP over the long term. THMFP was defined as THM formation after chlorination at 20 mg/L, at a pH of 6.5, a temperature of 26 ± 2°C, and a 3-day incubation time.

The study was run for 83 weeks. At the beginning of the study (weeks 0–21), both sets of columns achieved approximately 80% TOC and THMFP removal. These removals were primarily due to physical adsorption of the PM. In the final phase of the project (weeks 65–83), TOC removals averaged 24 and 27% in the GAC and O_3-GAC columns. During the same time period, THMFP removals averaged 24 and 30%, respectively, in the GAC and O_3-GAC columns. By this time, the GAC columns had each passed more than 50,000 bed volumes, and their adsorptive capacity was exhausted. In the absence of adsorption, TOC and THMFP removals during weeks 65–83 were deemed to be due to biological activity.

5.2. Bench-Scale Studies

A series of three bench-scale studies was performed to examine the impacts of ozone dose, water type, and attached versus suspended bacteria on the biological removal of DOC, THMFP, and HAAFP. The first study investigated the effect of ozone dose and biological treatment on PM control (122). THMFP and HAAFP were measured by chlorinating at 12 mg/L, at a pH of 7.5–8.0, and holding for 7 day at 20°C. Bench-scale biological treatment was performed in batch recycle tests (see Fig. 11.2). The water sample was circulated continuously through a bed of bioacclimated sand for 5 day. Oxygen was provided by applying a vacuum to the head space of the sample chamber and drawing in air through a water trap, which removed foreign bacteria. The 5-day contact time was considerably longer than the 5–10 min typical of pilot- or full-scale biological filters. As a result, estimates of biological removal in the batch test should be viewed as "ultimate" or "potential" removals, which will be higher than flow-through removals at equivalent organic matter concentrations, compositions, and ozone pretreatment.

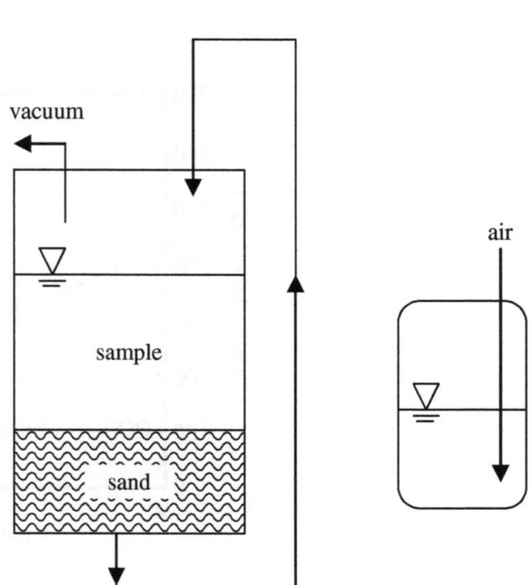

Fig. 11.2. Bench-scale biofiltration apparatus (120).

Water samples were ozonated at transferred doses of 0.5, 0.8, 1.1, 1.8, and 2.5 mg O_3/mg DOC. Biological treatment alone resulted in 13–14% removal of DOC. Ozonation followed by biological treatment yielded 20–30% DOC removals. Biological treatment alone yielded a 28% reduction in THMFP. Ozonation followed by biological treatment yielded a 40–47% reduction in THMFP. Ozonation followed by biotreatment yielded a 75–80% reduction in HAAFP. The reduction in THMFP with respect to ozone dose is shown in Fig.11.3. THMFP formation after biotreatment, but without ozonation, was approximately 220 μg/L. At the lowest transferred ozone dose of 0.8 mg O_3/mg DOC, followed by biotreatment, THMFP dropped to 160 μg/L. THMFP was not reduced appreciably at higher O_3/DOC ratios of 1.1, 1.8, and 2.5. The removal behavior of HAAFP through biotreatment as a function of applied ozone dose was similar to that observed for THMFP. These results imply that THMFP reduction through biotreatment is sensitive to the presence or absence of ozonation, but insensitive to the ozone dose after a minimum ozone dose ($\leq$0.8 mg/mg) is achieved. It is possible that the greatest reduction in PM molecular weight, and corresponding increase in PM bioavailability occurs when ozonation is introduced, and that further increases in ozone dose do not yield proportional decreases in PM molecular weight.

The second study examined DOC removal through ozonation and biofiltration as a function of water type (123). The first water was raw surface water from the Ohio River (ORW). The second was artificial water, produced using a solution of humic substances isolated and concentrated from ground water. The humic substances were mixed with dechlorinated tap water to the desired DOC concentration. Dechlorinated tap water was used to provide the noncarbonaceous nutrients and mineral matrix required for the growth of microorganisms. Humic compounds are one of the major categories of organic substances that make up DOC and which act as PM. The bench-scale biological filter was shown in Fig. 11.2. DOC removals reported for the artificial water are significantly higher than those observed for any of the previously described studies that used natural waters. It is possible

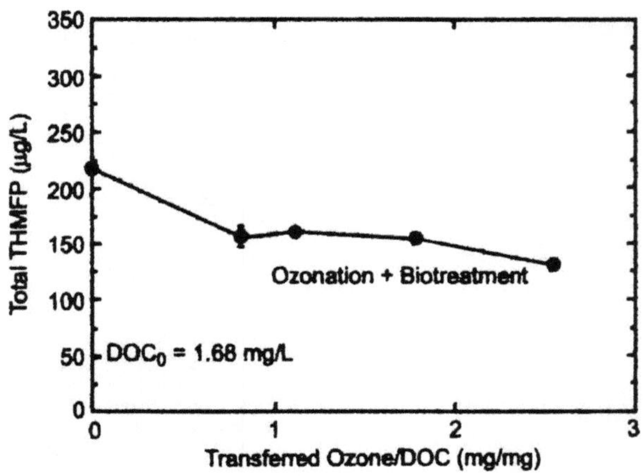

Fig. 11.3. Impact of ozone dose on THMFP removal (120).

that water with an organic fraction consisting only of humic substances would have a higher fraction of biodegradable components relative to natural water.

The third bench-scale study assessed the impact on PM removal of attached versus suspended bacteria (124). Three bench-scale bioreactors were filled with equal volumes of preozonated (0.7 mg O_3/mg DOC) ORW. The first reactor contained bio-acclimated sand. The second reactor contained only water and no sand. Reactor 3 was the same as reactor 2, except that the water contained mercuric chloride to suppress any biological activity. After 5 day of operation, DOC reductions in reactors 1, 2, and 3 were 23%, 10%, and 0%, respectively. These results imply that bacteria suspended in the water column can account for a significant fraction of PM reduction, but that bacteria attached to some type of fixed surface are necessary to achieve the full potential of the process.

5.3. *Pilot-Scale Study, Cincinnati, OH*

A year-long pilot-scale study (125) was carried out to assess: the impacts of filter disinfection, filter media and filter biomass on biological PM control (126), biological PM control during a filter cycle as a function of backwash disinfection (127), and the performance of biofilters with respect to depth (120). The plant schematic is shown in Fig. 11.4. Raw water was ozonated and then subjected to conventional treatment (alum coagulation, flocculation, and settling). After sedimentation, the water was distributed to eight filters, the design, operation, and performance of which are summarized in Table 11.2.

Of special interest during the study were the impacts of prechlorination and media type on PM removal in inert media filters. The filters used in the comparison were Filters 1, 2, 3, and 5. Filters 1–3 contained 20 in. of anthracite over 10 in. of sand, while Filter 5 contained 30 in. of sand. Filter 1 received chlorine in the influent and backwash water, Filter 2 received

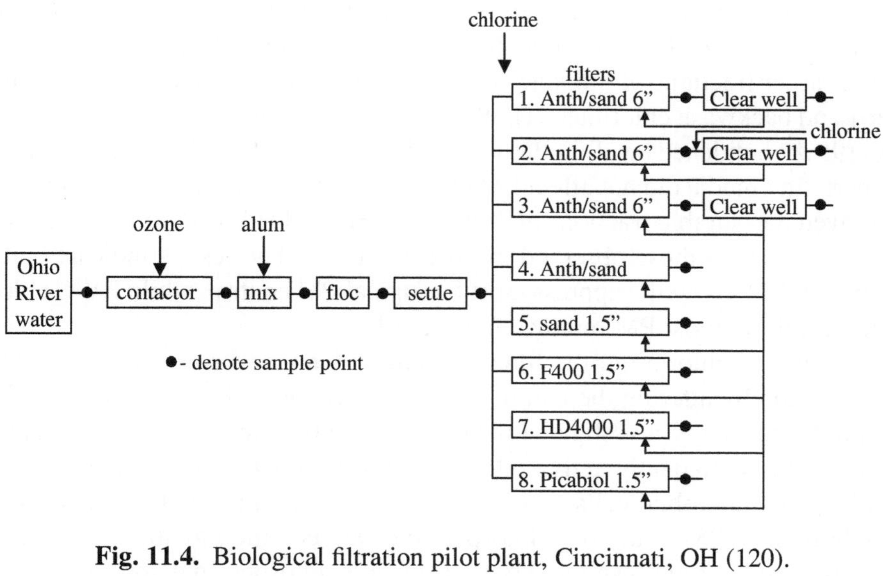

Fig. 11.4. Biological filtration pilot plant, Cincinnati, OH (120).

Table 11.2
Filter configuration, operation and performance, 12-month pilot-scale study (120)

Filter	Media	Chlorination Mean	Biomass (nmol lipid PO_4/g dry media)		TOC removal (%)[a]		THMFP removal (%)		HAAFP removal (%)	
			Mean	Std. Dev	Mean	Std. Dev	Mean	Std. Dev	Mean	Std. Dev
1	20" Anth./10" Sand[b]	Pre + BW	2.0	0.50	8.0	5.0	1.0	11	9	8
2	20" Anth./10" Sand	BW	6.0	0.60	16	9.0	13	6.0	28	7
3	20" Anth./10" Sand[c]	No	55	1.7	20	6.0	21	8.0	37	4
5	30" Sand	No	91	1.3	20	9.0	23	7.0		
6	26" GAC[d]/4" Sand	No	310	9.0	29	8.0	40	5.0		
7	26" GAC[e]/4" Sand	No	470	9.0	27	8.0	34	5.0		
8	26" GAC[f]/4" Sand	No	380	11	21	7.0	27	3.0		

[a]Filter influent TOC = 1.1–2.2 mg/L.
[b]Based on 12 months of data.
[c]Based on the last 10 months of data.
[d]Filtrasorb 400.
[e]Hydrodarco 4000.
[f]Picabiol.

chlorine only in the backwash water, and Filters 3 and 5 received no chlorine. Over the 1-year study period, all four filters were examined for TOC, THMFP, and HAAFP removals, which are summarized in Table 11.2.

THMFP and HAAFP were determined by chlorinating at 12–15 mg/L and incubating the samples for 7 day at 25°C. The nonchlorinated (NC) filters (Filters 3 and 5) removed equivalent or larger fractions of influent TOC, THMFP, and HAAFP than did prechlorinated (PC) Filter 1 and backwash chlorinated (BWC) Filter 2. There were no statistically significant differences (95% confidence level) in PM removals between Filters 3 and 5, implying that the choice of inert filter media did not affect PM removals. PM in Filter 1 were not significant, but Filter 2 removed measurable fractions of TOC, THMFP, and HAAFP. In fact, TOC removals in Filter 2 were close to those observed in Filters 3 and 5. The results indicate that PC and BWC combined (Filter 1) will suppress most biological PM control, while BWC chlorination alone (Filter 2) allows most PM removal to proceed.

The application of chlorine in the filter influent or filter backwash can affect biological substrate utilization by altering the nature and concentration of bacterial colonization. As a result, the pilot-scale study investigated the correlations between biomass development and PM removal in the biologically active pilot-scale filters (126). Biomass was defined as the total assemblage of microbial cells that have colonized the filter media and was quantified using phospholipids (128) as a proxy. Phospholipids are common to all viable bacterial cell membranes, but are broken down quickly in dead cells. Phospholipid concentrations during the pilot-scale study are summarized in Table 11.2. The biomass concentrations are the averages of triplicate samples collected from the tops of filters after 3 months of plant operation. The three GAC filters represented three significantly different types of commercially available activated carbon. None of the GAC filters were chlorinated. After this point in time, biomass levels at the tops of the filters did not change significantly throughout the remainder of the year-long pilot-plant run. The TOC and THMFP removal estimates are averages of samples collected between 155 and 330 day of plant operation. Based on analysis of breakthrough curves, the GAC filters were considered exhausted by this time, and as a result, any observed PM removal would have been due to biological activity.

The application of disinfectant in the filter influent or filter backwash water significantly suppressed biomass development relative to NC inert media filters. The PC and BWC dual media filters had biomass levels of 2.0 and 6.0 nmol PO_4/g dry media, respectively. In contrast, the NC sand and dual media filters developed 55–91 nmol PO_4/g dry media. All of the GAC filters accumulated more biomass than did the inert media filters. Biomass levels in the three GAC filters ranged from 310 to 470 nmol PO_4/g dry media. Increased biomass development in GAC filters was attributed to its porous structure and to the presence of adsorbed organic material. The large number of pore spaces protected biomass from fluid shear forces exerted during filter operation and backwashing. The adsorbed organic materials potentially provide extra utilizable substrate that is not available to bacteria on inert media. THMFP removals in the PC filter were not significantly different from zero. The THMFP fraction removed in the BWC filter was about an order of magnitude higher than the PC filter, despite similar biomass concentrations. THMFP removals in the NC dual media and sand filters were essentially identical, at 21–23%. THMFP removals in the three GAC filters were the highest of all the filters, ranging from 27 to 40%. With the exception of the PC filter, TOC removals did not appear to be correlated to biomass development. There appeared to be higher correlations between biomass development and THMFP removals, but differences in THMFP removals were not proportionate to differences in biomass concentrations.

In a final portion of the 1-year pilot-scale study, BWC and NC dual media filters were sampled for the control of PM over the course of several filter cycles (127). A filter cycle was the interval between startup following backwash to shutdown prior to the next backwash event. Filters were backwashed at 60 in. of headloss or every 48 h, whichever came first. Both filters received NC influent and were backwashed for 10 min at 50% bed expansion. The free chlorine residual in the backwash water for the BWC filter was about 1.0 mg/L. The filter effluents were sampled for TOC, THMFP, and HAAFP immediately prior to backwashing, as well as at selected times (1, 4, 12, 24, 48, and 72 h) during the next filter cycle. Biomass concentrations in the filter media were also evaluated. Statistically significant declines in biomass levels of approximately 25% were observed after backwashing in the BWC filter. The BWC filter biomass concentrations then increased steadily to prebackwash levels by the end of the filter cycle. No significant changes in biomass concentrations were observed in the NC filter with respect to backwashing. Backwashing with NC or BWC water had no detrimental effects on the biological control of TOC, THMFP, or HAAFP at any time during the three filter cycles in which backwash effects were examined. Figure 11.5 illustrates the impact of backwashing on HAAFP control during a filter cycle. The effluent HAAFP

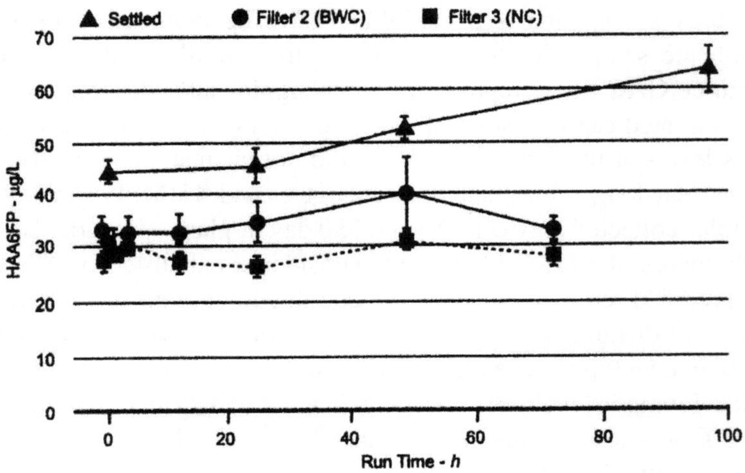

Fig. 11.5. AAFP control during a filter cycle, pilot-scale, Cincinnati, OH (120).

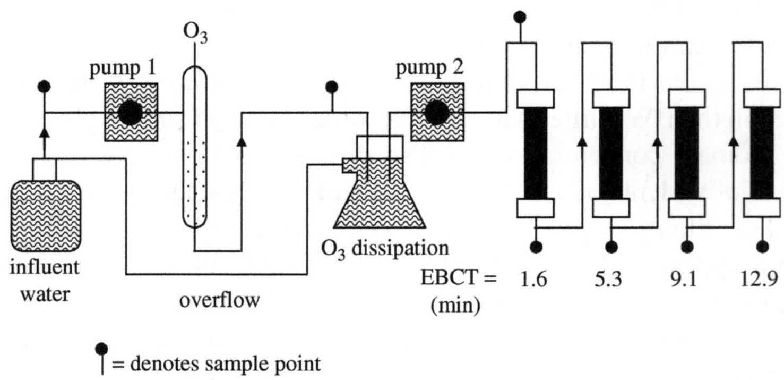

Fig. 11.6. Impact of contact time on biofiltration performance, segmented filter (120).

concentrations remained constant over the filter cycle, despite the increase in filter influent HAAFP levels. These results imply that HAAFP removal, as a fraction of the influent concentration, actually improved over the course of the filter cycle.

At the conclusion of the 1-year pilot-scale study, Filter 5 (30 in. sand, no PC or BWC) was cored along its full depth to examine the impact of filter depth on biological removal of DOC and THMFP. The core of biologically acclimated media was divided along its length into four sections, each representing approximately equal increments of contact time. The segmented filters were then run in series, using ORW that had been treated with potassium permanganate addition, followed by conventional treatment, and biological filtration. The pretreated water was then ozonated at the bench-scale prior to passage through the segmented filters. The experimental setup is shown in Fig. 11.6. The segmented filter was operated for 3 day, and sampling for PM removal was performed on each day. Total DOC removals across the segmented filters averaged 13%, with approximately 50% of the removal observed in the first 1.6 min of EBCT. THMFP removals across the segmented filters also averaged 13%. All of the observed THMFP removal occurred by 5.3 min of contact time.

After 3 days of sampling, the segmented filters were switched to parallel operation. This arrangement allowed water to directly enter each column segment without contact with previous filter sections. This setup yielded information about the capacity of the biomass in each section to remove compounds at concentrations not previously encountered by that filter section. The new arrangement was operated for 2 day and is detailed in Fig. 11.7 DOC removals in segments 1 and 2 were each less than 10%. No DOC removals were observed in segments 3 and 4, which comprised the lower half of the original filter. THMFP removals in segments 1 and 2 were each about 12% of the influent values. No THMFP removals were observed in columns 3 and 4. The results from the series phase of the study confirmed that the bulk of PM removal tends to occur in the top portion of a biological filter. This implies that the chemical composition of the PM reaching the lower portion of the biological filter may be fundamentally changed and that biomass in the lower level of the filter develops to utilize this remaining fraction. The results from the parallel phase of the study indicated that the biomass in the lower half of the filter might need a certain amount of acclimation time before it can effectively utilize a new type of substrate.

5.4. *Five-Month Pilot-Scale Study*

The year-long US EPA pilot-scale study did not evaluate the impact of preozonation or disinfectant type in an integrated pilot-scale environment. To fill in some of these gaps, a 5-month long pilot-scale biofiltration study was initiated at US EPA (120). This pilot plant was supplied with water from a local lake, the raw DOC of 5.8 mg/L of which was significantly higher than that of the ORW (1–2 mg/L). Filter design and performance are summarized in Table 11.3. All plant influent water received conventional pretreatment, consisting of alum coagulation, followed by flocculation, and sedimentation. Following sedimentation, some of

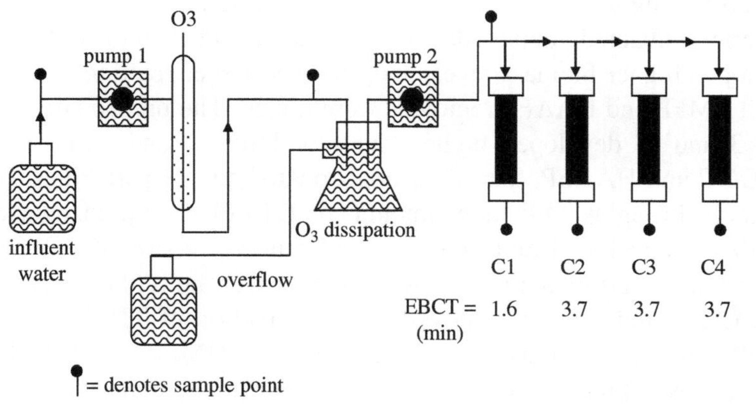

Fig. 11.7. Impact of contact time on biofiltration performance, parallel operation (120).

Table 11.3
Pilot-scale biological filters, a Cincinnati, OH (120)

Filter[c]	Preozonated	Backwash Disinfectant	Filter media	Biomass (nmol lipid PO_4/g dry media)	Percentage removals DOC[a]	THMFP	HAAFP
1	Yes	Chlorine	20" Anthracite/10" Sand	28	12	8	26
2	Yes	Chloramine	20" Anthracite/10" Sand	120	19	17	33
3	Yes	None	20" Anthracite/10" Sand	130	23	23	39
4	Yes	None	20" GAC[b]/10" Sand	288	42	53	62
5	Yes	None	30" Sand	96	23	21	39
7	No	None	20" Anthracite/10" Sand	66	8	20	27
8	No	Chlorine	20" Anthracite/10" Sand	16	0	24	23

[a]Average filter influent DOC = 2.8 mg/L.
[b]Filtrasorb 400.
[c]Loading rate for all filters = 2 gal/min/ft^2 (5 m/h).

the water was ozonated prior to filtration (Filters 1–5), while the remainder went directly to filtration (Filters 7 and 8). The average transferred ozone dose was 5.6 mg/L. None of the filters were PC. Filters 1 and 8 were backwashed with chlorinated water (1.6 mg/L), Filter 2 was backwashed with chloraminated water (2.1 mg/L), and the remaining filters received no disinfectant in the backwash. Once the plant reached steady-state operation, biomass development, and removals of DOC, THMFP, and HAAFP across the filters were measured.

Formation potentials were measured under uniform formation conditions (UFC). These consisted of chlorinating to yield 1 mg/L free chlorine residual after 24 h of holding time at pH 8. Preozonation impacted biomass development and DBPFP removal. Preozonated filters typically developed higher biomass levels and removed larger fractions of DOC. However, the impact on THMFP and HAAFP removals were mixed. The preozonated NC inert media filters (Filters 3 and 5) developed higher levels of biomass and removed more PM, as measured by DOC and HAAFP, than their nonozonated counterpart Filter 7. However, all three filters removed roughly equivalent amounts of THMFP. The preozonated, chlorinated, inert media Filter 1 developed more biomass and removed more DOC than did the non-ozonated, chlorinated, inert media filter (Filter 8). However, both filters removed equivalent fractions of HAAFP, and Filter 8 removed a larger fraction of THMFP. The presence or absence of BWC impacted biomass development, DOC, THMFP, and HAAFP removals in preozonated filters. NC, preozonated Filters 3 and 5 developed more biomass and removed more DOC, THMFP, and HAAFP than did BWC, preozonated Filter 1. BWC had less of an impact on nonozonated filters. Nonozonated, NC Filter 7 developed more biomass and removed more DOC than did nonozonated, chlorinated Filter 8. However, both filters removed equivalent amounts of THMFP and HAAFP.

The type of backwash disinfectant had an impact on biofiltration performance. Preozonated, chloraminated Filter 2 developed more biomass and removed more DOC, THMFP, and HAAFP than did preozonated, chlorinated Filter 1. Filter 2 developed as much biomass and removed as much DOC and HAAFP as preozonated, NC Filter 3. As in the previous pilot-scale study, the use of anthracite/sand versus sand only as a filter material had no impact on biological filtration performance. Preozonated, NC Filters 3 and 5 developed equivalent amounts of biomass and removed almost identical fractions of DOC, THMFP, and HAAFP. However, the use of exhausted GAC significantly improved biological filtration performance. Preozonated, NC Filter 4 (GAC) developed more biomass and removed roughly twice as much DOC, THMFP, and HAAFP than did Filters 3 and 5. All of the preozonated filters in the current study removed larger DOC fractions than did equivalent filters during the previous study. This was most likely due to the higher filter influent DOCs.

5.4.1. *Impact of Temperature*

Three biological filters were acclimated at 5, 20, and 35°C during a 73-day study in order to investigate biomass development and PM control as a function of temperature (120). For the first 44 day of the study, the filters were acclimated with ozonated ORW. During days 45–73, the filters were acclimated with an ozonated, isolated solution of NOM. The NOM was isolated by nanofiltration from a Florida ground water low in particulates and high in DOC. Prior to ozonation, the isolated NOM solution was diluted to a DOC concentration of 4.2 mg/L with dechlorinated tap water. The temperature in the 5 and 35°C filters was regulated using water-jacketed columns. The cooling or heating fluid was recirculated from the columns through constant temperature water baths. The temperature in the 20°C filter was maintained by the ambient internal building temperature. The tops of the filter media were sampled for biomass development on a regular basis. The filters were defined to have reached a biological steady-state when consecutive biomass samples did not vary by more than 20%. During the last 2 weeks of the study, the filters were sampled intensively for DOC and HAAFP control. Formation potentials were measured under uniform formation conditions. At the conclusion of the study, the filters were sacrificed and sampled for biomass development with respect to depth. Biomass profiles are summarized in Fig. 11.8, and PM control is summarized in Table 11.4. All filters were operated at 3.6 m/h loading rates. Top of filter biomass development was more extensive in the 20 and 35°C filters than in the 5°C unit. However, middle of filter biomass levels were higher at 5°C than at the two higher temperatures. DOC removal at 5°C was lower than at the two higher temperatures. There were no significant differences in DOC removals between 20 and 35°C. HAAFP removals at 5°C were about half the level reported for 20 and 35°C. HAAFP removals did not vary significantly between 20 and 35°C.

5.4.2. *Full-Scale Evaluation of Temperature Effects*

A study was performed to assess the impact of temperature on biomass growth and PM control in eight full-scale drinking water filters (120). Filters in these plants were sampled in

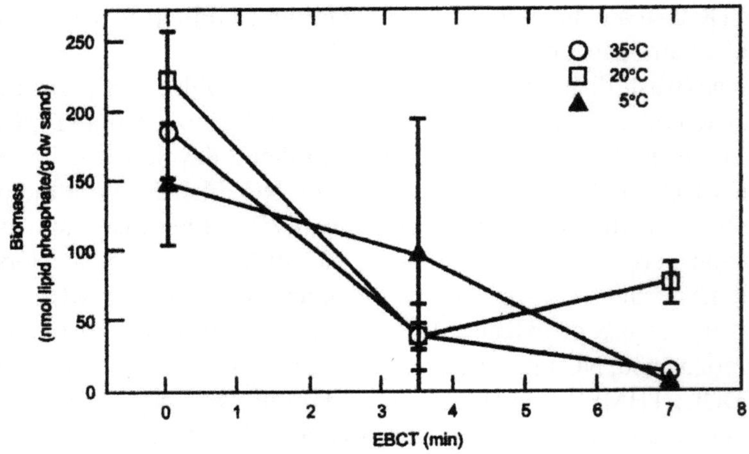

Fig. 11.8. Biomass development as a function of filter temperature, Cincinnati, OH (120).

Table 11.4
Precursor control as a function of temperature, Cincinnati, OH (120)

% Removal (mean, standard deviation)			
Parameter	5°C	20°C	35°C
DOC	15, 3	24, 3	24, 2
UFC HAA	14, 3	31, 3	34, 2

the summer and winter to capture the extremes in water temperature. All of the filters in these plants received preozonated influent. Filters in three of the plants contained exhausted GAC. Details on filter construction and performance are provided in Tables 11.5 and 11.6, respectively. The average winter and summer temperatures were 6.1 and 24°C, respectively. Top of filter biomass levels remained the same or decreased in all of the filters from winter to summer. On average, DOC removals decreased, while UFC THMFP and HAAFP removals did not increase significantly from winter to summer. The authors attributed these phenomena to increases in filter loading rates. On average, filter loading rates were 61% higher in the summer than in the winter. The corresponding decreases in EBCT may have mitigated the impacts of temperature increases on the rates of biological activity. As in the bench-scale temperature study, however, the data indicated that significant biological utilization of PM occurs even during the winter time.

5.5. Modeling Biological PM Control

A method was developed to predict DOC removal with respect to EBCT in biological filters (129). The model was based on a steady-state mass balance around a plug-flow reactor (PFR):

Table 11.5
Filter construction, full-scale temperature study (120)

Plant	GAC (in.)	Anthracite (in.)	Sand (in.)	Garnet (in.)	Chlorinated backwash	Filter loading rate	
						Winter	Summer
Celina, OH	–	–	18	–	Yes	4.6	3.9
Berea, OH	–	–	24	–	Yes	3.9	
Lake Bluff, IL	48	–	12	–	Yes	4.6	7.3
Somerset, NJ	43	–	9	4.5	Yes	3.7	4.9
Millwood, NY	–	24	12	–	No	2.9	4.4
Andover, MA	43	–	6	–	Yes	5.4	4.6
Santa Clarita, CA	–	72	–	–	No	9.8	15
Sylmar, CA	–	72	–	–	No	11	24

Table 11.6
Full-scale filter performance and biomass development as a function of temperature (120)

Plant	Winter					Summer				
	Temp (C)	Biomass (nmol PO_4)	% Removal		Temp (C)	Biomass (nmol PO_4)	% Removal			
			DOC	THMFP			HAAFP	DOC	THMFP	HAAFP
Celina	3	115	4	9	14	28	70	12	–	7
Berea	3	7	3	10	9					
Lake Bluff	2	127	48	8	30	21	78	30	32	39
Somerset	7	127	20	28	47	28	71	11	37	55
Millwood	5	31	16	–	3	22	31	2	26	23
Andover	7	106	25	49	81	26	47	28	40	67
S. Clarita	13	47	15	8	41	17	40	10	5	19
Sylmar	9	50	22	24	46	24	38	16	12	41

$$v_o \frac{\partial C_i}{\partial z} - \varepsilon E_L \frac{\partial C_i}{\partial z_2} + \frac{3}{r}(1 - \varepsilon) k_L (C_i - C_{ir}) = 0. \qquad (2)$$

The symbol z is depth within the filter (m), v_o is the filter loading rate (m/h), C_i is the bulk DOC concentration (mg/L), C_{ir} is the DOC concentration at the surface of the filter media (mg/L) where the biomass is attached, r is the radius of the filter media (m), c is the filter porosity (dimensionless), E_L is the substrate diffusivity in the bulk solution (m^2/s), and k_L is the overall mass transfer coefficient (m/s) that describes movement of the DOC from the bulk solution to the biomass colonies on the filter media surface. The first term in the above equation describes convection of DOC in the bulk solution. The second term describes diffusion of the DOC in the bulk solution. The third term describes transfer of DOC from the bulk solution to the biomass.

The hydrodynamic conditions in the filter were such that bulk diffusion could be neglected. As a result, the second term in the equation was dropped. DOC was divided into biodegradable (BDOC) and nonbiodegradable (NBDOC) fractions. The BDOC fraction, in turn, was divided into fast (FBDOC) and slowly (SBDOC) biodegradable fractions. The BDOC fraction was defined as the fraction of BDOC that could be biologically utilized in 5 day of contact with acclimated media. The FBDOC fraction was defined as that fraction of substrate that could be utilized in 3 min of EBCT with acclimated media. The SBDOC fraction was the difference between BDOC and FBDOC fractions. FBDOC and SBDOC utilization rates were described mathematically using Monod and first-order kinetics:

$$\frac{dC_1}{dt} = \frac{V_{max} X C_1}{K_{s1} + C_1}. \tag{3}$$

$$\frac{dC_2}{dt} = K_2 X C_2. \tag{4}$$

V_{max} (mg/mg cells·s) and K_{s1} (mg/L) are Monod kinetic coefficients, and K_2 (L/mg cells·s) is the first order kinetic coefficient. X is the concentration of biomass on the surface of the filter media (mg cells/L). The variation of biomass with respect to filter depth was expressed mathematically as:

$$X = A(1 + Be^{Fz}). \tag{5}$$

A, B, and F are regression coefficients, based on fitting the equation to an observed biomass profile. The two kinetic equations and the biomass profile equation were then combined with the mass balance to yield a first-order, linear, homogenous differential equation that could be solved numerically to predict bulk DOC concentration at any depth in the filter. The DOC utilization kinetic coefficients were assumed to be constant with respect to filter depth, and they were determined using a method originally developed by Rittman and McCarty (130, 131). Acclimated filter media were placed in small biological filters and exposed to influent water with varying concentrations of DOC. Utilization rates of DOC were observed by measuring DOC removals at varying influent DOC concentrations. The observed utilization rates in turn were used to estimate the kinetic parameters V, K_1, and K_2. The model was used to predict DOC utilization with respect to filter depth in sand-only biological filters that had been acclimated using a solution of ozonated ground water humic substances. A segmented filter arrangement was used, which permitted the measurement of biomass and DOC concentrations at various filter depths.

The observed DOC concentrations and modeling results are shown in Fig. 11.9. The individual points are the observed concentrations, and the solid line is the model prediction. The model was able to closely predict the biological utilization of DOC with respect to depth in the filter.

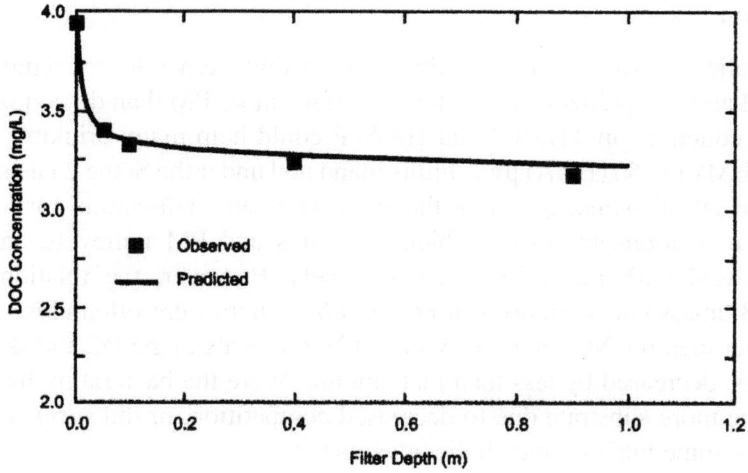

Fig. 11.9. Observed and predicted DOC utilization, 5 m/h (2.0 gal/min·ft^2) loading rate (120).

Table 11.7
Summary of model predictions, pilot-scale, Cincinnati, OH, 1996 (120)

Filter	Observed effluent DOC (mg/L)	95% confidence Interval (mg/L)	Predicted effluent DOC (mg/L)	Absolute deviation (mg/L)	Relative deviation (%)
1	2.3	0.7	2.1	−0.2	−8.7
2	2.1	0.4	1.9	−0.2	−9.5
3	2.0	0.7	1.8	−0.2	−10
5	2.1	0.7	1.8	−0.3	−14
7	2.5	0.9	2.7	+0.1	+6.0
8	2.7	0.7	2.7	0.0	0.0

The applicability of the model would be enhanced if it could accurately predict substrate utilization at the pilot-scale and in the presence or absence of ozonation and backwash disinfection. In order to address these issues, the model was evaluated during the 5-month pilot-scale study (132). Kinetic parameters were estimated for media acclimated with ozonated and nonozonated water. All of the filters listed in Table 11.3 were sacrificed at the end of the study in order to determine the biomass development with respect to filter depth. The model predictions and deviations from observed values are summarized in Table 11.7. The model predictions of effluent DOC were compared with the observed average effluent DOCs during the last 4 weeks of the pilot plant study, when DOC removals and biomass levels were at an approximate steady state. With one exception, model predictions fell within 10% of observed effluent DOCs.

5.6. Discussion

The data collected so far indicate that biologically active filters remove significant amounts of PM and that preozonated biofilters remove more PM than do nonozonated filters. The resulting reductions in THMFP and HAAFP could help many drinking water utilities meet the 80 (THM) and 60 (HAA) μg/L limits mandated under the Stage 2 Disinfection/DBPs Rule (133). The data also raise questions that indicate potential directions for future research. One area is the relationship between biomass levels and PM removals. In general, PM removals increased with increasing biomass levels. However, the relationship was not proportional. Biomass concentrations in PC or BWC filters were often close to an order of magnitude lower than for NC filters. However, PM removals in the PC and BWC filters, for some categories, decreased by less than that amount. Were the bacteria in disinfected filters able to consume more substrate due to decreased competition, or did surviving disinfectant exposure confer some intrinsic metabolic advantage?

Another direction for research would be the performance of biofilters under transient conditions. Only one paper (127) investigated PM control over the course of a single filter cycle. It would be worthwhile to examine biofiltration performance after extended periods of shutdown, rapid changes in loading rate, rapid changes in influent temperature or pH, and as a function of different filter backwashing methods (134–135).

6. DESIGN EXAMPLES

The basic principles for system design of pressure-driven membrane filtration include flux, recovery and flow balance (2).

Flux in filtration process is defined as the filtrate flow per unit of membrane filtration area, which is the membrane filtration system throughput. Calculation of flux (2) is shown in Eq. (6) as shown below:

$$J = \frac{Q_{\mathrm{p}}}{A_{\mathrm{m}}}, \tag{6}$$

where J is the water flux (m^3/m^2-h), Q_p is the filtrate flow produced by the membrane unit (m^3/h), A_m is the membrane surface area (m^2).

The recovery of a membrane unit is defined as the amount of feed flow that is converted to filtrate flow, expressed as a decimal percent (2), as shown in Eq. (7):

$$R = \frac{Q_{\mathrm{p}}}{Q_{\mathrm{f}}}, \tag{7}$$

where R is the recovery of the membrane unit (decimal percent), Q_p is the filtrate flow produced by the membrane unit (m^3/h), Q_f is the feed flow to the membrane unit (m^3/h).

A general flow balance that can be applied to all membrane filtration systems (2) is shown in Eq. (8):

$$Q_{\mathrm{f}} = Q_{\mathrm{c}} + Q_{\mathrm{p}}, \tag{8}$$

where Q_f is the feed flow to the membrane unit (m^3/h), Q_c is the concentrate flow from the membrane unit (m^3/h), Q_p is the filtrate flow produced by the membrane unit (m^3/h).

For the system operating without a concentration waste or "bleed" stream mode or cross-flow systems in which 100% of the concentrate is recirculated, the concentrate (i.e., bleed or reject) flow, Q_c, is zero. For the purpose of sizing a membrane filtration system, it may be desirable to account for the additional filtered water used for both backwashing and chemical cleaning in the determination of the filtrate flow, Q_p. Similarly, an estimate of the total required feed flow Q_f to the system should incorporate any raw water that may be used in these routine maintenance processes.

For the most widely used configuration as fiber membrane, the feed solution passes across the fibers or flows parallel to the fibers and the product water is collected at one or both ends of the bundle. The water flux can be determined from Eq. (9):

$$J = \frac{V_{\text{avg}} D}{4L}, \tag{9}$$

where J is the water flux (m^3/m^2-h), V_{avg} is the average water velocity (m/h), D is the tube diameter (m), L is the fiber length (m).

If the water velocity was constant at the exit value, the pressure drop could be calculated using Eq. (10):

$$\Delta p = \frac{1}{2}\frac{\mu L V_{\text{avg}}}{D^2}, \tag{10}$$

where Δp is the pressure drop (Pa), μ is the viscosity (Pa·s), V_{avg} is the average water velocity (m/s), L is the fiber length (m), D is the tube diameter (m).

Example 1

A hollow-fiber permeator with inside diameter 0.02 cm with a water flux of 8×10^{-6} m/s. The fiber length is 2 m.

(a) Determine the exit velocity of water
(b) Calculate the pressure drop

Solution

(a) Determine the exit velocity of water

Using Eq. (9) to determine the exit velocity of water

$$J = \frac{V_{\text{avg}} D}{4L},$$

$$V_{\text{avg}} = \frac{4JL}{D},$$

$$V_{avg} = \frac{4 \times (8 \times 10^{-6}\text{m/s}) \times (2\text{m})}{(0.02 \times 10^{-2}\text{m})},$$

$$V_{\text{avg}} = 0.32 \text{ m/s}.$$

The exit velocity of water is 0.32 m/s.

(b) Calculate the pressure drop
Using Eq. (10) to determine the pressure drop

$$\Delta p = \frac{1}{2}\frac{\mu L V_{\text{avg}}}{D^2}.$$

Assume: $\mu = 1 \times 10^{-3}$ Pa·s

$$\Delta p = \frac{1}{2}\frac{(1 \times 10^{-3}\text{Pa} \cdot \text{s}) \times (2\,\text{m}) \times (0.32\,\text{m/s})}{(0.02 \times 10^{-2}\text{m})^2},$$

$$\Delta p = 8 \times 10^3\text{Pa} = 0.08 \times 10^5\text{Pa} = 0.08 \text{ atm}.$$

The pressure drop is 8,000 Pa or 0.08 atm.

Example 2
Find the feed flow to the membrane unit if the water flux is designed as 5×10^{-6} m^3/m^2·h. The recovery of the membrane unit is set at 80% and the surface area of the membrane is 30 m^2.

Solution
1. Find the filtrate flow using Eq. (6)

$$J = \frac{Q_p}{A_m},$$

$$Q_p = \frac{A_m}{J},$$

$$Q_p = \frac{30\,\text{m}^2}{(5 \times 10^{-6}\text{m}^3/\text{m}^2 \cdot \text{s}) \times (3,600\,\text{s/h})},$$

$$Q_p = 1,666.67 \text{ m}^3/\text{s}.$$

2. Find the feed flow to the membrane using Eq. (7)

$$R = \frac{Q_p}{Q_f},$$

$$Q_f = \frac{Q_p}{R},$$

$$Q_f = \frac{1,666.67 \text{ m}^3/\text{s}}{0.08},$$

$$Q_f = 2,0833.38 \text{ m}^3/\text{s}.$$

The feed flow to the membrane unit is 2,0833.38 m^3/s.

7. NOMENCLATURE

A = Regression coefficient.
A_m = Membrane surface area (m^2).
B = Regression coefficient.
c = Filter porosity (dimensionless).
C_i = Bulk DOC concentration (mg/L).
C_{ir} = DOC concentration at the surface of the filter media (mg/L).
D = The tube diameter (m).
E_L = Substrate diffusivity in the bulk solution (m^2/s).
F = Regression coefficient.
J = Water flux ($m^3/m^2 \cdot h$).
k_L = Overall mass transfer coefficient (m/s).
K_2 = First order kinetic coefficient (L/mg cells·s).
K_{s1} = Monod kinetic coefficient (mg/L).
L = Fiber length (m).
Δp = Pressure drop (Pa).
Q_c = Concentrate flow from the membrane unit (m^3/h).
Q_p = Filtrate flow produced by the membrane unit (m^3/h).
Q_f = Feed flow to the membrane unit (m^3/h).
r = Radius of the filter media (m).
R = Recovery of the membrane unit (decimal percent).
V_{max} = Monod kinetic coefficient (mg/mg cells·s).
V_{avg} = Average water velocity (m/s).
X = Concentration of biomass on the surface of the filter media (mg cells/L).
z = Depth within the filter (m).
μ = Viscosity, (Pa·s).
v_o = Filter loading rate (m/h).

REFERENCES

1. National Drinking Water Clearinghouse (NDWC) (1999) Membrane Filtration, Tech brief #DWFSOM43, a National Drinking Water Clearinghouse Fact Sheet, Morgantown, WV, March
2. US EPA (2005) Membrane filtration guidance manual, EPA 815-R-06-009, US Environmental Protection Agency, Washington, DC
3. Jacangelo JG (1991) The Development of Membrane Technology, International report, Congress of the International Water Supply Association, Copenhagen
4. Jacangelo JG, Trussell RR, Watson M (1997) Role of membrane technology in drinking water treatment in the United States. Desalination 113:119
5. Oppenlander T (2003) Photochemical purification of water and air. Wiley-VCH, Germany
6. Richardson SD (1998) Drinking water disinfection by-products. In: Meyers RA (ed) The encyclopedia of environmental analysis & remediation, Vol 3. Wiley, NY
7. Servais P, Laurent P, Randon G (1993) Impact of biodegradable dissolved organic carbon (BDOC) on bacterial dynamics in the distribution system. In: Proceedings of AWWA Water Quality Technology Conference, Miami, FL

8. Amy GL, Chadik PA, Chowdhury ZK (1987) Developing models for predicting trihalomethane formation potential and kinetics. J Am Water Works Assoc 79:89
9. Killops SD (1986) Volatile ozonization products of aqueous humic material. Water Res 20:153
10. Glaze WH, Koga M, Cancilla D, Wang K, McGuire MJ, Liang S, Davis MK, Tate CH, Aieta EM (1989) Evaluation of ozonation by-products from two California surface waters. J Am Water Works Assoc 81:66
11. Coleman WE, Munch JW, Ringhand HP, Kaylor WH, Mitchell DE (1992) Ozonation/post-chlorination of humic acid: a model for predicting drinking water disinfection by-products. Ozone Sci Eng 14:51
12. Le Lacheur RM, Sonnenberg LB, Singer PC, Christman RF, Charles MJ (1993) Identification of carbonyl compounds in environmental samples. Environ Sci Technol 27:2745
13. Backlund P, Kronberg L, Pensar G, Tikkanen L (1985) Mutagenic activity in humic water and alum flocculated humic water treated with alternative disinfectants. Sci Total Environ 47:257
14. Matsuda H, Yamamori H, Sato T, Ose Y, Nagase H, Kito H, Sumida K (1992) Mutagenicity of ozonation products from humic substances and their components. Water Sci Technol 25:363
15. Weinberg HS, Glaze WH, Krasner SW, Sclimenti MJ (1993) Formation and removal of aldehydes in plants that use ozone. J Am Water Works Assoc 85:72
16. Kooij D (1992) Assimilable organic carbon as an indicator of bacterial regrowth. J Am Water Works Assoc 84:57
17. Melin ES, Édegaard H (2000) The effect of biofilters loading rate on the removal of organic ozonation by-products. Water Res 34:4464
18. US EPA (2000) Giardia: drinking water fact sheet. US Environmental Protection Agency, Washington, DC
19. Chen KC, Lin YF (1993) The relationship between denitrifying bacteria and methanogenic bacteria in a mixed culture system of acclimated sludges. Water Res 27:1749
20. Ergas SJ, Rheinheimer DE (2004) Drinking water denitrification using a membrane bioreactor. Water Res 38:3225
21. Nolan BT, Ruddy BC, Hitt KJ, Helsel DR (1997) Risk of nitrate in groundwaters of the United States – a national perspective. Environ Sci Technol 31:2229
22. Nuhoglu A, Pekdemir T, Yildiz E, Keskinler B, Akay G (2002) Drinking water denitrification by a membrane bio-reactor. Water Res 36:1155
23. Crespi M, Ramazzoth V (1991) Evidence that N-nitroso compounds contribute to the causation of certain human cancers. In: Bogardi I, Kuzelka RD (eds) NATO ASI Series, Volume G.30, Nitrate Contamination. Springer, Berlin, p 233
24. Winto EF (1997) Nitrate in drinking water. J Am Water Works Assoc 63:95
25. Shuval HI (1980) Infant methemoglobinemia and other health effects of nitrates in drinking-water. Prog Water Technol 12:1731
26. Mirvish SS (1977) N-nitroso compounds, nitrate, and nitrite: possible implications for the causation of human cancer. Prog Water Technol 8:195
27. Sakakibara Y, Araki K, Watanabe T, Kuroda M (1997) The denitrification and neutralization performance of an electrochemically activated biofilm reactor used to treat nitrate-contaminated groundwater. Water Sci Technol 36:61
28. Urbain V, Benoid R, Manem J (1996) Membrane bioreactor: a new treatment tool. J Am Water Works Assoc 88:75
29. Kapoor A, Viraraghavan T (1997) Nitrate removal from drinking water: review. J Environ Eng 123:371

30. Tiedje JM (1990) Ecology of denitrification and dissimilatory nitrate reduction to ammonia. In: Zehnder AJB (ed)Biology of anaerobic microorganisms. Wiley, New York, p 179
31. Knowles R (1982) Denitrificaton. Microbiological Rev 46:43
32. Rittmann BE, Huck PM (1989) Biological treatment of public water. CRC Critical Rev Envir Control 19:119
33. Hiscock KM, Lloyd JW, Lerner DN (1991) Review of natural and artificial denitrification of groundwater. Water Res 25:1099
34. Wang H, Qu J (2003) Combined bioelectrochemical and sulfur autotrophic denitrification for drinkingwater treatment. Water Res 37:3767
35. Zhang TC, David GL (1999) Sulfur:limestone autotrophic denitrification processes for treatment of nitrate-contaminated water: batch experiments. Water Res 33:599
36. Kurt M, Dunn IJ, Bource JR (1987) Biological denitrification of drinking water using autotrophic organisms with H_2 in a fluidized-bed biofilm reactor. Biotechnol Bioeng 29:493
37. Batchelor B, Lawrence AW (1978) Autotrophic denitrification using elemental sulfur. J Water Pollut Control Fed 50:1986
38. Rittmann BE, Snoeyink V (1984) Achieving biologically stable drinking water. J Am Water Works Assoc 76:106
39. Lee KC, Rittmann BE (2002) Applying a novel autohydrogenotrophic hollow-fiber membrane biofilm reactor for denitrification of drinking water. Water Res 36:2040
40. Flere JM, Zhang TC (1998) Sulfur-based autotrophic denitrification pond systems for in situ remediation of nitratecontaminated surface water. Water Sci Technol 38:15
41. Hoek JP, Hijnen WAM, Bennekom CA, Mijnarends BJ (1992) Optimization of the sulphur-limestone filtration process for nitrate removal from groundwater. J Water Suppl Res Technol-Aqua 41:209
42. Schippers JC, Kruithof JC, Mulder FG, Lieshout JW (1987) Removal of nitrate by slow sulphur limestone filtration. J Aqua 5:274
43. Buttiglieri G, Malpei F, Daverio E, Melchiori M, Nieman H and Ligthart J (2005) Denitrification of drinking water sources by advanced biological treatment using a membrane bioreactor. Desalination 178:211
44. National Drinking Water Clearinghouse (NDWC) (2000) Slow sand filtration, Tech brief #DWFSOM40, a National Drinking Water Clearinghouse Fact Sheet, Morgantown, WV, June
45. Vigneswaran S, Visvanathan C (1995) Water treatment processes: simple options. CRC, Boca Raton, FL
46. Clark RM, Clark DA (1995) Drinking water quality management. Technomic, Lancaster, PA
47. Hozalski RM (1996) Removal of biodegradable organic matter in drinking water biofilters: experimental studies and model development, Ph.D. dissertation. The Johns Hopkins University, Baltimore, MA
48. Cohen Y (2001) Biofiltration – the treatment of fluids by microorganisms immobilized into the filter bedding material: a review. Bioresour Technol 77:257
49. Hozalski RM, Bouwer EJ (2001) Non-steady state simulation of BOM removal in drinking water biofilters: Model development. Water Res 35:198
50. Kolot FB (1988) Principles, techniques and industrial applications. In: Robert E (ed) Immobilized microbial systems. Krieger, New York
51. Crope WA (1970) Attachment of marine bacteria to solid surfaces. In: Manly P (ed) Biological adhesion, Vol 73. Academic, New York

52. Cochet N, Lebeault JM, Vijayalakshmi A (1990) Physicochemical aspects of cell adsorption. In: Tyagi RD, Vembo K (eds) Wastewater treatment by immobilized cells. CRC, Boca Raton
53. Kosaric N, Blaszczyk R (1990) The morphology and electron microscopy of microbial aggregates. In: Tyagi RD, Vembo K (eds) Wastewater treatment by immobilized cells. CRC, Boca Raton
54. Wanner J (1994) Activated sludge: bulking and foaming control. In: A Technomic publication, CRC, Boca Raton
55. Wanner O, Gujer W (1984) Competition in biofilms. Water Sci Technol 17:27
56. Bickerstaff GF (1997) Immobilization of enzymes and cells. In: Bickerstar GF (ed) Immobilization of enzymes and cells. Humana, Clifton, UK
57. Tampion J, Tampion MD (1987) Immobilized cells: principles and application. Cambridge University Press, Cambridge
58. Iorio G, Calabro V (1995) Biotechnological applications of membrane systems in the agro-food industry. In: Ana, membrane technology: applications to industrial wastewater treatment, Kluwer, Dordrecht
59. Sutton PM, Mishra PN (1996) The membrane biological reactor for industrial wastewater treatment and bioremediation. In: Hickey RF, Smith GL (eds) Biotechnology in industrial waste treatment and bioremediation. CRC, Boca Raton
60. Gaeta SN (1995) The industrial development of polymeric membranes and membrane modules for reverse osmosis and ultrafiltration. In: Caetano A, Norberta PM, Drioli E, Muntau H (eds) Membrane technology: applications to industrial wastewater treatment. Kluwer, Dordrecht
61. Hozalski RM, Bouwer EJ, Goel S (1999) Removal of natural organic matter (nom) from drinking water supplies by ozone-biofiltration. Water Sci Technol 40:157
62. Urfer D, Huck PM, Booth S, Coffey BM (1997) Biological filtration for BOM and particle removal: a critical review. J Am Water Work Assoc 89:83
63. Huck PM, Anderson WB, Savage EA, Borstel RC, Daignault SA, Rector DW, Irvine GA, Williams DT (1989) Pilot scale evaluation of ozone and other drinking water disinfectants using mutagenicity testing. Ozone Sci Eng 11:245
64. Gagnon GA, Booth SDJ, Peldszus S, Mutti D, Smith F, Huck PM (1997) Carboxylic acids: formation and removal in full-scale plants. J Am Water Works Assoc 89:88
65. Urfer D, Huck PM (2001) Measurement of biomass activity in drinking water biofilters using a respirometric method. Water Res 35:1469
66. Hozalski RM, Bouwer EJ (2001) Non-steady state simulation of BOM removal in drinking water biofilters: applications and full-scale validation. Water Res 35:211
67. Persson F, Långmark J, Heinicke G, Hedberg T, Tobiason J, Stenström TA, Hermansson M (2005) Characterisation of the behaviour of particles in biofilters for pre-treatment of drinking water. Water Res 39:3791
68. Yavich AA, Lee KH, Chen KC, Pape L, Masten SJ (2004) Evaluation of biodegradability of NOM after ozonation. Water Res 38:2839
69. Fonseca AC, Summers RS, Hernandez MT (2001) Comparative measurements of microbial activity in drinking water biofilters. Water Res 35:3817
70. Waer MA (2006) Multiple barriers for a smelly situation. Am Water Work Assoc 32:3
71. Soares MIM, Belkin S, Abeliovich A (1988) Biological groundwater denitrification: laboratory studies. Water Sci Technol 20:189
72. Kappelhof JWNM, Hoek JP, Hijnen WAM (1992) Experience with fixed-bed denitrification using ethanol as substrate for nitrate removal from groundwater. Water Supply 10:91

73. Harremoes P, Jassen JLC, Kristensen GH (1980) Practical problems related to nitrogen bubble formations in fixed film reactors. Prog Water Technol 12:253
74. Hoek JP, Kappelhof JWNM, Schippers JC (1994) The use of vacuum deaeration in biological removal processes. Aqua 43:84
75. Woodbury BL, Dahab MF, Mihâltz P, Csikor Z (1998) Evaluation of reversible fixed-film static-bed bio-denitrification reactors. Water Sci Technol 38:311
76. Woodbury BL, Dahab MF (2001) Comparison of conventional and two-stage reversible flow, static-bed biodenitrification reactors. Water Res 35:1563
77. Min B, Evans PJ, Chu AK, Logan BE (2004) Perchlorate removal in sand and plastic media bioreactors. Water Res 38:47
78. Rittmann BE, Stilwell D, Ohashi A (2002) The transient-state, multiple-species biofilm model for biofiltration processes. Water Res 36:2342
79. Crawford G, Fernandez A, Shawwa A, Daigger G (2002) Competitive bidding and evaluation of membrane bioreactor equipment – three large case studies. In: Proceedings of WEF 75th Annual Conference 2nd Exposition, Chicago, IL
80. Yang W, Cicek N, Igl J (2006) State-of-the-art of membrane bioreactors: Worldwide research and commercial applications in North America. J Memb Sci 270:201–211
81. Cicek N, Franco JP, Suidan MT, Urbain V (1998) Using a membrane bioreactor to reclaim wastewater. J Am Water Works Assoc 90:105
82. Cicek N (2003) A review of membrane bioreactors and their potential application in the treatment of agricultural wastewater. Can Biosyst Eng 45(6):37
83. Buisson H, Cote P, Praderie M, Paillard H (1998) The use of immersed membranes for upgrading wastewater treatment plants. Water Sci Technol 37:89
84. Cote P, Buisson H, Pound C, Arakaki G (1997) Immersed membrane activated sludge for the reuse of municipal wastewater. Desalination 113:189
85. Rosenberger S, Kruger U, Witzig R, Manz W, Szewzyk U, Kraume M (2002) Performance of a bioreactor with submerged membranes for aerobic treatment of municipal waste water. Water Res 36:413
86. Cicek N, Winnen H, Suidan MT, Wrenn BE, Urbain V, Manem J (1998) Effectiveness of the membrane bioreactor in the biodegradation of high molecular weight compounds. Water Res 32:1553
87. Urbain V, Mobarry B, Silva V, Stahl DA, Rittmann BE, Manem J (1998) Integration of performance, molecular biology and modeling to describe the activated sludge process. Water Sci Technol 37:223
88. Visvanathan C, Aim RB, Parameshwaran K (2000) Membrane separation bioreactors for wastewater treatment. Crit Rev Env Sci Technol 30:1
89. Chang J, Manem J, Beaubien A (1993) Membrane bioprocesses for the denitrification of drinking water supplies. J Memb Sci 80:233
90. Li XY, Chu HP (2003) Membrane bioreactor for the drinking water treatment of polluted surface water supplies. Water Res 37:4781
91. McCleaf PR, Schroeder ED (1995) Denitrification using a membrane-immobilized biofilm. J Am Water Works Assoc 87:77
92. Mansell BO, Schroeder ED (1999) Biological denitrification in a continuous flow membrane reactor. Water Res 33:1845
93. Xing CH, Yamamoto K, Fukushi K (2006) Performance of an inclined-plate membrane bioreactor at zero excess sludge discharge. J Memb Sci 275:175–186

94. Crespo JG, Reis AM (2001) Treatment of aqueous media containing electrically charged compounds, PCT-WO 01/40118 A1, International Patent
95. Crespo JG, Velizarov S, Reis AM (2004) Membrane bioreactors for the removal of anionic micropollutants from drinking water. Curr Opin Biotechnol 15:463
96. Velizarov S, Matos C, Reis M, Crespo J (2005) Removal of inorganic charged micropollutants in an ion-exchange membrane bioreactor. Desalination 178:203
97. Wiśniewski J, Różańska A, Winnicki T (2005) Removal of troublesome anions from water by means of Donnan dialysis. Desalination 182:339
98. Velizarov S, Reis MA, Crespo JG (2002) Integrated transport and reaction in an ion exchange membrane bioreactor. Desalination 149:205
99. Velizarov S, Reis MA, Crespo JG (2003) Removal of trace mono-valent inorganic pollutants in an ion exchange membrane bioreactor: analysis of transport rate in a denitrification process. J Memb Sci 217:269
100. Chudyk WA, Snoeyink VL (1984) Bioregeneration of activateed carbon saturated with phenol. Environ Sci Technol 18:1
101. Speitel GE Jr., Lu CJ, Turakhia MH (1988) Biodegradation of synthetic organic chemicals in GAC beds. Water Sci Technol 20:463
102. Rice RG, Robson CM (1982) Biological activated carbon: enhanced aerobic biological activity in GAC systems. Ann Arbor Science, Ann Arbor, MI
103. Weber WJ, Pirbazari M, Melson GL (1978) Biological growth on activated carbon: an investigation by scanning electron microscopy. Environ Sci Technol 127:817
104. Suzuki Y, Mochidzuki K, Takeuchi Y, Yagishita Y, Fukuda T, Amakusa H, Abe H (1996) Biological activated carbon treatment of effluent water from wastewater treatment processes of plating industries. Sep Technol 6:147
105. Baker T, Dyer-Smith P, Boere J (1996) Optimizing ozone and GAC in potable water treatment. In: Proceedings of Asian Pacific Conference, Hong Kong, p 42
106. Jahangir MAQ (1994) Bioregeneration of granular activated carbon. Ph.D. thesis, The University of Birmingham, UK
107. Rodman CA (1973) Factors involved with biological regeneration of activated carbon adsorbers. J Water Pollut Control Fed 55:1168
108. Sirotkin AS, Koshkina LY, Ippolitov KG (2001) The BAC-process for treatment of wastewater containing non-ionogenic synthetic surfactants. Water Res 35:3265
109. Andrews GF, Tien C (1981) Bacterial film growth in adsorbent surfaces. AIChE J 27:396
110. Scholz M, Martin RJ (1997) Ecological equilibrium on biological activated carbon. Water Res 31:2959
111. Hoek JP, Hofman JAMH, Graveland A (1999) The use of biological activated carbon filtration for the removal of natural organic matter and organic micropollutants from water. Water Sci Technol 40:257
112. Nishijima W, Okada M (1998) Particle separation as a pretreatment of an advanced drinking water treatment process by ozonation and biological activated carbon. Water Sci Technol 37:117
113. Kim WH, Nishijima W, Shoto E, Okada M (1997) Competitive removal of dissolved organic carbon by adsorption and biodegradation on biological activated carbon. Water Sci Technol 35:147
114. Kameya T, Hada T, Urano K (1997) Changes of adsorption capacity and pore distribution of biological activated carbon on advanced water treatment. Water Sci Technol 35:155
115. Wang LK, Kopko SP (1997) City and Cape Coral Reverse Osmosis Water Treatment Facility, Technical report PB97–139547, US Department of Commerce, National Technical Information Service, Springfield, VA, Fluid Particle Separation J., 10, 144

116. Wang LK, Hung YT, Shammas NK (2005) Physicochemical treatment processes. Humana, Totowa, NJ, p 723
117. Wang LK, Hung YT, Shammas NK (2006) Advanced physicochemical treatment processes. Humana, Totowa, NJ, p 690
118. Wang LK, Hung YT, Shammas NK (2007) Advanced physicochemical treatment technologies. Humana, Totowa, NJ, p 710
119. Wang LK, Pereira NC, Hung YT, Shammas NK (2009) Biological treatment processes. Humana, Totowa, NJ, 818
120. US EPA (2001) Controlling disinfection by-products and microbial contaminants in drinking water, EPA/600/R-01/110, US Environmental Protection Agency, Washington, DC
121. US EPA (1982) Pilot-scale evaluation of biological activated carbon for the removal of THM precursors, EPA-600/S2–82–046. US Environmental Protection Agency, Washington, DC
122. Shukairy HM, Miltner RJ, Summers RS (1992) Control of disinfection by-products and biodegradable organic matter through biological treatment. Revue Des Sciences De L'Eau 5:1–15
123. Shukairy HM, Summers RS, Miltner RJ (1992) The impact of ozonation and biological treatment on disinfection by-products. In: Proceedings of 4th Drinking Water Workshop, Government of Quebec, Montreal, Canada, November
124. Miltner RJ, Shukairy HM, Summers RS (1992) Disinfection by-product formation and control by ozonation and biotreatment. J Am Water Works Assoc 84:53
125. Miltner RJ, Summers RS (1992) A pilot-scale study of biological treatment. In: Proceedings of American Water Works Association Annual Conference, Vancouver, British Columbia, Canada, June
126. Wang JZ, Summers RS, Miltner RJ (1995) Biofiltration performance: part 1, relationship to biomass. J Am Water Works Assoc 87:55
127. Miltner RJ, Summers RS, Wang JZ (1995) Biofiltration performance: part 2, effects of backwashing. J Am Water Works Assoc 87:64
128. Findlay RH, King GM, Watling L (1989) Efficacy of phospholipid analysis in determining microbial biomass in sediments. Appl Environ Microb 55:2888
29. Wang JZ, Summers RS (1995) Biomass growth and distribution in drinking water biofilters and its impact on NOM removal. In: Proceedings of American Water Works Association Annual Conference, Anaheim, CA, June
130. Rittman BE, McCarty PL (1980) Model of steady-state biofilm kinetics. Biotechnol Bioeng 22:2343
131. Rittman BE, McCarty PL (1980) Evaluation of steady-state biofilm kinetics. Biotechnol Bioeng 22:2359
132. Dugan NR, Summers RS (1997) A biomass based model to predict substrate utilization in field-scale drinking water biofilters. In: Proceedings of American Water Works Association Annual Conference, Atlanta, GA, June
133. AWWA (2005) Microfiltration and Ultrafiltration Membranes, Manual of Water Supply Practices M53, American Water Works Association, Denver, CO
134. Wang LK, Wang MHS, Suozzo T, Dixon RA, Wright TL, Sarraino S (2009) Chemical and biochemical technologies for environmental infrastructure sustainability. National Engineers Week Conference, Albany Marriott, Albany, NY. Feb. 5–6, 2009
135. Wang LK, Ivanov V, Tay JH, Hung YT (eds) (2010) Environmental biotechnology. Humana, Totowa, NJ

12
Desalination of Seawater by Thermal Distillation and Electrodialysis Technologies

Jiaping Paul Chen, Lawrence K. Wang, Lei Yang, and Yu-Ming Zheng

CONTENTS

Abstract Freshwater is one of the scarce resources in the world. In many countries, due to freshwater shortages, searching for freshwater resources has become extremely important and desalination is known to be an essential available solution for this. In this chapter, the thermal distillation of multistage flash distillation, multieffect distillation and vapor compression and electrodialysis processes for seawater desalination are presented. The working mechanisms, important issues in design, and the recent advances of the thermal distillation and electrodialysis processes for desalination are addressed. Reverse osmosis technology is also briefly presented. A comparison of these desalination processes is given. Other important issues, such as pre- and posttreatment process, energy consumption, and environmental aspects, are discussed.

Key Words Desalination • electrodialysis • energy consumption • thermal distillation

1. INTRODUCTION

Water is one of the scarce resources in the world. Naturally occurring freshwater sources are rain water, surface water, and ground water. After conventional treatment, most of the water can be directly used for various purposes. The demand for water to serve the world

From: *Handbook of Environmental Engineering, Volume 13: Membrane and Desalination Technologies*
Edited by: L.K. Wang et al., DOI: 10.1007/978-1-59745-278-6_12

continues to increase, however, freshwater supplies are finite, and it is becoming more difficult to develop them on a renewable basis. In addition, water pollution has been becoming increasingly serious as a result of which water sources are greatly affected.

Water has a great impact on the economic development of all countries. It is estimated that one-fifth of the world's population does not have access to safe drinking water. Water is particularly important in an arid region such as the Middle East where water is not only an economic issue, but also a key political issue (1). The scarcity of water and the high cost of its development have long been globally recognized. Water for direct human use could become more important than oil in the near future.

In many countries, as water storages have already been exploited and full utilization is reached, searching for water sources has become extremely important. The available solutions include reclamation of used water, recycling of used water, optimization of water usage, and desalination.

About 97% of water in the world is salty as its chief sources are sea and brackish water. Less than 1% of the world's water sources are considered potable. Desalination is a term used to describe the process, involving in the removal of dissolved mineral salts, organic substances, bacteria and viruses, and solids from seawater to obtain freshwater. The major function of the process is to remove essentially salt content or salinity of water. The salinity of water source is measured in terms of "total dissolved solids" (TDS), which is commonly reported in milligrams per liter (mg/L). Freshwater normally has less than 1,000 mg/L TDS. Raw water for desalination can be seawater from ocean or from shallow beach wells, brackish water (surface or aquifer), wastewater, and even water produced by oilfields. Water can be classified into four categories based on the salt content in it. They are

- recycled water with less than 1,500 mg/L TDS;
- slightly saline water with 1,000–3,000 mg/L TDS;
- moderately saline water with 3,000–10,000 mg/L TDS;
- highly saline water with over 10,000 mg/L TDS.

Brackish water normally refers to water with salinities between 1,000 and 10,000 mg/L TDS. The salinity of seawater is in the order of 35,000 mg/L TDS. The US Environmental Protection Agency (US EPA) and the World Health Organization (WHO) established a TDS guideline of 500 and 1,000 mg/L, respectively, for drinking water. In addition, silt density index (SDI) is another important parameter. SDI is a measure of the amount of 0.45-μm filter plugging that is caused by passing a sample of water through the filter for 15 min. It must be determined for water that is considered for reverse osmosis (RO) and electrodialysis (ED)/ electrodialysis reversal (EDR) treatment. SDI is also used for the evaluation of product water.

Total world capacity of potable water by desalination is approaching 30 million m^3/day, in some 15,000 plants (2). Half of these are in the Middle East, where water resource is extremely limited, whereas energy (oil) is widely available. Approximately 25% is located in Saudi Arabia. More than two-thirds of the world's desalting capacity is located in the arid, oil-rich areas of North Africa and Western Asia, or the Middle East (3). Figure 12.1 illustrates the distribution of desalination capacity in the world. About 10% of Israel's water is desalinated. In the UK, 150,000 m^3/day RO plant is proposed for the lower Thames estuary,

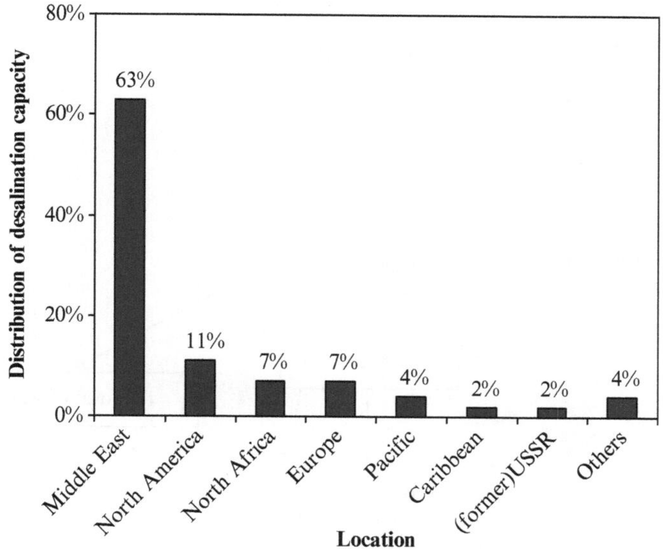

Fig. 12.1. Distribution of desalination capacity in the world.

utilizing brackish water. Around 800 desalination plants are operated in the USA, producing freshwater with a capacity of 225 MGD (about 1.4% of total water consumption). Most of them are in Florida, California, and Texas.

The general process of producing useable water from the water with high salt content is illustrated in Fig. 12.2. The desalination technologies are normally developed for treating large quantities of water (e.g. 100–1,000 MGD) at a central location. However, some have been recently adopted for small-scale water supply at home.

Modern desalination practices are broadly categorized into distillation or membrane processes. The distillation process involves phase changes and usually utilizes thermal energy and mechanical energy. It includes the multistage flash (MSF) distillation, the multieffect distillation (MED or ME), and vapor compression (VC). The membrane process involves no phase changes and the separation is mainly carried out using semipermeable membranes. This requires mechanical or electric energy. The commonly used techniques are RO and ED (EDR).

About 70% of the world's desalination capacity is dependent on the distilling process. However, in USA, the distribution is different; RO and ED (EDR) occupy around 80% of the market. The freshwater production capacity normally follows the descending sequence of MSF > RO > ED (EDR) > MED > VC. Table 12.1 gives the key factors in the different desalination operations. Other processes such as freezing and membrane distillation are seldom used. Of the aforementioned five main desalination technologies, MSF, MED, and VC thermal processes are generally used in the following applications:

- To treat highly saline waters (predominantly seawater).
- Where large volumes of product water are required.
- In locations where energy costs are low or where a waste heat source is available.

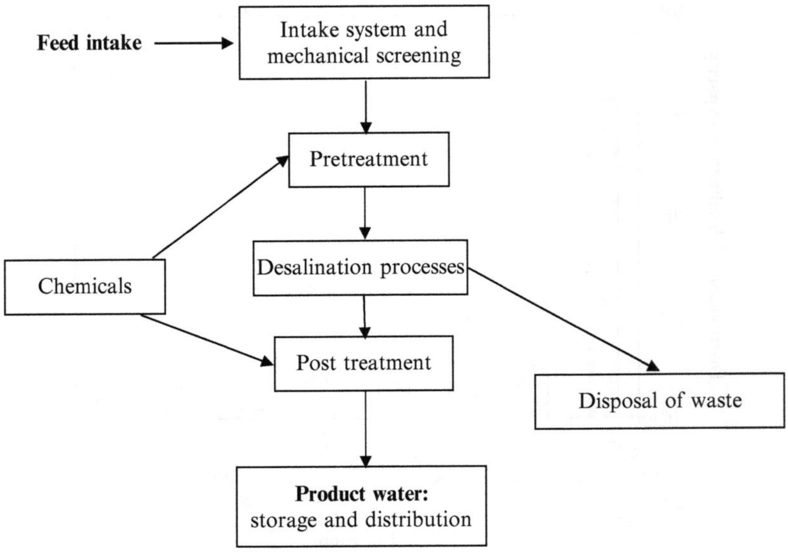

Fig. 12.2. General process for production of fresh water by desalination processes.

Membrane processes of RO and ED (EDR), on the contrary, are more favorable for treating brackish waters (under most conditions) or highly saline wastes for which energy costs are high or the flow rates are low.

In addition to the thermal and membrane processes, ion exchange (IX), freezing, and membrane distillation can be used for the removal of salt in water. IX is presented in another chapter of this handbook series. In the freezing process, saline water is frozen to form ice crystals under controlled conditions. Before the water is completely frozen, the mixture is usually rinsed to remove the salts in the remaining water or those adhering to the ice crystals. The ice is then melted to produce fresh water. Lower energy requirement, lower potential for corrosion, and lesser scaling or precipitation problems are the advantages; however, the disadvantage is the difficulty in operations (e.g. handling huge ice crystals). Membrane distillation process combines the use of both distillation and membranes (4, 5). Saline water is warmed to enhance vapor production, and this vapor is exposed to a membrane that can pass water vapor but not liquid water. After the vapor passes through the membrane, it is condensed on a cooler surface to produce freshwater. In the liquid form, the freshwater cannot pass back through the membrane; hence it is trapped and collected as the output of the plant. The advantages lie in its simplicity and the need for only small temperature differentials for its operation. However, when the process is run with low temperature differentials, large amounts of water must be used, which affects its overall energy efficiency. A recent development is adopting solar-powered membrane distillation technology for the production of pure water from brackish water (5). Solar energy is collected for the operation of membrane distillation system.

Before the raw water enters the aforementioned treatment systems, it is normally pretreated by various technologies, such as filtration, chemical coagulation, softening, and pH

Table 12.1
Comparison of major desalination technologies

Item	MSF	MED	ED/EDR	RO
Preferred water source	Seawater – brine	Seawater – brine	Brackish	Any
Final product salinity	5 mg/L < TDS < 50 mg/L	5 mg/L < TDS < 50 mg/L	TDS < 500 mg/L	TDS < 500 mg/L
Land usage	Large area	Large area	Small area	Small area
Capital costs	High	High	Medium	Medium
Operating temperature	High (~100°C)	Moderate (~70°C)	Not applicable	Not applicable
Energy cost	High	High	Moderate	Moderate
Feed-to-product ratio	7.0–12.0:1	6.0–10.0:1	1.2–2.0:1	2.0–2.5:1
Pre-treatment	Not required	Not required	Required	Required
Influent requirement	Not required	Not required	SDI < 12	SDI < 4
Quality of water produced	Pure	Pure	TDS of 350–500 mg/L	TDS of 20–50 mg/L
Susceptibility to scaling	Low	Low	Low	High
Bacterial contamination	Unlikely	Unlikely	Post-treatment needed	Possible
Waste	High salt content and corrosive	High salt content and corrosive	High salt content	Low salt content

adjustment. Selection of these technologies is dependent upon the water quality of the influent and the desalination process that is to be applied.

The cost of desalting saline water ranges from USD 0.4–2.2/m^3 of product water, which depends on the quality of the raw water and the technologies applied. Table 12.2 shows a comparison of capital and operational costs of desalination processes. The membrane technologies are cost effective for low-salinity sources (brackish water), whereas distillation-based technologies make sense only for higher salinity courses, such as seawater. The RO process can be regarded as the optimum choice for desalination of low-salinity water, regardless of plant capacity. The optimum choice for desalination of higher salinity feed depends on the required capacity. The MSF process would be the technology of choice for capacities more than 25,000 m^3/d, whereas the MED process should be chosen for capacities in the order of 10,000 m^3/d. VC is good for capacities in the order of 3,000 m^3/d (6). When a desalination plant with a larger capacity is built, the capital cost per unit of product water is lower. The daily operational cost, however, is not significantly affected by the size of the plant. It is advisable to build a larger plant instead of several smaller plants. Cost analysis must be conducted even before design is initiated. Computer models such as ARIMA model can be used (7).

Table 12.2
Capital and operational costs of desalination processes

Nature of technologies	Technologies	Capital cost (USD/m^3/d)	Operational cost (USD/m^3)
Thermal technologies	MSF	1,200–3,000	0.7–1.5 (with waste heat)
	MED	1,000–3,900	0.4–0.8 (with waste heat)
	VC	1,000–1,300	0.5–1.2
Membrane technologies	RO	500–1,200 (brackish water) 1,000–2,500 (seawater)	0.2–1.2 (brackish water) 0.2–1.7 (seawater)
	ED (EDR)	400–2,500	0.6–2.1

The cost for treating sea and brackish water by RO decreased by almost 50% from 1980 to 2005, because the cost of membrane materials dropped and prevention technologies of membrane fouling were well developed. For example, desalination of seawater cost approximately USD 1.5/m^3 production water in 1980, whereas the figure dropped to USD 0.8 in 2005. Over the last 50 years, the cost of freshwater production by MSF has decreased by an average of 44% per decade (8).

Large-scale desalination projects are extremely expensive and often require some degree of public financing. To build a 25-MGD plant, it costs 70–100 million USD. Thus, it becomes more common that contractual arrangements of selling/purchase water between governments and suppliers are made. The suppliers build and operate the plant. "Build, own, operate, and transfer" and "build, operate, and transfer" are two typical arrangements. The first one becomes more popular in developing countries, such as China.

Factors influencing selection of desalination technologies are many. The applicability of desalination is very site specific. Site-specific conditions will also determine the type of desalination technology selected (9). Financial issue is the first one to be solved. Both capital investment and cost of daily operation are essential. As desalination requires high energy input, energy requirement is an important issue. Selection of desalination process is highly dependent upon source water characteristics; the quality and the quantity of product water rely on them. Desalination plants should be close to the raw water source (e.g. sea) to obtain a steady flow of raw water and to be able to discharge its byproducts easily. Limited land space can be a major concern in the selection of desalination process. Desalination plants are usually big and their accommodation could be a main cost factor that needs to be considered. The plants that are built must have a large capacity in order to provide substantial amounts of water. Environmental factors and waste disposal options are also important. Wastewater with high salt content is produced. Handling the environmental problems is a great challenge and principles of pollution prevention can be implemented. Finally, operational and maintenance (O&M) issues must be considered. Training of operators is highly recommended.

Laws and regulations of federal, state, and local agencies play important roles in design and operation of desalination projects. "Water Desalination Act of 1996," "Pollution Prevention Act of 1990," and "Clean Water Act of 2002" regulate the projects at the federal level. In addition, government policies have a great impact on costs of desalination.

In this chapter, the thermal distillation and ED (EDR) processes for desalination are presented and the working mechanisms and recent advance of the same are illustrated. RO process is also briefly presented. A comparison of these desalination technologies is given. Other important issues, such as pre- and posttreatment processes, energy, and environmental aspects are discussed.

2. THERMAL DISTILLATION

2.1. *Introduction*

Distillation is a thermal process. It is the oldest and most commonly used method of desalination. A simple distillation process for desalination is illustrated in Fig. 12.3. The world's first desalination plant was operated in the mode of a multiple-effect distillation with a capacity of 60 m^3/d. This was installed in Curaçao, Netherlands Antilles, in 1928. Furthermore, commercial development of the seawater distillation units took place in the late 1950s (10). Distillation relies on the technology developed for industrial evaporation operation. MSF, MED, and VC processes have successfully led to the widespread use of distillation to desalinate seawater.

Product water recoveries by distillation range from 15 to 45% depending on the operation. The water is nearly mineral free, with very low TDS content. TDS of less than 25 mg/L is quite common and it normally ranges from 5 to 50 mg/L. However, this water must be subjected to proper posttreatment, as it is too corrosive to meet the

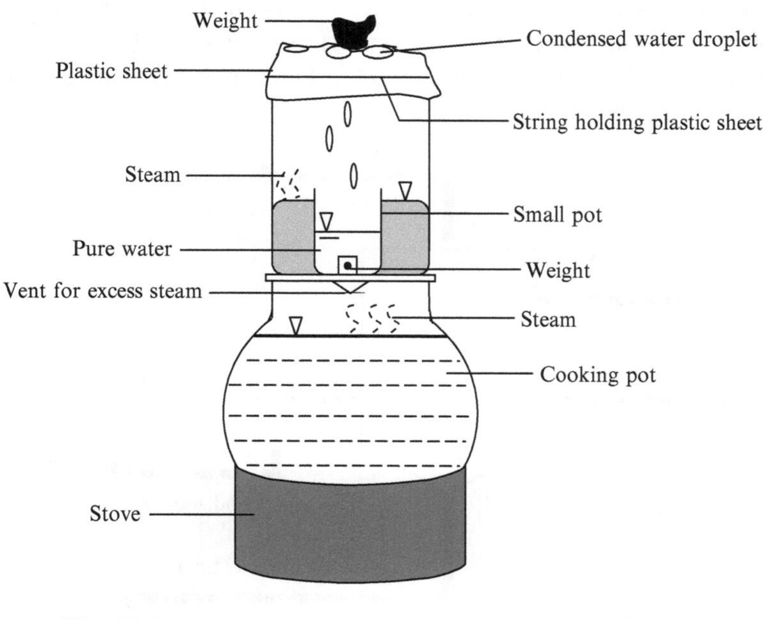

Fig. 12.3. Illustration of simple distillation for desalination.

corrosivity standards. The water can be stabilized by chemical treatment or by blending with other potable water.

2.2. *Working Mechanisms*

Distillation is a phase separation method. The saline water is heated to produce water vapor, which is then condensed to produce freshwater (Fig. 12.3). The distillation processes are generally operated on the principle of reducing the vapor pressure of water within the unit to permit boiling to occur at lower temperatures, without the use of additional heat. Distillation units routinely use designs that conserve as much thermal energy as possible by interchanging the heat of condensation and the heat of vaporization within the units. The major energy requirement in the distillation process is to provide the heat for vaporization to the saline feed.

The "distillation process" used for desalination can be analyzed as "evaporation," which is commonly adopted for concentration of substances in chemical and biochemical industries (11). The basic equation for solving for the capacity of an evaporator can be written as

$$q = U \times A \times \Delta T = U \times A \times (T_s - T_1), \quad (1)$$

where q (W) is the rate of heat transfer, U (W/m^2 K) is the overall heat-transfer coefficient, A (m^2) is the heat-transfer area, and ΔT (K) is the difference in temperature of the condensing steam (T_s, K) and that of the boiling liquid (T_1, K) in the evaporator.

The evaporation process can be determined by a heat and material balance shown in Fig. 12.4. The feed to the evaporator F (kg/h) has a solids content of X_F (mass fraction), temperature of T_F (K), and enthalpy of h_F (kJ/kg). The concentrated liquid leaving the evaporator (L, kg/h) has a solids content of X_L (mass fraction), temperature of T_L (K), and enthalpy of h_L. The vapor (V, kg/h) is given off as pure solvent having a solids content of $Y_v = 0$, temperature of T_V (K), and enthalpy (H_v). The vapor (V) is in equilibrium with the

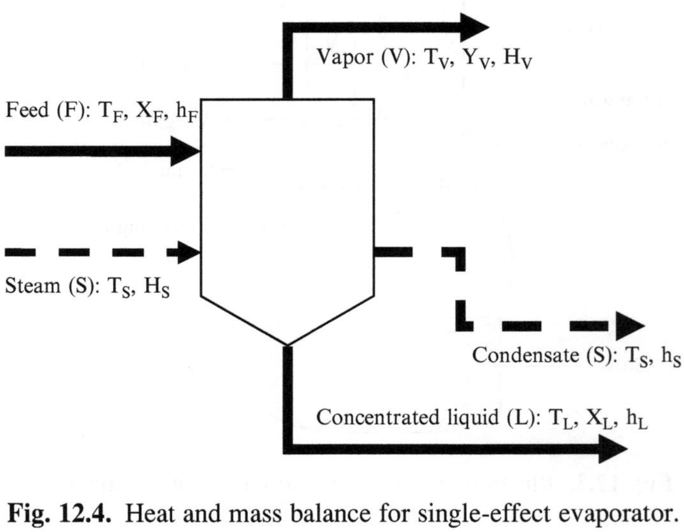

Fig. 12.4. Heat and mass balance for single-effect evaporator.

liquid (L) and both have the same temperature. The pressure (P_1) is the saturation vapor pressure of the liquid of composition (X_L) at its boiling point (T_L).

During the operation, a saturated steam (S, kg/h) must be added to the evaporator to provide heat to the feed solution. The steam has a temperature of T_s (K) and an enthalpy of H_s. The condensed steam leaving the system (S, kg/h) has a temperature of T_s (K) and an enthalpy of h_s. The steam gives off its latent heat (λ) as follows:

$$\lambda = H_s - h_s \tag{2}$$

The latent heat of steam at the saturation temperature (T_s) can be obtained from the steam tables. Since the system is operated at steady state, the rate of mass coming to the evaporator is equal to that leaving the evaporator. Thus,

$$F = L + V \tag{3}$$

The mass balance on the solute (solids) is

$$F(X_F) = L(X_L) \tag{4}$$

In the heat balance, the total heat entering the system is equal to the total heat leaving the system. Assuming no heat lost by radiation or concentration.

$$\begin{aligned}\text{Heat in feed} + \text{heat in steam} = \text{heat in concentrated liquid} + \text{heat in vapor}\\ + \text{heat in condensed steam}\end{aligned} \tag{5}$$

$$F \times h_F + S \times H_s = L \times h_L + V \times H_v + S \times h_s \tag{6}$$

Substituting Eq. (2) into Eq. (6) yields

$$F \times h_F + S \times \lambda = L \times h_L + V \times H_v. \tag{7}$$

The heat q transferred to the evaporator can be determined by

$$q = S \times (H_s - h_s) = S \times \lambda. \tag{8}$$

Example. A continuous evaporator is used for treatment of a salt solution that has a flow rate of 9,000 kg/h, 1% of salt, and a temperature of 37.8°C or 310.8 K. The concentrated liquid that leaves the evaporator has 1.5% of salt. The vapor space of the evaporator is at 1 atm, whereas the steam supplied is at 1.414 atm. The overall coefficient is 1,700 W/m^2 K. Determine the amounts of vapor and liquid leaving the system, the amount of steam added, and the rate of heat transfer.

Solution. According to the given conditions, F = 9,000 kg/h, X_F = 1%, X_L = 1.5%. Thus,

$$\begin{aligned}F &= L + V\\ 9{,}000 &= L + V\\ F \times X_F &= L \times X_L\\ 9{,}000 \times 0.01 &= L \times 0.015\end{aligned}$$

On solving these equations, L = 6,000 kg/h and V = 3,000 kg/L.

The operation produces freshwater of 3,000 kg/L (after condenser).
The boiling point of the solution in the evaporator is assumed to be that of water at 1 atm and 100°C (i.e. $T_L = 373.2$ K). The heat capacity of the feed is assumed to be that of water ($C_{PF} = 4.14$ kJ/kg K). H_v of water at 100°C (H_v) is 2,257 kJ/kg. The latent heat of steam at 1.414 atm is 2,230 kJ/kg. The saturation temperature [T_s] is 110°C. The enthalpy of feed is

$$h_F = C_{PF} \times (T_F - T_L), h_L = 0.$$

From Eq. (7), 9,000 × 4.14 × (37.8 − 100) + S × 2,230 = 6,000 × 0 + 3,000 × 2,257
S = 4,076 kg/h steam.
The rate of heat transfer (q) can be determined by

$$q = S \times \lambda = 4,076 \times 2,230 \times 1,000/3,600 = 2.52 \times 10^6 \text{ W}.$$

2.3. *Multistage Flash Distillation*

MSF distillation is currently the most common and simple technique that is in use. Since 1950s, it has been commercially used. The largest MSF plant, also the world's largest desalination plant, is the Al Jubail plant in Saudi Arabia. It was completed in 1985 and has a capacity of 208 MGD.

MSF is essentially a method whereby seawater is spontaneously boiled through a series of chambers, at progressively lower pressures. Figure 12.5 demonstrates the process. When the atmospheric pressure is lowered, the seawater will boil at a lower boiling temperature. As such, this will hasten the process of obtaining pure water. The water vapor given off is condensed into pure water at each chamber. The remaining seawater that is not evaporated is then moved on to the next chamber at a lower pressure. The sample process is subsequently repeated until the seawater leaves all the chambers.

A large amount of flashing brine is normally required in the MSF operation (10). As a result, 50–75% of the brine from the last stage is often mixed with the incoming feed, recirculated through the heat recovery sections of the brine heater, and flashed again through

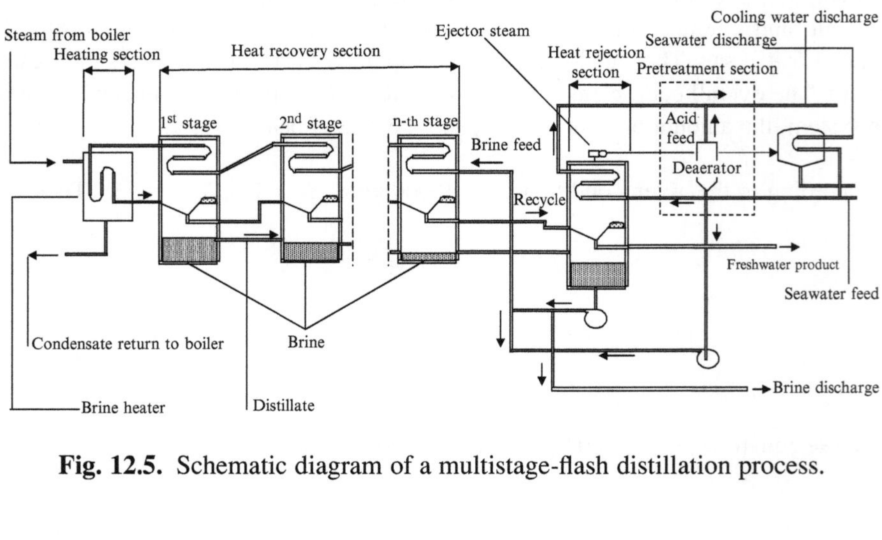

Fig. 12.5. Schematic diagram of a multistage-flash distillation process.

all of the subsequent stages. This "brine recycle" operation can reduce the amount of water-conditioning chemicals that must be added, which can significantly affect operational costs. This can also increase the salinity of the brine at the product end of the operation, raise the boiling point, and increase the danger of corrosion and scaling in the plant. In order to maintain a proper brine density in the system, a portion of the concentrated brine from the last stage is discharged to the ocean.

MSF is energy intensive and the cost of energy often constitutes a large portion of its water costs. An MSF plant also requires a large space area owing to its numerous chambers. The condenser tubes of the majority of operating units are made of Cu–Ni alloys (12). The corrosion of the alloys usually leads to tube failure and hence checking of the tubes is important during maintenance. Protective measures should also be adopted.

2.4. *Multieffect Distillation*

MED is actually a modification of the MSF process. Figure 12.6 demonstrates an MED process. Like MSF, seawater is heated into vapor and subsequently condensed to produce freshwater. The main difference lies in the arrangement of its chambers or pipes. MED usually operates on horizontal or vertical pipes. The steam is condensed on one side of the tube wall, whereas saline water is evaporated on the other side. This is similar to the VC process.

In the MED operation, some of the feed water is flash-evaporated, but most of the seawater is dispersed over an evaporator tube bundle and boiled. Steam then condenses to produce fresh water, and this process is continuously repeated. As the evaporation takes place in a vacuum, the sprayed seawater is able to reach boiling point even at low temperatures. As MED uses a progression of stages with ever-dwindling temperatures (60–70°C), it is comparatively smaller than MSF in terms of output. Unlike the MSF technique where water is

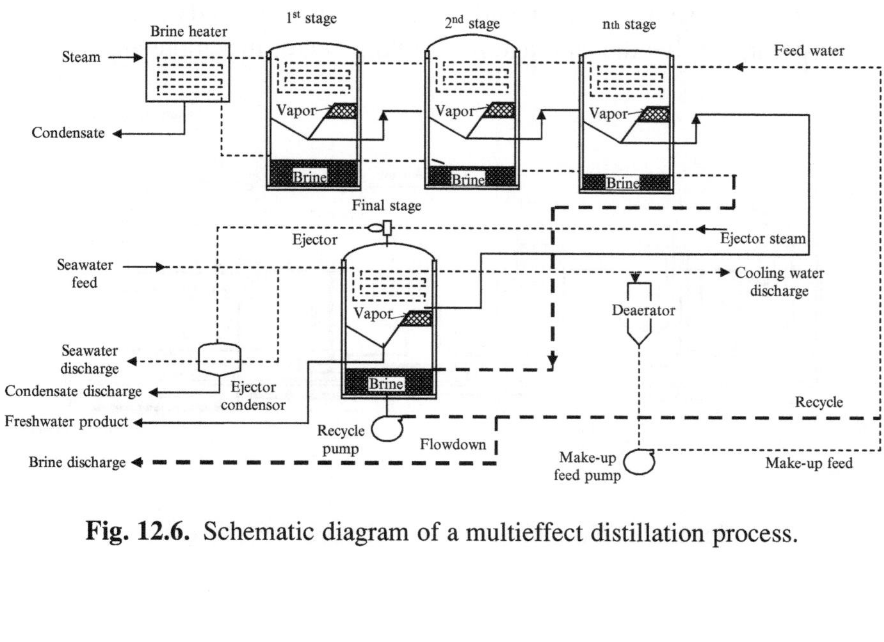

Fig. 12.6. Schematic diagram of a multieffect distillation process.

produced by turning heat into latent heat of evaporation, the MED technique uses latent heat to produce secondary latent heat in each chamber. This use of double condensing-evaporation heat transfer mechanisms is highly efficient and energy saving.

MED was developed earlier than MSF; however, it has not been extensively utilized for water production. A new type of low-temperature, horizontal-tube MED process has been successfully developed and used in the Caribbean. These plants appear to be very rugged, easy to operate, and economical, as they can be made of aluminum or other low-cost materials (10). The cost for O&M of MED is slightly higher than that for RO.

2.5. *Vapor Compression*

VC process uses mechanical energy rather than direct heat as a source of thermal energy. Figures 12.7 and 12.8 demonstrate the VC process. Water vapor is drawn from the evaporation chamber by a compressor. Except in the first stage, the vapor is condensed on the outsides of tubes in the same chambers. The heat of condensation is used to evaporate a film of saline water applied to the inside of the tubes within the evaporation chambers. VC normally has a low fresh water production capacity, i.e. <100 m^3/day, and is used at resorts and industrial sites (10).

2.6. *Solar Desalination*

Solar energy is always available and thus can be used to desalinate salt water. Its usage can be classified into two types: (a) direct usage: solar distillation and (b) indirect usage: energy from the sun rays is used to operate desalination processes.

Solar desalination is the simplest survival technique that is used to collect small quantities of drinkable water. It does not require any additional energy to run the operation. As shown in Fig. 12.9, a hole is dug in the ground, a sheet of plastic is placed on the floor, and a cup is placed at the bottom of the hole. Seawater is then added into the hole. The hole is covered with plastic sheet and a small stone is placed on the center of the sheet. Seawater evaporates from the hole and condenses on the underside of the plastic sheet and collects in the cup. This system can temporarily solve the shortage for drinking water in remote areas.

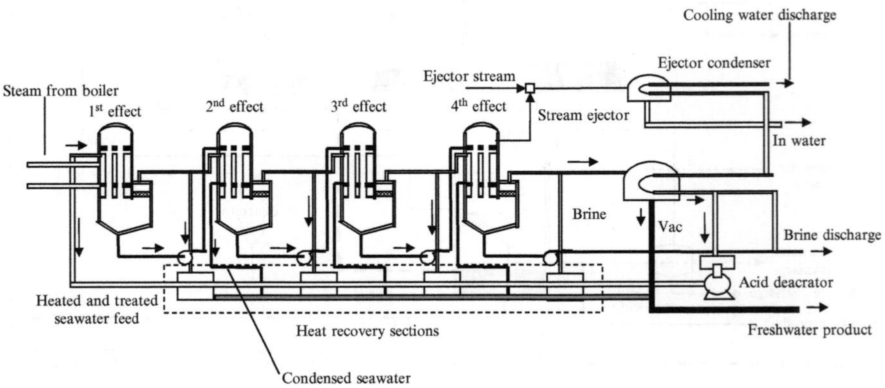

Fig. 12.7. Schematic of a vapor compression process.

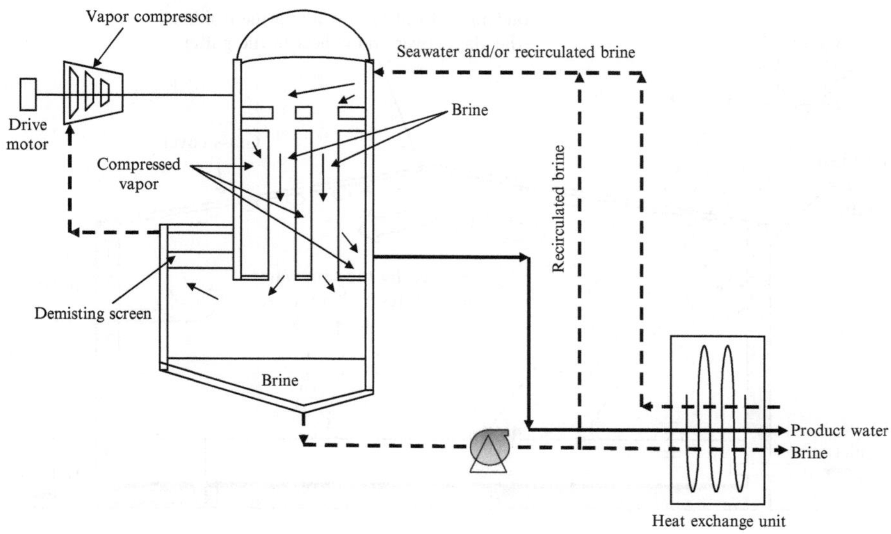

Fig. 12.8. Schematic diagram of a single-stage vapor compression process.

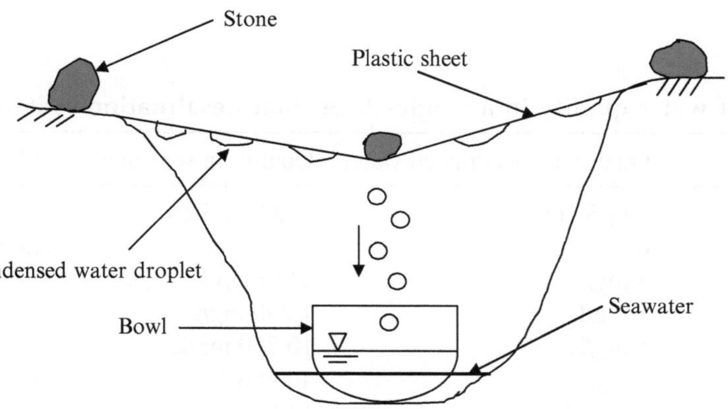

Fig. 12.9. Schematic diagram of simplest solar distillation for desalination.

Solar energy can be used to evaporate water from saltwater for household or small community water supplies by building sealed units covered by glass, as shown in Fig. 12.10. The water production rate ranges from 7 to 18 kg/m^2 day (13). The temperature of water from the outlets of collectors increases from 70°C at 9.00 A.M. to 95°C at 1.00 P.M., then decreases to 65°C at 4.00 P.M. (14). The quality of water before and after the solar desalination is given in Table 12.3. As shown, the product water is free of any impurities. This water, however, cannot be used as potable water. Mineral substances must be added.

Growth of algae on the underside of the units is normally observed. Effective sealing of the units, and regular checking and cleaning can efficiently control the problem. Small-scale

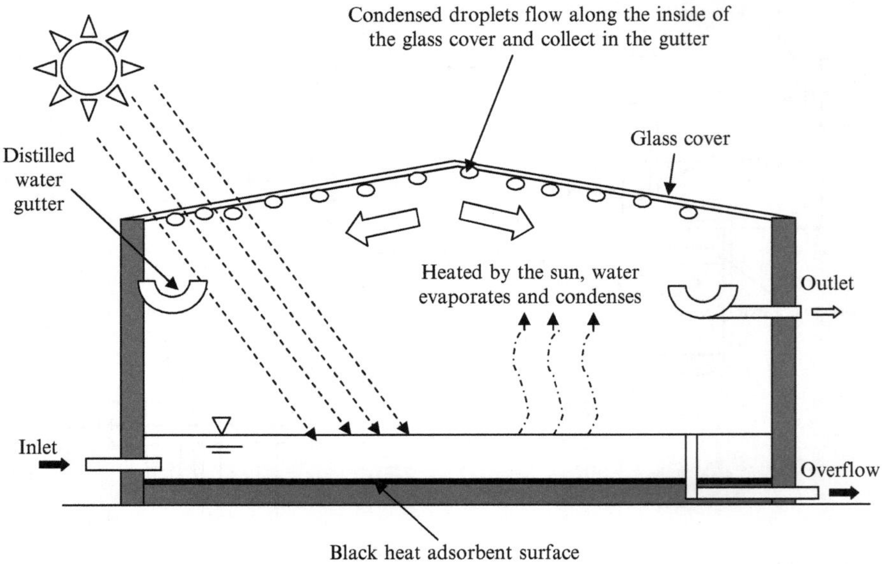

Fig. 12.10. Schematic diagram of solar desalination.

Table 12.3
Comparison of water quality in a single-stage solar desalination system

Parameter	Quality in desalinated water[a]	Quality in seawater[a]	US EPA/WHO standard
Conductivity	50 μS/cm	49,000 μS/cm	–/–
pH	6.8	8	6.5–8/–
Calcium	4 mg/L	411 mg/L	–/–
Magnesium	1 mg/L	1,290 mg/L	–
Sodium	5 mg/L	10,760 mg/L	–/200
Potassium	1 mg/L	400 mg/L	–/–
Bicarbonate	12 mg/L	142 mg/L	–/–
Carbonate	0	370 mg/L	–/–
Sulfate	0 mg/L	2,710 mg/L	250 mg/L/250 mg/L
Chloride	11 mg/L	19,350 mg/L	250 mg/L/250 mg/L
Nitrate	2 mg/L	120 mg/L	–/–
Total dissolved solids	30 mg/L	35,000 mg/L	500 mg/L/1,000 mg/L
Total hardness	15 mg/L as $CaCO_3$	4,300 as mg/L $CaCO_3$	–/–
Total alkalinity	10 mg/L as $CaCO_3$	2,150 as mg/L $CaCO_3$	–/–
Dissolved oxygen	7.6 mg/L	13 mg/L	–/–

[a]Ref. (14).

solar distillations are operated by families or small villages in many parts of the world. Solar-powered desalination plants are used in isolated parts of the world on a small scale, but the high capital cost, fragility, and complexity of operation have discouraged extensive use.

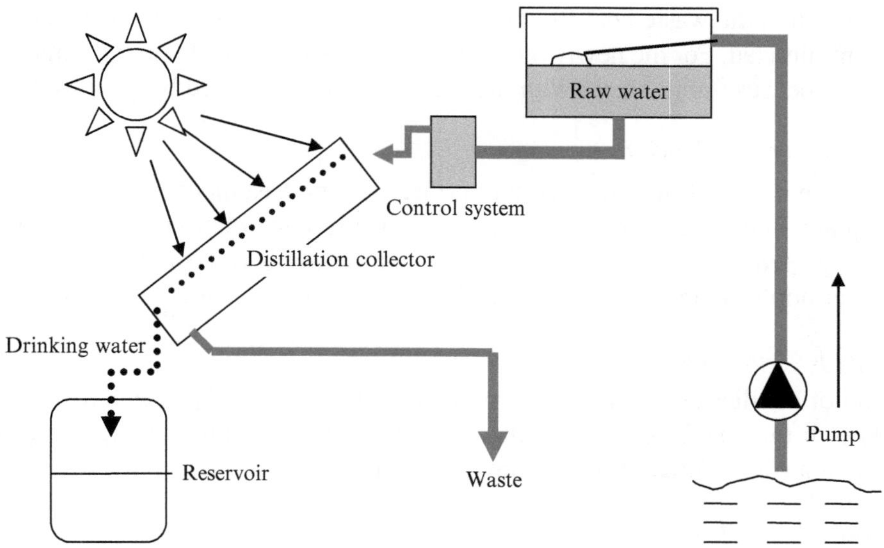

Fig. 12.11. Rosendahl system for desalination.

Figure 12.11 illustrates Rosendahl System for desalination. This system was developed by Wilfried Rosendahl of Germany in the1980s. The energy from sunlight increases the temperatures in the distillation collector to 80–90°C. About 50% of the raw water evaporates and condenses under the cool glass cover, which subsequently flows into a condensation channel and is channeled to the reservoir. The remaining untreated water (i.e. waste) that has a high salt content is collected and sent out for final disposal. As the system is exposed to sun light, the ultraviolet rays can cause a certain degree of disinfection for the freshwater. Its effect can be enhanced at the high temperature in the system. A water production rate of >2–7 L/d/m^2 of still area can be achieved on a day (15). Water production varies at different time. Normally, it reaches its maximum at 1.00 to 3.00 P.M. More water production is achieved when the water depth is 0.1–0.15 m (16).

In indirect usage, a solar collector is used to concentrate solar energy to produce high temperature in the thermal desalination process. Electrical energy, which is used to operate desalination processes, such as RO or ED, can also be produced by sunlight.

2.7. *Important Issues in Design (O&M)*

When a distillation plant for desalination is designed and operated, the following issues must be taken into consideration:

2.7.1. Thermal Discharge

A common problem resulting from the desalination operations is thermal discharge of liquids. Older high-temperature facilities produce brine at very high temperatures. Cooling towers, heat exchangers, or similar equipment must be considered in the design and construction so that the waste heat from the facilities can be properly handled and the water pollution is minimized. For the newly built distillation systems, the thermal discharge is not a major concern because brine water with high temperatures is cooled by the raw feed water.

2.7.2. *Materials for Construction*

The corrosive nature of high-temperature brines, acid pretreatments, and chemical scaling can cause plant failure. Presently, the only acceptable construction materials for wetted surfaces in high-temperature systems are an austenitic stainless steel. Anodized aluminum and many thermoplastic materials are acceptable for use in low-temperature systems.

2.7.3. *Energy for Operation*

Operation of distillation is mainly energy driven. The crude oil prices have increased by almost 100% in the last 2 years. It is, therefore, important to consider the energy cost for operations and a sound forecast of the same is essential.

2.7.4. *Characteristics of Raw Water*

The important parameters of raw water are TDS, salt contents, organic contents (TOC), and temperature. The availability of raw water is also important.

2.7.5. *Final Disposal of Waste Brine*

The waste brine has high temperatures and TDS. Thus, it must be properly treated before it is discharged to nearby waters. Cooling and IX are recommended for its treatment.

3. ELECTRODIALYSIS

3.1. *Introduction*

ED was commercially introduced in the early 1960s, about 10 years before the introduction of RO. The development of ED provides a cost-effective technology to remove salt from seawater. ED is also used in water and wastewater treatment, food processing, and chemical and pharmaceutical manufacturing.

In ED, ionic components are removed from aqueous solutions through the IX membranes using the driving force of an electric field. The membrane is selective and semipermeable and it only allows the passage of either the anions or the cations. The separation is because of the nature of charged substances rather than the difference in their sizes. In seawater desalination, the dissolution of salt leads to the formation of positively charged cations such as Na^+, Ca^{2+}, and Mg^{2+} and negatively charged anions such as Cl^-. The cations and the anions are attracted to the cathode and the anode, respectively, which are subsequently removed. As a result, the water after treatment carries less ionic substances. A typical electrodialysis plant operates at 50–90% conversion and this conversion value is determined by the brine recirculation rate. An ED system can operate over a wide pH range (pH 1.0–13.0) as well as a wide temperature range (up to 43°C). The cost for O&M of ED (EDR) is slightly higher than that for RO.

The drawback of this process is that the surfaces of membranes can be easily fouled. Thus, a modified process of EDR was developed. By intermittently reversing the polarity, the membranes' surfaces can be cleaned. Standard EDR operates for 15 min on each polarity. Electrodeionization (EDI) is a modification of conventional ED systems with IX resins installed in the ED stacks. The resins reduce the electrical resistance of the unit and the direct current potential splits water into hydrogen and hydroxyl ions that continuously regenerate a portion of the IX resins. EDI is particularly suited for the production of ultrapure water.

3.2. *Mechanisms*

The working mechanism of ED is demonstrated in Fig. 12.12. The IX membranes are arranged alternately with an anion-selective membrane followed by a cation-selective membrane. A cell is a space that is bounded by the two different membranes; it is arranged as anion exchange membrane, space, and cation exchange membrane. A unit cell or cell pair is made up of two cells. The ion-depleted cell is used for the passage of freshwater (the freshwater channel or desalted water channel), whereas the ion-concentrated cell is used for the passage of brine (the brine channel). The cells alternate between the fresh water channel and the brine channel. Each cell is connected by its own feed/product stream and brine stream. The unit cells and electrodes are arranged horizontally or vertically and the entire unit is called a stack. An ED system requires huge amounts of direct current (DC); a rectifier is usually used to transform the alternating current (AC) to DC.

The feed solution containing both positive and negative ions enters in a parallel path through all the cells, where an electric potential is maintained across the electrodes. This causes the ions to migrate to their respective electrodes. The positively charged cations in the

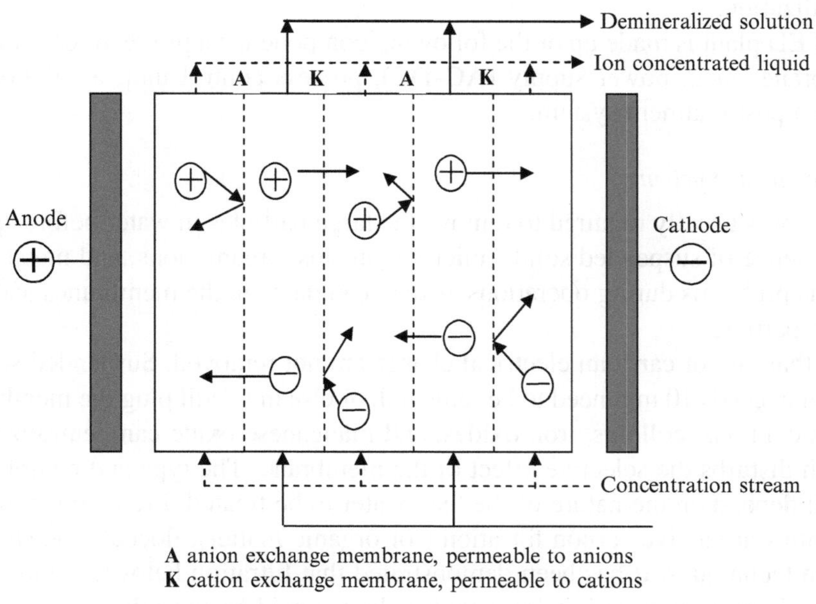

Fig. 12.12. Demonstration of mechanism of ED process.

solution will move toward the cathode and the negatively charged anions will move toward the anode. Cations can pass through the negatively charged cation exchange membrane but are retained by the positively charged anion exchange membrane. Similarly, anions pass through the positively charged anion exchange membrane but are retained by the negatively charged cation exchange membrane.

In the freshwater channel, the anions are attracted and diverted toward the anode, whereas the cations are attracted and diverted toward the cathode in a direction opposite to the movement of anions. The cations pass through the cation-selective membrane into the brine channel and are trapped in the brine channel as the next membrane is anion selective. Likewise, the anions pass through the anion-selective membrane into the brine channel but cannot pass through the next membrane because it is a cation-selective membrane. Consequently, the freshwater channel is depleted of both cations and anions through migration across the membranes, whereas the brine channel next to the freshwater channel becomes enriched with cations and anions. In other words, the alternating channels contain streams of dilute ion concentration and streams rich in ion concentration.

The quality of fresh water is dependent upon the following factors: (a) wastewater temperature, (b) amount of electrical current passed, (c) type and amount of ions, (d) permselectivity of the membrane, (e) fouling and scaling potential of the wastewater, (f) wastewater flowrates, and (g) number and configuration of stages.

3.3. *Important Issues in Design*

ED process may be operated in either a continuous or a batch mode as shown in Fig. 12.13. The units can be arranged either in parallel or in series in order to achieve the desired degree of demineralization.

A typical ED plant is made up of the following components: a pretreatment system for the feed, membrane stack, power supply (AC–DC), process control unit, a solution pumping system, and a posttreatment system.

3.3.1. *Pretreatment System*

Pretreatment is usually required to remove the large particles in water before applying the ED. The presence of suspended solids, microorganisms, organic ions, and metal oxides can cause serious problems during operations as they can damage the membranes and alter their chemical properties.

Particles that do not carry an electrical charge are not removed. Suspended solids with a diameter that exceeds 10 mm need to be removed, or else they will plug the membrane pores. Large organic anions, colloids, iron oxides, and manganese oxide can neutralize the membrane, which disturbs the selective effect of the membrane. The type and complexity of the pretreatment depend on the nature of the feed water to be treated. Pretreatment methods for these impurities are active carbon filtration (for organic matter), flocculation (for colloids), and filtration techniques. It has been demonstrated that filtration helps to remove the large, charged organic molecules and colloids such as humic acid because these molecules are too large to be removed through the membranes during ED.

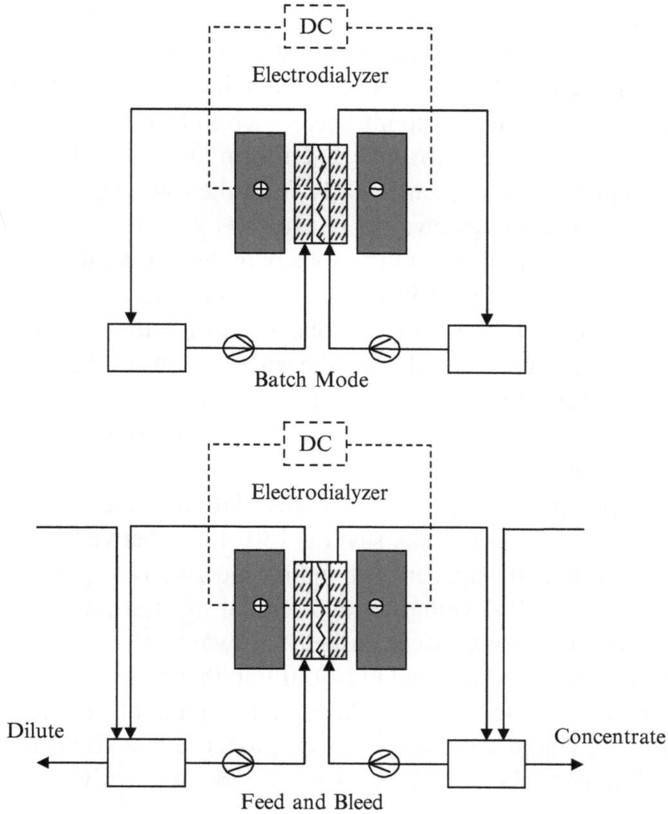

Fig. 12.13. Operation of electrodialysis process: (**a**) batch mode; (**b**) continuous mode.

The growth of microorganisms is another type of problem as it can cause fouling of the membrane. Chlorine should be used for disinfection them so that they cannot grow on the membrane. However, chlorine must be removed from the feed water before it enters the ED stack as chlorine can cause damage to the membranes. Sodium metahexaphosphate can be added to prevent fouling of the membrane surface by precipitation of sparingly soluble salts such as calcium sulfate.

After the pretreatment, the feed water can be pumped through the ED stack, which usually contains more than 100 membrane cell pairs, each with a membrane area of 1–2 m^2.

3.3.2. *Membrane Stack*

In a membrane stack, spacer gaskets separate the membrane sheets. A simple cell pair is made up of two membranes and two spacer gaskets. The spacers are turbulence-promoting support mesh used to create the channels through which the solutions flow. The holes in the spacer gaskets are aligned with holes in the membrane sheets to form the manifold channels that distribute the feed to the proper channels and remove the product and brine streams.

The distance between membranes (i.e. the cell thickness) needs to be as small as possible in order to minimize the electrical resistance and save energy consumption. The spacer maintains the shape of the membranes, controls the solution flow distribution in the stack, and reduces the film thickness on the membrane, which in turn limits the effects of concentration polarization. It is important to have a uniform flow distribution in the manifold channels to prevent internal leakage, particularly from the ion-concentrating cell to the ion-depleting cell. The feed water (wastewater) is usually retained for about 10–20 days in a single stack or stage. Both sheet flow and tortuous path design are used in ED.

In the sheet flow design, the solution flows in a relatively straight line from the entrance to the exit ports, which are located on opposite sides. Therefore, the supply ducts for the dilute and brine must be in line with the holes in the spacer gaskets, the membranes, and the electrode cell. An array of membranes is held between the electrodes and the process streams are kept separated. If the length-to-width ratio of the spacer is large enough, the entrance and exit ports can be located at the corners.

In the tortuous path design, a spacer gasket creates a labyrinthine flow and extends the path length within the cell. The solution takes several 180° turns between the entrance and exit ports located in the middle of the spacer. The main objective is to provide a long residence time for the solution in the cell despite the requirement of high linear velocity to limit the polarization effect. A tortuous path spacer gasket is shown next.

The end plate of the stack is a hard and plastic frame that contains the electrode compartment. The cathode is made of stainless and the anode is made of platinum-coated tantalum, niobium, or titanium. The whole setup is compressed together with bolts between the two end flow plates. The perimeter gaskets of the gasket spacers are tightly pressed into the membranes to form the cells.

The membranes of ED units are subject to fouling and scaling, and thus some pretreatment of the feed water is usually necessary. Precipitation of scale can be facilitated in the ED process by changes in pH that occur near the membranes as a result of the transport of H^+ and OH^- ions. However, as there is no flux of water through the membranes, ED can treat water with a higher level of suspended solids than RO. As nonionic solids, e.g. silica, are not concentrated by this process, these components are of less concern. Thus, inorganic acids such as H_2SO_4 can be fed to the concentrate stream to maintain a low pH and thus minimize scaling and fouling.

3.3.3. *Ion-Exchange Membrane*

In the ion-exchange membranes, the charged groups are attached to the polymer backbone of the membrane material. A cationic membrane with fixed negative groups will exclude the anions but is permeable to the cations. Similarly, an anionic membrane with fixed positive groups will exclude cations but is permeable to anions. Cation-selective membranes consist of sulfonated polystyrene, whereas anion-selective membranes consist of polystyrene with quaternary ammonia. There are two types of IX membranes: homogeneous membranes and heterogeneous membranes.

3.3.3.1 Homogeneous Membrane

Charged groups are uniformly distributed through the membrane matrix. The cation-selective membranes are often made up of crosslinked polystyrene (with divinyl benzene) that has been sulfonated to produce sulfonate that is attached to the polymer. The anion-selective membranes are made of crosslinked polystyrene containing quaternary ammonium groups.

3.3.3.2 Heterogeneous Membrane

The IX groups are contained in small domains distributed throughout an inert support matrix that provides the mechanical support. The simplest form has very finely powdered cation or anion exchange particles uniformly dispersed in polypropylene. A much finer heterogeneous dispersion of IX can be made from polyvinylchloride plastisol.

3.3.4. Solution Pumping System

The low-pressure pump helps to circulate the feed water through the stack. The pump must have enough power to overcome the resistance of water because it passes through the narrow passages. The energy required for the solution pumping system increases as the average salt concentration of the feed decreases. This can be significant when the feed concentration drops to a very low value. The pressure drop per stack varies from 15 to 30 psi for flow cells and 70 to 90 psi for tortuous cells. Depending on the extent of separation required, the water is pumped through 2–4 cells in series. Therefore, interstage pumps may be required.

3.3.5. Posttreatment

Posttreatment includes stabilizing the water and preparing it for distribution. Removal of unwanted gases such as hydrogen sulfide and adjustment of the pH are normally conducted. If the salt content in the water is still too high, other processes such as IX and RO can be used.

3.4. Electrodialysis Reversal

As the membranes used in the electrodialysis easily become fouled, efforts have been made to overcome the problem. EDR process was developed to help eliminate membrane fouling. In the early 1970s, an American company commercially introduced the EDR process for ED. EDR is a continuous self-cleaning ED (17, 18).

EDR was developed with ED technology as the basis. In the EDR process, the membrane polarity is reversed several times an hour as shown in Fig. 12.14. The brine stream becomes the product water stream and vice versa. As a result, the ions are now attracted in the reverse direction across the membrane stack. Immediately following the reversal of polarity and flow, the product water is dumped until the stack and lines are flushed out, and the desired water quality is restored. This flush takes about 1 or 2 min, and then the unit can resume producing water. The reversal process is useful in breaking up and flushing out scales, slimes, and other deposits in the cells before they can build up and create a problem. Flushing allows the unit to operate with fewer pretreatment chemicals and minimizes membrane fouling. The EDR systems operate with higher concentrations in the brine or concentrate the streams with less flow to waste (19).

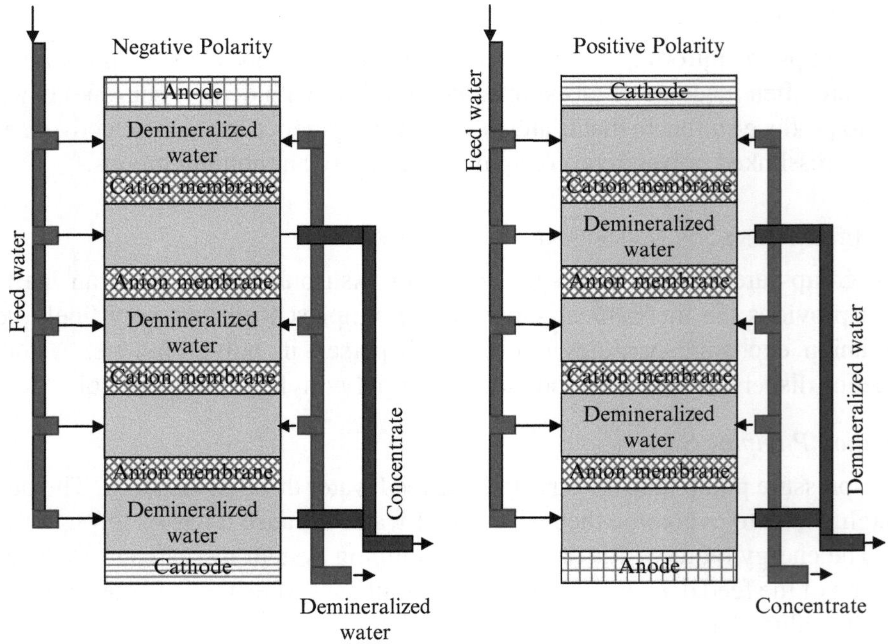

Fig. 12.14. Demonstration of EDR process.

The major advantage of EDR is its ability for cleaning colloidal particulates that are formed on the membrane during the operation, the mechanism of which can be better explained by Fig. 12.15. The particulates interact with water to form an effective negative charge at the surface of their bound water layer. The DC applied to ED stacks is a driving force that moves the negatively charged colloids toward the anion exchange membrane (20). When the particles reach the membrane surface, the electric field and electrostatic attraction to IX sites at the membrane surface tend to hold the deposit in place. The periodic DC power polarity reversal in an EDR system reverses the driving force for this deposition and tends to remove the deposit.

The EDR systems offer the following benefits:

- Automatic cleaning cycle every 10–20 min reduces scales, fouling, and slimes.
- Fresh water production is high.
- The systems can be either land based or mobile.
- The membrane life of the EDR systems ranges between 5 and10 years.
- The EDR membrane stacks can be easily cleaned.
- The systems are less sensitive than RO to particulates and metal oxides.
- The systems do not have silica limitation.
- The systems are stable and perform continuously.
- The systems can be operated with effective residuals of chlorine, chloramines, and chlorine dioxide to prevent biological fouling.
- Average free chlorine residuals up to 0.5 mg/L and chloramine (total chlorine) residuals up to 1.0 mg/L have no effect on membrane life.

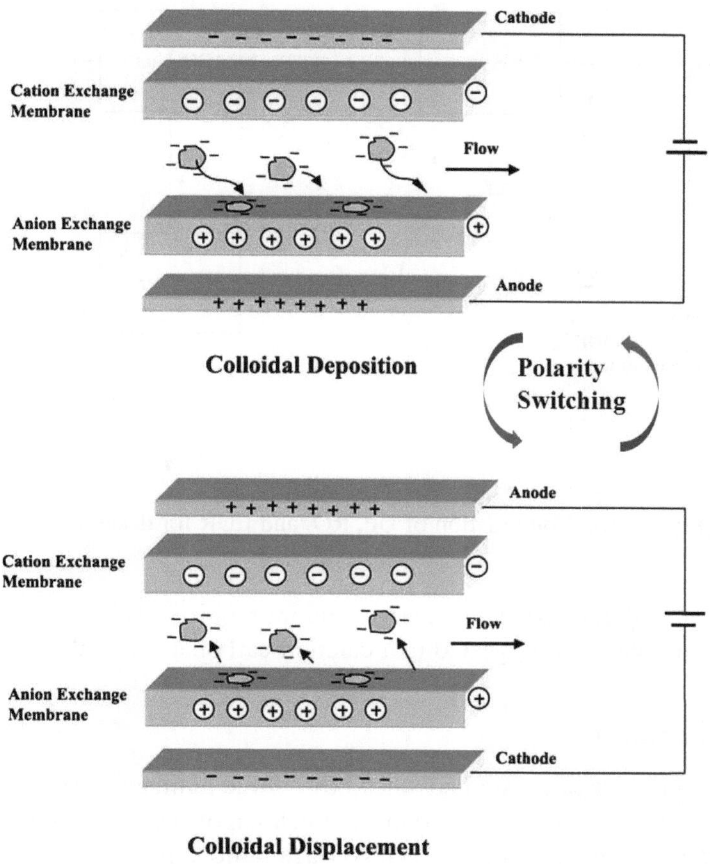

Fig. 12.15. Deposition and displacement of colloidal particulates during EDR operation.

The EDR system has a wide flow capacity ranging from 500 gal/d up to 1.6 million gal/d. By using multiple stages (stacks in series), the systems are designed to handle a wide range of treatment needs. It is able to reduce TDS with an efficiency of 50–95%.

The major advantage of EDR over RO is that it is not necessary to feed chemicals at remote water treatment sites. EDR and RO compete with one another on many desalting projects. EDR offers advantages over RO on some applications. It can be applied to reclaim the RO concentrate of approximately 8,000 mg/L TDS. The combined RO–EDR system in Fig. 12.16 can yield up to 96% of recovery (18). EDR is not affected as many feed water constituents as RO. The EDR can take raw water with turbidity up to 2 nephelometric turbidity unit (NTU), whereas the RO cannot function well if the turbidity is above 1 NTU. The oil & grease (O&G) of raw water to the EDR is up to 1 mg/L, whereas the raw water to RO must be O&G free.

Particles and microorganisms are not removed by the EDR (also ED). Hence, posttreatment of the product water becomes necessary. Filtration and disinfection by chlorination can

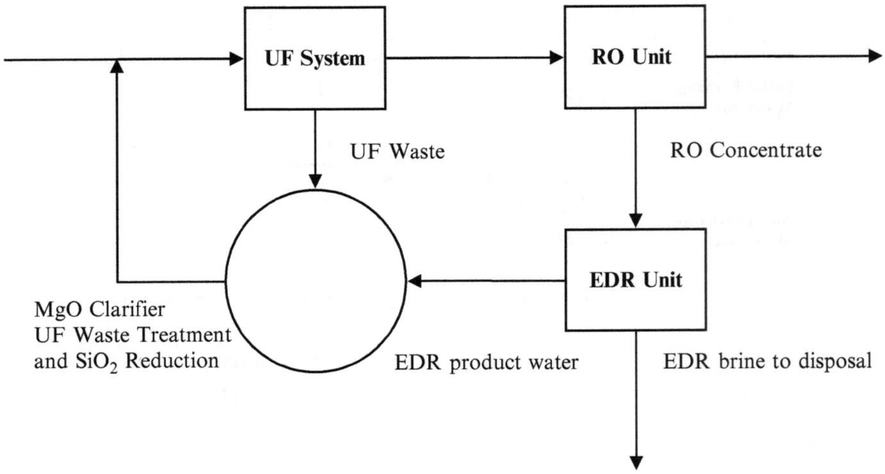

Fig. 12.16. Combination of UF, RO, and EDR for desalination.

be used. Remineralization of product water is seldom required. However, when recalcification is deemed desirable, it is suggested that calcium carbonate be added.

3.5. *Electrodeionization*

EDI theory and practice have been advanced by a large number of researchers throughout the world. EDI process was first described in a publication by a group of scientists at Argonne Labs in early 1955 as a method for removal of trace radioactive materials from water (21). EDI technology was first fully commercialized in early 1987 by a division of Millipore that is now part of US Filter Corporation (22). Several thousands commercial EDI devices are now manufactured by a number of companies, such as Ionpure, Ionics, GE, and Electropure, for the production of high-purity water at capacities ranging from <0.1 to >250 m^3/h.

EDI process is a continuous process, which has very high water recovery. RO/EDI systems may achieve overall water recoveries of greater than 90%. The electrical requirements are nominal. When 1 kWh of electricity is used to deionize 1,000 gallons of raw water, the conductivity can typically be reduced from 50 μS/cm in the feed to 0.1 μS/cm in the product water. As the EDI concentrate (or reject) stream contains only the feed water contaminants at 5–20 times higher concentration, it can usually be discharged without treatment, or used for another process. The process is particularly good for the production of ultrapure water.

The operation mechanism of EDI is shown in Fig. 12.17. The process uses a combination of ion-selective membranes and IX resins sandwiched between two electrodes (anode [+] and cathode [-]) under a DC voltage potential to remove ions from pretreated water. Compartment sets are called cell pairs and form the basic element in a module. Ion-selective membranes operate using the same principle and materials as IX resins. Anion-selective membranes are permeable to anions but not to cations, while cation-selective membranes are permeable to

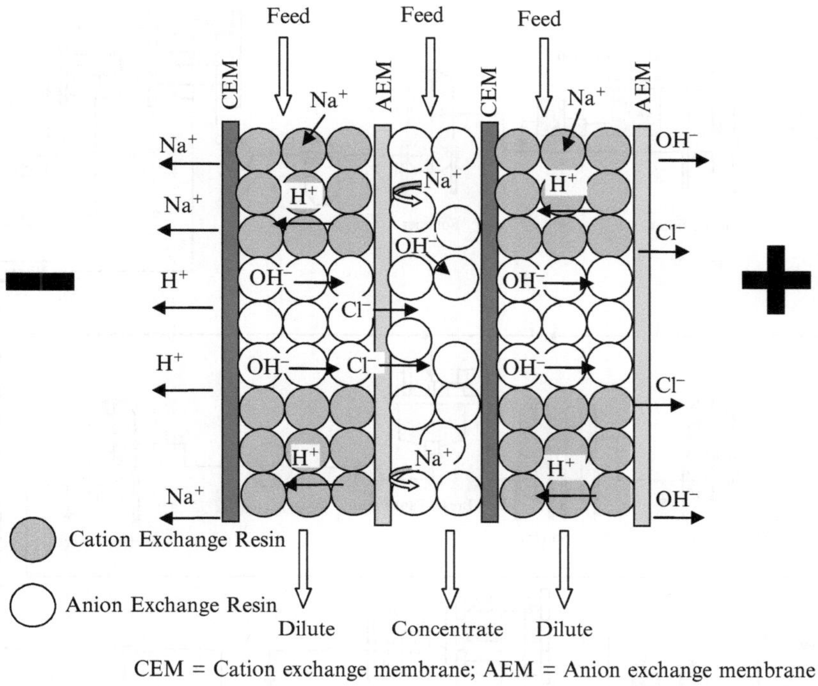

Fig. 12.17. Schematic diagram of electrodeionization for ultrapure water production.

cations but not to anions. The membranes are not water permeable. Two complete EDI processes are given in Fig. 12.18.

By spacing alternating layers of anion-selective and cation-selective membranes within a plate-and-frame module, a "stack" of parallel purifying and concentrating compartments is created. The "stack" of cell pairs is positioned between the two electrodes, which supply the DC potential to the module. Under the influence of the applied DC voltage potential, some of the ions across the respective membranes are captured by the IX resins. The remaining ions are transported across the membranes and migrated into the concentrating chambers. As a result, the salt content in the dilute compartments is very low (free of ions in most operations). The resins in the devices are continuously regenerated by the electric current; therefore, they do not become exhausted. This continuous electro-regeneration enables continuous EDI systems to produce pure water without chemical regeneration.

4. REVERSE OSMOSIS

RO process is one of the most important desalination processes. It is estimated that more than USD 10 billion will be spent for the construction of RO plants over the next 5 years and over USD 70 billion over the next 20 years. RO becomes a popular choice as the price of membranes has decreased significantly since 1980. For example, the relative cost of spiral-wound membranes has declined by more than 60% in the past 25 years (23).

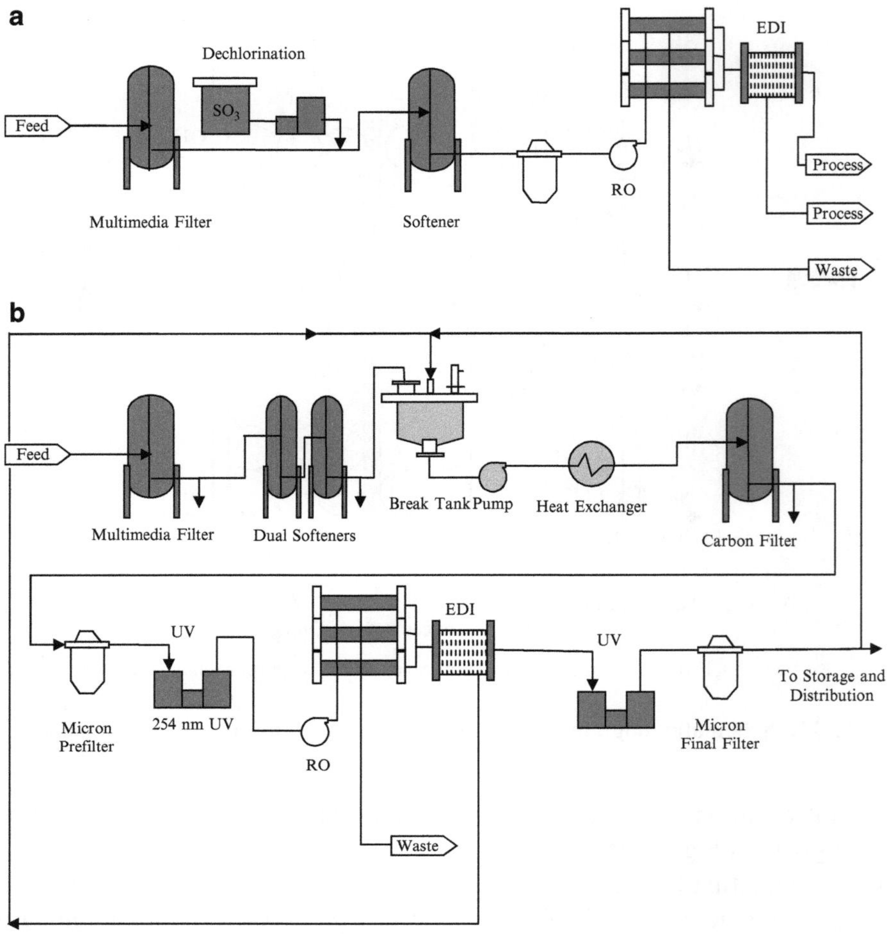

Fig. 12.18. Production of ultrapure water by EDI process: (**a**) pretreatment-single-pass RO/EDI; (**b**) USP purified water system with EDI.

In RO, saltwater on one side of a semipermeable membrane is subjected to a high pressure (24). Pure water diffuses through the membrane, leaving behind a more salty concentrate containing most of the dissolved organic and inorganic contaminants. The pressure ranges from 15 to 25 bar for brackish water and from 55 to 85 bar for seawater. Two of the most commercially successful RO configurations are spiral-wound and hollow fiber. Brackish water RO plants can typically recover 50–80% of the feed water, with 90–98% of salt rejection. For seawater, recovery rates vary from 20 to 40%, with 90–98% of salt rejection. The feed must have undergone pretreatment so that the solid and organic contents are below a certain level. The pretreatment usually consists of chemical coagulation, filtration (fine filtration, ultrafiltration or microfiltration), scaling control (e.g. softening), and acidification for regulating pH. Posttreatment of production water is needed for its distribution, which

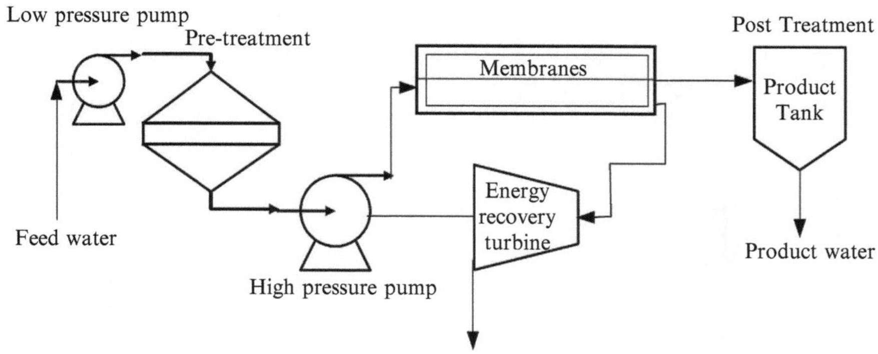

Fig. 12.19. Schematic presentation of a reverse osmosis desalination plant.

includes disinfection, removing gases (e.g. H_2S) and pH balancing. Interesting readers can refer to the literature (25–32) for details. A complete RO system is shown in Fig. 12.19.

5. ENERGY

In desalination, energy must be applied to drive the operations. The energy can be in the form of heat, pressure, and electricity. Its sources can be many, such as sun light. Desalination is very energyintensive, as illustrated in Fig. 12.20 (8). RO needs about 6 kWh of electricity per m^3 of water, whereas MSF and MED require heat at 70–130°C (25–200 kWh/m^3). A variety of low-temperature heat sources may be used, including solar energy. The choice of process generally depends on the relative economic values of freshwater and the particular energy source (e.g. fuel).

Traditionally, fossil fuels have been used for most desalination, which contributes to increased levels of greenhouse gases. The crude oil price has been dramatically increased in the last 2 years. Thus, the alternative energy sources must be sought.

Nuclear energy is becoming an important energy source as it is more costeffective (2), clean, and efficient. Nuclear energy is already being used for desalination and has the potential for much greater use.

Small- and medium-sized nuclear reactors are suitable for desalination, often with cogeneration of electricity using low-pressure steam from the turbine and hot sea water feed from the final cooling system. A nuclear plant may have desalination capacity of as high as 500,000 m^3/d.

Large-scale deployment of nuclear desalination on a commercial basis will depend primarily on economic factors. The United Nation's International Atomic Energy Agency has been fostering research and collaboration on the issue, and more than 20 countries are involved.

Instead of using the energy produced by nuclear reactions, the waste heat produced by the nuclear reactors can be directly used for running desalination operations. The reactors can be designed to couple with desalination systems. It can be expected to enable further cost

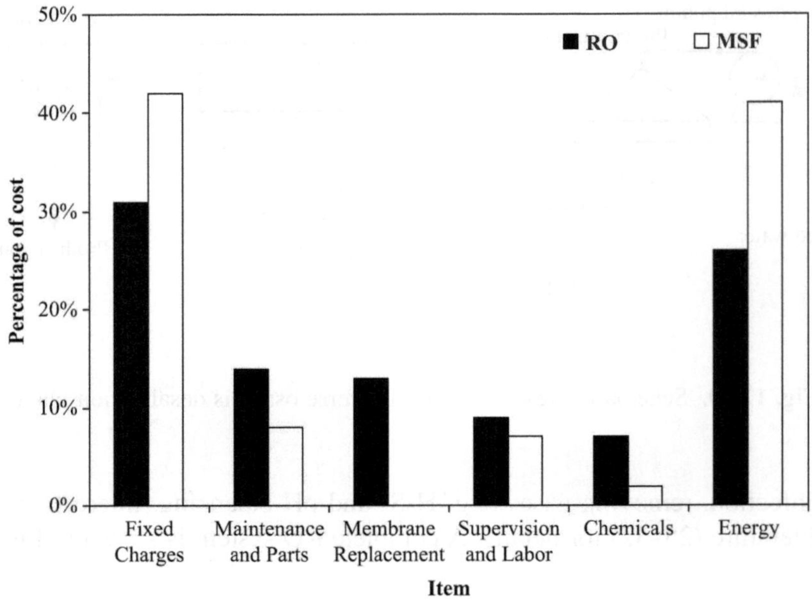

Fig. 12.20. Cost composition for representative RO and MSF seawater desalination plants.

reductions of nuclear desalination. Safety and reliability are of course key requirements. There are many successful cases.

In Japan, some 10 desalination facilities linked to pressurized water reactors operating for electricity production have yielded 1,000–3,000 m^3/d each of potable water. An MSF was initially employed, but an MED and an RO have been found to be more efficient. India has been engaged in desalination research since the 1970s and is about to set up a demonstration plant coupled to few nuclear power reactors at the Madras Atomic Power Station, Kalpakkam. This project will be a hybrid RO/MSF plant; RO and the MSF produce freshwater of 1,800 and 4,500 m^3/d, respectively. The plant will deliver water of 45,000 m^3/d by both kinds of desalination technologies. China is looking at the feasibility of a nuclear seawater desalination plant in the Yantai area producing 160,000 m^3/d by an MED process.

6. ENVIRONMENTAL ASPECT OF DESALINATION

Absolute environmental impacts of desalination plants and the respective processes are largely unknown owing to limited public attentions (8, 25). One impediment to the growing demand for desalting technologies is the environmental pollution that every desalination water treatment plant (DWTP) must face. Most environmental concerns that are raised relate to saltwater, air, and noise emissions.

Air emissions are a result of the desalination industry's heavy energy consumption and involve the commonly named pollutants carbon dioxide and sulfur dioxide. In light of their substantially higher energy consumption, thermal processes are inferior to membrane processes when it comes to air pollution. Noise pollution is very often associated with equipments such as high-pressure pumps in RO process, which produce noise during operations. Various technological measures can be utilized to reduce noise levels so that the risk to the operators can be minimized (26). The operations should be kept away from population centers.

The more important form of emission from desalination that raises environmental concerns comprises the discharge of concentrated saltwater after the desalination process is completed (27). The TDS of wastewater from a conventional DWTP is approximately 3–5 times higher than that in the feed; the flow rate is 20–50% of feed flow. The waste also contains chemicals used in the pretreatment of feed water, such as antiscalants, surfactants, and acid.

Major disposal alternatives for the water pollution are given in Table 12.4. Each method varies in complexity and costs. Sewer discharge is commonly the least complex and least

Table 12.4
List of disposal alternatives for water pollution in DWTP

Technology	Brief description
Ocean disposal	Ocean disposal is typically done several miles offshore with diffusers. Dilution by seawater can reduce the contents of the waste. The treatment plant outfall location may be affected by water quality rules and regulation.
Discharge to sewer and surface waters	Sewer discharge is the simplest; however, it depends on the ability of the wastewater treatment plant to accept high salinity discharge. Surface water discharge involves discharge to a point of outfall such as tidal lake or brackish canal. Dilution by surface waters can reduce the contents of wastes. The location of discharge and the pretreatment requirement are determined by state and regulatory agencies; water quality testing (e.g. bioassay toxicity) is required.
Land application	Irrigation is sometimes used for the waste streams with relatively lower salinity. Saline tolerant vegetation and habitat are required. This is usually determined by site-specific soil and drainage characteristics
Evaporation pond	Desalting concentrate is discharged to ponds and evaporated to dryness for final disposal. It is land intensive and requires relatively dry climates. Dry salt is the final waste product, which must be characterized and disposed of accordingly as solid waste. This is generally applicable only in arid environments and then usually in an inland location
Deep well injection	In deep well injection, the waste streams are injected to the ground and are permanently stored in the injection zone. It is very commonly used by the desalination water treatment plants
Zero discharge and industrial reuse	Zero discharge of wastes can be achieved by thermal processes, which greatly reduce or eliminate the waste stream. The thermal processes are energy intensive and very costly. Wastes must be characterized and disposed of accordingly. Marketable salts for general industrial uses can be recovered. This is still in the development stages, but past attempts indicate that recovery of marketable byproducts is economically difficult

costly. Zero discharge is the most complex and the thermal processes are costly. The salt produced can be reused by other industries; thus, the approach can be considered more environmental friendly according to the pollution prevention principles (28–32). It can help to improve public acceptance. The cost can be offset by beneficial use of byproducts (brine or specific salts).

7. RECENT ADVANCES OF THERMAL DISTILLATION AND ELECTRODIALYSIS FOR DESALINATION

The technology of thermal distillation and electrodialysis for desalination water continues to improve, driven by advances in technology, the need to reduce costs, and commercial competition. Other than the aforementioned MSF, MED, VC, and ED (EDR) processes, recently, a number of other distillation and electrodialysis processes have been developed to desalinate brine water. These processes have not achieved the level of commercial success that MSF, MED, VC, and ED (EDR) have, but they may become valuable under special circumstances or with further development. These important processes include membrane distillation and photovoltaic electrodialysis.

7.1. Membrane Distillation

Membrane distillation (MD) combines the use of both thermal distillation and membranes and was introduced commercially on a small scale in the 1980s. The process relies primarily upon thermal evaporation and the use of microporous hydrophobic membranes to pass vapor, which is then condensed to produce fresh water. MD operates on the principle of vapor–liquid equilibrium as a basis for molecular separation. The driving force of the process is supplied by the vapor pressure difference caused by temperature gradient imposed between the liquid–vapor interfaces (33).

MD yields highly purified permeate and separates contaminants from aqueous solutions, and has potential application in many areas of scientific and industrial interest. In particular, MD process can be used as a substitute for conventional desalination processes such as MSF, MED, and RO. The advantages of MD compared to that of these processes are as follows: (a) low operating temperature of a feed enables the utilization of waste heat as a preferable energy source; (b) lower vapor space required than traditional distillation, such as MSF and MED; (c) lower operating pressure than that required for pressure-driven membrane processes, such as RO; (d) almost a complete rejection of dissolved and nonvolatile species; and (e) performance not limited by high osmotic pressure or concentration polarization. Although the MD technology has the advantages that other membrane technologies do not have, one of the main obstacles impeding the implementation of MD is membrane fouling, which is caused by interactions between the membrane and the various components during the process (34).

Recently, the interest in using MD process for desalination is increasing worldwide due to these attractive features, especially when coupled with solar energy or utilizing low-grade heat source (35–37). Solar-powered desalination is an attractive and viable method for the production of fresh water in remote arid areas.

7.2. Photovoltaic Electrodialysis

As aforementioned, ED has proved its feasibility and high performance for desalination of brine water. ED desalination plants are well developed in industrial scale to provide big cities with freshwater. Small villages or settlements in remote rural areas without infrastructure do not profit from these techniques. The lack of energy source as well as a missing connection to the grid complicates the use of such techniques in these places. Furthermore, nowadays, the social concern about the environment is growing, and laws demanding environmental protection are being approved. These facts have led to the research and development of new environment-friendly technology for brine water desalination (38).

One of the most successful endeavors and technological achievements made was the use of photovoltaic (PV) system as power supply for an electrodialysis system for brine water desalination in remote areas (39). Generally, a PV system consists of PV modules, batteries, a regulator or controller, an inverter (dc–ac), and loads (charges). The PV modules can be connected in series and/or parallel and the solar radiation incident on the surface of the PV array is transformed into electric energy (direct current) by the PV modules. The generated electricity is passed to the regulator, which protects the batteries from overcharging or an excessive discharge. The batteries store energy that can be used as electrical back-up during periods of low solar insolation – for example, during cloudy weather or at night. The inverter transforms the direct current into alternating current (ac) for those devices that work with the latter. These systems can be used in remote sites for the self-sufficiency of electrical power in a reliable and autonomous way.

7.3. Future Study on Desalination Technologies

Researches are focused on the development and feasibility studies of desalination concepts that have not been fully explored and new desalination concepts for non-traditional desalination processes (40–41). Some of the areas considered for R&D activity are listed as follows: (a) New desalination concepts and feasibility studies; (b) coupling of desalination processes with nonconventional energy sources, such as solar energy; (c) new process design concepts of reported nonconventional desalination processes; (d) Development of new design concepts for process equipment; (e) Measurement and control systems; and (f) scale monitoring, controlling, and cleaning systems.

8. NOMENCLATURE

ΔT = Difference in temperature between the condensing steam (Ts) and the boiling liquid (T_L) in the evaporator (K)
A = Heat-transfer area (m^2)
C_{PF} = Heat capacity of the feed (kJ/kg K)
F = Feed to the evaporator (kg/h)
h_F = Enthalpy of the feed to the evaporator (kJ/kg)
h_L = Enthalpy of the concentrated liquid leaving the evaporator (kJ/kg)

h_S = Enthalpy of condensate (kJ/kg)
H_s = Enthalpy of steam (kJ/kg)
H_V = Enthalpy of vapor (kJ/kg)
L = Concentrated liquid leaving the evaporator (kg/h)
P_1 = Saturation vapor pressure of the liquid of composition (X_L) at its boiling point (T_L)
q = Rate of heat transfer (W)
S = Condensed steam leaving the system (kg/h)
T_L = Temperature of the concentrated liquid leaving the evaporator (K)
T_F = Temperature of the feed to the evaporator (K)
T_S = Temperature of the condensed steam leaving the system (K)
T_V = Temperature of the vapor (K)
U = Overall heat-transfer coefficient (W/m^2 K)
V = Vapor given off as pure solvent (kg/h)
X_F = Solid content of the feed to the evaporator (mass fraction)
X_L = Solid content of the concentrated liquid leaving the evaporator (mass fraction)
Y_V = Solid content of the vapor = 0
λ = Latent heat given off by steam (kJ/kg)

REFERENCES

1. Murakami M (1995) Managing water for peace in the Middle East: alternative strategies, United Nations University Press, Tokyo, New York, Paris
2. World Nuclear Association (2004) Nuclear desalination, London
3. Buros OK (2000) The ABCs of Desalting, International Desalination Association, Topsfield, MA
4. Alkaibi AM, Lior N (2004) Membrane-distillation desalination: status and potential. Desalination 171:111–131.
5. Ding ZW, Liu LY, El-Bourawi MS, Ma RY (2005) Analysis of a solar-powered membrane distillation system. Desalination 172:27–40.
6. Ettouney HM, El-Dessouky HT, Faibish RS, Gowin PJ (2002) Evaluating the economics of desalination. Chem Eng Prog 98:32–39.
7. Dore MHI (2005) Forecasting the economic costs of desalination technology. Desalination 172:207–214.
8. Ebensperger U, Isley P (2005) Review of the current state of desalination, Environmental Policy Program, Georgia State University, Atlanta, GA
9. Department of Natural Resources and Mines (2004) *Desalination in Queensland, Final report*, The State of Queensland, Australia
10. UNEP (1997) Source book of alternative technologies for freshwater augmentation in Latin America and the Caribbean, UNEP, International Environmental Technology Centre, Osaka/Shiga, Japan
11. Geankoplis CJ (2003) Transport processes and separation process principles. 4th ed., Pearson Education, Inc., pp 700–702.
12. El Din AMS, El-Dahshan ME, Haggag HH (2005) Carbon-induced corrosion of MSF condenser tubes in Arabian Gulf water. Desalination 172:215–226.
13. Mathioulakis E, Belessiotis V, Delyannis E (2007) Desalination by using alternative energy: Review and state-of-the-art. Desalination 203:346–365.

14. Joseph J, Saravanan R, Renganarayanan S (2005) Studies on a single-stage solar desalination system for domestic applications. Desalination 173:77–82.
15. Badran AA, Al-Hallaq AA, Eyal Salman IA, Odat MZ (2005) A solar still augmented with a flat-plate collector. Desalination 172:227–234.
16. Tripathi R, Tiwari GN (2005) Effect of water depth on internal heat and mass transfer for active solar distillation. Desalination 173:187–200.
17. US Department of Army (1986) Water desalination army TM 5–813–8, Washington, DC
18. Reahl ER (2004) Half a century of desalination with electrodialysis, Technical Paper, Ionics Inc.
19. Tsiakis P, Papageorgiou LG (2005) Optimal design of an electrodialysis brackish water desalination plant. Desalination 173:173–186.
20. Allison RP (1995) Electrodialysis reversal in water reuse applications. Desalination 103:11–18.
21. Walters WR, Weiser DW, Marek LJ (1955) Concentration of aqueous radioactive wastes. Indus Eng Chem 47:61–67.
22. Ganzi GC, Egozy Y, Giuffrida AJ, Jha A (1987) High purity water by electro-deionization. Ultrapure Water J **4**:3.
23. Semiat R (2000) Desalination: present and future. Water Int 25:54–65.
24. Kahraman N, Cengel YA, Wood B, Cerci Y (2004) Energy analysis of a combined RO, NF, and EDR desalination plant. Desalination 171:217–232.
25. Einav R, Lokiec R (2003) Environmental aspects of a desalination plant in Ashkelon. Desalination 156:79–85.
26. Crowl DA, Louvar JF (2002) Chemical process safety, fundamentals with applications. 2nd ed., Prentice Hall, NJ
27. Miller JE (2003) Review of water resources and desalination technologies, Sandia National Laboratories, SAND 2003–0880, Albuquerque, NM
28. Chen JP, Shen TT, Hung YT, Wang LK (2004) Pollution prevention, in handbook of industrial hazardous wastes treatment. 2nd ed., Marcel Dekker, NY, pp 971–1004.
29. AWWA (2006) Desalination of Seawater and Brackish Water. American Water Works Association, Denver, CO
30. Wang LK, Hung YT, Shammas NK (eds) (2006) Advanced physicochemical treatment processes. Humana Press Inc., Totowa, NJ, pp 549–580
31. California Coastal commission (2010) Seawater Desalination in California, http://www.coastal.ca.gov/desalrpt/dsynops.html.
32. (2006) Texas A&M University, 2nd Annual short course on the future of desalination in Texas, August 6–8, College Station, TX.
33. Al-Obaidani S, Curcio E, Macedonia F, Profio GD, Al-Hinai H, Drioli E (2008) Potential of membrane distillation in seawater desalination: thermal efficiency, sensitivity study and cost estimation. J Membrane Sci 323:85–98.
34. Wang J, Qu D, Tie M, Ren HJ, Peng XJ, Luan ZK (2008) Effect of coagulation pretreatment on membrane distillation process for desalination of recirculating cooling water. Sep Purif Technol 64:108–115.
35. Banat F, Jwaied N (2008) Economic evaluation of desalination by small-scale autonomous solar-powered membrane distillation units. Desalination 220:566–573.
36. Banat F, Jwaied N (2008) Exergy analysis of desalination by solar-powered membrane distillation units. Desalination 230:27–40.
37. Romel M, Koschikowski J, Wieghaus M (2007) Solar driven desalination systems based on membrane distillation. Solar Desalination for the 21st Century 247–257.

38. Ortiz JM, Exposito E, Gallud F, Garcia-Garcia V, Montiel V, Aldaz A (2008) Desalination of underground brackish waters using an electrodialysis system powered directly by photovoltaic energy. Solar Energy Mater Solar Cell. 92:1677–1688.
39. Ortiz JM, Exposito E, Gallud F, Garcia-Garcia V, Montiel V, Aldaz A. (2006) Photovoltaic electrodialysis system for brackish water desalination: Modeling of global process. J Membrane Sci 274:38–149.
40. Khawaji AD, Kutubkhanah IK, Wie JM (2008) Advances in seawater desalination technologies. Desalination 221:47–69.
41. Deng R, Xie L, Lin H, Liu J, Han W (2009) Integration of thermal energy and seawater desalination, Energy, in press, doi:10.1016/j.energy.2009.05.025.

13

Desalination of Seawater by Reverse Osmosis

Jiaping Paul Chen, Edward S. K. Chian, Ping-Xin Sheng, K. G. Nadeeshani Nanayakkara, Lawrence K. Wang, and Yen-Peng Ting

CONTENTS

Abstract Desalination allows the use of non-conventional water sources such as seawater for the production of potable water. Reverse osmosis (RO), one of the technologies for desalination, is becoming popular in the water industry. In this chapter, theory of RO process, plant configurations, and practical considerations related to the plant operation are addressed. Factors such as high permeate flux, high solute rejection, and mechanical and chemical stability govern the production of membranes for RO. Cellulose acetate membrane is popular due to the chlorine and fouling resistance. When it comes to rejection, thin film membranes are advantageous. Membranes are usually arranged in modules. Concentration polarization and compaction are two major limiting factors in the RO technology. Feed water must be pretreated using conventional and/or membrane filtration technologies in order to minimize membrane fouling. Reduction in permeate, pressure drop over the system, and decrease in rejection are the indications for the requirement of cleaning and regeneration of membranes. Chemical and/or physical methods can be used for the cleaning and regeneration of membranes. A case study and the recent developments are discussed in order to enhance the understanding of the process.

Key Words Desalination • fouling • membrane filtration • reverse osmosis.

From: *Handbook of Environmental Engineering, Volume 13: Membrane and Desalination Technologies*
Edited by: L.K. Wang et al., DOI: 10.1007/978-1-59745-278-6_13

1. INTRODUCTION

Demand for potable water is increasing because of the reduction in reliable fresh water sources and increase in global population. Therefore, it is necessary to search for technologies that can convert non-conventional water sources into fresh water. Seawater is one such abundant source which can be reached by most of the countries in the world. In order to utilize seawater as potable water, it is required to remove the high salinity. Desalination technologies are intended for the removal of dissolved salts that cannot be removed by conventional treatment processes. Thermal distillation technologies have been used on some ships for more than 100 years. Desalination was used on a limited scale for municipal water treatment in the late 1960s. On the basis of desalination, the past four decades can be divided into three phases: (1) 1950s were a time of discovery; (2) 1960s were concerned with research; and (3) 1970s and 1980s were the time of commercialization. In the beginning of the 1970s, the industry began to concentrate on commercially viable desalination applications and processes.

The first commercial plant for the production of potable water from a saline source using electrodialysis was put into operation in 1954. At that time, this process was not received favorably because of its inability to reduce dissolved solids to a desired extent. The first reverse osmosis (RO) water treatment plant was constructed in 1970s in Florida. Significant advances in membrane materials and technologies in the last three decades have greatly improved the cost effectiveness and performance capabilities of the processes. RO membrane processes are increasingly used worldwide to solve a variety of water treatment problems. In the desalination industry of USA, RO membrane technology is the most popular.

In this chapter, the process of RO and the mechanisms of membrane separation are illustrated. Membrane materials are presented. Several mathematical models are given for the membrane processes used in desalination. Pre- and posttreatment processes are discussed. Examples, a case study, and recent developments in the seawater RO desalination technology are elucidated.

2. MEMBRANE FILTRATION THEORY

2.1. *Osmosis and RO*

Osmosis is the phenomenon of water flow through a semipermeable membrane that blocks the transport of salts or other solutes through the membrane. It is applied to water purification and desalination, waste material treatment, and many other chemical and biochemical processes. When two aqueous solutions (or other solvents) are separated by a semipermeable membrane, water will flow from the side of low solute concentration to the side of high solute concentration. The flow is stopped, or even reversed by applying external pressure on the side of higher concentration. In such a case, the phenomenon is called RO, as shown in Fig. 13.1. If there are solute molecules only on one side of the system, then the pressure that stops the flow is called the osmotic pressure Π, which is given by the Van't Hoff equation:

$$\Pi = C_S RT. \tag{1}$$

Fig. 13.1. Schematic of reverse osmotic equilibrium (**a**) and reverse osmosis (RO) membrane (**b**).

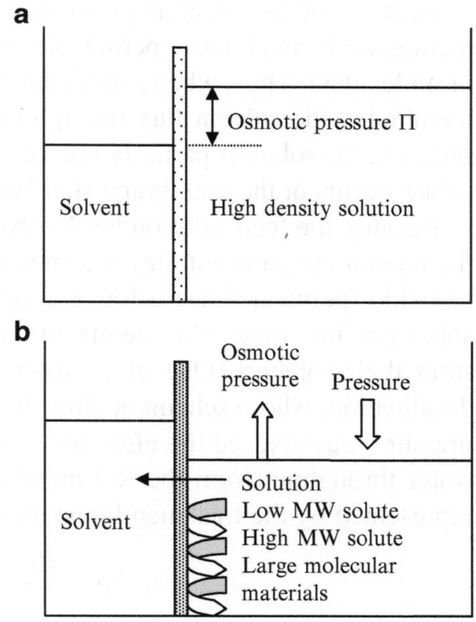

Here, Π is the osmotic pressure (force/length^2), C_S is the summation of molarities of all dissolved ions (moles/length^3), R is the ideal gas constant (force-length/mass-temperature), and T is the absolute temperature (K). The magnitude of the osmotic pressure, even in very dilute solution, is very large.

For instance, with a solute of 0.01 M at room temperature (298 K), the osmotic pressure would be

$$\Pi = C_S RT = \left(\frac{0.010 \text{ mol}}{1 \text{ L}}\right) \times \left(\frac{0.0821 \text{ L atm}}{\text{mol K}}\right) \times 298 \text{ K} = 0.24 \text{ atm.}$$

This pressure is sufficient to support a column of water 8.1-ft high.

For applications of the RO of seawater, an approximate expression is (88):

$$\Pi = 1.12T \sum \bar{m}_i, \tag{2}$$

where Π is in psia, T is in K, and $\sum \bar{m}_i$ is the summation of molarities of all dissolved ions and nonionic species in the solution (M).

RO membrane has an anisotropic (also called asymmetric) structure: it consists of an extremely thin skin layer (100–500-nm thick) with very fine microporous texture (1) and on the top of it a much thicker, spongy, supporting sublayer with much larger pore size. In the case of cellulose acetate (CA) RO membranes of the Loeb–Sourirajan type (2), for example, the pores of the skin layer have been estimated, based on results of titrated water diffusion experiments, to be less than 0.8 nm (8 Å) in diameter (3), and the pore size of the substructure is in the range of 100–400 nm (4).

Because of its peculiar physicochemical properties and porous texture (5, 6), an RO membrane is much more permeable to solvent molecules than to solute particles (ions and/or molecules). Thus, when a sufficiently pressurized solution (feed solution) is applied to the membrane, the solvent flux through the membrane is usually much greater than the solute flux, i.e., the solute is partially rejected. It is believed that the separation between solvent and solute occurs at the membrane skin layer (7).

Because the feed solution loses solvent, the resulting solution on the high-pressure side of the membrane (concentrate) becomes more concentrated, whereas the solution on the opposite side (permeate) has a lower concentration of solute. The process is used to separate solutes of low molecular weight (usually molecules and/or ions with diameters <100 nm) from their solvents. One of its most important applications is water (seawater included) desalination, whose schematic flow diagram is shown in Fig. 13.2. In this case, the operating pressure must exceed the effective osmotic pressure of the feed solution in order to get any water throughput from the RO membrane. The resultant hydrostatic pressure difference is represented by the transmembrane pressure difference Δp:

$$\Delta p = \frac{P_{\text{feed}} + P_{\text{concentrate}}}{2} - P_{\text{permeate}}, \tag{3}$$

where P_{feed}, $P_{\text{concentrate}}$, and P_{permeate} are the hydrostatic pressure of the feed, concentrate, and permeate flows, respectively.

The effectiveness of the RO process for water desalination largely depends on the permeability and selectivity of the membrane. Characteristics that define the membrane performance can be quantified for a set of given operating conditions by determining the permeate flux through the membrane and the ratio between the permeate flow rate and the effective area of the membrane.

The solute rejection R of each solute of interest is defined by

$$R = \frac{C_s^{\text{I}} - C_s^{\text{II}}}{C_s^{\text{I}}} = 1 - \frac{C_s^{\text{II}}}{C_s^{\text{I}}}, \tag{4}$$

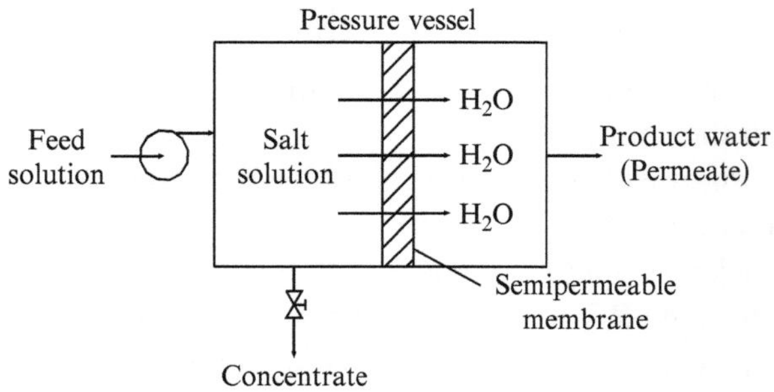

Fig. 13.2. Simplified RO flow diagram.

where C_s^{I} is the solute concentration at the feed solution – membrane interface and C_s^{II} is the solute concentration in the permeate. The desalination ratio D_r is defined by

$$D_r = \frac{C_s^{I}}{C_s^{II}}. \tag{5}$$

Some important requirements for an RO membrane to be used in water desalination are as follows:

- High permeate flux.
- High solute rejection.
- Mechanical strength to withstand elevated operating pressures.
- Chemical stability for a wide range of temperature and pH.
- Operating capability over a broad range of ionic species present in the feed solution.
- Resistance to biological attack.

2.2. *Membranes*

A wide variety of membrane types, sizes, construction options, and techniques are available for removing salt from seawater. There are basically two types of commercial membranes used in RO applications to date: CA and thin film (TF). The former is considered an integral membrane and the latter a composite membrane. Membrane manufacturing, operating conditions, and performances differ significantly for each group of polymeric material.

2.2.1. CA Membranes

The CA membrane originally consists of cellulose diacetate polymer (2). Current CA membrane is usually made from a blend of cellulose diacetate and triacetate. A TF acetone-based solution of CA polymer with swelling additives is first cast onto a non-woven polyester fabric. Part of the solvent evaporates during casting. After the casting step, the membrane is immersed into a cold waterbath to remove the remaining acetone and other leachable compounds. Following the cold bath, an annealing step is carried out by immersing the membrane into a hot waterbath at a temperature of 60–90°C. The annealing step improves the semipermeability of the membrane with a decrease of water transport and a significant decrease of salt passage. The resulting cellulose membrane has an asymmetric structure with a dense surface layer of about 0.1–0.2 μm that rejects the salt. The substructure of the membrane film is spongy and porous, and has high water permeability. Semipermeability of a CA membrane is largely dependent on the temperature and duration of the annealing step.

The superior chlorine and fouling resistance of CA makes it the membrane of choice for many applications. In desalination, applications are mainly limited for brackish water because of the compressibility of the membrane itself under the high pressures required for operation.

2.2.2. TF Membranes

Two distinct steps are used to manufacture TF composite (TFC) polyamide (PA) membranes. First, a support layer is formed by casting polysulfone solution onto a non-woven polyester fabric. The support layer is very porous and hence does not reject the salt. In the following separate manufacturing step, a semipermeable membrane skin is formed on the polysulfone substrate by interfacial polymerization of monomers containing amine and carboxylic acid chloride functional groups.

In the 1970s, Cadotte prepared a composite membrane that consisted of a thin layer of PA formed in situ by condensation of branched polyethyleneimine and 2,4-diisocyanate on a porous polysulfone membrane (8). The effect of surface modification on the performance of an aromatic PA–TFC membrane (FT-30) has been studied (9). After soaking in solutions of various concentrations of hydrofluoric acid (HF), fluorosililic acid, and their mixtures at controlled temperature for various times, the membrane was rinsed with deionized (DI) water and tested for their RO performances. For membranes soaked in 15 wt.% HF solution for 7 days, the flux increased from 3.5 to 18 L/m^2 h, whereas NaCl separation increased from 94.5 to 95.3% under the same operating pressure of 250 psig. For membrane soaked in 15 wt. % HF solution for 4 days, X-ray photoelectron spectroscopy showed that the fluorine ratio on the membrane surface increased from 0 to 0.012 on the fresh membrane. The value increased to 0.044 for the membrane immersed in 15 wt.% HF solution for 75 days. Thus, surface chemical modification contributed to the dramatic increase of the flux as well as the thinning of the selective skin layer, whereas NaCl separation remained practically unchanged.

The discovery of TF membranes is a breakthrough in achieving flows and rejections suitable for seawater desalination. In some cases, three-layer configurations are used for extra durability and performance. Some characteristics of the two membranes are listed in Table 13.1.

Thin film composite membranes are subject to continuous development. Membranes with chlorine resistance, low energy, and low fouling have been produced. Recently, about 60% of seawater recovery has been reported for some extra high-rejection TF membranes. RO systems were originally designed using flat CA or hollow fiber membranes that were retrofitted with TFC membranes of polyurea and PA. The TFC PA membranes have many advantages in comparison with CA or hollow fiber membranes. These include wider chemical and physical tolerance ranges, higher silt density index (SDI) tolerance, feasibility of operation at higher temperatures, and higher pH tolerance. These make pretreatment easier and less expensive, and operation easily controlled. Furthermore, the introduction of the TFC PA

Table 13.1
Characteristics of cellulose acetate and thin-film membranes

Parameters	Cellulose acetate	Thin film
Temperature	Up to 30°C	Up to 50–90°C
Operating pressure (psi)	Up to 450 typically; up to 800 occasionally	100–1,200
pH	3.0–9.0	1.0–12.0
Chlorine tolerance	Fair	Poor
Oxidization tolerance	Good	Poor
Rejection (%)	Good	Excellent

membrane helps in better combating biofouling and raising the effective plant availability in view of the higher resistance of the spiral-wound configuration to mechanical blocking and of the PA polymer to biodegradation. Particularly in seawater desalination, higher RO process efficiency and recovery rates are realized by the development of the very high-rejection TFC membranes, which are resistant to compaction at high applied pressures and enable brine recovery in two-stage systems. Consequently, operation and project cost have decreased significantly due to the application of TFC PA membranes.

2.3. *Membrane Filtration Theory*

2.3.1. *Membrane Transport*

Although the transport in the solution circulating in the space between the membranes is important, it is the ion and water transport in the membranes that determine the performance of membrane process. Theories used to characterize transport in the skin layer of membranes have been reviewed (10–16). Mechanistic and mathematical models proposed can be divided into three types: irreversible thermodynamics models; nonporous or homogeneous membrane models (such as the solution-diffusion [SD], SD-imperfection, and extended SD models); and pore models (such as the finely porous, preferential sorption capillary flow [PSCF] and surface force-pore flow models [SFPF]). Some of these descriptions rely on relatively simple concepts, whereas others are more complex and require sophisticated solution techniques. Models that adequately describe the performance of RO membranes are very important because these are needed in the design of RO processes. Models that predict separation characteristics also minimize the number of experiments that must be performed to describe a particular system. The transport models focus on the top thin skin of asymmetric membranes or the top thin skin layer of composite membranes because these determine fluxes and selectivities of most membranes (17). Also, most of the membrane models assume equilibrium (or near equilibrium) or steady-state conditions in the membrane.

2.3.2. *Irreversible Thermodynamics Models*

Irreversible thermodynamics lead to the following expressions for solvent and solute flow:

$$J_w = L_p \cdot (\Delta P - \Delta \Pi), \tag{6}$$

$$J_s = C_s(1 - \sigma)J_w + D\varepsilon\Delta C. \tag{7}$$

ΔP and $\Delta \Pi$ are transmembrane pressure and osmotic pressure across membrane, respectively; C_s and ΔC are average concentration of solute and solute concentration difference across membrane, respectively; D is the diffusivity and ε is the porosity of the membrane; L_p is a coefficient; and σ is a measure of the solute–water coupling within the membrane and may often be treated as 1. The theory then characterizes the membrane by the parameters L_p and $D\varepsilon$, which may be measured by experiments other than RO and then used to quantify the performance of the membrane. A particular strength of this approach is its extension to multicomponent systems in which it can be used to predict membrane behavior. However, it does not elucidate the actual transfer mechanisms within the membrane.

2.3.3. Homogeneous Models

The homogeneous models assume that the membrane is nonporous. The extent of separation of the solvent and solute by the membrane depends on the difference in the solubilities and diffusivities of the solvent and solute in the membrane phase. In other words, when the solvent and solute molecules come into contact with the membrane surface, the solvent and the solute molecules will dissolve in the surface layer. The solubilities of the solute molecules depend on their respective solute distribution coefficient between the solution and the membrane phase. Once in the membrane phase, the solvent and solute molecules will diffuse across the membrane. The flux of these molecules depends on the transmembrane pressure difference and their respective concentration difference across the membrane (collectively known as the molecular species' individual chemical potential gradient).

The most commonly applied homogeneous model is the SD model. The model gives the following relationships for flux of solvent:

$$J_w = \frac{P_w}{\Delta x}(\Delta P - \Delta \Pi) = A_w(\Delta P - \Delta \Pi). \tag{8}$$

Here, $A_w = P_w/\Delta x$, P_w is the solvent permeability, Δx is the thickness of the membrane, and A_w is the solvent permeability constant. Although J_w depends mainly on the difference between ΔP and $\Delta \Pi$, the solute flux, J_s, depends mainly on the solute concentration difference between the feed and permeate sides of the membrane as shown in Eq. (9):

$$J_s = \frac{D_s K_s (C_s^{\mathrm{I}} - C_s^{\mathrm{II}})}{\Delta x} = A_s(C_s^{\mathrm{I}} - C_s^{\mathrm{II}}). \tag{9}$$

Here, $A_s = D_s K_s/\Delta x$, D_s is the diffusivity of the solute, K_s is the solute distribution coefficient between the solution and membrane phase, C_s^{I} is the concentration of the solute in the feed solution on the membrane surface, C_s^{II} is the concentration of the solute in the permeate solution, and A_s is the solute permeability constant.

A comparison between the aforementioned two equations shows that the flow of the solvent and the flow of the solute through the membrane do not affect each other and are indeed separate flows affected only by their own chemical potential gradient as described in the SD model. The intrinsic membrane solute rejection for this model is given by

$$R = \left[1 + \left(\frac{\rho_w A_s}{A_w}\right)\left(\frac{1}{\Delta P - \Delta \Pi}\right)\right]^{-1}, \tag{10}$$

where R is the solute rejection and ρ_w is the density of the solvent.

Example

A RO membrane to be used at 25°C for a 2.5 kg/m^3 NaCl feed solution has a water permeability constant of $A_w = 5.32 \times 10^{-4}$ kg/s m^2 and a solute (NaCl) permeability constant of $A_s = 4.89 \times 10^{-7}$ m/s. ΔP is 27.2 atm. Calculate the water flux and solute flux through the membrane, the solute rejection R, and C_s^{II} of the product solution.

SOLUTION

In the feed solution, $C_s^I = 2.5\,\text{kg/m}^3$ (2.5 g/L), then substituting this into Eq. (1)

$$\Pi_1 = C_s^I RT = \left(\frac{2 \times 2.5\ \text{g/L}}{58.45\ \text{g/mol}}\right) \times \left(\frac{0.0821\ \text{L atm}}{\text{mol K}}\right) \times 298\ \text{K} = 2.09\ \text{atm}.$$

Because the product solution C_s^{II} is unknown, a value of $C_s^{II} = 0.1$ g NaCl/L is assumed, then $\Pi_2 = 0.08$ atm and DP $= P_1 - P_2 = 2.09 - 0.08 = 2.01$ atm. Also, because permeate is quite dilute, a value of $\rho_w = 997\ \text{kg/m}^3$ is assumed.

Substituting into Eq. (8),

$$J_w = A_w(\Delta P - \Delta\Pi) = 5.32 \times 10^{-4}(27.20 - 2.01) = 1.341 \times 10^{-2}\ \text{kg H}_2\text{O/s m}^2.$$

For calculation of R, substituting into Eq. (10),

$$R = \left[1 + \left(\frac{\rho_w A_s}{A_w}\right)\left(\frac{1}{\Delta P - \Delta\Pi}\right)\right]^{-1} = \left[1 + \left(\frac{997 \times 4.89 \times 10^{-7}}{5.32 \times 10^{-4}}\right)\left(\frac{1}{27.20 - 2.01}\right)\right]^{-1}$$

$$= 0.965.$$

Using this value of R in Eq. (4),

$$R = 0.965 = 1 - \frac{C_s^{II}}{C_s^I} = 1 - \frac{C_s^{II}}{2.5}.$$

Solving $C_s^{II} = 0.0875\,\text{kg/m}^3$ for the product solution. This is not close enough to the assumed value of $C_s^{II} = 0.10\,\text{kg/m}^3$. At least one more iteration should be conducted until the solved C_s^{II} is very close to another assumed value of C_s^{II}, and Π_2 will not change significantly on another trial. Only for the purpose of illustration, it is assumed that the final value of C_s^{II} is $0.175\ \text{kg/m}^3$.

Substituting into Eq. (9),

$$J_s = A_s(C_s^I - C_s^{II}) = 4.89 \times 10^{-7}(2.5 - 0.175) = 1.137 \times 10^{-6}\text{kg H}_2\text{O/s m}^2.$$

As transmembrane pressure increases, Eq. (8) suggests that the solvent flux will increase proportionally. Solute flux (Eq. (9)) will rise less rapidly, giving an improvement in solute rejection, as is observed experimentally. Similarly, the model correctly predicts the decrease in rejection that occurs with increased solvent recovery. As a consequence of increased solvent recovery, the concentration of retained solute increases and by Eq. (9), a greater passage of solute (low rejection) is expected. A transport model based on the SD model incorporating concentration polarization has been successfully used to interpret the experimental results and predict rejection over a range of operating conditions (18). However, the SD model is limited to membranes with low water content, and for many RO membranes and solutes, particularly organics, the SD model does not adequately describe water or solute flux (12, 13). Possible causes for these deviations include imperfections in the membrane barrier layer, pore flow (convection effects), and solute–solvent–membrane interactions (12, 13).

2.3.4. Solution-Diffusion-Imperfection Model

Although the SD model assumes the absence of pores on the membrane surface layer, there will inevitably be imperfections (pores) on the surface of the membrane. These pores can provide pathways by which the solvent and solute molecules can diffuse through the membrane. The SD-imperfection model extends the SD model by accounting for the existence of these pores and the flow through them.

This model gives the solvent flux as

$$J_w = \frac{P_w}{\Delta x}(\Delta P - \Delta \Pi) + \frac{P_3}{\Delta x}\Delta P, \tag{11}$$

where $P_3/\Delta x$ is a coupling coefficient. Comparing Eqs. (8) and (11), it is clear that the second term in Eq. (11) accounts for solvent flowthrough the imperfections in the membrane surfaces, in addition to diffusion as accounted for by the first term in the same equation.

Existence of pore flow is also accounted for in the second term of the extended solute flux equation:

$$J_s = \frac{P_2(C_s^{\mathrm{I}} - C_s^{\mathrm{II}})}{\Delta x} + \frac{P_3}{\Delta x}\Delta P C_s^{\mathrm{I}}, \tag{12}$$

where $P_2/\Delta x$ is a solute permeability coefficient equivalent to $P_s/\Delta x$ in Eq. (9).

The corresponding intrinsic membrane solute rejection for this model is

$$R = \left[1 + \left(\frac{P_2}{P_w}\right)\left(\frac{1}{\Delta P - \Delta \Pi}\right) + \left(\frac{P_3}{P_w}\right)\left(\frac{\Delta P}{\Delta P - \Delta \Pi}\right)\right]^{-1}. \tag{13}$$

Extended SD model is mainly used for organic solutes by an introduction of the pressure-dependent term (19). However, the model has two major disadvantages: it contains three parameters that must be determined by nonlinear regression in order to characterize the membrane system, and the parameters describing the system are usually functions of both feed concentration and pressure (12).

2.3.5. Preferential Sorption Capillary Flow Model

An early pore model was the PSCF model proposed by Sourirajan. In contrast to the SD model, the membrane is assumed to be microporous (1, 20). Therefore, membrane surface properties and transport phenomena in the membrane pore determine the mechanism for the separation of solute and solvent molecules. According to the model, on contact of the feed solution with the membrane, chemical properties of the barrier layer allow for the preferential sorption of the solvent molecules and subsequently the formation of a solvent layer, on the surfaces and in the pores of the membrane. The solute molecules, on the contrary, are rejected by the membrane and, therefore, cannot form any surface layer. The solvent molecules in the solvent layer are then forced, under hydrostatic pressure, through the membrane capillary pores to the permeate side.

Solvent flux, J_w, as given by this model is

$$J_w = A\{\Delta P - [\Pi(C_s^{I}) - \Pi(C_s^{II})]\}, \tag{14}$$

where A is the pure solvent permeability constant and $\Pi(C_s)$ is the osmotic pressure of a solution with solute mole fraction C_s. Unlike solvent flux, the applied pressure does not influence the solute flux:

$$J_s = \frac{K_s D_s}{\Delta x}(C_s^{I} - C_s^{II}). \tag{15}$$

A comparison between Eqs. (9) and (15) shows that they are identical. Thus in the PSCF model, the transport of solute molecules takes place only by diffusion through the membrane surface layer and not partly by pore flow as suggested by the SD-imperfection model. The model has been utilized to analyze the transport of a large number of solutes and membranes (20).

2.3.6. *Finely Porous Model*

The finely porous model assumes that the transport of solvent takes place by viscous flow through uniform membrane pores and that the transport of solute occurs by both diffusion and convection in these pores. It can provide valuable insight into parameters such as pore size, solute–membrane interaction (friction parameter), and solute distribution coefficient that affect solute transport (12). However, for some solute systems such as dilute organics, the model does not adequately describe decreases in water flux compared with the pure water flux unless a correction is made in the pore size; it is usually necessary to reduce the pore size in order for the measured and predicted water fluxes to agree. This disadvantage limits the applicability of the finely porous models for water flux prediction in these systems.

2.3.7. *Surface Force Pore Flow Model*

The SFPF model is a two-dimensional extension of the finely porous model (20, 21). The finely porous model considers only axial solute concentration gradients, whereas the SFPF model recognizes that the solute concentration in an RO membrane pore may be a function of radial and axial positions (15, 17, 20). The model gives excellent predictions of solute separation for a wide range of inorganics and organics under varying operating conditions (22–24). However, this model does not adequately predict the water flux ratio for some dilute organics that cause substantial decreases in water flux. The pore radius must be reduced in order to force the predicted and measured water flux ratio into agreement for these systems.

2.4. *Concentration Polarization*

This phenomenon describes the increase in retained solids or solute concentration at the membrane surface. According to a simple stagnant film theory, there will be a concentration gradient over the thickness of the stagnant film. Thus, the concentration will be the highest on the membrane surface and will decrease to bulk concentration in the well-mixed liquid. This will cause the diffusion of solids in the direction opposite to the main flux and, therefore, a reduction in the overall flux. Other negative results include an increase in solute flux resulting from the increased concentration gradient across the membrane, and scaling or particle scaling induced by the precipitation of the solute. Consequently, the separation properties of membrane are reduced. The extent of concentration polarization can be reduced by promoting good mixing of the bulk feed solution with the solution near the membrane wall. Mixing can be enhanced through membrane module optimization of turbulence promoters, spacer placement, hollow fiber diameter, and so on, or by simply increasing axial velocity to promote turbulent flow (16, 17, 25, 26).

Concentration polarization complicates the modeling of membrane systems because it is very difficult to determine experimentally the solute wall concentration, which is used in most RO transport models, and not the bulk concentration. For very high feed flow rates, enough mixing near the membrane surface occurs so that the wall concentration can be assumed to be equal to the bulk concentration. Similarly, at low permeation rates, the extent of concentration polarization is likely to be limited. Thus, for example, Du Pont hollow fiber permeators, because of the intrinsic low permeability of the membrane, can be operated under essentially backmixed conditions on the shell (feed) side (27). The presence of a mesh in spiral-wound modules promotes feed side mixing and minimizes the effect of concentration polarization on the membrane surface. However, at lower feed flow rates, the difference between the wall and bulk concentration can be substantial and so the wall concentration must be calculated.

The simplest relationship for the membrane surface concentration (C_{m}) is provided by

$$J_w = k \cdot \ln \frac{(C_{\mathrm{m}} - C_{\mathrm{s}}^{\mathrm{II}})}{(C_{\mathrm{b}} - C_{\mathrm{s}}^{\mathrm{II}})}, \tag{16}$$

where k is the mass transfer coefficient, dependent on the extent of cross flow and physical properties prevailing on the surface of the membrane; C_{b} is the concentration of solute in bulk feed; and C_{m} is the concentration of solute on the membrane surface.

More complete solutions that account for diffusion effects in the axial direction and loss of solvent along the flow channel have been provided (28, 29). All of them broadly show a greater level of polarization than that predicted by Eq. 4(16) and would encourage the use of effective mixing on the feed side of an RO membrane.

2.5. *Compaction*

Compaction is another phenomenon in membrane mass transport. When a polymeric membrane is put under pressure, the polymers are slightly reorganized and the structure is changed. This results in lowered volume porosity, increased membrane resistance, a dense structure with smaller pores, and eventually, a lowered flux. For example, after long-term compaction of the asymmetric CA membranes, the selectivity was found to be unchanged but the flux decreased (30). The flux loss could be attributed to a compaction of the porous sublayer and the unchanged skin. It was concluded that under all circumstances, compaction results in a notable membrane permeability loss. Thus the correct application of pressure is very important.

3. MEMBRANE MODULES AND PLANT CONFIGURATION

3.1. Membrane Modules

Basically, the RO process is a modular scale process wherein scale-up beyond a certain size is not economically advantageous (3). The modules are packaging devices of the membranes and their supports, designed in such a way that the pressurized feed solution can be applied to the surface of the membrane skin layer and the permeate can be collected without being contaminated by the concentrate.

At present, there are basically four RO module designs commercially available, namely, tubular, hollow fiber, spiral-wound, and flat configuration. For RO applications, the most commonly used configurations are hollow fiber and spiral-wound modules. Some commercial RO membranes and corresponding modules are summarized in Table 13.2.

3.1.1. Hollow Fiber

This configuration consists of a very large number of hollow membrane fibers, which are asymmetric in structure and as fine as a human hair (Fig. 13.3). The membrane fibers are assembled into a cylindrical bundle evenly spaced about a central feed distributor tube. The ends of the fibers are epoxy sealed to form a sheet-like permeate tube. In some cases, half-fold bundles are used. The hollow fiber membrane bundle is contained in a cylindrical housing or shell. The assembly is called a permeator. During operation, the pressurized feed water enters the permeator feed end, passes through the central distributor tube and through the tube wall,

Table 13.2
Commercially available RO membranes and modules

Membrane material	Module type	Company
Asymmetric aramid	Spiral-wound	Du Pont (USA)
Asymmetric cellulose acetate	Spiral-wound	Toray (Japan)
	Spiral-wound	Desalination system (USA)
	Plate and frame	DDS (Denmark)
	Tubular	Paterson Candy (UK)
Asymmetric poly(acrylonitrile)	Spiral-wound	Sumitomo (Japan)
Cellulose acetate	Hollow fiber	Toyobo (Japan)
	Hollow fiber	Osmonics (USA)
	Spiral-wound	Hydranautics/Nitto Denko (USA/Japan)
Composite aramid	Spiral-wound	Du Pont (USA)
Composite cellulose acetate	Spiral-wound	Du Pont (USA)
Composite PEC	Spiral-wound	Toray (Japan)
Composite polyamide	Hollow fiber	Filmtec/DOW (USA)
	Spiral-wound	Fluid system/UOP
	Spiral-wound	Desalination system (USA)
	Plate and frame	DDS (Denmark)
	Tubular	Paterson Candy (UK)
	Spiral-wound	Hydranautics/Nitto Denko (USA/Japan)

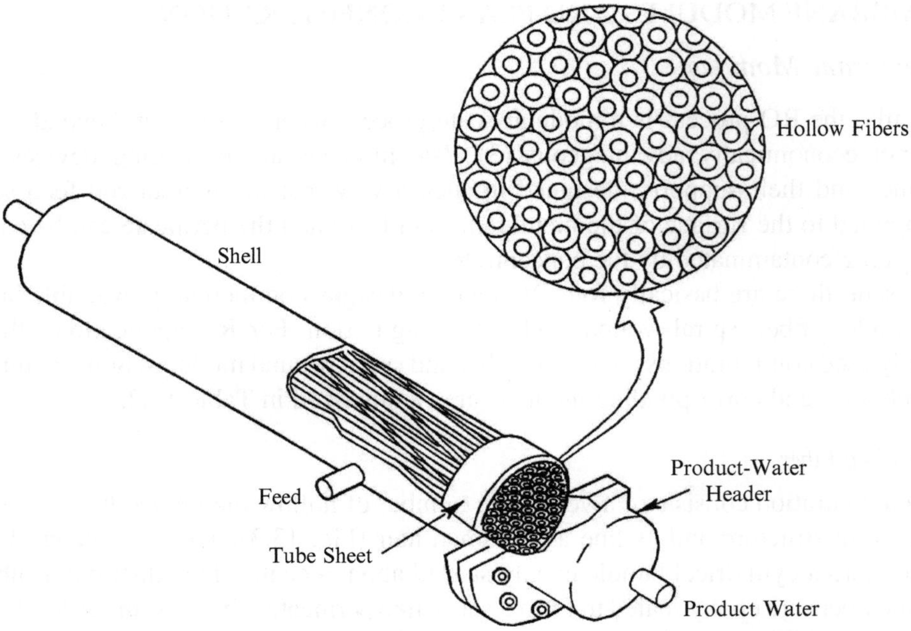

Fig. 13.3. Schematic drawing of a hollow fiber module.

and flows radially around the fiber bundle toward the outer permeator pressure shell. Water permeates through the outside wall of the fibers into the hollow fiber bore and is removed from the tube sheet end of the fiber bundle.

An increase in permeate flux will increase the delivery rate of ions to the membrane surface and increases salt concentration on the membrane surface, whereas an increase in feed flow increases turbulence and reduces the thickness of the high concentration layer near the membrane surface. For a hollow fiber module, the permeate water flow per unit area of membrane is generally low, and therefore, the concentration polarization is not high on the membrane surface. Furthermore, the feed water flow must exceed a minimum reject value in order to minimize concentration polarization and maintain even flow distribution through the fiber bundle. Typically, a single hollow fiber permeator can be operated at up to half recovery and meet the minimum reject flow required.

Systems based on this configuration are extremely compact and require relatively little space because of its extremely high packing density. Its primary disadvantage is that the configuration is highly susceptible to fouling and is difficult to clean effectively because of the very small spacing between fibers in the bundle. Thus, extensive pretreatment is required even on relatively clear feed waters.

3.1.2. *Spiral-Wound*

The spiral-wound configuration (Fig. 13.4) consists of several flat membranes separated by the permeate collector channel materials to form a leaf. The edges of the leaf are sealed on

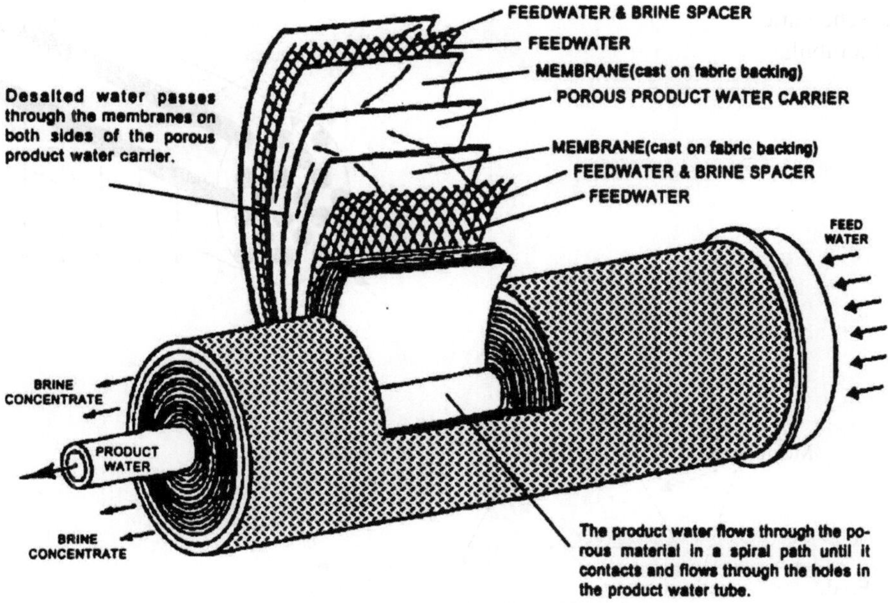

Fig. 13.4. Schematic drawing of a spiral-wound module.

three sides with the fourth side left open for the permeate to exit. A number of these permeate assembles or leaves are wound around a central plastic permeate tube. This tube is perforated to collect the permeate from the multiple leaf assemblies.

Spiral systems are usually staged with three to six membrane elements connected in series in a pressure tube. The feed stream from the first element becomes the feed to the following element, and so on, for each element within the pressure tube. The feed flows through the element in a straight axial path from the feed end to the opposite end, running parallel to the membrane surface. The turbulence induced by the feed channel spacer helps to reduce concentration polarization. The brine stream from the last element exits the pressure tube as waste. The permeate from each element enters the permeate collector tube and exits the vessel as a common permeate stream. A single pressure vessel with four to six membrane elements connected in series can be operated at up to half recovery under normal design conditions.

The spiral configuration makes better use of space than tubular or flat-sheet types. Furthermore, it has low manufacturing cost and can be cleaned both chemically and hydraulically with relative ease. Extensive pretreatment is necessary for highly turbid feed waters because the quite small flow passages are subject to clogging.

3.1.3. Tubular Module

The tubular configuration (Fig. 13.5) is conceptually the simplest of all the configurations. In this configuration, the membrane is either cast on the inner surface of, or placed within, a porous tube and sealed into place. During operation, pressurized feed water is circulated

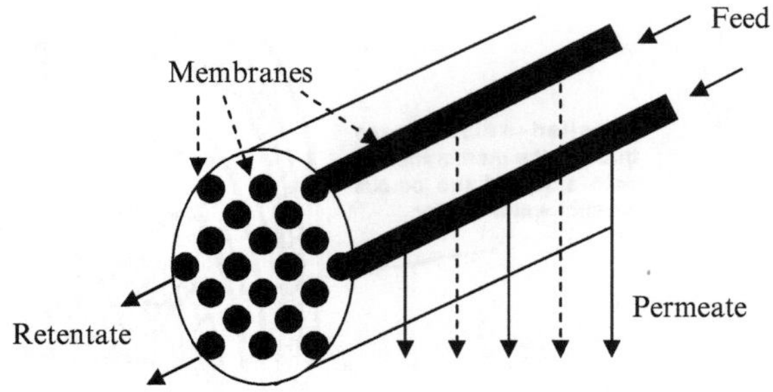

Fig. 13.5. Schematic drawing of a tubular module.

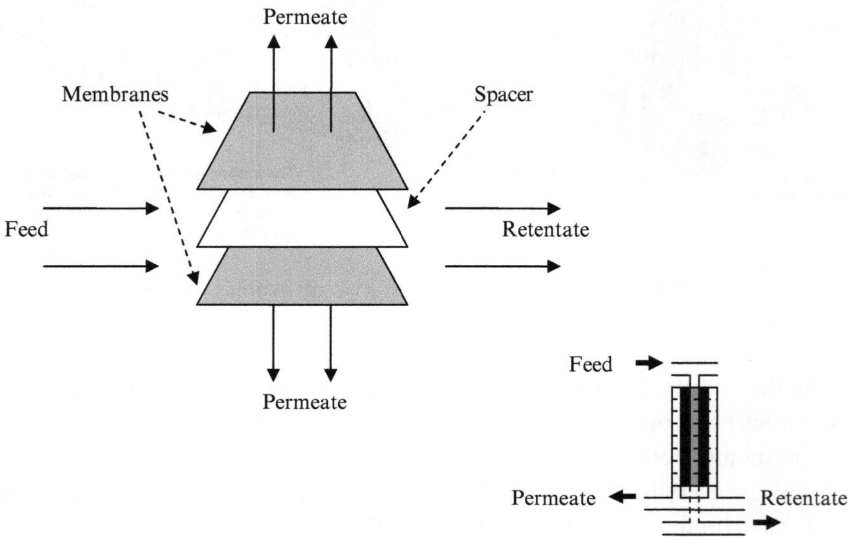

Fig. 13.6. Schematic drawing of a flat configuration module.

through tubes in series or parallel. Purified water passes through the membrane, through the porous tube, and drops off into a convenient collection receptacle for removal from the system.

The tubular system has the advantage that it can be operated on extremely turbid feed waters and can be cleaned either mechanically or hydraulically with ease. It has the disadvantage of very high capital costs when compared with either the spiral or the hollow fine fiber for water treatment applications.

3.1.4. Flat Configuration

In the plate and frame design (Fig. 13.6), membranes are attached to both sides of a rigid plate. The plate may be constructed of solid plastic with grooved channels on the surface, porous fiberglass materials, or reinforced porous paper. The plate/frame membrane units are placed in a pressurized vessel in a configuration that allows feed water to contact all sides of the plates. Feed water under pressure is fed though the vessel and concentrated brine is removed from the other end.

Each plate in the vessel is at low pressure so that water passes through the membrane which is in contact with the high-pressure brine and is collected in the porous media. Fresh water is collected from all the plates. This design is useful for chemical process applications but, because of its complexity, is very expensive to operate for large-scale operations.

3.2. *Plant Configuration of Membrane Modules*

3.2.1. Plant Configuration

A typical RO plant configuration generally includes raw water pumps, pretreatment, membrane units, disinfection units, storage, and distribution elements. For desalination application, membrane units are often configured into a multistage continuous process. In such a system, the concentration of the feed stream increases gradually along the length of several stages of membrane modules arranged in series (Fig. 13.7). The feed reaches its final concentration only at the last stage. Variations in water temperature and fouling degree of the feed water are compensated by adjustments of the feed pressure. In some case, concentrate recirculation is used to achieve a high enough system recovery. Part of the concentrate is directed back to the feed water side of the module. The recycled concentrate mixes with the feed water and will be treated once more. Other design considerations for membrane units are discussed next.

Feed Water Pressure

The feed pressure is determined by the flux, the energy loss in the system, and the osmotic pressure in the membrane system. The required feed pressure will increase when the membrane elements become contaminated with time. A feed pump that enables a higher flow than the theoretically required flow will then be preferred to keep the feed pressure continual. Generally, a feed pump that increases the feed pressure by 25% will be satisfactory in practice. Typical feed pressures range from 5 to 24 bars. RO systems rated at less than

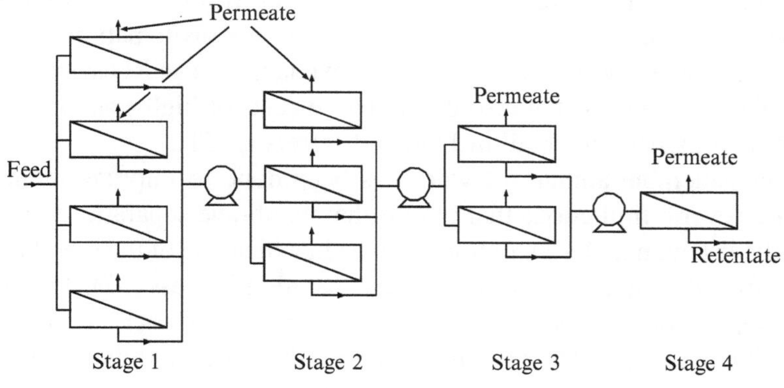

Fig. 13.7. Schematic flow-diagram of a multistage continuous RO process.

17 bars are classified as low-pressure units, whereas those operating at more than 24 bars are classified as high-pressure units. Although high operating pressure can cause noise, vibration, and corrosion problems, these systems are more effective. More recently, new models have been developed to perform well at low pressures.

Membrane Type and Pore Size

Two types of membrane modules are usually selected for RO systems: spiral wound and hollow fiber. The spiral-wound unit clogs less frequently, whereas the hollow fiber membrane has much greater surface area per unit volume. In the early days, membranes were usually made of CA. These days, membranes are also made of aromatic PA and TF polymer composites. Different materials will have distinct characteristics, such as hydraulic resistance, pH range, temperature range, chlorine tolerance, and biodegradation tolerance.

Pretreatment Requirements

Pretreatment requirements consist of the following major factors: (1) influent suspended solid concentration; (2) ionic size of the contaminants; and (3) membrane type. Other characteristics such as contaminant concentration, water temperature, and presence of competing ions are also important. Pretreatment is most commonly used to prevent fouling of the membrane during operation. Typical pretreatment for RO includes particle removal by filtration, sequestering hardness ions by precipitation, and pH control to prevent clogging.

Product Conversion Rate

This is the ratio of reject water to finished water and it depends on several factors, mainly ionic charge and ionic size. The higher the ionic charge and the larger the ionic size of the contaminant, the more easily the ion is removed and the more finished water is recovered relative to the amount of rejected water.

Energy Saving

Application of a pressure exchanger can be used to win back energy from the concentrate flow released under high pressure. The concentrate flow from the membranes is directed through the pressure exchanger, where it directly transfers energy to part of the incoming feed water with maximum affectivity. Use of a pressure exchanger in an RO system saves not only more than half the total energy, but also the purchase costs of high-pressure pumps. An RO design software is illustrated in detail in Judd and Jefferson (31).

Although a considerable amount of work has been done to improve membrane performance, there are several limitations that inhibit the membrane separation processes. Therefore, engineers have begun to design hybrid systems in which one of the membrane separation processes is combined with a more conventional separation process (32). The cost associated with a hybrid separation process is often lower than that of each individual separation process. According to General des Eaux and USFilter, there are a number of ways to combine distillation process and various membrane separation processes in hybrid desalination systems. For example, the energy consumption is reduced by feeding the cooling water from the distillation process to an RO unit; ultrafiltration (UF) and microfiltration (MF) are used to pretreat feed stream; and the upstream of a distillation process is softened by nanofiltration (NF) to reduce scale-forming ions, then the productivity of the distillation process is increased by operating at high temperature.

3.2.2. Energy Saving

Seawater RO desalination is an energy intensive process. Therefore, it is important to make use of energy saving methodologies. Energy saving methodologies are employed mainly to recover the energy carried over by the brine leaving the RO system at high pressure. For a seawater RO system operating at 55–69 bar with 15–40% conversion, the brine leaves the RO modules at a pressure of 51–65 bar. When a hydraulic turbine is used to convert the brine stream to power, a major part of the energy of the high-pressure brine can be recovered. Without energy recovery, RO for seawater conversion consumes 35–45 kWh/1,000 gal depending on feed salinity and other operating conditions. With an energy recovery turbine (ERT), the above energy consumption can be reduced to 20–25 kWh/1,000 gal (20).

Consequently, the most important part in any RO plant, as far as the energy consumption is concerned, is the energy recovery system. Basically, there are two different approaches for energy recovery. First, the energy recovery technologies are commonly referred to as the positive displacement systems such as Energy Recovery, Inc.'s Pressure Exchanger, Desalco's Work Exchanger Energy Recovery system, Siemag's system, and RO Kinetic's System. All of these technologies use the principle of positive displacement to transfer the energy contained in the reject stream directly into a new stream (Fig. 13.8). Positive displacement pumps typically offer lower flow rate and higher discharge pressure capabilities than centrifugal pumps. The efficiencies of all these devices can be quantified as the hydraulic energy out divided by the hydraulic energy in. Most of these devices achieve relatively similar net energy transfer efficiencies between 91 and 96%. The energy saving is achieved by reducing the volumetric output of the high-pressure pump.

These constituents provide the onset of membrane fouling (33):

- Particulate deposition (colloidal fouling).
- Adsorption of organic molecules (organic fouling).
- Inorganic deposits (scaling).
- Microbial adhesion and growth (biofilm formation).

A pressure exchange system from Siemag was applied in a 5,000 m^3/day seawater desalination plant INALSA I, Spain. The results proved that this system contributed to reduce the energy demand for RO units in the range of 25–30% when compared with plants equipped

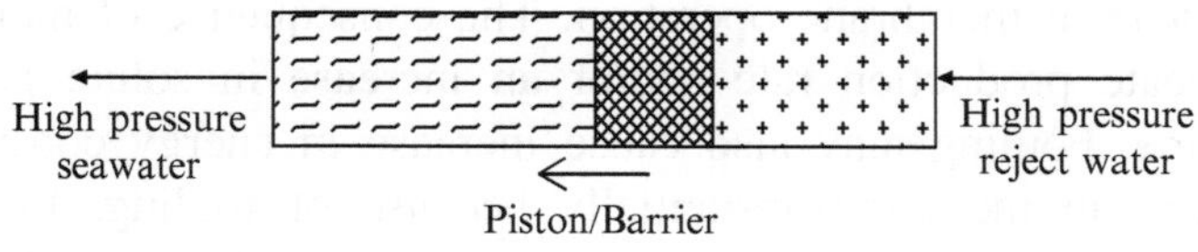

Fig. 13.8. Simple schematic of energy recovery with positive displacement device.

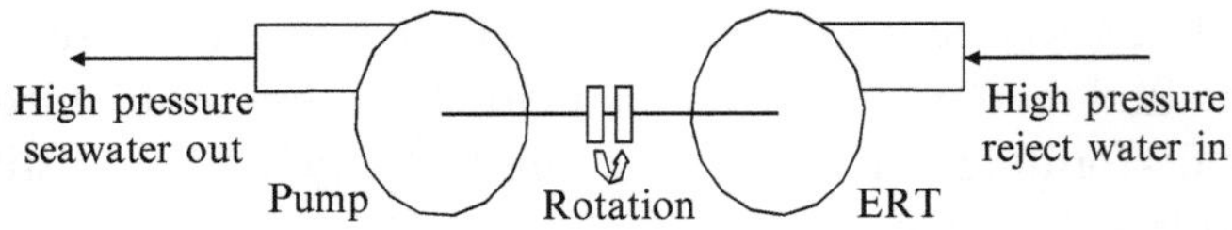

Fig. 13.9. Simple schematic of energy recovery with energy recovery turbine and pump.

with usual energy recovery devices based on turbines. The direct transfer of the fluid pressure from the concentrate to the fluid pressure in the feed stream with an efficiency of approximately 98% allows for short amortization times for the system itself if installed in new plants as well as for retrofitting of existing plants (34).

Another energy recovery technology is to convert the hydraulic energy in the reject stream into rotational energy, in the form of mechanical shaft power, by an ERT. In an RO system, this rotational mechanical energy must then be transferred into another pumping device that pressurizes the incoming stream (Fig. 13.9). The efficiency of centrifugal ERT is typically in the range of 80–88% in converting hydraulic energy into rotational mechanical power which is converted back to hydraulic energy. Therefore, when the efficiency losses of the pump are accounted for, the real net energy transfer efficiencies are about 35–75% for these devices even when operating at their optimum efficiency design points.

4. PRETREATMENT AND CLEANING OF MEMBRANE

4.1. *Mechanisms of Membrane Fouling*

The pretreatment system that is used is highly dependent on the feed water quality. A complete and exact analysis of source water is thus important for the design of a pretreatment system and the entire RO system because this often determines the type and size of the pretreatment. The source of water may contain various concentrations of suspended solids and dissolved matter. Suspended solids may consist of inorganic particles, colloids, and biological debris including bacteria and algae. Dissolved matter may consist of highly soluble salts, such as chlorides, and sparingly soluble salts, such as carbonates, sulfates, and silica. During operation, the concentration of suspended particles and dissolved ions increases with a decreasing volume of feed water. The settled suspended particles on the membrane surface may block feed channels and increase pressure drop across the system. The precipitated soluble salts from the concentrate stream may form a scale on the membrane surface and thus decrease water permeability through the RO membranes. The process of forming a deposited layer on the membrane surface is called membrane fouling. Membrane fouling remains the most difficult obstacle in membrane operation. The consequence of membrane fouling is reduction of permeate production rate and/or an increase in solute passage across the membrane with time. Fouling may also cause increase in energy consumption as transmembrane pressure can increase substantially because of fouling. In addition, fouling increases downtime and may shorten membrane life span. There is also a high level of biological activity resulting from biological treatment.

Colloidal fouling is caused by the accumulation of particles and macromolecules on, in, and near a membrane. Materials accumulated on the membrane surface form an additional layer of resistance to permeation. Early work on colloidal fouling of RO membrane used to treat feed water indicated that subparticles of 5 μm contribute more substantially to fouling than larger particles (35). It was postulated that particles are subjected to higher velocity and shear force on the membrane surface as particle size increases. Therefore, larger particles tend to be swept away in bulk flow rather than deposits on the membrane surface. In addition to surface deposition, some particles may be small enough to penetrate and remain within the pores of the membrane (36). In a study on the influence of colloidal fouling on salt rejection of TFC RO, a sharp decrease in permeate flux and significant decline in salt rejection with increasing concentration factor were observed under conditions in which colloidal fouling occurred (37).

The adsorption of organic matter on membrane surfaces is detrimental to permeate flux and affects the salt rejection of membranes. The negatively charged functional groups on organic foulants have an affinity for the charged membrane surface, thereby forming a permeate-resistant layer. Organic foulants also interact with colloidal foulants. Polyphenolic compounds, proteins, and polysaccharides bind colloids and particles and increase their cohesion to the membrane surface (35). In addition, the biochemical interaction between organics and microorganisms promote biofilm formation and growth. Insufficient knowledge on the composition of dissolved organics in water makes the control of organic fouling difficult (38).

Inorganic fouling is caused by the deposition of iron, silica, aluminum, calcium, phosphorus, and sulfate. The fouling mechanism on the membrane surface can be caused by the concentration polarization effect. A concentrated boundary layer is created on the separation surface as product water passes through the membrane. Within this boundary layer, the high concentration causes the salts to precipitate and suspended solids initiate the deposition on the membrane surface, thereby leading to scaling and fouling (36). Scale deposition on and into RO membranes impairs the hydrodynamic conditions of the feed flow. When fouling conditions are not controlled properly, scaling becomes a self-sustaining phenomenon. Under severe concentration polarization, channeling, failure of RO performance, and damage to membrane surface occur.

Biofouling is the term given to the adhesion of microorganisms and growth of biofilm on the membrane surface. In addition to the detrimental effects of increased transmembrane pressure and decreased permeate flux, biofouling may cause chemical degradation of the membrane material. This could result from direct enzymatic biodegradation of the membrane surface or by generation of extreme local pH that may hydrolyze the membrane polymer (39). Such fouling can significantly reduce the membrane lifetime.

The ratio of carbon/nitrogen/phosphorus has important influences on the rate of biofilm development (40). It has been reported that membranes that suffered severe biofouling were found to contain a high percentage (typically >60%) of organics. Laboratory characterization of membrane biofilms found that a typical biofilm contains

- 90% moisture,
- 50% total organic matter,

- up to 40% humic substances,
- low inorganic content, and
- high microbiological counts including bacteria and fungi.

Biofouling is perhaps the most difficult fouling threat to overcome. No system operator can afford to ignore the potential for biofouling. Biological organisms include bacteria, algae, and fungi. Of these, bacteria cause the majority of problems in membrane treatment systems for a variety of reasons. Many bacteria can easily adapt to the environment inside the membrane treatment system. With the organic compounds concentrated on the membrane surface, the bacteria multiply rapidly and are rejected by the membrane. These bacteria have a number of defense mechanisms. For example, some have fimbriae, which stick out from all sides of the cell. These allow the bacteria to attach themselves, and remain attached, to the surface of the membrane or feed space. In addition, bacteria may secrete a mucous capsule, or slime, which coats the cell and protects them from any harsh elements entering their environment.

The susceptibility of membranes to biological fouling is significantly dependent on the membrane composition. CA membranes are easily attacked by bacteria. Disinfection of feed water is required for such membranes. PA membranes are also susceptible to bacterial attack, but TFC membranes are generally quite resistant. Chlorination can be used for CA membranes, but must be followed by dechlorination for PA and other membranes, usually by the addition of sodium metabisulfate. In ultrapure water production, sterility is often maintained by ultraviolet (UV) sterilizers (41).

A recent study has shown that some RO membranes are more prone to biofouling than others. The study involved a bioadhesion assay which utilizes a model bacterium, SW 8, known to adhere to membranes. Examination of the bacteria adhered to the membranes using optical microscopy revealed that membranes that are less susceptible to bioadhesion are hydrophilic in nature (39). Further investigation carried out on RO simulators that consisted of flat sheet membranes to simulate spiral-wound module revealed biofilm characteristics under field emission scanning electron microscope. Microorganisms covered the surface of all types of RO membranes used in the experiment to a density of about 2.25×10^8 cells/cm^2 (39). Bacteria of different shapes (mostly rod shaped) with dimensions between 1 and 3 μm were observed. The organisms appeared to excrete extracellular polymeric substances (EPS).

Understanding membrane fouling creates the foundation for researchers to devise an approach to counter or minimize fouling in order to maintain high RO performance. One strategy to alleviate membrane fouling is feed pretreatment to reduce or remove fouling constituents in the feed. A proper membrane cleaning and regeneration protocol to remove foulants periodically from the membrane surface is also essential in maximizing RO efficiency.

4.2. *Feed Pretreatment*

The success of RO process is highly dependent on appropriate feed pretreatment. Pretreatment must be effective in reducing RO fouling potential in a reliable and consistent manner. Feed pretreatment continues to be extensively studied. In the development of a pretreatment program, the focus is on removing as many fouling constituents in the feed water as possible before RO processing. The feed waters are pretreated to extend the lifetime of the membrane, prevent fouling of the membranes, and maintain performances (i.e., salt rejection and recovery) of the system (42).

There are some important indicators related to membrane fouling and feed water pretreatment. Among them, turbidity and SDI are often used in the RO industry as the indicators of suspended particles. The recommended limitation values are 1 NTU and 4 for turbidity and SDI, respectively. Continuous operation of an RO system with feed water, which has turbidity or SDI values near the limits of these values, may result in significant membrane fouling. For long-term reliable operation of the RO unit, the average values of turbidity and SDI in the feed water should not exceed 0.5 NTU and 2.5 SDI units, respectively. Langelier Saturation Index (LSI) and the saturation ratios are the indicators of saturation levels of sparingly soluble salts in the concentrate stream. The LSI provides an indication of the calcium carbonate saturation. Negative LSI values indicate that the water is undersaturated with respect to calcium carbonate. Undersaturated water has a tendency to remove existing calcium carbonate protective coatings in pipeline and equipment. Positive LSI values indicate that the water is supersaturated with respect to calcium carbonate, and scaling may occur. The LSI was originally developed by Langelier for potable water of low salinity. For the RO application for desalination of high-salinity water (e.g., seawater), the LSI is an approximate indicator only. The saturation ratio is the ratio of the actual concentration of ions in the concentrate stream to the theoretical solubilities of the salts at given conditions of temperature and ionic strength. These ratios are applicable mainly to sparingly soluble sulfates of calcium, barium, and strontium.

4.2.1. Conventional Pretreatment

RO desalination was pioneered using conventional treatment processes to upgrade product water quality. Generally, the conventional pretreatment process for RO application usually includes many interdependent pretreatment stages such as disinfection, coagulation/flocculation, sedimentation/filtration, dechlorination, and scale control.

4.2.2. Disinfection

The goal of disinfection is to prevent and/or control the colonization of microorganisms within the membrane modules and other parts of the RO desalination process from the intake to the discharge. Chlorination, sodium hypochlorite, formaldehyde, iodine, ozone, copper sulfate, and UV light have been used in disinfection. On the choice of the disinfectant to use, it is best, in theory, to alternate between different disinfectants in case a resistant strain of bacteria develops. A better reason for opting to use different chemical agents is for their individual characteristics. For example, one may be more aggressive at attacking the biofilm by limiting the food supply, whereas another may prevent the growth of bacteria in RO systems through creating 'unfriendly' living environment (e.g., chlorine). Disinfection is also partially accomplished by using conventional water pretreatment processes such as coagulation/flocculation, sedimentation, and filtration.

4.2.3. Coagulation/Flocculation

The process of coagulation and flocculation is applied to separate suspended solids (colloids) in water when small particles aggregate to form larger particles, which settle down as sediments. This is usually achieved by the addition of chemicals such as alum, ion salts, and high molecular weight polymers. For example, aluminum sulfate and ferric chloride as coagulants were examined for use in seawater (43). Results showed that the best clarification was found using aluminum sulfate at dosages of 20–30 mg/L with an anionic polyelectrolyte flocculant aid at 0.025–0.1 mg/L. Another study showed that seawater turbidity could be effectively removed by all tested coagulants of aluminum sulfate, ferric chloride, and electrolytic aluminum. Ferric chloride was recommended in the pretreatment of seawater because less of it remains in the treated seawater than aluminum over a wide pH range. Similarly, electrolytic aluminum was found to be ideal for turbidity removal of brackish water (44).

Although coagulation/flocculation is a very effective pretreatment for removing colloidal and suspended matter, the process is expensive as chemical dosage is required. The process is also difficult to operate because optimum dosage depends on influent quality. Furthermore, the coagulation–clarification process generates solid waste that requires additional handling and disposal. In cases where overdosage occurs, high metal salt content of the pretreatment water may result in metal hydroxide precipitation on the subsequent RO membranes (45).

4.2.4. Sedimentation and Filtration

In sedimentation, suspended solids in the water are removed by gravitational settling. The settling rate of the particles depends on their size and density as well as the density of the water. Filtration is the most common pretreatment to reduce membrane fouling by particular matters. The most common media filters are silica sand and crushed coal (anthracite). The selection of filter design depends very much on the required quality of effluent.

4.2.5. Dechlorination

Because of high sensitivity of polymide membranes to chlorine, dechlorination is usually achieved by the addition of sodium bisulfite ($NaHSO_3$). The problem related with this approach is that the solution itself becomes an incubator of bacteria, causing biofouling of the membranes. In some cases, granular activated carbon is used to remove chlorine, but it can also serve as an incubator of bacteria because of its porous structure and nutrient-rich environment. Recently, UV treatment has become an increasingly popular dechlorination technology with none of the aforementioned drawbacks. UV light produces photochemical reactions between 180 and 400 nm that dissociate free chlorine to form hydrochloric acid. The peak wavelengths for dissociation of free chlorine range from 180 to 200 nm, whereas the peak wavelengths for dissociation of combined chlorine range from 245 to 365 nm. The UV dosage required for dechlorination depends on the adsorption of UV in the water, total chlorine level, ratio of free vs. combined chlorine, background level of organics, and target reduction concentrations.

4.2.6. *Scale Control*

Scaling potential salts often encountered in RO are calcium carbonate, calcium sulfate, silica, calcium fluoride, and strontium sulfate. Calcium carbonate scaling can be controlled by the addition of acid such as sulfuric acid. Sodium hexametaphosphate (SHMP) is commonly used to inhibit calcium sulfate. Silica is usually controlled either by decreasing the recovery ratio or by chemical pretreatment. In seawaters and most brackish waters, the concentration of Sr^{2+} is quite low compared with that of Ca^{2+}. In situations where there is a potential for both calcium-based and strontium-based scales formation, the amount of $CaSO_4/CaCO_3$ scale formed is so large compared with that of $SrSO_4/SrCO_3$ scale that the latter is not a significant factor in scale inhibition.

The efficiency of SHMP as a scale inhibitor in preventing calcium sulfate scale growth has been evaluated (46). It was shown that 5 ppm of SHMP resulted in a significant increase in calcium sulfate solubility. However, increasing the SHMP dosage to 20 ppm did not result in a corresponding increase in the solubility. The effectiveness of a polyelectrolyte-based antiscalant, commercially known as AF-400, was compared with that of SHMP for the control of calcium sulfate scale (47). It was shown that the induction period increased with the increase of AF-400 dose level. At 2 ppm, the crystallization of calcium sulfate was completely inhibited for at least 2 h. At 0.5 ppm, the induction period achieved with AF-400 was about twice as long as that with SHMP.

The effectiveness of four scale inhibitors (ethylene maleic anhydride, acrylate, acrylic acid, and phosphonate) in modifying crystal growth of $CaCO_3$ was studied. The presence of these inhibitors in ppm quantities significantly altered crystal growth of $CaCO_3$; the normal rhombohedral form was changed to a needle-like structure. The inhibitor was believed to have inhibited the crystal growth by adsorbing on to the fast-growing crystal face. In another study, hydroxyethylidene diphosphonate was found to be most effective against $CaCO_3$ scale among four phosphonate-based scale inhibitors (48).

Another pilot-scale study showed that the conventional SHMP + H_2SO_4 inhibitor outperformed a so-called advanced antiscalant, consisting of a polyacrylate and hydroxyethylidene diphosphonate, for the treatment of brackish water (49). The study was carried out in two identical hollow fine fiber RO units under the same feed pressure (27.6 bar) and product water recovery (70%). The pumping energy and scale control treatment costs (as USD/m^3 of water produced) were the same for the two treatment units. However, in terms of salt rejection, product total dissolved solids (TDS), and output, the performance of the advanced antiscalant was inferior. After 3,000 h, the salt rejection of the unit with the advanced antiscalant was about 81 vs. 94% for the unit with H_2SO_4 + SHMP. Similarly, at 3,000 h, the former unit produced 20% less water of unacceptable quality (620 ppm), whereas the latter unit produced more water of excellent quality (180 ppm).

An example of a complete pretreatment flow scheme for seawater RO plant is shown in Fig. 13.10 in which water is controlled for pH, scale, particulates, and biological fouling.

4.2.7. *Membrane Pretreatment*

Conventional pretreatment methods have been used in RO desalination plants for many years. This system is less expensive than the membrane separation system, although the

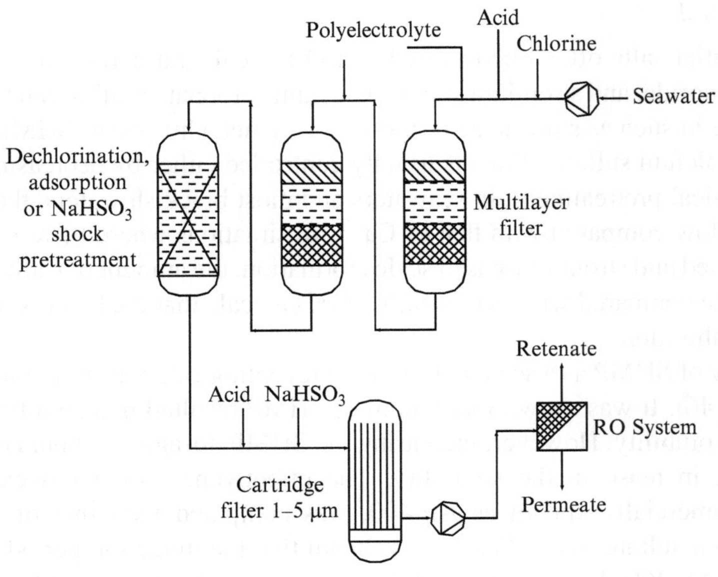

Fig. 13.10. Pretreatment flow scheme for seawater RO plant.

quality of the filtrate is inferior and highly inconsistent. In comparison with conventional pretreatment, MF and UF membrane processes can consistently produce filtrate of a better quality. The membrane separation system is also able to remove viral or microbial contaminants and hence reduce the need for disinfection. Micropollutants such as agricultural chemicals can be removed by the membrane separation, while the conventional pretreatment system is unable to remove them. The ease of operation of the membrane pretreatment processes is also a major advantage. Moreover, membrane pretreatment systems generally require less space and fewer chemicals compared with a conventional pretreatment system, and RO membrane replacement frequency can be reduced (50). Together with improved maintenance procedures such as filtrate backwashing and air scouring, complete flux and pressure recoveries are achievable with minimal use of chemicals. The RO desalination pretreatment systems that utilize MF and UF membranes have started to take the place of conventional pretreatment in public water supply systems and the benefits of these systems are being felt.

The performance of conventional media-filtration and membrane filtration techniques was studied in an RO desalination process (51). For the membrane filtration techniques, two UF and one MF pilot plants were used in the experimental study. During the experiments, the SDI of the filtrate samples was regularly measured to quantify the performance of pretreatment systems in rejecting colloidal particles. Other important parameters that were measured include filtrate flux, transmembrane pressure, total suspended solids, colloidal silica, and total organic carbon. The results showed that the quality of the filtrate produced by the conventional media-filtration technique was inferior and highly inconsistent. SDI of the filtrate varied from 2.8 to 3.8 and spikes as high as 6.3 were frequently observed. Membrane pretreatment produced filtrate of a better quality; SDI of the filtrate produced was consistently less than 3, a prerequisite for proper operation of an RO desalination plant.

The competitiveness of UF pretreatment over conventional pretreatment was assessed by evaluating the impact on RO hydraulic performances (52). The study showed that UF provided permeate water with high and constant quality, resulting in a higher reliability of the RO process than with a conventional pretreatment. The SDI of surface seawater was reduced from 13–25 to less than 1, whereas the conventional pretreatment failed to reduce it to less than 2.5. The combination of UF with precoagulation at low dose helped in controlling UF membrane fouling and providing filtered water in steady state conditions. The performance of RO membranes, downstream UF, exceeded the usual operating conditions encountered in seawater desalination. The combined effect of higher recovery and higher flux rate significantly reduced the RO plant costs.

A study was performed to investigate MF and UF pretreatment performances in treating a broad range of water for RO desalination (53). The results of 2-year testing showed that MF gave the best solution for the pretreatment of raw water to produce infiltration water. The effective mean filtrate production of the pilot plant is higher than that for the other systems that were tested, and the amount of chemicals needed in the process is much lower owing to the effectiveness of the air backwash system. The quality of the filtrate produced with UF membranes was slightly better compared with that of the filtrate produced with MF membranes. A demonstration experiment using 35 pilot-scale MF and UF plants showed no significant difference in membrane filtrate quality and energy consumption between MF and UF membranes. No coliforms were found in the filtrates (44).

MF and UF processes are capable of reducing biofouling tendency in RO membrane as they pose a physical barrier to these microorganisms. MF is capable of removing protozoa (approximately 10 μm), coliforms (approximately 1 μm), and cysts (approximately 0.1 μm). The pore size of UF membranes is smaller and thus can further remove viruses (approximately 0.01–0.1 μm) (54). It has been shown that the indicator organisms *Escherichia coli* and spores of sulfite-reducing *Clostridia*, present in conventionally treated water, could be reduced by UF to values below the detection limit (55). The application of membranes thus reduces the dosage of aggressive chemicals such as chlorine and ozone, thereby minimizing or avoiding the production of undesirable disinfection byproducts. The interest in using membranes as part of the disinfection process has intensified with the emergence of chlorine-resistant pathogens. Chlorine-resistant *Cryptosporidium parvum* has been reported to cause outbreak of diarrhea epidemic in USA and UK. Water supply authorities are looking to UF and MF application to act as an absolute barrier to *Cryptosporidium oocysts*, which range from 4 to 6 μm in size (56).

The use of MF membrane as pretreatment also allows the use of TFC RO membrane instead of CA RO membrane. TFC membranes are anisotropic membranes with a porous sublayer supporting a very thin top layer ranging from 0.1 to 0.5 μm in thickness. Such membranes have higher flux compared to thicker isotropic membranes, as flux is inversely proportional to the thickness of the membrane (57). TFC membranes have higher solute rejection capability and can be operated at lower pressure (58).

Chlorination of feed water prior to membrane pretreatment may extend membrane run times between cleaning. Over 90 h of MF operation was achieved with prechlorinated feed water compared with 42-h operation reported when feed water was not chlorinated (59). Similar observations were reported with dosage of chloramine prior to MF pretreatment (53). It was speculated that preoxidation resulting from chlorination altered the chemistry of EPS produced by the microorganisms in the feed water. This could weaken the attachment of the EPS on the membrane and thus offset the detrimental effect on the membrane flux. However, care must be taken to verify the compatibility of the membrane with chlorination as some membranes are not tolerant to the aggressive action of chlorine.

4.3. *Membrane Cleaning and Regeneration*

Fouling is almost an inevitable consequence of the nature of the RO process itself even when good pretreatment is employed. Furthermore, the high operating pressures compress the membrane and cause further loss of membrane properties. During operation, symptoms of fouling would include one or all of the following conditions:

(a) Decrease in normalized water flow by 10–15% from startup (reference) conditions.
(b) Increase in pressure drop over a stage or the system by 10–15%.
(c) Significant decrease in salt rejection over time.

The challenge is, therefore, to reduce and control fouling sufficiently to minimize the rate of RO flux decline and prolong membrane lifetime. This can be accomplished by a combination of good feed pretreatment as aforementioned and membrane cleaning programs.

Periodic cleaning is essential to improve the performance and to prolong the life of RO membranes. Cleaning methods used for RO membrane restoration could be broadly categorized into two types: chemical and physical.

4.3.1. Chemical Cleaning Methods

Most membrane manufacturers recommend chemical methods for membrane cleaning and regeneration. Chemical cleaning methods depend on chemical reactions to weaken the cohesion forces between the foulants, and the adhesion forces between the foulants and the membrane surface. Chemical reactions involved in cleaning include hydrolysis, peptization, saponification, solubilization, dispersion, chelation, sequestering, and suspending. The choice of chemical cleaning agents not only depends on the type of foulants present in the membrane system but also on the chemical resistance of the membrane material and the whole system. Caution must be taken in applying these chemicals as the aggressive nature of these chemicals may adversely affect the system or membrane integrity when not applied properly. This is especially true for aromatic PA membranes, which are less robust than CA and polysulfone membranes. For example, some cationic and nonionic surfactants may adsorb onto PA membranes and cause a flux decrease. PA membranes are also not resistant to the strong oxidizing actions of disinfectants such as hydrogen peroxide and hypochlorite. Acid used to clean the membrane system must be rinsed out thoroughly before application of hypochlorite for disinfection, because hypochlorite at low pH can cause corrosion in stainless steel. The pH of the chemical cleaning solution must fall within the tolerable range of 1.0–13.0 for polysulfone membranes and 3.0–8.0 for CA membranes (60). Generally, chemicals used for cleaning membranes should ideally posses the following desirable properties:

(a) Loosen and dissolve foulants from membrane surface.
(b) Keep foulants in dispersed and soluble form.
(c) Avoid fresh fouling.
(d) Do not cause damage to membrane material.
(e) Easily rinsed off after cleaning.
(f) Chemically stable before, during, and after use.
(g) Cost effective.

Table 13.3 summarizes common fouling factors and cleaning and control methods. The cleaning procedures may vary depending on the situation, but the basic steps of chemical cleaning are as follows:

(a) Make up the cleaning solution according to the manufacturer's instructions.
(b) Perform a low-pressure flush with permeate water or cleaning solution to displace the solution in the vessels, the process water can be dumped to drain until the cleaning solution has filled the vessels.
(c) Recycle the solution through the elements and back to the tank. During operation, part of the cleaning solution is dumped to drain before returning back to the RO cleaning tank. Readjust the pH to that of the target when it changes more than 0.5 pH units.
(d) An optional soak and recirculation sequence can be used. The soak time can be from 1 h to overnight depending on the manufacturer's recommendations. Additional recirculation of cleaning solutions at a high flow will help to displace the foulants from the membrane.
(e) A low pressure cleaning rinse with permeate water is required to remove all traces of chemical from the cleaning skid and the RO skid.

Once all the stages of a strain are cleaned, the RO membrane can be placed back into service. The RO permeate quality usually becomes stable after a few hours to a few days, especially after the cleanings in high pH solutions.

Frequent and appropriate cleaning procedures are vital in order to maintain the performance of RO membranes. Although the procedures may be similar, in many cases, the cleaning solution concentrations, volumes of solution, and stages of cleaning will differ. The frequency of cleaning will be determined by the rate of fouling. The main factors that influence the effectiveness of membrane cleaning are as follows (33):

- Type of cleaning chemicals used.
- Cleaning solution volume (it must be appropriate for the size of the plant).
- Contact time (soaking of membranes in solution can aid in cleaning).
- Temperature (cleaning chemicals are most effective at warm temperatures).
- Design of the cleaning circuit and operating parameters.

Although chemical cleaning agents are categorized with respect to their action on different types of foulants, their combined effects are much more complex. These cleaning agents may interfere with the cleaning effects of each other. Some may provide efficient control over particular foulants while adversely affecting fouling control of another foulants. For example,

Table 13.3
Summary of fouling factors and methods to clean and control

Fouling factors	Effects on RO performances	Methods for cleaning and controlling fouling
Scales: Precipitation of soluble salts (minerals), for example, $CaCO_3$, $CaSO_4$, $MgCO_3$, $MgSO_4$, $BaSO_4$, $SrSO_4$, SiO_2, $Fe(OH)_3$	Major loss of salt rejection	Clean with citric acid or EDTA-based solution
	Moderate increase in differential pressure	Clean silicate-based foulants with ammonium bifluoride solutions
	Slight to moderate flow loss	Reduce recovery
	Effects usually occur in last stage or element	Adjust pH Use antiscalant
Colloids: Agglomeration of suspended matter on the membrane surface, for example, clay or silt particles, often organic and inorganic constituents	Significant increase in differential pressure	Clean with chelants or detergents at high pH
	Moderate flow loss	Clean silicate-based foulants with ammonium bifluoride solutions
	Moderate loss of rejection	Improve prefiltration
	Effects usually occur in first stage or element	Increase feed flow Reduce recovery
Organics: Attachment of organics, for example, fats, oils, grease, hydrocarbon compounds, synthetic coagulants, and flocculants	Rapid and significant flow loss	Difficult to clean
	Stable or moderate increase in salt rejection	Prefiltrate with granular active carbon
	Generally, no or slight increase in differential pressure	High pH soaks effective for some organic types
Biological: Formation of bio-growth on membrane surface, for example, sulfur reducing bacteria, mycobacterium	Rapid and major loss of production	Clean with commercial detergents or EDTA-based solutions at high pH
	Moderate but rapidly increasing differential pressure	Add solution bisulfate; chlorinate with or without activated carbon filtration
	Slight to moderate increase in passage	
	Effects occur slowly	

some cationic polymers, although effective in silica scale inhibition, reduce calcium carbonate scale inhibition by 30–40% (61). Humic and fulvic acid act as good calcium scale inhibitors but they promote biofouling in the membrane system.

A membrane restoration program was conducted at the site of a large (>3 MGD) RO brackish water desalting plant (62). The procedure applied consisted of low-pressure flushing with acidic (pH 4.0) solution of 2% citric acid with ammonia, followed by alkaline (pH 10.5) flush with NaOH solution. Before and after applying each cleaning solution, the elements were thoroughly flushed with product water. After this a tannic acid solution (100 ppm) was applied at the pressure of 300 psi for hollow fiber elements and at 400 psi for spiral-wound elements. The average salt passage of hollow fiber elements was about 14% before cleaning. Following restoration, it was reduced to 6–10% and was maintained at that level for several thousand operating hours by applying continuous online treatment. The reduction of salt passage was accompanied by a decrease of about 10% in productivity. Similar treatment results were observed with the PA, spiral-wound elements, which had a salt passage in the range of 14–24%. As a result of the treatment, an average decrease in salt passage of 40%, accompanied by a tolerable (approximately 10%) decrease of productivity, was obtained. The treated spiral-wound elements were reinstalled in the desalting unit and operated for over 1,000 h with no significant increase in salt passage.

The 28,800-gal/day RO system consisting of five cellulose triacetate hollow fine fiber permeators in a 3/2 array was cleaned using a combination of FLOCLEAN 103, a low pH liquid formulation for removal of carbonates and metal hydroxides, and FLOCLEAN 107, a neutral pH liquid formulation designed to remove organic, silt, and other particulates from CA membranes. It was found that differential pressure of the membrane system could be reduced by 42% and the permeate flow was increased by 6.4%. The TFC membrane can withstand the higher pH of this formulation without concern about hydrolytic degradation, and a high pH is generally conducive to the removal of colloidal soils. Another RO system equipped with the TFC membrane was cleaned by FLOCLEAN 411, a high pH formulation designed to remove organics, silt, and other particulate deposits from PA, polysulfone, and TFC membranes. It resulted in an excellent improvement in permeate flow. After cleaning, the membrane produced desired recovery at a lower feed pressure. The differential pressure was reduced by 16.7% and the permeate flow was increased by 23.8% (63). Another pilot-scale study demonstrated the outperformance of some commercial cleaners over the membrane manufacturer's recommended recipe of 2% citric acid at pH 4.0 for cleaning seawater RO membranes (64). For example, percentage improvement in nominal product flow achieved with Floclean 403 and 411 from Pfizer was 11.6–30.8% higher than that with citric acid.

Systems with chlorination prior to the membrane processes were easier to clean because of the presence of chloramines, which act as a disinfectant and reduce the tendency of biofouling. However, membranes in such a system are more susceptible to structural damage because of prolonged exposure to the aggressive action of the combined chlorine. CA membranes are reported to become more brittle when chlorine dosage was in the range of 15–20 mg/L (65). It was also suggested that certain organochlorine derivatives might modify the molecular structure of the membranes, thus resulting in flux decline and decreased salt rejecting efficiency.

4.3.2. *Physical Cleaning Methods*

Physical cleaning methods depend on mechanical forces to dislodge and remove foulants from the membrane surface. Physical methods used include forward flushing, backward flushing, backwashing, vibrations, air sparge, and CO_2 back permeation.

Forward Flushing

The purpose of a forward flush is the removal of a constructed layer of contaminants on the membrane through the creation of turbulence. A high hydraulic pressure gradient is in order during forward flush. When forward flush is applied in a membrane, the barrier that is responsible for dead-end management is opened. At the same time, the membrane temporarily performs cross-flow filtration, without the production of permeate.

Backward Flushing

Permeate is always used for a backward flush and the backward flush pressure is about 2.5 times more than the production pressure. The pressure on the permeate side of the membrane is higher than the pressure within the membranes. The pores of a membrane are flushed inside out, resulting from the higher pressure on the permeate side of the membrane than that within the membranes.

Air/Water Flush

Air is added to the forward flush to form air bubbles, causing turbulence which removes fouling from the membrane surface. The benefit of the air flush over the forward flush is that it uses a smaller pumping capacity during the cleaning process.

Permeate Back Pressure

The pressure added on the product side reverses the normal water transport across the membrane. At the same time, the materials dislodged from the surface are carried away by the feed to brine flush simultaneously.

Vibration

The permeator is vibrated by a pneumatic hammer device attached to the pressure vessel. The materials dislodged from the surface are carried away by the feed to brine flush simultaneously.

CO_2 Back Permeation

This method is suitable for hollow fiber configuration. A high-pressure CO_2 gas is introduced into product side through the internal fiber bores and removed through the fiber walls. The foulants lifted from the surface are carried away by a flush stream.

Seven RO cleaning methods were compared for cleaning a hollow fiber: simple forward flush, reverse flow, permeate back pressure, vibration, air drain and water refill, air sparge, and CO_2 back permeation (66). The results showed that the reverse flow technique (in which a permeator flush stream was periodically switched from a feed-to-brine to a brine-to-feed flow direction) was the most effective of the methods tested. The cleaning effectiveness of all the cleaning methods could be improved by an average of 55% with the addition of surfactants to the cleaning flush stream. Three brackish RO units using TFC membranes were studied (67).

The literature review reveals that even though some of the above physical methods, such as forward flushing and reverse flushing, seem to be economically very attractive, physical methods are not implemented widely in the RO industry. Only forward flushing with permeate water is used between cleanings, when more than one chemical cleaning agent is applied. MF and UF membranes used in pretreatment to RO are more frequently cleaned by physical cleaning and less frequently by chemical cleaning. Chemical cleaning may be required, as physical cleaning (e.g., backwashing) does not restore the flux to the initial value and, therefore, the flux would gradually decline to a level requiring a chemical wash, although chemical wash would not necessarily restore the original flux either. A study showed that the MF/UF water recovery (in several stages) could be restored between 80 and 90% by backwashing for every 40 min and chemically cleaning every 6 months (61). In another case, a gas backwash was performed periodically to remove the particle fouling and restored the pressure of a pretreatment MF. Consequently, the period between backwash cycles varied typically from 5 to 20 min with different source water. By a rapid flow of gas through the membrane wall (from inside to outside), the solids were removed from the outer surface of the membrane. However, a gradual accumulation of foulants still occurred over time. A periodic chemical cleaning of the membranes was performed every 2–7 days. This comprised an acid and caustic clean every 2–7 days. It was found that the mean recovery during the whole period of testing was 80–83% (53).

5. CASE STUDY

The Cape Coral water treatment system is one of the largest reuse systems for residential use in USA (66, 67). The total system includes two RO plants with a total production capability of 15 MGD. The design features of the RO facilities are listed in Table 13.4. Well water is used as the feed water with quality listed in Table 13.5. The water treatment system treats the raw well water to meet the drinking water standards of both the US Environmental Protection Agency and the Florida Department of Environmental Protection. The unit operations and unit processes include acidification, scale prevention, cartridge filtration, RO, degasification, disinfection, and neutralization (66, 67).

5.1. *Acidification and Scale Prevention for Pretreatment*

With the application of sulfuric acid for acidification, and a scale inhibitor (Flocon-100 Antiscalant) for scale prevention, certain impurities in raw well water can be kept in liquid form prior to the subsequent water treatment processes. Sulfuric acid feed systems are designed to apply a 98% solution directly to the raw well water through a metering system that includes standby metering pumps. Complete dispersion of scale inhibitor (Flocon-100) is accomplished in very little time (in seconds) and with appropriate velocity gradients. The

Table 13.4
Design features of the Cape Coral water treatment system

Plant data	Plant 1	Plant 2
Production capability	6 MGD	9 MGD
Raw water supply	Lower Hawthorn aquifer	Lower Hawthorn aquifer
	Brackish water	Brackish water
	10 wells (approximately 700-ft deep)	12 wells (approximately 700-ft deep)
Production unit data		
Number of production units	10	8
Operating RO pressure (psi)	250	160
Recovery (%)	75	85
Pressure vessel array	10 (first stage)	16 (first stage)
	7 (second stage)	8 (second stage)
	4 (third stage)	
Number of elements/vessel	4	7
Number of elements/production unit	84	168
Membrane element data		
Manufacturer	Fluid systems	Hydranautics
Model number	8021 LP	8040-LSY-CPA2
Configuration	Spiral-wound	Spiral-wound
Type	Thin film composite	Thin film composite polyamide
Minimum salt rejection (%)	97.5	99

Table 13.5
Water quality in the Cape Coral water treatment system

Parameters	Raw water	RO reject	Plant effluent
pH	7.7	6.8	8.8
Total hardness as calcium carbonate (ppm)	500	2,500	80
Calcium hardness as calcium carbonate (ppm)	190	850	30
Magnesium hardness as calcium carbonate (ppm)	310	1,650	50
Conductivity (ms/cm)	2,400	9,000	575
Total dissolved solids (ppm)	1,534	6,390	313
Chlorides (ppm)	600	2,550	150
Fluorides (ppm)	2	1.6	0.7
Residual chlorine (ppm)			1.82
Turbidity (NTU)			0.66

velocity gradient is related to the amount of energy imparted to the water during mixing of water and scale inhibitor (67).

5.2. *Cartridge Filters for Prefiltration*

After pretreatment by acidification and scale prevention, the raw well water becomes the feed water. The feed water then passes through a series of cartridge filters for removal of suspended particulates, such as microbes, dirt, sand, and silt, and turbidity. Cartridge filtration is considered as an emerging technology suitable for removing suspended matter from water. These cartridge filters are similar to the swimming pool filters and are mechanically simple for easy operation.

The cartridge filters use ceramic or polypropylene microporous filter elements that are packed into pressurized housings. They are operated by the physical process of straining water through porous filter elements and can exclude particles down to less than 1 μm in size. The ease of operation and maintenance of cartridge filters makes them very attractive for small water systems with treatment capacity equal to or below 15 MGD (67).

5.3. *Reverse Osmosis*

In the Cape Coral RO operations, spiral membranes are used in all 18 production units (Table 13.4). Ten of these production units operate at 234 psi. Eight of the newer generation spiral membranes operate at 160 psi. About 12.6% of the cartridge filter effluent becomes the "bypass stream" which bypasses the RO units and goes to the posttreatment facilities directly. The remaining about 87.4% of the cartridge filter effluent becomes the "feed water" which is fed to the RO units for RO membrane filtration and subsequent posttreatment. The average pH of the cartridge filter effluent or the feed water (after pretreatment and cartridge filtration) is 5.9 (67).

The capability of RO for removal of 97–99% of salts, color, bacteria, virus, colloidal suspensions, and other toxic organic and inorganic substances has been demonstrated. The RO units at Cape Coral water treatment system are mainly used to remove chlorides (salts), total hardness, and many other undesirable substances. The average chlorides are reduced from 600 to 62 ppm. The product water is then blended with influent feed water, bringing the chlorides up to 150 ppm. It is important to note that finished water is blended with 10–15% of influent feed water. Normal drinking water should have no more than 250 ppm of chlorides as mandated by the Florida Department of Environmental Regulation. Therefore, the chloride concentration at the level of 150 ppm is acceptable. Total hardness can be reduced from 500 ppm (as calcium carbonate) to 80 ppm (as calcium carbonate) after blending.

After RO treatment, between 15 and 25% of the influent feed water becomes the RO reject, which is discharged into a receiving salt water lake. The average characteristics of the RO reject or brine concentrate are listed in Table 13.5. This highly concentrated RO reject is discharged from the RO system by way of the brine line, and is also called "waste stream." The brine concentrate, treated with chlorine and oxygen to maintain environmental standards, although highly concentrated with impurities, sustains a concentration lower than that of the salt water lake it empties into. The brine is evenly diffused throughout the lake using an intricate series of submersed spray nozzles. The brine or RO reject is neither drinkable nor suitable for irrigation because of the high salt and impurity concentration (67).

After RO treatment, between 75 and 85% of the feed water becomes the RO product water which has low concentrations of total hardness, conductivity, TDS, chlorides, and fluorides. The RO product water is also called "permeate stream."

5.4. Neutralization and Posttreatment

The RO product water (i.e., the permeate stream) and the bypass stream are both further treated with an air stream by degasification for removing hydrogen sulfide and carbon dioxide gases at low pH. The principal design considerations for degasification process (also known as the "aeration" process) are air-to-water ratio, number of degasifiers, and hydraulic loading rate. Removal efficiency of degasification improves with increases in air-to-water ratios and increasing number of degasification units.

Finally, chlorine (10 ppm) and sodium hydroxide (40 ppm) are dosed as the disinfectant and neutralizing agent, respectively. Chlorine is an excellent disinfectant and oxidant. It provides a stable residual for the water distribution system. Average free residual chlorine and average plant effluent turbidity are 1.82 ppm and 0.66 NTU, respectively. The City's RO Water Treatment Facility has been designed to use liquid sodium hydroxide for finished water neutralization. The plant feeds a premixed 50% solution containing 6.38 lb of sodium hydroxide/gal.

The average characteristics of the finished water (i.e. the plant effluent) after the post-treatment of degasification, chlorination, and neutralization are listed in Table 13.5. However, it should be noted that the TDS removal efficiency of RO is very high. Average TDS concentration of the RO produced water (i.e., permeate stream) is only 111 ppm. The higher TDS concentration (313 ppm average) in the finished water (i.e., plant effluent) is mainly contributed by the bypass stream as well as the chemical addition in the posttreatment steps (67). Recent RO and NF developments can be found in the literature (67–69).

5.5. Total Water Production Cost and Grand Total Costs

In 1996, the average water production cost was estimated to be USD 0.73/1,000 gal of water produced (Table 13.6). The cost includes chemical treatment, operating labor, power consumption, and miscellaneous plant administration cost. Typical chemicals required include 130 ppm of sulfuric acid, 3 ppm of scale inhibitor (Flocon-100), 10 ppm of chlorine, and 40 ppm of sodium hydroxide. Normally, between 10 and 20 full-time employees (Chief operators, technician, analyst, supervisor, mechanic, and electricians) are needed to work on site at the plant (67).

The grand total cost includes water production, RO membrane replacement, and capital investment cost, and it is estimated to be USD 1.25/1,000 gal of water produced.

Table 13.6
Water production cost (1996) (65)

Items	Cost (USD/1,000 gal)
Sulfuric acid (130 ppm)	0.03
Scale inhibitor (3 ppm [Flocon-100])	0.03
Chlorine (10 ppm)	0.01
Sodium hydroxide (40 ppm)	0.03
Electricity power	0.34
Labor cost	0.19
Miscellaneous	0.1
Total	0.73

6. RECENT ADVANCES IN RO TECHNOLOGY FOR SEAWATER DESALINATION

Desalination is becoming popular because of the decrease in reliable water sources and increase in the global population. The total desalination capacity of the world is estimated to be 40 million m^3/day (70), whereas the seawater desalination accounts for 24.5 million m^3/day (71). As discussed earlier, RO technology is one of the widely using seawater desalination technologies. In order to meet the demand, RO technology has to be improved in many aspects, and deficiencies in the technology must be minimized. Recent developments in the field of RO desalination of seawater to avoid the deficiencies are discussed in this section.

About 70% of the operating cost of seawater RO desalination is the energy cost (72). On the contrary, energy is as important as water and is becoming scarce. Therefore, energy recovery devices and alternative energy sources receive the attention of the scientific community. Energy recovery devices discussed earlier in this chapter are modified further in recent studies. A novel "fluid switcher" is introduced to the energy recovery devices. It is said that the hydraulic recovery efficiency is increased to 76.83% after the introduction of the "fluid switcher" (73). Alternative renewable energy sources are evaluated for the seawater RO desalination, especially for the small communities in rural areas where electricity is not available. A plant with capacity of 20 m^3/day was evaluated for the possibilities of using photovoltaic (PV) array as the energy source. Modeling study showed that the solar-driven plant is a favorable energy option compared to a fully diesel-driven or diesel-assisted PV-RO plant (74). In another study, solar organic rankine cycle engines are proposed for a plant with a capacity of 15 m^3/day (75). Wave energy is another option proposed for small-scale seawater RO desalination plants. RO plant with a pressure exchanger intensifier and with specific energy consumption less than 2 kWh/m^3 was recognized as the promising location where the wave energy can be utilized (76). Possibility of use of the energy from fuel cells (77), PV energy (78), and hydrostatic pressure (79) was evaluated as well.

Attention has been paid to improving the recovery of the RO membrane. A high recovery seawater RO desalination system has been developed in Japan. The system is based on two major concepts: (1) pressure resistance increment and (2) control of the concentration polarization. RO module with increased pressure resistance (from 7 to 8.2 MPa) was incorporated in the system. The system was tested for two different situations: (1) for standard seawater with TDS of 35,000 mg/L and (2) for Middle East seawater with TDS of 43,000 mg/L. The tests were carried out in Japan in a facility with the capacity of 50,000 m^3/day for standard seawater and in Saudi Arabian RO plant with the capacity of 14,000 m^3/day for Middle East seawater. It was found that the recovery of standard seawater can be increased from 40 to 60% for standard seawater and for the Middle East seawater, the recovery was increased from 35 to 50% successfully (80).

Fouling control is another concern of improvement. Biofouling is a major problem in seawater RO desalination. Intermittent chlorine injection was found to be one of the effective biofouling control strategies. However, it is required to use high chlorine resistant membranes. Cellulose triacetate membranes are recommended for use with direct chlorine sterilization, with intermittent injection of chlorine due to the membrane's high chlorine resistance. The process was optimized (81, 82). In another study, membrane surface was modified with nanosilver to mitigate biofouling (83). Effective pretreatment is important in fouling control. Conventional pretreatment and UF/MF membrane filtration pretreatment were discussed earlier in this chapter. Although UF and MF are cost effective compared to double filtration steps, it is costlier compared to single filtration. On the contrary, it is argued that the utilization of UF/MF only shifts the fouling from RO to upstream UF/MF membranes. It was found that dissolved air flotation (Aqua DAF™) prior to UF membrane filtration improved the UF flux by 60% and the cost was optimized. It is said that this arrangement has improved the flux of RO and reduced the fouling (84). Flotation, a well-known option in wastewater treatment, can be used in seawater RO desalination as a pretreatment step. This technology will effectively remove oil, grease, and algae (85).

Because of the low ionization product of boric acid at pH lower than 10, seawater desalination by RO cannot effectively remove boron from the feed water. Electrodialytic boron removal is proposed recently to avoid the abovementioned problem. Two different feed water samples were used: (1) assuming 40% water recovery and (2) a higher rejection. TDS and boron concentration of the first feed water were 604 and 2.25 mg/L, respectively. For the second feed water, TDS and boron concentration were 400 and 1.3 mg/L, respectively. Electrodialytic unit with AMX and CMX Neosepta (Tokuyama Co.) membranes was used for the boron removal experiments. Membrane to membrane distance was 0.4 mm. Boron concentration was reduced to 0.4 mg/L using the above system, while the energy consumption of the first feed water was 0.237 kWh/m^3 and that of the second feed water was 0.186 kWh/m^3 (86).

Apart from the technological concerns over seawater desalination, environmental concerns are arising. During the process, concentrate and chemicals are released to the marine environment. On the contrary, energy demand is very high and air pollutants are emitted. Therefore, environmental impact assessment should be conducted for each desalination project (71). In Israel, the issue of seawater desalination related environmental pollution is addressed seriously. Guidelines for pretreatment, discharge composition, marine monitoring, etc., are adopted (87).

7. NOMENCLATURE

ΔC = Solute concentration difference across membrane
$\Delta \Pi$ = Osmotic pressure across membrane (atm)
ρ_w = The density of the solvent (kg/m^3)
Δp = The transmembrane pressure difference
Δx = The thickness of the membrane
ΔP = Transmembrane pressure
σ = A measure of the solute–water coupling within the membrane, and may often be treated as 1
Π = Osmotic pressure (force/length^2) (psia or atm)
ε = Porosity of the membrane
$\sum \bar{m}_i$ = The summation of molarities of all dissolved ions and nonionic species in the solution, M
A_s = The solute permeability constant = $D_s K_s / \Delta x$
A_w = The solvent permeability constant = $P_W / \Delta x$ (m/s)
C_S = Average concentration of solute = summation of molarities of all dissolved ions, moles/length^3
C_s^{I} = The concentration of the solute in the feed solution on the membrane surface (kg/m^3)
C_s^{II} = The concentration of the solute in the permeate solution (kg/m^3)
D = Diffusivity
D_s = Diffusivity of the solute
J_s = Solute flux (kg/s m^2)
J_w = Solvent flux (kg/s m^2)
K_s = The solute distribution coefficient between the solution and membrane phase
L_p = A coefficient
$P_{concentrate}$ = The hydrostatic pressure of the concentrate
P_{feed} = The hydrostatic pressure of the feed
$P_{permeate}$ = The hydrostatic pressure of the permeate flow
P_w = The solvent permeability
R = The ideal gas constant, force-length/mass-temperature or the solute rejection value, dimensionless
R = The solute rejection value, dimensionless
T = The absolute temperature (K)

REFERENCES

1. Sourirajan S (1970) Reverse osmosis. Academic Press, New York
2. Loeb S, Sourirajan S (1960) Sea water demineralization by means of a semipermeable membrane. UCLA engineering report 60–60, University of California, Los Angeles
3. Podall HE (1972) Reverse osmosis. In: Li NN (ed) Recent developments in separation science, vol II. CRC Press, Cleveland, pp 171–203
4. Sun-tak H, Kammermeyer K (1976) Membrane in separations. In: Weissberger A (ed) Techniques of chemistry, Wiley, New York, 15(4):247–248

5. Matsuura T, Blais P, Pageau L, Sourirajan S (1977) Parameters for prediction of reverse osmosis performance of aromatic polyamide-hydrazide (1:1) copolymer membranes. Ind Eng Chem Process Des Dev 16:361–372
6. Sourirajan S (1980) Reverse osmosis–a new field of applied chemistry and chemical engineering. In: Plenary lecture at ACS symposium on synthetic membranes and their applications, San Francisco, CA
7. Howe ED (1974) Fundamentals of water desalination. Marcel Dekker, New York
8. Rozelle LT, Cadotte JE, Cobian KE, Kopp CV (1977) Nonpolysaccharide membranes for reverse osmosis: NS-100 membranes. In: Sourirajan S (ed) Reverse osmosis and synthetic membranes, theory-technology-engineering. National Research Council, Canada, pp 249–312
9. Mukherjee D, Kulkami A, Gill WN (1994) Flux enhancement of reverse osmosis membranes by chemical surface modification. J Memb Sci 97:231–249
10. Noble RD, Stern SA (1995) Membrane separations technology: principles and applications. Elsevier, New York
11. Jonsson G (1980) Overview of theories for water and solute transport in UF/RO membranes. Desalination 35:21–28
12. Soltanieh M, Gill W (1981) Review of reverse osmosis membranes and transport models. Chem Eng Commun 12:279–287
13. Mazid M (1984) Mechanisms of transport through reverse osmosis membranes. Sep Sci Technol 19:357–364
14. Pusch W (1986) Measurement techniques of transport through membranes. Desalination 59: 105–115
15. Dickson J (1998) Fundamental aspects of reverse osmosis. In: Parekh B (ed) Reverse osmosis technology. Marcel Dekker, New York, pp 1–51
16. Rautenbach R, Albrecht R (1989) Membrane processes. Wiley, New York
17. Bhattacharyya D, Williams M (1992) Theory – reverse osmosis. In: Ho W, Sirkar K (eds) Membrane handbook. Van Nostrand Reinhold, New York, pp 269–280
18. Lee S, Lueptow RM (2001) Reverse osmosis filtration for space mission wastewater: membrane properties and operating conditions. J Memb Sci 182:77–90
19. Burghoff H, Lee K, Pusch W (1980) Characterization of transport across cellulose acetate membranes in the presence of strong solute-membrane interactions. J Appl Polym Sci 25:323–329
20. Sourirajanand S, Matsuura T (1995) Reverse osmosis/ultrafiltration principles. National Research Council of Canada, Ottawa
21. Matsuura T, Sourirajan S (1981) Reverse osmosis transport through capillary pores under the influence of surface forces. Ind Eng Chem Process Des Dev 20:273–279
22. Mehdizadeh H, Dickson J, Eriksson P (1989) Temperature effects on the performance of thin-film composite, aromatic polyamide membranes. Ind Eng Chem Res 28:814–819
23. Bhattacharyya D, Jevtitch M, Schrodt J, Fairweather G (1986) Prediction of membrane separation characteristics by pore distribution measurements and surface force-pore flow model. Chem Eng Commun 42:111–123
24. Jevtitch M (1986) Reverse osmosis membrane separation characteristics of various organics: prediction of separation by surface force-pore flow model and solute surface concentration by finite element method. Dissertation, D. Bhattacharyya, Director, Department of Chemical Engineering, University of Kentucky, Lexington, Kentucky
25. Matthiasson E, Sivik B (1980) Concentration polarization and fouling. Desalination 35:59–65

26. Gekas V, Hallstrom B (1987) Mass transfer in the membrane concentration polarization layer under turbulent cross flow. J Memb Sci 30:153–161
27. Gill WN, Matsumoto MR, Gill AL, Lee YT (1988) Flow patterns in radial flow hollow fiber reverse osmosis systems. Desalination 68:11–28
28. Kimura S, Sourirajan S (1968) Mass transfer coefficients for use in reverse osmosis process design. Ind Eng Chem Process Des Dev 7:539–547
29. Sirkar KK, Rao GH (1983) Additivity between donnan salt and ion-exchanged salt in the specific conductance of membranes. Desalination 48:25–31
30. Jonsson G (1978) The influence of pressure in the compaction of asymmetric cellulose acetate membranes. In: Proceedings of the 6th international symposium in fresh water from sea, Athens
31. Judd S, Jefferson B (2003) Membrane for industrial wastewater recovery and re-use. Elsevier Advanced Technology, Oxford
32. Matsuura T (2001) Progress in membrane science and technology for seawater desalination a review. Desalination 134:47–54
33. Dudley LY (1998) Membrane autopsies for reversing fouling in reverse osmosis. Membr Technol 95:9–12
34. Geisler P, Krumm W, Peters TA (2001) Reduction of the energy demand for seawater RO with the pressure exchange system PES. Desalination 135:205–210
35. Winfield BA (1979) A Study of the factors affecting the rate of fouling of reverse osmosis membranes treating secondary sewage effluent. Water Res 13:565–569
36. Weisner MR, Aptel P (1996) Mass transport and permeate flux and fouling in pressure-driven processes. In: Odendaal PE, Wiesner MR, Mallevialle J (eds) Water treatment membrane processes. McGraw-Hill Company, New York, pp 4.1–4.30
37. Lee S, Cho J, Elimelech M (2004) Influence of colloidal fouling and feed water recovery on salt rejection of RO and NF membranes. Desalination 160:1–12
38. Winter H (1987) Control of organic fouling at two seawater reverse osmosis plants. Desalination 66:319–325
39. Sadr Ghayeni SB, Beatson PJ, Schneider RP, Fane AG (1998) Water reclamation from municipal wastewater using combined microfiltration-reverse osmosis (ME-RO): preliminary performance data and microbiological aspects of system operation. Desalination 116:65–80
40. Baker JS, Dudley LY (1998) Biofouling in membrane systems – a review. Desalination 118:81–90
41. Wang LK (2006) Innovative ultraviolet, ion exchange, membrane and flotation technologies for water and wastewater treatment, National Engineers Week Seminar, Training Manual, National Association of Professional Engineers and Practicing Institute of Engineers, Albony, 12–14 February 2006
42. Bou-Hamad S, Abdel-Jawad M, Ebrahim S, Al-Mansour A, Al-Hijji A (1997) Performance evaluation of three different pretreatment systems for seawater reverse osmosis technique. Desalination 110:85–92
43. Adin A, Klein-banay C (1986) Pretreatment of seawater by flocculation and settling for particulates removal. Desalination 58:227–241
44. Taniguchi Y (1997) An overview of pretreatment technology for reverse osmosis desalination plants in Japan. Desalination 110:21–36
45. Ebrahim S (1994) Cleaning and regeneration of membranes in desalination and wastewater applications: state-of-the-art. Desalination 96:225–238
46. Yeatts LB, Lantz PM, Marshall WL (1974) Calcium sulfate solubility in brackish water concentrates and applications to reverse osmosis processes; polyphosphate additives. Desalination 15:177–192

47. Amjad Z (1985) Applications of antiscalants to control calcium sulfate scaling in reverse osmosis systems. Desalination 54:263–276
48. Reddy MM, Nancollas GH (1973) Calcite crystal growth inhibition by phosphonates. Desalination 12:61–73
49. Butt FH, Rahman F, Baduruthamal U (1995) Pilot plant evaluation of advanced vs. conventional scale inhibitors for RO desalination. Desalination 103:189–198
50. Pearce GK (2008) UF/MF pre-treatment to RO in seawater and wastewater reuse applications: a comparison of energy costs. Desalination 222:66–73
51. Chua KT, Hawlader MNA, Malekb A (2003) Pretreatment of seawater: results of pilot trials in Singapore. Desalination 159:225–243
52. Brehant A, Bonnelyeb V, Perez M (2002) Comparison of MF/UF pretreatment with conventional filtration prior to RO membranes for surface seawater desalination. Desalination 144:353–360
53. Van Houtte E, Verbauwhede J, Vanlerberghe F, Demunter S, Cabooter J (1998) Treating different types of raw water with micro- and ultrafiltration for further desalination using reverse osmosis. Desalination 117:49–60
54. Fane AG (1996) Membranes for water production and wastewater reuse. Desalination 106:1–9
55. Kruithof JC, Schippers JC, Kamp PC, Folmer HC, Hofman JAMH (1998) Integrated multi-objective membrane systems for surface water treatment: pretreatment of reverse osmosis by conventional treatment and ultrafiltration. Desalination 117:37–48
56. Hills P, Padley MB, Powell NI, Gallegher PM (1998) Effects of backwash conditions on out-to-in membrane microfiltration. Desalination 118:197–204
57. Aptel P, Buckley CA (1996) Categories of membrane operations. In: Odendaal PE, Wiesner MR, Mallevialle J (eds) Water treatment membrane processes. McGraw-Hill, New York, pp 2.1–2.24
58. Buckley CA, Hurt QE (1996) Membrane applications: a contaminant-based perspective. In: Odendaal PE, Wiesner MR, Mallevialle J (eds) Water treatment membrane processes. McGraw-Hill, New York, pp 3.1–3.24
59. Jolis D, Hirano RA, Pitt PA, Müller A, Mamais D (1996) Assessment of tertiary treatment technology for water reclamation in San Francisco, California. Water Sci Technol 33:181–192
60. Trägårdh G (1989) Membrane cleaning. Desalination 71:325–335
61. Sheikholeslami R (1999) Fouling mitigation in membrane processes. Desalination 123:45–53
62. Wilf M, Glueckstern P (1985) Restoration of commercial reverse osmosis membranes under field conditions. Desalination 54:343–350
63. Graham SI, Reitz RL, Hickman CE (1989) Improving reverse osmosis performance by periodic cleaning. Desalination 74:113–124
64. Ebrahim S, El-Dessouky H (1994) Evaluation of commercial cleaning agents for seawater reverse osmosis membranes. Desalination 99:169–188
65. Ridgway HF, Justice CA, Whittaker C, Argo DG, Olson BH (1984) Biofilm fouling of RO membranes – its nature and effect on treatment of water reuse. J Am Water Works Assoc 76:94–102
66. Johnson J, Leahy M (1982) Development of new cleaning techniques for reverse osmosis membranes. OWRT Contract 14-340-001-8519, Office of Water Research and Technology, Washington
67. Wang LK, Kopko SP (1987) City of Cape Coral reverse osmosis water treatment facility. Technical report No. NTIS-PB97-139547. US Department of Commerce, National Technical Information Service, Springfield, VA.
68. Benko K, Pellegrino J, Price MK (2006) Measurement of water permeation kinetics across reverse osmosis and nanofiltration membranes-apparatus development. J Memb Sci 270:187–195

69. AWWA (2006) Desalination of seawater and brackish water. American Water Works Association, Denver
70. Khawaji AD, Kutubkhanan IK, Wie JM (2008) Advances in seawater desalination technologies. Desalination 221:47–69
71. Lattemann S, Hopner T (2008) Environmental impact and impact assessment of seawater desalination. Desalination 220:1–15
72. Akgul D, Cakmakci M, Kayaalp N, Koyuncu I (2008) Cost analysis of seawater desalination with reverse osmosis in Turkey. Desalination 220:123–131
73. Sun J, Wang Y, Xu S, Wang S (2008) Energy recovery device with a fluid switcher for seawater reverse osmosis system. Chin J Chem Eng 16(2):329–332
74. Helal AM, Al-Malek SA, Al-Katheeri ES (2008) Economic feasibility of alternative designs of a PV-RO desalination unit for remote areas in the United Arab Emirates. Desalination 221:1–16
75. Bruno JC, Lopez-Villada J, Letelier E, Romera S, Coronas A (2008) Modelling and optimization of solar organic rankine cycle engines for reverse osmosis desalination. Appl Therm Eng 28:2212–2226
76. Folley M, Suarez BP, Whittaker T (2008) An autonomous wave-powered desalination system. Desalination 220:412–421
77. Singh R (2008) Sustainable fuel cell integrated membrane desalination systems. Desalination 227:14–33
78. Gandhidasan P, Al-Mojel SA (2009) Effect of feed pressure on the performance of the photovoltaic powered reverse osmosis seawater desalination system. Renewable Energy 34(12): 2824–2830
79. Charcosset C, Falconet C, Combe M (2009) Hydrostatic pressure plants for desalination via reverse osmosis. Renewable Energy 34(12):2878–2882
80. Fujiwara N, Matsuyama H (2008) High recovery system in seawater reverse osmosis plants. J Appl Polym Sci 108:3403–3410
81. Fujiwara N, Matsuyama H (2008) Optimization of the intermittent chlorine rejection (ICI) method for seawater desalination RO plants. Desalination 229:231–244
82. Fujiwara N, Matsuyama H (2008) Elimination of biological fouling in seawater reverse osmosis desalination plants. Desalination 227:295–305
83. Yang HL, Lin JC, Huang C (2009) Application of nanosilver surface modification to RO membrane and spacer for mitigating biofouling in seawater desalination. Water Res 43(15): 3777–3786
84. Bonnelye V, Guey L, Del Castillo J (2008) UF/MF as RO pre-treatment: the real benefit. Desalination 222:59–65
85. Peleka EN, Matis KA (2008) Application of flotation as a pretreatment process during desalination. Desalination 222:1–8
86. Turek M, Bandura B, Dydo P (2008) Electrodialytic boron removal from SWRO permeate. Desalination 223:17–22
87. Safrai I, Zask A (2008) Reverse osmosis desalination plants – marine environmentalist regulator point of view. Desalination 220:72–84
88. Seader JD, Henley EJ (1998) Separation Process Principles. John Wiley & Sons, New York 920p.

14

Membrane Technologies for Point-of-Use and Point-of-Entry Applications

Puangrat Kajitvichyanukul, Yung-Tse Hung, and Lawrence K. Wang

Contents

Abstract Point-of-use (POU) system is the treatment process aimed to treat only water intended for direct consumption (drinking and cooking), typically at a single tap or limited number of taps. Point-of-entry (POE) treatment devices are typically installed to treat all water entering a single home, business, school, or facility. Reverse osmosis (RO) is recognized by the industry as one of the top POU and POE treatment technologies. This chapter describes the advantages and limitations in using RO for POU and POE applications. Types and configurations of reverse osmosis, and installation, operation and maintenance, and testing of RO are also included.

Key Words Point-of-use • POU • point-of-entry • POE • ultrafiltration • microfiltration • nanofiltration • reverse osmosis • oil–water • membrane

1. INTRODUCTION

Potable water with a good quality that is chemically and microbiologically safe for consumption is necessary for humans. In general, the central drinking water treatment plants receive raw water from either surface water or groundwater and produce high-quality water using various types of physical, chemical, and biological treatment processes before distribution to consumers. While central treatment plants treat all water to the same level, plants that treat only a portion of the total flow for specific consumption are needed. Two types of

From: *Handbook of Environmental Engineering, Volume 13: Membrane and Desalination Technologies*
Edited by: L.K. Wang et al., DOI: 10.1007/978-1-59745-278-6_14

water treatment devices designed for this purpose include point-of-use (POU) and point-of-entry (POE). POU devices are designed to treat only the water intended for direct consumption (drinking and cooking), typically at a single tap or limited number of taps, as shown in Fig. 14.1 (1). POE treatment devices are typically installed to treat all water entering a single home, business, school, or facility, as presented in Fig. 14.2 (1). POE systems, or wholehouse systems, are usually installed near the water meter. (Water meters are usually located in the basement of a house. In warm weather climates, the water meter may be installed in the garage or outside the house.)

POU or POE treatment devices may be an option for public water systems where central treatment is not affordable. The plumbed-in units with separate faucets are the most widely

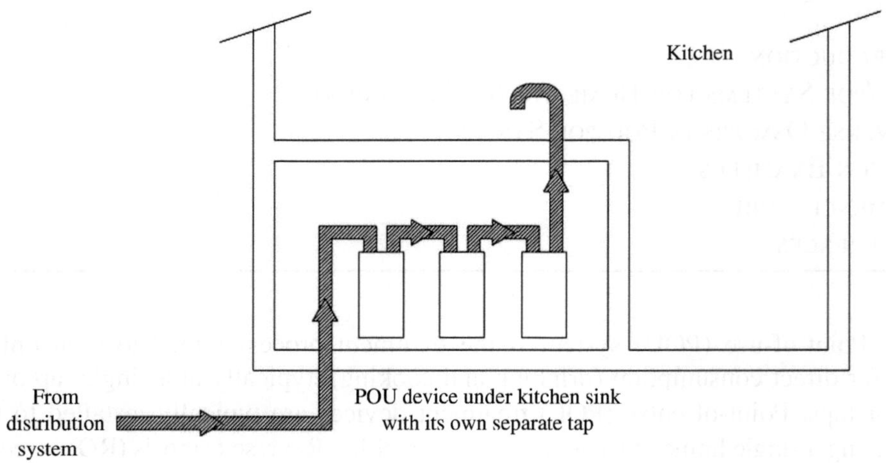

Fig. 14.1. Typical of point-of-use (POU) installation (1).

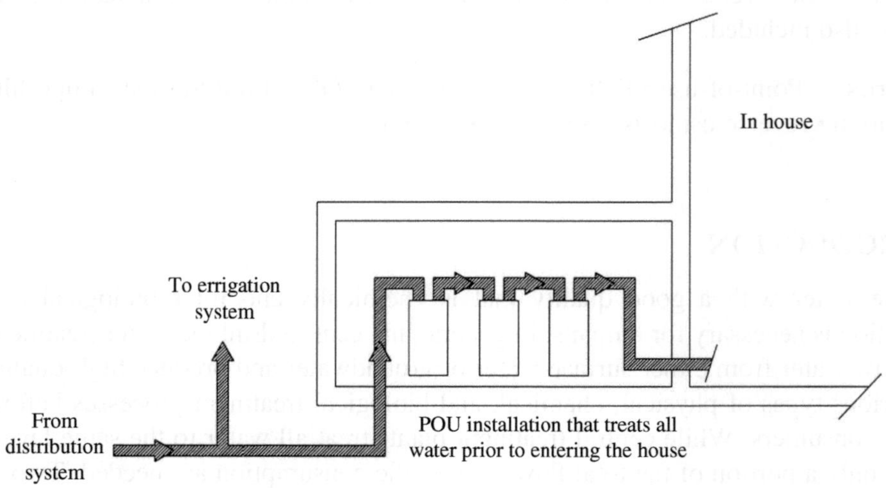

Fig. 14.2. Typical of point-of-entry (POE) installation (1).

used application of POU device. Such units are typically installed under the kitchen sink so as to provide convenient use for drinking and cooking water. Separate faucets allow for the use of untreated water for washing and cleaning, thus helping to reduce operating costs of the treatment device (1).

2. POU/POE SYSTEMS FOR HOME WATER TREATMENT

2.1. *Types of POU/POE Systems*

Several different types of POU/POE are applied to treat tap water with the required quality for home use. Details of each type of POU/POE systems are provided below.

Personal water bottle: This type of product consists of a bottle and a filter. The filter may be integrated with the push/pull cap of the filter bottle or may be integrated with a straw.

Pour-through: In pour-through products, gravity causes water to drip through a pitcher, which is usually stored in the refrigerator. They typically have a lower capacity (i.e., can filter fewer gallons) than other types of systems.

Faucet mount or faucet-attached unit: This type of filter is attached on an existing kitchen sink faucet. A diverter is usually used to direct water through the system when treated drinking water is desired.

Counter-top manual fill: This system is usually placed on a counter and filled by pouring water into the system and activating it for a batch of water. (A manual fill distiller is usually considered to be a Counter-Top Manual Fill.)

Counter-top connected to sink faucet or faucet-connected counter-top units: This product is usually placed on a counter and connected by tubing to an existing kitchen sink faucet. The treated water dispenses out of a return tube from the kitchen faucet, or the treated water is dispensed from a spout in the system.

Plumbed-in: This type of system is usually installed under a sink and requires a permanent connection to an existing water pipe. The filtered water is dispensed through an existing sink faucet.

Plumbed-in to separate tap: This system is so-called plumbed-in units with separate faucets for the POU device. This product is installed in the same manner as the plumbed-in systems (above). However, the filtered water is dispensed through an auxiliary faucet mounted next to the kitchen sink.

2.2. *Technologies in POU/POE Systems*

Technologies applied in POU/POE include adsorptive media, ion exchange (IX), granular activated carbon (GAC), and reverse osmosis (RO). Details of each system in POU/POE are explained as follows:

Adsorptive media: Adsorptive media includes activated alumina (AA), granular ferric hydroxide (GFH), or other specialty iron-based media. The widely used adsorptive media is AA, a heat-treated, hydrated aluminum oxide. Iron-based media is typically generated in a proprietary process and may consist of granules of ferric oxide or ferric hydroxide, activated alumina coated with iron, or natural minerals impregnated with a substantial quantity of ferric hydroxide. Activated alumina is used for a wide range of adsorbent as a selective adsorbent for many chemicals including arsenic (in an oxidized state) and fluoride, and for selenium removal. Activated alumina is listed as a Small System Compliance Technology (SSCT) for arsenic.

Ion exchange: Ion exchange is a process in which ions are exchanged between a solution and an ion exchanger, an insoluble solid or gel. IX can consist of anion exchanger (AX) or cation exchanger (CX). IX achieves the selective removal of charged inorganic species from water using an ion-specific resin. The cation exchanger (CX) is used for positively charged cations and the anion exchanger (AX) is used for negatively charged anions. Ion exchange is a method widely used in POU/POE system to produce soft water. This is accomplished by exchanging calcium (Ca^{2+}) and magnesium (Mg^{2+}) cations against sodium (Na^{+}) or hydrogen (H^{+}) cations. As water containing undesired ions passes through a column of resin media, charged ions on the resin surface are exchanged with the undesired ions in the water. In a large centralized treatment system, the resin is regenerated and a regenerant waste stream is discharged. For POU/POE units, the resin is replaced periodically as opposed to regenerating. Ion exchange is classified as an SSCT for uranium and can also be used to remove arsenic.

Granular activated carbon: GAC is an established technology that works through absorption of problematic compounds. GAC is a highly porous material with a very large surface area. Chemical pollutants are attracted to and held by the surface of GAC. This unit is primarily used to remove taste and odor as well as some harmful contaminants. GAC is potentially useful in small system applications for the removal of organic chemicals in the POU/POE system. The capacity of GAC to adsorb organic chemicals depends on the chemical properties of the organic chemicals. GAC has the added benefit of improving esthetics (taste, odor, and color) of water and is sometimes included in POU/POE applications for improved esthetics. In addition, this unit is best suited for the removal of organic compounds and is selected as an SSCT for synthetic organic contaminants (organic chemicals, e.g., pesticides and herbicides).

Reverse Osmosis: In water filter terms, RO (or hyper-filtration) is the process of filtering water under pressure through a semi-permeable membrane, allowing water to pass through but rejecting other particles such as bacteria, toxins, salts, and anything bigger than around 150 Da in size. RO can remove contaminants as small as a molecule and is listed as an SSCT for arsenic, copper, lead, fluoride, radium, and uranium.

The applicability of the applied technologies in the POU/POE system is shown in Table 14.1 (1).

2.3. *Selection of POU/POE Technologies*

The top two POU/POE treatment technologies recognized by the industry are adsorption and RO (1). Adsorption uses filtration media to bind the arsenic to the adsorbent chemically. Advantages include low cost, no requirement of chemicals or regeneration, safe handling, and removal of other heavy metals such as copper. However, it requires certain contact time; competing ions can reduce capacity and performance decreases with pH. RO also has its advantages and disadvantages. RO can remove dissolved contaminants as well as taste and odor. Other challenges for RO include a high cost if arsenic is the primary target contaminant;

Table 14.1
Applicability of POU/POE treatment technologies

Treatment technology	Contaminant							
	Arsenic	Copper	Lead	Fluoride	Nitrate	SOCs	Radium	Uranium
Activated alumina (AA)	SSCT			UI				X
Distillation	X	X	X		SSCT		?	?
Granular activated carbon (GAC)						SSCT		
Anion exchange (AX)	X				SFI			
Cation exchange (CX)		SSCT	SSCT				SSCT	
Reverse osmosis (RO)	SSCT	SSCT	SSCT	SSCT	SFI		SSCT	SSCT
Other adsorption media[a]	X							
Treatment technology	Contaminant							
	Antimony	Barium	Beryllium	Cadmium	Chromium	Selenium	Thallium	
Anion exchange (AX)	SSCT				SSCT	SSCT		
Cation exchange (CX)		SSCT	SSCT	SSCT			SSCT	
Reverse osmosis (RO)	SSCT	SSCT	SSCT	SSCT	SSCT	SSCT	SSCT	

Note: SSCT = treatment technology has been identified by EPA as an SSCT (*Federal Register*, Volume 63, No. 151, August 6, 1998). SFI = Treatment technology has been suggested to receive further investigation for the listed contaminant; anion exchange for nitrates is not currently recommended. UI = Under investigation. X = Treatment technology can remove the noted contaminant but it is not listed as an SSCT in the Federal Register or in a rule. ? = treatment technology is questionable for the listed contaminant.

[a]Other adsorption media such as silica, fluoride, phosphate, sulfate, dissolved iron and manganese.

it is not suitable for the POE systems due to corrosion and affordability, wastes 3–5 gallons of water for every treated gallon, and requires routine maintenance.

Although some POU/POE technologies are capable of removing microbial contaminants, VOCs, or radon, POU/POE devices should not be used for achieving compliance with these contaminant rules. VOCs and radon are both volatile and present an inhalation or contact exposure risk at untreated taps (e.g., showerheads). Thus, POU/POE system is not listed as a compliance technology for any MCL or treatment technique requirement for a microbial contaminant or indicator of a microbial contaminant by US EPA as prohibited by the Safe Drinking Water Act (SDWA). Therefore, POU/POE devices at a single kitchen tap will not sufficiently protect the public from these risks.

To select an appropriate POU or POE system for household water treatment, tap water has to be analyzed by the local health department or a reputable laboratory. This step will help toward solving a suspected water quality problem. A water analysis is also used to determine the most appropriate solution to the problem. For example, the problem of arsenic contamination in water might be identified and the proper POU/POE system will be selected. To obtain the quality of water entering the POU/POE system, state or local health officials can interpret water analysis results. Some laboratories may also provide this service. The appropriate POU/POE system will then be selected based on the characteristics of tap water and the quality of treated water for household use. The selection of specific types of POU/POE technologies may be restricted by site-specific water quality issues. The presence of high concentrations of competing contaminants or foulants can significantly reduce the removal efficiencies of these devices, making water quality testing and pilot testing important steps in selecting a POU/POE technology. The water quality parameters and competing ions that may reduce the efficiency of POU and POE devices are shown in Table 14.2 (1).

To select an appropriate POU/POE technology, the following criteria have to be considered:

- The case of use of the technology
- The microbiological quality of water
- System costs
- Environmental sustainability
- Socio-cultural acceptability
- The required operation and maintenance

Table 14.2
Water quality parameters and competing ions that may reduce the efficiency of POU/POE

Technology	Water quality parameter of concern	Issue
Ion exchange	Iron, manganese, copper	Fouling, competing ions
Adsorptive media	Silica, fluoride, phosphate, sulfate, dissolved iron and manganese	Interfering/competing ions
Reverse osmosis	Hardness, iron, manganese	Fouling
Granular activated carbon	Organics, multiple SOCs or VOCs present	Competing ions
Aeration	Hardness, iron, manganese	Fouling, scaling

Available technologies for POU/POE applications used in developing, transition, and industrialized countries are summarized in Table 14.3 (2).

Some POU/POE treatment methods are described in detail below.

(1) Solar or UV irradiation

Solar radiation is a simple technology for destroying pathogenic microorganism and, consequently, improving the microbiological quality of water. This method can be used effectively for small volumes of water. According to several researches (12–14), solar water disinfection (SODIS) was applied for treating water of small volume (<10 L) and with relatively low turbidity (<30 NTU). The low-turbidity water was filled in clear PET bottle (1–2 L), aerated with air, and exposed to full sunlight for 5 h. This system highly depends on sunlight for disinfection quality.

Table 14.3
Overview of available POU/POE water treatment technologies for developing/industrialized/ transition countries

Water treatment system	Type of supply	Estimated costs		Source
		Investment USD	Operational USD[a]	
Developing countries				
Boiling with fuel	POU	Cook pot	Depends on fuel price	Sobsey (3)
Solar disinfection	POU	Plastic bottles	None	Sobsey (3)
UV disinfection with lamps	POU/POE	100–300	10–100	Sobsey (3)
Free chlorine	POU/POE	2–8 (vessel)	1–3	Sobsey (3)
Biosand filters	POU/POE	10–20	None	Kaiser et al. (4)
Ceramic filters	POU/POE	10–25	None-10	Clasen et al. (5)
Coagulation, filtration, chlorination	POU/POE	5–10	140–220	Sobsey (3)
Transition and industrialized countries				
Activated carbon filtration	Faucet-mounted	25–50	25–50	WSC (6); Ecosoft (7)
	Under a sink	50–300	50–100	
	POE	500–800	n.a.	
Microfiltration[b]	POU/POE	3	12	LifeStraw (8); Li and Chu (9); Pillay (10)
Ultrafiltration	POU/POE	40	None	LifeStraw (8); Homespring (11)
	POE	2,700–3,000	n.a.	
Reverse osmosis	Single tap	300–600	80–120	WSC (6)

n.a. data not available, *POE* point-of-entry, *POU* point-of-use.
[a]Operational costs for POU/POE systems are given for a family of four.
[b]Ceramic filters are considered separately.

UV irradiation with lamps is another disinfection technique using a similar mechanism in destroying pathogenic microorganisms. This method is well known for its performance in inactivating two waterborne, chlorine-resistant protozoans, *Cryptosporidium parvum* oocysts and *Giardia lamblia* cysts. To obtain high performance of this method, low turbidity of water is required.

(2) Chemical treatment
Among several chemicals, free chlorine is the most widely used in POU/POE system. This chemical can destroy various types of waterborne pathogens except *Cryptosporidium parvum* oocysts and the Mycobacteria species (15). The taste and odor problems cause low socio-cultural acceptation of this method.

(3) Filtration
Several types of filtration have been used to remove various types of pathogens. These methods are biosand filtration and fiber or fabric filtration (2). Biosand filter is the sand filter with a bioactive layer that can eliminate pathogens such as bacteria and protozoa. This filtration process has relatively low performance in destroying viruses.
Fiber or fabric filters are commonly used in the removal of large waterborne pathogens, such as free-swimming larval forms of Schistosomes and Faciola species, guinea worm larvae within their intermediate crustacean host, and bacterial pathogens associated with relatively large copepods and other zooplanktons in water, such as the bacterium *Vibrio cholerae* (2, 3, 16, 17). This filter has limitation in removing viruses, bacteria, and smaller protozoan parasites (17).

(4) Membrane filtration Membrane filtration is the most widely used treatment method for drinking water, especially in POU/POE system. This method provides superb advantage as an absolute barrier for pathogens and suspended solids or turbidity. In comparison to other conventional water treatment methods, this process can treat water in one step without the addition of chemicals. Membrane systems used in POU/POE system include ultrafiltration (UF), microfiltration (MF), nanofiltration (NF), and RO. Summary of available POU/POE water treatment technologies for developing and transition countries is presented in Table 14.4. Among these membranes, RO is the most recognized system at the industrial level (1). Details of fundamental concept, type, composition, operation and maintenance, and design of RO are described in the next section.

3. REVERSE OSMOSIS IN POU/POE SYSTEM

3.1. Application of Reverse Osmosis in POU/POE

Reverse osmosis is one type of membrane filtration that is widely used in drinking water treatment. The RO water treatment method has been used comprehensively to convert brackish or seawater to drinking water (18–21), to clean up wastewater (22–24), and to recover and treat water from industrial processes (25–29). Nowadays, RO is becoming more popular as the household water treatment method for the direct consumption of water, such as for drinking and cooking purposes. Owing to the increasing concern about contaminants that affect human health, as well as non-hazardous chemicals that affect the taste, odor, or color of drinking water, RO has become a common method for the treatment of household drinking water supplies.

Table 14.4
Available membrane systems for decentralized application in developing/transition countries and their properties

Membrane system	Capacity (L/day)	Pre treatment (posttreatment)	Feed water quality	System investment costs (USD)
RO, POU	145–340	MF, activated carbon, multistage	Tap water	ca. 400
RO, SSS	1,000	UF	Brackish water	High
	720,000	MF/UF	Sea or brackish water	ca. 10,000
FO, POU	3 (9 L/system)[a]	None	Surface water	30
UF POU	20–30, 18,000 L/system[a]	Chlorine	Surface/ groundwater	ca. 40
UF POE	25,000	Active carbon	Tap or groundwater	2,700–3,000
UF SSS	60,000	None		141,000
	240,000	None		178,000
	1,000[b]	– Coarse filter – Microfilter – Security filter	Surface water	n.a.
	≤6,000	Different configuration	Surface/ groundwater	Depending on layout
	10 m^3/day	None	Surface water	n.a.
	48,000	Multistage MF, UV	Brackish water	ca. 26,000
MF, POU/ POE	100 L/system[a]	None	Surface water	49.95
	700 L/system[a]			3
	ca. 5,000 L/system[a]	None	Surface water	10–25
	10,000 L/system[a]			150–300
	100–750 L/system[a]	MF, active carbon		200–400
MF, SSS	10,000	None	Surface/ groundwater	ca. 1,000–2,000
	n.a.	None	Surface water	n.a.
	≤5,760,000	Different configurations	Surface/ groundwater	Depending on layout

n.a. data not available. The literature references for this table are discussed in the text, and the data on costs and capacity for the market products were obtained from the producer. *MF* microfiltration, *RO* reverse osmosis, *SSS* small-scale systems, *UF* ultrafiltration.
[a]Capacity per system service life.
[b]Capacity per membrane module.

The advantages of RO over other POU/POE systems are as follows:

- RO is relatively insensitive to flow and total dissolved solids (TDS) level. In addition, it can remove a variety of ions and metals and very fine suspended particles such as asbestos that may be found in water. It is also effective in removing some detergents; some taste, color, and odor-producing chemicals, certain organic contaminants; and some pesticides.
- RO is suitable for small systems with a high degree of seasonal fluctuation in water demand.
- Water treated by RO does not adversely affect people on sodium-restricted diets.

Some of the limitations of RO are as follows:

- Reverse osmosis units are slow and produce more wastewater. A little more than 1 gallon of potable water is produced every 6 h, and 3–5 gallons of wastewater are generated in that time.
- High pressure (and the associated electrical energy costs) is required to operate the unit.
- High level of pretreatment is required in some cases.
- RO requires routine maintenance.

The main purpose of RO treatment in POU/POE system is generally to reduce the levels of naturally occurring substances that cause water supplies to be unhealthy or unappealing (foul tastes, smells, or colors) and the substances that have contaminated the water supply, resulting in possible adverse health effects or decreased desirability. Before installing the RO and other treatment units, the water has to be tested to identify the bacteria and level of minerals that are present. The water test analysis and interpretation will provide information about both naturally occurring substances and those resulting from human activity. The intended use of the water (drinking only, drinking and cooking, laundry, or all household uses) will also help in determining the extent of treatment needed and the type of system to select. However, treatment of contaminated water supplies should be considered only as a temporary solution. The best solution is to remove the source(s) of contamination and/or obtain a new water supply.

For the POU/POE system, RO is listed as an SSCT by USEPA for copper, lead, fluoride, antimony, barium, beryllium, cadmium, chromium, selenium, and thallium. In addition, RO is also suggested to receive further investigation for its potential application for nitrate removal. In fact, RO is an effective method for reducing the concentration of total dissolved solids, including a variety of ions and metals and very fine suspended particles such as asbestos that may be found in water. It is also effective in removing some detergents; some taste, color, and odor-producing chemicals; certain organic contaminants; and some pesticides. List of the impurities that can be reduced by RO is shown in Table 14.5 (30). However, the rate of reduction of each specific compound will depend on the RO membrane type and the system's operating conditions.

Various major contaminants can be removed by RO in POU/POE units. Details and the requirements in pollutant removal for each contaminant are provided below.

Hardness: Hardness of water is a measure of the amount of minerals, primarily calcium and magnesium, it contains. Water that contains more than 200 mg/L or 200 parts per million (ppm) as calcium carbonate ($CaCO_3$), or 12 grains per gallon, is considered to be hard and may cause plumbing and laundry staining.

Total dissolved solids: The TDS test measures the total amount of dissolved minerals in water. The solids can be iron, chlorides, sulfates, calcium, or other minerals found on the

Table 14.5
List of impurities that can be removed by reverse osmosis (adapted from Kneen et al. (30))

Ions and metals		Organic chemicals
Arsenic (As)	Aluminum (Al)	Benzene
Barium (Ba)	Bicarbonate (HCO_3^-)	Carbon tetrachloride
Cadmium (Cd)	Calcium (Ca)	Dichlorobenzene
Carbonate (CO_3^{2-})	Chloride (Cl)	MTBE
Chromium (Cr)	Copper (Cu)	Toluene
Fluoride (F)	Iron (Fe)	Trichloroethylene
Lead (Pb)	Magnesium (Mg)	Total trihalomethanes (THMs)
Manganese (Mn)	Mercury (Hg)	Pesticides
Nitrate/nitrite (NO_3^-/NO_2^-)	Potassium (K)	1,2,4-trichlorobenzene
Radium (Ra)	Selenium (Se)	2,4-D
Silver (Ag)	Sodium (Na)	Atrazine
Sulfate (SO_4^{2-})	Zinc (Zn)	Endrin
Particles		Heptachlor
Asbestos	Protazoan cysts	Lindane
		Pentachlorophenol

earth's surface. The dissolved minerals can produce an unpleasant taste or appearance and can contribute to scale deposits on pipe walls. The only effective means of reducing TDS is by using RO.

Arsenic (AS): Arsenic (As) is a naturally occurring contaminant found in many groundwaters in two forms (valences or oxidation states): pentavalent arsenic (also known as As(V), As^{5+}, or arsenate) and trivalent arsenic (also known as As(III), As^{3+}, or arsenite). In some occasion, a combination of both can be found in groundwater. Although both forms of arsenic are potentially harmful to human health, trivalent arsenic is considered more harmful than pentavalent arsenic. It is important for consumers to determine which type of arsenic is present in their water prior to selecting a water treatment device. When pentavalent arsenic is found in water, either reverse osmosis or distillation can be selected. RO is considered the effective treatment in POU/POE systems to remove this contaminant species.

Iron: Iron is frequently found in drinking water. It is introduced from iron pipes in the water distribution system. Iron in drinking water can give a rusty color to laundered clothes and may affect taste. In the presence of hydrogen sulfide, iron causes a sediment to form that may give the water a blackish color. Iron exists in natural groundwater in the soluble (ferrous) state but when exposed to oxygen, it is converted into the insoluble (ferric) state with its characteristic reddish brown or rusty color. If allowed to stand long enough, this rusty sediment will usually settle to the bottom of a container; however, it is difficult to use this type of settling to remove the iron. RO is also an effective treatment method to remove iron from drinking water.

Chloride: Chlorides in groundwater can be naturally occurring in deep aquifers or caused by pollution from seawater, brine, or industrial or domestic wastes. Chloride concentration above 250 mg/L can produce a distinct taste in drinking water. Although National Sanitation Foundation (NSF) International has not certified any water treatment products for reduction of chlorides, reverse osmosis and distillation may be effective.

Sulfates: Sulfates in groundwater are caused by natural deposits of magnesium sulfate, calcium sulfate, or sodium sulfate. Concentrations should be below 250 ppm. Higher concentrations are undesirable because of their laxative effects.

The impurities that cannot be efficiently removed by RO include dissolved gases and most volatile and semi-volatile organic contaminants including some pesticides and solvents. Although RO membranes can remove all microorganisms, RO units are not recommended for treatment of bacteria and other microscopic organisms. It is currently recommended that only microbiologically safe (i.e. coliform negative) water be fed into RO systems.

In POU/POE applications, many factors play a role in the removal efficiency of RO. It is noted that not all compounds can be efficiently removed by this process. The efficiency with which membranes reject the contaminant molecules depends on the pollutant concentration and chemical properties of the pollutant. Membrane type, water pressure, and operating conditions will also affect the degree of pollutant removal. The efficiency of removal of various contaminants can be varied; homeowners should evaluate this when considering using RO for home treatment. In addition, other home treatment methods may be better for a specific contaminant. RO systems are normally used to treat only drinking and cooking water supplies. It is not appropriate when larger supplies are being treated. Source of water supply entering the RO process must be potable. The dugout water, the water supplies contaminated by coliform bacteria, and other surface supplies that have not been municipally treated are not recommended to be treated using RO units.

3.2. Fundamental Concept of Reverse Osmosis

Reverse osmosis is based on the process of osmosis. Osmosis involves the selective movement of water from one side of a membrane to the other. To make the process work, pressure is applied to the contaminated water, forcing water through the membrane. To understand how RO works, it is important to know about osmosis, which is the important issue in the membrane process.

Osmosis is a physical phenomenon that has been extensively studied by scientists in various disciplines of science and engineering. Conventionally, osmosis is defined as the net movement of water across a selectively permeable membrane driven by a difference in osmotic pressure across the membrane (31). A selectively permeable membrane allows the passage of water, but rejects solute molecules or ions.

In fact, osmosis is the transport of water across a selectively permeable membrane from a region of higher water chemical potential to a region of lower water chemical potential. It is driven by a difference in solute concentrations across the membrane that allows passage of water, but rejects most solute molecules or ions. Osmotic pressure (π) is the pressure which, if applied to the more concentrated solution, would prevent transport of water across the membrane (31). In RO process, hydraulic pressure is applied in the opposite direction of the osmotic pressure gradient. In addition, osmotic pressure is also used in forward osmosis

(FO) and pressure-retarded osmosis (PRO). In FO process, the osmotic pressure differential ($\Delta\pi$) across the membrane is used. PRO can be viewed as an intermediate process between FO and RO, where hydraulic pressure is applied in the opposite direction of the osmotic pressure gradient (similar to RO). However, the net water flux is still in the direction of the concentrated draw solution (similar to FO).

Forward osmosis has been used to treat industrial wastewaters (at bench scale) (32–34), to concentrate landfill leachate (at pilot and full scale) (35–37), and to treat liquid foods in the food industry (at bench scale) (38–46). FO is also being evaluated for reclaiming wastewater for potable reuse in life support systems (at demonstration scale) (47–50), for desalinating seawater (51, 52), and for purifying water in emergency relief situations (53). PRO, a closely related process, has been tested and evaluated since the 1960s as a potential process for power generation (54–65).

The general equation describing water transport in RO, FO, and PRO is as follow (31):

$$J_W = A(\sigma\Delta\pi - \Delta P), \tag{1}$$

where Jw is the water flux, A is the water permeability constant of the membrane, σ is the reflection coefficient, and ΔP is the applied pressure.

For FO, ΔP is zero; for RO, $\Delta P > \Delta\pi$; and for PRO, $\Delta\pi > \Delta P$. The flux directions of the permeating water in FO, PRO, and RO are illustrated in Fig. 14.3 (31).

Flux directions and driving forces for the three processes were characterized in the early 1980s by Lee et al. (56). The FO point, PRO zone, and RO zone, along with the flux reversal point, are illustrated in Fig. 14.4 (56).

From Fig. 14.4, FO takes place when the hydraulic pressure difference is zero, the PRO zone is where the applied pressure difference is between zero and the flux reversal point, and the RO zone is where the applied pressure difference is greater than the osmotic pressure difference.

In general, the osmotic pressure of the dissolved chemical causes pure water to pass through the membrane from the dilute to the more concentrated solution, as shown in

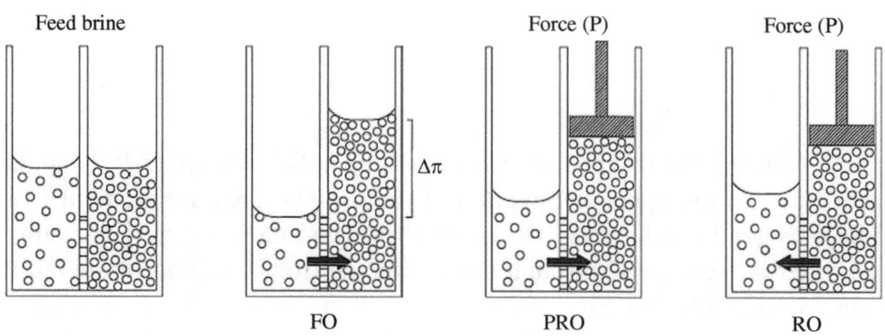

Fig. 14.3. Solvent flows in FO, PRO, and RO (31). *Note*: For FO, ΔP is approximately zero, and water diffuses to the more saline side of the membrane. For PRO, water diffuses to the more saline liquid that is under positive pressure ($\Delta\pi > \Delta P$). For RO, water diffuses to the less saline side due to hydraulic pressure ($\Delta P > \Delta\pi$).

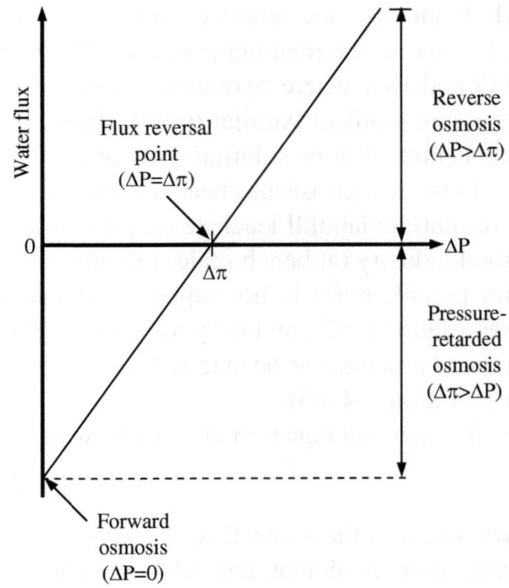

Fig. 14.4. Direction and magnitude of water flux as a function of applied pressure in FO, PRO, and RO (adapted from Lee et al. (56)).

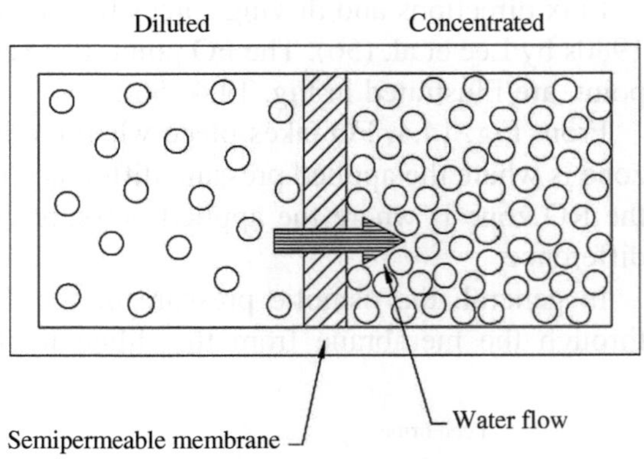

Fig. 14.5. Water across the membrane from dilute to the concentrated solution in osmosis (adapted from Kneen et al. (30)).

Fig. 14.5 (30). To make the RO process work, the hydraulic pressure (ΔP) is applied to the contaminated water in the opposite direction of the osmotic pressure gradient ($\Delta\pi$), where the hydraulic pressure is higher than the osmotic pressure ($\Delta P > \Delta\pi$). Under this circumstance, pure water is "squeezed" through the membrane from the concentrated to the dilute side as shown in Fig. 14.6 (30). Since contaminants do not move with the water as it moves across the membrane, pure water collects on the other side of the membrane. The purified water that accumulates on one side of the membrane can then be used or stored. This purified water is cleaner and healthier drinking water. The rejected impurities on the concentrated side of the membrane are washed away in a stream of wastewater and not accumulated as on a traditional filter.

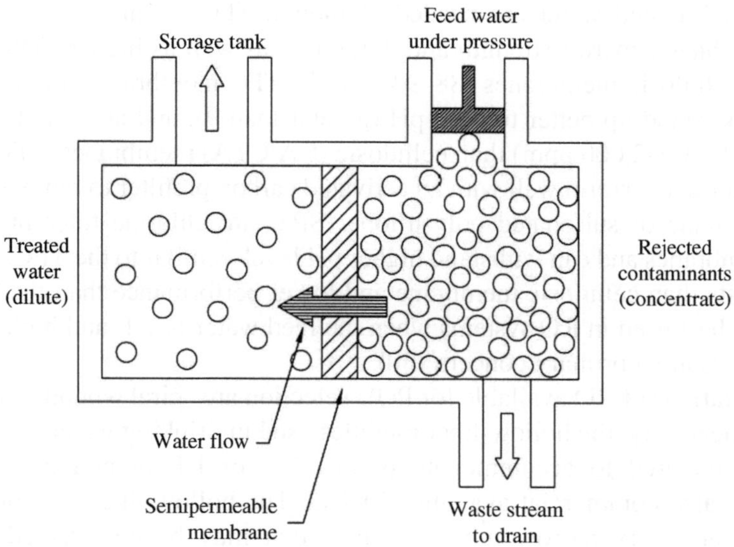

Fig. 14.6. Water across the membrane from the concentrated solution into the more dilute solution in reverse osmosis (30).

3.3. *Types and Configuration of Reverse Osmosis*

Normally, the central part of the reverse osmosis for POU/POE system is the module which is a pressurized container that houses the semipermeable membrane. Several square feet of this membrane are arranged to form the module. There are two types of membranes, one for chlorinated supplies and another for nonchlorinated supplies. It is important that the proper membrane is used. Chlorinated water can damage the membrane used for nonchlorinated water.

Reverse osmosis membranes are generally manufactured from cellulosic types (cellulose acetate, cellulose triacetate, or blends) or polyamide materials (and their respective derivatives). Very thin membranes are made from these synthetic fibers. However, there are various advantages and disadvantages associated with each. The advantages of cellulosic membranes are their lower cost and ability to tolerate chlorine, which curtails the growth of microorganisms in the system. However, cellulose membranes are susceptible to biodegradation and must be operated within a relatively narrow pH range of about 4–8; however, they do have some resistance to continuous low-level oxidant exposure (66). In general, for example, chlorine doses of 0.5 mg/L or less may control biodegradation as well as biological fouling without damaging the membrane. Polyamide (PA) membranes, by contrast, can be used under a wide range of pH conditions and are not subject to biodegradation (66). Although PA membranes have very limited tolerance for the presence of strong oxidants, they are compatible with weaker oxidants such as chloramines. PA membranes require significantly less pressure to operate and have become the predominant material used for RO applications (66).

The membrane made from cellulose acetate, cellulose triacetate, or blends can produce an adequate capacity for most households, but thin-film composites (TFC) should be used if large volumes of treated water are needed. Although TFC membranes are initially more expensive, they have superior strength and durability, as well as higher TDS rejection rates (>95%), than cellulosic membranes (88–94%) (14). TFC membranes are more resistant to microbial attack, stand up better to high pH (greater than 9), and are able to handle higher levels of TDS (1, 500–2,000 ppm) than cellulosic (CA/CTA) membranes. TFC deteriorate in chlorinated water but perform well with an activated carbon prefilter to remove chlorine (30).

Membranes made of sulfonated polysulfone (SPS) are chlorine tolerant like cellulosic (CA/CTA) membranes and can withstand higher pH levels similar to the TFC. However, they have higher costs than cellulosic membrane and lower performance than the TFC. The SPS membranes are best used in RO systems when the feed water is soft and high in pH or when high nitrate levels are a primary concern (30).

Two configurations of RO available for POU selection are spiral wound around a tube and hollow fibers. Generally, the hollow-fiber modules used in drinking water treatment applications are manufactured to accommodate porous MF or UF membranes; however, this configuration is also commercial available for RO. The hollow-fiber membranes are comprised of long and very narrow tubes of fibers that may be constructed using various membrane materials. These fibers may be bundled in one of several different arrangements (66). In one common configuration used by many manufacturers, the fibers are bundled together longitudinally, potted in a resin on both ends, and encased in a pressure vessel that is included as a part of the hollow-fiber module. These modules are typically mounted vertically, although horizontal mounting may also be utilized. One alternate configuration is similar to spiral-wound modules in that both are inserted into pressure vessels that are independent of the module itself (66). These modules are mounted horizontally.

A typical commercially available hollow-fiber module may consist of several hundred to over 10,000 fibers. A symmetric hollow fiber with two different operation modes, inside-out and outside-in modes, are shown in Fig. 14.7 (66). When a hollow-fiber module is operated in

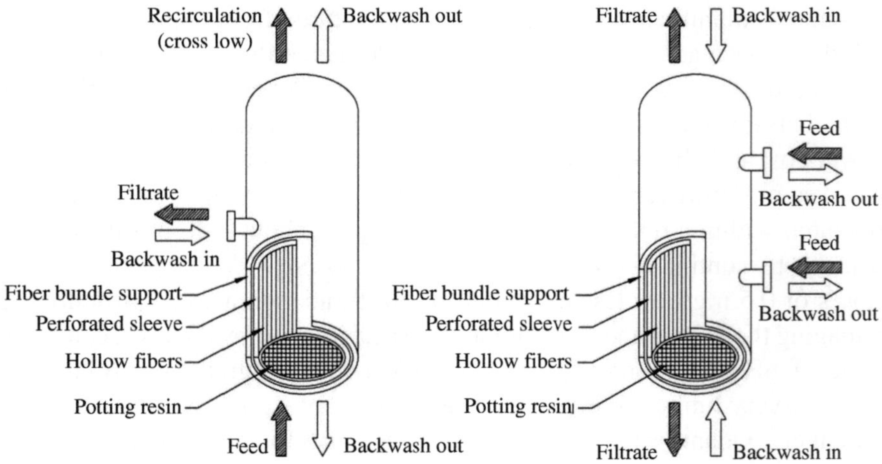

Fig. 14.7. A symmetric hollow-fiber with two different modes of operation; inside-out and outside-in modes (66).

an inside-out mode, the pressurized feed water may enter the fiber lumen at either end of the module, while the filtrate exits through a filtrate port located at the center or end of the module. In outside-in mode, the feed water typically enters the module through an inlet port located in the center and is filtered into the fiber lumen, where the filtrate collects prior to exiting through a port at one end of the module (66).

Spiral-wound modules are the most widely used configuration in RO process, with the configuration designed for the use of semipermeable membranes. The basic arrangement of a spiral-wound module is the flat membrane sheets called a "leaf" wound around a central perforated tube. One leaf consists of two membrane sheets placed back to back and separated by a fabric spacer called a permeate carrier. The layers of the leaf are glued along three edges, while the unglued edge is sealed around the perforated central tube. A diagram of a spiral-wound element is shown in Fig. 14.8 (66).

For the operation of the spiral-wound modules (66), the feed water enters the spacer channels at the end of the spiral-wound element in a path parallel to the central tube. As the feed water flows across the membrane surface through the spacers, a portion permeates through either of the two surrounding membrane layers and into the permeate carrier, leaving behind any dissolved and particulate contaminants that are rejected by the semipermeable membrane. The filtered water in the permeate carrier travels spirally inward around the element toward the central collector tube, while the water in the feed spacer that does not permeate through the membrane layer continues to flow across the membrane surface, becoming increasingly concentrated in rejected contaminants. This concentrate stream exits the element parallel to the central tube through the opposite end from which the feed water entered (66).

In the selection of configuration of RO units, the advantage and disadvantage of each type have to be concerned. The hollow fibers provide a tremendous surface area for water treatment

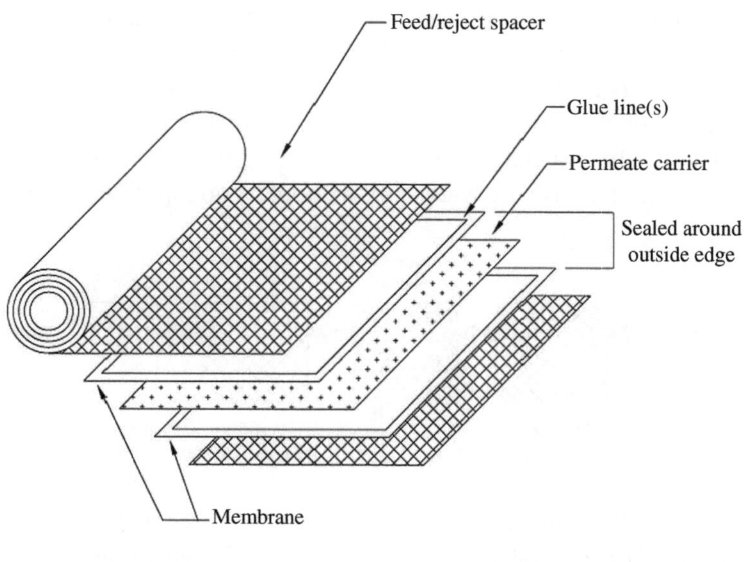

Fig. 14.8. A diagram of a spiral-wound element in RO units (66).

inside a compact cylindrical element. As they have greater surface area, they provide greater capacity. However, they are more easily clogged than the spiral-wound membranes commonly used in home RO systems. With more appropriate design for RO process, the spiral-wound module receives more attention from many manufacturers as the major configuration for RO units in POU/POE system.

3.4. Components and Installation of Reverse Osmosis in POU/POE Units

In installation of RO in POU/POE units, many components are required. Generally, the feed water with proper characteristics enter into and through a membrane, while contaminant ions are rejected and discharged in a concentrated waste stream, and the treated water has to be pressurized and kept in the storage tank to produce an adequate flow when the tap is open. Some RO membranes are sensitive to chlorine, a consideration for systems that have centralized chlorination installed. Thus, various types of units are required for one RO system. A typical home reverse osmosis treatment system is shown in Fig. 14.9 (21). RO systems consist of the booster pump, pretreatment unit, reverse osmosis membrane unit, posttreatment unit (21), pressurized treated-water storage container for the treated water, and distribution elements (a separate delivery tap for the treated water supply). These units are able to process practically any desired quantity or quality of water by configuring units sequentially to reprocess waste brine from the earlier stages of the process.

The feed water entering the RO system should be bacteriologically safe. RO units will remove virtually all microorganisms but they are not recommended for this use because of the possibility of contamination through pinhole leaks or deterioration due to bacterial growth. Details of each unit in RO system are provided below:

Booster pump: The line pressure of the water system in a home usually supplies the pressure for RO. In some installations, a small booster pump may be required to increase the water pressure for the membrane process. For RO in POU/POE units, water pressure must be maintained at 40–70 pounds per square inch to keep a driving force across the membrane. In fact, the driving force is furnished by both the pressure and the concentration differentials across the thin layer. For the feed water, the pressure effect is the most important, and for dissolved mineral ions, the concentration difference is the most important. Consequently, increases in pressure will increase the water flux without a corresponding increase in salt flux.

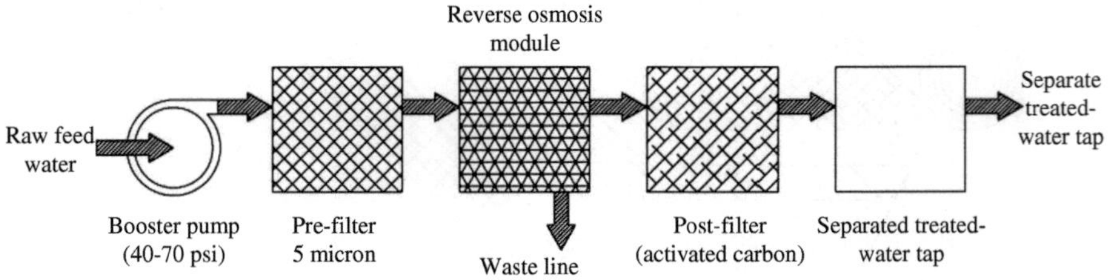

Fig. 14.9. A typical home RO system includes (1) a booster pump (2) a pretreatment unit (3) a reverse osmosis membrane unit, (4) a posttreatment unit (5) a pressurized treated-water storage container, and (6) a separate treated-water tap (21).

The booster pump is installed with the major functions in raising and maintaining the water pressure at the optimum level to ensure the highest rejection rate and maximum water production.

Pretreatment: A pretreatment unit is required for RO in POU/POE system to obtain the desired quantity or quality of water. Most reverse osmosis units require pretreatment to remove all manganese, iron, hardness minerals and the TDS. High levels of water hardness tend to reduce membrane efficacy and result in more frequent replacement of the RO membrane. Also, high levels of iron, manganese, and aluminum can also cause membrane fouling. Thus, many treatment methods such as filter and activated carbon can be used for this purpose.

Filtration is a widely used pretreatment unit for RO in POU/POE system to remove tiny particles of suspended dirt and sediment to prevent membrane fouling. The prefilter is sometimes referred to as a sediment filter. It removes small suspended particles to extend the life of the membrane. Fine grain filtration may be required if suspended solids are a problem. It is important to select a unit which is capable of producing a sufficient amount of treated water. Additionally, the activated carbon prefilter can be installed. This process can remove organic materials and any dissolved gases not treated by the RO membrane.

Generally, the upper TDS required for an RO unit is limited to approximately 1,500 mg/L. In addition, some membrane units are damaged by chlorine, and others by bacterial growth. In this situation, a carbon prefilter is also recommended. A GAC prefilter is typically used to remove chlorine, which can damage some types of RO membranes. When choosing a pretreatment unit for RO process, the characteristic of feed water has to be considered to determine the appropriate pretreatment method, and the upper limit for TDS and other contaminants has to be identified to operate the pretreatment unit properly.

Reverse osmosis membrane unit: The central part of the reverse osmosis system is the module, which is a pressurized container that houses the semipermeable membrane. Several kinds of reverse osmosis membranes are available. The most common materials are cellulose acetate or polyamide resins. Mixtures or variations of these materials are also used. Each product has certain advantages and limitations and these needs to be considered carefully (see Sect. 3.3 Types and configuration of reverse osmosis for more details). In general, the time needed to produce 1 gallon of water ranges from 2 to 7 h. Typically, the capacity of RO membranes for home systems is rated in the range of 24–100 gallons per day. This treated water comes out much slower than water from a regular tap, so a tank is used to store the treated water. This treated water is often used only for drinking and cooking.

Posttreatment unit: The major function of a posttreatment unit is to remove any desirable taste and odor as well as any residual organics from the treated water. The typical unit of posttreatment is a carbon filter. A post-carbon filter removes test and odor to give water a final polish prior to delivery. However, if a carbon filter is used as a part of the prefiltration step, postfiltration is normally eliminated. A standard carbon filter positioned after the storage tank removes compounds that cause unpleasant taste and odors, including those from the tank or plastic tubing, just before water is dispensed.

An additional prolonged contact carbon filter (PCCF) can be installed to remove high levels of organic chemicals such as trihalomethanes, volatile organic chemicals, and chloramines. This unit is normally placed between the RO membrane and the storage tank. Combining an activated carbon filter with RO expands the range of chemicals the system can remove. The carbon filter is required to change regularly. After the posttreatment step, the purified water is passed into a small pressure tank for storage until used at the tap.

Pressurized treated-water storage container: As the RO units supply treated water at very low rates (around 40%), a water storage container of 2–5 gallons is typically required for a POU/POE application. These units are pressurized to produce an adequate flow when the tap is open. Minimum pressure is required to deliver water to the water storage tank. The storage tank, tubing, and dispensing faucet should be made of plastic, stainless steel, or other nontoxic materials. The pressurized storage tank will hold the purified drinking and cooking water ready for use.

Separate treated-water tap: A separate delivery tap for the treated water is used for drinking and cooking purposes.

Other accessories: In the RO process, the units are operated by the use of pressure-sensitive switches, check valves, or flexible bladders. Shut-off valves are necessary to conserve water during low-use periods. Monitoring gauges or servicing lights are used for monitoring and maintenance processes.

Generally, the RO unit is normally located beneath the kitchen sink, since it is used to treat water for drinking and cooking purposes. The typical RO installation in POU/POE devices is shown in Fig. 14.10. UV disinfection is optional for this process.

The RO system in POU/POE has been identified in both the arsenic and radionuclide rules as an SSCT for arsenic, uranium, and radium, as shown in Table 14.4. RO is also listed as an SSCT by USEPA for copper, lead, fluoride, antimony, barium, beryllium, cadmium, chromium, selenium, and thallium. More suggestions in investigation of the potential application of RO for nitrate removal are recommend by USEPA.

3.5. Operation and Maintenance of Reverse Osmosis in POU/POE

Some of the factors that should be considered in operating an RO unit include (a) type and initial concentrations of the contaminants entering the process, (b) the water supply rate, to determine the produced quantity of water to meet normal daily drinking and cooking

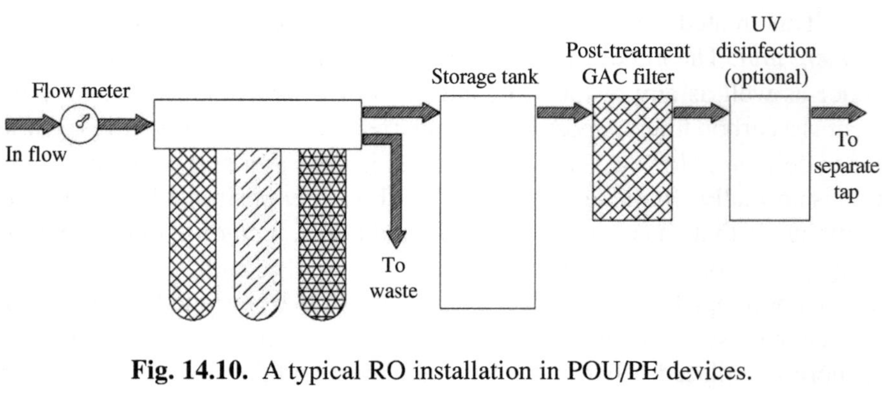

Fig. 14.10. A typical RO installation in POU/PE devices.

requirements, (c) the water pressure to meet the supply and rejection rates (this value has to be considered to determine whether the RO unit will be operated on the normal operating pressure of a home water system or using a booster pump), (d) the rejection rate or the percentage of contaminants to be removed by the membrane, and (e) the monitoring or testing method of the system's performance. Details of the major factors in operating an RO system in POU/POE are provided below.

3.5.1. Types and Initial Concentrations of the Contaminants in Feed Water

Types and initial concentrations of the contaminant entering the RO process are the crucial factors that determine the quality of the purified water. Overall water quality affects the efficiency of an RO system and its ability to remove specific contaminants. In operating an RO unit, it is required that the water source entering the RO units must not contain more than 10 grains of hardness (0.2 mg/L of iron and less than 0.25 mg/L of manganese). The TDS should be about 500 ppm. The higher the TDS, the greater the osmotic pressure and the lower the recovery rate of treated water (Table 14.6).

Moreover, the initial concentration of contaminants is also affected depending on the rejection rate. Rejection rates need to be high enough to reduce the contaminant level in the untreated water to a safe level. For example, if the initial concentration of nitrate is relatively high as 120 mg/L, 90% of nitrate removal would still leave 10% or 12 mg/L of nitrate in the treated water. This value is greater than the maximum contaminant level (MCL), which is 10 mg/L for nitrate allowed in drinking water. Thus, the rejection rate of nitrate at 10% from this RO unit is not an appropriate value for purified water. Water with very high levels of nitrates (such as 100 mg/L) would remain near or above the health standard levels even after treatment. However, high concentrations of nitrate levels are not expected in feed water and indicate unusual problems that require special investigation and handling. The NSF recommends that special designs be used for RO units where the NO_3^-N level exceeds 40 mg/L (nitrates as used in this part refers to nitrate-nitrogen or NO_3^-N).

3.5.2. Water Pressure and Temperature

The standard at which manufacturers rate RO system performance is 77°F, 60 pounds per square inch (psi), and TDS at 500 ppm (30). Membrane productivity is very sensitive to changes in water temperature. As water temperature increases, water flux increases due to the higher diffusion rate of water through membrane. It is reported that the amount of treated

Table 14.6
Best available technologies (include RO) for radionuclides (40 CFR 141.66(g), 142.65(a))

Contaminant	Best available technology
Combined radium-226 and radium-228	Ion exchange, reverse osmosis, lime softening
Uranium	Ion exchange, reverse osmosis, lime softening, coagulation/filtration
Beta particle and photon radioactivity	Ion exchange, reverse osmosis

water produced decreases 1–2% for every decrease in degree below the standard temperature of 77°F. An RO system supplied with well water at a temperature of 60°F produces only three quarters of the volume it would produce at 77°F (30). Increased feed water temperature also results in lower salt rejection or higher salt passage due to a higher diffusion rate for salt through the membrane.

As described earlier about the mechanism of RO process, the major factor controlling the efficiency in contaminant removal is water pressure. RO can work properly if there is enough water pressure to overcome the osmotic pressure. Moreover, the RO system must work against back pressure created in the storage tank as it fills with water and compresses the air in the tank. The net water pressure at the RO membrane can be calculated by subtracting back pressure and osmotic pressure from the feed line pressure. If the net water pressure at the membrane is lower than 15 psi, treated water production is less efficient and contaminant rejection rates are lower (30).

Water pressure is also a function of the water flux across the membrane. Water pressure increases in direct relationship to increases in water flux across the membrane. In addition, increased water pressure results in increased salt rejection. If the water pressure is not enough, auxiliary pumps or booster pumps can be added to the RO system to raise and maintain the pressure. Consequently, the quality and quantity of water produced can be achieved. High-quality RO systems have valves that shut off the flow whenever storage tank pressure reaches two-thirds of the feed pressure; at this point, low net water pressure can result in low rejection rates.

3.5.3. *Rejection Rate*

Most reverse osmosis systems operate in the 75–80% recovery efficiency range. For example, this means that a unit operating at 75% recovery efficiency will produce 75 gallons of pure treated water and 25 gallons of wastewater from 100 gallons of raw feed water.

To determine the efficiency of RO unit in contaminant removal, the term rejection percentage (% rejection) is always used. The rejection percentage is the percentage of a particular contaminant that does not move through, or is rejected by, the membrane. RO membranes are rated for their ability to reject compounds from contaminated water. A rejection rate is calculated for each specific ion or contaminant as well as for reduction of TDS. The rejection percentage for nitrate can be as high as 90%. However, the initial concentration of contaminants entering the process has to be considered to determine the efficiency of the RO process (see above, Types and initial concentrations of the contaminants in feed water). It is important to know the specific requirements for water quality when selecting the RO system for drinking and cooking purposes. For example, high rejection rates are essential when high nitrates or lead concentrations in the water must be brought below the EPA maximum contaminant or action levels (30). Moreover, when water contains more than one contaminant, the rejection rate for each contaminant may be reduced or one of the contaminants may be reduced in preference to the other contaminant. For example, cases have been reported where water supplies containing either high TDS levels or high sulfates in combination with nitrates show no decrease in nitrates after treatment.

3.5.4. Recovery Rate

As the feed water enters the RO system, only part of the water comes out as treated water. Part of the water fed into the system is used to wash away the rejected compounds and goes down the drain as waste. If not properly designed, only little treated water will be obtained from large quantities of water fed to the RO systems. Generally, most home RO systems are designed for 20–30% recovery (i.e., 2–3 gallons of treated water are produced for every 10 gallons fed into the system). In fact, RO systems for household drinking water treatment can operate at higher recovery rates, but doing so may shorten membrane life.

The recovery rate, or efficiency, of the RO system can be calculated by dividing the volume of treated water produced by the volume of water fed into the system:

$$\%\ \text{Recovery} = \frac{\text{Volume of Treated Water Produced}}{\text{Volume of Feed Water Used}} \tag{2}$$

As described previously, reverse osmosis can occur when the natural osmotic flow between a dilute solution and a concentrated solution is reversed through the application of water pressure. If the recovery rate is increased when water pressure remains constant, the salts in the residual feed become more concentrated and the natural osmotic pressure will increase until it is as high as the applied feed pressure. The maximum recovery rate usually depends on the limiting osmotic pressure, the concentration of salts present in the feed water, and the tendency of salts to precipitate on the membrane.

The device in controlling the recovery rate is the flow regulator, which must be properly adjusted on the amount of rejection stream. If the flow is slow, the recovery rate is high. In addition, RO membranes are easily fouled if concentrated impurities are not washed away quick enough. If the flow is too fast, the recovery rate is low and too much water goes down the drain.

3.5.5. Maintenance of an RO System

An RO system must be well maintained to ensure reliable performance. The major component of RO in POU/POE system is not only the RO membrane itself but also other combined units such as pre- and post-device treatment. Maintenance of the whole system is important for the RO unit to work in the proper manner. The device might require continuous operation to maintain peak membrane performance.

The first step to ensure that the system can work properly is to install and operate according to the manufacturer's instructions. After installation, retest both the feed water (prior to treatment) and the treated water at a state-certified laboratory to ensure the system is working properly and removing the contaminants. The testing is required for both the untreated and treated water annually. More frequent testing (quarterly or semi-annually) is suggested if high levels of contaminants are present in the raw water. Results of the water quality from the testing will help in determining the performance of the RO system, the replacement of components, and the required maintenance. The maintenance, cleaning, and part replacement should be performed in accordance with the manufacturer's recommendations.

Characteristics of feed water to RO system have to be monitored to ensure the quantity and quality of the purified water. The efficiency of a reverse osmosis treatment unit can be affected by the presence of turbidity, iron, manganese, hydrogen sulfide, and chlorine in water. High levels of particulate matter in the water such as iron can foul or clog the membrane, decreasing water flow. Because RO devices are point-of-use units, it may be necessary for a point-of-entry device to be installed to treat these problems if they exist. Many systems now have a built-in continuous monitor that indicates a high TDS level, a sign that the system is not operating properly. It may also be necessary to test regularly for specific health-related contaminants such as nitrates or lead.

Replacement of RO components in POU/POE is a routine maintenance process. Pre- or postfilters must be replaced once a year or more often, depending on the volume of water fed through the system and the quality of the feed water. Customers have to follow the manufacturer's instructions for the maintenance procedure in each part of the RO system. For RO maintenance, some systems have built-in monitors such as conductivity meters, pressure gauges, and other devices, which can be used to detect the RO operating problems. However, many RO units are not provided with these. Maintenance in accordance with the manufacturer's recommendations is required for all parts of RO units, as described above.

3.6. Fouling and Cleaning of Reverse Osmosis

Membrane fouling is classified as the most serious problem encountered in RO systems. It can reduce the performance of RO tremendously. Clogged RO membranes, filters, or flow controls will decrease water flow and the system's performance. If fouling is detected in the early stages, the membrane can often be cleaned and regenerated. Completely clogged or torn RO membranes must be replaced. The cleaning procedure varies depending on the type of membrane and fouling.

Generally, fouling refers to the deposition of undesirable material on the membrane, which leads to several negative effects including reduced membrane fluxes, decreased permeate flow, increased differential pressure, increased pressure drops, and decreased salt rejection, necessitating periodic cleanings and causing risks of mechanical and chemical damage. All these effects result in the deterioration of plant performance and increased treatment costs (67, 68).

Foulings are classified into four categories: scaling, colloidal fouling, organic fouling, and biological fouling. Scaling occurs when the concentration of salts present in the concentrate exceeds the concentration of the solubility product. The fouling formed by deposits from small amounts of relatively inert suspended solids, which typically are silica, silicates, and previously precipitated scales, is called colloidal fouling. Metallic oxides also play a critical role in the cohesion of the scale formed or the fouling layer (69, 70). Divalent cations (e.g., Ca^{+2}) were also proved to have a dramatic effect on fouling (71).

Organic fouling occurs when surface water contains residual material from the decay of organic matter with a high percentage of multi-functional organic acids (e.g., humic acids). Precipitated organic acids are the most common form of organic fouling and are easily dissolved at elevated pH. These compounds are measured as DOC (dissolved organic carbon). Amphoteric and hydrophobic organic matters are primary organic foulants for polyamide nanofiltration membranes. These two species displayed different interaction adsorption with membranes and caused different degree of organic fouling. With an accelerating deterioration of water quality, especially for a micropolluted water source, organic matters and/or microorganisms have become the main foulant in membrane systems.

The last type of fouling, biological fouling or biofouling, results from bacteria that pass the pretreatment, attach to the membrane or spacer components, and form a foulant layer. Biofouling arises from microbial biofilms which are ubiquitous in water systems of any nonsterile unit (72). Undesired effects of biofouling include energy loss in cooling towers and heat exchangers, decreased water quality and possible contamination of drinking water with pathogenic bacteria, plugging of filter systems, corrosion or biodeterioration of equipment or pipeline, and degradation of product quality in paper manufacturing (72). The presence of organic and inorganic compounds serving as sources of energy for bacteria in drinking water results in the multiplication of microorganisms in drinking water distribution systems (regrowth) (73). Bacteria growing attached onto surfaces exposed to drinking water remain in the distribution system and may form the main part of the amount of metabolically active biomass present, in particular when drinking water contains a disinfectant residual (74). Release of bacteria from the walls of the pipes and reservoirs and from sediments may be responsible for the increase in numbers of bacteria, including coliforms, in drinking water during distribution (75–77). Several factors affect the formation of biofouling such as nutrient concentration in water system. In nonsterile water systems, microorganisms attach to any kind of surface and form biofilms which reach the plateau growth phase within a short time. Biomass accumulates on the reactor surfaces, and the nutrient-depleted effluent water allows microbial growth, including the development of biofilms, only to a lesser extent. The lower the nutrient concentration, the less the biofilm formed (72).

Several techniques are used to prevent the fouling problems. Graham et al. (70) suggested that periodic cleaning of a membrane to remove foulant accumulations is essential to maintain quality and quantity of product water and to maximize membrane life. Scaling is prevented by operating at a relatively low recovery of 85% and hydrochloric acid (HC1) dosing to the feed (67). To prevent colloidal fouling, a Modified Fouling Index (MFI) lower than 1 s/L^2 is proposed (67). Hong and Elimelech (71) indicated that the removal of divalent cations in the cleaning process could reverse fouling. In their work, EDTA was used as the strong chelating agent. It was demonstrated that this chemical is an effective agent in removing the natural organic matter (NOM)-calcium fouling layer and restoring permeate flux. However, for complex fouling, which often includes multiform foulants or interaction with membranes, little applied information and few cleaning techniques are available regarding cleaning solutions and procedures to remove these complex foulants from a RO system. In preventing organic fouling, control of organic specie with the value of dissolved organic carbon (DOC) lower than 1 mg C/L is proposed (67).

To determine the scaling problem, calculation of the scaling potential is provided in ASTM standards, which are (1) ASTM D3739-94 (2003): Standard Practice for Calculation and Adjustment of the Langelier Saturation Index for Reverse Osmosis, (2) ASTM D4582-91 (2001): Standard Practice for Calculation and Adjustment of the Stiff and Davis Stability Index for Reverse Osmosis, and (3) ASTM D4692-01: Standard Practice for Calculation and

Adjustment of Sulfate Scaling Salts ($CaSO_4$, $SrSO_4$, and $BaSO_4$) for Reverse Osmosis. In the specified method, the ion products, IP_c, of the considered salt in the concentrate stream and the solubility product, K_{sp}, of that salt under conditions in the concentrate stream have to be compared. Generally, scale-control measures are not needed if $IP_c < K_{sp}$.

The cleaning chemicals used to prevent fouling can be divided into acid and alkaline cleaners. Acid cleaners are used to remove inorganic precipitates. Examples of acid cleaners are HCl and phosphoric acid (H_3PO_4). Sulfuric acid is not recommended for cleaning purpose due to the risk of calcium sulfate precipitation. Alkaline cleaners are used to remove organic fouling including biological matter. The major alkaline cleaner is sodium hydroxide (NaOH). RO permeate can be used for preparing cleaning solutions. However, prefiltered raw water may be used.

For biofouling, controlling the microbiological activity using surface area or bacteria-contaminated water as a feed source in the system design process as follows can prevent the accumulation of microorganisms. As the sunlight is essential for the microorganism growth, the components of the pre-treatment system such as pipes, filters, and retention tanks should be covered to avoid the sunlight exposure. It is also suggested that the stand-by devices with large surfaces such as sand or cartridge filters should be avoided. If they are not avoidable, drains should be installed to allow discharge of the sanitization chemicals. The most common way to eliminate microorganisms from an RO system is the application of antimicrobial chemicals. This practice is practical when mechanical cleaning is impracticable and costly due to equipment downtime. The most widely used biocides are chlorine and aldehyde. However, these biocides often fail to disinfect surfaces of water systems successfully. It is well known that microorganisms living in biofilms are more resistant and thus require much higher concentrations of biocides to be disinfected (78). Prevention of biofouling is frequently achieved by continuous dosage of biocides. However, in many cases, permanent dosage of biocides is undesirable for environmental reasons. Biocides and their toxic by-products contaminate sewage water, where they still act as unwanted antimicrobial chemicals. Furthermore, biocides require special attention in terms of storage and handling in order to meet legal regulations (72).

Chlorine is the important biocide used for the different types of RO membranes. Continuous chlorination can be used with cellulose membranes to protect the system from biofouling and eliminate the particle trapping slime. However, chlorine and other oxidizing disinfectants are harmful to some type of RO membranes such as TFC. If the TFC membrane is used in RO units, an activated carbon must be in place to remove the oxidizing chemicals before the chlorinated water reaches the TFC membrane (30).

Other techniques to prevent biofouling are periodic testing and monitoring and using activated carbon filter. Microorganisms, dead or alive, can clog RO membranes. If coliform bacteria or other special contaminants are a known or suspected problem, periodic testing is recommended. For the application of activated carbon (AC) prefilters, it is recommended that this pretreatment process should be used with chlorinated water. Application with nonchlorinated water supplies is not recommended because they provide a place for microorganisms to multiply and lead to increased biofouling of the RO membrane surface (30). It is important to replace carbon filters periodically following the manufacturer's instructions, especially after an extended shutdown period during which microorganisms can flourish.

3.7. Membrane Testing for Point-of-Use and Point-of-Entry Purposes

Membrane testing procedure for RO unit selection can be divided into three categories: screening test, application test, and pilot test (79–83). In most cases, the computer program from manufacturers can predict the system performance. However, the testing is recommended (80–82).

The screening test is aimed to select the appropriate membrane for the desired separation, and obtained important design parameters included membrane flux and rejection properties of the membrane. The test is performed by mounting a small piece of flat sheet membrane and exposing it to the test solution. The method is fast and inexpensive. The application test is performed to obtain the scale-up data for an RO process including permeate flux and permeate quality as a function of operating pressure and system recovery. In determining the operating pressure in application test, the feed pressure is adjusted until the desired permeate quality and permeate flux rate are obtained. Sufficient feed flow should be maintained to ensure a low recovery rate (<5%) as the membrane flux rate is increased. In determining the recovery rate (permeate flow/feed flow), the test is performed by directing the permeate stream into a container while returning the concentrate stream to the feed tank. Both the permeate flow and permeate quality are monitored during the test. The test is stopped when the permeate flow rate declines to a low value or when the permeate quality declines below acceptable limits. The recovery rate is calculated by subtracting the remaining feed volume from the original feed volume and then dividing by the original feed volume. A pilot test is usually run in the field in a continuous mode. The objective is to confirm the system design and fine-tune operating parameters (80–82).

Testing parameters to measure the performance of RO system include conductivity, oxidation–reduction potential/redox (ORP), and pH. Electrical conductivity is the most convenient method for testing permeate water quality and RO performance. Generally, pure water is a poor electrical conductor. Conductivity in water is represented by the amount of ionized substances or salts dissolved in water. Water with high TDS leads to increase in electrical conductivity. Thus, the electrical conductivity in permeate water can indicate the performance of an RO process. ORP and pH are important parameters in measuring the success and useful life of an RO membrane. ORP is also used to determine the activity of oxidizers such as chlorine, bromine, ozone, and hydrogen peroxide. Thus, a high ORP reading indicates a need for pretreatment to remove these oxidizers. In addition, a low ORP reading may indicate biofouling of the membrane. The measurement of pH is used in predicting the membrane life and the scaling potential of feed water. As described earlier, scaling could be formed with high amount of calcium and high pH. For silicon-based compounds, a low pH is required in scale forming. Optimum pH is also required to monitor the performance of an RO process.

4. DESIGN EXAMPLES

4.1. Design of RO in POU and POE Applications

The basic principles for system design of pressure-driven membrane filtration include flux, recovery, and flow balance (66, 83).

Flux in filtration process is defined as the filtrate flow per unit of membrane filtration area, which is the membrane filtration system throughput. Calculation of flux is shown in Eq. (3) as shown below (66):

$$J = \frac{Q_{\mathrm{p}}}{A_{\mathrm{m}}}, \tag{3}$$

where J = water flux ($\mathrm{m^3/m^2\,h}$), Q_{p} = filtrate flow produced by the membrane unit ($\mathrm{m^3/h}$), and A_{m} = wetted surface area of membrane a ($\mathrm{m^2}$).

The recovery of a membrane unit is defined as the amount of feed flow that is converted to filtrate flow, expressed as a decimal percent, as shown in Eq. (4) (66):

$$R = \frac{Q_{\mathrm{p}}}{Q_{\mathrm{f}}} \tag{4}$$

Here, R = recovery of the membrane unit (decimal percent), Q_{p} = filtrate flow produced by the membrane unit ($\mathrm{m^3/h}$), and Q_{f} = feed flow to the membrane unit ($\mathrm{m^3/h}$).

A general flow balance that can be applied to all membrane filtration systems is shown in Eq. (5) (66):

$$Q_{\mathrm{f}} = Q_{\mathrm{c}} + Q_{\mathrm{p}} \tag{5}$$

Here, Q_{f} = feed flow to the membrane unit ($\mathrm{m^3/h}$), Q_{c} = concentrate flow from the membrane unit ($\mathrm{m^3/h}$), and Q_{p} = filtrate flow produced by the membrane unit ($\mathrm{m^3/h}$).

For the system operating without a concentration waste or “bleed” stream mode or cross-flow systems in which 100% of the concentrate is recirculated, the concentrate (i.e., bleed or reject) flow, Q_{c}, is zero. In determining of the filtrate flow, Q_{p}, for the purpose of sizing a membrane filtration system, the additional filtered water for both backwashing and chemical cleaning should be included. Similarly, an estimate of the total required feed flow Q_{f} to the system should incorporate any raw water that may be used in these routine maintenance processes.

In the RO process, the driving force for the transport of water is a pressure gradient across the membrane. The resulting osmotic pressure gradient against the transport of water from the feed to the filtrate side of the membrane has to be considered. Typically, the osmotic pressure gradient is approximated from the concentration of TDS on the feed and filtrate sides of the membrane. The corrected driving force across the semipermeable membrane in an RO process is termed the net driving pressure (NDP) and can be calculated by Eq. (6). In this equation, NDP is in a simplified form by combining the respective parameters associated with two terms, the transmembrane differential pressure (ΔP) and the transmembrane osmotic pressure differential ($\Delta\pi$).

$$\mathrm{NDP} = \Delta P - \Delta\pi \tag{6}$$

Here, NDP = net driving pressure (psi), ΔP = transmembrane differential pressure (psi), and $\Delta\pi$ = transmembrane osmotic pressure differential (psi).

The performance of an RO system is defined by the feed flow to the membrane unit (Q_f) and its salt passage. The feed flow through the RO is directly proportional to the wetted surface area of the membrane (A_m) multiplied by the NDP and the membrane permeability coefficient (k). The water permeation equation can be described by the following equation:

$$Q_f = (A_m)(k)(NDP) = (A_m)(k)(\Delta P - \Delta\pi) \tag{7}$$

Here, Q_f = feed flow to the membrane unit (m^3/h), A_m = wetted surface area of membrane (m^2), k = membrane permeability coefficient ($psi^{-1}\ h^{-1}$), NDP = net driving pressure (psi), ΔP = transmembrane differential pressure (psi), and $\Delta\pi$ = transmembrane osmotic pressure differential (psi).

In addition, salt passage has to be considered in an RO system, as the salt flux (S_A) is proportional to the salt concentration difference between both sides of the membrane. The salt flux across the membrane can be explained by Eq. (8):

$$S_A = k'(C_f - C_P), \tag{8}$$

where S_A = salt flux (mg/m^2 h), k' = salt diffusion coefficient (m^2 h), C_f = concentration of solid in feed (mg/L), and C_P = concentration of solid in permeate (mg/L).

It is important to note that the flux associated with an RO system is typically referenced to a temperature of 25°C (298 K). It is necessary to have the temperature correction factor (dimensionless) for the purposes of assessing operational performance. The calculation of temperature correction factor (TCF) for RO systems is shown in Eq. (9):

$$\mathrm{TCF} = \exp\left[U \times \left(\frac{1}{T+273} - \frac{1}{298}\right)\right], \tag{9}$$

where TCF = temperature correction factor (dimensionless), T = water temperature (°C), and U = membrane-specific manufacturer-supplied constant ($1/K$).

As the RO process is normally rated to produce purified water at 60 psi (or 4.1 bars), the pressure correction factor (PCF, dimensionless) is necessary in some systems that operate at any pressure other than 60 psi. PCF can be determined as follows:

$$\mathrm{PCF} = P/60, \tag{10}$$

where PCF = pressure correction factor (dimensionless) and P = line pressure (psi).

Typically, the RO process produces the purified flow at 60 psi (4.1 bars) operating pressure, 25°C (77°F) operating temperature, and 500 ppm TDS. Once the TCF and PCF have been determined, the flux normalized to 25°C and 60 psi can be calculated according to Eq. (14):

$$J_{25} = J_T(\mathrm{PCF})(\mathrm{TCF}) \tag{11}$$

Here, J_{25} = normalized flux at 25°C (gfd), J_T = actual flux at temperature T (gfd), PCF = pressure correction factor (dimensionless), and TCF = temperature correction factor (dimensionless).

For the performance comparison of RO (and NF) with MF and UF, the term J_{25} normalized flux may not be used as the MF, UF, and MCF systems are typically referenced to a temperature of 20°C.

To identify changes in productivity that is attributable to fouling of membrane, the flux is determined as follows:

$$M_{25} = \frac{J_{25}}{\text{NPD}} \tag{12}$$

Here, M_{25} = temperature- and pressure-normalized flux (gfd/psi), J_{25} = normalized flux at 25°C (gfd), and NDP = net driving pressure (psi).

Rejection rate (*Re*) is a fraction of the total contaminant mass/concentration rejected by the membrane. It is also determined by the following Eq. (13):

$$Re = 1 - (C_p/C_f), \tag{13}$$

where Re = rejection rate (%), C_f = concentration of solid in feed (mg/L), and C_P = concentration of solid in permeate (mg/L).

In determining the rejection rate (*Re*), the conductivity measurement can be performed. Conductivity in the RO testing procedure could be replaced by hardness, alkalinity, nitrate, phosphate, silica, etc. (measured in ppm or mg/L). Rejection rates less than 95% may indicate that the membrane should be replaced.

Volume recovery, %: Volume recovery (R_v) is the fraction of the total feed volume of the test water that can be recovered as clean permeate:

$$R_v = V_p/V_f, \tag{14}$$

where R_v = volume recovery, V_F = volume of feed, L, and V_p = volume of filtrate, L.

4.2. *Calculations*

Example 1

Find the rejection percentage of RO units after the consumer had run the RO system for 20 min and found that the hardness of the RO water is 5 ppm. Let the tap water hardness be 80 ppm.

Solution

Find the rejection percentage from

$$Re = 1 - (C_p/C_f)$$

Here, C_f = feed average concentration = 80 ppm and C_P = filtrate concentration = 5 ppm.

Therefore,

$$Re = 1 - (5/80)$$
$$Re = 1 - 0.0625$$
$$Re = 0.9375 = 93.75\%$$

Thus, membrane hardness rejection is 93.75%. Rejection rates less than 95% may indicate that the membrane should be replaced.

Example 2

Find the rejection flow and the concentration of solute in rejection from the RO unit, if the initial concentration of solute in the feed is 300 mg/L. Let the solute concentration in permeate be 50 mg/L and assume that the RO unit produces 50 m^3/day of clean water. The recovery of the membrane unit is 95%.

Solution

(1) Write the mass balance diagram for the RO unit.

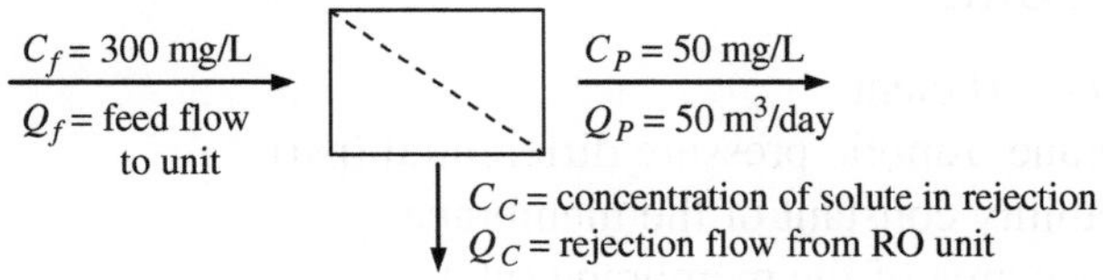

(2) Find Q_f feed flow to RO unit

From Eq. (4)

$$R = \frac{Q_p}{Q_f}, \tag{4}$$

where R = 95%, Q_p = 50 m^3/day, and Q_f = feed flow to the membrane unit. Therefore,

$$Q_f = \frac{Q_p}{R} = \frac{50\,m^3/day}{0.95}$$
$$Q_f = 52.63\ m^3/day$$

Feed flow to RO unit is 52.63 m^3/day.

(3) Find Q_C, rejection flow from the RO unit, from the flow balance.

From Eq. (5),

$$Q_f = Q_c + Q_p \tag{5}$$
$$Q_c = Q_f - Q_p$$
$$Q_f = 52.63\,m^3/day$$

Here, Q_c = concentrate flow from the membrane unit, Q_p = 50 m^3/day, Q_C = 2.63–50 m^3/day, and Q_C = 2.63 m^3/day.

(4) Find C_C, concentration of solute in rejection from the mass balance.

From mass balance diagram,

$$C_f Q_f = C_C Q_C + C_P Q_P$$
$$C_C Q_C = C_f Q_f - C_P Q_P$$
$$C_c = \frac{C_f Q_f - C_p Q_p}{Q_c}$$

$$C_c = \frac{(300\,\mathrm{mg/L}) \times (52.63\,\mathrm{m^3/day}) - (50\,\mathrm{mg/L}) \times (50\,\mathrm{m^3/day})}{(2.63\,\mathrm{m^3/day})}$$

$$C_c = 5{,}052.85\,\mathrm{mg/L}$$

The concentration of solute in rejection flow is 5,052.85 mg/L.

5. NOMENCLATURE

σ = The reflection coefficient
$\Delta\pi$ = Transmembrane osmotic pressure differential (psi)
A = Water permeability constant of the membrane
A_m = Wetted surface area of the membrane (m^2)
C_f = Concentration of solid in feed (mg/L)
C_P = Concentration of solid in permeate (mg/L)
k = Membrane permeability coefficient ($psi^{-1}\ h^{-1}$)
k' = Salt diffusion coefficient (m^2 h)
J = Water flux (m^3/m^2 h)
J_{25} = Normalized flux at 25°C (gfd)
J_T = Actual flux at temperature T (gfd)
M_{25} = Temperature- and pressure-normalized flux (gfd/psi)
MF = Microfiltration
NDP = Net driving pressure (psi)
ΔP = Transmembrane differential pressure (psi)
P = Line pressure (psi)
PCF = Pressure correction factor (dimensionless)
Q_c = Concentrate flow from the membrane unit (m^3/h)
Q_f = Feed flow to the membrane unit (m^3/h)
Q_p = Filtrate flow produced by the membrane unit (m^3/h)
R = Recovery of the membrane unit (decimal percent)
Re = Rejection rate (%)
R_v = Volume recovery
S_A = Salt flux (mg/m^2 h)
T = Water temperature (°C)
TCF = Temperature correction factor (dimensionless)
U = Membrane-specific manufacturer-supplied constant (1/K)
V_F = Volume of feed, L
V_p = Volume of filtrate, L

REFERENCES

1. U.S. EPA (2009) *Point-of-use or point-of-entry treatment options for small drinking water system*, U.S. Environmental Protection Agency, Washington, DC. EPA 815-R-06-010

2. Pronk W, Zurbrügg C, Swartz C, Pronk W (2008) Decentralized systems for potable water and the potential of membrane technology. Water Res, doi:10.1016/j.watres.2008.10.030
3. Sobsey MD (2002) Managing water in the home: accelerated health gains from improved water supply. Water, sanitation and health. Department of Protection of the Human Environment, World Health Organization, Geneva, pp 1–70
4. Kaiser N, Liang K, Maertens M, Snider R (2007) BSF Evaluation Report: Summary of All Laboratory and Field Studies. Centre for Affordable Water and Sanitation Technology, Calgary, Alberta, Canada, http://www.cawst.org
5. Clasen T, Brown J, Suntura O, Collin S (2004) Safe household water treatment and storage using ceramic drip filters: a randomised controlled trial in Bolivia. Water Sci Technol. 50(1):111–115
6. WSC (2007) Water Systems Council, Wellcare Information sheets, Water Treatment. http://www.watersystemscouncil.org/wellcare/infosheets.cfm (accessed on Dec 10, 2008)
7. Ecosoft (2007) Health and beauty filters. http://nashavoda.com.ua/en/main/ (accessed on Dec 10, 2008)
8. LifeStraw (2008) Vestergaard Frandsen. http://www.vestergaard-frandsen.com/lifestraw.htm (accessed on Dec 10, 2008)
9. Li XY, Chu HP (2003) Membrane bioreactor for the drinking water treatment of polluted surface water supplies. Water Res 37:4781–4791
10. Pillay VL (2006) Durban Institute of Technology (DIT), personal communication
11. Homespring (2007) GE Water & Process Technology. http://www.homespring.com (accessed on Dec 10, 2008)
12. Wegelin M, Canonica S, Mechsner K, Fleischmann T, Pesaro F, Metzler A (1994) Solar water disinfection: scope of the process and analysis of radiation experiments. J Water Supply Res Technol-Aqua 43:154–169
13. Reed RH, Mani SK, Meyer V (2000) Solar photo-oxidative disinfection of drinking water: preliminary field observations. Lett Appl Microbiol 30:432–436
14. Mintz E, Bartram J, Lochery P, Wegelin M (2001) Not just a drop in the bucket: expanding access to point-of-use water treatment systems. Am J Public Health 91:1565–1570
15. Clasen T, Bastable A (2003) Faecal contamination of drinking water during collection and household storage: the need to extend protection to the point of use. J Water Health 1:109–115
16. Huq A, Xu B, Chowdhury MAR, Islam MS, Montilla R, Colwell RR (1996) A simple filtration method to remove plankton-associated Vibrio cholerae in raw water supplies in developing countries. Appl Environ Microbiol 62:2508–2512
17. Sobsey MD, Stauber CE, Casanova LM, Brown JM, Elliott MA (2008) Point of use household drinking water filtration: a practical, effective solution for providing sustained access to safe drinking water in the developing world. Environ Sci Technol 42:4261–4267
18. Mohamed ES, Papadakis G, Mathioulakis E, Belessiotis V (2005) The effect of hydraulic energy recovery in a small **sea water reverse osmosis** desalination system; experimental and economical evaluation. Desalination 184:241–246
19. Atikol U, Aybar HS (2005) Estimation of **water** production cost in the feasibility analysis of RO systems. Desalination 184:253–258
20. Afonso MD, Jaber JO, Mohsen MS (2004) Brackish groundwater treatment by **reverse osmosis** in Jordan. Desalination 164:157–171
21. Van der Bruggen B (2003) Desalination by distillation and by **reverse osmosis** – trends towards the future. Membr Technol 2:6–9

22. Madaeni SS, Koocheki S (2006) Application of taguchi method in the optimization of **wastewater** treatment using spiral-wound **reverse osmosis** element. Chem Eng J 119:37–44
23. López-Ramírez JA, Oviedo MDC, Alonso JMQ (2006) Comparative studies of **reverse osmosis** membranes for **wastewater** reclamation. Desalination 191:137–147
24. Suthanthararajan R, Ravindranath E, Chits K, Umamaheswari B, Ramesh T, Rajamam S (2004) Membrane application for recovery and reuse of water from treated tannery **wastewater**. Desalination 164:151–156
25. Kim I-C, Lee K-H (2006) Dyeing process **wastewater** treatment using fouling resistant nanofiltration and **reverse osmosis** membranes. Desalination 192:246–251
26. Jung Y-J, Kiso Y, Yamada T, Shibata T, Lee T-G (2006) Chemical cleaning of **reverse osmosis** membranes used for treating **wastewater** from a rolling mill process. Desalination 190:181–188
27. Lee J-W, Kwon T-O, Moon I-S (2006) Performance of polyamide **reverse osmosis** membranes for steel **wastewater** reuse. Desalination 189:309–322
28. Bódalo A, Gómez JL, Gómez E, Hidalgo AM, Alemán A (2005) Viability study of different **reverse osmosis** membranes for application in the tertiary treatment of wastes from the tanning industry. Desalination 180:277–284
29. Into M, Jönsson A-S, Lengdén G (2004) Reuse of industrial **wastewater** following treatment with **reverse osmosis**. J Membr Sci 242:21–25
30. Kneen B, Lemley A, Wagenet L (1995) Water treatment notes: reverse osmosis treatment of drinking water, Cornell Cooperative Extension, FACT SHEET 4
31. Cath TY, Childress AE, Elimelech M (2006) Forward osmosis: principles, applications, and recent developments. J Membr Sci 281:70–87
32. Votta F, Barnett SM, Anderson DK (1974) Concentration of industrial waste by direct osmosis: completion report, Providence, RI
33. Anderson DK (1977) Concentration of Dilute Industrial Wastes by Direct Osmosis, University of Rhode Island, Providence
34. Holloway RW, Cath TY, Dennett KE, Childress AE (2005) Forward osmosis for concentration of anaerobic digester centrate, in: Proceedings of the AWWA membrane technology conference and exposition, Phoenix, AZ
35. Beaudry EG, Herron JR (1997) Direct osmosis for concentrating wastewater, in: Proceedings of the 27th international conference on environmental systems, Lake Tahoe, NV
36. York RJ, Thiel RS, Beaudry EG (1999) Full-scale experience of direct osmosis concentration applied to leachate management, in: Margherita di Pula S. (ed) Proceedings of the seventh international waste management and landfill symposium, Cagliari, Sardinia, Italy
37. Osmotek Inc (2003) Landfill leachate treatment. (http://www.rimnetics.com/osmotek.htm, Available: 14 November 2006)
38. Beaudry EG, Lampi KA (1990) Membrane technology for directs osmosis concentration of fruit juices. Food Technol 44:121
39. Dova MI, Petrotos KB, Lazarides HN (2007) On the direct osmotic concentration of liquid foods. Part I: Impact of process parameters on process performance. J Food Eng 78(2):422–430
40. Dova MI, Petrotos KB, Lazarides HN (2007) On the direct osmotic concentration of liquid foods: Part II. Development of a generalized model. J Food Eng 78(2):431–437
41. Jiao B, Cassano A, Drioli E (2004) Recent advances on membrane processes for the concentration of fruit juices: a review. J Food Eng 63:303–324

42. Petrotos KB, Quantick PC, Petropakis H (1998) A study of the direct osmotic concentration of tomato juice in tubular membrane-module configuration. I. The effect of certain basic process parameters on the process performance. J Membr Sci 150:99–110
43. Petrotos KB, Quantick PC, Petropakis H (1999) Direct osmotic concentration of tomato juice in tubular membrane-module configuration. II. The effect of using clarified tomato juice on the process performance. J Membr Sci 160:171–177
44. Petrotos KB, Lazarides HN (2001) Osmotic concentration of liquid foods. J Food Eng 49:201–206
45. Popper K, Camirand WM, Nury F, Stanley WL (1966) Dialyzer concentrates beverages. Food Eng. 38:102–104
46. Wrolstad RE, McDaniel MR, Durst RW, Micheals N, Lampi KA, Beaudry EG (1993) Composition and sensory characterization of red raspberry juice concentrated by direct-osmosis or evaporation. J Food Sci 58:633–637
47. Beaudry EG, Herron JR, Peterson SW (1999) Direct osmosis concentration of waste water: final report, Osmotek Inc., Corvallis, OR
48. Cath TY, Gormly S, Beaudry EG, Adams VD, Childress AE (2005) Membrane contactor processes for wastewater reclamation in space. I. Direct osmotic concentration as pretreatment for reverse osmosis. J Membr Sci 257:85–98
49. Cath TY, Adams VD, Childress AE (2005) Membrane contactor processes for wastewater reclamation in space. II. Combined direct osmosis, osmotic distillation, and membrane distillation for treatment of metabolic wastewater. J Membr Sci 257:111–119
50. Flynn M, Fisher J, Borchers B (1998) An evaluation of potential Mars transit vehicle water treatment systems, NASA Ames Research Center, Moffett Field, CA
51. Kravath RE, Davis JA (1975) Desalination of seawater by direct osmosis. Desalination 16:151–155
52. McCutcheon JR, McGinnis RL, Elimelech M (2005) A novel ammonia–carbon dioxide forward (direct) osmosis desalination process. Desalination 174:1–11
53. Cohen D (2004) Mixing moves osmosis technology forward, in: Chemical Processing magazine (http://www.chemicalprocessing.com/ articles/2004/346.html, Avaliable: 14 November 2006)
54. Aaberg RJ (2003) Osmotic power – a new and powerful renewable energy source, ReFocus 4:48–50
55. Jellinek HHG, Masuda H (1981) Osmo-power: theory and performance of an osmo-power pilot plant. Ocean Eng 8:103–128
56. Lee KL, Baker RW, Lonsdale HK (1981) Membranes for power generation by pressure-retarded osmosis. J Membr Sci 8:141–171
57. Loeb S (1975) Osmotic power plants. Science 189:654–655
58. Loeb S (1976) Production of energy from concentrated brines by pressureretarded osmosis. I. Preliminary technical and economic correlations. J Membr Sci 1:49–63
59. Loeb S (1998) Energy production at the Dead Sea by pressure-retarded osmosis: challenge or chimera. Desalination 120:247–262
60. Loeb S (2001) One hundred and thirty benign and renewable megawatts from Great Salt Lake. The possibilities of hydroelectric power by pressure retarded osmosis. Desalination 141:85–91
61. Loeb S (2002) Large-scale power production by pressure-retarded osmosis using river water and sea water passing through spiral modules. Desalination 143:115–122
62. Mehta GD (1982) Further results on the performance of present-day osmotic membranes in various osmotic regions. J Membr Sci 10:3–19

63. Seppälä A, Lampinen MJ (1999) Thermodynamic optimizing of pressureretarded osmosis power generation systems. J Membr Sci 161:115–138
64. Wick GL (1978) Energy from salinity gradients. Energy 3:95–100
65. Mehta GD, Loeb S (1978) Internal polarization in the porous substructure of a semi-permeable membrane under pressure-retarded osmosis. J Membr Sci 4:261–265
66. U.S. EPA (2005) Membrane filtration guidance manual, EPA 815-R-06-009, Office of Water
67. Nederlof MM, Kxuithof JC, Herman JAMH, de Koning M, van der Hoek J-P, Bonne PAC (1998) Integrated multi-objective membrane systems application of reverse osmosis at the Amsterdam Water Supply. Desalination 119:263–273
68. Boerlage SFE, Kennedy MD, Bonne PAC, Galjaard NG, Schippers JC (1997) Prediction of flux decline in membrane systems due to particulate fouling. Desalination 113:231–233
69. Butt FH, Rahman F, Baduruthamal U (1995) Identification of scale deposits through membrane autopsy. Desalination 101:219–230
70. Graham SI, Reitz RL, Hickman CE (1989) Improving reverse osmosis performance through periodic cleaning. Desalination 74:113–124
71. Hong S, Elimelech M (1997) Chemical and physical aspects of natural organic matter (NOM) fouling of nanofiltration membranes. J Membr Sci 132:159–181
72. Griebe T, Flemming H-C (1998) Biocide-free antifouling strategy to protect RO membranes from biofouling. Desalination 118:153–156
73. van der Kooij D, Veenendaal HR, Baars-Lorist C, van der Klift DW, Drost YC (1995) Biofilm formation on surfaces of glass and Teflon exposed to treated water. Water Res 29:1655–1662
74. Donlan RM, Pipes WO (1988) Selected drinking water characteristics and attached microbial population density. J Am War Works Assoc 80:70–76
75. LeChevallier MW, Babcock TM, Lee RG (1987) Examination and characterization of distribution system biofilms. Appl Environ Microbiol 53:2714–2724
76. van der Wende E, Characklis WG, Smith DB (1989) Biofilms and bacterial drinking water quality. Water Res. 23:1313–1322
77. Van der Kooij D (1992) Assimilable organic carbon as an indicator of bacterial regrowth. J Am Water Works Assoc 84:57–65
78. Srinivasan R, Stewart PS, Griebe T, Chen C-I, Xu X (1995) Biofilm parameters influencing biocide efficacy. Biotechnol Bioeng 46:553–560
79. The Dow Chemical Company (2006) Liquid separations. (http://www.dow.com/liquidseps/service/lm_feas.htm
80. Wang LK, Wang MHS, Suozzo T, Dixon RA, Wright TL, Sarraino S (2009) Chemical and Biochemical Technologies for Environmental Infrastructure Sustainability. 2009 National Engineers Week Conference, Albany Marriott, Albany, NY. Feb. 5–6
81. Andrew R (2009) POU and POE standards in Canada. Water Conditioning Purif 51(9):6–58
82. Andrew R (2007) Point of entry systems and the NSF/ANSI standards. Water Conditioning Purif 49(10):6–88
83. Wolfe C (2009) Water purifiers keep army moving. Water Conditioning Purif 51(8):44–45

15
Membrane Technologies for Oil–Water Separation

Puangrat Kajitvichyanukul, Yung-Tse Hung, and Lawrence K. Wang

CONTENTS

Abstract Oily wastewater treatment can be classified into two categories; primary and secondary treatment systems. The primary treatment is employed to separate floatable oils from water and emulsified oil. Secondary treatment system is aimed to treat or break emulsified oil and, then, remove oil from water. This chapter mainly describes use of classical membrane technologies, which are ultrafiltration, microfiltration, nanofiltration, and reverse osmosis, for oil–water separation. Advances in membrane technology such as membrane modification, development of inorganic membrane, and improvement of hydrophilicity of membrane for oil water separation are also discussed as well.

Key Words Ultrafiltration • microfiltration • nanofiltration • reverse osmosis • oil–water • membrane

1. INTRODUCTION

Oil and grease term contains a wide range of contaminants, which may include, but not limited to fatty acids, surfactants, petroleum hydrocarbons, phenolic compounds, animal and vegetable oils, etc. Oil and grease concentrations in wastewater, as recommended by US Environmental Protection Agency (USEPA), are not determined as the presence of specific compounds, but measured by their extractability employing a particular solvent. Oil and

From: *Handbook of Environmental Engineering, Volume 13: Membrane and Desalination Technologies*
Edited by: L.K. Wang et al., DOI: 10.1007/978-1-59745-278-6_15

grease (O&G) is a common pollutant frequently found in effluent of a wide range of industries. Many industries such as steel, aluminum, food, textile, leather, petrochemical, and metal finishing were reported as sources of high concentrations of oil and grease in their wastewaters as shown in Table 15.1 (1).

Guerin et al. (2), have reported average annual oily wastewater loadings from mining operation of about 140 kL/day. Clearing of spills, leaks and overflows with floor wash down are accounted for 60% of the operations' oily wastewater. Composition of oily wastewater from mining operation reported by site locations is shown in Table 15.2.

"Produced water" is also the single largest volume of oily wastewater generated by the oil and gas industry (3). This oily wastewater comes from crude oil and natural gas production, which contain both soluble and insoluble (oil droplets not removed prior to physical separation) petroleum fractions at variable concentrations. The major hydrocarbon groups present in produced water include alkanes, alkenes, alkynes, aromatics, polynuclear aromatics, and complex hydrocarbon compounds containing oxygen, nitrogen, and sulfur. Produced water

Table 15.1
Oil and grease concentrations in wastewater of selected industries

Type of industry	Oil and grease concentration (mg/L)
Palm oil industry	4,000
Food processing	3,000–4,000
Mining operation	3,000–23,000
Metal finishing	100–5,000
Steel-rolling mill	7,200
Aluminum rolling	5,000–50,000
Oil drilling	7–1,300
Crude oil tank ballast	3–72
Petroleum refinery	16–3,200
Can production forming	200,000

Adapted from Patterson (1).

Table 15.2
Oily wastewater stream compositions in mining operation from selected site locations

Site locations	Free oil	Emulsified oil	Grease	Diesel or gasoline
Ore crusher houses	/	/	/	–
Plant workshop	/	/	/	–
Drill and shovels	/	/	/	/
Workshop	/	/	/	/
Primary crusher	–	/	/	–

from gas production operations generally contains higher levels of Benzene, Toluene, Ethylbenzene, and Xylene (BTEX) than those generated from oil production (4). Chapelle et al. (5) has reported that relatively water soluble light aromatics of BTEX comprise only 2–3% of most crude oils as a whole. Stephenson (6) reported mean concentrations of benzene ranging from 5.8 to 12.2 mg/L and 1.3 to 8.7 mg/L for gas and oil production, respectively.

Other major source of oily wastewater is food processing industry. Oily and fatty materials are produced mainly from slaughtering, cleaning, and by-product processing (1). The oily vegetable extraction is also the source of oily waste. Ahmad et al. have reported oil and grease concentrations from palm oil processing, which produced as high as 4,000 mg/L of oil resulting in high biochemical oxygen demand (BOD) and chemical oxygen demand (COD) contents.

In general, oily wastewater found in the industry comes from many sources including floor wash, machine coolants, alkaline/acid cleaners, and spills from manufacturing process as shown in Table 15.3. Nature of oily wastewater is varied due to their production source. Oil from spill is mainly free and emulsified oils, while oil from alkaline and acid cleaning process is normally highly emulsified due to presence of surfactants. Mixtures of various types of oil could be found in wastewater generated from floor wash. These oily wastewaters can be present in both free and emulsified forms stabilized by dirt, debris, and solvents. Oils from petroleum refining or oil from drilling activity are frequently found in both free and emulsified oils making it more difficult in later oily–water separation process.

Table 15.3
Sources of oily wastes from industries

Source	Industries	Nature
Alkaline and acid cleaners	Metal fabrication, iron and steel, metal finishing, industrial laundries	Normally highly emulsified due to surfactants; difficult to treat
Floor washes	All industries	Mixture of various types of oils from spills of hydraulic and cutting fluids, oil mists from spraying/coating, etc.; Can be present in both free and emulsified forms stabilized by dirt and debris, and solvents
Machine coolants	Metals manufacturing, machining	Normally emulsified and difficult to treat
Vegetable and animal fats splitting, refining, rendering	Edible oil, detergent manufacture, fish processing, textile (wool scouring), leather (hide processing), tank car washing	Both free and emulsified oils; difficulty of treatment varies
Petroleum oils	Petroleum refining, petroleum drilling	Both free and emulsified oils; difficulty of treatment varies

Adapted from Cheryan and Rajagopolan (7).

2. FUNDAMENTAL KNOWLEDGE OF OIL WATER SEPARATION

2.1. Oil Properties

Several forms of oil and grease presented in wastewater are free, dispersed, or emulsified (8). Droplet size of oil is a major factor for their classifications. Free oil has a droplet size greater than 150 μm. Dispersed oil is characterized by droplet sizes ranging from 120 to 150 μm. Oily water, with droplet size less than 20 μm are classified as emulsified oil.

Tabakin et al. (9) have proposed more specific oil categories based on its physical forms in wastewater. It includes:

- Free oil: the rapidly rising oil to the surface of wastewater under quiescent conditions.
- Mechanical dispersions: fine droplets that are stabilized by electrical charges or other forces, but not through the influence of surface active agents.
- Chemically stabilized emulsions: oil droplets that are stabilized by the surface active agents at the oil–water interface.
- Dissolved oil: very fine droplets (typically less than 5 μm) that are truly soluble species in the chemical sense.
- Oil-wet solids: oil that adhered to the surface of particulate material in the wastewater.

Additionally, oil can be classified by its density. If the ratio of the density of oil to receiving water is greater than 1.0, oil will not float and thus, with a density less than 1.0, oil will float. The density is also shown in terms of the American petroleum institute (API) gravity. As shown in Fig. 15.1, with higher densities than receiving water (above the line), oil sinks; with lower densities (below the line), it initially floats.

In oily wastewater treatment, two types of oils are mainly considered to be removed; free and emulsified oils. Free oil has lower specific gravity than water. It can rapidly rise to the surface, and thus it is called “floatable oil.” It is generally considered to have droplets size larger than 250 μm in diameter.

Emulsified oils are often resistant to get separated from water because the droplets are either resistant to rise to the surface or they rise so slowly that they cannot be removed with most oil water separators.

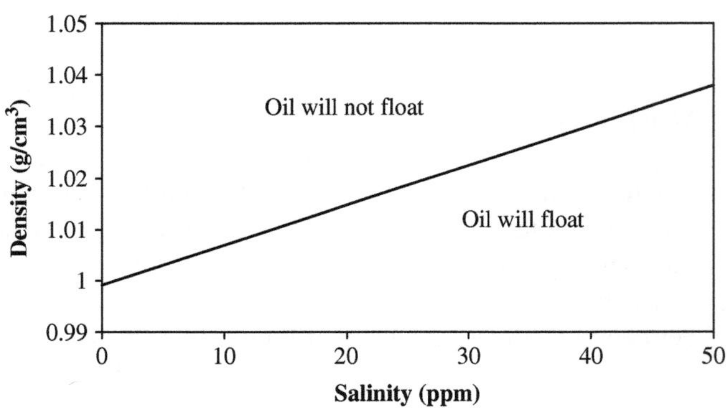

Fig. 15.1. The relationship between density and salinity at a temperature of 15°C.

Oil properties are the crucial factors for separation and removal from water. Compounds in the oil, such as resins, asphaltenes, and waxes, contribute to the formation of stable emulsions (10, 11). Nickel porphyrin in seawater is a stabilizing component in oil (12). Changes in density and viscosity, the formation of stable emulsions, and the dispersion of oil and emulsified oil droplets, all play important roles in inhibiting an effective separation of both oil and emulsion droplets in water, and water droplets from an emulsion (13).

Temperature is also one factor that causes changes in oil properties. For example, most crude oils and refined products have specific gravities between 0.80–0.98 g/cm^3. The density of both water and oil can be lowered with the increasing temperature as shown in Table 15.4. Thus, the net effect on the buoyancy force is very restricted for separation of oil droplets from water.

2.2. *Emulsion*

Emulsions are defined as a colloidal suspension of a liquid within another liquid with droplet sizes typically less than 20 μm. The emulsions contain the colloid particles, which scatter light causing the cloudiness; they are either dispersion of oil droplets in water or of water droplets in oil, in which the internal phase is oil and water, respectively (see Fig. 15.2).

When emulsions are formed, their droplet structures are very constant. However, when the agitating is continuously precede, the size of emulsion increases due to the droplet fusion. Eventually, the emulsion will separate into the basic components as oil and water. The rate of droplet fusion can be decreased with the addition of surfactant and appropriate controlled

Table 15.4
Changes in density of oil as a function of temperature

Temperature (°C)	Water density	Fuel oil density	Density difference
10	1.024	0.880	0.144
30	1.020	0.870	0.150
50	1.012	0.855	0.157

Adapted from Nordvik et al. (13).

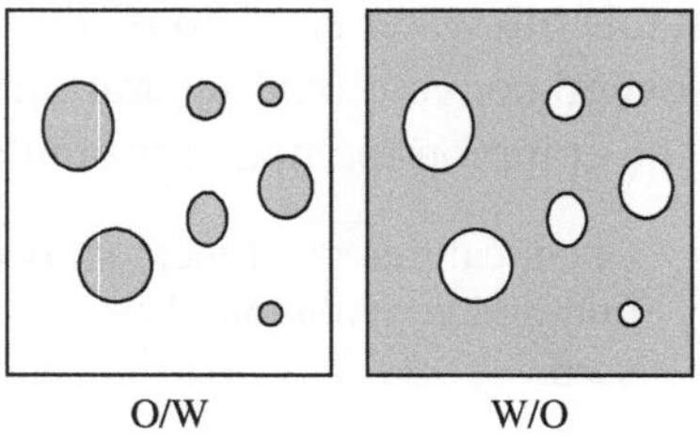

Fig. 15.2. Different types of emulsions as oil droplets in water and water droplets in oil. *White* and *gray* areas represent water and oil, respectively (14).

conditions. Besides the widely used terms, emulsified oils are the oil droplets that are reduced in size to such a degree that the oil's normal electrical repulsion of the water molecule is overcome due to its minute size.

Emulsions can also form in several ways as follows (15):

- Energy: Pumping, mixing, and other activities of adding energy to a wastewater may accidentally or purposefully, mechanically emulsify oil.
- Emulsifiers: certain organic chemicals such as soaps and detergents can lower specific gravity of oil causing it more difficult in separating oil from water by gravity. This process is widely used in industrial application to dilute oil with water without stratification.
- Heat: some organic chemicals can become emulsified at high or low temperatures.

Generally, when oil and one or more surfactants are agitated with a mixture of water under controlled conditions, a cloudy mixture, or emulsion, will be formed. In addition, a transparent solution so called microemulsion also occurs (2–6). Microemulsion is obtained when two immiscible liquids are stirred, either droplets of oil in water or droplets of water in oil depending on the dispersed phase, where the situation related mainly to the formulation and to a lesser degree to the droplets of water in oil. The droplets that are very fine, less than 10 nm, are usually named miniemulsions.

The microemulsion is one type of emulsion in which the droplets are much smaller than the usual emulsions. Their structures have characteristic dimensions between 5–100 nm. Thus, the aggregates of this size are poor scatters of visible light. Consequently, these solutions are optically clear when the light can pass through the solution without much scattering. They are dynamic to form, and disintegrate and reform in fractions of milliseconds. Microemulsions can easily form with an increase of the temperature. In this condition, the molecular motion increases and deteriorates the directional interactions. The addition of alcohol also accelerates microemulsion formation because the molecules get inserted in between surfactant molecules and push them apart, with a corresponding reduction in polar interactions and rigidity. Microemulsions are thermodynamically stable solutions. When these components are brought together, they can stay stable for a long period until the ingredients are intact. To disturb the stability of these microemulsions, a high amount of surfactant under the controlled temperature is required.

Wastewaters containing emulsions are found in several industries. Oil emulsions are usually mixtures of several components, each with a specific function. Emulsified oily wastewater can contain oil (mineral, vegetable, or synthetic), fatty acids, emulsifiers (anionic and nonionic surfactants), corrosion inhibitors (amines), bactericides, and other chemicals. Two broad categories of emulsified oil presented in industrial wastewater are unstable and stable oil–water emulsions. The unstable oil–water emulsion can be readily removed by using several conventional separation processes such as chemical, electrical, or physical processes. The common methods for emulsion breaking are:

- Chemical treatment method: two methods usually applied for breaking emulsion include acidification and coagulation. The combination of pH adjustment and coagulant addition is also widely used.

- Flotation system: this process is functioned by increased differences in specific gravity between the oil and water by blowing fine air bubbles through the oily wastewater. The most common floatation system is dissolved air flotation (DAF) process.
- Filtration: Membrane filtration is the common process which is widely used in many industries. The examples of membrane filtration include microfiltration (MF), ultrafiltration (UF), nanofiltration (NF), reverse osmosis (RO), and the newly developed technology such as hydrophobic membrane.
- Hydrocyclone: Hydrocyclone is a physical method used to break emulsion and remove oil from water. This method increases force for phase separation which also can handle solids in the influent.
- Electrical process: Electrofloatation and Electrocoagulation are oil destabilization techniques that utilize electricity as the major unit for oil and water separation.

For removing stable oil–water emulsion, several conventional methods are not effective. For example, a very long residence time is required for emulsion to rise onto the top to enable gravity separation. In several cases, chemical addition cannot effectively break the stable oil–water emulsion, especially when the oil droplets are finely dispersed and have a low concentration. These conventional methods were reported for their capabilities in reducing oil concentration to hardly 1% by volume of the total wastewater and they cannot efficiently remove oil droplets below 10 μm size (16, 17). Several membrane techniques were suggested to be used to remove the emulsion in the micron size from oily wastewater (18–26). These promising membrane based-methods include ultrafiltration (18, 19), microfiltration (20, 21), membrane distillation (22), dehydration of oil emulsion by prevaporation (23), and reverse osmosis (24). Many emerging techniques such as ion-exchange membrane (25), hybrid-modified resin, and activated carbon are recently developed and exhibit promising results to be used in the near future (26).

2.3. *Coalescence in Oil Water Separation*

Coalescence in oil–water separation is usually defined as the process by which two or more droplets or particles merge during contact to form a single daughter droplet (or bubble). In membrane process, the coalescence is controlled by the collision frequency between droplet–droplet and droplet–interface, which is determined by the transport of oil droplets out of the bulk solution to the membrane, and by film drainage between adjacent droplets or droplet–membrane as the governing process for coalescence to occur (27). In the coalescence process, there will be a greater chance for droplet collision and droplet coalescence especially at higher oil droplet concentrations. Several techniques such as crossflow, intermittency, pulsatile flow, backflush, and flow reversal operations were developed to modify the process fluid dynamics in an attempt to reduce flux decline by promoting turbulence in the feed channel (28–35).

In application of membrane coalescence, the crossflow technique is frequently selected. The effect of crossflow should be to modify the drop sizes approaching the membrane and also shear thin any oil film wetting the membrane. Lefferts (36) has found that membrane surface roughness and oil-droplet interception at the membrane surface can promote emulsion coalescence. In crossflow process, the membrane-feed interactions are proposed as follows (27). At low crossflow velocities, the movement of oil droplets is mainly dominated by Brownian diffusion and electrostatic effects. Droplet coalescence may occur in localized places at the surface of the membrane (depending on the local roughness) as well as inside of the membrane pores. The coalescence of droplets at the membrane surface results in the formation of a thin oil layer that would reduce the total number of available pores for permeate flow, resulting in lower permeate flux. When steady state is reached, the thin oil layer slowly permeates through the pores via surface flow due to the hydrophobic nature of both oil droplets and membrane pores. However, during operation at higher crossflow velocities, coalescence at the membrane surface may be reduced. The effect of crossflow would be to keep the largest droplets away from the surface by shear-induced diffusion or lateral migration. The droplets reaching the surface would be the smallest in the distribution and would be able to penetrate into the pores without substantial film formation. In this scenario, oil film forming at the surface of the membrane is diminished and there is less reduction of the total number of available pores, resulting in higher permeate fluxes. At high crossflow velocities, the amount of flux decline is also smaller. This is because the interfacial oil film is not as pronounced during operation as it is at small crossflow velocities, and steady state permeate flow is reached without significant flux decline (27). From Hong's work, the membrane coalescence under crossflow and intermittent permeate operation is found to be the promising method to be a potentially novel process for the treatment of oily wastewater (27).

3. MEMBRANE TECHNOLOGY FOR OIL WATER SEPARATION

Oily wastewater treatment can be classified into two categories; primary and secondary treatment systems. The primary treatment is employed to separate floatable oils from water and emulsified oil. Free oil is readily removed in this step by normal mechanical separation processes, such as gravity flotation or skimming. Secondary treatment system is aimed to treat or break emulsified oil and, then, remove oil from water. An unstable oil–water emulsion can be destabilized in this step by acid cracking and the oil is subsequently separated by chemical flocculation or coagulation, followed by air flotation, centrifugation, or filtration. Oil emulsions have, however, become more and more stable, and a number of them are now almost impossible to crack by chemical means (37). For the time being, regulations imposed on oily–water pollution have become tighter and the conventional treatment methods cannot cope with the problem anymore. The membrane separation process has been one of the candidates for oil–water separation (38–43). The obvious advantages of membrane process would be lower capital cost, the absence of chemical addition, and the subsequent generation of oily sludge. Concentrated oil from oily wastewater could be reused and handled with accumulated straight oils and, subsequently, disposed of into existing incinerators.

Membrane processes are widely used in oil–water separation. In general, membrane is classified into two groups; pressure-driven membrane and electrical membrane, known as electrodialysis. The most applicable process for oily wastewater removal is the former type. The pressure-driven membrane applications include microfiltration, ultrafiltration, nanofiltration, and reverse osmosis. All of them are categorized by the molecular weight or particle size cut-off of the membrane as shown in Table 15.5. Possible guidelines for membrane-process

Table 15.5
Types of pressure-driven membrane processes as categorized by size cut-off range

Membrane processes	Size cut-off range (μm)
Microfiltration (MF)	0.05–1.5
Ultrafiltration (UF)	0.002–0.05
Nanofiltration (NF)	0.0005–0.007
Reverse osmosis (RO)	0.0001–0.003

Adapted from Casey (35).

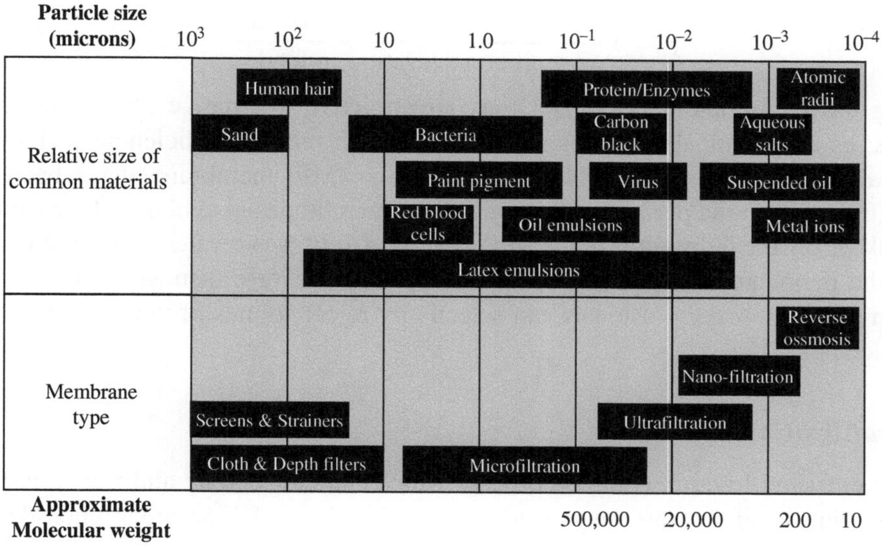

Fig. 15.3. Separation technologies based on droplet sizes.

selection from various oil droplet size categories have been developed (14) as shown in Fig. 15.3.

Membranes are useful for stable emulsion removal particularly water-soluble oily wastes (7). This process is generally regarded as polishing technologies for emulsified oils. The low effluent concentrations of oil less than 5 mg/L could be achieved by this technique (2). Membrane processes have several advantages as described as follows (7):

- Applicable for wide range of industries.
- A barrier to rejected components. The quality of treated water is relatively uniform.
- Extraneous chemicals are not needed.
- It can be used in-process to allow recycling of selected waste streams within a plant.
- Concentrates up to 40–70% oil and solids can be obtained by UF or MF.
- Membrane equipment has a smaller footprint.

- Energy costs are lower compared to thermal treatments.
- The treatment plant can be highly automated and does not require highly skilled operators.

The development of membrane technologies in the last 30 years has embodied applications in the processing of emulsions. Both microfiltration and ultrafiltration have been used for concentrating emulsions (43–45). Ideally, the continuous phase is driven through the membrane by the application of a transmembrane pressure as the dispersed phase is retained at the surface by a sieving mechanism (43). However, because the liquid droplets are deformable, they can still be squeezed into smaller pores if the applied pressure exceeds the dispersed phase capillary pressure (44). This would then result in the breakthrough of dispersed phase as well as permeate contamination.

The separation efficiency (f) for the separation of o/w emulsions is given by (43):

$$f = 1 - \frac{\text{oil concentration in permeate}}{\text{oil concentration in feed}}$$

Among the membrane processes, UF is an attractive alternative for the treatment of emulsions, as no chemical additives are required, high removal efficiencies achieved, and low capital and operating costs (44). Macrofiltration (MF) membranes have been used to recover surfactants in the permeate. Microfiltration membranes give higher fluxes, but pose a higher risk of oil breakthrough. If the salt content of oily wastewater is too high for direct reuse of the permeate in the plant, it can be treated by reverse osmosis and nanofiltration (7, 46). In addition, reverse osmosis can selectively reject solutes of the same size order as water molecules.

3.1. *Ultrafiltration*

UF is a pressure driven membrane process which can concentrate and fractionate macromolecular solutes and separate suspended species from water. UF provides a non-destructive separation, which can be performed without any phase transition (47). Because the osmotic pressure exerted by the high molecular weight solutes can be negligible, this process is operated at relatively low pressure in range of 0.69–6.91 atm (70–700 kPa) (48, 49). The concentrate may contain up to 50% of oil. The oily concentrate can be further separated by centrifugation.

UF membranes, which have tight pores, have been selected in most applications to ensure steady permeate quality. The flux frequently declines with time; this has been attributed to surfactant adsorption on the pore walls, to the buildup of a polarized layer of concentrated emulsion at the membrane surface and also pore plugging by oil droplets (44, 45). However, this technique has two drawbacks; concentration polarization and membrane fouling (50). As a filtration process, membranes used are truly porous and separation is a physical process requiring elevated pressure to achieve passage of fluid through the filter. The water and other low molecular weight solutes pass through the membrane pores. The larger molecules or aggregates are rejected. The successful ultrafiltration separation performance is usually obtained when discrete and stable emulsion particles of oil, larger than the membrane pore size, are maintained. However, the mechanism of separation of oil and water in ultrafiltration involves not only the size differences among various solute and solvent molecules under consideration, but also adsorption and surface charge characteristics of membranes (51). Thus, the ultrafiltration membranes are made from several types of polymers such as cellulosic esters, polyamides, polysulfone, polyacrylonitrile (PAN), and polyvinylidene difluoride (PVDF) to obtain the specific characteristics. The composite ultrafiltration membranes from blended polymers are also developed for oil and water separation purpose, such as the combination of PVDF and polymethyl methacrylate (52), the blended polyurea and polyurethane ultrafiltration over macroporous polysulfone substrate (53), and cellulose propionate (CP) polymer using dimethyl acetamide solvent (51). The ultrafiltration performances in oil water separation of each types of polymer are also investigated. Bodzek and Konieczny (48) investigated the ultrafiltration performances of PAN and polyvinyl chloride membranes for three oil emulsions. Over 90% of COD rejection and over 95% of oil retention were attained by the ultrafiltration process. The inorganic membrane ultrafiltration is also employed in oil water separation. Results from laboratory and pilot plant tests for oily and latex wastewater have shown that an ultrafiltration treatment process is efficient and economically attractive. In addition, a high percentage removal of suspended solids and COD using ultrafiltration to treat the mineral oil water emulsions and oily wastewater from oil product refining processes were also reported (54). From continuing investigation from both research and real applications, ultrafiltration becomes the most appropriate process for oil–water separation from industries.

3.2. *Microfiltration*

Microfiltration is one type of membrane filtration that is widely used in oil–water separation. It is generally defined as the pressure-driven flow of a suspension containing colloidal or fine particles with dimensions within the size range of 0.02–10 μm through a membrane. The microfiltration process is carried out in two types of configurations: dead-end and cross-flow. Cross-flow microfiltration which is a downstream solid–liquid separation process in which a suspension flows tangentially under moderate pressure through a microporous membrane is of particular interest in processes that demand continuous or semicontinuous operation; however, use in wastewater treatment applications has remained limited as a result of low long-term permeate fluxes due to membrane fouling (55).

Microfiltration membranes give higher fluxes, but pose a higher risk of oil breakthrough. In this system, purified fluid passes through the membrane as permeate, whereas the retained aggregates form a polarization layer and/or a thin cake on the membrane surface. In the operation, the serious problem occurred when the accumulation of aggregates was close to the membrane surface due to its permselectivity, the so-called concentration polarization and/or cake buildup. The filtration resistance increases and the permeate flux (productivity) decays continuously as the thickness of the polarisation layer or cake increases until a steady-state is reached. However, since the feed flows tangentially to the membrane surface in crossflow microfiltration, the accumulation of aggregates is limited and the pressure drop is lower in comparison to the conventional filtration process. This phenomenon is called membrane fouling. Koltuniewiez et al. (56) investigated the microfiltration fouling for the separation of oil-in-water emulsions by using three types of microfiltration membranes. They performed simultaneous experiments in both dead-end and cross-flow modes at various pressures, and in the latter case at different cross-flow velocities. They reported that after an initial phase of pore blocking, the buildup of a layer of droplets and the resultant switch to cake filtration dominates the performance (57). Because of membrane fouling, microfiltration has to be stopped frequently for membrane cleaning to restore membrane permeability. Membrane cleaning, as well as other measures for fouling control, significantly increases the cost and complexity of the processes.

Recently, there are several reports using microfiltration (MF) for oil and water separation. It was reported that the membranes made of highly hydrophilic and solvent resistant materials should be applied to oily water treatment (56, 57). For a membrane study in rejecting oil droplets (58), it was found that oil droplets are completely retained and the continuous phase is permeated. However, because the oil droplets are deformable, depending on the applied pressure, they can be squeezed through the pores and contaminate the permeate. Nazzal and Wiesner (59) also investigated the effects of transmembrane pressure and membrane pore size. If transmembrane pressures were below a critical pressure, emulsion rejection could be maximized. Conversely, if the membranes were used as coalescers, the applied pressure has to be above the critical pressure, enabling the oil droplets to wet the membranes and initiate coalescence. An important difference between membrane coalescers and rejecting membranes is the potential flux, which could differ by an order of magnitude in favour of coalescers.

In the application of microfiltration on coalescence process in the stable oil-in-water emulsions (60), the membranes are also used to enlarge oil droplets. In this manner, the membrane pore size had a large influence on the coalescence of oil droplets and that membrane orientation is also important. Also, imposed shear rates inside the membrane pores play a key role during membrane coalescence, where sufficient in-pore shear rates are required to coalesce oil droplets, although higher shear rates may reverse the process.

3.3. Nanofiltration

Nanofiltration is used most commonly when the objective of the water/wastewater treatment is to separate both target components, and non-target constituents. Nanofiltration is capable of concentrating sugars, bacteria, proteins, particles, and other constituents that have a molecular weight greater than 1,000 Da unlike RO, where almost complete rejection of ions is seen.

As types of membrane filtration frequently used in oil–water separation are ultrafiltration and microfiltration, only some works are applied nanofiltration in this area. Peng et al. (61) used nanofiltration for the improvement of water management in oil sands operation. It was reported that nanofiltration is an effective process for this application in both for selective water softening and for potential discharge requirements. Proper application of membrane technologies can result in better reuse of process waters, reduction of raw water intake by replacement with nanofiltration produced water, and the improved possibility of meeting discharge water quality criteria by reduction of toxicity properties in process-affected waters.

Specifically, nanofiltration is effective in removing divalent ions of Ca^{2+}, Mg^{2+}, and $SO_4{}^{2-}$, which are responsible for high water hardness, with average rejection efficiency of more than 95%. Nanofiltration may provide an efficient approach for water softening of selective water streams that could be preferable to wider scale application of lime-softening technology. In addition, nanofiltration is effective in the removal of total organic carbon. This should be sufficient to remove acute toxicity associated with these materials in process-affected waters. For the fouling consideration, the fouling in this membrane type is reversible. However, the membrane permeate flux could be fully recovered (61).

3.4. Reverse Osmosis

Reverse osmosis is one type of membrane filtration that has been applied for treating a wide variety of industrial effluents with the objective to separate oil from water. Sridhar et al. (62) used the pilot-scale reverse osmosis to determine the feasibility of the process for treating the wastewater from a vegetable oil industry. The system was found to be very suitable for treating oil effluents having total dissolve solid (TDS) concentrations upto 52,215 mg/L. Reverse osmosis has been found to be a very promising separation process for treatment of oil industry effluent and water recovery due to the high fluxes obtained alongside with the significant rejection of TDS, COD, BOD, and color.

Application of reverse osmosis units in conditions when the feed may be contaminated with crude oil and fuel oil spillages with feedwater comprising NaC1/water solutions of 2,000–35,000 mg/L concentration is also reported (63). In this work, hydrocarbons retained in solution in water were not found to exhibit damaging effects on the performance of the reverse osmosis membranes. However, it was reported that diesel contamination possibly posed harmful effects with a capacity to reduce membrane fluxes to zero if present in large concentrations for even short periods of time. Hexane, which is one of lighter crude oil fraction can also cause serious deterioration of the performance of reverse osmosis membranes when in contact, in pure or emulsified form. The damage is worse in more concentrated hydrocarbon mixtures and at longer exposure times. The damaging effects of the hydrocarbon contaminants are also dependent on types of membrane. Hodgkless et al. (63) reported that the brackish water membrane suffers substantial reductions in flux and roughly proportionate increases in salt passage. While the seawater membrane undergoes larger deleterious effects on salt passage and much lower reductions in permeate flux, it appears that the damage caused by exposure to hydrocarbons is difficult to reverse.

3.5. Integrated Membrane System

3.5.1. Combination of Ultrafiltration/Reverse Osmosis Processes

Generally, direct application of reverse osmosis for processing of oily wastewater would require some physical and chemical pretreatment, depending on the used module configuration. The simplest pretreatment would be for a tubular module, and can be done with an "equalization tank," from which the free-floating oil and the settleable solids are removed (64). Application of ultrafiltration as a pretreatment stage before reverse osmosis should

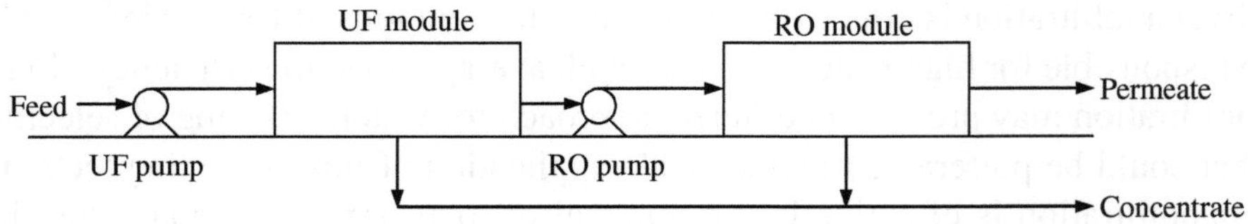

Fig. 15.4. Flow schematic of a unit combination of UF/RO processes (64).

remove nearly all the oil and some COD. An example of Flow schematic of a unit combination of UF/RO processes is shown in Fig. 15.4.

Lin and Lan (65) investigated the treatment of waste drawing oil which is a high-strength waste oil–water emulsion commonly used in the cable and wire industries. Semi-batch ultrafiltration and reverse osmosis processes along with prefiltration by a microfilter were employed to treat the waste oil–water emulsion. Drawing oil or cutting machine oil is a typical oil–water emulsion which is commonly used in the precision machining and cable and wire industries. The oil–water emulsion serves the purposes of lubrication, cooling, surface cleaning, and corrosion prevention in the manufacturing process. Depending on specific applications, the oil emulsion can consist of up to 97% water, the rest being a complex aqueous mixture which comprises different kinds of oils – mineral, animal, vegetable and synthetic, alcohols, sequestrants, and surfactants. The test results show that the ultrafiltration treatment is very effective in reducing the COD and copper concentrations and in improving the turbidity, but it is relatively ineffective in reducing the conductivity of the ultrafiltration permeate. In conjunction with the reverse osmosis treatment, the final permeate quality is found to be excellent with over 99% improvements in the COD, copper, conductivity, and turbidity (65).

3.5.2. Combination of Ultrafiltration/Nanofiltration Processes

Combination of nanofiltration with other types of membrane is also used in treating industrial effluent. Generally, the nanofiltration process is carried out under lower operating pressures than reverse osmosis and with lower molecular weight cut-offs (MWCO) than ultrafiltration. Thus the combination of nanofiltration with either reverse osmosis or ultrafiltration membrane is possible with the use of nanofiltration as either a primary or secondary treatment. The application of the ultrafiltration process as a pretreatment step for industrial effluents replaces the conventional methods of pretreatment for nanofiltration, resulting in longer membrane life, and the ability to operate the nanofiltration system at higher flux rates. Karakulski and Morawski (66) investigated the integrated membrane system based on ultrafiltration and nanofiltration in purification of effluents from a cable factory. The application of ultrafiltration membranes with MWCO of 100 kDa results in the complete rejection of suspended solids and the retention of oil and lubricants at a level of 99%. The resulting ultrafiltration permeate was further purified by nanofiltration to achieve the retention of remaining pollutants (oil and lubricants) and to reduce the content of copper ions. As a result of the nanofiltration process, the content of organic compounds in the permeate determined as total organic carbon was below 1 mg/dm^3, and the rejection of copper ions obtained.

The quality of permeate obtained by the application of an ultrafiltration and nanofiltration integrated membrane system allows the direct reuse of treated water for fresh emulsion top up.

3.6. Membrane Bioreactor

Membrane bioreactor is usually referred to as a bioreactor integrated to a membrane module system. This technology has been developed over 30 years ago (67) and it is mostly applied to treat industrial wastewater, domestic wastewater, and specific municipal wastewater, where a few small work appears in treatment of water (68).

In general, membrane bioreactors consist of two compartments, the biological unit and the membrane module (69). The biodegradation of the compounds occurs in the biological unit, while the physical separation of the treated water from mixed liquor takes place in the membrane module. Membrane bioreactor systems can be classified into two groups: integrated or submerged membrane bioreactors, and external or recirculated membrane bioreactors (69). The first group, integrated membrane bioreactors, involves the outer skin membranes that are internal to the bioreactor as shown in Fig. 15.5. The driving forces across the membrane is achieved by pressurizing the bioreactor or creating negative pressure on the permeate side (69–72). A diffuser is provided for mixing and facilitating the filtration surface scouring. The second type, external or recirculated membrane bioreactors, involves of the recirculation of the mixed liquor through a membrane module that is outside the bioreactor (72). The driving force is the pressure created by high cross-flow velocity along the membrane surface (73, 74). A schematic of this system is shown in Fig. 15.6.

Types and configurations of membrane used in membrane bioreactor applications include tubular, plate and frame, rotary disk, hollow fiber, organic (polyethylene, polyethersulfone, polysulfone, polyolefin, etc.), metallic, and inorganic (ceramic) microfiltration and ultrafiltration membranes (75). The pore size of membranes used ranged from 0.01 to 0.4 μm (69).

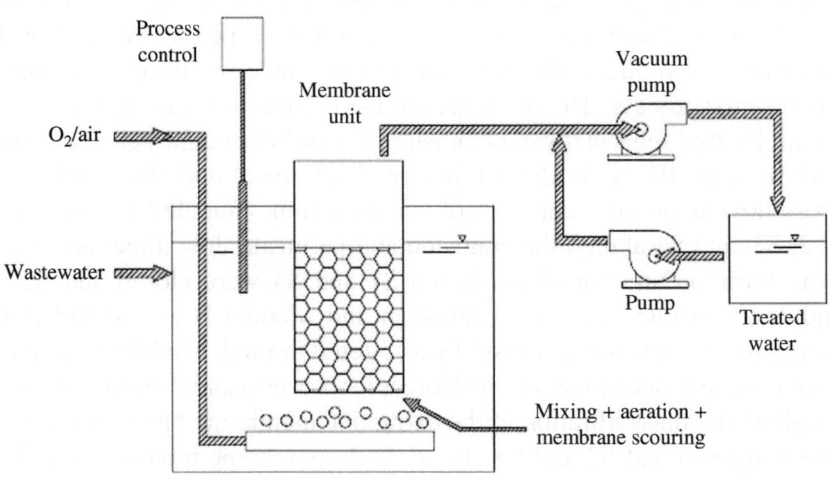

Fig. 15.5. Schematic of integrated membrane bioreactors (69).

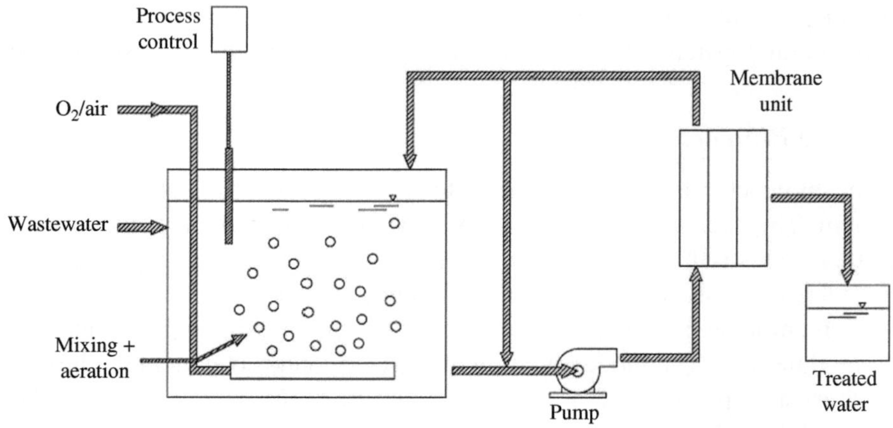

Fig. 15.6. Schematic of external membrane bioreactors (69).

The fluxes obtained ranged from 0.05 to 10 m/d (m^3 m^{-2} d^{-1}), strongly depending on the configuration and membrane material (69).

An example of application of membrane bioreactor in separation of oil from water is shown below (76). The conceptual extractive membrane bioreactor process is a hybrid process combining liquid–liquid extraction, cross-flow microfiltration and a side-loop membrane bioreactor. Two hydrophilic ceramic membranes were used for oil–water emulsion separation and biomass separation. These are operated alternately for each purpose in opposing directions. Thus, one filtration serves as backflush of the other and vice versa, reducing membrane fouling. In this work, when oil–water emulsion is separated by hydrophilic membrane filtration, water passes through the membrane and the big oil droplets are retained on the membrane surface by a sieving mechanism. The deformable oil droplets cannot be squeezed into pores unless the applied pressure is higher than the capillary pressures. A schematic diagram of the device used in this work is shown in Fig. 15.7 (76).

It was designed to simulate the forward and the reverse filtration in the extractive membrane bioreactor process. The oil water emulsion filtration was the reverse filtration in this experiment. Diethyl sebacate/water emulsion, 2% (w/w), and the permeate were stored in Vessels I and II, respectively. Both liquids were well mixed and circulated by centrifugal pumps in cross-flow along different sides of one membrane module (A or B). Four solenoid valves S1, S2, S3, and S4 altered the emulsion and permeate flow directions and controlled the membrane filtration and backflush. When S1 and S4 were closed and S2 and S3 are opened, Module A operates as reverse filtration and Module B as forward filtration. The diethyl sebacate/water emulsion in Vessel I was filtered through Module A, entered Vessel II (the bioreactor in the extractive membrane bioreactor process) and then went back to Vessel I, through Module B (biomass filtration in the extractive membrane bioreactor process). When S1 and S4 were opened and S2 and S3 closed, both membrane modules were backflushed. The backflush length was set at 2 s. After 5 h the functions of Modules A and B were switched, providing membrane cleaning. The results of this study showed that the ceramic

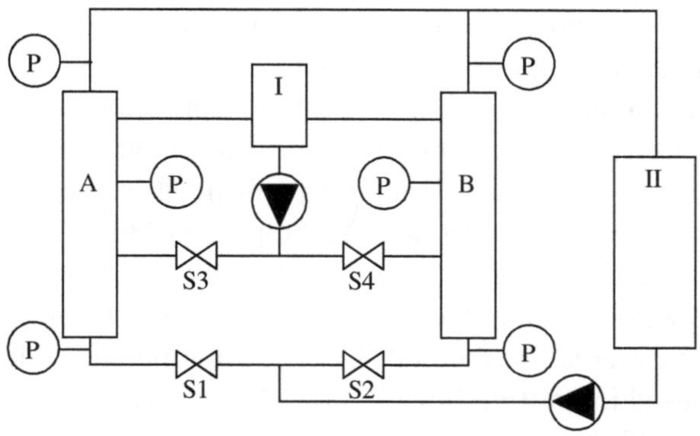

Fig. 15.7. Schematic of dual-membrane bioreactor (76).

hydrophilic microfiltration membrane could effectively separate the diethyl sebacate/water emulsion. Diethyl sebacate concentrations in the permeate were lowest when the transmembrane pressure was low and when the membrane was frequently backflushed. Membrane fouling during oil–water emulsion filtration is mainly due to adsorption of oil on membrane structure, which causes modification of the critical surface tension. This process exhibits a good example to apply membrane bioreactor as an effective oil–water separation method (76).

4. ADVANCES IN MEMBRANE TECHNOLOGY

4.1. Modification of Membrane

The most serious limitation of membranes is the continuous decline of flow through the membrane due to fouling. Permeation rates of macromolecular solutions through membranes are much smaller than pure water permeation rates and decrease rapidly as the operation continues. Flux decline has a significant negative impact on the feasibility and economics of the membrane process. Recently, many works have focused on membrane modification with a fundamental approach to reduce membrane fouling. Numerous attempts have been undertaken to modify membrane materials aimed at producing membranes less susceptible to fouling. One technique widely used for this purpose is blending of the original polymer with polymers to achieve the required properties (77, 78). For example, to enhance the membrane hydrophilicity, the polyether sulfonamide) is blended with the original polymer to obtain a modified poly(etherimide) ultrafiltration membrane (79). To change the adsorption and permeation properties of the membrane, the addition of polymeric layers on the active surfaces of membranes is performed (80). Adsorbed hydrophilic polymers were assessed as a possible means of reducing protein fouling in microfiltration and ultrafiltration membranes (81–85). Polysulfone ultrafiltration and microfiltration membranes were modified by preadsorption of water-soluble polymers (85). The preadsorbed polymers cause the narrowing of pore sizes on the membrane surface. In addition, the desirable anti-fouling effect of the preadsorbed polymers was reported to depend at times on the porosity of the membrane, being less effective for the more porous membranes.

Development of membrane materials by irradiation process is also the new technique. Commercial ultrafiltration and microfiltration membranes were modified via plasma deposition of acrylic acid onto their surfaces, leading to enhanced solute rejection, while the flux decline pattern of bleach plant effluents was not improved (86). A photochemical grafting procedure was also used to increase the hydrophilicity of polysulfone. The surface of polyethersulfone hollow-fibre membranes was modified via grafting of polyethylene glycol (PEG) polymer using an irradiation method (87). Low temperature plasma-induced surface modification has been used in grafting functional groups onto PAN and polysulfone, enhancing their hydrophilicity (88). Surface modification of preformed polysulfone membranes was also attempted through heterogeneous reactions (89–91). Control of the reaction conditions was very difficult and the microporous structure could be damaged.

Less work focused on modification of hydrophobic membrane. It is much less known and scarcely investigated as to how a more hydrophobic membrane surface will perform in treating aqueous solutions such as oil–water emulsions that tend to be foulants. An attractive and simple method of membrane modification in this area is to add or blend a surface modifying macromolecular additive into the polymer casting solution (92). The modified membrane could have more favorable characteristics in treating oil–water emulsions than the unmodified membrane. The surface active additives in polymer materials can migrate to the surface and change the surface properties of the materials. A macromolecular additive is surface active if it migrates to the surface of the phase in which it is dispersed. Its structure consists of low and high polar components. The driving force for spontaneous surface migration is the tendency to minimize interfacial energy, which is universal to all condensed phase of matter. However, the kinetics of this phenomenon is slower in a polymeric system than in low viscosity liquid systems. In the work of Hamza et al. (92), the employed macromolecules is copolymeric in nature and is used to modified ultrafiltration membranes in treating oily aqueous solutions. Polyethersulfone ultrafiltration membranes are modified utilizing surface modifying macromolecules. Casting solutions are prepared containing various amounts of surface modifying macromolecules as additives. Ultrafiltration membranes are prepared using the phase inversion technique. It is reported that the modified polyethersulfone membranes in this work have a superior performance, reflected in their higher flux when treating oil–water emulsions than the control unmodified membranes. The oil gel layer resistance of the membranes decreased with an increase in the surface modifying macromolecules.

4.2. *Improving of Hydrophilicity of Membrane for Oil Water Separation*

Interaction of membranes with water can be divided into two types, attractive or repulsive response to water. A hydrophilic membrane is the type of membrane that exhibits an affinity for water. It has a high surface tension value and can form hydrogen bonds with water. This interaction of water is possible via the material composition of the membrane with the suitable characteristics such as surface chemistry to interact with water.

In hydrophilic membrane, the surface chemistry of membrane is required to bind to water better than other materials. The particles such as colloids, proteins, clays, and oily particles have hydrophobic interaction with water. For the purpose of oil water separation, hydrophilic membrane separation will be possible only when the oil content is less than 50% in which the water will be the suspension fluid and the oil will be the globule. Oil and other particles in the waste stream tend to agglomerate and cause fouling in the membrane process. To prevent fouling, the membrane requires a surface chemistry that prefers binding to water better than other materials. The modification of the membrane to have high hydrophilic properties to prevent the fouling is brought to the ultrafiltration membrane as this type of membrane is widely used in separation of oil and water for industrial influent. The superb application for minimizing oily wastewater streams of ultrafiltration has been limited due to the membrane-fouling problem.

The commercial membrane with high hydrophilic properties has been done by Filtration Solution, Inc. The new types of membrane, superphilic membrane, have been developed. They are made of a chemically modified PAN polymer. To avoid fouling by oils, the membrane was engineered to be extremely hydrophilic as compared with conventional PVDF and polysulfone (PS) membranes which are oleophilic (i.e. oil attracting). The improvement in membrane hydrophilicity is quantified by measuring water-membrane-air contact angles for the various membrane types. A smaller contact angle correlates with a more hydrophilic surface and less fouling by free oils and concentrated oily emulsions. A higher contact angle represents a high hydrophobic surface with the repulsive response for water. The fouling problem with free oil can occur with this type of membrane and cause the difficulty in membrane cleaning. Hydrophilicity property is mainly dependent on type of membrane materials. The value of hydrophilicity can be measured by placing a drop of water on the surface of a material and then measuring the contact angle. The contact angle of ceramic membrane is more than 30° and for other polymeric membrane materials is in the range of 44–112°. In details, the contact angles of polytetrafluoroethylene (PTFE), polypropylne, PVDF, and hydrophilic polysulfone are 112, 108, 66, and 44°, respectively.

Modification of ultrafiltration membranes to have high hydrophilic properties to reduce the fouling problem has been done in many research works. A hydrophilic polymer additive is often blended with the membrane-forming polymer to obtain hydrophilic membranes. The polymer additive may be polyvinylpyrrolidone (PVP) (93, 94), PEG (95), and so on. Ochoa and Masuelli (96) made membranes with different degrees of hydrophilicity of polyvinylidene fluoride (PVDF) and polymethyl methacrylate (PMMA). A higher hydrophilic character with an increase of PMMA in the casting solution and the appearance of larger macrovoids in the porous substructure were found without a substantial modification of the selective surface structure. Other hydrophilic polymers that can improve hydrophilicity of hydrophobic materials include polymethyacrylate (PMA), polyvinylacetate (PMAc), cellulose acetate (CA), etc.

Surface modification of a hydrophobic membrane material by graft polymerization was also widely utilized. Sulfonation of a hydrophobic polymer was also used to render the polymer hydrophilic. For example, Shen et al. (97) sulfonated polyetherimide to obtain a hydrophilic polymer for preparation of ultrafiltration membranes. In addition, a hydrophilic hollow-fiber UF membrane for oil–water separation was developed from a new dope containing cellulose/monohydrate *N*-methylmorpholine-N-oxide (NMMO·H_2O)/polyethylene glycol (PEG400) using a dry–wet spinning technique (98). It is reported that the cellulose hollow-fiber ultrafiltration membrane can resist fouling and is tolerant to a wide pH range 1–14 in treating oily wastewater.

4.3. Development of Inorganic Membrane

Recently, inorganic membranes have been developed with the purpose on the treatment of waste emulsion. However, there is still no report on industrial application using this type of organic membranes. The main reason may be the low oil rejection efficiency, the liability to membrane fouling and high costs.

The well known inorganic membrane is ceramic membrane. Ceramic membranes have been widely used owing to their numerous advantages, which are stability at high temperatures and pressure resistance, good chemical stability, high mechanical resistance, long life, and good defouling properties (99). Research on ceramic membranes was initially directed towards the preparation of alumina membranes. Nowadays, many porous membrane materials that have been applied to develop the ceramic membrane include alumina, zirconia, titania, and silica (99). Among them zirconia is a very interesting material for inorganic membranes. The outstanding qualities of zirconia membranes are their high chemical resistance that allows steam sterilization and cleaning procedures in the pH range 0–14 (100), good pure water permeability, high membrane flux in separation and filtration (101–103) due to their specific surface properties (104), and high thermal stability (101) which is very attractive for catalytic membrane reactors at high temperature.

Many commercial α-Al_2O_3 microfiltration membranes have been successfully prepared by particles sintering (105, 106). The main process is to first prepare a suspension of fine particles, then to deposit the particles on a porous support by a slip-casting method or dip-coating method, and heating the coated support to partly sinter the particles. Synthesis of alumina microfiltration membrane was also reported by Hyun and Kim (107). It was reported that the effective average pore size of alumina top layers *f* should be 0.16 nm. The reported flux and rejection efficiency is about 50 L/h m^2 (LMH) and 98%.

Several research works have been focused on the synthesis of zirconia microfiltration membrane and its application in oily wastewater. Yang et al. (99) had prepared a new kind of ceramic MF membrane, which uses zirconia as the top layer; it was found that this kind of membrane shows good performance in treating the waste oily emulsion. This kind of membrane seems to be a good candidate for the treatment of waste rolling emulsion. In synthesis of zirconia microfiltration membrane, the effective pore size of zirconia top layer should less than 0.07 mm, which provides the flux and rejection efficiency as 20 L/h m^2 and 98%, respectively. Wang et al. (108) used the zirconia microfiltration membrane in treating waste rolling emulsion. It was reported that the pretreatment of waste emulsion can improve both the flux and the oil rejection efficiency.

5. DESIGN EXAMPLES

Three terms usually used in designing of membrane unit include permeate flux, rejection, and volume recovery. The description of each term is defined as follows.

Permeate flux, L/h m^2: Permeate flux is the volume of permeate generated from a membrane surface area within a given time as shown in Eq. (1):

$$J = \frac{V}{F \times t} \tag{1}$$

where J = permeate flux, L/h m^2, V = volume of permeate, L, F surface area, m^2, t = time, h.

Rejection, %: Rejection (R) is a fraction of the total contaminant mass/concentration rejected by the membrane. It is also determined by the following Eq. (2):

$$R = 1 - (C_{\mathrm{p}}/C_{\mathrm{r}}) \tag{2}$$

where R = observed rejection coefficient, C_{p} = bulk solute concentration in the permeate, mg/L, C_{r} the retained bulk solute concentration, mg/L.

Volume recovery, %: Volume recovery (R) is the fraction of the total feed volume of the test water that can be recovered as the clean permeate:

$$R_{\mathrm{v}} = V_{\mathrm{P}}/V_{\mathrm{F}} \tag{3}$$

where R_{v} volume recovery, V_{F} = volume of feed, L, V_{p} volume of permeate, L.

The concentration polarization ratio ($C_{\mathrm{membrane}}/C_{\mathrm{bulk}}$) depends on the flux, rejection, and mass transfer coefficient and is estimated from the following equation:

$$\frac{C_{\mathrm{membrane}}}{C_{\mathrm{bulk}}} = \frac{\exp\left(\frac{J}{k}\right)}{R_{\mathrm{O}} + (1 - R_{\mathrm{O}})\exp\left(\frac{J}{k}\right)} \tag{4}$$

where R_{O} observed rejection which becomes equal to 1 when the solute is completely retained by the membrane, k = the mass transfer coefficient, J permeate flux, L/hm^2.

The value of mass transfer coefficient, k, is determined by using R_{O} and J data taken at different pressures, but constant feed concentration and feed rate to plot a straight line of best fit between $\ln((1 - R_{\mathrm{O}})J/R_{\mathrm{O}})$ versus J of slope $1/k$ and intercept $\ln(Dk/\delta)$ where D is diffusion coefficient and δ represents boundary layer thickness.

Some design information of the widely used ultrafiltration for oil and water separation are provided in this section. The permeate flux (J) is an important parameter in the design and economic feasibility analysis of the UF separation process. Hydrodynamics of membrane modules have an important effect on the mass transfer, separation, and fouling behavior of membrane systems.

Generally, the pure solvent transporting through porous ultrafiltration membranes is directly proportional to the applied transmembrane pressure (ΔP). The Kozeny–Carman and Hagen–Poiseuille equations describe the convection flow (J_0) as follows (109):

$$J_0 = \frac{\Delta P}{\eta R_{\mathrm{m}}} \tag{5}$$

where J_0 = convection flow, L/h m^2, ΔP applied transmembrane pressure, kPa, η = solvent viscosity, mPa/s, R_m intrinsic resistance of clean membrane (m^{-1}).

A permeate flux declines in the presence of solute due to membrane fouling (110–115). A decrease in flux is a result of several phenomenons including adsorption of macromolecules to membrane surface involving pore blocking, concentration polarization, and formation of a gel-like cake layer within membrane pores (110). Several models have been used to describe solute fouling, among them are hydraulic resistance, osmotic pressure, gel polarization, and film models (111, 114).

The permeate flux (J) to the applied pressure is related to the fouling resistance as described by Darcy's law (30):

$$J = \frac{\Delta P}{\eta(R'_{\mathrm{m}} + R_{\mathrm{P}})} \tag{6}$$

where J permeate flux, L/h m^2, ΔP applied transmembrane pressure, kPa, η solvent viscosity, mPa/s, R'_m intrinsic membrane resistance (m^{-1}), R_P polarization layer resistance (m^{-1}).

In this equation, R'_m ($=R_m + R_f$) is the intrinsic membrane resistance that includes the fouling layer resistance (R_f) due to specific membrane-solute interactions. The polarization layer resistance, R_P, consists of two resistances: R_g due to gel-polarized layer and R_b due to associate boundary layer. The intrinsic membrane resistance is unaffected by operating parameters whereas the polarization layer resistance is a function of applied pressure. When R_p is negligible, the filtrate flux is given by (109):

$$J = \frac{\Delta P}{\eta R'_{\mathrm{m}}} = \frac{\Delta P}{\eta(R_{\mathrm{m}} + R_{\mathrm{f}})} \tag{7}$$

where J permeate flux, L/h m^2, ΔP applied transmembrane pressure, kPa, η solvent viscosity, mPa/s, R_f fouling layer resistance (m^{-1}), R_m intrinsic resistance of the clean membrane (m^{-1}), R'_m intrinsic membrane resistance (m^{-1}).

To determine the relative degree of purification in a given ultrafiltration process or to estimate the period of ultrafiltration processing required to achieve a certain degree of separation or purification, the ultrafiltration process must be mathematically modeled (46).

During ultrafiltration membrane operation, there is a volume rejection. Thus, ultrafiltration data can be presented in terms of volume concentration ratio (VCR) or concentration factor (CF) as shown by the following equation:

$$\mathrm{VCR} = \mathrm{CF} = \frac{V_0}{V_{\mathrm{r}}} = 1 + \frac{V_{\mathrm{P}}}{V_{\mathrm{r}}} \tag{8}$$

where VCR volume-to-concentration ratio, CF concentration factor, V_0 initial feed tank volume (m^3), V_p volume of permeate (m^3), V_r volume of retained (m^3).

The material balance at any time during ultrafiltration operation is given by

$$\log(\mathrm{SCR}) = \log\left(\frac{C_{\mathrm{r}}}{C_0}\right) = R\log(\mathrm{CF}) \tag{9}$$

where SCR solute concentration ratio, R observed rejection coefficient, CF concentration factor, C_0 initial solute concentration in the feed (mg/L), C_r the retained bulk solute concentration (mg/L).

This equation allows the calculation of rejection using only retentive data.

5.1. *Example 1*

Membrane technology is used to remove oil particles from industrial wastewater. The initial feed tank volume is 8,640 m^3/day and after treatment, the retained volume is required to be 50% of the initial volume entering to the basin. If the observed rejected coefficient is 0.02, find solute concentration ratio.

Solution:

1. Find concentration factor

$$\text{Concentration factor} = \text{CF} = \frac{V_0}{V_r}$$

$$V_0 \; 8,640\,\text{m}^3/\text{day}; V_r \; 50\% \; V_0 \; 0.5 \; V_0$$

$$\text{CF} = \frac{V_0}{0.5 \; V_0} = 2$$

Concentration factor 2

2. Find solute concentration ratio
   Solute concentration ratio, SCR, could be find from

$$\log(\text{SCR}) = R \log(\text{CF})$$

$$\log(\text{SCR}) = 0.02 \log(2) = 0.06$$

SCR 1.15

   Solute concentration ratio is 1.15.

5.2. *Example 2*

The permeate flux of the membrane is 20 L/h m^2. Calculate the required surface area for the membrane if the volume of permeate are 10 L, 20 L, and 50 L, consequently. Let the membrane operating time is 3 h.

Solution: Find the surface area of membrane using the following equation:

$$J = \frac{V}{F \times t} \tag{1}$$

1. If volume of permeate is 10 L, the surface area can be calculated as follow:

$$F = \frac{V}{J \times t}$$

$$F = \frac{10\ \text{L}}{20\ \text{L/h} \cdot \text{m}^2 \times 3\ \text{h}}$$

$$F = 1.6 \times 10^{-1}\ \text{m}^2$$

Surface area of membrane is $1.6 \times 10^{-1}\ \text{m}^2$.

2. If volume of permeate is 20 L, the surface area can be calculated as follow:

$$F = \frac{20\ \text{L}}{20\ \text{L/h} \cdot \text{m}^2 \times 3\ \text{h}}$$

$$F\ 3.3 \times 10^{-1}\ \text{m}^2$$

Surface area of membrane is $3.3 \times 10^{-1}\ \text{m}^{-2}$.

3. If volume of permeate is 50 L, the surface area can be calculated as follow:

$$F = \frac{50\ \text{L}}{20\ \text{L/h} \cdot \text{m}^2 \times 3\ \text{h}}$$

$$F\ 8.3 \times 10^{-1}\ \text{m}^2$$

Surface area of membrane is $8.3 \times 10^{-1}\ \text{m}^2$.

It is noted that when the surface area is direct proportional with volume of filtrate when the permeate flux is constant. The increasing of filtrate volume required the higher surface area of membrane.

6. NOMENCLATURE

C_O = Initial solute concentration in the feed, mg/L
C_r = Retained solute concentration, mg/L
C_p = Bulk solute concentration in the permeate, mg/L
C_r = The retained bulk solute concentration, mg/L
CF = Concentration factor
F = Surface area, m^2
η = Solvent viscosity, mPa/s
J = Permeate flux, L/h .m^2
J_O = Initial permeate flux, L/h m^2
k = The mass transfer coefficient (k)
ΔP = Applied transmembrane pressure, kPa
R = Observed rejection coefficient
R_O = Observed rejection which becomes equal to 1 when the solute is completely retained by the membrane
R_f = Fouling layer resistance, L/m
R_m = Intrinsic resistance of clean membrane, L/m
R'_m = Intrinsic membrane resistance, L/m
R_P = Polarization layer resistance, L/m
R_v = Volume recovery

SCR = Solute concentration ratio
t = Time, h
V = Volume of permeate, L
V_0 = Initial feed tank volume, m^3
V_F = Volume of feed, L
V_r = Volume of retained, m^3
V_p = Volume of permeate, L
VCR = Volume-to-concentration ratio

REFERENCES

1. Patterson JW (1985) Industrial wastewater treatment technology, 2nd edn. Butterworth Publishers, Stoneham, MA
2. Guerin TF (2002) Heavy equipment maintenance wastes and environmental management in the mining industry. J Environ Manage 66:185–199
3. Gilbert T, Telleza GT, Nirmalakhandanb N, Gardea-Torresdey JL (2002) Performance evaluation of an activated sludge system for removing petroleum hydrocarbons from oilfield produced water. Adv Environ Res 6:455–470
4. Fillo JP, Koraido SM, Evans JM (1992) Sources, characteristics, and management of produced water from natural gas production and storage operations. Plenum Press, New York
5. Chapelle FH (1993) Ground-water microbiology and geochemistry. John Wiley and Sons, Inc, Somerset, NJ
6. Stephenson MT (1992) Components of produced water: a compilation of industry studies. J Petrol Technol 44:548–603
7. Cheryan M, Rajagopalan N (1998) Membrane processing of oily streams. Wastewater treatment and waste reduction. J Membr Sci 151:13–28
8. Rhee CH, Martyn PC, Kremer JG (1987) Removal of oil and grease in the hydrocarbon processing industry. In: Proceedings of the 42nd purdue industrial waste conference, West Lafayette, IN, Lewis Publishers, Chelsea, MI, p 143
9. Tabakin RB, Trattner R, Cheremisinoff PN (1978) Oil/water separation: The options available. Part 1 and Part 2. Water Sew Works 74–77
10. Lewis A, Singsaas I, Johannesen BO, Nordvik AB (1995) Key factors that control the efficiency of oil spill mechanical recovery method. MSRC Technical Report Series 95–038, Marine Spill Response Corporation, Washington, DC, p 54
11. Bobra M (1992) A study of water-in-oil emulsification. Report EE-132, Environment Canada, Environment Protection Directorate, Ottawa, Canada, p 63
12. Lee R (1995) Isolation and identification of compounds and mixtures which promote and stabilize water-in-oil emulsions. MSRC Technical Report Series 95–002, Marine Spill Response Corporation, Washington, DC, p 62
13. Nordvik AB, Simmons JL, Bitting KR, Lewis A, Storm-Kristiansen T (1996) Oil and water separation in marine oil spill clean-up operations. Spill Sci Technol Bull 3:107–122
14. Burguera JL, Burguera M (2004) Analytical applications of organized assemblies for on-line spectrometric determinations: present and future. Talanta 64:1099–1108
15. Roberts AE (2000) Water quality control handbook. McGraw-Hill, New York, USA

16. Benito JM, Rios G, Ortea E, Fernhdez E, Cambiella A, Pazos C, Coca J (2002) Design and construction of a modular pilot plant for the treatment of oil-containing wastewaters. Desalination 147:5–10
17. Abdessemed D, Nezzal G, Aim RB (2000) Coagulation-adsorption-ultrafiltration for wastewater treatment and reuse. Desalination 131:307–314
18. Chakrabarty B, Ghoshal AK, Purkait MK (2008) Ultrafiltration of stable oil-in-water emulsion by polysulfone membrane. J Memb Sci 325:427–437
19. Matos M, Lobo A, Fernández E, Benito JM, Pazos C, Coca J (2008) Recycling of oily ultrafiltration permeates to reformulate O/W emulsions. Colloids Surf A 331:8–15
20. Ohya H, Kim JJ, Chinen A, Aihara M, Semenova SI, Negishi Y, Mori O, Yasuda M (1998) Effects of pore size on separation mechanisms of microfiltration of oily water using porous glass tubular membrane. J Memb Sci 145:1–14
21. Al-Obeidani SKS, Al-Hinai H, Goosen MFA, Sablani S, Taniguchi Y, Okamura H (2008) Chemical cleaning of oil contaminated polyethylene hollow fiber microfiltration membranes. J Memb Sci 307:299–308
22. Gryta M, Karakulski K (1999) The application of membrane distillation for the concentration of oil–water emulsions. Desalination 121:23–29
23. Deng S, Sourirajan S, Chan K, Farnand B, Okada T, Matsuura T (1991) Dehydration of oil–water emulsion by pervaporation using porous hydrophilic membranes. J Colloid Interface Sci 141:218–225
24. Mohammadi T, Kazemimoghadam M, Saadabadi M (2003) Modeling of membrane fouling and flux decline in reverse osmosis during separation of oil in water emulsions. Desalination 157:369–375
25. Ata N, Yazicigil Z, Oztekin Y (2008) The electrochemical investigation of salts partition with ion exchange membranes. J Hazard Mater 160:154–160
26. Zhou Y-B, Tang X-Y, Hu X-M, Fritschi S, Lu J (2008) Emulsified oily wastewater treatment using a hybrid-modified resin and activated carbon system. Sep Purif Technol 63:400–406
27. Hong AC, Fane AG, Burford RP (2002) The effects of intermittent permeate flow and crossflow on membrane coalescence of oil-in-water emulsions. Desalination 144:185–191
28. Li HY, Bertram CD, Wiley DE (1998) Mechanisms by which pulsatile flow affects cross-flow microfiltration. AIChE J 44:1950–1961
29. Gupta BB, Blanpain P, Jaffrin MY (1992) Permeate flux enhancement by pressure and flow pulsations in microfiltration with mineral membranes. J Memb Sci 70:257–266
30. Belfort G, Davis RH, Zydney AL (1994) The behavior of suspensions and macromolecular solutions in crossflow microfiltration – review. J Memb Sci 96:1–58
31. Winzeler HB, Belfort G (1993) Enhanced performance for pressure-driven membrane processes: the argument for fluid instabilities. J Memb Sci 80:35–47
32. Jonsson AS (1993) Influence of shear rate on the flux during ultrafiltration of colloidal substances. J Memb Sci 79:93–99
33. Rodgers VGJ, Sparks RE (1993) Effects of solution properties on polarization redevelopment and flux in pressure pulsed ultrafiltration. J Memb Sci 78:163–180
34. Spiazzi E, Lenoir J, Grangeon A (1993) A new generator of unsteady-state flow regime in tubular membranes as an anti-fouling technique: a hydrodynamic approach. J Memb Sci 80:49–57
35. Bertram CD, Hoogland MR, Li H, Odell RA, Fane AG (1993) Flux enhancement in crossflow microfiltration using a collapsible-tube pulsation generator. J Memb Sci 84:279–292

36. Lefferts AG (1997) Membranes as separators of dispersed emulsion phases, in chemical engineering. Netherlands University, Twentia
37. Ohya H, Kim JJ, Chinen A, Aihara M, Semenova SI, Negishi Y, Mori O, Yasuda M (1998) Effects of pore size on separation mechanisms of microfiltration of oily water, using porous glass tubular membrane. J Memb Sci 145:1–14
38. Goldsmith RL, Roberts DA, Burre DL (1974) Ultrafiltration of soluble oil wastes. J Water Pollut Control Fed 46:2183–2192
39. Bailey PA (1977) The treatment of waste emulsified oils by ultrafiltration. Filtr Sep 14:53–55
40. Goldsmith RL, Hossain S (1973) Oil water separation by ultrafiltration. In: Proceedings of the joint conference on prevention and control of oil spills, Washington, DC, pp 441–456
41. Bansal IK (1976) Concentration of oily and latex wastewaters using ultrafiltration inorganic membranes. Ind Water Eng 13:6–11
42. Priest W (1978) Treatment of waste oil emulsions by ultrafiltration. Water Waste Treat March: 42–43
43. Hlavacek M (1995) Break-up of oil-in-water emulsions induced by permeation through a microfiltration membrane. J Memb Sci 102:1–7
44. Lee S, Aurelle Y, Roques H (1984) Concentration polarization, membrane fouling and cleaning in ultrafiltration of soluble oil. J Memb Sci 19:23–38
45. Lipp P, Lee CH, Fane AG, Fell CJD (1988) A fundamental study of the ultrafiltration of oil–water emulsions. J Memb Sci 36:161–177
46. Cheryan M (1998) Ultrafiltration and microfiltration handbook. Technomic, Lancaster, PA
47. Gryta M, Karakulski K, Morawski AW (2001) Purification of oily wastewater by hybrid UF/MD. Water Res 35:3665–3669
48. Bodzek M, Konieczny K (1992) The use of ultrafiltration membranes made of various polymers in the treatment of oil emulsion wastewaters. Waste Manage 12:75–84
49. Karakulski K, Kozlowski A, Morawski AW (1995) Purification of oily wastewater by ultrafiltration. Sep Technol 5:197–205
50. Lindau J, Jonsson AS (1994) Cleaning of ultrafiltration membranes after treatment of oily waste water. J Memb Sci 87:71–78
51. Khan S, Ghosh AK, Ramachandhran V, Bellare J, Hanra MS, Trivedi MK, Misra BM (2000) Synthesis and characterization of low molecular weight cut off ultrafiltration membranes from cellulose propionate polymer. Desalination 128:57–66
52. Suzane P, Klaus VP (1992) Preparation and performance of asymmetric polyetherimide membranes for isopropanol dehydration by pervaporation. J Memb Sci 109:165–172
53. Hvid KB (1990) Preparation and characterization of a new ultrafiltration membrane. J Memb Sci 53:189–202
54. Gorzka Z, Kazmierczak M, Socha A, Jednoralska AM (1992) Utilization and treatment of exhausted cooling oil-emulsions. Waste Manage 12:345–348
55. Freeman SDN, Morin OJ (1995) Recent developments in membrane water reuse projects. Desalination 103:19–30
56. Koltuniewicz AB, Field RW, Arnot TC (1995) Cross-flow and dead-end microfiltration of oily-water emulsion. Part 1: experimental study and analysis of flux decline. J Memb Sci 102:193–207
57. Gallagher P (1992) Membrane removal processes for oil–water emulsion treatment. Water Waste Treat 42–43
58. Farnand BA, Sawatzky H, Poirier MA (1985) An evaluation of the use of porous membranes for the dewatering of wellhead bitumen/water/mineral emulsions. Sep Sci Technol 20:193–203

59. Nazzal FF, Wiesner MR (1996) Microfiltration of oil-in-water emulsion. Water Environ Res 68:1187–1191
60. Hong A, Fane AG, Burford R (2003) Factors affecting membrane coalescence of stable oil-in-water emulsions. J Memb Sci 222:19–39
61. Peng H, Volchek K, MacKinnon M, Wong WP, Brown CE (2004) Application of nanofiltration to water management options for oil sands operations. Desalination 170:137–150
62. Sridhar S, Kale A, Khan AA (2002) Reverse osmosis of edible vegetable oil industry effluent. J Memb Sci 205:83–90
63. Hodgkless T, Hanbury WT, Law GB, Al-Ghasham TY (2001) Effect of hydrocarbon contaminants on the performance of RO membranes. Desalination 138:283–289
64. Sarakulski KJ, Kozfowskib A, Morawskia AW (1995) Purification of oily wastewater by ultrafiltration. Sep Technol 5:197–205
65. Lin SH, Lan WJ (1998) Waste oil/water emulsion treatment by membrane processes. J Hazard Mater 59:189–199
66. Karakulski K, Morawski AW (2002) Treatment of spent emulsion from a cable factory by an integrated UF/NF membrane system. Desalination 149:163–167
67. Crawford G, Fernandez, A, Shawwa A, Daigger G (2002) Competitive bidding and evaluation of membrane bioreactor equipment – three large case studies. In: Proceedings of WEF 75th annual conference 2nd exposition, Chicago, IL
68. Yang W, Cicek N, Igl J (2005) State-of-the-art of membrane bioreactors: Worldwide research and commercial applications in North America. J Memb Sci 270(1–2):201–211
69. Cicek N (2003) A review of membrane bioreactors and their potential application in the treatment of agricultural wastewater. Can Biosyst Eng 45:6.37–6.49
70. Buisson H, Cote P, Praderie M, Paillard H (1998) The use of immersed membranes for upgrading wastewater treatment plants. Water Sci Technol 37:89–95
71. Cote P, Buisson H, Pound C, Arakaki G (1997) Immersed membrane activated sludge for the reuse of municipal wastewater. Desalination 113:189–196
72. Rosenberger S, Kruger U, Witzig R, Manz W, Szewzyk U, Kraume M (2002) Performance of a bioreactor with submerged membranes for aerobic treatment of municipal waste water. Water Res 36:413–420
73. Cicek N, Winnen H, Suidan MT, Wrenn BE, Urbain V, Manem J (1998) Effectiveness of the membrane bioreactor in the biodegradation of high molecular weight compounds. Water Res 32:1553–1563
74. Urbain V, Mobarry B, Silva V, Stahl DA, Rittmann BE, Manem J (1998) Integration of performance, molecular biology and modeling to describe the activated sludge process. Water Sci Technol 37:223–229
75. Visvanathan C, Aim RB, Parameshwaran K (2000) Membrane separation bioreactors for wastewater treatment. Crit Rev Env Sci Technol 30:1–48
76. Liu W, Howell JA, Arnot TC, Scott JA (2001) A novel extractive membrane bioreactor for treating biorefractory organic pollutants in the presence of high concentrations of inorganics: application to a synthetic acidic effluent containing high concentrations of chlorophenol and salt. J Memb Sci 181:127–140
77. Nunes SP, Peinemann KV (1992) Ultrafiltration membranes from PVDF/PMMA blends. J Memb Sci 73:25–35

78. Uragami T, Fujimoto M, Sugihara M (1980) Studies on syntheses and permeabilities of special polymer membranes. 28. Permeation characteristics and structure of interpolymer membranes from poly (vinylidene fluoride) and poly (styrene sulfonic acid). Desalination 34:311–323
79. Blicke C, Peinemann KV, Nunes SP (1993) Ultrafiltration membranes from poty(ether sulfonamide)/poly(ether imide) blends. J Memb Sci 79:83–91
80. Zeman LJ (1983) Adsorption effects in rejection of macromolecules by ultrafiltration membranes. J Memb Sci 15:214–230
81. Hanemaaijer JH, Robbertsen T, van den Boomgaard Th, Gunnik JW (1989) Fouling of ultrafiltration membranes. The role of protein adsorption and salt precipitation. J Memb Sci 40:199–217
82. Sivik B, Wahlgren M, Miezis Y (1990) A rheological screening method for membrane modifying polymers. Desalination 77:181–193
83. Kim KJ, Fane AG, Fell CJD (1988) The performance of ultrafiltration membranes pretreated by polymers. Desalination 70:229–249
84. Brink LES, Romijn DJ (1990) Reducing the protein fouling of polysulfone surfaces and polysulfone ultrafiltration membranes; Optimization of the type of presorbed layer. Desalination 78:209–233
85. Brink LES, Elbers SJG, Robbertson T, Both P (1993) The anti-fouling action of polymers preadsorbed on ultrafiltration and microfiltration membranes. J Memb Sci 76:281–291
86. Cho DL, Ekengren O (1993) Composite membrane formed by plasma-polymerized acrylic acid for ultrafiltration of bleach effluent. J Appl Polym Sci 47:2125–2133
87. Mok S, Worsfold DJ, Fouda A, Matsuura T (1994) Surface modification of polyethersulfone hollow-fibre membranes by g-ray irradiation. J Appl Polym Sci 51:193–199
88. Ulbricht M, Belfort G (1996) Surface modification of ultrafiltration membranes by low temperature plasma, II. Graft polymerization onto polyacrylonitfile and polysulfone. J Memb Sci 111:193–215
89. Higuchi A, Iwata N, Nakagawa T (1988) Surface-modified polysulfone hollow fibers. J Appl Polym Sci 36:1753–1767
90. Breithach L, Henke E, Staude E (1991) Heterogeneous functionalizing of polysulfone membranes. Angew Makromol Chem 184:183–196
91. Guiver MD, Black E, Tam CM, Deslandes Y (1993) Functionalized polysulfone membranes by heterogeneous lithiation. J Appl Polym Sci 48:1597–1606
92. Hamza A, Pham VA, Matsuura T, Santerre JR (1997) Development of membranes with low surface energy to reduce the fouling in ultrafiltration applications. J Memb Sci 131:217–227
93. Xu ZL, Chung TS, Huang Y (1999) Effect of polyvinyl pyrrolidone molecular weights on morphology, oil/water separation, mechanical and thermal properties of polyetherimide/polyvinyl pyrrolidone hollow fiber membranes. J Appl Polym Sci 74:2220–2233
94. Marchese J, Ponce M, Ochoa NA, Prádanos P, Palacio L, Hernández A (2003) Fouling behaviour of polyethersulfone UF membranes made with different PVP. J Membr Sci 211:1–11
95. Xu ZL, Sichuan T, Loha KC, Lima BC (1999) Polymeric asymmetric membranes made from polyetherimide/polybenzimidazole/poly(ethylene glycol) (PEI/PBI/ PEG) for oil–surfactant–water separation. J Memb Sci 158:41–53
96. Ochoa NA, Masuelli M, Marchese J (2003) Effect of hydrophilicity on fouling of an emulsified oil wastewater with PVDF/PMMA membranes. J Memb Sci 226:203–211
97. Shen LQ, Xu ZK, Liu ZM, Xu YY (2003) Ultrafiltration hollow fiber membranes of sulfonated polyetherimide/polyetherimide blends: preparation, morphologies and anti-fouling properties. J Memb Sci 218:279–293

98. Li HJ, Cao YM, Qin JJ, Jie XM, Wang TH, Liu JH, Yuan Q (2006) Development and characterization of anti-fouling cellulose hollow fiber UF membranes for oil–water separation. J Memb Sci 279:328–335
99. Yang C, Zhang G, Xu N, Shi J (1998) Preparation and application in oil–water separation of Zro_2/Al_2O_3 MF membrane. J Memb Sci 142:235–243
100. Bhave RR (1991) Inorganic membranes synthesis, characteristics and applications. Van Nostrand Reinhold, New York
101. Larbot A, Fabre JP, Guizard C, Cot L, Gillot J (1989) New inorganic ultrafiltration membranes: titania and zirconia membranes. J Am Ceram Soc 72:257
102. Higgins RJ, Bishop BA, Goldsmith RL (1994) Reclamation of waste lubricating oil using ceramic membranes. In: Proceeding of the third international conference on inorganic membranes, Worcester, MA, pp 447–463
103. Guizard C, Rambault D, Urhing D, Dufour J, Cot L (1994) Deasphalting of a long residue using ultrafiltration inorganic membranes. In: Proceeding of the Third International Conference on Inorganic Membranes, Massachusetts, Worcester, pp 345–354
104. Randon J, de Lira HL, Paterson R (1994) Improved separations using surface modification of ceramic membranes. In: Proceeding of the third international conference on inorganic membranes, Massachusetts, Worcester, pp 429–435
105. Atkinson A, Segal DL (1993) Process for making ceramic membrane filters. Patent Appl, UK 2,273,672
106. van't Veen WH, Engel AJG, Bonekamp BC, Veringa HJ, Terpstra RA (1988) Composite ceramic micropermeable membrane, process and apparatus for producing such membrane. Eur Patent 0,320,033
107. Hyun SH, Kim GT (1997) Synthesis of ceramic microfiltration membranes for oil/water separation. Sep Sci Technol 32:2927–2943
108. Wang P, Xu N, Shi J (2000) A pilot study of the treatment of waste rolling emulsion using zirconia microfiltration membranes. J Memb Sci 173:159–166
109. Marchese J, Ochoa NA, Pagliero C, Almandoz C (2000) Pilot-scale ultrafiltration of an emulsified oil wastewater. Environ Sci Technol 34:2990–2996
110. Ko MK, Pellegrino JJ (1992) Determination of osmotic pressure and fouling resistance and their effects of performance of ultrafiltration membranes. J Memb Sci 74:141–157
111. Pradanos P, Hernandez A, Calvo JI, Tejerina F (1996) Mechanisms of protein fouling in cross-flow UF through an asymmetric inorganic membrane. J Memb Sci 114:115–126
112. Wu CJ, Li A, Li L, Zhang L, Wang H, Qi XH, Zhang QX (2008) Treatment of oily water by a poly (vinyl alcohol) ultrafiltration membrane. Desalination 225:312–321
113. Subhi AJ, Anne N (2008) An experimental evaluation of reverse osmosis membrane performance in oily water. Desalination 228:287–294
114. Wang LK, Wang MHS, Suozzo T, Dixon RA, Wright TL, Sarraino S (2009) Chemical and biochemical technologies for environmental infrastructure sustainability. 2009 National Engineers Week Conference. Albany Marriott, Albany, NY, Feb. 5–6
115. Koch Membrane Systems, Inc. Tubular membranes handle extremely difficult oily wastewater at Metaldyne. Pollution Equipment http://www.pollutionequipment-digital.com/pollutionequipment/200908/?pg=25#pg25 (2010)

16

Gas-Sparged Ultrafiltration: Recent Trends, Applications and Future Challenges

Kaustubha Mohanty and Raja Ghosh

CONTENTS

Abstract Ultrafiltration is a pressure-driven membrane technique whose applications are wide ranging: protein fractionation to wastewater treatment. The performance of ultrafiltration is limited by concentration polarization and subsequent fouling. Gas sparging i.e. introduction of gas bubbles along with the feed has been shown to be effective in reducing concentration polarization and thus controlling fouling. This chapter reviews the recent developments in gas-sparged ultrafiltration. The review focuses on the basics of ultrafiltration and two-phase flow hydrodynamics, the use of gas bubbles for enhancing permeate flux and its applications in bioseparation and wastewater treatment. Some practical issues and future challenges are also discussed.

Key Words Ultrafiltration • gas sparging • concentration polarization • fouling • flux enhancement • protein fractionation • membrane bioreactor.

From: *Handbook of Environmental Engineering, Volume 13: Membrane and Desalination Technologies*
Edited by: L.K. Wang et al., DOI: 10.1007/978-1-59745-278-6_16

1. INTRODUCTION

Membrane technology due to its advantages over traditional separation processes is increasingly drawing the attention of researchers in its various fields of application. Recent novel developments in membrane technology have made it possible to achieve many of the separation objectives economically and efficiently, in a flexible manner. Examples of successful innovations that have led to outstanding results in large-scale commercial processes can be found in the areas of water treatment, wastewater treatment, membrane bioreactor. (MBRs) and bioseparation engineering (1). Brackish and seawater desalination using reverse osmosis membranes and the treatment of surface water using micro- and ultrafiltration membranes are important applications of membrane technology (2). In the area of wastewater treatment, ultra- and microfiltration membranes are increasingly being used, particularly as components of MBRs (3–5). Membrane processes are being extensively used in the area of bioseparation engineering (6). Ultrafiltration is increasingly being used for the purification of recombinant protein products and blood-plasma derived products. Compared to other bioseparation methods such as chromatography, ultrafiltration offers several advantages namely higher productivity, lower capital investment, ease of translation to large-scale commercial production and easy equipment cleaning (7).

Ultrafiltration is adversely affected by concentration polarization and subsequent fouling (1–2). Various system hydrodynamics improvement approaches have been investigated for controlling concentration polarization. These include vortex mixing (8, 9), use of baffles and tube inserts (10), pulsatile flow of feed (11), and gas sparging (1, 12, 13). Gas sparging has been shown to be the most cost-effective approach among these. When gas bubbles are introduced as the second phase into the membrane module along with the liquid feed, they are reasonably benign towards the membrane and are easily separated from the retentate stream. Gas bubbles reduce concentration polarization by increasing the solute mass transfer coefficient. This is mainly due to bubble-induced secondary flow, which promotes local mixing near the membrane surface. Bubbles also physically remove accumulated particles and macrosolutes from the membrane surface by a scouring action. In some hybrid membrane devices such as MBRs, gas bubbles are required for an additional purpose i.e. oxygen supply.

Gas sparging can be used in both modes of membrane filtration i.e. positive pressurization driven permeation and suction driven permeation. The first mode is mainly used in ultrafiltration processes where higher transmembrane pressures (TMPs) are required. The second mode where the TMP is restricted to one bar is more commonly used for microfiltration. Due to the various advantages resulting from , there has been a recent upsurge of interest in its use in membrane filtration. This chapter reviews the state-of-the-art of this promising technique. We start with a short discussion on the basics of ultrafiltration. The fundamental principles and concepts of two-phase flow are then briefly discussed. This is followed by a detailed discussion on the use of gas bubbles in different membrane modules viz. tubular, hollow fibre and flat-sheet, and the permeate mechanisms involved in each case. Specific applications of gas sparged ultrafiltration in areas ranging from bioseparations to waste water treatment are dealt with next. The chapter concludes with discussions on practical issues involved and future challenges.

2. ULTRAFILTRATION BASICS

Ultrafiltration (UF) fits between nanofiltration and microfiltration in the filtration spectrum and involves separation of species ranging from about 1 to 100 nm in size, or about 500 to 500,000 kg/kg mole in molecular weight. Ultrafiltration is basically used for the concentration, diafiltration, clarification and fractionation of macromolecules. Separation primarily depends on species size, but several other factors are also known to affect separation (6, 14). Ultrafiltration is usually carried out in the cross-flow mode where the feed flows parallel to the membrane surface, thereby being split into two product streams. The stream that goes through the membrane is called permeate or the filtrate, while that remaining on the feed side is called the retentate. The volume, type and amount of species present in the permeate depend on the characteristics of the membrane, the operating conditions, and the quality of feed. The composition of the retentate depends on what goes out along with the permeate.

Ultrafiltration is used to separate large molecules like proteins, DNA and starch, colloidal dispersions such as clays, paints, pigments and latex particles and fine suspensions such as bacteria and virus. The osmotic backpressure in ultrafiltration is typically low. Hence the applied pressure is mostly utilized for generating permeate unlike in reverse osmosis where a high percentage of the applied pressure is used for overcoming osmotic pressure. The permeate flux of a membrane is defined as the amount of permeate produced per unit area of membrane surface per unit time. The permeate flux is directly related to the productivity of an ultrafiltration process. Different parameters that affect the permeate flux are: the properties of the membrane, the properties of the feed, the TMP and the system hydrodynamics.

2.1. *Applications of Ultrafiltration*

Ultrafiltration was initially developed for concentrating the macromolecules such as proteins. Its current range of applications includes water and wastewater treatment, food processing, chemical processing and bioprocessing. Ultrafiltration is now playing a major role in making the downstream processing of proteins and nucleic acids cost effective. Current downstream processing cost for such products using conventional separation technology is in the range of 60–80% to total production cost. Hence there is significant room for cost cutting.

In the water industry there are two major applications of ultrafiltration: production of process water for the manufacturing industries and production of potable water for human consumption. In the pharmaceutical and biotechnology industries there is a need for ultrapure water for a range of applications: preparation of tissue culture and fermentation media, buffer solutions, analytical solvents, drug and intravenous solutions etc. Ultrafiltration is extensively being used for producing ultrapure water. Ultrafiltration is able to remove particulate matter from water without any use of additives.

The manufacturing and service sectors produce large quantities of wastewater. Municipal and industrial wastewaters are increasingly being treated in MBRs which use ultrafiltration membranes. The MBR based effluent treatment plant is smaller and the effluent is free from pathogens (5). Current applications of MBR include water recycling in buildings (15–16), municipal wastewater treatment for small communities (17–20), industrial wastewater treatment (21–23) and landfill leachate treatment (24).

2.2. *Advantages and Limitations of Ultrafiltration*

The major advantages of ultrafiltration over other competing separation techniques are (6, 14):

1. Low equipment and operating costs
2. High throughput of product
3. Relative ease of scale-up
4. Ease of equipment cleaning and sanitization
5. No chemical additives required
6. No changes in pH and ionic strength, an advantage in
7. Operation at ambient temperature

Some of the limitations of ultrafiltration are (14):

1. Osmotic pressure resulting from concentration polarization (though not as severe as in reverse osmosis)
2. Susceptibility to membrane fouling
3. Low upper feed solids content limit
4. Limitations with processing viscous feed

Concentration polarization is a physical phenomenon related to solute mass transfer and refers to an exponentially increasing solute concentration profile near the membrane surface. Concentration polarization is usually restricted within the hydrodynamic boundary layer on the feed side. In some cases the osmotic back pressure resulting from such elevated solute concentration at the membrane surface can diminish the applied TMP. However, the magnitude of osmotic pressure in most ultrafiltration processes tends to be low. Concentration polarization mainly causes its detrimental effect by leading to severe fouling. Membrane fouling while being linked to concentration polarization is governed by the adsorption of macromolecular solutes on the membranes mainly by physicochemical interactions.

2.3. *Fouling Control: The Need for Gas Sparging*

Fouling which refers to the adsorption and deposition of feed components on and within the membrane is the main reason for decrease in permeate flux and hence productivity in an ultrafiltration process. In some cases, fouling can adversely affect product quality. For a given membrane-feed system, fouling is influenced by concentration polarization, i.e. the more severe the concentration polarization, the more severe is the fouling. Hence controlling concentration polarization helps in reducing fouling. Concentration polarization and subsequent fouling can be reduced by improving the hydrodynamic conditions near the membrane surface. Various techniques, such as use of pulsatile flow, generation of secondary flow (e.g. Dean and Taylor vortices), feed flow obstruction using helical inserts and baffles and the use of corrugated membrane surfaces have been shown to be effective at promoting local mixing near the membrane (8–11). However, these approaches are not very cost effective since these involve significant modifications to the membrane module and some cases are very energy intensive. Concentration polarization can also be controlled by solid–liquid two-phase flow (25–27). However, the added solids can damage the membrane and more importantly these need to be removed from the retentate stream. Gas sparging, i.e., injecting gas (e.g. air, nitrogen) in the form of bubbles into the feed stream has proved to be an effective, simple and cost-effective technique for controlling concentration polarization (1, 12–13). Gas sparging works by formation of bubble induced secondary flow, physical displacement of the mass transfer boundary layer, scouring of the membrane surface and pressure pulsing caused by gas slugs (28). The contributions of these mechanisms in controlling concentration polarization depend of the type of membrane module and their mode of operation. The basic and applied aspects of gas-sparged ultrafiltration particularly the use of gas bubbles for permeate flux enhancement are discussed in detail in the subsequent sections.

3. FUNDAMENTALS OF GAS–LIQUID TWO-PHASE FLOW

In order to use gas sparging in an effective manner it is necessary to understand gas–liquid two-phase flow behaviour. This section provides a brief overview of gas–liquid two-phase flow.

3.1. Bubbles

When gas is injected into a stationary liquid, gas bubbles are formed and these move upward due to buoyancy. The size of the bubbles depends on the way the gas is introduced and on the gas flow rate. The bubble motion generates a secondary flow behind the bubble, i.e., the wake region. The strength and extent of wake region depends on the shape and size of the bubble. Based on their shape, bubbles in stationary liquids are classified into spherical, ellipsoidal and spherical-cap bubbles. The shape of the bubbles can be anticipated empirically, as shown in Fig. 16.1 (29). As the bubble size increases, the shape of the bubble changes from spherical to ellipsoidal and eventually to spherical-cap.

Spherical bubbles are small and their size is usually less than 1 mm. Since there is no boundary layer separation around such bubbles, there is no wake. Ellipsoidal bubbles are typically of 1.5–15 mm size. There is boundary layer separation around such bubbles and hence these are followed by helical vortex wakes. Spherical-cap bubbles are usually big bubbles (>15 mm) and their primary wake is about 4.5 times the bubble volume (30). These big bubbles can generate strong secondary flow effect.

The terminal rise velocity of spherical bubbles can be approximated by Stokes law ($Re < 1$),

$$U_b = \frac{d_e^2 g(\rho_1 - \rho_g)}{18\mu_1}. \tag{1}$$

Ellipsoidal bubbles do not obey Stokes law and their rise velocity in water is approximately 0.24 m/s (31). This rise velocity decreases as viscosity of the liquid increases. The rise velocity of spherical-cap bubbles can be approximated by,

$$U_b = 0.71(gd_e)^{0.5}. \tag{2}$$

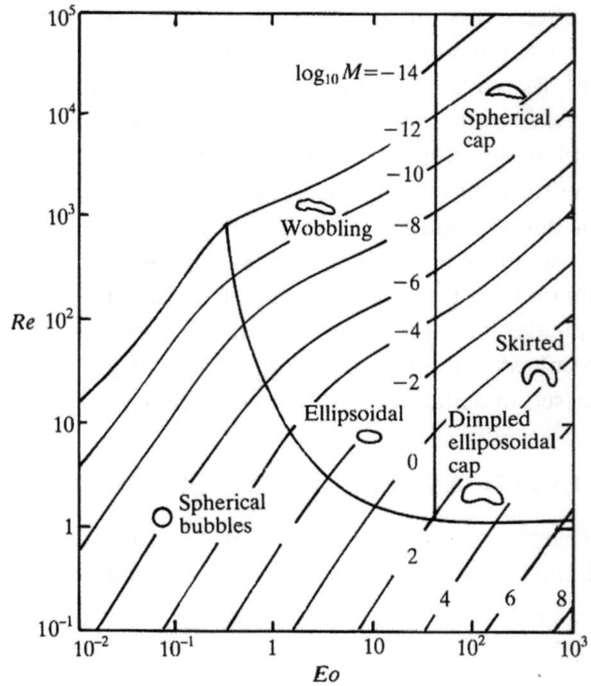

Fig. 16.1. Bubble characteristics chart (after Clift et al. (29)).

3.2. *Two-Phase Flow Patterns*

In gas–liquid flow the two phases can adopt various geometric configurations, these being known as flow patterns or flow regimes (32). The common flow patterns for a vertical upward flow are shown in Fig. 16.2. The various regimes are:

Bubbly flow or bubble flow. Here the gas phase is distributed in discrete bubbles within a liquid continuum.

Slug (or plug) flow. In this type of flow some of the gas bubbles have nearly the same cross-section as that of the channel and move along in the characteristic bullet-shaped Taylor bubbles. The Taylor bubbles are separated by liquid slugs which may or may not contain smaller spherical gas bubbles.

Churn flow. If the velocity of the two-phase flow mixture flowing in a slug flow in a channel is increased, the flow becomes unstable resulting in "churning" or oscillatory flow.

Annular flow. In this type of flow the liquid flows in the form of a thin film along the channel walls, while the gas flows along the centre.

4. GAS-SPARGING INSIDE MEMBRANE MODULES

Different types of membrane modules (e.g. tubular, flat sheet, hollow fibre) are used in ultrafiltration processes. The effect of gas sparging inside these membrane modules depend primarily on the dimensions of the flow path and the bubbles. In tubular and hollow-fibre

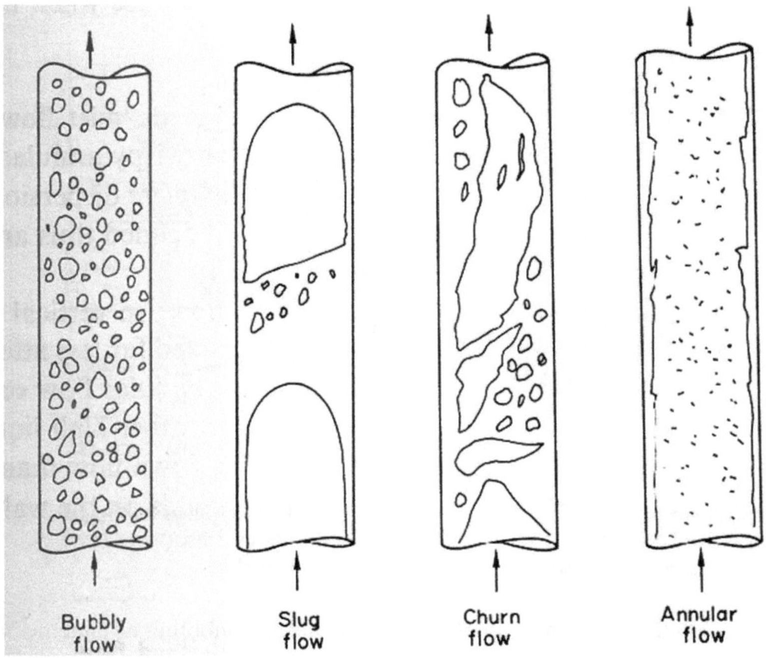

Fig. 16.2. Gas–liquid two-phase flow patterns (adapted from Hewitt and Hall-Taylor (32)).

membranes, slug flow is preferred. Figure 16.3 shows various membrane/bubbling arrangements (1) (Fig. 16.3). A comparison of flux enhancement by using different membrane modules, operating conditions and test media are summarized in Table 16.1.

4.1. Gas-Sparging in Tubular Membrane Modules

Effect of gas sparging in tubular membrane modules has been extensively studied by many researchers. Gas sparging has been examined over a wide range of operating conditions (such as TMP, gas flow rate, and feed concentration) and using different test media such as dextrans, therapeutic proteins (e.g. IgG, human serum albumin), other proteins (such as lysozyme, bovine serum albumin) and cells (yeast) (1, 7, 12, 13). The effect of membrane orientation has also been studied.

The use of gas sparging for permeate flux enhancement was first reported by Cui (12). He used a vertical tubular membrane module (12.7 mm i.d., PVDF, MWCO: 100 kDa) with colored dextran solution (MW: 87 kDa) as the test media. Figure 16.4 shows that gas sparging even at a low gas flow rate (ratio $V_g/V_l = 0.07$) increased the flux significantly. However, higher gas flow rates did not seem to increase permeate flux any further. These initial observations by Cui (12) were later confirmed by many researchers (13, 27, 33–37). The flux enhancement was found to depend on the type of membrane module and its orientation, and operating parameters like TMP, feed concentration, gas flow rate, liquid cross-flow velocity.

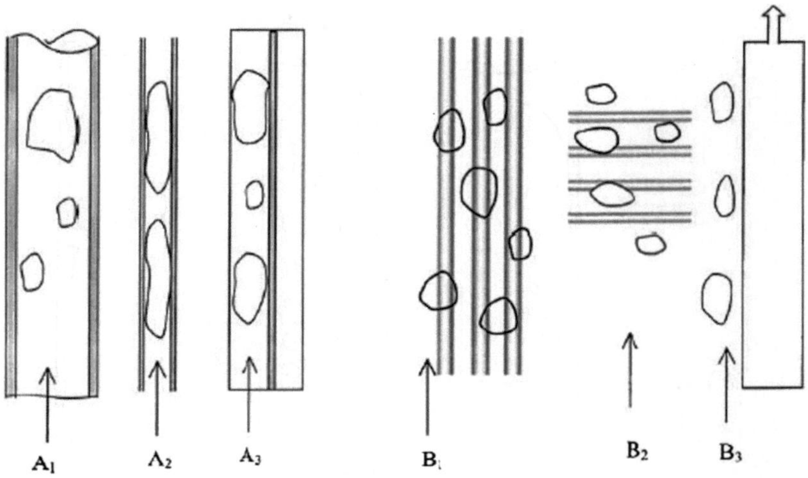

Fig. 16.3. Schematic representation of various membrane/bubbling arrangements (1).

Cui and Wright (37) studied the flux enhancement in downwards crossflow condition using dextran as the test media. They reported that flux increases up to 320% with gas sparging. They attributed the significant increase in permeate flux to the disruption of the concentration polarization layer and improved mass transfer. The best results were obtained at higher TMPs, lower liquid cross-flow velocities and higher feed concentrations.

Membrane orientation has been found to have a significant effect on flux enhancement (12, 38). Figure 16.5 compares the flux enhancement for vertical and horizontal membranes. It can be seen that flux enhancement is higher for vertical upward flow than the horizontal flow for the same range of operating parameters. Cheng et al. (39) have studied the effect of the mounting angle of the membrane and reported an optimum inclination angle of around 50° that has resulted in maximum flux enhancement (Fig. 16.6). Table 16.2 summarizes the effect of inclination angle in two-phase ultrafiltration (40). The flux enhancement due to membrane inclination increases with increasing gas velocity, feed concentration, and TMP, while it decreases with increasing liquid velocity.

Effect of liquid Reynolds number on the enhancement of permeate flux by gas sparging was reported by Cui and Wright (37). The flux enhancement was found to be higher in the laminar flow region (Fig. 16.7) and not very significant in the turbulent regime. The observed enhancement was believed to be caused by improved distribution of bubbles over the membrane surface at lower liquid flow rates.

4.2. Gas-Sparging in Hollow Fibre Membrane Modules

Hollow fibre membrane modules are of great interest due to their compactness (41). Their application was earlier limited to water and wastewater treatment. Of late they are increasingly

Table 16.1
Comparison of flux enhancement by using different membrane modules, operating conditions and test media

Test media	Membrane module & operating conditions	Flux enhancement (%)	References
Coloured dextran MW: 87 kDa	Vertical tubular module d_i: 12.7 mm V_g/V_l: 0.7	70–250	(12)
Dextran MW: 260 kDa	Vertical tubular module V_g: 0–0.35 m/s V_l: 0.14–0.78 m/s	320	(37)
Dextran Blue dextran BSA	Vertical tubular module V_g: 0–0.13 m/s V_l: 0.13–0.39 m/s	60 113 91	(13)
Dextran MW: 83 kDa	Hollow fibre modules V_g: 0.006–0.014 m/s V_l: 0.097–0.18 m/s d_i: 0.2mm	20–50	(27)
HSA MW: 66 kDa	V_g: 0.007–0.017 m/s V_l: 0.12; 0.21 m/s d_i: 0.5 mm	10–60	
Clay suspensions d_p: 1 μm	Hollow fibre module V_g: 0–0.6 m/s	155	(45)
Clay suspensions d_p: 1 μm	Hollow fibre module V_g: 0–1 m/s d_i: 0.93 mm	110	(43)
Bentonite suspensions	Tubular module V_g: 0–2.5 m/s V_l: 0.31–1.57 m/s d_i: 15 mm	200	(33)
HSA & IgG mixture	Tubular module V_g: 30–150 mL/min V_l: 0.25–1 L/min	70	(34)
BSA, Lysozyme, HSA, IgG	Rectangular flat-sheet tangential flow module V_g: 0–200 mL/min	50	(51)
Yeast suspension	Flat-sheet ceramic module V_g: 0–10 L/min V_l: 3–16.66 L/min	100	(53)

being used for bioseparation. The first reported use of gas sparging to enhance flux in a hollow fibre membrane module was by Lee at al. (42). In their study of *E. coli* suspension ultrafiltration using 1 mm inner-diameter PS hollow fibres, 65% increase in permeate flux under the stabilized operating conditions was observed.

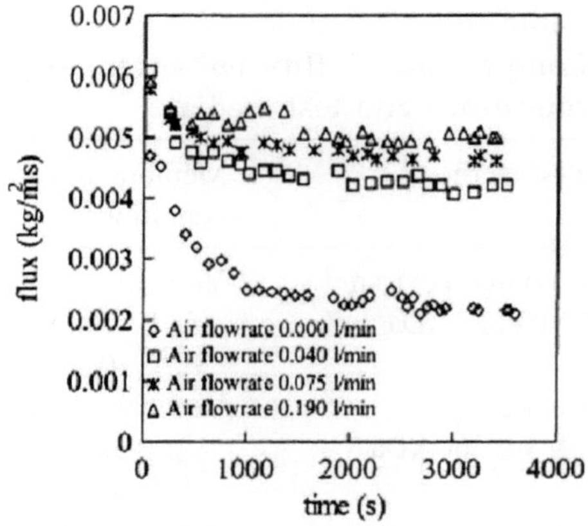

Fig. 16.4. Effect of gas sparging on permeate flux (vertical tube UF of dextran, $V_l = 0.61$ L/min, $P = 0.75$ bar, $C_o = 1.2$ wt%, MW = 87 kDa) (12).

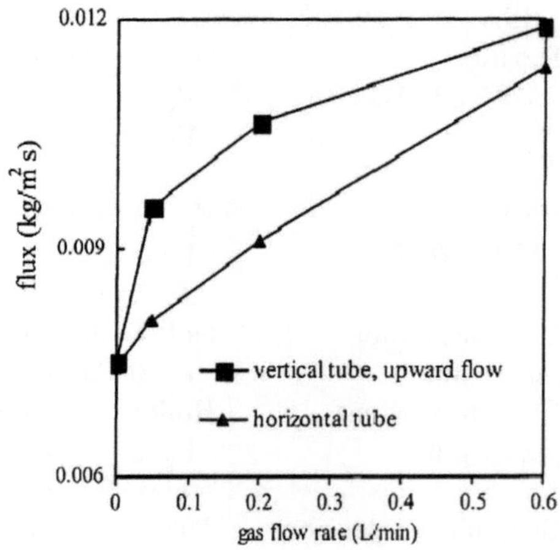

Fig. 16.5. Effect of tube orientation on gas-sparged UF (12).

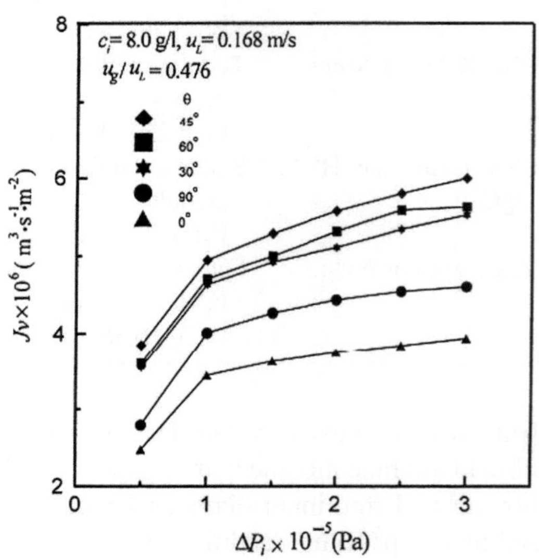

Fig. 16.6. Effect of permeate flux for gas-sparged UF at various inclination angles ($V_g = 0.08$ m/s, $V_l = 0.168$ m/s) (39).

Table 16.2
Optimal inclination angles of membrane in two-phase ultrafiltration

ΔP_i (kPa)	V_g (m/s)				
	$C_o = 4.0$ kg/m^3		$C_o = 8.0$ kg/m^3	$C_o = 16.0$ kg/m^3	
	0.08	0.12	0.08	0.08	0.12
$V_l = 0.168$ m/s					
100	50°	52°	50°	50°	52°
200	51°	51°	51°	51°	53°
300	51°	51°	51°	51°	53°
	$C_o = 4.0$ kg/m^3		$C_o = 8.0$ kg/m^3	$C_o = 16.0$ kg/m^3	
	0.08	0.16	0.16	0.08	0.16
$V_l = 0.672$ m/s					
100	48°	46°	48°	48°	48°
200	47°	47°	48°	47°	47°
300	48°	48°	48°	47°	48°

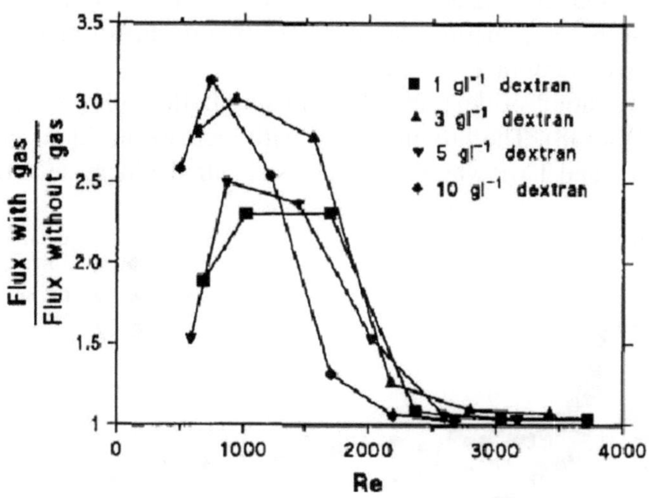

Fig. 16.7. Permeate flux enhancement vs. liquid Reynolds number at fixed gas rate ($V_g = 0.018$ m/s, TMP = 1.0 bar) (37).

Bellara et al. (27) studied the ultrafiltration of two different macromolecules (dextran and albumin) using commercial hollow fibre cartridges (PS, MWCO: 30 and 200 kDa). These researchers used a solenoid valve for injecting the bubbles to ensure all fibres received gas bubbles at a controlled frequency. Flux enhancements reported were 20–50% for dextran and 10–60% for albumin at the operating conditions examined.

Cabassud et al. (43–44) studied the filtration of clay particle suspension (0.01 μm particle size) using hollow fibre ultrafiltration membranes with the aim of studying the potentialities of such processes for drinking water production. They observed increase in flux by up to 110% (at a superficial gas velocity of 1 m/s) and 60% (at a superficial gas velocity of 0.1 m/s). Figure 16.8 shows the effect of gas velocity on the permeate flux. It can be seen that when the air velocity is higher, the permeate flux is maintained at a higher level. Permeate flux was reported to be dependent on both gas and liquid velocity.

In related work (45–46), Cabassud and coworkers analyzed the effects of bubbling on cake structure (porosity, thickness and specific resistance) using cake filtration model. New experimental results on ultrafiltration of river water demonstrated that gas sparging would be more effective with feed waters having significant fouling ability (see Fig. 16.9). They demonstrated experimentally that cake deposition is inhibited by gas–liquid two-phase flow. Hydrodynamics parameters such as shear stress and Reynolds number were characterized and correlated to flux enhancement. These workers suggested that mixing or turbulence near the membrane surface seems to control the flux enhancement whereas the cake structure seems to be linked either to mixing or to the flow intermittence.

Bellara et al. (47) studied the fractionation of BSA and lysozyme using gas-sparged ultrafiltration in hollow fibre membrane modules. Pilot-plant scale polysulphone hollow fibres (MWCO: 150 and 200 kDa) were used in this study. The results obtained (see Table 16.3) indicated that gas sparging was an effective means of disrupting concentration polarization to improve membrane selectivity.

Gas-sparged ultrafiltration of dextran solution using hollow fibre membranes was examined by Cheng and Wu (48). The flux increased with increase in liquid superficial velocity, gas superficial velocity, and TMP whereas it decreased with the increase in feed concentration.

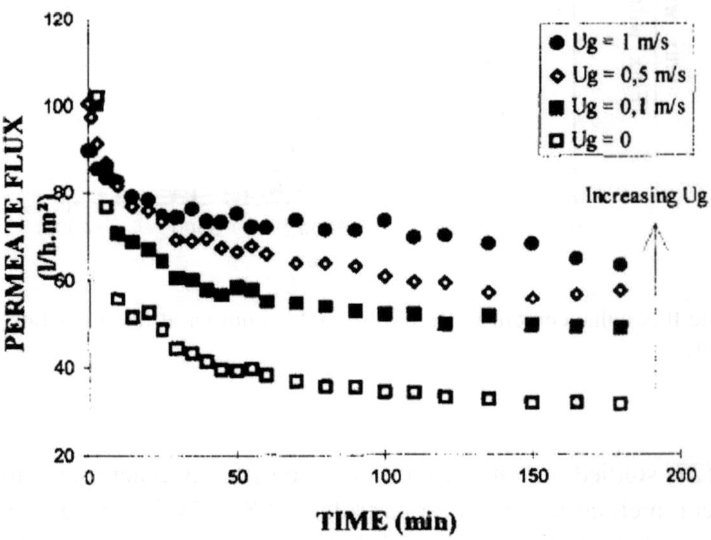

Fig. 16.8. Effect of air velocity on permeate flux (clay concentration = 0.93 g/L, TMP = 0.6 bar, $V_l =$ 0.5 m/s, V_g = 0–1 m/s) (43).

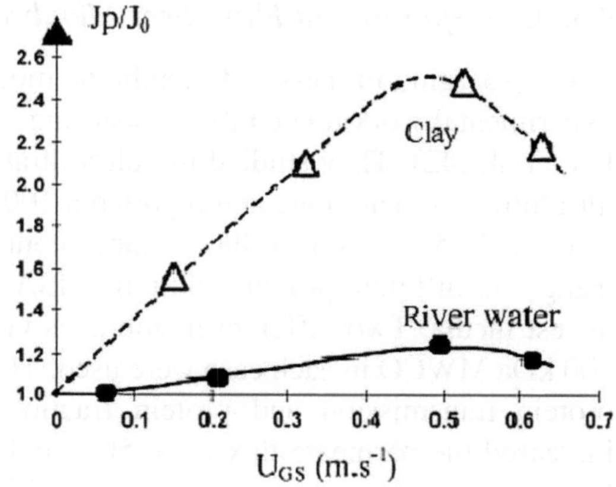

Fig. 16.9. Variation of permeate flux with the gas velocity for clay suspension and river water (V_l = 0.7 m/s, TMP = 1.0 bar, T = 20°C) (46).

Table 16.3
Effect of gas sparging on BSA/LYS fractionation using hollow fibre membrane module

Gas bubbling freq (Hz)	TMP (kPa)	Permeate flux (L/m^2 h)	S_{LYS}	S_{BSA}	ψ	Lys purity (%)
0	25	70.5	84.9	2.4	35	97.3
0.1		75.9	81.7	1.0	82	98.8
0.2		78.0	77.8	0.7	111	99.1
0	75	73.8	88.4	3.8	23	95.9
0.1		79.8	75.3	0.8	94	98.9
0.2		82.5	78.0	0.8	98	99.0

Conditions: 1 g/L BSA + 1 g/L LYS, 100 mM NaCl + 20 mM phosphate buffer, liquid circulation rate = 1 L/min

The maximum flux enhancement reported for the range of operating conditions examined was 160%. This study showed once again that flux enhancement by gas sparging is particularly significant when the system is operated at a lower liquid velocity.

In an experimental study, Smith and Cui (49) reported an increase in flux of up to 102% in gas-sparged dextran ultrafiltration using hollow fibre membranes (MWCO: 50 kDa). It was found that flux enhancement was higher with continuous gas sparging when compared with intermittent gas sparging. Flux enhancement was found to rely heavily on the thinning of the mass transfer boundary layer as successive gas slugs moved along the membrane. Smith and Cui (50) developed a simplified boundary layer theory based model for the prediction of permeate flux in gas-sparged enhanced ultrafiltration processes. The model points out that neglecting the suction velocity leads to inaccuracy in flux predictions and hence should be included in any widely applicable model.

As mentioned earlier in this section, flux enhancement due to gas sparging inside hollow fibres is significantly lower than in tubular membranes. This is basically attributed to the presence of high shear rates in hollow fibre membranes. As the shear rates and hence mass transfer coefficients are already quite high, further increase becomes less significant (27).

4.3. Gas-Sparging in Flat-sheet Membrane Modules

Gas sparging in flat-sheet membrane modules has been less extensively studied. The first experimental work related to gas sparging in a flat-sheet membrane module was reported by Lee et al. (42). They studied the ultrafiltration of a bacterial suspension using 300 kDa PS ultrafiltration membrane and reported a 100% increase in permeate flux.

Li et al. (51) examined flux enhancement by gas sparging in flat-sheet membrane modules using four different protein solutions (HSA, BSA, human immunoglobulin G and lysozyme) as test media. Two different membranes viz., polysulphone and polyether sulphone having 100 kDa MWCO in each case were used. The effects of gas sparging on permeate flux, single protein transmission and protein fractionation were extensively studied. Gas sparging increased the permeate flux by 7–50% and as with other types of membrane modules, was found to be more effective at lower liquid feed flow rates. Protein transmissions in single protein ultrafiltration were reduced by gas sparging. Protein transmission was found to be lower with PS membranes than with PES membranes, this being attributed to protein fouling (see Table 16.4). However, protein fractionation was enhanced in binary protein ultrafiltration. Gas sparging affected the transmission of different proteins to different extents and hence increased selectivity.

Ghosh et al. (52) examined the effect of gas sparging on fractionation of proteins BSA and lysozyme using a flat-sheet PS membrane having 100 kDa MWCO. Though the enhancement of permeate flux was reported to be only 10%, gas sparging enhanced the selectivity of protein fractionation very significantly. Nearly complete separation of BSA and lysozyme with gas-sparged ultrafiltration was reported. This was attributed to the disruption of concentration polarization layer and enhanced mass transfer due to bubble-induced secondary flow.

Mercier-Bonin et al. (53) conducted experiments on yeast-cell ultrafiltration using a ceramic flat-sheet membrane. They also studied the effect of membrane orientation and found that horizontally installed membrane modules were more efficient in terms of flux enhancement than vertically installed membrane modules. The maximum flux enhancement

Table 16.4
Effect of gas sparging on protein fractionation

Protein (g/L)	Membrane	Flux (kg/m^2h)		Transmission (%)			Selectivity	
		NG[a]	G[b]	NG[a]		G[b]	NG[a]	G[b]
HSA, 4.5 + IgG, 1.0	PES 100	25.8	31.3	18.5	(HSA)	5.6	3.7[c]	9.5[c]
				68.3	(IgG)	53.1		
HSA, 4.5 + IgG, 1.0	PS 100	22.5	24.7	5.5	(HSA)	1.7	4.25[c]	8.9[c]
				23.4	(IgG)	15.2		
BSA, 2.0 + Lys, 2.0	PES 100	45.5	53.3	38.46	(BSA)	6.04	2.47	14.83
				95.27	(Lys)	89.57		

[a]Without gas sparging
[b]With gas sparging
[c]Reversed selectivity, the ratio of IgG and HSA transmission

was observed when particle deposition was more severe (i.e. at high feed concentrations and low liquid flow rates). This clearly indicated that the mechanism behind such enhancement was the disruption of the fouling layer.

In a recent study, Cheng and Lin (54) reported the flux behavior of an inclined flat-sheet membrane module of 100 kDa MWCO using dextran T500 as the test media. It was found that an increase in cross-flow velocity increased the permeate flux and decreased the time required to reach the steady state. Under the experimental operating conditions, examined maximum flux enhancement was observed when the membrane inclination was 180°, i.e. with a horizontally installed module with the feed below the membrane. At this angle the gas slugs were the closest to the membrane surface and thereby disrupted the concentration boundary layer more effectively. The effect of inclination was more pronounced when the system was operated at a low liquid flow rate.

4.4. Mechanisms of Flux Enhancement

It is clear from the experimental results discussed in the previous sub-sections that flux enhancement was more significant when the concentration polarization was more severe. Generally, the gas bubbles disrupted the concentration polarization layer thereby enhancing the permeate flux. However, the differences in the performance of gas sparging in tubular and hollow fibre membranes suggest that the exact mechanisms could be somehow different (27). Two main mechanisms for flux enhancement have been identified:

1. *Bubble induced secondary flow* – The secondary flow is generated by the buoyancy driven bubble motion. The wake following bubble promotes local mixing in and near the mass transfer boundary layer, thereby decreasing concentration polarization (27). This is similar to that observed in gas-sparged enhancement of heat transfer (55–56).
2. *Physical displacement of the concentration polarization layer* – Taylor bubbles in tubular membranes disrupt the concentration polarization layer by physically eroding them, leading to high mass transfer rates. With hollow fibre membranes, the thickness of the liquid film between the membrane wall and a gas slug is usually lower than the calculated boundary layer thickness.

In addition to these two mechanisms, other factors such as pressure pulsing caused by passing gas slugs, increase in superficial cross-flow velocity and vibration (in the case of hollow fibre membranes) are also responsible for flux enhancement (1).

5. GAS-SPARGING IN SUBMERGED MEMBRANE SYSTEMS

In aerobic MBRs gas sparging is utilized for biological oxidation and enhancement of membrane filtration. The advantages of MBRs in municipal wastewater treatment include excellent quality of treated water, small treatment plant footprint, reduced sludge production and better process reliability. Submerged MBRs are operated in the suction mode and hence the need for a pressurized module is eliminated. However control of fouling is challenging since the biomass concentration in the feed is high. The nature and extent of fouling are strongly influenced by three factors: biomass characteristics, operating conditions and

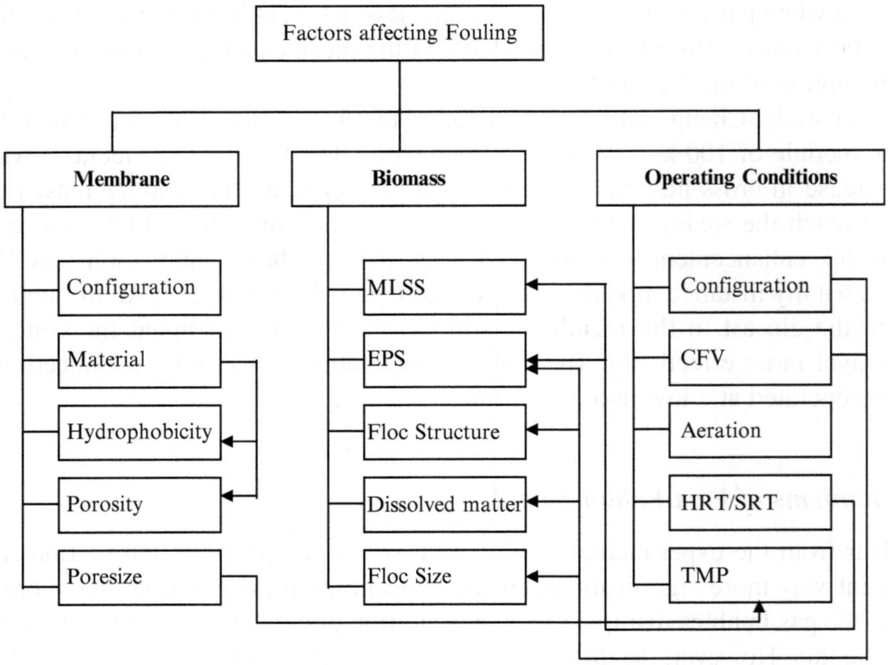

Fig. 16.10. Factors influencing membrane fouling in membrane bioreactor (MBR) process (adapted from Chang et al. (57)).

membrane characteristics (see Fig. 16.10, 57). Gas sparging helps to operate a MBR at a low fouling condition.

Most submerged membrane systems are based on hollow fibre or flat-sheet membranes. Flat-sheets are usually oriented vertical, while the hollow fibres could be mounted either vertically or horizontally. The feed is present "outside" the membrane module and the permeate is drawn inside by suction.

5.1. Submerged Flat-Sheet Systems

The membrane surface in flat-sheet systems is more accessible to well directed bubbles and hence the aeration is better utilized. However, flat-sheet systems have limited membrane packing density, typically 25% lower than that of hollow fibre systems. Moreover vigorous backwashing is not feasible since the membrane is not supported on the feed side. The flat-sheet based MBRs have been successfully commercialized by Kubota (Japan). Figure 16.11 shows a schematic representation of the Kubota system where a membrane is submerged in an activated sludge sewage treatment tank. By using this arrangement, the energy required for aeration is utilized for reducing fouling (58). Gunder and Krauth (59) compared a flat-plate system (Kubota, 80 m^2 membrane area) with a hollow fibre system (Zenon, 83.4 m^2 membrane area) under similar operating conditions. The results showed that both systems achieved excellent treatment quality but the hydraulic performance of flat-sheet system was

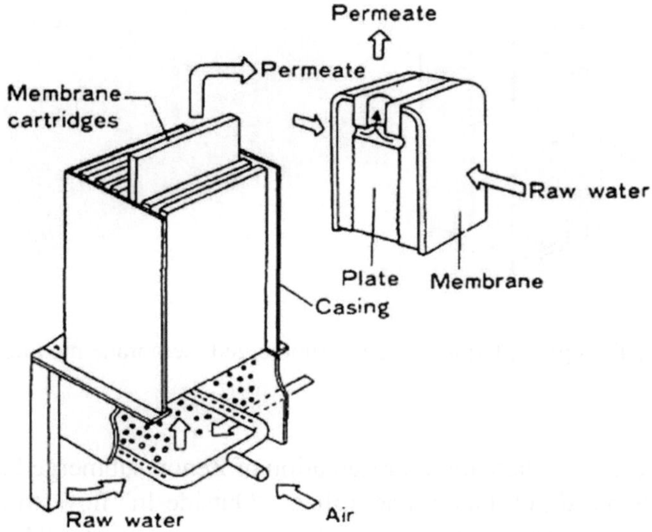

Fig. 16.11. A schematic diagram of the Kubota membrane unit operation and panel Construction (58).

slightly better than that of hollow fibre system. Better bubble distribution in well defined flat channels and the loss of driving force in the hollow fibre system due to lumen pressure drop were suggested as plausible reasons.

5.2. *Submerged Hollow Fibre Membranes*

Unlike flat-sheet submerged membrane systems, hollow fibre submerged membrane systems have been extensively studied. The relative ease of fabrication, high surface-to-volume ratio, their rigid self-supporting and back washable nature properties make hollow fibres attractive (1, 60). Moreover, incorporating hollow fibres in the MBRs can reduce energy requirement for the membrane separation (61). A typical device of this kind consists of a bundle of parallel polymeric hollow fibres submerged in a tank containing the liquid feed. The concept of submerged hollow fibre system was introduced in Japan in mid-1980s, following a patent by Tajima and Yamamoto (62). The patent describes a filter which incorporates hollow fibres in a vessel with intermittent air bubbling around the fibres to remove entrapped solid particles. Yamamoto et al. (63) are the first to report the use of submerged hollow fibres in a MBR for wastewater treatment. In this reactor, the role of gas sparging was limited to aeration and mixing and induces liquid flow. In a later work, Yamamoto et al. (64) added a jet aerating device to provide intermittent vigorous aeration to increase the efficiency. In early 1990s, Futamura et al. (65) has studied the effect of continuous bubbling in submerged hollow fibre MBRs by using both vertical and horizontal fibres. In a later work, they suggested the use of vertically aligned hollow fibres with vigorous bubbling as a more efficient way to control fouling. During this period, Zenon (Canada) were developing a submerged hollow fibre system with gas sparging for wastewater treatment. Several patents were taken based on the concept of hollow fibres with gas sparging (66–67).

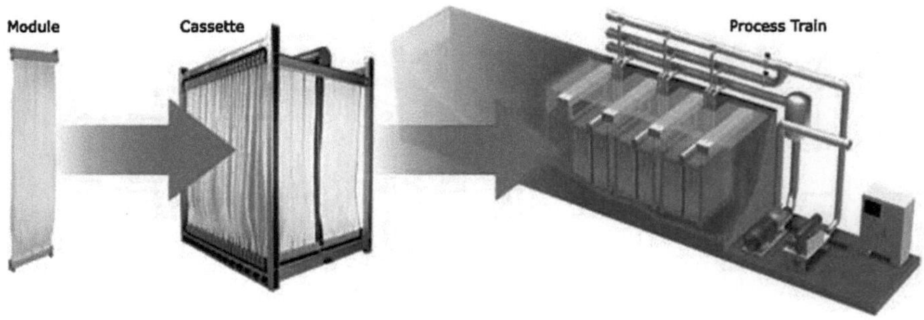

Fig. 16.12. Schematic representation of Zenon submerged membrane module (ZeeWeed 500).

Figure 16.12 describes the schematic representation of Zenon submerged membrane module (ZeeWeed 500). ZeeWeed 500 membrane utilizes "Outside-In" flow, through a reinforced, hollow-fiber membrane that has nominal and absolute pore sizes of 0.04 and 0.1 μm respectively which excludes particulate matter including solids, bacteria, pathogens and certain viruses. Zenon recommends some specific requirements like the need for vigorous coarse bubbling and the advantage of loose fibres in the ZeeWeed 500 system. Mitsubishi has also developed a submerged hollow fibre (Sterapore series) MBR with bubble injection to control fouling.

Leonard et al. (68) studied the impact of gas–liquid two-phase flow on both permeate flux and oxygen transfer rate in a submerged MBR. The flux enhancement was found to last for an extended period even though the biomass concentration increased. This was due to the permanent effect of the unsteadiness generated by the two-phase flow (see Fig. 16.13). Lee et al. (69) studied the flux enhancement in a submerged membrane system fitted with a stainless steel prefilter and reported that this system produced a constant permeate flux for 50 days of operation. The prefilter helped in reducing the cake resistance of the membrane.

The effect of the alignment of submerged hollow fibres has been looked at. Horizontal alignment provided transverse flow for the bubbles whereas vertical alignment provided axial flow for the bubbles. For single-phase liquid flow it has been reported that transverse flow across hollow fibres gave better mass transfer than external axial flow (70).

Bouhabila et al. (71) has demonstrated that the combined use of gas sparging and periodic backwashing increased permeate flux significantly without seriously increasing fouling resistance. Gas sparging alone proved to be less efficient due to its limited effects on internal fouling. It was reported that under optimal conditions for backwashing (15 s backwash every 5 min) along with gas sparging, the fouling resistance decreased by 3.5 fold. In a recent work, a similar finding was reported by Posch and Schiewer (72).

Guibert et al. (73) has proposed a new technique of "air cycling" with air injected sequentially on either side of a group of hollow fibre bundles instead of being injected continuously. Under this strategy, the local bubbling intensity can be increased to a higher value than for continuous bubbling and due to the density difference of aerated and unaerated column of water. The efficiency of the process depends on the cycle frequency.

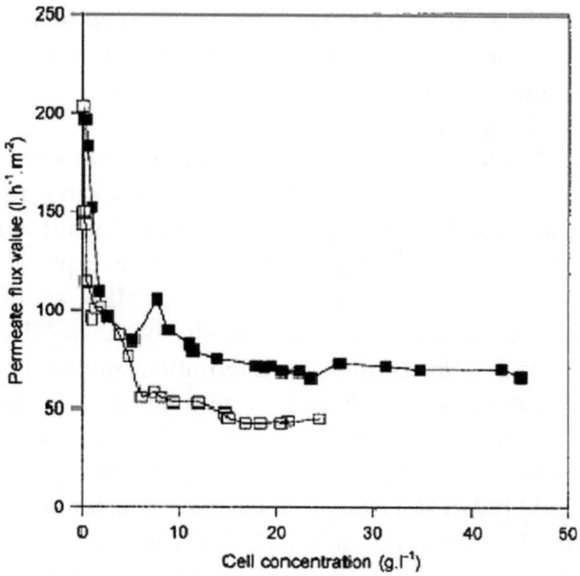

Fig. 16.13. Permeate fluxes during growth of *A. eutrophus* on phenol in a classical MBR (*open square*), or a two-phase unsteady slug flow bioreactor (*filled square*) as a function of the total cell concentration (68).

In a recent study, Ghosh (74) has demonstrated the effect of gas sparging to enhance membrane permeability in suction-driven submerged hollow fibre ultrafiltration of macromolecular solutions using two different module types. It was found that the effectiveness of gas sparging increased with increase in feed concentration. Depending on the operating conditions and on the module design, gas sparging enhanced effective hydraulic permeability by as much as 115%.

In addition to the above mentioned, there are some very interesting reports in the literature on the effect of fibre size and flexibility on effectiveness of gas sparging (75–78). Since, these are based on microfiltration membranes they are beyond the scope of this chapter. The authors would strongly recommend these papers for further insight in to the various aspects of submerged hollow fibre MBR systems in general.

6. APPLICATIONS OF GAS-SPARGING

The application of gas-sparged ultrafiltration is widespread and ranges from water treatment to protein bioseparation. Some of these are discussed in this section.

6.1. Water and Wastewater Treatment

Membranes are being used for water and wastewater treatment since the early 1960s (4). However, most of the rapid developments in the field of MBRs have taken place in the last 10–15 years. MBRs started as small-scale units for niche applications. Currently used large

MBRs can handle flows in excess of 10,000 m^3/day. Submerged MBRs are ideal for effluent treatment applications and are being used for treating a wide range of effluents: sewage, brewery waste, dairy waste, seafood waste, vegetable processing waste, bakery and canteen waste, pharmaceutical and paper effluents (79).

There are more than 2,200 MBR installations in operation or under construction worldwide (80). There are four main manufacturers providing MBRs to North America. They are Zenon Environmental Inc. (Canada), USFilter (USA), Kubota (Japan) and Mitsubishi-Rayon (Japan). Most of the MBRs use submerged hollow fibres except for the Kubota systems which uses the flat-sheet membranes. Table 16.5 outlines MBR applications in the area of domestic and municipal wastewater treatment. Majority of these MBRs employ controlled flux, intermittent suction and backwashing. The intermittent suction provides more opportunity for the bubbling to remove fouling during the zero flux periods. However, this strategy requires optimization of the on/off periods (1, 81). Application of gas sparging in a submerged tubular membrane system treating domestic and municipal wastewater was studied by Chang and Judd (82–83). Detailed description is already discussed in Sect. 5.2.

Hybrid processes using UF with coagulation (or PAC) are increasingly being looked at for water-treatment type applications. In these hybrid systems, the UF membranes are used to remove suspended solids and microorganisms, while PAC is responsible for forming a "reaction zone" where adsorption, biodegradation and coagulation can be carried out simultaneously (1). In these systems, gas sparging can serve three purposes: creating turbulence for

Table 16.5
MBR applications in the area of domestic and municipal wastewater treatment

Membrane configuration	Size of operation	Treatment success	Country of application	Reference
Plate and frame external	Full-scale average ~125 m^3/day	Effluent COD <5 mg/L	Japan	(24)
Hollow fibre submerged	Pilot-scale <1.5 m^3/day	Effluent COD <10 mg/L	Japan	(92)
Tubular external	Pilot-scale 360–840 m^3/day	Effluent TC <12 mg/L	The Netherlands	(93)
Tubular external	Pilot-scale 2.4–4.8 m^3/day	COD removal >94%	France	(19)
Tubular external	Bench-scale 0.16 m^3/day	COD removal >98%	USA	(94)
Hollow fibre submerged	Pilot-scale 2.6–5.0 m^3/day	COD removal >96.5%	Canada and France	(95)
Cartridge-disc external	Pilot-scale 48 m^3/day	Effluent COD <5 mg/L	Korea	(96)
Hollow fibre submerged	Pilot-scale 6–9 m^3/day	COD removal >95%	Germany	(97)
Hollow fibre submerged	Pilot-scale 46–47 m^3/day	COD removal >93%	The Netherlands	(98)
Plate and frame submerged	Pilot-scale 48–72 m^3/day	COD removal >91%	The Netherlands	(98)

fouling control, mixing for adsorption and oxygenation for biodegradation. Various strategies applied to MBRs viz., bubbling, intermittent suction and backwashing can also be used effectively to optimize the hybrid membrane process performance (1).

6.2. *Protein Fractionation and Concentration*

Ultrafiltration is widely used for protein desalting and concentration. In recent years, protein fractionation by UF has generated considerable interest in terms of its applicability for bioseparation of recombinant protein and plasma products (6). The advantages of UF compared to other methods like chromatography are: high throughput of products, lower capital investment and ease of scale-up, while maintaining product purity under ambient conditions. Efficiency of bioseparation depends on protein–protein and protein-membrane interactions as well as on operating conditions. Low selectivity and severe decline in permeate flux have been identified as the major challenges with ultrafiltration. Low selectivity results from concentration polarization, while decline in permeate flux mainly results from membrane fouling. Both these problems can be minimized by improving the system hydrodynamics. Gas sparging has been found to be an attractive way by which system hydrodynamics could be improved in a cost effective manner.

Ghosh et al. (52) has studied the effect of bubbles on fractionation of BSA (67 kDa) and lysozyme (14.1 kDa) using a flat-sheet membrane (MWCO 100 kDa). Under the optimal operating conditions of gas-sparged UF complete separation of these two proteins was achieved. It can be seen from Table 16.6; about 18-fold increase in selectivity was achieved simply by sparging a small amount of gas. Further increase in sparging did not have much effect. Similar effects were observed in hollow fibre modules (47).

Fractionation of HSA (67 kDa) and IgG (167 kDa) by gas-sparged UF using a tubular PVDF membrane (MWCO 100 kDa) was reported by Li et al. (7). Complete separation of the

Table 16.6
Effect of gas sparging on the fractionation of BSA and lysozyme

Gas flowrate ($\times 10^6$ m^3/s)	Permeate flux ($\times 10^2$ kg/m^2s)	S_a for BSA	S_a for lysozyme	Selectivity (ψ)
0	0.93	0.078	0.846	10.9
1.67	0.96	0.004	0.776	194.0
3.33	0.97	0.004	0.785	196.3
6.67	1.00	0.004	0.739	184.8
[a]0	0.75	0.059	0.780	13.2
[a]0.83	0.78	Undetectable	0.724	Nearly complete separation
[a]1.67	0.82	Undetectable	0.710	Nearly complete separation

Operating conditions: feed = 2 kg/m^3 BSA + 2 kg/m^3 lysozyme, buffer = 20 mM phosphate. pH = 7.0, PS 100 kDa membrane, TMP = 0.5 bar, $V_l = 8.33 \times 10^{-6}$ m^3/s, $u_l = 0.074$ m/s

[a]TMP = 0.3 bar, all other operating conditions are same as above

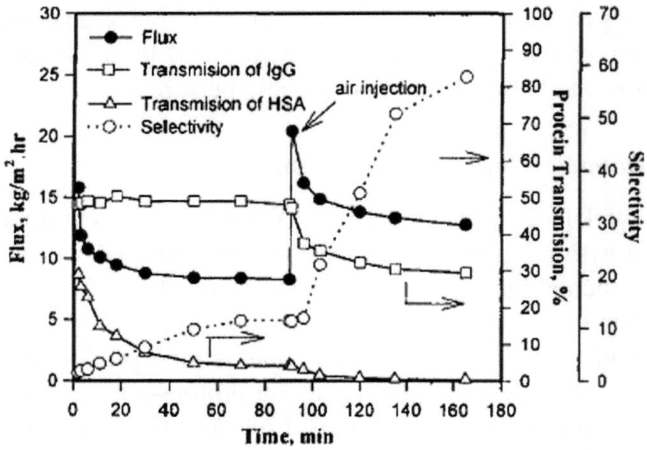

Fig. 16.14. Effect of gas sparging on HSA/IgG fractionation (7).

two proteins was achieved with IgG going thorough the membrane and HSA being rejected. Figure 16.14 shows that injecting bubble increases permeate flux and reduces protein transmission. The injection of air reduced HSA transmission to zero resulting in selective transmission of IgG only. The gas sparging has an on–off effect on increasing permeate flux and selectivity, i.e. significant enhancement on protein fractionation can be achieved under a relatively low air flow rate which also reduces the potential damage to proteins due to foaming.

6.3. Membrane Cleaning

As mentioned earlier, membrane fouling is the major problem in UF systems. Periodic cleaning must be incorporated into a membrane operation as a preventive measure against severe and irreversible fouling. Membranes could be cleaned by various physical, chemical and physicochemical methods (84–85). Gas sparging has been proved to be an efficient method for cleaning membranes. The gas is usually sparged on the feed side of the module sometimes in combination with backwash or reversed permeate flow (86–87). The use of gas sparging with backwash for cleaning hollow fibre membranes was investigated by Serra et al. (87). It was found that the cake layer was instantaneously lifted off by the reversed permeate flux and the injection of air improved material removal and reduced the volume of concentrated foulant to be flushed. Hence the backwash time was reduced and its efficiency improved. Verbeck et al. (88) used the airflush technique which involved an intermittent two-phase flush through the module to dislodge and remove the cake from the membranes. They found that airflush was a much better cleaning method than single-phase liquid forward flow at the same velocity.

7. PRACTICAL ISSUES AND FUTURE CHALLENGES

The use of gas sparging in membrane processes for controlling concentration polarization and subsequent fouling in order to enhance permeate flux is now widespread. However, there are some practical issues and challenges which remain unaddressed and these constrain wider usage. One major concern is the tendency of foaming in gas sparged processes. Excessive foaming and bursting of bubbles can damage proteins and in some cases cells during bio-separation. Foaming and bubble damage are closely related. Clarkson et al. (89–90) studied protein denaturation during foaming using a bubble column. They found that damage to the protein is mainly due to surface denaturation at the gas–liquid interface. The degree of denaturation was found to correlate directly with the interfacial exposure area. Shear stress does not contribute significantly to the degree of damage and neither does oxidation. When the gas flowrate is low, the foaming is less and the damage due to it is also less significant. However, with high gas flowrates, foaming can cause severe problems (90). In a related work, Clarkson et al. (91) investigated the conditions under which protein damage due to foaming can be reduced. They concluded that protein damage could be reduced by operating at optimal ionic strength and pH, and to a lesser extent, by the addition of sugars. There are some other possible mechanisms of bubble-induced damage to proteins such as high shear rate resulting from bursting bubbles and aggregation due to the change of protein structure (1).

Gas distribution is another challenge in gas-sparged UF. It has to be ensured that injected bubbles influence the entire membrane surface area in a module. Most conventional membrane modules are therefore not suitable for gas-sparged UF and special modules need to be designed. Gas sparging is also difficult to implement in multi-pass operations and cannot be used in spiral wound modules.

Gas-sparged UF in MBRs for the treatment of municipal as well as industrial wastewater is the most promising application of this technique. Gas sparging (with or without backwashing) has been successful in controlling concentration polarization and cake deposition. However, there are some important issues that need more attention viz. optimization of the energy need for bubbling (e.g. intermittent or cyclic, 73, 81), and optimization of bubble size and frequency for a given application.

8. CONCLUSIONS

Gas sparging has been successfully applied to ultrafiltration to enhance its performance. The bubbles are responsible for creating a secondary flow which enhances mass transfer and hence the permeate flux. Bubbles also contribute towards fouling control by eroding the mass transfer layer, physical scouring action, pressure pulsing and vibration. Gas sparging is found to be effective even at modest gas flowrates. The flux enhancement is more significant at lower liquid flowrates, higher solute/particle concentrations and higher operating pressures.

Most current applications of gas sparging are in the area of submerged MBRs for water and municipal wastewater treatment. With controlled bubbling and better module design, the use of gas sparging could be extended to food and biotechnology processes. In the case of MBRs, cake deposition and fouling is minimized by gas sparging. Further improvement in efficiency can be achieved by use in combination with backwashing. Gas sparging increases efficiency of protein fractionation by controlling concentration polarization. This results in high selectivity of separation and sustained operation due to low fouling (99–104).

With most UF systems the gas is sparged internally, i.e. within the module, while with submerged systems the gas is sparged externally. With tubular and hollow fibre membranes, Taylor bubbles or gas slugs are found to be most effective. With submerged systems larger bubbles are found to be more effective as they have larger and stronger wake regions which create stronger secondary flows and promote local mixing. Larger bubbles are also easier to separate from the retentate and cause less foaming. Submerged systems are also influenced by fibre size, orientation and looseness. Use of intermittent bubbling was found to be more effective than continuous bubbling and this also reduced energy requirement.

9. ACRONYMS

BSA	Bovine serum albumin
HSA	Human serum albumin
IgG	Human immunoglobulin G
MBR	Membrane bioreactor
MWCO	Molecular weight cut off
PAC	Powdered activated carbon
TMP	Transmembrane pressure
UF	Ultrafiltration

10. NOMENCLATURE

C = Empirical parameter in (16.3), dimensionless
C_o = Feed concentration, kg/m^3
d_i = Inside diameter of membrane, mm
d_p = Mean particle diameter, μm
Fr = Froude number, dimensionless
P = Pressure of the system, bar
T = Temperature of the process, °C
u_l = Superficial liquid velocity within flow channel, m/s
U_s = Slug rising velocity, m/s
V_b = Bubble volume, mL
V_g = Gas flowrate, m/s
V_l = Liquid flowrate, m/s

REFERENCES

1. Cui ZF, Chang S, Fane AG (2003) The use of gas bubbling to enhance membrane processes. J Memb Sci 221:1–35
2. Al-Bastaki N, Abbas A (2001) Use of fluid instabilities to enhance membrane performance: a review. Desalination 136:255–262
3. Brindle K, Stephenson T (1996) The application of membrane biological reactors for the treatment of wastewaters. Biotechnol Bioeng 49:601–610

4. Viswanathan C, Aim RB, Parameswaran K (2000) Membrane separation bioreactors for wastewater treatment. Crit Rev Environ Sci Technol 30(1): 1–48
5. Cicek N (2003) A review of membrane bioreactors and their potential application in the treatment of agricultural wastewater. Can Biosyst Eng 45:37–49
6. Ghosh R (2003) Protein bioseparation using ultrafiltration: theory, applications and new developments. Imperial College Press/World Scientific, London
7. Li QY, Cui ZF, Pepper DS (1997) Fractionation of HSA and IgG by gas-sparged ultrafiltration. J Memb Sci 136:181–190
8. Belfort G, Pimbley JM, Greiner A, Chung KY (1993) Diagnosis of membrane fouling using a rotating annular filter: 1. cell culture media. J Memb Sci 77:1–22
9. Bellhouse BJ, Sobey IJ, Alani S, DeBlois BM (1994) Enhanced filtration using flat membranes and standing vortex waves. Bioseparation 4:127
10. Najarian S, Bellhouse BJ (1996) Enhanced microfiltration of bovine blood using a tubular membrane with screw-threaded insert and oscillatory flow. J Memb Sci 112:249–261
11. Wang YY, Howell JA, Field RW, Wu DX (1994) Simulation of cross-flow filtration for baffled tubular channels and pulsatile flow. J Memb Sci 95:243–258
12. Cui ZF (1993) Experimental investigation on enhancement of cross flow ultrafiltration with air sparging. In: Patterson R (ed) Effective membrane processes – new perspectives. Mechanical Engineering, London
13. Cui ZF, Wright KIT (1994) Gas-liquid two phase flow ultrafiltration of BSA and dextran solution. J Memb Sci 90:183
14. Cheryan M (1998) Ultrafiltration and microfiltration handbook. Technomic, Lancaster
15. Kimura S (1991) Japan's aqua renaissance 90 project. Water Sci Technol 23:1573–1582
16. Yokomizo T (1994) Ultrafiltration membrane technology for regeneration of building wastewaters for reuse. Desalination 98:319–326
17. Buisson H, Cote P, Praderie M, Paillard H (1998) The use of immersed membranes for upgrading wastewater treatment plants. Water Sci Technol 37:89–95
18. Cote P, Buisson H, Pound C, Arakaki G (1997) Immersed membrane activated sludge for the reuse of municipal wastewater. Desalination 113:189–196
19. Fan XJ, Urbain V, Qian Y, Manem J (1996) Nitrification and mass balance with a membrane bioreactor for municipal wastewater treatment. Water Sci Technol 34:129–136
20. Irwin J (1990) On-site wastewater reclamation and recycling. Water Environ Technol 2:90–91
21. Berube PR, Hall ER (2001) Fate and removal kinetics of contaminants contained in evaporator condensate during treatment for reuse using a high-temperature membrane bioreactor. J Pulp Paper Sci 27:41–45
22. Fan YB, Wang JS, Jiang ZC (1998) Test of membrane bioreactor for wastewater treatment of a petrochemical complex. J Environ Sci 10:269–275
23. Scholzy W, Fuchs W (2000) Treatment of oil contaminated wastewater in a membrane bioreactor. Water Res 34:3621–3629
24. Manem JAS (1996) Membrane bioreactors. In: Mallevialle J, Odendaal PE, Wiesner MR (eds) Water treatment membrane processes. McGraw Hill, New York
25. Bixler HJ, Rappe GC (1970) Increasing the mass-transport rate across ultrafiltration membranes. US Patent 3 541 006. 17 Nov 1970
26. Van der Waal MJ, van der Velden PM, Koning J, Smolders CA, van Swaay WPM (1977) Use of fluidized beds as turbulence promoters in tubular membrane systems. Desalination 22:465–483

27. Bellara SR, Cui ZF, Pepper DS (1996) Gas sparging to enhance permeate flux in ultrafiltration using hollow fibre membranes. J Memb Sci 121:175–184
28. Taha T, Cui ZF (2002) Hydrodynamic analysis of upward slug flow in tubular membranes. Desalination 145:179–182
29. Clift R, Grace JR, Weber ME (1978) Bubbles, drops and particles. Academic, New York
30. Miyahara T, Tsuchiya K, Fan LS (1988) Wake properties of a single gas bubble in a three-dimensional liquid–solid fluidized bed. Int J Multiphase Flow 41:749–763
31. Whalley PB (1987) Boiling condensation and gas–liquid flow. Clarendon, Oxford
32. Hewitt GF, Hall-Taylor NS (1970) Annular two-phase flow. Pergamon, Oxford
33. Mercier M, Fonade C, Lafforgue-Delorme C (1997) How slug–flow can enhance the ultrafiltration flux in mineral tubular membranes. J Memb Sci 128(1):103–113
34. Li QY, Cui ZF, Pepper DS (1997) Effect of bubble size and frequency on the permeate flux of gas sparged ultrafiltration with tubular membranes. Chem Eng J 67:71–75
35. Ghosh R, Cui ZF (1999) Mass transfer in gas sparged ultrafiltration: upward slug–flow in tubular membranes. J Memb Sci 162:91–103
36. Cui ZF, Bellara SR, Homewood P (1997) Airlift crossflow membrane filtration – a feasibility study with dextran ultrafiltration. J Memb Sci 128:83–91
37. Cui ZF, Wright KLT (1996) Flux enhancement with gas sparging in downwards crossflow ultrafiltration: performance and mechanism. J Memb Sci 117:109–116
38. Cui ZF, Ghosh R, Yu J, Luan SD (2001) An experimental study of flux enhancement with air sparging in a horizontal tubular membrane module In: Proceedings of the sixth world congress on chemical engineering, Melbourne
39. Cheng TW, Yeh HM, Wu JH (1999) Effects of gas slugs and inclination angle on the ultrafiltration flux in tubular membrane module. J Memb Sci 158:223–234
40. Cheng TW (2002) Influence of inclination on gas-sparged cross-flow ultrafiltration through an inorganic tubular membrane. J Memb Sci 196:103–110
41. Kulkarni SS, Funk EW, Li NN (1992) Ultrafiltration: membranes. In: Ho WS, Sirkar KK (eds) Membrane handbook. Van Nostrand Reinhold, NewYork
42. Lee C, Chang W, Ju Y (1993) Air slugs entrapped cross-flow filtration of bacteria suspension. Biotechnol Bioeng 41:525–530
43. Cabassud C, Laborie S, Lainé JM (1997) How slug–flow can improve ultrafiltration flux in organic hollow fibres. J Memb Sci 128:93–101
44. Laborie S, Cabassud C, Durand-Bourlier L, Laine JM (1998) Fouling control by air sparging inside hollow fibre membranes – effects on energy consumption. Desalination 118:189–196
45. Laborie S, Cabassud C, Durand-Bourlier L, Lainé JM (1997) Flux enhancement by a continuous tangential gas flow in ultrafiltration hollow fibres for drinking water production: effects of slug–flow on cake structure. Filtr Sep 34:887–891
46. Cabassud C, Laborie S, Durand-Bourlier L, Laine JM (2001) Air sparging in ultrafiltration hollow fibers: relationship between flux enhancement, cake characteristics and hydrodynamic parameters. J Memb Sci 181:57–69
47. Bellara SR, Cui ZF, Pepper DS (1997) Fractionation of BSA and lysozyme using gas sparged ultrafiltration in hollow fibre membrane modules. Biotechnol Prog 13:869–872
48. Cheng TW, Wu JG (2003) Quantitative flux analysis of gas-liquid two-phase ultrafiltration. Sep Sci Technol 38:817–835
49. Smith SR, Cui ZF (2004) Gas-slug enhanced hollow fibre ultrafiltration – an experimental study. J Memb Sci 242:117–128

50. Smith SR, Cui ZF (2004) Analysis of developing laminar pipe flow – an application to gas slug enhanced hollow fibre ultrafiltration. Chem Eng Sci 59:5975–5986
51. Li QY, Cui ZF, Pepper DS (1998) Enhancement of ultrafiltration by gas sparging with flat sheet membrane modules. Sep Purif Technol 14:79–83
52. Ghosh R, Li QY, Cui ZF (1998) Fractionation of BSA and lysozyme using ultrafiltration: effect of gas sparging. AIChE J 44:61–67
53. Mercier-Bonin M, Lagane C, Fonade C (2000) Influence of a gas/liquid two-phase flow on the ultrafiltration and microfiltration performances: case of a ceramic flat sheet membrane. J Memb Sci 180:93–102
54. Cheng T, Lin C (2004) A study on cross-flow ultrafiltration with various membrane orientations. Sep Purif Technol 39:13–22
55. Kenning DBR, Kao YS (1972) Convective heat transfer to water containing bubbles: enhancement not dependent on thermocapillary. Int J Heat Mass Transf 15:1709–1717
56. Tokuhiro AT, Lykoudis PS (1994) Natural convection heat transfer from a vertical plate. 1. Enhancement with gas injection. Int J Heat Mass Transf 37:997
57. Chang IS, Clech PL, Jefferson B, Judd S (2002) Membrane fouling in membrane bioreactors for wastewater treatment. J Environ Eng 128(11):1018–1029
58. Churchouse S (1997) Membrane bioreactors for wastewater treatment – operating experiences with the Kubota submerged membrane activated sludge process. Memb Technol 83:5–9
59. Gunder B, Krauth K (1998) Replacement of secondary clarification by membrane separation – results with plate and hollow fibre modules. Water Sci Technol 38:383–393
60. Calabro V, Curcio S, Iorio G (2002) A theoretical analysis of transport phenomena in a hollow fibre membrane bioreactor with immobilized biocatalyst. J Memb Sci 206:217–241
61. van Dijk L, Roncken GCG (1997) Membrane bioreactors for wastewater treatment: the state of the art and new developments. Water Sci Technol 35(10):35–41
62. Tajima F, Yamamoto T (1988) Apparatus for filtering water containing radioactive substances in nuclear power plants, Toshiba, US Patent 4 756 875
63. Yamamoto K, Hiasa M, Mahmood T, Matsuo T (1989) Direct solid–liquid separation using hollow fiber membrane in an activated sludge aeration tank. Water Sci Technol 21:43–54
64. Chiemchaisri C, Wong YK, Urase T, Yamamoto K (1992) Organic stabilization and nitrogen removal in membrane separation bioreactor for domestic wastewater treatment. Water Sci Technol 25:231–240
65. Futamura O, Katoh M, Takeuchi K (1994) Organic waste water treatment by activated sludge process using integrated type membrane separation. Desalination 98:17–25
66. Cote PL, Smith BM, Deutschmann AA, Rodrigues CF, Pedersen SK (1994) Frameless array of hollow fiber membranes and method of maintaining clean fiber surfaces while filtering a substrate to withdraw a permeate. PCT WO 94/11094
67. Mahendran M, Pedersen SK, Henshaw WJ, Behmann H, Rodrigues CF (1997) Vertical skein of hollow fiber membranes and method of maintaining clean fiber surfaces. PCT WO 97/06880
68. Leonard D, Mercier-Bonin M, Lindley ND, Lafforgue C (1998) Novel membrane bioreactor with gas/liquid two-phase flow for high-performance degradation of phenol. Biotechnol Prog 14:680–688
69. Lee SM, Jung JY, Chung YC (2001) Novel method for enhancing permeate flux of submerged membrane system in two-phase anaerobic reactor. Water Res 35:471–477

70. Futselaar H, Zoontjes RJC, Reith T, Rácz IG (1993) Economics comparison of transverse and longitudinal flow hollow fibre membrane modules for reverse osmosis and ultrafiltration. Desalination 90:345–361
71. Bouhabila EH, Ben Aim R, Buisson H (2001) Fouling characterization in membrane bioreactors. Sep Purif Technol 1(22–23):123–132
72. Posch C, Schiewer S (2005) Critical flux aspect of air sparging and backflushing on membrane bioreactors. Desalination 175:61–71
73. Guibert D, Ben Aim R, Rabie H, Cote P (2002) Aeration performance of immersed hollow-fiber membranes in a bentonite suspension. Desalination 148:395–400
74. Ghosh R (2006) Enhancement of membrane permeability by gas sparging in submerged hollow fibre ultrafiltration of macromolecular solutions: role of module design. J Memb Sci 274:73–82
75. Chang S, Fane AG (2001) The effect of fibre diameter on filtration and flux distribution – relevance to submerged hollow fibre modules. J Memb Sci 184:221–231
76. Chang S, Fane AG, Vigneswaran S (2002) Experimental assessment of filtration of biomass with model axial and transverse fibres. Chem Eng J 87:121–127
77. Chang S, Fane AG, Vigneswaran S (2002) Modelling and optimisation of submerged hollow fibre membrane modules. AIChE J 48:2203–2212
78. Field RW, Wu D, Howell JA, Gupta BB (1995) Critical flux concept for microfiltration fouling. J Memb Sci 100:259–272
79. Churchouse S, Wildgoose D (1999) Membrane bioreactors progress from the laboratory to full-scale use. Memb Technol 111:4–8
80. Yang W, Cicek N, Ilg J (2006) State-of-the-art of membrane bioreactors: worldwide research and commercial applications in North America. J Memb Sci 270:201–211
81. Chua HC, Arnot TC, Howell JA (2002) Controlling fouling in membrane bioreactors operated with a variable throughput. Desalination 149:225–229
82. Chang IS, Judd SJ (2002) Air sparging of a submerged MBR for municipal wastewater treatment. Process Biochem 37:915–920
83. Chang IS, Judd SJ (2003) Domestic wastewater treatment by a submerged MBR (membrane bioreactor) with enhanced air sparging. Water Sci Technol 47(12):149–154
84. Ebrahim S (1994) Cleaning and regeneration of membranes in desalination and wastewater applications: State-of-the-art. Desalination 96:225–238
85. Paul Chen J, Kim SL, Ting YP (2003) Optimization of membrane physical and chemical cleaning by a statistically designed approach. J Memb Sci 219:27–45
86. Tanaka T, Itoh H, Nakanishi K, Kume T, Matsumo R (1995) Crossflow filtration of Baker's yeast with periodical stopping of permeation flow and bubbling. Biotechnol Bioeng 47:404–410
87. Serra C, Durand-Bourlier L, Clifton MJ, Moulin P, Rouch JC, Aptel P (1999) Use of air sparging to improve backwashing efficiency in hollow-fibre modules, J Memb Sci 161:95–113
88. Verbeck J, Worm G, Futselaar H, van Dijk JC (2000) Combined air–water flush in dead-end ultrafiltration. In: Proceedings of the IWA Conference on Drinking and Industrial Water Production, Paris, pp 655–663
89. Clarkson JR, Cui ZF, Darton RC (1999) Protein denaturation in foam. I. Mechanism study. J Colloid Interface Sci 215:323–332
90. Clarkson JR, Cui ZF, Darton RC (1999) Protein denaturation in foam. II. Surface activity and conformational change. J Colloid Interface Sci 215:333–338
91. Clarkson JR, Cui ZF, Darton RC (2000) Effect of solution conditions on protein damage in foam. Biochem Eng J 2:107–114

92. Chiemchaisri C, Yamamoto K, Vigneswaram S (1993) Household membrane bioreactor in domestic wastewater treatment. Water Sci Technol 27(1):171–178
93. Muller EB, Stouthamer AH, Verseveld HW, Eikelboom DH (1995) Aerobic domestic wastewater treatment in a pilot plant with complete sludge retention by cross-flow filtration. Water Res 29:1179–1189
94. Cicek N, Franco JP, Suidan MT, Urbain V (1998) Using a membrane bioreactor to reclaim wastewater. J Am Water Works Assoc 90(11):105–113
95. Cote P, Buisson H, Praderie M (1998) Immersed membranes activated sludge process applied to the treatment of municipal wastewater. Water Sci Technol 38(4–5):437–442
96. Ahn KH, Cha HY, Song KG (1999) Retorfitting municipal sewage treatment plants using an innovative membrane–bioreactor system. Desalination 124:279–286
97. Rosenberger S, Kruger U, Witzig R, Manz W, Szewzyk U, Kraume M (2002) Performance of a bioreactor with submerged membranes for aerobic treatment of municipal wastewater. Water Res 36:413–420
98. Roest Van de HF, Lawrence DP, Van Bentem AGN (2002) Membrane bioreactors for municipal wastewater treatment. STOWA Report. IWA, London
99. Psoch C, Schiewer S (2006) Anti-fouling application of air sparging and backflushing for MBR. J Memb Sci 283(2):273–280
100. Psoch C, Schiewer S (2009) Air sparged membrane bioreactor for performance increase and less fouling. Water & Environmental Research Center, University of Alaska, Fairbanks, Alaska
101. Wang LK, Wang MHS (2008) Development and applications of membrane bioreactor technologies. In: National engineers week conference, Albany Marriott, Albany, NY, Feb 2008
102. Levesque S, Wallis-Large C, Hemken B, Bontrager S, Kreuzwiesner S (2009) Plan ahead with MBRs. Water Environ Fed 21(1):34–37
103. Wang LK, Shammas NK, Hung YT (eds) (2009) Advanced biological treatment processes. Humana, Totowa, pp 129–156
104. Wang LK, Ivanov V, Tay JH, Hung YT (eds) (2010) Environmental biotechnology. Humana, Totowa

Index

From: *Handbook of Environmental Engineering, Volume 13: Membrane and Desalination Technologies*
Edited by: L.K. Wang et al., DOI: 10.1007/978-1-59745-278-6 © Springer Science+Business Media, LLC 2011

MIX
Papier aus verantwortungsvollen Quellen
Paper from responsible sources
FSC® C105338

If you have any concerns about our products,
you can contact us on
ProductSafety@springernature.com

In case Publisher is established outside the EU,
the EU authorized representative is:
Springer Nature Customer Service Center GmbH
Europaplatz 3, 69115 Heidelberg, Germany

Printed by Libri Plureos GmbH
in Hamburg, Germany